Outdoor Recreation

David Huddart • Tim Stott

Outdoor Recreation

Environmental Impacts and Management

David Huddart
Liverpool John Moores University
Liverpool, UK

Tim Stott
Liverpool John Moores University
Liverpool, UK

ISBN 978-3-319-97757-7   ISBN 978-3-319-97758-4 (eBook)
https://doi.org/10.1007/978-3-319-97758-4

Library of Congress Control Number: 2018955843

© The Editor(s) (if applicable) and The Author(s), under exclusive licence to Springer Nature Switzerland AG 2019
This work is subject to copyright. All rights are solely and exclusively licensed by the Publisher, whether the whole or part of the material is concerned, specifically the rights of translation, reprinting, reuse of illustrations, recitation, broadcasting, reproduction on microfilms or in any other physical way, and transmission or information storage and retrieval, electronic adaptation, computer software, or by similar or dissimilar methodology now known or hereafter developed.
The use of general descriptive names, registered names, trademarks, service marks, etc. in this publication does not imply, even in the absence of a specific statement, that such names are exempt from the relevant protective laws and regulations and therefore free for general use.
The publisher, the authors and the editors are safe to assume that the advice and information in this book are believed to be true and accurate at the date of publication. Neither the publisher nor the authors or the editors give a warranty, express or implied, with respect to the material contained herein or for any errors or omissions that may have been made. The publisher remains neutral with regard to jurisdictional claims in published maps and institutional affiliations.

Cover illustration: MITO images GmbH / Alamy Stock Photo

This Palgrave Macmillan imprint is published by the registered company Springer Nature Switzerland AG
The registered company address is: Gewerbestrasse 11, 6330 Cham, Switzerland

Acknowledgements

We would like to acknowledge the generations of outdoor education students at Liverpool John Moores University for their help in developing our interest in recreation ecology and management. We thank our wives, Silvia and Debbie, for their patience and understanding during the long period of time that it took to write this book.

Contents

1	**Introduction to Outdoor Recreation and Recreation Ecology**		1
1.1	Outdoor Recreation		1
1.2	Recreation Impacts and Recreation Ecology		1
1.3	The Recreation Management System in the USA		2
	1.3.1	US National Park Service	3
		1.3.1.1 The Acts Establishing the National Park Service	4
		1.3.1.2 The National Parks Omnibus Management Act of 1998	4
		1.3.1.3 The Management Policies and Guidelines	4
		1.3.1.4 Visitor Carrying Capacity Decision-Making	6
		1.3.1.5 Visitor Perceptions of Resource Conditions	8
		1.3.1.6 Monitoring Programme Capabilities	8
	1.3.2	Management Considerations	8
		1.3.2.1 Evaluating Impact Acceptability	8
		1.3.2.2 Site Management of Recreation Sites	9
		1.3.2.3 Visitor Management	11
		1.3.2.4 Construct and Manage Facilities	11
		1.3.2.5 Educational Practices	11
		1.3.2.6 Leave No Trace Programmes and Information	12
	1.3.3	Other Environmental Impacts That Are Not Recreation-Induced	13
References			14
2	**Recreational Walking**		17
2.1	Introduction		17
2.2	Devotional Trails		18
2.3	Formal and Informal Trails		18
2.4	Global Perspective on the Numbers Involved		20
2.5	Trampling		21
2.6	Footpath Erosion		22
2.7	Trampling Impacts on Vegetation		23
2.8	Trampling Impacts on Soils		27
2.9	Trampling Impacts on Water Quality		28

2.10	Ways of Assessing Trampling Patterns Caused by Recreational Walking...............................	29
	2.10.1 Analytical, Descriptive Field Survey.............	29
	2.10.2 Experimental Trampling......................	30
2.11	Hiking Pole Impacts................................	30
	2.11.1 Vegetation Impacts...........................	31
	2.11.2 Soil Impacts................................	31
	2.11.3 Rock Impacts...............................	31
2.12	Summary Related to Impacts of Recreational Walking......	31
2.13	Effects on Wildlife.................................	32
	2.13.1 Flight and Behaviour Change...................	33
2.14	Impacts Are Not Always Negative......................	36
2.15	Management Implications for Recreational Walking-Induced Change in the Landscape....................	37
2.16	Path Wear and Deterioration..........................	39
2.17	Techniques for Managing the Footpath Surface...........	40
	2.17.1 Creating More Resistant Footpath Surfaces........	40
	2.17.1.1 Geotextiles........................	40
	2.17.1.2 Three-Dimensional Nettings............	40
	2.17.1.3 Chemical Binders....................	41
	2.17.1.4 Mulch Mats........................	41
	2.17.1.5 Mesh Elements......................	41
	2.17.2 Surface Glues...............................	41
	2.17.3 Surface Moulding............................	42
	2.17.4 Aggregate Paths.............................	42
	2.17.5 "Floated" Aggregate Paths.....................	42
	2.17.6 Boardwalk..................................	43
2.18	Vegetation Reinstatement............................	43
	2.18.1 Transplanting...............................	44
	2.18.2 Seeding.....................................	44
2.19	The Trampling Impact on Blanket Peat and Other Organic-Rich Soils.................................	44
2.20	The Three Peaks Project: Background to the Project (Yorkshire Dales, UK)...............................	45
	2.20.1 Chemical Consolidation of Soil..................	45
	2.20.2 Aggregate Path Construction....................	45
	2.20.3 Temporary Boardwalks........................	46
	2.20.4 Stone Pitching...............................	46
	2.20.5 Mechanised Path Construction Using Subsoil.......	47
2.21	Ecological Trials....................................	47
	2.21.1 Reinforcement of Existing Vegetation.............	47
	2.21.2 Restoration of Severely Damaged Peat Soils........	47
	2.21.3 Revegetating Mineral Soils.....................	47
	2.21.4 Revegetating Aggregate Path Surfaces............	48
	2.21.5 Conclusions Related to the Three Peaks Project.....	48
References..		49

3 Mountain Marathons, Adventure Racing, and Mountain Tours ... 55
- 3.1 Definitions ... 55
  - 3.1.1 Mountain Marathon ... 55
  - 3.1.2 Adventure or Expedition Racing ... 56
  - 3.1.3 High Mountain Tours ... 56
- 3.2 History, Diversity, and Participation Numbers ... 59
  - 3.2.1 Mountain Marathons ... 59
  - 3.2.2 Adventure Racing ... 60
  - 3.2.3 High Mountain Tours ... 61
- 3.3 Safety and Legal Regulation ... 64
- 3.4 Environmental Impact, Management, and Education ... 65
  - 3.4.1 Research Needs ... 65
  - 3.4.2 The National Three Peaks Challenge ... 67
  - 3.4.3 The Yorkshire Three Peaks ... 69
  - 3.4.4 Management Approaches to Minimise Damage ... 69
- References ... 70

4 Recreational Climbing and Scrambling ... 73
- 4.1 Introduction ... 73
- 4.2 Types of Climbing ... 74
- 4.3 Numbers Involved in Climbing ... 76
- 4.4 How Do Climbers Affect Crag Ecosystems? ... 77
- 4.5 Effects of Climbing on Cliff Vegetation and Other Biota ... 78
  - 4.5.1 Traditional Summer Climber Impacts on the Vegetated Crag Environment ... 78
  - 4.5.2 Impacts on Gastropods ... 84
  - 4.5.3 Climbing Effects on Bird Populations, Particularly Raptors ... 85
  - 4.5.4 Unusual Potential Impacts on Mammals ... 87
  - 4.5.5 Overview of the Climbing Impacts on Cliff Biotic Diversity ... 88
  - 4.5.6 Damage to the Rock ... 88
    - 4.5.6.1 Chalk ... 89
    - 4.5.6.2 Effects of Protection ... 89
    - 4.5.6.3 Rock Damage by Ropes ... 92
    - 4.5.6.4 Rock Polish ... 92
- 4.6 Bouldering and Its Environmental Impacts ... 93
- 4.7 Winter Climbing and Its Environmental Impacts ... 93
- 4.8 Management of Climbing ... 96
  - 4.8.1 Management Plans ... 96
  - 4.8.2 Memorandums of Understanding or Agreements ... 98
  - 4.8.3 Liaison Groups ... 98
  - 4.8.4 Closures ... 98
  - 4.8.5 Seasonal Restrictions ... 98
  - 4.8.6 "Remote Areas" ... 99
  - 4.8.7 Rerouting ... 100
  - 4.8.8 Use of a Star System in Guidebooks ... 100
  - 4.8.9 Booking or Permit System ... 100

		4.8.10	Outreach and Education........................ 100

                         4.8.10.1 Provision of Information................ 101

References... 106

5 Gorge Walking, Canyoneering, or Canyoning 111
- 5.1 Introduction 111
- 5.2 Numbers of Participants............................... 113
- 5.3 Case Studies of Three Types of Canyoning............... 114
  - 5.3.1 Gorge Walking or Gill Scrambling in the UK 114
    - 5.3.1.1 What Are the Needs of the Gorge Walker?... 117
    - 5.3.1.2 How Can We Control Any Erosion in Gorges?............................... 120
  - 5.3.2 Canyoning in the Greater Blue Mountains World Heritage Area, Australia........................ 122
  - 5.3.3 Canyoneering in Utah and Arizona 123
- 5.4 General Management Issues for Canyoning and Canyoneering.. 127
  - 5.4.1 Banning the Activity Completely.................. 127
  - 5.4.2 Restoration and Clean-Up Projects 127
  - 5.4.3 Code of Conduct Issues and Ethics Guidelines for Canyoneers and Canyoners 128
- 5.5 Education... 128

References... 129

6 Off-Road and All-Terrain Vehicles, Including Snowmobiling 131
- 6.1 Definitions 131
  - 6.1.1 Off-Road Vehicles (ORVs)....................... 131
  - 6.1.2 All-Terrain Vehicles (ATVs) 132
  - 6.1.3 Off-Road Motorcycles: Motocross and Enduro Motorcycling 133
  - 6.1.4 Snowmobiles 134
- 6.2 History, Diversity, and Participation Numbers 134
  - 6.2.1 ORVs 134
  - 6.2.2 ATVs 135
  - 6.2.3 Snowmobiles 137
- 6.3 Safety and Legal Regulation 138
  - 6.3.1 ORVs 138
  - 6.3.2 ATVs 139
  - 6.3.3 Snowmobiles 141
- 6.4 Environmental Impact 141
  - 6.4.1 ORVs, ATVs, and Motorcycles 142
    - 6.4.1.1 Damage to Soil and Vegetation........... 142
    - 6.4.1.2 Pollution and Noise..................... 145
    - 6.4.1.3 Wildlife 145
  - 6.4.2 Snowmobiles 146
    - 6.4.2.1 Damage to Soil and Vegetation........... 146
    - 6.4.2.2 Pollution and Noise..................... 147
    - 6.4.2.3 Wildlife 148

	6.5	Management and Education	150
		6.5.1 Introduction	150
		6.5.2 Legal Controls	150
		6.5.3 Managing ORV Use and Experiences	152
		6.5.4 Education and ORV Users	153
	6.6	Future Trends	154
	References		160

7 Mountain Biking ... 163

- 7.1 Definitions ... 163
  - 7.1.1 Mountain Biking (MTB) ... 163
  - 7.1.2 Fatbikes ... 164
  - 7.1.3 BMX Bikes ... 164
- 7.2 Participation Numbers ... 164
- 7.3 History, Designs, and Disciplines with MTB ... 168
  - 7.3.1 History ... 168
  - 7.3.2 Designs ... 168
  - 7.3.3 MTB Disciplines ... 169
- 7.4 Environmental Impact ... 171
  - 7.4.1 Damage to Soil and Vegetation ... 171
  - 7.4.2 Impacts of Mountain Biking on Wildlife ... 174
- 7.5 Management and Education ... 176
  - 7.5.1 Impacts to Vegetation: Management Implications ... 176
  - 7.5.2 Impacts to Soils: Management Implications ... 176
  - 7.5.3 Impacts to Water Resources: Management Implications ... 177
  - 7.5.4 Managing and Educating Mountain Bikers ... 178
  - 7.5.5 Forest-Based Mountain Biking: The UK Experience ... 179
  - 7.5.6 Future Research and Management Implications for Mountain Biking ... 182
- References ... 183

8 Camping, Wild Camping, Snow Holing, and Bothies ... 187

- 8.1 Definitions ... 187
  - 8.1.1 Camping ... 187
  - 8.1.2 Snow Caves, Quinzhees, and Igloos ... 190
  - 8.1.3 Bothies ... 194
- 8.2 Participation Numbers ... 194
- 8.3 Environmental Impact ... 198
  - 8.3.1 Damage to Soil and Vegetation ... 198
  - 8.3.2 Impacts of Camping, Snow Holing, and Bothying on Water Resources ... 201
  - 8.3.3 Impacts of Camping, Snow Holing, and Bothying on Wildlife ... 202
- 8.4 Management and Education ... 203
  - 8.4.1 Managing the Impacts of Camping, Snow Holing, and Bothying on Vegetation and Soils ... 203
  - 8.4.2 Managing the Impacts of Camping, Snow Holing, and Bothying on Water Resources ... 206

	8.4.3	Managing the Impacts of Camping, Snow Holing, and Bothying on Wildlife............................ 209

References... 213

9 Horseback Riding 215
9.1 Introduction 215
    9.1.1 Recreational Horse Riding......................... 216
9.2 Numbers Involved in Horseback Riding 217
9.3 Horseback Riding: Biophysical and Social Impacts 219
    9.3.1 Biophysical Impacts Caused by Horseback Riding.. 219
    9.3.2 Trail Proliferation.............................. 223
    9.3.3 Field Experiments on the Impacts 223
9.4 Impact on Water Crossings and Rivers 225
9.5 Impacts of Horse Manure 226
9.6 Horses As Agents of Weed Spreading 227
9.7 Horses As Agents of the Spread of Non-native or Alien Species.. 232
    9.7.1 Summary Related to Impact of Horses on Weed and Alien Plant Spreading 233
9.8 Damage to Campsites by Horses 234
9.9 Impact of Packstock Grazing 234
9.10 Impacts on Animals 235
    9.10.1 Impacts on Breeding Birds........................ 236
9.11 Potential Spread of Pathogens by Horses 236
9.12 Why Horse Riding Impacts Are of Particular Concern in Australian Ecosystems................................... 236
9.13 Social Impacts of Horse Riding 238
9.14 Management of Horseback Riding Impacts 238
9.15 What Are Other Management Options and How Acceptable Are They? 240
    9.15.1 Prohibit Horseback Riding........................ 240
    9.15.2 Unrestricted Open Access 241
    9.15.3 Managing Horseback Riding Commercial Operators 242
    9.15.4 Horseback Riders' Reaction to Management 243
    9.15.5 Scientific Monitoring Over an Extended Time Period 243
    9.15.6 Packstock Management Strategies.................. 244
References... 245

10 Geocaching, Letterboxing, and Orienteering................. 249
10.1 Introduction 249
10.2 Numbers of Participants................................. 252
10.3 Geocaching 252
    10.3.1 Types of Geocaches 253
10.4 Geocaching: Environmental Impacts...................... 254
10.5 Geocaching: Educational Benefits and the Promotion of Environmental Stewardship............................... 255

	10.6	Management of Geocaching and Letterboxing	256
	10.7	Orienteering: Environmental Impacts	256
		10.7.1 Disturbance to Vegetation.	257
		10.7.2 Disturbance to Mammals	257
		10.7.2.1 Disturbance of Elk and Deer in Sweden .	258
		10.7.2.2 Disturbance of Deer in the New Forest in Southern England.	259
		10.7.3 Disturbance of Birds.	259
		10.7.3.1 Disturbance of Nesting Birds in Drumore Wood, Aberfoyle, Scotland	260
		10.7.3.2 Discussion of Research into the Disturbance of Birds by Orienteers	261
	10.8	A Comparison Between Orienteering and Letterboxing Impacts.	262
	References.	263
11	**Skiing, Snowboarding, and Snowshoeing**.	267
	11.1	Definitions	267
		11.1.1 Alpine Skiing	267
		11.1.2 Nordic Skiing	267
		11.1.3 Telemark.	268
		11.1.4 Ski Mountaineering	269
		11.1.5 Snowboarding.	269
		11.1.6 Snowshoeing.	270
	11.2	Snow Sport Competition	273
	11.3	Participation Numbers	273
	11.4	Environmental Impact	278
		11.4.1 Impacts of Snow Sport on Soil and Vegetation	278
		11.4.2 Impacts of Snow Sport on Wildlife	288
		11.4.3 Impacts of Snow Sport on Water Resources.	289
	11.5	Management and Education.	291
		11.5.1 Effects of Climate Change on Snow Sport.	291
		11.5.2 Do Ski Resorts Need to Become "Greener" for Tourism to Become Sustainable?.	293
		11.5.3 Concluding Comments.	294
	References.	295
12	**Caving**	299
	12.1	Introduction and Numbers Involved in Caving	299
	12.2	How Can Caves Be Damaged?	303
	12.3	Cave Fauna and Flora.	307
		12.3.1 Zones in Caves	308
		12.3.2 Bats in Caves	309
	12.4	Management Strategies to Conserve Caves.	310
		12.4.1 Potential Strategies.	310
		12.4.2 Federal Cave Management and the National Park Service	311
		12.4.3 NPS Cooperative Relationships	312
		12.4.4 Canadian National Parks	313

	12.4.5	Access Agreements and Physical Barriers 313
	12.4.6	Secret Conservation 314
	12.4.7	Zero Access................................. 314
	12.4.8	Restricted Access 314
	12.4.9	Periodical Access 315
	12.4.10	Booking..................................... 315
	12.4.11	Gating 315
	12.4.12	Sacrificial Caves 316
	12.4.13	Endurance Conservation..................... 318
	12.4.14	Artificial Obstacles 318
	12.4.15	Zoning Off................................... 318
	12.4.16	Formation Repair Work....................... 319
	12.4.17	Exploration Policy 319
	12.4.18	Cave Adoption Schemes 319
	12.4.19	Cave Fauna Management 319
	12.4.20	Artificial Bat Caves.......................... 321
	12.4.21	Management of Lampenflora in Show Caves..... 322
12.5	Education... 322
	12.5.1	Cave Conservation and Responsible Caving Practices 322
	12.5.2	Alternatives to Caving to Take Pressure Off the Caves..................................... 324
12.6	Cave Art Teaching and Experimental Archaeology........ 325
12.7	Minimum Impact Caving Techniques for Fauna Developed in Tasmania 326
References... 327

13 Water Sports and Water-Based Recreation 331

13.1	Definitions 331
	13.1.1	Non-motorised Water Sports................... 331
		13.1.1.1 Canoeing......................... 331
		13.1.1.2 Kayaking......................... 335
		13.1.1.3 Rafting/White-Water Rafting 335
		13.1.1.4 Rowing 335
		13.1.1.5 Sailing/Yachting 335
		13.1.1.6 Windsurfing....................... 335
		13.1.1.7 Surfing........................... 337
	13.1.2	Motorised Water Sports...................... 338
13.2	Participation Numbers 338
13.3	Environmental Impact 341
	13.3.1	Physical Impacts of Water Sports 344
	13.3.2	Biological Impacts of Water Sports 346
		13.3.2.1 Water Quality and Micro-organisms.... 346
		13.3.2.2 Impacts on Plants and the Spread of Invasive Species 348
		13.3.2.3 Disturbance to Wildlife.............. 350
	13.3.3	Chemical Impacts of Water Sports 352
		13.3.3.1 Heavy Metals 352
		13.3.3.2 Motorboat Engine Products and Bi-products 353

	13.4	Management and Education................................. 353
		13.4.1 Managing Physical Impacts 353
		13.4.2 Managing Biological/Water Quality Impacts...... 355
		13.4.3 Managing for Climate Change 355
		13.4.4 Concluding Comments...................... 356
	References... 357
14	**Recreational Scuba Diving and Snorkelling**................. 361
	14.1	Introduction 361
	14.2	Types of Recreational Diving........................... 361
		14.2.1 Recreational Scuba Diving.................... 362
		14.2.2 Technical Diving............................ 362
		14.2.3 Wreck Diving 362
		14.2.4 Snorkelling................................ 362
		14.2.5 Swimming with Cetaceans 363
	14.3	Benefits of Dive Tourism 363
	14.4	Estimates of the Numbers Involved in These Recreation Pursuits... 364
	14.5	Direct and Indirect Impacts of Recreational Diving 364
		14.5.1 Direct Trampling by Reef Walking.............. 364
		14.5.2 Effects of Pontoons......................... 365
		14.5.3 Fish Feeding and Pontoons.................... 366
		14.5.4 Direct Impact by Mooring Installation........... 366
		14.5.5 Direct Impact by Anchor Damage 366
		14.5.6 Other Boating Impacts 368
	14.6	Direct Scuba Diving Impacts........................... 368
		14.6.1 Direct Impact by Divers (See Fig. 14.3 and Table 14.2) 368
		14.6.2 Effects of Diver Behaviour.................... 369
		14.6.3 Impacts from Sediments Raised by Divers........ 373
		14.6.4 Impacts on the Corals........................ 374
		14.6.4.1 Coral Species and Physical Damage 375
		14.6.4.2 Effects of Sunscreens and Insect Repellents on Coral Growth........... 377
	14.7	Effects of Recreational Diving on Fish Communities...... 378
	14.8	Impact of Recreation Divers on Kelp Forests 380
	14.9	Management of Scuba Diving and Snorkelling........... 381
		14.9.1 Introduction 381
		14.9.2 Management of the Divers' Physical Environment ... 381
		14.9.3 The Development of Artificial Reefs and Marine Sculpture or Art Trails 382
		14.9.4 Modification of Diver Behaviour 384
		14.9.5 Charging Fees or Differential Fees.............. 387
		14.9.6 Recreation Divers As Part of the Conservation Effort...................................... 388
		14.9.7 Activity Standards, Park Rules and Regulations................................. 389
		14.9.8 Other Damaging Impacts on Coral Reefs......... 389
	References... 390

15 Recreational Fishing ... 395
- 15.1 Introduction ... 395
- 15.2 Definition of Recreational Fishing ... 396
  - 15.2.1 Types of Recreational Fishing ... 396
- 15.3 Numbers of Recreation Fishers ... 396
- 15.4 Direct Impacts of Recreational Fishing ... 398
  - 15.4.1 Effects on Fish Stocks ... 398
  - 15.4.2 Large Species Range ... 399
  - 15.4.3 Endangered Fish Species and Trophy Fishing ... 400
  - 15.4.4 Size Selection and Fish Community Structure ... 400
  - 15.4.5 Consequences of High Exploitation Rates and Selectivity ... 401
- 15.5 Direct Consequences ... 401
  - 15.5.1 Depensation Instead of Compensation ... 401
  - 15.5.2 Truncation of Age and Size Structure ... 401
  - 15.5.3 Loss of Genetic Diversity ... 402
  - 15.5.4 Evolutionary Changes Due to Selective Angling ... 403
  - 15.5.5 Discards or By-Catch from Recreational Fishing ... 404
  - 15.5.6 Catch-and-Release Impacts ... 404
- 15.6 Impact of Invasive, Non-native Species ... 407
- 15.7 Indirect Impacts of Recreational Fishing ... 409
  - 15.7.1 Disturbance of Habitats ... 409
    - 15.7.1.1 Walking Tracks (See Burgin 2017) ... 409
    - 15.7.1.2 Impacts of Off-Road Vehicles ... 410
    - 15.7.1.3 Wading Associated with Instream Angling ... 410
  - 15.7.2 Disturbance of Wildlife ... 410
- 15.8 Plastic in Various Forms ... 410
- 15.9 Fishing Gear Loss ... 411
- 15.10 Boat Strike and Boat Traffic Impacts ... 412
- 15.11 Nutrient Input ... 413
- 15.12 Exotic Species of Bait and Bait Gathering Effects ... 413
- 15.13 Pathogen Transmission ... 414
- 15.14 Inadvertent Overland Dispersal of Non-native Plants ... 414
- 15.15 Management of Recreational Fishing Impacts ... 415
- 15.16 Education Related to Recreational Fishing Impacts ... 419
  - 15.16.1 Trade Sector and Recreation Fisheries Conservation ... 419
  - 15.16.2 Mandatory Education Programmes ... 419
  - 15.16.3 Best Practices Guidelines for Catch-and-Release in General and, As an Example, Guidelines for Striped Bass Catch-and-Release ... 420
    - 15.16.3.1 Techniques to Increase Survival of Released Fish ... 420
    - 15.16.3.2 Terminal Tackle Type ... 420
    - 15.16.3.3 Playing Time ... 420
    - 15.16.3.4 Landing and Handling Techniques ... 420
    - 15.16.3.5 Unhooking Techniques ... 421
    - 15.16.3.6 Release Methods ... 421

	15.17	Behavioural Response of Anglers to Management Actions 422
	15.18	Voluntary Codes of Practice, Codes of Conduct, and Angler's Codes of Ethics (Like the Fly Fishers Code of Ethics, 2002).. 423
	References.. 425

16 Expeditions 429
   16.1 Definitions 429
   16.2 History of Overseas Expeditions 430
   16.3 Participation Numbers 432
   16.4 Environmental Impact 433
      16.4.1 The Impact of Movement and Access 433
      16.4.2 Expedition Campsites......................... 437
      16.4.3 Impact on Local Communities 439
      16.4.4 Impact of Expedition Fieldwork................ 440
   16.5 Management and Education........................... 440
   References.. 445

17 Overall Summary 447
   17.1 Introduction 447
   17.2 Numbers Involved in Outdoor Recreation, the Relative Importance of the Various Activities, and Likely Future Trends.. 447
   17.3 Research into the Recreation Impacts (Recreation Ecology) and the Land Management of These Impacts..... 450
   References.. 457

Index 459

About the Authors

 **David Huddart** is Emeritus Professor of Quaternary Geology and Environmental Education at Liverpool John Moores University. He has spent many years teaching landscape interpretation, recreation ecology, and outdoor and environmental education. He was the course leader for the postgraduate diploma in outdoor education and the undergraduate BEd and BSc in science and outdoor education and outdoor education degree courses. He has supervised undergraduate and PhD student dissertations in many aspects of recreation ecology. He has researched in many parts of the world including Svalbard, Iceland, Greenland, Alaska, Mexico, Guatemala, and the United Kingdom.

 **Tim Stott** is Professor of Physical Geography and Outdoor Education at Liverpool John Moores University where he has been responsible for leading and teaching on the outdoor education programmes for 25 years. He has travelled widely, skiing, trekking, cycling, and canoeing, and has carried out fieldwork for his research in Iceland, Svalbard, Greenland, European Alps, British Columbia, China, Peru, Bolivia, and the Caucasus, as well as the United Kingdom.

List of Figures

Fig. 2.1	(A) Lanzarote Geopark, Montaña Corona, demarcation of path by large boulders and information sign.	19
Fig. 2.2	Excessive wide footpath in clay-rich soil caused by human and probably horse, trampling pressure, Pennine Way (UK).	23
Fig. 2.3	(A) Braiding of footpaths, Snowdon, North Wales. (B) Eroded and braided, informal footpath, Clwydian Hills, Denbighshire Moors, North Wales.	24
Fig. 2.4	Deep incision caused by trampling pressure, Peak District.	28
Fig. 2.5	Sign asking for no access up the volcano flank because of the highly erodible coarse volcanic ash, Lanzarote Geopark, Montana Corona.	38
Fig. 2.6	Geotextile. (Three Peaks project).	41
Fig. 2.7	Boardwalk across extremely rough aa lava, Timanfaya National Park, Lanzarote.	43
Fig. 2.8	Stone pitching in Snowdonia, North Wales.	46
Fig. 3.1	Competitor on a cycling leg of an adventure race in the English Lake District.	56
Fig. 3.2	The start of the high mountain tour at the end of the eighteenth century: contemporary portrait of Horace-Bénédict de Saussure on Mont Blanc in 1787.	57
Fig. 4.1	Big wall climbing, El Capitan southeast face, Yosemite.	74
Fig. 4.2	Erosion at base of crags, Harrison's Rocks, South East England.	75
Fig. 4.3	El Capitan (Yosemite) SE face closures due to peregrine nesting areas.	88
Fig. 4.4	Rope damage to Sandstone, Zion National Park, Utah.	89
Fig. 4.5	Nut placement.	89
Fig. 4.6	(A) Idwal Slabs (Snowdonia): erosion at base of crag, oblique and vertical jointing, and weathering holes. (B) Weathering pits in dolerite, Idwal Slabs, Snowdonia.	91
Fig. 4.7	Snowdon lily (*Lloydia serotina*), Clogwyn Du'r Arddu, Snowdonia.	96
Fig. 4.8	Landscape Arch, Arches National Park, Utah. Closure of all climbing routes in this national park on arches.	99

Fig. 4.9	Signage at Harrisons Rocks, South East England: information related to climbing etiquette.	102
Fig. 5.1	Gill scrambling often includes climbing or descending waterfalls and rapids.	112
Fig. 5.2	Canyoneering, Arches National Park, Utah.	112
Fig. 5.3	Abseiling down a waterfall close to the heavily vegetated rock walls.	114
Fig. 5.4	(A) Sea plantain (*Plantago maritimus*). (B) Scarce Turf Moss (*Rhytidiadelphus subpinnatus*). (C) Starry Saxifrage (*Saxifraga stellaris*).	115
Fig. 5.5	Nantcol Gorge (North Wales).	119
Fig. 5.6	Padley Gorge.	120
Fig. 5.7	Grand Canyon with many tributary canyons hidden from the main canyon rim.	124
Fig. 6.1	A Ford Bronco dune bashing.	132
Fig. 6.2	A motocross rider coming off a jump.	133
Fig. 6.3	(A) A snowmobile. (B) Snowmobile being used by reindeer herders.	134
Fig. 6.4	(A) Negative environmental effects caused by a motorcycle to a portion of the Los Padres National Forest, California. (B) Land Rover Series III mud plugging. (C) A Jeep Grand Cherokee, in action, drives through a watercourse.	143
Fig. 6.5	(A) Surveying an eight-year-old vehicular track in Gipsdalen, Svalbard, to measure the eroded cross-section area. (B) An eight-year-old vehicular track in Gipsdalen, Svalbard.	144
Fig. 6.6	Average soundscape power (normalised watts/kHz) and 95% confidence intervals within ten 1-kHz frequency intervals summarised for snowmobile noise and natural quiet identified from 59,598 sound recordings acquired in Kenai National Wildlife Refuge, Alaska.	149
Fig. 6.7	Out of control: the impact of ORVs and roads on wildlife and habitat in Florida's national forests: an educational publication by Defenders of Wildlife.	154
Fig. 7.1	A full-suspension mountain bike.	164
Fig. 7.2	Participation in gateway outdoor activities.	167
Fig. 7.3	Example of a North Shore board way at Llandegla mountain bike centre, North Wales.	173
Fig. 7.4	(A) Damage to moorland vegetation due to the passage of mountain bikes on a moorland in NE Wales in winter. (B) Mountain bikes crossing watercourses can release fine sediment which can result in siltation on stream beds. (C) Mountain bikes passing through waterlogged flushes in upland areas can cause compaction, remove binding vegetation, and release fine sediments. (D) Mountain bikes passing through upland moorland flatten vegetation.	174

List of Figures

Fig. 7.5	(A) Trail guide for Innerleithen, part of the 7stanes mountain biking suite of trails developed and managed by Forestry Commission Scotland. (B) Mountain bike centres and bases in Wales which offer a variety of ride experiences.	181
Fig. 8.1	(A) A hooped bivouac bag used for lightweight survival camping. (B) Using a tarpaulin to create an overnight shelter in Yorkshire, UK. (C) A tarpaulin used as a lightweight overnight shelter in a woodland. The campers sleep on the ground. (D) A hammock with tarpaulin used as a lightweight overnight shelter in a woodland. The camper sleeps off the ground. (E) A mountain tent used for a camp at 2000 m by Castle Creek Glacier in the Cariboo Mountains, British Columbia. (F) Typical family camping tent on a campsite in Anglesey, Wales. (G) A campsite in Switzerland in summer showing a range of typical tent designs.	189
Fig. 8.2	(A) A 5-berth UK touring caravan on a campsite in Wales. (B) A typical touring caravan park at Strathyre in central Scotland. (C) A typical static caravan park in North Wales.	191
Fig. 8.3	(A) Excavating snow caves at Garbh Uist Beag, Cairngorms, Scotland. (B) A snow cave is constructed by excavating snow so that the tunnel entrance is below the main space to retain warm air. (C) Inside a snow cave in the Cairngorms, Scotland. (D) Cooking equipment inside a snow cave.	192
Fig. 8.4	(A) A quinzhee (or quinzee) is a snow shelter that is made from a large pile of loose snow which is shaped then hollowed. (B) Cutting blocks to make the base of an igloo, Cairngorms, Scotland. (C) Construction of an igloo nearing completion. (D) A finished igloo. The person standing on top is demonstrating the structural strength of the igloo.	193
Fig. 8.5	Ryvoan bothy in the Spey Valley, Scotland.	194
Fig. 8.6	Participation in gateway outdoor activities.	198
Fig. 8.7	(A) Damage to grass by trampling around the entrance to a family tent on a commercial campsite in Switzerland. (B) Damage to natural vegetation by trampling around tents at a mountain training camp in SW Greenland. (C) Building open fires for cooking is a common activity when camping.	200
Fig. 8.8	(A) Example of a visitor sign used by Maine Lakes Environmental Association, USA, to control the spread of invasive aquatic species. (B) Visitor sign in the Eastern USA alerting water users to the spread of the invasive species, Eurasian Watermilfoil.	206
Fig. 8.9	(A) A degraded section of the Tasmanian Overland Track (2005). (B) A renovated section of the Tasmanian Overland Track (2005). (C) Hardening of campsites at selected overnight nodes (foreground) on the Tasmanian Overland Track. (D) Camping platform near a overnight stay cabin on the Tasmanian Overland Track.	207

Fig. 8.10	"Where to Go in the Great Outdoors". Mountaineering Council of Scotland Advisory Leaflet.	208
Fig. 8.11	The Cairngorm Poo Project provides visitors with these bottles to bring back their human waste in winter.	209
Fig. 8.12	The Mountain Bothies Association's Bothy Code.	210
Fig. 9.1	Pony trekking, Brecon Beacons.	216
Fig. 9.2	(A) Pioneering pack mule and miner on route to mine at Mount Lowe, California. (B) Packstock on Baja California (Mexico) mountain trail.	218
Fig. 9.3	The change in soil depth from the baseline micro-topography across 5–100 cm of the treatment transects cross-sectional profile after various intensities of horse trampling at study site DE3 in D'Entrecasteaux National Park, Western Australia.	224
Fig. 9.4	Horse trekking through river, Cochamó, Chile.	226
Fig. 10.1	(A) Geocache location in Delaware State Park woods. (B) Geocaching in Delaware State Park.	250
Fig. 10.2	Small official geocache in Delaware State Park woods.	250
Fig. 10.3	Wheatear (*Oenanthe oenanthe*).	261
Fig. 11.1	(A) An Alpine ski boot and bindings. (B) An Alpine ski is shaped to enhance its turning capability. (C) Alpine ski technique with skis parallel. (D) Drag lift (also called a poma tow) at Chamrousse 1650, France. (E) Chairlift and ski resort infrastructure at Les Deux Alpes, France.	268
Fig. 11.2	(A) Nordic ski boot and binding. (B) A Nordic ski is longer and narrower than an Alpine ski. (C) Nordic skiing at the Plateau D'Alsace, Chamrousse, France.	269
Fig. 11.3	(A) Telemark boot and binding. (B) A telemark ski is shaped like an Alpine ski. (C) Telemark ski technique.	270
Fig. 11.4	(A) Ski mountaineering boot and binding. (B) Ski mountaineering boots, bindings, and skis are lighter than Alpine equivalents. (C) A ski mountaineering party in northern Norway skinning uphill. (D) Ski mountaineers make zigzag tracks up steep slopes. (E) Ski mountaineering ski technique for the descent is basically the same as for Alpine skiing, though the snow conditions can be variable. (F) Ski mountaineering descent in Kvaloya, near Tromso, Norway. (G) Ski mountaineers sometimes have to walk in carrying their skis (on approach to the Komna plateau, Slovenia). (H) Ski mountaineers reaching a summit in Norway.	271
Fig. 11.5	(A) Snowboarders use soft boots which they strap into bindings. (B) Snowboarding on a piste. (C) Snowboarders in a play park.	272

Fig. 11.6	Modern snowshoes made from plastic have bindings, heel lift, and small spikes on the underside for grip on hard snow	273
Fig. 11.7	Number of people who ski in Europe as of 2016, by country (in 1000)	282
Fig. 11.8	(A) End-of-season skiing at Val Thorens, France, (B) Skiing on thin snow cover at Lecht Ski Resort in Scotland	284
Fig. 11.9	Reduction in the number of ski days and the percentage closure of ski resorts in various regions as a function of temperature increase.	292
Fig. 12.1	*Homo naledi* bones from Rising Star Cave, part of the Cradle of Humankind World Heritage site about 50 km north-west of Johannesburg	301
Fig. 12.2	(A) Long Churn Cave in soluble limestone: (B) Lava tube and lake, Jameos del Agua, Lanzarote. (C) Glacier cave in Iceland	302
Fig. 12.3	Great Douk dig, vertically from the collapsed cave entrance section to try and connect the active streamway from the cave to the River Greta to the west.	304
Fig. 12.4	(A) Blind albino crab (*Munidopsis polymorpha*), Jameos del Agua lake, Malpais de la Corona, Lanzarote. (B) *Zospeum tholussum*, a microscopic cave snail completely blind with a translucent shell, from the Lukina Jama-Trojama cave system (Velebit Mountains, Croatia). (C) Devil's Hole (*Cyprinodon diabolis*)	307
Fig. 12.5	(A) Healthy little brown bats, Aeolus Cave or Dorset Bat Cave in the Taconic Mountains in East Dorset, Vermont. (B) Bat roosting in cave with white nose syndrome, Greeley Mine. Vermont	310
Fig. 12.6	(A) Gating at Agen Allwedd, Llangattock, South Wales. (B) Bat Gate at the entrance of Skeleton Cave, Oregon; lava tube on the northern flank of Newberry volcano	316
Fig. 12.7	(A) Great Douk Cave entrance and Great Douk Pot (collapse section of the cave). (B) Long Churn passage (Yorkshire Dales), active streamway, phreatic upper passage with vadose trench incised showing several water levels. (C) Great Douk sacrificial cave.	317
Fig. 12.8	Zoning off of stalagmites, Matienzo Caves, Spain	318
Fig. 12.9	Waitomo Glowworm Caves, New Zealand	320
Fig. 12.10	Green Glow Caves, New Zealand	321
Fig. 12.11	Swan Mine, Mendips	325
Fig. 12.12	Replica of lionesses painting from Chauvet Cave, Ardeche, in the Moravian Museum, Brno	326

Fig. 13.1	(A) Tranquil open canoe journey on the River Stour, Southern England. (B) Open canoe camping. The canoe is ideal for transporting heavy loads. (C) The author solo paddling an open canoe on Grade II water of the River Dee, North Wales. (D) The open canoe can be rigged for sailing. (E) The US National Canoe Poling Championships on the Meramec River, Missouri, USA.	334
Fig. 13.2	(A) Recreational kayaker descends a rapid on the upper Afon Tryweryn, North Wales. (B) Recreational kayaker with the kayak almost totally under the water. (C) A recreational white-water kayaker using a high-volume kayak to descent "big water". (D) White-water rafting on the upper Afon Tryweryn, North Wales. (E) White-water rafting on a wave on the Durance River, France. (F) A typical "beach" on the Durance River, France.	336
Fig. 13.3	(A) Dinghy sailing in Liverpool Marina. (B) Yachts racing on the Mersey Estuary near Liverpool. (C) Cruising boats in Liverpool Marina. (D) Windsurfing in the inland sea at Valley on Anglesey in North Wales. (E) Surfing at Bundoran, Northern Ireland.	337
Fig. 13.4	Small powerboat.	338
Fig. 13.5	Main impacts of recreation-water resource impacts	345
Fig. 13.6	The impacts of boats on plants.	348
Fig. 13.7	Likely influences and impacts of powerboating activities on fishes and their habitats and the likely time frame over which the impacts may occur	351
Fig. 14.1	Technical diver using mixed gases to dive to 60 m	362
Fig. 14.2	Diving platform or pontoon, Agincourt Reef, Great Barrier Reef, Queensland.	366
Fig. 14.3	Scuba diving impacts on coral reefs.	368
Fig. 14.4	Stony coral: table coral of genus Acropora at French Frigate Shoals, North-western Hawaiian Islands.	369
Fig. 14.5	Density of fishes (per 20 m^2) (mean ± SE) at Picãozinho and Quebra Quilhas	379
Fig. 14.6	Materials commonly used for artificial reefs	383
Fig. 14.7	Reef balls, Lake Pontchartrain Basin.	384
Fig. 14.8	(A) Artificial reef sculptures by Jason deCaires Taylor. (B) Reef sculptures, Playa del Carmen, Lanzarote	385
Fig. 14.9	Green Fins information leaflet	387
Fig. 15.1	Fly fishing on River Sava, Bohinjka, Slovenia.	397
Fig. 15.2	Trophy fishing for striped blue marlin, caught off Cabo San Lucas, Baja California.	397
Fig. 15.3	Round goby (*Neogobius melanostomus*)	408
Fig. 15.4	Northern snakehead (*Channa argus*)	409
Fig. 15.5	Gizzard shad with VHSv, a deadly infectious disease which causes bleeding	414

Fig. 16.1	These six men, the northern party of Captain Scott's last expedition, stand outside the entrance to the snow hole in which they have just spent the 1911–1912 Antarctic Winter in darkness	431
Fig. 16.2	The number of planned and executed expeditions on the Royal Geographical Society Expeditions Database, 1964–2018	433
Fig. 16.3	(A) Expedition footwear trampling experiment on tundra in Svalbard. (B) Crampons used by mountaineering expeditions can leave scratches on rock, which on popular routes can leave a permanent scar	435
Fig. 16.4	(A) Bringing in supplies by boat for the 2009 British Exploring Society expedition to Tasermiut Fjord, SW Greenland. (B) Supplies for the 2009 British Exploring Society expedition to Tasermiut Fjord, SW Greenland, were brought on this raft. (C) Loading supplies onto the Langoysund in Longyearbyen for the 2001 British Exploring Society expedition in Svalbard. (D) Small inflatable zodiacs with an outboard engine are popular for use on expeditions to moving equipment and people. (E) Cruise ships started to visit Longyearbyen, Svalbard, in the 1990s	436
Fig. 16.5	(A) Small expedition camp site in Zara Valley, Ladakh. (B) Large British Exploring Society Expedition base camp site in Gipsdalen, Svalbard. (C) British Exploring Society Expedition base camp site on a storm beach at Brucebyen on Isfjord, Svalbard, in 2001. (D) British Exploring Society Expedition campsite on dry river bed in Ladakh in 2013	437
Fig. 16.6	River in SW Greenland where a group of 12 young expeditioners assisted a professor in sampling the river for four weeks	441

List of Tables

Table 1.1	Core management strategies and actions for minimising or avoiding resource and social impacts in wilderness settings.	7
Table 2.1	Growth in popularity of recreational walking in the USA between 1982–1983 and 2005–2009 (numbers in millions)	20
Table 2.2	Estimated growth in hiking and primitive area visits in the USA to 2060 (figures in millions)	21
Table 2.3	Changes in plant physiognomy of two grasses with trampling	24
Table 2.4	Changes in plant physiognomy from Beer head (Devon)	25
Table 2.5	Changes in onset of flowering of two herbaceous plants with trampling	25
Table 2.6	Resilient species to trampling	25
Table 3.1	Outdoor participation by activity in the USA, 2006–2016 (The Outdoor Foundation, 2017, p. 8).	62
Table 3.2	The EXCEDO hiking difficulty scale based on the classification of the Swiss Alpine Club	64
Table 3.3	Example of adventure race event rules from Marmot Dark Mountains	65
Table 3.4	General reviews, recent Australian research, and activity-specific issues/impacts associated with activities that are often part of adventure races	66
Table 3.5	The National Three Peaks Challenge: Ben Nevis, Scafell Pike, and Snowdon—for or against?	68
Table 3.6	Three Peaks Challenge Code of Practice	68
Table 3.7	The Annual Yorkshire Three Peaks Challenge: environmental concerns	69
Table 4.1	Simulated abseils (all trees were of maturity class 4).	80
Table 4.2	Mean number of taxa per plot in three climbed and unclimbed areas	81
Table 4.3	Birds at risk from climbing disturbance in the UK	85
Table 4.4	Southern Sandstone Code of Practice	104

Table 5.1	Species growing in Lake District Gorges (based on over 200 plant species living in gorges)	116
Table 6.1	Trends in number of people ages 16 and older participating in recreation activities by historic period in the USA, 1982–2001 (Source: Cordell 2012, p. 33)	136
Table 6.2	Trends in number of people ages 16 and older participating in recreation activities in the USA, 1999–2001 and 2005–2009 for activities with between 25 and 49 million participants from 2005 through 2009 (Source: Cordell 2012, p. 37)	136
Table 6.3	Trends in number of people ages 16 and older participating in recreation activities in the USA, 1999–2001 and 2005–2009 for activities with fewer than 15 million participants from 2005 through 2009 (Source: Cordell 2012, p. 40)	137
Table 6.4	Mean and total annual days for activities adding more than 100 million participation days between 1999–2001 and 2005–2009 (Source: Cordell 2012, p. 42)	138
Table 6.5	ATV exposure and riding behaviours of pupils in Iowa schools from 2010 to 2013 ($n = 4320$)	140
Table 6.6	Outdoor recreation activities for 2008 by participants, participation rate, days, and days per participant	155
Table 6.7	Projected motorised off-road participation and use (off-road driving) by adult US residents, 2008–2060, by Resources Planning Act (RPA) scenario and related climate futures	156
Table 6.8	Projected motorised snow activity participation and use (snowmobiling) by adult US residents, 2008–2060, by Resources Planning Act (RPA) scenario and related climate futures	156
Table 6.9	Changes in total outdoor recreation participants between 2008 and 2060 across all activities and scenarios	157
Table 6.10	Changes in total outdoor recreation days between 2008 and 2060 across all activities and scenarios	158
Table 7.1	Outdoor participation by activity (ages 6+) in the USA, 2006–2016 (The Outdoor Foundation 2017, p. 8)	165
Table 7.2	Trends in number of people ages 16 and older participating in recreation activities in the USA, 1999–2001 and 2005–2009 for activities with between 25 and 49 million participants from 2005 through 2009 (Source: Cordell 2012, p. 37)	167
Table 7.3	Average monthly participation in sport and recreation in England, October 2012–2013	168

Table 7.4	Official IMBA "Mountain Bike Rules of the Trail" in which the IMBA considers that "every mountain biker should know and live by…"	172
Table 7.5	Mountain biking in England: venues listed on the Forestry Commission website	179
Table 8.1	A proposed camping spectrum	188
Table 8.2	Trends in number of people of ages 16 and older participating in recreation activities by historic period in the USA, 1982–2001 (Source: Cordell 2012, p. 33)	195
Table 8.3	Trends in number of people of ages 16 and older participating in recreation activities in the USA, 1999–2001 and 2005–2009 for activities with between 25 and 49 million participants from 2005 through 2009 (Source: Cordell 2012, p. 37)	196
Table 8.4	Outdoor participation by activity (ages 6+) in the USA, 2006–2016 (The Outdoor Foundation 2017, p. 8)	197
Table 8.5	Vegetation and soil conditions on 29 campsites and undisturbed control sites at Delaware Water Gap National Recreation Area, 1986	201
Table 8.6	Spotting of animals by villagers of the surrounding areas before and after the start of camping/rafting	204
Table 9.1	Numbers involved in horseback riding in the USA	219
Table 9.2	Suggested environmental impacts of horse riding on natural ecosystems	220
Table 9.3	Variables that can be used to monitor horse riding impacts	237
Table 9.4	Management strategies for horse riding in protected areas	241
Table 11.1	Outdoor participation by activity (ages 6+) in the USA, 2006–2016 (The Outdoor Foundation 2017, p. 8)	275
Table 11.2	Trends in number of people of ages 16 and older participating in recreation activities in the USA, 1999–2001 and 2005–2009 for activities with fewer than 15 million participants from 2005 through 2009 (Source: Cordell 2012, p.40)	277
Table 11.3	Percentage of participants and population, rations of percentages, and statistical test results for the activity group snow skiing or boarding	278
Table 11.4	Changes in total outdoor recreation participants between 2008 and 2060 across all activities and scenarios (Source: Bowker et al. 2012, p. 28)	279
Table 11.5	Changes in total outdoor recreation days between 2008 and 2060 across all activities and scenarios (Source: Bowker et al. 2012, p. 29)	280

Table 11.6	Once a week participation in funded sports (16 years and over)	281
Table 13.1	List of potential water-based activities which take place on the water surface	332
Table 13.2	Outdoor participation by activity (ages 6+) in the USA, 2006–2016 (The Outdoor Foundation 2017, p. 8)	340
Table 13.3	Trends in number of people of ages 16 and older participating in recreation activities in the USA, 1999–2001 and 2005–2009 for activities with fewer than 15 million participants from 2005 through 2009 (Source: Cordell 2012, p.40)	341
Table 13.4	Changes in total outdoor recreation participants between 2008 and 2060 across all activities and scenarios (Source: Bowker et al. 2012, p. 28)	342
Table 13.5	Changes in total outdoor recreation days between 2008 and 2060 across all activities and scenarios (Source: Bowker et al. 2012, p. 29)	343
Table 13.6	Once a week participation in funded sports (16 years and over)	344
Table 13.7	Coliform populations for various classes of campsites. University of Minnesota Boundary Waters Canoe Area Campsite Study, 1970	346
Table 13.8	Breeding densities (pairs/10 km channel) of three common species of English waterbirds in used and disused canals	350
Table 13.9	Summary of the major findings relating to recreational motorboat activities	354
Table 14.1	Estimates for scuba diving and snorkelling (participants 16+) in the USA	364
Table 14.2	Diver-induced damage compared	369
Table 15.1	Comparative catches of species shared by recreation and commercial fishers in various Australian studies	399
Table 16.1	Types of expeditions	430
Table 16.2	Checklists provided by British Ecological Society/Young Explorers' Trust	442
Table 17.1	Outdoor recreation activity numbers and total number of participation days in the USA; in millions of participants and days	448
Table 17.2	American participation in certain outdoor recreation activities, figures in % of population participating	449
Table 17.3	Future estimated demand for some outdoor activities between 2020 and 2040 as a percentage of 1987 demand	449
Table 17.4	Future participation in selected outdoor recreation activities up to 2060: numbers in millions and % increase based on three possible scenarios outlined in that report	450

Introduction to Outdoor Recreation and Recreation Ecology

Chapter Summary

Here we define outdoor recreation, recreation impacts, and recreation ecology which are at the core of this book. The management of wilderness recreation in the USA is discussed as an example of how a recreation management system for protected areas operates, in particular how the National Park Service (NPS) has been controlled by legislative acts, policies, and guidelines. Strategies for defining visitor carrying capacity and monitoring visitor impacts are discussed along with site and visitor management techniques and strategies for educating the recreationists. The NPS originally employed the Visitor Experience and Resource Protection and decision-making framework for evaluating the visitor carrying capacity limits and currently uses the Visitor Use Management planning framework and decision-making framework which are described in this chapter.

The overall aim of this book is to examine the environmental impacts of a range of outdoor activities, to understand how these can be minimised by various management approaches, and to see if education of the recreationists is one approach that can be successful in reducing the environmental impacts.

1.1 Outdoor Recreation

This refers to outdoor leisure activities which take place in natural or at least semi-natural locations in the countryside. When that recreation involves excitement, real or perceived risk, or physically challenging situations, it is referred to as adventure recreation and is carried out for physical or social, goal-related benefits by individuals or groups. The benefits are predominantly physical, such as physical health, but they can also be mentally, emotionally, and spiritually rewarding. At the same time, the physical activity often gives pleasurable appreciation of the environment where the participant finds peace in nature, relaxes, and enjoys life. The activities are often used as a means of educational and team-building goals.

1.2 Recreation Impacts and Recreation Ecology

In this book we define a recreational environmental impact as any undesirable, negative, visitor-related, biophysical change to natural resources, such as vegetation, wildlife, soils, and water, which can be an agent of change in the natural landscape. How acceptable such visitor impacts and changes are is debatable and based on value judgements, but the effective land management actions either to avoid or more likely to minimise

© The Author(s) 2019
D. Huddart, T. Stott, *Outdoor Recreation*, https://doi.org/10.1007/978-3-319-97758-4_1

the environmental impacts should depend on scientific knowledge related to soil science, geomorphology, ecology, and hydrology and the numbers and types of impact caused by the recreationists.

This branch of science is called recreation ecology, defined as the study of ecological changes associated with visitor activities, including the role of influential factors, both in natural and semi-natural areas (Liddle 1997; Monz et al. 2010; Hammitt et al. 2015). As a field of study, it seeks to develop a deeper understanding of the role and function of all the factors that influence visitor impacts and thus the probability of selecting sustainable management remedial actions. The best visitor impact management is seen to arise from collaborations between recreation ecologists, social scientists, experienced recreation land managers, recreation stakeholders, and the recreationists themselves. There are many outdoor recreation activities that contribute to these ecological changes, and we will consider here what we consider to be some of the most important. There are many gaps though, and some topics are considered in more depth than others. However, some of these gaps are covered in Liddle (1997) and Buckley (2004, 2006), such as the impacts on terrestrial wildlife and birds and the impacts and management of tourist engagements with cetaceans (Higham and Lusseau 2004). We consider the individual outdoor recreation activities and the environmental impacts that they can cause and attempt to outline the numbers involved in each activity, but this is an extremely difficult task. Then the management and educational options are considered in order to minimise the recreational environmental impacts for each activity. The difficulty in estimating the numbers involved in these activities and in different world regions can be illustrated, for example, by the work of Balmford et al. (2015) who tried to estimate the global number of visits per year to protected areas, which they considered to be about 8 B visits/year, of which over 80% were in Europe (3.8 B) and North America (3.3 B). However, given the confidence intervals for the global total (5.4–18.5 B/year), there could be wide discrepancies in the assumed figure, but it is considered implausible that the figure could be fewer than 5 B/year. This is because there were several conservative aspects to the calculations, for example, the exclusion of ~40,000 very small sites and the incomplete nature of the world database for protected areas. This seems to be supported by three national estimates: the 2.5 B visit days/year to US protected areas in 1996 (Eagles et al. 2000), the over 1 B visits/year (although many of these were cultural) to Chinese national parks in 2006 (Ma et al. 2009), and the 3.2–3.9 B visits/year to all British "ecosystems" in 2010 (Sen et al. 2012). As these figures are generally for several years ago, today's figures are likely to have increased markedly. Similar imprecise figures for individual activities are found worldwide, although there are relatively accurate figures for the USA, UK, Australia, and parts of Europe. We give estimates for each outdoor recreation activity in every chapter.

As an example of how the recreation management system for protected areas is operated in one part of the world, we will look in detail at the USA, because this is well documented by legislation and by a system that caters for large numbers and is relatively well organised throughout the country.

1.3 The Recreation Management System in the USA

The National Wilderness Preservation System (NWPS) and the four federal management agencies—the Bureau of Land Management, the Fish and Wildlife Service, the Forest Service, and the National Park Service (NPS)—manage 765 wilderness areas, across some 110 million acres of protected lands. The NPS alone accommodates over 330 million visitors per year, up from 275 million in 2008, which presents managers with major management challenges. These visitor figures have grown considerably since the 1964 Wilderness Act, and currently the wilderness area represents about 5% of the USA. An increasing number of visitors inevitably contribute negative effects to fragile natural, historical, and cultural resources. Such visitation-related resource impacts can degrade natural conditions and processes and the quality of recreation experiences. However, although greater physical numbers must mean more negative impacts, it is also visitor behaviour and the spatial management of these visitors that are keys to minimising impacts.

The term wilderness is defined as an area where the earth and community of life are not changed by man and where man himself is a visitor who does not remain and as an area of undeveloped federal land which retains its primeval character and influence without permanent improvements or human habitation, which is protected and managed so as to preserve its natural conditions. This type of land appears to have been affected mainly by the forces of nature, and the imprint of man's activities is mostly unnoticeable, and there are outstanding opportunities for solitude and an unconfined type of recreation. Mostly these wilderness areas are at least 5000 acres in size which makes their preservation and use practicable in an unimpaired condition; but these areas also contain ecological, geological, or other features of scientific, educational, scenic, or historical value. However, there are no specific criteria in law, although the Wilderness Act prohibits commercial activities, motorised access and roads, structures, and facilities, but there are exceptions in Section 4d of that act (Hoover 2014). According to the NPS Management Policies, the fundamental purpose of the national park system, established by the Organic Act and reaffirmed by the General Authorities Act, as amended, begins with a mandate to conserve park resources and values. The fundamental purpose of all parks also includes providing for the enjoyment of park resources and values by the people of the USA (NPS 2006, section 1.4.3). However, what might appear to be a dual mandate, visitation and resource protection, is clarified to reveal the primacy of resource protection. The Management Policies acknowledge that some resource degradation is an inevitable consequence of visitation but directs managers to ensure that any adverse impacts are the minimum necessary, unavoidable, cannot be further mitigated, and do not constitute impairment of, or to detract from, the original park resources and values (NPS 2006).

Four federal land management agencies are responsible for the stewardship of these wilderness, protected lands: the NPS (~44 million acres), the Forest Service (~36 million acres), the Fish and Wildlife Service (~21 million acres), and the Bureau of Land Management (~9 million acres) (Marion et al. 2016). These lands are largely in Alaska, California, Arizona, Idaho, and Washington (in total nearly 80% in these areas). The professional stewardship of these lands is to maintain their wilderness character which requires objective information about internal and external threats. Recreation is one of the main internal threats, despite being recognised as one of the core traditional uses of wilderness.

1.3.1 US National Park Service

We now describe how the NPS in the USA is organised as the increasing popularity of the national park system presents substantial management challenges. Too many visitors may cause unacceptable impacts to fragile natural and cultural resources and may also cause overcrowding and other social impacts, which can also degrade the quality of visitor experiences. Questions that are often posed are: how many visitors can ultimately be accommodated in a park or related area? And, how much resource and social impact should be allowed? These and related questions are commonly referred to as visitor capacity (Manning 1999; Stankey and Manning 1986; Shelby and Heberlein 1986; Graefe et al. 1984). Sustaining any type of long-term natural resource monitoring programme over time can be exceptionally difficult for management agencies due to changing personnel, management priorities, growth in numbers, and relatively low budgets. Initially the legislative and management intent regarding visitor impact monitoring and its role in balancing visitor use and resource protection objectives are described and reviewed.

Legislative mandates challenge managers to develop and implement management policies, strategies, and actions that permit recreation without compromising ecological and aesthetic integrity. Furthermore, managers are often forced to engage in this balancing act under the close scrutiny of the public, competing interest groups, and even the courts. Managers can no longer afford to adopt a wait-and-see attitude, or rely on subjective impressions of deterioration in resource conditions. Professional land management increasingly requires the collection and use of scientifically valid research and monitoring data. Such data should describe the nature and severity of visitor impacts and the relationships between controlling visitor use and biophysical factors. These relationships are complex and not always intuitive. A reliable information base is therefore essential to managers seeking to develop, implement, and gauge the success of visitor and resource management programmes.

Current legislation and agency documents establish mandates for monitoring (Marion 1991), but managers who make proactive decisions should be prepared to prove the viability of their strategies, or risk public disapproval, or even legal action against the agency. Survey and monitoring programmes provide the means for such demonstrations.

1.3.1.1 The Acts Establishing the National Park Service

The National Park Service Organic Act of 1916 established the Service, directing it to "promote and regulate the use [of parks] to conserve the scenery and the natural and historic objects and the wildlife therein and to provide for the enjoyment of the same in such manner and by such means as will leave them unimpaired for the enjoyment of future generations."

These provisions were supplemented and clarified by Congress through enactment of the General Authorities Act in 1970 and through a 1978 amendment expanding Redwood National Park (16 USC 1a-1): "the protection, management, and administration of these areas shall be conducted in light of the high public value and integrity of the National Park System and shall not be exercised in derogation of the values and purposes for which these various areas have been established."

Congress intended park visitation to be contingent upon the NPS's ability to preserve park environments in an unimpaired condition. However, unimpaired does not mean unaltered or unchanged because recreation activity, no matter how infrequent, will cause impacts for a period of time. What constitutes an impaired resource is ultimately a management decision and is a judgement. If interpreted too strictly, the legal mandate of unimpaired preservation may not be achievable, yet it provides a useful goal for managers in reconciling these two competing objectives.

1.3.1.2 The National Parks Omnibus Management Act of 1998

This act established a framework for fully integrating natural resource monitoring and other science activities into the management processes of the National Park System. The act charges the Secretary of the Interior to "develop a program of inventory and monitoring of National Park System resources to establish baseline information and to provide information on the long-term trends in the condition of National Park System resources."

Thus, relative to visitor use, park managers must evaluate the types and extents of resource impacts associated with recreation activities and determine to what extent they are unacceptable and constitute impairment. Further, managers must seek to avoid, or limit, any form of resource impact, including those judged to fall short of impairment. Visitor impact monitoring programmes can assist managers in making objective evaluations of impact acceptability and impairment and in selecting effective impact management practices by providing quantitative documentation of the types and extent of recreation-related impacts to natural resources.

1.3.1.3 The Management Policies and Guidelines

Authority to implement congressional legislation is delegated to agencies, which identify and interpret all relevant laws and formulate administrative policies to guide their implementation. A document titled Management Policies (NPS 2006) describes these policies to provide more specific direction to management decision-making. For example, relative to the need for balancing visitor use and resource impacts, the NPS Management Policies state that "The fundamental purpose of the national park system, established by the Organic Act and reaffirmed by the General Authorities Act, as amended, begins with a mandate to conserve park resources and values. This mandate is independent of the separate prohibition on impairment, and so applies all the time, with respect to all park resources and values, even when there is no risk that any park resources or values may be impaired. NPS managers must always seek ways to avoid, or to minimize to the greatest degree practicable, adverse impacts on park resources and values."

Congress, recognising that the enjoyment by future generations of the national parks can be ensured only if the superb quality of park resources and values is left unimpaired, has provided that when there is a conflict between conserving resources and values and providing for enjoyment of them, conservation is to be predominant. This is how courts have consistently interpreted the Organic Act, in decisions that variously describe it as making resource protection the primary goal, or resource protection the overarching concern.

The impairment that is prohibited by the Organic Act and the General Authorities Act is an impact that, in the professional judgement of the responsible NPS manager, would harm the integrity of park resources or values, including the opportunities that otherwise would be present for the enjoyment of those resources or values. Whether an impact meets this definition depends on the particular resources and values that would be affected; the severity, duration, and timing of the impact; the direct and indirect effects of the impact; and the cumulative effects of the impact in question and other impacts.

Impacts may affect park resources or values and still be within the limits of the discretionary authority conferred by the Organic Act. In these situations, the Service will ensure that the impacts are unavoidable and cannot be further mitigated. Even when they fall far short of impairment, unacceptable impacts can rapidly lead to impairment and must be avoided. When a use is mandated by law but causes unacceptable impacts on park resources or values, the Service will take appropriate management actions to avoid or mitigate the adverse effects.

Natural systems in the national park system, and the human influences upon them, will be monitored to detect change. The Service will use the results of monitoring and research to understand the detected change and to develop appropriate management actions.

Further, the Service will identify, acquire, and interpret the needed inventory, monitoring, and research, including applicable traditional knowledge, to obtain information and data that will help park managers accomplish park management objectives provided for in law and planning documents. It will define, assemble, and synthesise comprehensive baseline inventory data describing the natural resources under its stewardship and identify the processes that influence those resources, and it will use qualitative and quantitative techniques to monitor key aspects of resources and processes at regular intervals. It will analyse the resulting information to detect or predict changes, including interrelationships with visitor carrying capacities that may require management intervention, and to provide reference points for comparison with other environments and time frames. It will use the resulting information to maintain, and, where necessary, restore, the integrity of natural systems.

Although numerous reasons for implementing a visitor impact monitoring programme are described in the following sections, the actual value of these programmes is entirely dependent upon the park staff who manage them. Programmes developed with little regard to data quality assurance, or operated in isolation from resource protection decision-making, will be short-lived. In contrast, programmes that provide managers with relevant and reliable information necessary for developing and evaluating resource protection actions can be of significant value. Only through the development and implementation of professionally managed and scientifically defensible, monitoring programmes can we hope to provide legitimate answers to the often asked question, "Are we loving our parks to death?"

The NPS has implemented a strategy designed to institutionalise natural resource inventory and monitoring on a programmatic basis throughout the agency. A service-wide NPS Inventory and Monitoring Program has been implemented to ensure that park units with significant natural resources possess the resource information needed for effective, science-based, managerial decision-making and resource protection. A key component of this effort is the organisation of park units into 32 eco-regional networks to conduct long-term monitoring for key indicators of change, or vital signs. These vital signs are measurable, early warning signals that indicate changes that could impair the long-term health of natural systems. Early detection of potential problems allows park managers to take steps to restore ecological health of park resources before serious damage can happen.

1.3.1.4 Visitor Carrying Capacity Decision-Making

Decisions regarding impact acceptability and the selection of actions needed to prevent resource impairment frequently fall into the domain of visitor carrying capacity decision-making. The 1978 National Parks and Recreation Act requires the NPS to determine carrying capacities for each park as part of the process of developing a general management plan, which must include identification of, and implementation commitments for, visitor carrying capacities for all areas of the park and determine whether park visitation patterns are consistent with social and ecological carrying capacities.

The NPS originally employed the Visitor Experience and Resource Protection (VERP) planning and decision-making framework (National Park Service 1997) for formal evaluations of the acceptability of visitor impacts and for establishing carrying capacity limits on visitation. Visitor impact monitoring programmes provided an essential component of such efforts. VERP and other similar frameworks (e.g. Limits of Acceptable Change), evolved from, and largely replaced, management approaches based on the more traditional carrying capacity model (Stankey et al. 1985). Under these newer frameworks, numerical standards were set for individual biophysical or social condition indicators. These limits defined the critical boundary between acceptable and unacceptable changes in resource or social conditions and against which future conditions could be compared through periodic monitoring. VERP was an adaptive management process where periodic monitoring was conducted to compare actual conditions to quantitatively defined standards of quality. If standards were exceeded, an evaluation was conducted to identify those factors that managers could effectively manipulate to improve conditions for the indicators with unacceptable conditions. For example, if a standard for the individual or aggregate size of recreation sites was exceeded, managers might consider implementing one or more site management or educational actions. If the next cycle of monitoring also found sub-standard conditions, more restrictive actions like fencing or area closures could be considered. These frameworks were incorporated into many protected area planning documents, although staffing and funding levels frequently challenged and prevented management from sustaining their effective use (Farrell and Marion 2002; Manning 2007). In response Farrell and Marion (2002) proposed the Protected Areas Visitor Impact Management (PAVIM) framework which was thought to increase efficiency through greater reliance on expert panels of managers, scientists, and knowledgeable stakeholders. Visitor carrying capacity is the type and level of visitor use that can be accommodated while sustaining the desired resource and visitor experience conditions in the park. By identifying and staying within carrying capacities, park uses that may unacceptably impact the resources and values for which the parks were established can be prevented. Additional guidance on visitor carrying capacity decision-making is provided in the NPS Management Policies (2006): more recently the six US federal agencies (Bureau of Land Management, the Forest Service, the US Army Corps of Engineers, the Fish and Wildlife Service, the NPS, and the National Oceanic and Atmospheric Administration) formed an Interagency Visitor Use Management Council (IVUMC) to "increase awareness of commitment to proactive, professional and science-based visitor use management on federally-managed lands and waters." They defined Visitor Use Management (VUM) as the "proactive and adaptive process for managing characteristics of visitor use and managerial setting using a variety of strategies and tools to achieve and maintain desired resource conditions and visitor experience." Marion (2016) describes this VUM planning framework and decision-making process and reviews five key visitor impact management strategies in further detail, but the overall aim is to provide consistent guidance for federal land management agencies.

The VUM framework which is currently in use includes 4 core elements (Build the Foundation, Define Visitor Use Management Direction, Identify Management Strategies, and Implement, Monitor, Evaluate, and Adjust) and 14 steps. The steps that fall under Build the Foundation are as follows: (1) clarify project purpose and need; (2) review the area's purpose, along with related legislative, policy, and management direction; (3) assess existing information and current conditions and identify project issues and opportunities; and (4) develop the project strategy. The outcome of these four steps will be to understand what needs to be done and how to organise the planning effort. The steps that fall under Define Visitor Use Management Direction are as follows: (5) define desired conditions for resources, recreation opportunities, and visitor experiences for the project area(s); (6) define appropriate visitor uses, facilities, and services based on desired conditions; and (7) select indicators and establish thresholds. The outcome of these three steps will be to describe the conditions to be achieved or maintained and how conditions will be tracked over time. The steps that fall under Identify Management Strategies are as follows: (8) compare and document the differences between existing and desired conditions, and, for visitor use-related impacts, clarify the specific links to visitor use characteristics; (9) identify VUM strategies and actions to achieve desired conditions; (10) where necessary, identify visitor capacities and strategies to manage use levels within capacities; and (11) develop a monitoring strategy. The outcome of these steps will be a decision on strategies to manage visitor use to achieve or maintain desired conditions. Finally, the steps that fall under Implement, Monitor, Evaluate, and Adjust are as follows: (12) implement management actions; (13) conduct and document ongoing monitoring, and evaluate the effectiveness of management actions in achieving desired conditions; and (14) adjust management actions if needed to achieve desired conditions and document the rationale. The outcome of these steps will be to implement management actions and adjust based on lessons learned. There are five core management strategies and actions for minimising or avoiding resource and social impacts in wilderness settings which are outlined in Table 1.1 which is adapted from Cole et al. (1987) and Marion (2003, 2016). These will be discussed

Table 1.1 Core management strategies and actions for minimising or avoiding resource and social impacts in wilderness settings

Core strategies	Management actions
1. Manage use levels	Redistribute, discourage, limit use; redistribute or reduce use during times of peak use, in high use locations, or when impact potential is high
2. Modify the location of use	Concentrate use on sustainable, expansion-resistant trails and campsites to limit the total area of impact; disperse use on durable substrates at levels that prevent the formation of trails and campsites; encourage or require visitors to camp out of sight or a minimum distance from trails and campsites; restrict certain types of use to specific locations (e.g. horses on horse trails only)
3. Increase resource resistance	Construct, reconstruct, or maintain impact-resistant trails and campsites
4. Modify visitor behaviour	Persuasive communication, interpretation, and/or education; encourage Leave No Trace practices when on a trail or camping regulation and enforcement: prohibit or require certain practices and equipment when on a trail or camping
5. Close and rehabilitate the resource	Close and rehabilitate unnecessary or less sustainable trail sections and campsites

briefly later in this chapter, and examples will be discussed in the later recreation activity chapters.

1.3.1.5 Visitor Perceptions of Resource Conditions

Visitors to protected areas are aware of resource conditions along trails and at recreation sites, just like the managers (Lucas 1979; Marion and Lime 1986; Vaske et al. 1982). Legislative mandates set high standards when they direct managers to keep protected natural areas unimpaired and human impacts substantially unnoticeable. Seeing trails and recreation sites, particularly those in degraded condition, reminds visitors that others have gone before them. In remote areas even the presence of trails and recreation sites reduces perceived naturalness and can diminish opportunities for solitude. In accessible and popular areas, the proliferation and deterioration of trails and recreation sites present a used appearance, in contrast to the ideal of a pristine natural environment (Leung and Marion 2002).

Degraded resource conditions on trails and recreation sites can have significant utilitarian, safety, and experiential consequences for visitors (Leung and Marion 2002). Trails serve a vital transportation function in protected natural areas, and their degradation greatly diminishes their utility for visitors and land managers. For example, excessive tread erosion or muddiness can render trails difficult and unpleasant to use. Such conditions can also threaten visitor or packstock safety and prevent or slow rescues, possibly increasing agency liability. Impacts associated with certain types of uses, such as linear rutting from bikes or vehicles, or muddy hoof prints from horses, can also exacerbate conflicts between recreationists.

Visitors spend most of their time within protected natural areas on trails and recreation sites, so their perceptions of the area and its naturalness are strongly influenced by trail and site conditions. Visitors are sensitive to overt effects of other visitors, such as the presence of litter, horse manure, and visually obtrusive examples of impacts, such as tree root exposure, tree felling, and soil erosion.

1.3.1.6 Monitoring Programme Capabilities

Visitor impact monitoring programmes can be of significant value when providing managers with reliable information necessary for establishing and evaluating resource protection policies, strategies, and actions. When implemented properly and with periodic reassessments, these programmes produce a database with significant benefits to protected area managers. Data from the first application of impact assessment methods developed for a long-term monitoring programme can objectively document the types and extent of recreation-related resource impacts. Such work also provides information needed to select appropriate biophysical indicators and formulate realistic standards, as required in VERP, LAC, or VUM planning and decision-making frameworks.

Reapplication of impact assessment protocols as part of a monitoring programme provides an essential mechanism for periodically evaluating resource conditions in relation to standards. Visitor impact monitoring programmes provide an objective record of impacts, even though individual managers might change jobs and roles or move to a new area. A monitoring programme can identify and evaluate trends when data are compared between present and past resource assessments. It may detect deteriorating conditions before severe, or irreversible, changes occur, allowing time to implement corrective actions. Analysis of monitoring data can reveal insights into relationships with causal or noncausal yet influential factors. For example, the trampling and loss of vegetation or soils may be greatly reduced by shifting trails to more resistant and resilient vegetation types or topographic alignments, instead of more contentious limitations on use. Following the implementation of corrective actions, monitoring programmes can evaluate their efficacy.

1.3.2 Management Considerations

1.3.2.1 Evaluating Impact Acceptability

An important first step in management decision-making is evaluating and determining the acceptability of recreation impacts. Such judgements can be made by examining the results of scientific studies, agency planning, and management guidance documents and with appropriate public input.

Firstly, managers might consider the management zone and associated objectives where visitor impacts are occurring. Impacts occurring in pristine areas where preservation values are paramount are less acceptable than when located in areas that are intensively developed and managed for heavy recreation use. Secondly, managers might consider environmental and cultural factors. Visitor impacts occurring within rare, sensitive, or fragile communities of flora or fauna, or cultural resources, are less acceptable than when located in areas that lack such characteristics. Within such sensitive areas, managers may also consider the specific locations where recreation activity is concentrated, their proximity to rare plants or cultural resources, and actual threat potentials. Finally, managers should consider use-related factors. Impacts that can be easily avoided are less acceptable, such as when three informal trails in close proximity to each other access a recreation site that could be accessed by a single trail. Similarly, are three viewpoints present when one would suffice? Is visitor behaviour a factor? Could minimum *Leave No Trace* practices be communicated and adopted by visitors to reduce their per capita impacts? Some impacts are desirable to visitors and facilitate visitor use. For example, the lack of vegetation on trails and recreation sites attracts and spatially concentrates visitor use and trampling and facilitates their use by visitors. A trail or climb that lacks vegetation is simply easier to use. A careful consideration of these and other relevant factors (e.g. visitor safety) can assist managers in making what are inherently value-laden decisions regarding the acceptability of visitor impacts. The acceptability of visitor impacts, in turn, guides decisions about the need for and selection of appropriate and effective management interventions.

For visitor impacts found to be within acceptable limits, managers may continue existing management actions and monitoring. As previously noted, visitation to parks is an important mandate, and some degree of degradation is an inevitable consequence. Formal trails and recreation sites are never sufficient by themselves for sustaining all forms of park visitation. Dispersed off-trail traffic that can lead to the development of informal trails and recreation sites is necessary for accessing and using less visited locations, like rock climbing or fishing sites. Some degree of visitor impact associated with dispersed use activities is inevitable. The challenge is to avoid those visitor-associated impacts that can be avoided and minimise those that cannot. Hence defining, monitoring, and managing resources and visitors to avoid exceeding acceptable limits of change is a key challenge for park planning and management activities.

Managers should also consider the costs to visitors in reduced opportunities, or experiential quality, associated with alternative management actions under consideration. Frequent dialogue with recreation representatives can aid in the selection of the most effective practices that have the least cost to visitors. Fortunately, managers have some powerful site and visitor impact management strategies and actions available for avoiding or minimising such impacts. The following sections review some site and visitor management options.

1.3.2.2 Site Management of Recreation Sites

Recreation site management actions fall into three general categories: close and restore sites, redesign sites, and install facilities.

Close and Restore Recreation Sites

Recreation sites that represent avoidable impacts—resulting from illegal uses, poor design, or threatened rare or sensitive resources—should be considered for closure and restoration. Consideration needs to be given to how closures will affect recreation opportunities, including what alternative sites are available, if visitors will find them acceptable, and how information about closures and alternatives will be communicated. Will visitors migrate from a closed area to other areas? This is because closures are only effective when there is a shift in visitation from high-value/sensitive areas to lower-value/resistant areas.

Working directly with the affected recreation groups to evaluate problems and alternatives, and implement appropriate solutions, is always preferable and provides the most effective outcome. Such groups may also provide volunteers to assist with site closure and restoration work as we will see in some of the later chapters.

Communication Strategies

There needs to be an evaluation of existing communication, including printed literature, signs, and personal communication. If there are rare or sensitive resources existing in an area and they are being adversely affected by recreation activities, this needs to be communicated to the visitors. Visitors must be asked to remain on the formal trails and sites whenever possible to protect sensitive resources from damage. The objective is to reduce casual or unnecessary off-trail traffic, recognising that off-trail activity may be essential for visitors engaged in activities, such as nature study, climbing, and fishing.

Visitors must easily be able to distinguish between formal trails/viewpoints and informal trails/viewpoints, and clear marking with paint blazes or signs can help visitors make distinctions. Maps should be included in printed materials and on trailhead signs showing the presence and locations of formal recreation sites (e.g. designated viewpoints) along trails. Otherwise, visitors may venture off-trail in search of viewpoints before they encounter the formal sites.

Actions to Close Sites

A variety of options for closing sites can be considered. Vegetative restoration actions only should be considered after sites have been effectively closed to visitor traffic, and the need for relocating formal trails well-away from sites that may otherwise be difficult to close should be considered.

When all recreation site access trails are to be closed, actions found to be effective include the placement of logs across informal trails at their junctions with formal trails, use of symbolic "No-hiking" prompter signs, and dispersal of organic leaf litter, rocks, and light brushing to naturalise and hide informal trails (Hockett et al. 2010; Wimpey and Marion 2011).

If needed, a post to sign the direction and distance to the closest formal viewpoint site should be installed, and low symbolic fencing, or high barrier fencing, should be considered to block site access, or large felled trees or rocks should be moved onto the site to deter use. Heavy brushing with materials that a single person cannot easily remove can be effective but may also shift traffic around them. A Blue Ridge Parkway (North Carolina and Virginia) study evaluating the effectiveness of heavy brushing work to close informal trails in rare plant communities reported that visitors dismantled 12 of 14 brushings within two months (Johnson et al. 1987). The two successful brushings also failed to stop hikers, instead diverting them into rare plant habitats and creating new trails. The investigators stressed that managers need to focus on addressing the causes for the off-trail traffic, that is, why does a visitor desire to access a particular location? In this study contributing factors included lack of adequate signing to direct visitors to official trails and sites, confusion about formal trail locations and destinations, and a desire to explore or pick blueberries.

A difficult task is to try and restore soils and vegetation to pre-use conditions through planting of native trees, shrubs, and herbs/grasses, although we will see such techniques in a later chapter on footpath restoration.

Sometimes physical barriers are necessary to prevent visitor access. Low borders of rocks or logs can sometimes communicate management intent to deter access and are less visually obtrusive than high barriers. Higher barriers include scree walls of native rock and low or high fencing that physically block access and provide indisputable evidence of management intent. A study at Acadia NP (Maine) found low symbolic post and rope fencing to be substantially more effective than signs alone and deterred nearly all off-trail traffic (Park et al. 2008). Fencing can include rope or chain strung through wooden or steel posts and various types of manufactured fencing. Tall fencing is a highly effective solution, but even these must have clear signage to prevent passage over or around the fence, and fences should be terminated at locations that prevent informal trails from developing around the ends.

High fencing can also present problems to visitors who must venture off-trail to pursue their recreation activities. Finally, barriers of all types can be temporary, altering visitor distribution patterns until vegetative recovery occurs, or permanent. However, it is critical that management actions effectively address the original cause for a recreation site's creation; otherwise, it may simply reappear. Deterring continued access and use of recreation sites will initiate natural recovery of recreation sites, but even limited or low levels of traffic can prevent or retard recovery (Cole and Spildie 2007; Cole 1992; Leung and Marion 2002). Unassisted recovery rates are extremely slow, particularly when soils are thin and dry, and a recreation site created in a single year can require many years to recover (Cole and Spildie 2007; Therrell et al. 2006). Recovery of native vegetation may not even be possible if native soils are not first restored or if non-native species become established on the site. Ecological restoration efforts should begin with evaluations of substrates on the site and comparisons to adjacent off-site substrates. Factors that affect restoration success include season of year for plantings; use of seed, transplants, or greenhouse stock; and soil type and preparation, fertilisation, and watering (Cole and Spildie 2007, Therrell et al. 2006).

Redesign Recreation Sites

For those sites remaining open to visitor use, several effective site management actions can aid managers in restricting their size and reducing resource impacts. First, consider site access, selecting a single well-designed trail, preferably with durable, rocky substrates and low gradients. If needed, redesign and reconstruct the access trail to make it sustainable, adding durable substrates and rock borders or even fencing if needed to restrict off-trail hiking. Second, evaluate the recreation site's uses, size, and expansion potential, proximity to sensitive resources, substrates, and safety: where needed, restrict site size and expansion potential with border stones, logs or felled trees, scree walls, or fencing. If at all possible, do not use unnatural materials. When possible, retain site portions that have the best viewpoints.

1.3.2.3 Visitor Management

There are several visitor management options to avoid or minimise impacts by altering behaviour through educational messages or regulations. For example, educational messages on trailhead displays can inform all visitors of the presence of rare plants and the need to stay on formal trails and recreation sites to prevent their trampling (Cole et al. 1997; Marion and Reid 2007). Regulations could limit the type or amount of visitation or even prohibit off-trail hiking in certain areas. Unfortunately, studies reveal that education or regulations would need to eliminate nearly all trampling from environments with low trampling resistance, or resilience (ability to recover), to achieve any substantial recovery on well-established recreation sites or trails (Cole 1987; Leung and Marion 2002). This is quite challenging given that education-induced behaviour change is voluntary and the effectiveness of regulatory approaches depends on the frequency of patrolling by agency law enforcement officers (Marion and Reid 2007). Integrating visitor management and site management actions provides the best opportunities for achieving success.

1.3.2.4 Construct and Manage Facilities

Visitor impacts can also be avoided or minimised through the construction of facilities to attract and concentrate use to a specific intended location. A new, well-designed and constructed formal trail can replace several poorly located and rapidly degrading, informal trails. Replacing numerous informal viewpoint sites with a few carefully selected and developed formal sites can also greatly reduce the total area of trampling disturbance within sensitive resource areas. Within these features, managers can also construct facilities that increase site durability, attract and concentrate use, and discourage or prevent site expansion and off-site activity.

1.3.2.5 Educational Practices

Visitor education efforts should be to increase the spatial containment of visitor trampling on a limited number of small, well-defined, and resistant locations (Cole 1992; Leung and Marion 1999; Marion and Farrell 2002). For example, research has shown that effectively worded and communicated *Leave No Trace* messages can reduce off-trail hiking rates (Marion and Reid 2007). However, compliance can be further enhanced by improving trailblazing and constructing trail borders or fencing. In high-use areas, educational messages can ask visitors to concentrate traffic on formal trails and recreation sites. In low-use areas, messages can ask visitors to restrict traffic to trampling-resistant natural surfaces (e.g. rock) or on the bare substrates of well-established form.

Educational messages for climbers and hikers where there are extensive climbing and walking routes can be communicated by signs placed at a trailhead or cliff-site location, hiking or climbing guides and pamphlets, agency or climbing organisation websites, retail stores that sell climbing gear, climbing walls and guiding/instructional services, climbing clubs and organisations, or personal communication through park staff or volunteer trail or climbing stewards. Educational signs placed in backcountry, wilderness settings are generally limited to those deemed critical to resource protection efforts and are generally for short-term use, giving restoration and vegetative recovery a chance to take hold.

An effective but inexpensive educational method is to install small prompter signs, such as a symbolic sign showing a Vibram boot print, with a red slash symbol superimposed. These can be screwed onto a log pulled across an informal trail or site or mounted on a short well-anchored stake. For trails or sites that are difficult to close, an effective technique is to cover initial portions with peat moss and/or organic litter and jute netting and install a restoration sign. Visitors who see an earnest attempt to restore a trail or site damaged by foot traffic will usually be less willing to ruin that effort by walking on it. An effectively worded educational sign should clearly define the appropriate behaviour, educate visitors about how their personal actions contribute to resource impacts, and provide a compelling rationale (Vande Kamp et al. 1994; Winter 2006). For example, a number of studies have demonstrated the efficacy of a sign with message wording like this: Please Do Not Leave Designated Trails to Preserve Sensitive Vegetation (Cialdini 1996; Cialdini et al. 2006; Johnson and Swearingen 1992). However, even the most effectively worded signs are ineffective if visitors cannot easily distinguish between official and visitor-created trails or recreation sites.

Some sensitive resources are widespread and difficult to avoid during site selection. If present, there should be a consideration of site designs and actions to protect these resources. For example, the higher level of protection offered by fencing may be justified to protect sensitive resources adjacent to some sites or rocks placed to move traffic patterns around a small patch of plants. Improving site substrates by extracting rocks or stumps to provide smoother walking surfaces attracts visitors to a site and keeps them there. Similarly, adding large rocks or logs to the trail margins can help, although the purist will object to the unnatural form.

1.3.2.6 Leave No Trace Programmes and Information

The *Leave No Trace* programme develops and promotes low-impact outdoor practices and ethics designed to make all outdoor activities more sustainable. This US national non-profit programme has staff and numerous collaborators who develop the best available low-impact practices for a diverse array of environments and recreation activities, including materials and messaging that specifically target climbing, hiking, and fishing (see www.LNT.org and Marion 2014). Management staff should assemble and actively promote these practices, which have been formally adopted by most protected area land managers, including numerous private organisations. For example, the Leave No Trace Center for Outdoor Ethics has developed a comprehensive 25-page Skills and Ethics booklet on rock climbing that could be sold in visitor centres and used as a resource for educational efforts. The ethical component of the *Leave No Trace* programme encourages visitors to become more actively involved in planning, management, and stewardship efforts for the areas they visit. Climbers and hikers at both Great Falls Park

(Virginia) and Carderock Recreation Area (Maryland) have actively participated in past stewardship projects held at both parks and have been actively involved in park planning efforts and in some of the cliff and trail ecological studies. However, including the best available *Leave No Trace* practices in climbing guidebooks for an area is perhaps the best way to inform climbers.

Other effective communication methods for climbers for *Leave No Trace* practices include the development and free distribution of a climber's pamphlet, a *Leave No Trace* bulletin board panel, and a separate climber's page on the park's website. Climbing-related information would include (1) basic park-wide climbing guidance, (2) information about cliff-associated rare plant communities and species with visitor impact management concerns, (3) low-impact climbing practices, and (4) where necessary, guidance for specific cliffs, climbing areas, or climbs within the park (e.g. how to locate and use a reduced network on climbing area access trails).

Finally, it is thought that there is good reason to believe that rock climbers will be receptive to such information and will be compliant with educational messaging. A survey of rock climbers at Shenandoah NP (Virginia) revealed substantial support and receptivity for the provision of information on low-impact climbing practices, required use of designated access trails, closing cliffs during critical wildlife seasons, and closing climbing routes with sensitive rare plants (Lawson et al. 2006). Interestingly, findings also revealed that climbers were relatively unaware of visitation-related impacts to cliff-associated, rare flora and fauna. Such educational information needs to be communicated more effectively to all cliff visitors, including hikers. All visitors need to become better informed about the resource impacts associated with their recreational activities, for if they are, they can modify their behaviour to avoid or minimise these impacts. Doing so will also protect their continued access to the areas they cherish and avoid future restrictions and regulations that may be imposed on visitors.

1.3.3 Other Environmental Impacts That Are Not Recreation-Induced

Apart from the human recreational environmental impacts that are the focus of this book, there are many geological, geomorphological, and climatic impacts that can cause major changes to the environment. These can range from rockfall events like in Yosemite National Park (USA) which have far bigger effects on the landscape than rock climbing in that area to tsunami due to the effects of earthquakes, caused by fault movements along subduction zones which can cause coastal flooding and erosion of coral reefs. Other climatically controlled environmental change can be caused by hurricanes and typhoons, but the biggest effect is being caused by climate change caused by human impact on the atmosphere since the industrial revolution which is having, for example, a major impact on the world's coral reefs. Other human-induced environmental changes can be caused by deforestation, agricultural practices which can cause soil erosion, acidification of the world's lakes and oceans, and the use of plastics and chemicals in various forms which can cause pollution and problems for wildlife. Many of these important current and future environmental impacts are caused by human-induced climate change which needs to be controlled now. The point worth emphasising is that although recreational environmental impacts are relatively minor compared to the major earth processes, they do have an important and wide-ranging role to play in environmental change and a recreationist's perception of the landscape. This will be discussed in the subsequent chapters where the impacts caused by what are considered to be the most important recreation activities are outlined. There are sections in each chapter discussing the numbers involved in each activity, there is a description of the ways in which management of the activities can take place, and there is a discussion of the role of education of both land managers and recreationists in the process of minimising change to the environment.

References

Balmford, A., Green, J. M. H., Anderson, M., Beresford, J., Huang, C., Naido, R., et al. (2015). Walk on the wild side: Estimating the global magnitude of visits to protected areas. *PLoS Biology, 13*(2), e1002074. https://doi.org/10.1371/journal.pbio.1002074.

Buckley, R. (Ed.). (2004). *Environmental Impacts of Ecotourism*. Wallingford: CABI, 389pp.

Buckley, R. (2006). *Adventure Tourism*. Wallingford: CABI, 515pp.

Cialdini, R. B. (1996). Activating and aligning two kinds of norms in persuasive communications. *Journal of Interpretation Research, 1*, 3–10.

Cialdini, R. B., Demaine, L. J., Sagarin, B. J., Barrett, D. W., Rhoads, K., & Winter, P. L. (2006). Managing social norms for persuasive impact. *Social Influence, 1*, 3–15.

Cole, D. N. (1987). Effects of three seasons of experimental trampling on five montane forest communities and a grassland in western Montana. *Biological Conservation, 40*, 219–244.

Cole, D. N. (1992). Modeling wilderness campsites: Factors that influence amount of impact. *Environmental Management, 16*, 255–264.

Cole, D. N., Petersen, M. E., & Lucas, R. C. (1987). *Managing Wilderness Recreation Use: Common Problems and Potential Solutions*. USDA Forest Service General Technical Report INT-230, Intermountain Research Station, Ogden, UT, 60pp.

Cole, D. N., & Spildie, D. R. (2007). *Vegetation and Soil Restoration on Highly Impacted Campsites in the Eagle Cap Wilderness, Oregon*. USDA Forest Service, General Technical Report RMRS-GTR-185, Rocky Mountain Research Station, Fort Collins, CO, 26pp.

Cole, D. N., Watson, A. E., Hall, T. E., et al. (1997). *High-Use Destinations in Wilderness: Social and Biophysical Impacts, Visitor Responses, and Management Options*. USDA Forest Service Research Paper INT-RP-496, Intermountain Research Station, Ogden, UT, 30pp.

Eagles, P. F. J., McLean, D., & Stabler, M. J. (2000). Estimating the tourism volume and value in protected areas in Canada and the USA. *George Wright Forum, 17*, 62–76.

Farrell, T. A., & Marion, J. L. (2002). The Protected Areas Visitor Impact Management (PAVIM) framework: A simplified process for making management decisions. *Journal of Sustainable Tourism, 10*, 31–51.

Graefe, A. R., Vaske, J. J., & Kuss, F. R. (1984). Social carrying capacity: An integration and synthesis of twenty years of research. *Leisure Sciences, 6*, 395–431.

Hammitt, W. E., Cole, D. N., & Monz, C. A. (2015). *Wildland Recreation: Ecology and Management* (3rd ed.p. 313). Chichester: Wiley Blackwell.

Higham, J., & Lusseau, D. (2004). Ecological impacts and management of tourist engagements with Cetaceans. In R. Buckley (Ed.), *Environmental Impacts of Ecotourism* (pp. 171–186). Wallingford, Oxfordshire and Cambridge, MA: CABI Publishing.

Hockett, K., Clark, A., Leung, Y.-F., Marion, J. L., & Park, L. (2010). *Deterring Off-trail Hiking in Protected Natural Areas: Evaluating Options with Surveys and Unobtrusive Observation*. Final Research Report, Virginia Technical College of Natural Resources and Environment, Blacksburg, VA, 191pp.

Hoover, K. (2014). *Wilderness: Overview and Statistics*. Congressional Research Service Report 7-5700, RL 31447, 17pp. Retrieved from www.crs.gov.

Johnson, B. R., Bratton, S. P., & Firth, I. (1987). *The Feasibility of Using Brushing to Deter Visitor Use of Unofficial Trails at Craggy Gardens, Blue Ridge Parkway, North Carolina (CPSU Report No. 43)*. National Park Service Cooperative Studies Unit, Institute of Ecology, University of Georgia, Athens, GA.

Johnson, D. R., & Swearingen, T. C. (1992). The effectiveness of selected trailside sign texts in deterring off-trail hiking at Paradise Meadows, Mt. Rainier National Park. In H. H. Christensen, D. R. Johnson, & M. H. Brookes (Technical Coordinators), *Vandalism: Research, Prevention and Social Policy* (pp. 103–120). Portland, OR: USDA Forest Service, Pacific Northwest Research Station.

Lawson, S., Wood, K., Hockett, K., Bullock, S., Kiser, B., & Moldovanyi, A. (2006). *Social Science Research on Recreational Use and Users of Shenandoah National Park's Rock Outcrops and Cliffs*. Virginia Tech College of Natural Resources and Environment, Department of Forest Resources and Environmental Conservation, Blacksburg, VA. 161pp.

Leung, Y.-F., & Marion, J. L. (1999). Spatial strategies for managing visitor impacts in national parks. *Journal of Park Recreation and Administration, 17*, 20–38.

Leung, Y.-F., & Marion, J. L. (2002). Recreation impacts and management in wilderness: A state-of-knowledge review. In *Wilderness Science in a Time of Change Conference 1999* (Vol. 5, pp. 23–48). USDA Forest Service RMRS-P-15-VOL-5, Rocky Mountain Research Station, Fort Collins, CO.

Liddle, M. J. (1997). *Recreation Ecology: The Ecological Impact of Outdoor Recreation and Ecotourism*. London: Chapman and Hall.

Lucas, R. C. (1979). Perceptions of non-motorized recreational impacts: A review of research findings. In R. Ittner, D. R. Potter, J. K. Agee, & S. Anschell (Eds.), *Recreational Impacts on Wildlands: Conference Proceedings* (pp. 24–31). USDA Forest Service R-6-001-1979. Seattle, WA: Pacific Northwest Forest and Range Experiment Station and USDI National Park Service.

Ma, X.-L., Ryan, C., & Bao, J.-G. (2009). Chinese national parks: Differences, resource use and tourism product portfolios. *Tourism Management, 30*, 21–30.

Manning, R. E. (1999). *Studies in Outdoor Recreation: Search and Research for Satisfaction* (2nd ed.). Corvallis, OR: Oregon State University Press.

Manning, R. E. (2007). *Parks and Carrying Capacity: Commons without Tragedy* (2nd ed.). Washington, DC: Island Press, 328pp.

References

Marion, J. L. (1991). *Developing a Natural Resource Inventory and Monitoring Program for Visitor Impacts on Recreation Sites: A Procedural Manual*. USDI National Park Service, National Research Report NPS/NRVT/NRR-91/06, Denver, CO, 59pp.

Marion, J. L. (2003). *Camping Impact Management on the Appalachian National Scenic Trail*. Appalachian Trail Conference, Harper's Ferry, WV, 109pp.

Marion, J. L. (2014). *Leave No Trace in the Outdoors*. Mechanicsburg, PA: Stackpole Books, 128pp.

Marion, J. L. (2016). A review and synthesis of recreation ecology research supporting carrying capacity and visitor use management decisionmaking. *Journal of Forestry, 114*, 339–351.

Marion, J. L., & Farrell, T. (2002). Management practices that concentrate visitor activities: Camping impact management at Isle Royale National Park, USA. *Journal of Environmental Management, 66*, 201–212.

Marion, J. L., Leung, Y.-F., Eagleston, H., & Burroughs, K. (2016). A review and synthesis of recreation ecology research findings on visitor impacts to wilderness and protected natural areas. *Journal of Forestry, 114*, 352–362.

Marion, J. L., & Lime, D. W. (1986). Recreational resource impacts: Visitor perceptions and management responses. In D. L. Kulhavy & R. N. Conner (Eds.), *Wilderness and Natural Areas in the Eastern United States: A Management Challenge* (pp. 229–235). Nacogdoches, TX: Stephen F. Austin State University, School of Forestry.

Marion, J. L., & Reid, S. E. (2007). Minimising visitor impacts to protected areas: The efficacy of low impact education programmes. *Journal of Sustainable Tourism, 15*, 5–27.

Monz, C. A., Cole, D. N., Leung, Y.-F., & Marion, J. L. (2010). Sustaining visitor use in protected areas: Future opportunities in recreation ecology research based in the USA experience. *Environmental Management, 45*, 551–562.

National Park Service. (1997). *A Summary of the Visitor Experience and Resource Protection (VERP) Framework*. USDI National Park Service, Denver Service Control Publication NPS D-1214, Denver, CO, 35pp.

National Park Service. (2006). *Management Policies*. Washington, DC: USDI National Park Service.

Park, L. O., Marion, J. L., Manning, R. E., Lawson, S. R., & Jacobi, C. (2008). Managing visitor impacts in parks: A multi-method study of the effectiveness of alternative management practices. *Journal of Parks and Recreation Administration, 26*, 97–121.

Sen, A., Harwood, A. R., Bateman, I. J, Munday, P., Crowe, A., Brander, L., et al. (2012). *Economic Assessments of the Recreation Value of Ecosystems in Great Britain*. CSERG Working Paper of the University of East Anglia, Norwich, 33pp.

Shelby, B., & Heberlein, T. A. (1986). *Carrying Capacity in Recreation Settings*. Corvallis, OR: Oregon State University Press.

Stankey, G. H., Cole, R. C., Lucas, R. C., Petersen, M. E., & Frissell, S. S. (1985). *The Limit of Acceptable Change (LAC) System for Wilderness Planning*. USDA Forest Service, General Technical Report INT-176, Intermountain Research Station, Ogden, UT, 37pp.

Stankey, G. H., & Manning, R. E. (1986). *Carrying Capacity of Recreation Settings. The President's Commission on Americans Outdoors: A Literature Review* (pp. 47–57). Washington, DC: U.S. Government Printing Office.

Therrell, L., Cole, D., Claassen, V., Ryan, C., & Davies, M. A. (2006). *Wilderness and Backcountry Site Restoration Guide*. Technical Report 0623-2815-MTDC. USDA Forest Service, Missoula Technology and Development Center, Missoula, MT, 394pp.

Vande Kamp, M., Johnson, D., & Swearingen, T. (1994). *Deterring Minor Acts of Noncompliance: A Literature Review*. Technical Report NPS/PNRUN/NRTR-92/08. Cooperative Park Studies Unit, College of Forest Resources, AR-10, University of Washington.

Vaske, J. J., Graefe, A. R., & Dempster, A. (1982). Social and environmental influences on perceived crowding. In F. E. Botler (Ed.), *Proceedings of the Third Annual Conference of the Wilderness Psychology Group* (pp. 211–227). Morgantown, WV: West Virginia University, Division of Forestry.

Wimpey, J., & Marion, J. L. (2011). A spatial exploration of informal trail networks within Great Falls Park, VA. *Journal of Environmental Management, 92*, 1012–1022.

Winter, P. I. (2006). The impact of normative message types on off-trail hiking. *Journal of Interpretation Research, 11*, 34–52.

Recreational Walking

Chapter Summary

The various categories of recreational walking are described in this chapter, including devotional trails and formal and informal trails, and the numbers involved in the activity are estimated. Trampling impacts on vegetation, soil, and water quality, and the resultant footpath erosion are discussed and evaluated, including the impacts of hiking poles. The ways to assess the trampling patterns caused are summarised, including experimental trampling. The effects of recreational walking, including dog walking, on wildlife, especially on birds, through flight and behaviour changes, are evaluated. Techniques for managing the footpath surface are described such as creating more resistant surfaces, such as using geotextiles, surface nettings, chemical binders and surface glues, mulch mats, aggregate paths, and boardwalks. Vegetation reinstatement using transplanting techniques and seeding is evaluated, and an example of management experiments are described and evaluated from the Three Peaks project (Yorkshire Dales, UK).

2.1 Introduction

There are several ways of categorising recreation walkers. For example, the typologies of walkers used by Natural England, who fund and maintain the English National Trails, were as follows: they categorised 33% of the trail users as dedicated users (completing the trail in one go or in different sections over time) (Edwards 2007), and the remaining 67% were split into three categories: (a) Amblers (people on trips of one hour or less), accounting for 6% of users; (b) Ramblers (who typically walk between one and four hours), accounting for 44% of users; and (c) Scramblers (full-day walkers and anyone walking over four hours), accounting for 50% of users. The latter category included hiking, hiking to a summit, backpacking, backpacking to a summit, mountain climbing, and rock climbing. However, the link between scrambling, mountain climbing, and rock climbing demonstrates a conceptualisation of certain forms of walking with other activities that cannot be considered walking, although all the three terms do require walking as part of the preparatory phase of the recreational activity. We will consider scrambling in the chapter on gorge walking and canyoning and rock climbing in a separate chapter. It has been suggested that hillwalking in exotic places has been "redefined as trekking" (Beedie and Hudson 2003). In the USA, the UK, and Canada, hiking refers to walking outdoors on a trail or off-trail for recreational purposes. The terms rambling and fell-walking are used as a synonym in the UK and Northern England, respectively. In Australia bushwalking refers to both off-trail and on-trail hiking in natural areas, but bushwacking is used specifically for hiking through dense forest where the vegetation needs to be wacked or slashed using a machete in order to progress. In New Zealand a long, vigorous walk or hike is called tramping. Backpacking is where multi-day hikes take place, with the hikers carrying their food supplies and overnight camping equipment. This can be part of long-distance hikes along National Trails in England and Wales, the Kungsleden in Sweden, and the National Recreational Trail (NRT) system in the USA. The latter are often called thru' trails and thru'-hiking entails hiking the entire length of a long-distance trail from one terminus to the other terminus within the span of one hiking season. The numbers attempting this type of hike have greatly increased, and the number of thru' hikers on the Pacific Crest Trail has been restricted to 50/day starting at the monument at the Mexican border. In a similar way reflecting the popularity of these types of hikes, the Appalachian Trail Conservancy has begun to charge for a thru'-hiking permit to pass through the Great Smoky Mountains National Park. The US NRTs were first designated as part of the National Trails System Act in 1968, and the first was inaugurated in 1971. They have to be over 100 miles long, and there are nearly 1300 NRTs, in all states, with over 26,000 miles of trails. The National Trails include the 1200-mile Ice Age National Science Trail established in 1980 which traces the last ice sheet's edge through the middle of the country, the 2190-mile Appalachian Trail, and the 2600-mile Pacific Crest Trail from Mexico to Canada. There are 19 National Historic Trails within the National Trails System which are long-distance trails and historic in nature where the aim is to explore the rich history and cultural diversity, for example, the Iditarod National Historic Trail (Alaska) and 11 National Scenic Trails to explore the American natural landscapes, for example, the Pacific Northwest National Scenic Trail and the Florida National Scenic Trail (americantrails.org). In the UK there are 13 National Trails in England since the first was established as the Pennine Way in 1965, with over 3500 km of path network, 2 in Wales, and 4 in Scotland, with 745 km of trail.

lems. Croagh Patrick in Co. Mayo (Republic of Ireland) is absolutely unique in its cultural status as it has international significance as a religious and spiritual summit, and its popularity is a direct consequence of this status. The mountain is named after St Patrick who is reputed to have spent 40 days and nights on the summit in 441 AD and is the location from where he reputedly banished snakes from Ireland. However, long before the arrival of Christianity, the summit of Croagh Patrick was occupied by Iron Age hill forts, ramparts, and dwellings. The whole of the summit area and its surroundings have significant archaeological remains, while a number of archaeological features, ranging from ancient walls to dwellings and religious stations, are found all over the mountain and adjacent to the main path. The religious status of the mountain also has an influence on the nature of footpath restoration works that may be acceptable on this mountain which is extremely heavily eroded, the worst in Ireland. The annual pilgrimage on Reek Sunday sees an estimated 30,000 individuals ascending the mountain in one day. Many of these individuals see this as a purely religious pilgrimage and regard the challenge of ascending a worn and difficult path as part of the pilgrimage, often wearing totally unsuitable clothing, some crawling or going part of the way on hands and knees. They would not want to see the challenge reduced by over-engineered paths. This pilgrimage is one of Ireland's most significant cultural events and is reputed to date back at least 1500 years and is one of the most extraordinary surviving pilgrimages in Western Europe. Any path works would need to ensure that the sanctity and religious significance of the route are fully appreciated, and the views of the pilgrims and their representatives would be a factor in determining the path works that were acceptable here (Jones 2013). A similar trail across northern Spain is the El Camino de Santiago from France to Santiago de Compostela, stretching 800 km and attracting approximately 200,000 pilgrims per year.

2.2 Devotional Trails

There are some mountains and trails which are special because they inspire and attract spiritual and devotional pilgrims and as such create extra prob-

2.3 Formal and Informal Trails

Trails are either created as formal trails by various land management organisations and as we will see can be built from varying types of mate-

2.3 Formal and Informal Trails

Fig. 2.1 (A) Lanzarote Geopark, Montaña Corona, demarcation of path by large boulders and information sign. The path surface is the local volcanic ash and is an example of a formal footpath. Photo by David Huddart. (B) Single formal trail up to the volcano flanks but then multiple informal, highly eroded trails in the volcanic ash, Lanzarote Geopark. Photo by David Huddart

rials depending on the local conditions (Fig. 2.1A), or they appear as informal, visitor-created networks which are totally unplanned (Fig. 2.1B).

An example of the juxtaposition of these two types of trail and the problems the informal trails can cause can be taken from Cadillac Mountain in Maine (Monz et al. 2010). Here there is high off-trail use as visitors disperse away from the formal trail system despite the adequate formal trail network that consists of two summit access routes that connect with a circular summit trail and formal observation sites. A total of 335 informal trails form a network that was mapped within the summit study area, and these trails lead to substantial vegetation cover loss, soil exposure, and erosion. These informal trail impacts are important as the subalpine/alpine vegetation communities on these mountain summits are typically fragile, spatially restricted, and rare (Ketchledge et al. 1985). Due to the fact that these informal trails were not planned or constructed, they are usually poorly located with regard to the terrain and resource protection needs. The proliferation over time means that there has been habitat fragmentation and impacts on the sensitive ecological communities. From the visitor's point of view, this has created a visually scarred landscape, especially above the tree line.

2.4 Global Perspective on the Numbers Involved

It is clear there are many ways in which walking participation in the UK is reported, either for tourism, recreation, or as an undifferentiated activity. Most data is based on surveys of relatively small and varied samples, and methodologies differ depending on the organisation and context. From the Sport England Active People Survey (2014), it was estimated that 23.313 million people (over 14 years of age) took part in recreational walking which was 53.8% of the population, whilst in Scotland this was between 79 and 88% of the population each year between 2004 and 2014, although 34% of that figure walked under two miles and only 8% walked over eight miles. In fact to illustrate the difficulties of interpreting these figures for 2013–2014, hill-walking and mountaineering were classed together and embraced just 2% of the population. The estimated annual recreational walking visits nevertheless reached over 300 million. Looking at the UK in the context of other European countries is similarly problematic. A study by Bell et al. (2007) compared the participation in walking as a recreation activity in six countries, and a consultation by TNS (2008) provides a figure for the UK from a similar time frame. The UK at 63% (at least once a year) is equal to Denmark but demonstrates less walking in population terms than Finland (68% of the population), Holland (74%), Norway (84%), and a similar ballpark figure for the USA (67%). Bell's comparison drew from secondary data taken from national surveys, and it is not clear whether all figures included the whole population (it states, e.g., that the Danish data only covered the adult population, whilst the Finnish data related to the "whole population"). Whilst these figures vary widely, both in their results and method of compilation, they do provide a rough benchmark on the importance of recreational walking. What can be seen from figures over time for the USA in Table 2.1 is the growth in popularity of this recreational activity, and it is easily the number one active outdoor pursuit worldwide.

The number of days for day hiking was estimated by Bowker et al. (2012) to be 1835 million, and for backpacking and primitive camping, the figure was estimated to be 1239 million. The figures continue to rise, and an estimate of the percentage of the US population taking part in

Table 2.1 Growth in popularity of recreational walking in the USA between 1982–1983 and 2005–2009 (numbers in millions) (adapted from Cordell 2012)

Activity	1982–1983	1994–1995	1999–2001	% change over that period	2005–2009	% of US population	% change 2001–2009
Day hiking	24.3	53.6	69.1	+44.8	79.7	33.9	+15.4
Walk for pleasure	91.0	138.5	175.6	+83.7	200.0	85.0	+13.9
Primitive camping	17.3	27.7	30.8	+13.5	34.2	14.5	+3.2
Backpacking	8.7	17.0	21.5	+12.8	23.2	9.9	+7.9

Table 2.2 Estimated growth in hiking and primitive area visits in the USA to 2060 (figures in millions) (adapted from Bowker et al. 2012)

Activity	Mean number	Mean number of total days	Range	% increase from 2008
Hiking	134.4	3330	117–150	45–82
Backpacking	141.0	1909	120–152	26–57

hiking/backpacking rose from 34% in 2012 to 47% in 2017. Bowker et al. (2012) had a range of estimates based on various parameters for the growth of these activities up to 2060 (Table 2.2).

Aggregating national figures of walking would be useful to gain a global perspective, but an exhaustive search for reports by organisations, including the World Tourism Organization and World Health Organization, suggests that no overall figures exist.

In terms of trail walking, Natural England (2015) synthesised research and monitoring from a number of National Trails: people counters, anecdotal estimates from the observations of their wardens, and baseline data from their "Monitor of Engagement with the Natural Environment Survey (MENE)." This provided them with a range of 63 million to 140 million people passing through areas intersected by a national trail. The large variation is due to an upper limit of people being surveyed within a range of 500 m from a trail. These estimates encompass any visitor, from those walking very small distances to the few that complete the long-distance trail. It is possible to estimate demand based on visits to locations and national scale surveys on activity participation, although the fragmented nature of available data suggests a low degree of precision.

2.5 Trampling

This is the direct disturbance that results in the bruising, crushing, and breaking of plant tissue, and, as we can see from the growth figures since the 1970s and the projected figures for the next 40 years, the trampling pressure from recreational walking is likely to grow considerably. The management of the erosion caused has been one of the most serious problems facing recreation resource managers and continues to be so. The paradox is one of trying to protect ecosystems whilst providing for their recreation use as much as possible. In the UK up until the 1970s, there was not much concern about the ecological effects of recreation, and planners placed a low priority on this issue. This was partly because the total acreage of land that had become degraded was fairly minimal, except for a few well-publicised examples of vegetation loss around the Cairngorm ski lift in Scotland and the loss of stability of sand dune ecosystems. Few areas appeared to be under immediate threat. The outdoor recreation boom was only just getting under way, but there was a very rapid growth in the late 1960s and early 1970s. There was also little scientific information about the rates and directions of changes in natural ecosystems caused by recreation. Any reasoned discussion about the possible changes was hindered by a lack of quantitative data, so there was little for recreation resource managers to work with. At the same time, there were often divergent interests between conservationists and ecologists and the general public and the land managers. Conservationists were often trying to ensure the survival of a rare plant association, or even a single species, whilst the general public was unaware of the potential recreation impacts, was not concerned about them, or was not educated about them. The land managers were caught between the two groups. However, by the 1970s, there was a change because of the outdoor recreation boom which was part of a return-to-nature movement, because ecological change caused by this boom became much more obvious and research started to give land managers some information as to how to tackle the problems in an applied manner. Recreation ecology and recreation land management had become important in planning the countryside. It became clear that certain types of habitat were more vulnerable to recreation trampling: coastal systems such as sand dune and salt marshes, especially those in the early stages of the seral succession and on an unstable substrata; mountain habitats where the growth capacity and self-recovery were reduced by the climate and the mountain soils were thin, nutrient deficient, and often wet; systems with shallow soils, such as chalk grasslands, those with nutrient-deficient soils such as lowland heaths, and those with very wet soils like blanket bogs and fens; and those ecosystems which have taken a long time to develop, for example, caves, or have developed in different environmental conditions than those currently. In different habitats too, the numbers of recreation walkers will be different before major environmental damage results, although generally the more walkers, the more damage occurs. For example, it was estimated that 7500 walkers/season moving from a concrete path onto mature salt marsh caused complete vegetation loss, whereas the same figure over yellow sand dunes caused complete elimination of marram, sea couch grass, and prickly saltwort and considerable erosion. It has to be realised though that other factors are involved apart from the numbers involved, such as the slope of the terrain, the soil characteristics, the type of footwear used, visitor behaviour, and the climate of the area where trampling takes place.

Yet erosion is part of the natural world, and there is no doubt that geomorphic processes on hillslopes, within the soil profile and in all the other surface sub-environments, cause far more erosion in total than recreational walking. Nevertheless trails are often an important sediment source, even if they only represent a small fragment of the landscape, and sometimes, as Cole (2004) suggests, it is difficult to separate the impacts of recreational walking on trails from impacts caused by trail construction and maintenance and the impacts that would occur in the absence of walking, from erosion by rain channelled down a trail tread. Nevertheless, for example, Ramos-Scharrón et al. (2014) found that on the island of St. Croix in the US Virgin Islands, sediment production measured with sediment traps on 12 trail segments varied in annual erosion rates from 0.6 to 81 $Mgha^-yr$. The lower figure was from abandoned trails with a dense vegetation cover, whilst the higher rates were associated with de-vegetated trails immediately following construction or restoration. Annual trail erosion rates were one to three orders of magnitude higher than measured surface erosion rates on undisturbed hillslopes that had not been trampled. So the problem here is how to minimise erosion caused by both recreational walking and any remedial attempts to counteract the damage caused. This is because all recreational walking is going to create changes in the natural ecosystems where it takes place. The aim must be to be able to control these changes when they get out of hand or preferably before they reach that stage and where prevention is better than cure. The avoidance of future erosion must be a strategy to use if at all possible. This may be attempted by discouraging the use of an eroded path for a time to allow natural revegetation, by the fencing off of a section of path and the provision of an alternative. However, as has been found in the Peak District (Northern England), this approach is difficult to apply if the path is a public right of way. It is not possible to divert or close a public right of way, without due legal procedure under the Highways Act (1980). It is not possible to do this unless that path is not needed for public use, and if a path is so well used that it is becoming eroded, then it is obvious that this path is needed for public use.

2.6 Footpath Erosion

The extent of footpath erosion in popular mountain areas in Scotland was mapped by Grieve et al. (1995) from air photos, and interestingly the mean length of eroded footpaths overall was found to be less than half that of large natural gullies. We cannot aim to manage these natural gullies, and it has been claimed by some that we have gone too far in terms of "erosion control" as the impassioned pleas by Hayes (1997) suggest that a footpath is the line of least resistance in a way that complements the land, whereas his so-called footways are a planned, engineered, and pre-eminently artificial imposition on the landscape, which are ugly to look at and often difficult to walk on. He argued that there was an unfortunate lack of recognition for the intrinsic beauty of a naturally eroded path, and many would agree with some of his views.

Trails though are generally regarded as an essential facility in parks and recreation areas, providing access to areas with no roads, offering recreation opportunities, and protecting resources by concentrating visitor-use impacts on resistant tread surfaces. However, much ecological change as we have seen assessed on trails is associated with their construction or rehabilitation and is considered unavoidable (Birchard and Proudman 2000). The principal challenge for trail providers is therefore to prevent post-construction degradation from both recreation use and natural processes such as rainfall and water runoff and to minimise the damage potentially caused by any remedial work on the trail network.

Trampling too is part of a set of processes in natural systems which are caused by animals such as rabbits, sheep, feral horses, and deer which cause paths to be created. Yet when recreation management in many popular areas has been lacking, or applied too late, there can be a series of changes that involve vegetation deterioration and erosion of the vegetation cover, soil changes, and soil erosion. However, there can be a whole spectrum of ecosystem changes that are classed under the umbrella term of trampling pressure which can range from the loss of a single attractive flowering species, the creation of permanent puddles in areas of locally concentrated pressure, such as gateways to multiple, braided, deeply eroded trails, and major gullying on thick, peat soils (Figs. 2.2 and 2.3).

In the last 20 years though, there has been a considerable amount of research into trampling pressure both on vegetation and soils, and there have been several reviews related to the impacts of recreational walking. Before we outline what the effects of trampling pressure are on vegetation and soils, the detailed reviews and summaries of the research literature can be found in Kuss and Graefe (1985), Cole (2004), Cole et al. (1987), Liddle (1975, 1997), Leung and Marion (1996, 2000), Marion and Leung (2004), Pescott and Stewart (2014), Hammitt et al. (2015), Marion (2016), and Marion et al. (2016) and the extensive literature used therein.

2.7 Trampling Impacts on Vegetation

Vegetation becomes bruised by light trampling pressure where there are broken rigid stems and a decrease in vegetation height. This is likely to represent a drop in primary production of the ecosystem, although after very low trampling intensity, there is a stimulation of plant growth,

Fig. 2.2 Excessive wide footpath in clay-rich soil caused by human and probably horse, trampling pressure, Pennine Way (UK). Photo by David Huddart

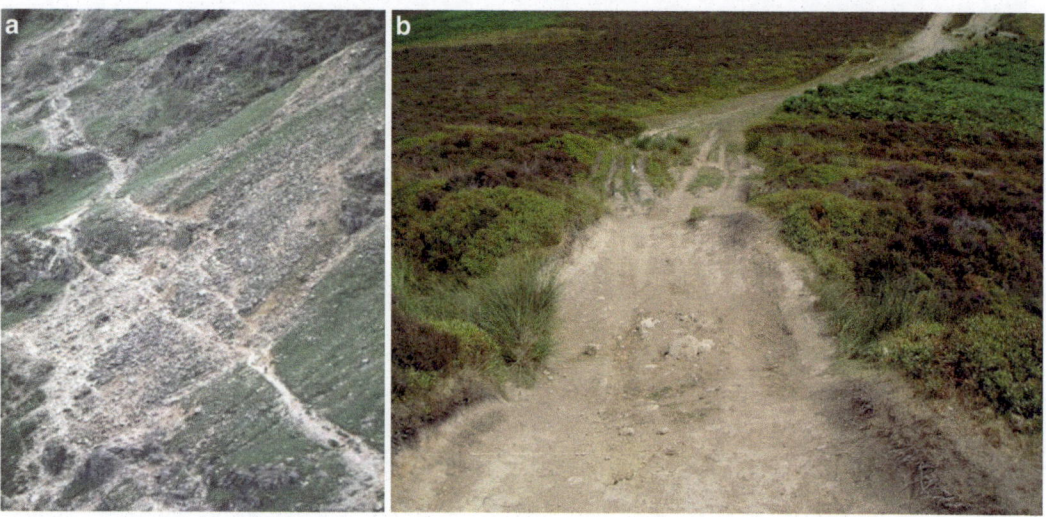

Fig. 2.3 (A) Braiding of footpaths, Snowdon, North Wales. Photo by David Huddart. (B) Eroded and braided, informal footpath, Clwydian Hills, Denbighshire Moors, North Wales. Photo by David Huddart

Table 2.3 Changes in plant physiognomy of two grasses with trampling (after Trew 1973)

Character average	*Lolium perenne* trampled	*Lolium perenne* untrampled	*Festuca rubra* trampled	*Festuca rubra* untrampled	Rye grass and red fescue
Height above ground (cm)	7.35	23.76	5.1	9.72	
Number of seedpods	12.61	21.00	13.00	18.00	
Length of seed head (cm)	3.1	8.6	1.75	3.2	
Length of leaves (cm)	2.1	5.3			

which probably marginally increases production. Very light use may only cause a slight decrease in total vegetation cover, although there may be increases in branching and a decline in the incidence of flowering. Damage sufficient to cause small patches of bare ground will, in many types of habitat, result in invasion by trample-resistant species, such as *Lolium perenne* and *Plantago major* (Speight 1973). Heavier use can reduce or even eliminate plant cover, with only the most trample-tolerant species surviving in gaps between stones or in similar places where direct wear is low. Generally though there is a loss in vigour in the plants, and if a single species is examined, the crested dog's tail grass (*Cynosurus cristatus*) from Dovedale (Peak District), it shows a 50% reduction in stem length in transects taken along the line of a grass-covered path. Similar changes can be seen in Tables 2.3 and 2.4 from Beer head (Devon) where there are also changes in the number of seedpods, the length of the seed head, and the length of the leaves. There are also changes in the onset of flowering time of herbaceous plants as can be seen from Table 2.5 from Porteynon Bay in Gower (South Wales). The trampling damage must at early stages increase the percentage of organic matter on the trampled areas, for example, the movement of 8000 people across chalk grassland in Southern England produced an accumulation of broken and dead ground vegetation sufficient to double the weight of leaf litter.

Trampling on upland vegetation, such as *Calluna vulgaris* (heather) and *Vaccinium myrtillus* (bilberry), has been found to make them vulnerable to desiccation and winter frost browning (Watson 1984; Cole et al. 1987). Of even more relevance to blanket bog and its restoration is the

2.7 Trampling Impacts on Vegetation

Table 2.4 Changes in plant physiognomy from Beer head (Devon) (after Price 1987)

Average height of grass	On path (cm)	Adjacent to path (cm)
Agropyron repens couch grass	10	60
Phleum pratense (timothy)	15	60
Agropyron repens on pathway		
Average number of seedpods	16	20
Average length of seedpods	4 mm	9 mm
Average length of leaves	4	17

Table 2.5 Changes in onset of flowering of two herbaceous plants with trampling (after Trew 1973)

Intensity of trampling (people/year)	Onset of flowering: *Galium verum* (lady's bedstraw)	Onset of flowering: *Veronica arvensis* (wall speedwell)
2000	22 June	14 June
1000	7 June	2 June
500	29 May	26 May

Table 2.6 Resilient species to trampling (after Huxley 1970; Price 1987)

Common bent grass	*Agrostis tenuis*	Resistant species after Huxley
Crested dog's tail grass	*Cynosurus cristatus*	
Meadow fescue	*Festuca pratense*	
Rye grass	*Lolium perenne*	
Annual meadow grass	*Poa annua*	
Meadow grass	*Poa pratensis*	
White clover	*Trifolium repens*	
Couch grass	*Agropyron repens*	Resistant to trampling in middle of path, after Price
Timothy	*Phleum pratense*	
Crested dog's tail grass	*Cynosurus cristatus*	
Sheep's fescue	*Festuca ovina*	
Glaucous sedge	*Carex flacca*	
Yarrow	*Achillea millefolium*	Flourishing at path margin
Hoary plantain	*Plantago media*	
Dwarf thistle	*Cirsium acaulon*	
Restharrow	*Ononis repens*	

impact of recreation on lower plants, particularly sphagnum mosses and lichens. Borcard and Matthey (1995) found that as low a figure as ten minutes of experimental trampling repeated only three times a year for three years almost destroyed the cover of *Sphagnum recurvum* and *S. fuscum*.

There are effects on flowering with many species failing to flower, and the abundance of flowering heads of grasses recorded a drop in frequency along a grass-covered path in Dovedale (Table 2.3).

Some plant species are resistant to trampling, and this plant resistance allows these opportunistic and aggressive plants to become more dominant. The species that are intolerant to trampling rapidly die out, and these are generally the species most characteristic of the habitat, for example, wild thyme on chalk grasslands in Southern England. The result is that with intermediate trampling pressure, there is some vegetation cover loss and some compositional changes. The resilient species start to dominate, and this can be seen in Table 2.6 where grasses are important. Research to monitor the changes taking place under increasingly severe recreational walking may identify stages at which species disappear and ultimately at what intensity the vegetation cover is removed completely. The loss of indicator species can show the general levels of wear and tear before the situation becomes too evident to the average recreationist, and it is then that appropriate management decisions can be taken. Certain growth forms become predominant, and the resistant species often possess several of them. These are the plants that are procumbent or trailing rather than erect; they have protective thorns or prickles; the stems are flexible rather than rigid or woody; the leaves are in a basal rosette rather than any other arrangement; the leaves are flexible and able to fold under pressure rather than fracture; growth is possible from intercalary (inserted between others) as well as apical meristems; seasonal regrowth depends on hidden cryptophytic buds, concealed beneath the soil surface rather than on aerial parts; reproduction is possible by means of suckers, stolons (shoots from the base of a plant, rooting, and budding at the nodes), or corms, in addition to seeding or runners and finally rapid growth rate. At the same time, it is possible to identify certain plant characteristics that make a species prone to trampling damage. Perennial species that are woody and have an erect growth form seem particularly vulnerable, including the dwarf shrubs such as heather (Calluna species) and tree seedlings.

In fact systematically reviewing and a meta-analysis of all the experimental trampling pressure evidence led Pescott and Stewart (2014) to conclude that the effects of life-forms were more important than the intensity of the trampling experienced in causing erosion. This is illustrated from an example from Cole (1995a) in an experimental trampling study on an alpine grass and sedge turf, where 500 passes by a walker reduced plant cover by 40%, whereas the same trampling level in a subalpine forest with a herb and fern understory reduced cover by 97%. Studies have also shown that these differences in morphology and trampling resistance can be strongly correlated with sunlight intensity (Liddle 1997; Cole and Monz 2002). Nonwoody, shade-tolerant species require large leaf surfaces supported by strong rigid stems which are easily crushed, whereas in contrast sun-loving plants, especially grasses, can obtain the necessary sunlight with small or narrow leaves and flexible stems. Plant resilience which is the capacity of vegetation to recover from trampling pressure is well known from grasses and is attributed to leaf durability, fast growth rates, and stem flexibility (Sun and Liddle 1993).

As the trampling pressure increases beyond moderate levels, plant cover and biomass are reduced. Damage and leaf removal means that plants are unable to produce sugars and store carbohydrates in roots which slows or halts flowering and seed production and reduces plant growth in subsequent years (Liddle 1997). This is exacerbated by root damage via soil compaction and root death eventually. For example, tree root damage in the California redwoods can show a decrease in root branching and in the number of feeder rootlets and root exposure and damage, which can lead to tree death. The soil compaction caused by excessive trampling eventually inhibits seed germination as the roots cannot penetrate the soil. Seedling mortality is caused through desiccation and drought which are accentuated by the destruction of vegetation cover and the microclimatic changes which increase the surface soil temperatures. Injured plants trying to grow in full sun are particularly prone to desiccation and death. The high evaporation rates and high temperatures inhibit the plants becoming established, and crusts which form on the surface of bare, fine-grained soils inhibit seedling emergence. These crusts act too as a seal to water penetration which creates moisture stress in the upper soil surface layer and for the seedlings.

The trampling injury to plants causes many problems as outlined by Kuss and Graefe (1985), for example, a disordered use of energy which is reflected in abnormal cellular activity, injurious physiological processes, and impaired vital functions of the injured plants. The vital processes such as root formation; photosynthesis and assimilation; translocation of water, nutrients, and photosynthate; respiration; energy metabolism; and transpiration are impaired as a result of trampling injury. The symptoms of this impairment are similar to those caused by plant disease and nutritional disorders, and signs of stressed plants are reduced vigour and stunting, loss of photosynthetic surfaces, aborted or reduced flower counts, lessened fruit and seed production, defoliation, wilting and curling of leaves, dieback, and reduced biomass. Overall with time before complete vegetation loss, there is a decrease in species diversity and an increase in invasive species which are non-native to the area.

In fact though, what has been noted in many studies is that the trampling pressure and recreational walking use can increase with little further impact to the vegetation if the users stay on the well-established trails (Cole 1995a; Leung and Marion 2000; Monz et al. 2010). The impact-use relationship is usually curvilinear, but for more resistant and resilient grasses and sedges, they can withstand prolonged low to medium impacts, but eventually the cover is eroded in high-impact use.

In addition to physical changes in vegetation cover and in soil erosion resulting from trampling, there can be other effects, such as reductions in the cover of orchids as a result of picking by tourists (Bratton 1985) and of dead wood for campsite fires (Bratton et al. 1979).

The responses of vegetation to trampling have been reported to be affected by trampling intensity (number of human trampling passes; e.g. Cole et al. 1987; Cole 1995a), frequency (trampling passes per time period (Cole and Monz 2002), distribution (whether trampling passes are dispersed or clumped for a particular trampling frequency (Gallet et al. 2004)), season (Gallet and Roz'e 2002), weather (Gallet and Roz'e 2001), habitat (Liddle 1975), species (Gallet et al. 2004), Raunkiaer life-form (i.e. perennating bud position) and growth form (Cole 1995b; Pescott and Stewart 2014), and soil type (Talbot et al. 2003). Pescott and Stewart (2014) considered variation in trampling intensity, vegetation resistance, recovery time, Raunkiaer life-form of the community dominant, and broad habitat type as potential reasons for heterogeneity in experimental results across primary studies.

The results from Pescott and Stewart's (2014) analysis suggest that the initial resistance of a plant community, and the length of the recovery period, may be better predictors of vegetation resilience than the intensity of trampling undergone; that is, intrinsic properties of vegetation appear to be some of the most important determinants of resilience, with the magnitude of the actual disturbance explaining much less of the community response.

The results support a situation where particular plant functional traits are likely to be more important than projected intensity of use when considering the siting of recreation activities involving human trampling. This somewhat surprising result has important management ramifications because it suggests that even relatively low-intensity trampling could be as damaging as high-intensity trampling in certain plant communities. Thus, trampling may sometimes be unsustainable for vulnerable vegetation, potentially creating conflict between even relatively limited access and plant species, or community-focused conservation objectives.

These results confirmed the importance of Raunkiaer life-form as documented by Cole (1995a) across the 188 trials investigated, suggesting that hemicryptophytes and geophytes will be more resilient to trampling impacts relative to other life-forms. In contrast, chamaephyte-dominated vegetation did not show a main effect of recovery; indeed, chamaephyte-dominated communities have been shown to die back after trampling disturbance, despite initially high resistance (Cole 1995a; Cole and Monz 2002). Sites of conservation importance dominated by phanerophytes, chamaephytes, helophytes, or therophytes should not experience regular trampling disturbance if damaging impacts are to be avoided. Trying to reduce trampling intensities may not be effective where adverse impacts are already occurring, although it has been found that there is a negative relationship between initial resistance and resilience for chamaephyte-dominated vegetation (i.e. high initial impacts may be followed by some recovery). Conversely, the current evidence base suggests that vegetation dominated by hemicryptophytes and geophytes, life-forms with more protection for their perennating buds (Kent 2012), recover to a greater extent than vegetation dominated by other life-forms. They could therefore potentially be trampled more intensively, provided monitoring is undertaken to provide early warning of deterioration or unsustainable use.

2.8 Trampling Impacts on Soils

Soil erosion can result in aquatic system disturbance, excessively muddy trails, widening of trails, tread incision (Fig. 2.4), and braided or multiple trails and can lead to the creation of undesired trails (Hammitt et al. 2015; Marion et al. 1993). Unlike disturbed vegetation and compacted soil, soil erosion is the only trail degradation indicator, relatively speaking, that does not recover naturally over time. A study of 106 US National Park Service units found that almost

Fig. 2.4 Deep incision caused by trampling pressure, Peak District. Photo by David Huddart

50% of all park managers indicated that soil erosion on trails was a problem in many or most areas of the backcountry. Trail widening was cited by 31% of park managers, and 29% rated the formation of braided or multiple trails and the creation of undesired trails as serious problems (Marion et al. 1993).

Low levels of trampling pressure initially create increased organic material to be released to the soil because of dead plant leaves, broken grass stems and break-up of needles, and the pulverisation of the organic matter. However, fairly quickly this trampling also allows the disappearance of the soil litter layer by wind or water erosion, or it can decompose into the underlying organic soil layer. So soil break-up of the A_L and A_F layers takes place relatively quickly in the initial stages of path formation, but the break-up of soil layers deeper than the A horizons appears to be unusual, except on organic, peat soils. Instead soil compaction gradually takes place with use, and there is an increase in the soil's moisture content/unit volume of pore space. The mechanical forces of foot pressure cause the soil grains to rearrange and pack together more tightly, which increases the soil density and decreases water and air permeability, and roots can find it difficult to penetrate into the soil, and seeds find it more difficult to germinate. On flat surfaces compaction of the soil can lead to puddling and increased muddiness of the soil surface, especially where there are organic or clay components in the soil. This can lead to impeded drainage and gleyed soils which destroy the plant roots which might have helped re-establish the vegetation cover. It can also cause trail widening and the creation of secondary trails when users try to circumnavigate the puddled areas.

The continued loss of the soil litter layer can only lead to decreased nutrient cycling and a reduced population of organisms responsible for the recycling processes, changes in soil organism populations, and plant death. In addition there is a decrease in pore space abundance with compaction, and this leads to a decrease in the numbers of larger soil organisms. The impedance of drainage can lead to increased runoff on sloping surfaces and accelerates the soil erosion whilst at the same time decreasing the infiltration rates into the soil. For example, in Switzerland in woodland, it was found that on footpaths there was an 80% decrease in infiltration. The end products of high trampling are soil erosion, gullying, and often sediment fan deposition outside the trail boundaries, especially on steeper, relatively unvegetated slopes. Wind erosion can cause soil loss where the substrates are dry and loose and lack a vegetation cover, like in sand dunes, but water erosion is much more common, especially in mountain areas with steep slopes and in areas with high rainfall. Amount of use can be an important factor in soil erosion, but probably more important is the type of use, and horses and all-terrain vehicles are the worst culprits as we will see in later chapters of this book.

Soil erosion loss is important ecologically because the processes of new soil creation can be extremely slow and the sediment can have an effect on turbidity, sedimentation, and the populations in freshwater ecosystems. This of course is an argument for footpath management to control this soil erosion and against the argument of people such as Hayes (1997), quoted earlier in this chapter who have defended natural erosion.

2.9 Trampling Impacts on Water Quality

The impacts of concern here are all secondary impacts that can be caused by hiking on trails, with or without dogs, or by using packstock animals (see Chap. 9). There seems to be a correlation between increased humans and packstock animals and an increase in harmful bacteria in water and a degradation in water quality in wilderness areas (Derlet and Carlson 2006; Kellogg et al. 2012; Clow et al. 2013). For example, Derlet and Carlson (2006) in the Sierra Nevada found that 12 out of 15 backcountry sites sampled with packstock traffic gave high levels of coliform bacteria, and Kellogg et al. (2012) found high faecal hand contamination amongst tested wilderness hikers. Meanwhile Reed and Rasnake (2016) found elevated *Escherichia coli* and other coliform bacteria in springs and streams near Appalachian Trail shelters in the Great Smoky Mountains National Park, particularly during the summer months. Other biological impacts could be the introduction or spread of protozoa like *Giardia lamblia*. Chemical impacts are mainly related to nutrient influx to streams and lakes which can lead to lowered dissolved oxygen rates but which can include impacts from soap, sunscreen, food particles, and human and animal wastes (Ursem et al. 2009). Nutrient loading in lakes can contribute to algal bloom increases and decreased water quality (Hammitt et al. 2015). Most of these water quality issues are local in nature, and usually it is impossible to directly link the water quality to a specific trail impact.

An indirect effect too could be on river bank erosion as Madej et al. (1994) linked a 27% increase in channel changes, including bank erosion in the Merced River in Yosemite National Park to the effects of heavy human traffic. The changes in the river since 1920 documented on detailed maps correlate with a dramatic increase in tourists since that date, and there were over 3.5 million visitors to that national park in 1991, and it was estimated that 90% of them visited the upper Yosemite valley and the upper river study reach. Over 1000 campsites were within 500 m of that reach too. Trampling can damage or destroy the riparian vegetation close to the river and reduce the bank stability. The causes of the increased bank erosion were evaluated, but there were no changes in precipitation, flood peaks, or sediment load. However, the degree of channel widening was inversely correlated with bank stability ratings, and these were associated with high visitor use areas. Hence Madej et al. (1994) suggested that human trampling and flow constriction by bridges, locally aggravated by bank revetment placements, were likely to be the most probable causes of the erosion problems. In reality how much was caused by human trampling is still debateable.

2.10 Ways of Assessing Trampling Patterns Caused by Recreational Walking

2.10.1 Analytical, Descriptive Field Survey

There are several ways of assessing the trampling patterns caused, and these have been developed and summarised by Liddle (1975), Bayfield et al. 1988, Cole and Bayfield (1993), Cole (1991, 2004), Leung et al. (2011), Marion and Cole (1996), Marion et al. (2011), and Hammitt et al. (2015) in their Chapter 11. There is analytical, descriptive field survey where the vegetation and soil parameters are measured for the trail and the adjacent areas to assess the current conditions. So here the impact conditions for trampled and untrampled areas can be measured over large areas rapidly and with minimal expertise. The assumptions here are that the whole area was homogeneous before, that the adjacent taller vegetation is undisturbed by trampling, and that there has been no overall environmental change since the trail was introduced. The adjacent areas are the undisturbed control sites which provide an estimate of the change which has resulted from use, but the control sites are never perfect replicates of the pre-existing conditions. The parameters that are measured can be the width and depth of the path; the vegetation cover to estimate the plant frequency, usually using a quadrat; and the plant height and soil parameters such as soil compaction (using an impact penetrometer), organic content, grain size, pH, and water content. It would be possible to measure plant biomass, but as this is destructive of the vegetation, it is not often applied. It is useful if there are estimates of the number of tramples in an area, but these are not usually available, but sometimes there are walker counts, automatic counters using an interrupted beam or tramplometers (Bayfield 1971), using lengths of wire soldered to pins.

2.10.2 Experimental Trampling

An alternative approach would be to use experimental trampling, although approaches using heavy weights or rollers have been totally unrealistic. It is common though to subject an untrampled area to a known amount of trampling and record the changes. It is then possible to use a method to show the vulnerability of different habitats which can be compared (Liddle 1975). This technique uses the number of passes the vegetation could withstand before the vegetation is reduced to a 50% cover: the trample dosage (TD50). So what are recorded on a previously untrampled area are the vegetation cover at intervals and the number of passes. The percentage of live vegetation remaining at the various measurements (y) is then regressed on the log of the number of passes (x) to produce an equation which may be used to estimate the number of passes which reduced the vegetation to 50% cover. This has proved to be a useful concept of the vulnerability to abrasion, but it does not take into account longer-term changes resulting from either the regenerative ability of the vegetation or responses to change mechanisms other than direct trampling, for example, climate factors. As Cole (2004) suggests, the maximum insight can be obtained by using several approaches simultaneously, as was used by Marion and Cole (1996).

All the trampling results can be treated graphically, with regression lines and correlation coefficients. The percentage cover reduction (Cole 1978) across the trail can be measured by $CR = (C_2 - C_1) 100 / C_2$ where C_1 is the percentage cover in quadrat 1 close to the path and C_2 is the percentage cover in quadrat 2 away from the path. The change in species composition can be quantified by a coefficient of floristic dissimilarity where the larger values indicate greater vegetation alongside trails (Cole 1978), a change in species diversity can be found using a diversity index (Liddle 1975), or the Shannon diversity index and the relative abundance of species in the point quadrats can be worked out by the number of touches of species A x100/total touches of all species and the frequency in percent by the number of points of As occurrence/total number of points measured.

A study of experimental trampling can be taken from the Cairngorms (Scotland) which looked at the vegetation damage since the opening up of the Cairngorm plateau by the introduction of ski chairlifts and roads (Pryor 1986). The *Juncus trifidus* heath was one of the main vegetation types assessed because it is widespread but confined only to the Cairngorm Mountains. Thus the trampling of this vegetation type was important for its continued existence and survival. An experimental approach was adopted, and ten sites at each of six different abundant vegetation types were used. For each site a plot was divided into two parallel sections, one to be trampled on and the other as a control. Each plot and vegetation type was subjected to the same degree of trampling, 300 tramples in the same direction on the same day. Four quadrats were used to measure percentage vegetation cover in each plot. The results were that the most resistant was the alpine Nardus grass and the open *Juncus trifidus* heath. Medium resistant was the Rhacomitrium heath and the Empetrum-Vaccinium heath, whilst the least resistant was the heather (Calluna) and the lichen-rich, dwarf Calluna heath. So the most resistant species were the tussock species which produced a large number of tillers over the winter which emerge in the spring causing a rejuvenation of the plant in trampled areas. Other results were that there was generally a size reduction of the tussock species; moderate trampling actually increased the number of smaller and younger individuals, and the effects of trampling were only evident when disturbance was high.

2.11 Hiking Pole Impacts

With the great increase in use of hiking poles over the last 20 years, there is some evidence that there are impacts on the vegetation and soils (Marion et al. 2000).

2.11.1 Vegetation Impacts

Trailside vegetation can be damaged from the swinging action of trekking poles, particularly from contact with the baskets, which can get caught in low-growing plants. One North Carolina hiker noted in an email to the Appalachian Trail Conference that "the ground was becoming "torn up" by spiked walking poles. On the uphill side of the trail, moss and wild flowers were torn from their bedding. On the downside of the trail, parts of the trail were also torn away." The potential consequences of such damage include a reduction or loss of vegetation cover, change in vegetation composition, and trail widening. It was also noted that trail maintainers generally trim only higher, overhanging vegetation that is unaffected by trekking pole use.

2.11.2 Soil Impacts

A number of soil impacts could result from repeated contact and penetration by trekking pole tips. In wet or loose soils, pole tips can penetrate up to two inches and leave holes half an inch in diameter. These holes are often V-shaped, wider at the top due to the swing of the upper pole once the tip is embedded in soil. Under some conditions it was seen that soil was lifted by pole tips and dropped onto the ground surface. In a letter to the editor of Appalachian Trailway News, a Virginia hiker observed that trekking pole use has become nearly universal and that "These things are tearing up the trail on each side of the footpath. Some places look like they have been freshly plowed." Potential soil impacts from such disturbance include the loss of organic litter and exposure of soil and increased erosional rates and muddiness. Research is needed to document if, or to what extent, pole use could increase erosional rates. Muddiness could develop following rainfall, when surface water runoff fills the holes created by pole tips. The increased water and soil contact in areas with high densities of holes could turn trail sides to mud, as often occurs on horse trails when water fills hoof prints. Trails that are outsloped for water drainage would not prevent such muddiness, and water bars and drainage dips would prevent muddiness only on the downhill sides of trails.

2.11.3 Rock Impacts

The carbide tips on trekking poles leave visually obvious white scratch marks on rock surfaces and also damage lichens. A hiker in Maine related in an ATN letter that "the scratching is so pronounced on granite surfaces that it is sometimes easier to follow where the poles have been than to locate a white blaze. …the scratching is something I vividly remember from my hike, so remarking about it is justified." In an opinion letter to *Backpacker* magazine's website, a hiker in the Adirondack reported that "I was upset to see all the rocks had little white marks on them. Not just a rock here or there, but *all* the rocks on the trail were chipped by hundreds of people …. It got to the point where I could not concentrate on anything else but these thousands of little white gashes in the rocks I was stepping on. It really left a bad taste in my mouth and a grim look to the future." How serious the impacts of trail pole use are is debatable, but with the dramatic increase in use, they are clearly having some impact.

In some parts of the world, there have been suggestions that hiking poles should be banned as in Hong Kong country parks where rapid erosion has partly been attributed to the effects of the extensive use of hiking poles (South China Morning Post 2018, www.scmp.com/lifestyle/health-wellness/article/2140427/could-hikin).

2.12 Summary Related to Impacts of Recreational Walking

As a summary here, Cole (2004) offers five key generalisations regarding the impacts of recreational walking which are:

1. Impact is inevitable with repetitive use. Numerous studies have shown that even very low levels of repetitive use cause impact.

Therefore, avoiding impact is not an option unless all recreation use is curtailed as even extremely low levels of repetitive use cause impacts. Managers must decide on acceptable levels of impact and then implement actions capable of keeping use to these levels.
2. Impact occurs rapidly, while recovery occurs more slowly. This underscores the importance of proactive management, since it is much easier to avoid impact than to restore impacted sites. It also suggests that relatively pristine places should receive substantial management attention, in contrast to the common situation of focusing most resources in the impacted places. Finally, it indicates that restoration of sites (periodically closing damaged sites, to allow recovery, before reopening them to use) is likely to be ineffective.
3. In many situations, impact increases more as a result of new places being disturbed than from the deterioration of places that have been disturbed for a long time. This also emphasises the need to be attentive to relatively pristine places and to focus attention on the spatial distribution of use. It suggests that periodic inventories of all impacted sites are often more important than monitoring change on a sample of established sites.
4. Magnitude of impact is a function of frequency of use, the type and behaviour of use, season of use, environmental conditions, and the spatial distribution of use. Therefore, the primary management tools involve manipulation of these factors.
5. The relationship between amount of use and amount of impact is usually curvilinear (asymptotic), although this now is debated (as we will discuss later). This has numerous management implications and is also fundamental to many minimum-impact educational messages. It suggests that it is best to concentrate use and impact in popular places and to disperse use and impact in relatively pristine places.

2.13 Effects on Wildlife

Trails, and the presence of visitors, can also impact wildlife, fragment wildlife habitat, and cause avoidance behaviour in some animals and attraction behaviour in others seeking to obtain human food (Hellmund 1998; Knight and Cole 1991). While most impacts are limited to a linear disturbance corridor, some impacts, such as alterations in surface water flow, introduction of invasive plants, and disturbance of wildlife, can extend considerably further into natural landscapes (Kasworm and Monley 1990; Tyser and Worley 1992). Even localised disturbance can harm rare or endangered species or damage sensitive resources, particularly in environments with slow recovery rates. Animal life is disturbed by trampling, and most species decline in numbers or move somewhere else as most animals are unsettled by noise, rapid movement, and the nearness of people. However, some species are more sensitive than others, usually the commoner species the least so. In the breeding season, nearly all are very sensitive, and the decline of the Little Tern in Europe correlates with the recreation use of its once-inaccessible breeding beaches. It has been shown that in Scotland 5% of greylag geese do not return to the nest once flushed and duck nests are also vulnerable. In breeding areas that were visited three times a week, all the nests were predated; weekly visits meant a 60% loss, whilst in the nests with no visits, there was only a 10% loss. Birdwatchers are possibly responsible, but recreation walkers must be too.

Recreation disturbances often cause birds to spend considerable amounts from their energy reserves which is likely to be detrimental to their breeding success and migration, and it is thought that recreation can have significant negative effects at the individual, population, and community levels.

Human recreation disturbances are known to cause reduction of reproductive success and nesting failures (Bolduc and Guillemette 2003; Finney et al. 2005; McGowan and Simons 2006; Steven et al. 2011; Whitfield and Rae 2014). As explained by Holm and Laursen (2009), disturbances caused by hikers alone can negatively affect territorial densities of certain birds, causing effective habitat loss for breeding. However, such impacts seem to be species-specific, and for some birds, human use of recreation trails has no apparent effect on nest survival (Smith-Castro and Rodewald 2010).

Many studies have correlated higher pedestrian or vehicle traffic in wildlife refuges to reduced species richness and abundance of birds (Fernández-Juricic 2000; Burger et al. 2004; Marcum 2005; Steven et al. 2011). However, not all bird species are equally affected by human recreation disturbances. Certain bird species can tolerate greater degrees of disturbances (Marcum 2005; Gill 2007; Cardoni et al. 2008). The response of birds to human recreation disturbances may depend on the nature or type of disturbance, as well as on the distance from the disturbance (Fernández-Juricic 2000; Pease et al. 2005; Ruddock and Whitfield 2007). For instance, contrary to the general belief that motorised, nature-based tourism activities cause greater disturbances to birds (Stolen 2003; Schlacher et al. 2013), non-motorised recreation such as hiking and biking that often involve close encounters with birds have been found to cause more severe negative effects on a wide range of bird species (Buckley 2004; Pease et al. 2005). In the case of species that are sensitive to human recreation disturbances, continuous exposure to such disturbances may ultimately lead to permanent avoidance of habitats and changes in regular behavioural patterns.

2.13.1 Flight and Behaviour Change

A very substantial amount of research has focused upon measuring how and when walkers disturb wildlife through approaching them, and/or causing noise, which triggers, in essence, an anti-predator response of escape ("flight"). Within this literature there is once again, however, a very heavy focus upon birdlife (for reviews see Sidaway 1990; Taylor et al. 2005), which itself focuses substantially upon ground-nesting birds (for a "systematic review," see Showler et al. 2010) and disturbance by dogs accompanying walkers. Indeed, in their review of the disturbance impacts of dogs, Taylor et al. (2005) conclude that there is very little relevant research that has focused on the effects of dogs on animal groups other than birds. The central concern is that disturbance can cause birds, and other animals, to flee from cover or nests which can impact on their energy balances, feeding behaviour, and the vulnerability of young, eggs, or fledglings (Dahlgren and Korschgen 1992; Fox and Madsen 1997; Rasmussen and Simpson 2010). Each of these potentially affects not only individuals but also populations through affecting breeding success and can thus be a particular concern for endangered or vulnerable species of conservation interest. Considerable attention has been given to flight responses of water birds (see, e.g., Carney and Sydeman 1999; Nisbet 2000) but much less to forest bird species. Searches relating to the recreation disturbance of 35 "woodland bird" species found in the UK (as defined by Amar et al. 2006) identified very few studies (Ibanez-Alamo and Soler 2010; Luka and Hrsak 2005; Fernández-Juricic et al. 2001a, b; Fernández-Juricic and Tellería 2000; Fernández-Juricic 2000; Müller et al. 2006). Most of these studies were conducted in Europe, and five relate to empirical work in urban woodlands in Madrid (Spain), and conclusions from these studies are useful. Human disturbance was found to negatively influence the number of bird species, their persistence, and guild density (Fernández-Juricic 2000), along with blackbird feeding strategies, habitat selection, and abundance (Fernández-Juricic and Tellería 2000). However, various factors affect an animal's tolerance of disturbance and subsequent likelihood of flight, particularly the surrounding habitat structure and composition (Fernández-Juricic 2000). In essence, alert distances and individual "buffer zones" vary with the presence of "escape cover" such as shrub and tree cover. This effect is reported in the wider literature (e.g. Langston et al. 2007). Interestingly, Fernández-Juricic et al. (2001b) noted that blackbird buffer distances were greater in "highly visited" parks, which the authors related to habituation. Studies relating to other birds associated with woodlands in the UK include black grouse (*Tetrao tetrix*) and capercaillie (*Tetrao urogallus*). Baines and Richardson (2007), for example, report that "The disturbance regimes imposed had no discernible impact upon black grouse population dynamics," although one study by Patthey et al. (2008) revealed a considerable impact of skiing on black grouse populations in the European Alps. An earlier study of red grouse (Picozzi 1971) similarly showed no negative breeding impact, stating that grouse bred no worse on study areas on moors where people had unrestricted access, and grouse bags showed no evidence of a decline associated with public access agreements. Newton et al. (1981) investigated the potential impacts of recreation walkers on merlin (*Falco columbarius*) in the Peak District National Park (UK). Their conclusion was that it was "unlikely" to have caused the sharp decline in merlins during the 1950s but that it could possibly slow re-colonisation. Other studies of merlin (e.g. by Meek 1988) similarly suggest little negative impact by recreation, instead focusing on general habitat degradation by agriculture and pollution as the most likely causes of decline. In contrast, studies of capercaillie suggest a negative impact by recreation activity (Summers et al. 2004, 2007; Theil et al. 2011) where attention was drawn to the birds' avoidance of woodland areas near tracks and suggested a causal connection between this and recreation use. Although counts of recreation visitors in these studies are very low, the authors find a statistically significant difference between capercaillie use of wooded areas adjacent to tracks classified as "high" and "low" human use. Extrapolation from total track length led these authors to assert reduced woodland "carrying capacity" as the species avoids using between 21 and 41% of the two forests studied. Studies of forest bird disturbance by walkers and dogs beyond the UK reveal some useful findings. In their study of 90 peri-urban (urban fringe) woodlands north of Sydney, Banks and Bryant (2007) identified a substantial, although seemingly short-term, effect of dogs on native birds, especially ground-nesters. They found that dog walking caused a 41% reduction in the numbers of bird individuals detected and a 35% reduction in species richness compared with areas where dogs were prohibited, but they suggested that the long-term impacts may be small. Nevertheless they argue against access by dog walkers to sensitive conservation areas. In the UK, a high proportion of walkers using woods and forests are accompanied by dogs: Taylor et al. (2005) assert a figure of up to 50% in lowland areas, with fewer in upland areas This can serve to increase (in some cases dramatically) the scale of disturbance (or "sphere of influence," Taylor et al. 2005). The impact of dogs has received widespread attention, although again primarily in relation to ground-nesting birds, although Miller et al. (2001) illustrated increased disturbance of mule deer by dogs and in non-forest environments.

In contrast, Gutzwiller et al. (1998) found little evidence that intrusion altered vertical distributions of four passerines that nest, forage, sing, and seek refuge in subalpine forest. The minimal effects they observed indicate that the species studied were able to tolerate low levels of intrusion. Similarly, in their study of nesting northern cardinals in riparian forests in Ohio (USA), Smith-Castro and Rodewald (2010) found no association between nest survival and the tendency of birds to flush. On balance, the available evidence does not indicate significant negative impacts on forest birds following "flight" responses to walking, including no clear long-term or population-level impacts.

However, responses to human intrusion are well documented for a range of species. These include elevated heart rate (Weimerskirch et al. 2002), increased alarming or defensive behaviours (Andersen et al. 1996; Reby et al. 1999), and ultimately the avoidance of high-risk areas, either completely or by using them for limited periods only (Gill et al. 1996). Disturbance by people can also increase the risk of predation (Anderson 1988). Consequently, in areas where levels of human activity are high, repeated disturbance by visitors can lead to a reduction in the survival or reproductive success of individuals (Goodrich and Berger 1994; Burger et al. 1995). Ground-nesting birds, such as waders (*Charadriidae* spp.), are thought to be particularly at risk from human disturbance. When approached, birds often flush from nests, leaving eggs and chicks exposed to possible chilling or predation and imposing an energetic cost on the adults (Nudds and Bryant 2000; Bolduc and Guillemette 2003). For example, human disturbance is thought to significantly reduce the chick-rearing ability of African black oystercatchers *Haematopus moquini*, which breed on the coasts of South Africa at the height of the summer tourist season; breeding success outside protected areas was approximately one third of that on reserves (Leseberg et al. 2000). Similarly, human disturbance was found to interrupt incubation and reduce chick foraging time in New Zealand dotterels *Charadrius obscurus* (Lord et al. 1997, 2001). However, other studies have found no evidence of an adverse effect of human disturbance on bird populations (Gill et al. 2001; Verboven et al. 2001). Nevertheless responses to transient human disturbance seem well known, and they are predicted to lead to population-level impacts on some bird species (Hill et al. 1997). Local wildlife does not seem to become habituated to continued disturbance because the effects of dogs seem to occur even where dog walking is frequent.

Data collected over 13 years was used to investigate the impact of recreation disturbance on the distribution and reproductive performance of golden plovers breeding in close proximity to the Pennine Way (UK), an intensively used long-distance footpath (Finney et al. 2005). Importantly, the Pennine Way was resurfaced in 1994 to prevent further erosion of the surrounding vegetation. Finney et al. (2005) were therefore able to examine if the response of golden plovers to recreation disturbance was influenced by changes in the intensity and extent of human activity resulting from the resurfacing work. Before the Pennine Way was resurfaced, golden plovers avoided areas within 200 m of the footpath during the chick-rearing period. At this time over 30% of people strayed from the footpath, and the movement of people across the moorland was therefore widespread and unpredictable. Following resurfacing, over 96% of walkers remained on the Pennine Way, which significantly reduced the impact of recreation disturbance on golden plover distribution; golden plovers only avoided areas within 50 m of the footpath at this time. Despite the clear behavioural responses of golden plovers to the presence of visitors, there was no detectable impact of disturbance on reproductive performance.

These findings are consistent with those from an earlier study (Yalden and Yalden 1989), which used the alarm-calling behaviour of adult birds to estimate the sensitivity of golden plovers to visitor disturbance. They found that the average distance at which adult birds began alarm-calling in response to an approaching human was approximately 200 m during the chick-rearing period. This suggests that for breeding waders, similar behavioural studies could be used to indicate the distances from sources of disturbance over which habitat occupancy is likely to be reduced. For example, response distances include 75 m for common sandpipers *Actitis hypoleucos* (Yalden 1992), 100 m for New Zealand dotterels (Lord et al. 2001), and in excess of 1 km for both curlew, *Numenius arquata*, and redshank, *Tringa totanus* (Yalden and Yalden 1989). These figures suggest that the impact of increased recreation activity on breeding waders will vary depending on the sensitivity of the particular species concerned. For golden plovers, the avoidance of areas within 200 m of a footpath is unlikely to be a serious threat in places where a single footpath crosses a large area of suitable habitat. However, human disturbance may become a problem in areas where there is a network of footpaths.

Alwis et al. (2016) found that heavy use of nature trails for recreation has affected the bird species that occur on the Kudawa nature trail in the Sinharaja World Heritage Forest. Under high levels of disturbance, bird communities avoided edge habitats and flushed far into the forest (up to 150 m). Certain bird species seem to tolerate greater degrees of recreation disturbances. The sensitivity of individual bird species to visitor recreation disturbances varies with the stratum/layer of the rain forest usually occupied by these bird species. Effects of recreation disturbances were more profoundly felt by birds occupying understory and sub-canopy layers of the forest near the nature trail. Accordingly, the following species ashy-headed, laughing thrush, dark-fronted babbler, spot-winged thrush, Tickell's blue flycatcher, brown-breasted flycatcher, greater flameback, Malabar trogon, orange-billed babbler, Sri Lanka scimitar-babbler, and yellow-browed bulbul avoid habitat edges along the jungle trail under increased visitor activity. Bird species occupying the canopy and higher layers of the forest are more tolerant to recreation disturbances. A forest bird, the Sri Lanka blue magpie did not show noticeable signs of avoidance behaviour under human presence. Instead, they seem to be attracted towards small- to medium-sized visitor groups. This was evident by the high number of Sri Lanka blue magpies being recorded under low and moderate levels of recreation disturbances and their numbers showing a positive correlation with disturbance level, although the relationship was not statistically significant. This suggests a possible habituation of the Sri Lanka blue magpie population ranging around the Kudawa nature trail to low and moderate levels of recreation disturbances or human presence. In fact, during their field studies, it was often observed that a group of Sri Lanka blue magpies was perching near the trail in anticipation of food, when visitor groups were present. Similar behavioural observations were made on ground-occupying Sri Lanka junglefowl and visitors feeding both species were a common observation. Such visitor behaviour along with exposure to recurring recreation disturbances can alter the normal behaviour of birds and induce habituation to human presence.

Few studies have attempted to assess the impacts of flight responses to walking on forest species other than birds. Some studies show, for example, that human presence on foot can in some circumstances disturb wild deer. Langbein and Putnam (1998) and Recarte et al. (1998) studied disturbance of British park deer, although came to different conclusions. The former reported significant immediate behavioural responses of deer to human presence, but these had no long-term impacts (such as on body weights or overwinter mortality), whilst Recarte et al. (1998) reported less disturbance and concluded that level of disturbance response was related to surrounding habitat and habituation. Other UK deer research includes Ward et al. (2004), who found that wild roe deer (*Capreolus capreolus*) did not flee from or otherwise change their behaviour, when disturbed by night-time ecological survey. They were found, however, to avoid paths and roads even at night when human activity was very low. In a US study, Miller et al. (2001) reported that for all species, area of influence, flush distance, distance moved, and alert distance (for mule deer) were greater when activities occurred off-trail versus on-trail and that for mule deer, the presence of a dog resulted in a greater area of influence, alert and flush distance, and distance moved than when a pedestrian was alone. Studies by de Boer et al. (2004) and Marini et al. (2008) highlight a number of factors affecting the flight responses of wild deer. The structure of surrounding habitat is repeatedly identified as a major factor. In the only study of disturbance of squirrels by recreation identified in this review, Gutzwiller and Riffell (2008) concluded that abundance of red squirrels at foot sites in the USA did not differ significantly from that at other control sites during their experiments. Although immediate/short-term behaviour change may be apparent, the limited available evidence shows little or no long-term negative impacts upon forest mammals following "flight" caused by walking in woodlands.

2.14 Impacts Are Not Always Negative

A study of salamanders actually identified a beneficial relationship between trail presence and species success, noting that trails result in more microhabitats for salamanders around them (Davis 2007). However, other analyses of human disturbance of reptiles describe some significant negative impacts, for example, the removal and accelerated decay of woody debris vital for skinks (Hecnar and M'Closkey 1998).

Invertebrates must be affected by walking trampling pressure, and an example can be given from sand dunes at Dundrum (Northern Ireland) where Buchanan (1976) found the following: (a) A lowering of total animal numbers and the number of species (diversity was lower) occurred. Mites and Collembola (springtails) which form the bulk of the terrestrial soil fauna are reduced by around 90% in the fixed yellow dunes by a trampling pressure of just over 13 people/m/hour. 1500 people/year seemed to be insufficient to cause a reduction in invertebrate numbers in Calluna heath or fixed dunes but caused a 45% decrease in mite populations in bracken heath, whilst the springtails were reduced by 65%. (b) The roundworms and threadworms increased significantly with trampling, but they decreased in the Calluna and bracken heath. However, some doubts were expressed about the validity of these results because of the extraction method used. (c) An increase in scavenging species occurred because of the deaths of many animals, and there was an increase in species associated with ephemeral habitats, for example, the tiger beetle probably requires the presence of bare, sandy paths to be kept open by trampling in order to survive in heathland localities.

2.15 Management Implications for Recreational Walking-Induced Change in the Landscape

The evidence presented so far, systematically accumulated across field observations and measurements and high-quality experimental studies, suggests that vulnerable vegetation of conservation value should not be trampled, irrespective of the projected intensity of use: even moderate disturbance can have significant effects on plant communities. Simple indicators such as life-form of the dominant community may be useful for rapid assessments of a community's vulnerability to recreation pressure. The evidence is clear that positive management to conserve the physical, biological, and aesthetic qualities of the countryside is needed, and this implies control of the walker's behaviour, the management of vegetation, and improvements in trail surfaces in many cases. The management approaches are many, but they involve either raising the capacity or reducing use, and usually a combination of elements of both. The alternative is to do nothing and accept deterioration, but this is not a viable alternative for recreation land managers who are looking after national parks and wilderness areas. What is important is a consideration of the recreation carrying capacity which is the level of recreation an area can sustain without an unacceptable degree of deterioration of the character and quality of the land resource in terms of vegetation, soils, and the recreation experience. However there are various types of carrying capacity and not just one. The ecological carrying capacity is the maximum level of recreational walking that can be accommodated before there is a decline in the ecological value, assessed from an ecological viewpoint. There has been criticism of this definition because it fails to take sufficient account of any acceptable change away from the desired state. It relies on three conditions which have to be accepted: there is a most desirable state, there is a degree of change away from this which is only just acceptable, and both of these are matters of judgement. In a country park close to urban areas, a rye grass cover may be acceptable, but in a protected nature reserve, the loss of a single species may be looked at as too much to pay for a certain use level. There are also problems because a number of factors can bring about a change in ecological carrying capacity such as wet weather, steep slopes, and the kind of use. Thus this concept is never likely to be a useful applied principle where the number of people which bring about whatever threshold of change is used to define it can be predicted for a wide range of situations: the concept seems specific to particular sites and local circumstances. However, productive vegetation generally has a higher ecological carrying capacity than vegetation on poorer soils, and most amenity ecosystems are unproductive and are so because most of the fertile soils are under cultivation. The physical capacity is the maximum level of use that a site can accommodate spatially and is never likely to apply on recreation trails, but the perceptual carrying capacity, which is the use level above which there is a loss of enjoyment because of overcrowding, may well occur. Again it varies between different people, but it is likely to be well below the physical capacity.

What we can say from all this is that carrying capacity depends on the management policy and is the maximum intensity of recreational walking use an area will continue to support under a particular management regime without inducing permanent change in the biotic environment maintained by that management. As Wagar (1964) put it, "the final definitions of recreational carrying capacity must be of an administrative nature." The recreation manager has effectively four attitudes to adopt: (a) ensure that recreational walking exerts a minimal modifying influence on the ecosystem, (b) attempt to retain the essential character of the ecosystem but otherwise accept some changes from the walking, (c) replace those elements of the ecosystem which are more susceptible to trampling pressure by components which are more resilient, and (d) ignore ecological changes resulting from the trampling pressure. However, in most cases with suitable management, the planned use of an area for recreational walking is perfectly compatible with the maintenance of this ecological value. The degradation from overuse must be prevented, and walkers must be excluded from using certain paths or discouraged from doing so.

Zonation of recreational walking into high- and low-intensity usage, with "honeypots" located away from vulnerable vegetation, may be a more effective conservation strategy than encouraging moderately intensive but more widespread walking usage. This is especially given that occasional use can result in the development of informal path networks which may subsequently encourage further disturbance (Roovers et al. 2004). The siting and development of new trails, nature trails, car parks, toilets, and information centres can be used to dictate the recreational walking use pattern and help implement a zoning policy.

The potential strategies to manage recreation walkers and their experiences are many and have been discussed in many publications, but excellent reviews can be found in Cole et al. (1987), Leung and Marion (2000), Manning and Anderson (2012), Hammitt et al. (2015), and Marion (2016). The many elements usually embrace trying either to raise capacity or reduce the use of an area. Reducing the use of the entire area may be attempted by limiting the number of walkers; limiting the length of stay; charging increased fees either to enter the recreation area or to park in reduced car parks, which will reduce access; encouraging the use of other areas by developing better facilities there; and advertising extensively these new areas or existing alternatives. Reducing the use of the areas where problems occur can be attempted by informing the recreational walkers of the problems of these areas (Figs. 2.1A and 2.5) and/or the advantages of alternative areas, limiting the numbers to these

Fig. 2.5 Sign asking for no access up the volcano flank because of the highly erodible coarse volcanic ash, Lanzarote Geopark, Montana Corona. Photo by David Huddart

areas or limiting the stay length, charging differential fees, prohibiting use, making access more difficult, and eliminating facilities.

A modification of the use location within problem areas can be attempted by locating facilities on durable sites, making trails more durable, supplying information on these walking trails, and banning off-trail walking. A key effort must be made to reduce off-trail hiking, and methods to try and achieve this have been developed and evaluated by Clark and Leung (2007) and Hockett et al. (2010, 2017). It might be possible to modify the timing of use by encouraging use outside peak use periods, by charging higher fees for use during high-use or high-impact periods or lowering fees for use during low-use or low-impact periods, and by discouraging or banning use when there is high-impact potential. It might be possible to change visitor behaviour by information and education and to modify their expectations. The resistance of the trails can be strengthened in various ways, and the problem areas can be monitored, maintained, and/or rehabilitated.

Soils vary greatly in their resistance to wear. Surfaces with a high proportion of coarse particles (rocks or stones) are generally least affected by recreation use, and clay and peat soils are most affected in that they show the greatest erosion of soil material and changes in soil structure. Most soils have lower resistance to wear under wet ground conditions because water acts as a lubricant and allows soil particles to rub against each other and soil compaction to occur. Peat and clay soils are particularly vulnerable to wear under wet conditions and often require special protection, such as improved drainage, more durable surfacing, or reduced levels of use.

The wear-resistant properties of vegetation vary from species to species. However, the types of wear exhibited by different communities reflect not only the durability of the vegetation but also the extent to which the vegetation structure limits or modifies recreation movement. For example, stands of rushes are not easily walked through, so use and wear tend to be confined to meandering routes between clumps. In contrast short grassland has high resistance to wear but low resistance to movement, resulting in extensive scuffing with only a few areas of localised severe wear. Bog and fen have poor wear resistance but high impedance to movement, with the result that there tend to be fewer but heavily worn paths.

What we can see here is that certain environments are prone to recreational walking pressure: peat soils, upland blanket bog, and water-logged soils. We will look at a case study from the Three Peaks project (English Pennines) in Sect. 2.20, which has been particularly badly affected by trampling pressure and many remedial approaches have been developed to try and counteract severe erosion problems in that area.

2.16 Path Wear and Deterioration

The factors that influence path deterioration are many and include soil and vegetation characteristics and the type and intensity of use. There are, however, some relatively straightforward relationships between site characteristics and path wear that are of fairly general applicability and which have important implications for path design and management. Much pioneering work for remedial work on footpaths was carried out on paths in the Cairngorms, in other parts of Scotland, and in the Yorkshire Dales (see later section) by Bayfield and Miller (1988) and Bayfield et al. (1990, 1991a, b) but subsequently confirmed by studies in other types of terrain (review by Liddle 1989). In the Cairngorms investigations, the effects of site factors were determined by examining path sample data where site characteristics other than the major variable were similar, for example, samples that varied in surface wetness but were otherwise similar in terrain type, level of use, and soil and vegetation types. The data showed that irrespective of the level of use, path width increased with the surface wetness and roughness and also with the angle of slope along the path. More surprising was that the roughness of adjacent ground decreased path width; and paths were narrowest where they passed through the roughest terrain (provided the path surface remained of the same quality). Although fairly obvious, these relationships can be of considerable importance for path design and management. Clearly for minimum path deterioration, the path surface needs to be dry and smooth (minimum values for wetness and roughness) and it should not have steep slopes where possible. The effect of adjacent ground roughness implies that paths should be sited through broken terrain rather than smooth ground, if a choice exists. Furthermore, deterioration might be limited on existing routes by increasing the roughness of ground adjacent to paths by placing rocks or logs or planting tussocky or otherwise coarse vegetation as obstacles to help reduce path spread. A further important observation concerned the relative impacts of uphill and downhill walking. Downhill walkers were found to take more steps and to have a greater impact with each step than uphill walkers. They also tended to deviate more from the centre of the path than walkers going uphill, possibly because when walking uphill, the field of view (and choice) is relatively restricted. The differences increased with slope steepness. These observations imply that it may be possible to reduce the impacts of use if the direction of travel can be manipulated so that users go mainly uphill on steep slopes and downhill on gentle slopes. On circular walks, for example, there is usually an optional direction for use.

2.17 Techniques for Managing the Footpath Surface

2.17.1 Creating More Resistant Footpath Surfaces

Land managers can apply more sustainable construction and maintenance techniques to increase the ability of their trails to resist impact by adding types of stonework, such as stone flags laid across peat, aggregates, borders, and boardwalks, or adding drainage features, like water bars, drains, treads that dip away from the slope, and ditches. Much pioneering work took place in the UK and was published in handbooks and research reports by Agate (1983), Bayfield and Aitken (1992), Davies and Loxham (1996), Barlow and Thomas (1998), Backshall et al. (2001), the National Trust for Scotland (2003), and a whole series of reports published by the Institute of Terrestrial Ecology, Banchory, Scotland, published by Bayfield and his co-workers, such as Bayfield and McGowan (1986) and Bayfield and Miller (1988). In the USA there were publications addressing similar issues, for example, by Hingston (1982), USDA Forest Service (1985), Birchard and Proudman (2000), Marion and Olive (2006), and Hesselbarth et al. (2007).

2.17.1.1 Geotextiles

These are water-permeable textile materials (fabrics,) used as an underlay to conserve gravel on trails and stabilise erodible surfaces. The textile allows water to pass through it but keeps soil layers from mixing and breaking down. There are three main types of geotextiles used for erosion control: (1) nettings laid on the surface mainly to trap sediment and slow surface runoff (Fig. 2.6), (2) partly buried three-dimensional nettings intended to provide some shallow subsurface stability and sediment trapping, and (3) subsurface cellular webs, which provide deeper surface stability.

Surface Nettings

Geojute is an example of this category which is a jute-based, open-weave netting with a mesh size of 1–2 cm. The netting is rolled out and pegged in place with long wire staples. It has the advantages of stretching and closely fitting irregular ground contours (particularly after being wetted) and is also biodegradable. The jute retains some of the moisture from rain and surface flow and rots to provide surface organic material, both features which may be of minor benefit to establishing vegetation. As the netting is merely pegged to the surface, no special soil preparation is required.

2.17.1.2 Three-Dimensional Nettings

Enkarnat types of geotextile are made from two or more layers of fine and coarse grade net, tacked together to provide both reinforcing and soil-holding abilities. In use they are laid on the ground, and soil is worked into the upper layer. The geotextile is thus placed at or just below the

Fig. 2.6 Geotextile. Originally covered with aggregate, but this has been eroded as the geotextile was not laid deep enough so that there was an adequate aggregate fill. It should have trapped sediment and slowed surface runoff (Three Peaks project). Photo by David Huddart

surface. These materials are stronger than jute netting but not biodegradable. Unless laid very carefully, they do not fit as closely to the surface as jute, and sometimes surface runoff can wash out soil from under the netting, leaving the netting suspended, visually intrusive, and ineffective. Geocells and geogrids are three-dimensional webs providing a network of cells resembling a honeycomb (e.g. Armater). The webs are placed on the slope to be protected, pegged down at intervals with stakes, and the cells filled with soil. This is a substantial reinforcement technique that can provide subsurface stabilisation to about 10 cm or more and some surface protection by reducing the runoff velocity and runoff and sediment trapping. Due to its high cost and heavy earthmoving requirement, it is not often justified for erosion control at recreation sites but has been effective in increasing the resistance and load-bearing capacity of wet tread substrates. Other effective products include drainage mats (Polnet) and turf reinforcement mats, like Pyramat (Monlux and Vachowski 2000; Meyer 2002; Marion and Leung 2004; Groenier et al. 2008).

2.17.1.3 Chemical Binders

There are a number of soil stabilisers using different chemicals that have been developed to increase the adhesion of soils, improving moisture resistance and bearing and shear strength of the soil (Bergmann 1995; Meyer 2002).

2.17.1.4 Mulch Mats

Mats, such as Greenfix, are sheets of lightweight netting enclosing a mulch layer of straw, coir, or other organic materials. The mats are sometimes pre-sown with appropriate grass seed or may be laid on top of sown slopes. They are pegged down with wire staples and provide protection from raindrops, reduce runoff velocities, and trap sediment. The mulch can have the disadvantage in some situations of stifling the growth of vegetation, although in the longer term, decomposition adds valuable organic matter to the soil. Mulch mats are bulky and relatively costly and, like geowebs, not often justified at small-scale recreation sites.

2.17.1.5 Mesh Elements

This technique involves reinforcing soils by mixing them with small pieces of plastic mesh. The mesh acts as a root substitute to strengthen the soil and to some extent traps sediment and reduces runoff velocities. The main drawback of this technique is the problem of satisfactorily mixing the mesh elements into the soil to be protected. In commercial practice soils and mesh are often pre-mixed off site, but this substantially increases the cost.

2.17.2 Surface Glues

Surface glues, or soil stabilisers, are usually applied to soils as a component of hydroseed mixtures or sprayed on after manual seeding or planting. They form a porous skin on the surface to prevent soil particles washing away and are mainly intended for short-term protection. As in the case of geotextiles, there are a large number of products on the market. All appear to be more or less efficient as glues, although there is very little comparative information on the effectiveness of different formulations. In selecting a product, important considerations will be cost and the method of application. Powder glues are much easier to apply to small areas than emulsions, which tend to clog small sprayer equipment. All types can be used in hydroseeding. Although each manufacturer claims low toxicity, some products are more toxic to plants than others. Bayfield and McGowan (1990) compared the toxicity of a small range of glues on both grasses and bryophytes and found use had a stimulating effect on the growth of some test species. It was not clear if this effect was due to release of nutrients or some hormonal effect of the glues.

2.17.3 Surface Moulding

This technique consists of cutting horizontal ledges or grooves in slopes instead of dressing them to a flat profile. This approach is fairly common in the USA and is also used in the tropics but appears to have been rarely tried in the UK. In the USA, serrated "steps" are recommended, between 15 cm and 1.3 m high. In Malaysia, grooves 5 cm deep and 25 cm between centres are cut in the dressed surface. Both methods help trap waterborne sediment, reduce runoff velocities, and increase slope permeability. Trials by the Malaysian Thai Development Company indicate about 25–75% better germination of grasses and legumes on grooved slopes, probably because of better permeability and moisture retention than on ungrooved slopes.

2.17.4 Aggregate Paths

Aggregate paths are mainly suitable for use on mineral soils, where there is firm subgrade material. Simple excavation of a path base is followed by infilling with an appropriate fill material. No reinforcing is required although a filter geotextile may help to prevent the surface becoming clogged with fines. Angular quarry graded material is the most suitable fill since it will pack down firm and solid. Often, though, material from local sources, such as borrow pits or streams, will be used. These materials may be satisfactory, but problems can arise if there is too high a proportion of rounded gravel or too much or too little clay present: these combinations make poor path surfaces. A solution can be to mix in a proportion of angular material or clay to improve the properties of the mixture. Mixing can, however, be difficult and time-consuming, particularly at remote sites. A variation is to form a wearing surface with stones or rock ("cobbling"). This can be very durable but has to be carefully laid and consolidated, as protruding angular material can be both unpleasant to walk on and even dangerous. Effective drainage is important to prevent scour and loss of finer surfacing material. It should be possible to build in a camber or shedding slope, but in practice this appears to be difficult to achieve and sustain.

2.17.5 "Floated" Aggregate Paths

Aggregate paths can be rafted or floated on soft ground or peat by laying them on geotextiles, to prevent them from sinking. The earliest such use of geotextiles in path construction was with Terram. Although this material is a good filter textile, it lacks tensile strength, and some of the early trials were not successful, as the resultant paths tended to slump and sag. In recent years there has been recognition that over wet amorphous peat, slabs need to be laid onto a base of chestnut paling, about 1.5 m wide, bound with polypropylene cord. In a few locations, slabs have been laid onto a base of aggregate, but this appears to be overspecification and should not be required at most sites. Except at very wet locations, the evidence is that direct laying onto bare peat is satisfactory as there have been few instances of sinking or loose slabs.

2.17.6 Boardwalk

Boardwalk is a well-established technique for creating an acceptable walking surface across difficult ground, usually wet and peat-covered, although also very rocky ground (Fig. 2.7). Possibly the oldest documented path in Britain, the Bronze Age "Sweet Track" through a bog in Somerset is a type of boardwalk, made from split logs. Boards for the surface can be laid either across or along the route. In either case they normally rest on bearers to tie the structure together and to help spread the load. Many boardwalks are built on site, particularly where the ground is uneven or the route circuitous, but prefabricated sections or rafts are sometimes delivered to site by vehicle or helicopter. Boardwalk is relatively more temporary than most path surfaces in that it can be lifted and removed fairly easily without leaving much damage. Sections need to be checked regularly for wear and tear since, more than most path techniques, it manifestly implies a public safety obligation on the manager. For similar reasons it should be covered with chicken wire or tar and gravel to counteract any slippery surface. Boardwalk is an excellent solution for paths through wet sites of high ecological interest as it avoids any interference with lines of drainage. It can also be successful in woodland and coastal settings. Although it does not offer a pleasant walking experience over long distances, it has nevertheless been used in quantity in some remote areas such as the Pennine Way in the Cheviots, where there is a shortage of local materials for alternatives.

2.18 Vegetation Reinstatement

Given time, almost any bare soil surface will be revegetated by natural colonisation. Studies of bulldozed hill roads in the Cairngorms show that even at almost 1000 m, sites are probably recolonised within five years, and some in two or three. Even these rates are, however, unacceptably slow for most recreation sites, where the aim is to reinstate bare or damaged surfaces as rapidly as possible, to minimise both erosion and visual intrusion. In the past it has been acceptable to merely provide some kind of vegetation cover, in fact any kind of vegetation cover. Recently though planners and managers have begun to demand higher standards of revegetation, involving use of native species and landscaping schemes that try to blend damaged areas to the surrounding ground.

Reinstatement aims to use appropriate native species, to use local or native strains which are similar to the impacted ground, to create ecologically diverse stands of vegetation, to prepare surfaces for planting that blend with surrounding

Fig. 2.7 Boardwalk across extremely rough aa lava, Timanfaya National Park, Lanzarote. Photo by David Huddart

landforms, and to integrate engineering, vegetation reinstatement, and landscaping management of damaged sites.

2.18.1 Transplanting

The techniques described here are all ways of reusing existing plant resources. They include movement of intact vegetation in the form of turves or larger "clods," dividing clumps of vegetation, bare root transplants, and taking cuttings. All transplanting techniques have the advantages of providing a greater or lesser degree of instant cover and of being able to use local material (where available) appropriate to the site. All the techniques also have the disadvantage of having a relatively high labour requirement and are usually more costly than corresponding seeding methods.

2.18.2 Seeding

Seeding is probably the most widely used method of reinstating damaged ground at recreation sites, but the issues include the selection of appropriate seed mixtures, methods of sowing, and the use of species as nurse cover, or as permanent contributors to vegetation.

There are three main methods: drilling by direct placement of seeds in the soil, broadcasting (dry spreading seeds), and hydroseeding (spreading seeds in a water slurry, usually with other ingredients, such as fertilisers and tackifiers). Broadcasting sowing is the most commonly used method and can involve hand-sowing or the use of various types of backpack or tractor-mounted, seed spreaders. Care needs to be taken to ensure even spreading of seed. With very small seeds, it is useful to mix the seed with sand or fine sawdust to make it easier to see gaps or dense patches. The ground to be sown is best raked and roughly levelled to form a seedbed prior to sowing. After seeding, rolling or some other method of light compaction will help partially bury seeds and keep the surface moist. Hydroseeding is a relatively large-scale operation like drilling but is better suited to sloping and rough ground. Little surface preparation is needed, although establishment will be improved if the surface is raked and levelled. Hydroseed slurries are sprayed on to areas to be seeded from a vehicle-mounted nozzle or using extension hoses. The slurry typically includes seeds, soluble and slow-release fertilisers, and peat or woodpulp mulch and tackifier (soil stabiliser or glue). Most of the ingredients are insoluble and have to be kept in suspension by agitation prior to spraying. Although this is essentially a large-scale technique requiring special equipment, it has been quite widely used at ski resorts such as Cairngorm and has been tested on parts of the Three Peaks footpath system.

Overall though the site management actions should remain as unnoticeable as possible, and they should be visually and ecologically less obtrusive to natural conditions than the walkers' impacts that prompted the remedial action (Marion and Sober 1987).

2.19 The Trampling Impact on Blanket Peat and Other Organic-Rich Soils

Much of the published information indicates that there have been progressive declines in the condition of footpaths, particularly in the uplands, and that in some places there has been a proliferation of routes. For example, early observations on the popular, intensively used Pennine Way long-distance footpath in Northern England in 1971 and 1983 (Bayfield 1985) showed that bare and trampled widths had about doubled over this period. Later monitoring in 1988 showed that the width of bare ground had increased by 300–900%, resulting in an average width of 7–8 m of bare peat but extending to 70 m in places and an estimated peat loss of 10 mm/yr. (Porter 1990).

The removal of surface vegetation and exposure of bare peat soils by trampling also heighten the risk of soil erosion through the natural weathering by wind and water. In an upland environment such as a blanket bog, the linear nature of the paths on a gradient may localise water runoff, rapid transport of sediment, and soil deposition in watercourses, and change species composition, with extensive gully erosion, up to several metres deep (Morgan 1995; Grieve 2001).

2.20 The Three Peaks Project: Background to the Project (Yorkshire Dales, UK)

In 1986 a survey of the area's 65 km of footpaths found that 21 km were severely damaged and a further 21 km needed immediate remedial work to avoid them deteriorating to a similar state (Smith 1987). The average path width was found to be 11.4 m, whilst in north of Pen-y-ghent at Black Dub Moss, the path was 150 m wide. The paths had been getting measurably worse, and something had to be done in terms of management as the path erosion rivalled some of the more famous areas of the Pennine Way, with deep mud gulleys in places and paths on the flanks of the fells as ugly scars visible from a long distance. What was attempted in 1987 was that the Yorkshire Dales National Park set up the Three Peaks project where the aim was to repair the paths using techniques which would be ecologically and visually acceptable. This was an undertaking which would take some time since little work of this type had been carried out elsewhere, especially on deep peat soils. Initially the project was set up to run until 1992 and was funded by the Countryside Commission, the Sports Council, and the Nature Conservancy Council. These fragile uplands could not withstand the levels of use without remedial work because the alternative was destruction and permanent loss of a unique upland habitat. The measured use levels were obtained by mechanical counter-stiles after an initial visitor survey in 1985 had established that Ingleborough had 150,000/yr., Pen-y-ghent 60,000/yr., and Whernside 40,000/yr. The techniques used were based around a whole series of both engineering and ecological trials to try and establish what was likely to succeed best and the potential value of path construction techniques on paths that were already severely damaged and where no other form of intervention seemed possible (Bayfield 1987).

2.20.1 Chemical Consolidation of Soil

Between 1986 and 1988, engineering trials on organic peat soils with a clay fraction of under 5% were attempted using a chemical consolidator of weak, clay soils called "Solidry." The chemical was supposed to stick the soil particles together, giving the soil an improved structure and so increasing its strength. Although the manufacturer suggested an application rate of 1% by weight, the chemical was tested at rates of 1%, 1.5%, and 2% on 20-m path lengths at Churn Milk Hole, on the southern flank of Pen-y-ghent. Each section had been rolled and resulted in a surface strong enough to carry a truck weighing four tonnes. The surface remained intact for three months until the hard winter of 1987 when freeze-thaw transformed the whole stretch into a quagmire. An application of 4% was tried, but this proved no more successful, and so the entire path was covered in bituminised geotextile matting and also on some stretches with limestone chippings. So this chemical consolidation was not a success, especially as a later trial at High Birkwith at lower elevation also failed.

2.20.2 Aggregate Path Construction

Traditional methods of path construction on organic soils involve the excavation of the weak, load-bearing soils. However, this technique has considerable drawbacks: the impact on the site hydrology is marked especially if peat pipes are encountered, and it may have localised impact on the adjacent plant communities. The disposal of unwanted soil is often problematic, and narrow paths can create a serious barrier to the use of machinery for future path building and maintenance. Therefore alternative methods for aggregate path construction were examined. The most extensively used repair technique has been the use of matting covered by stone chippings. Two different types of mat have been used: (a) a plastic mesh alongside the Ribblehead Viaduct and (b) geotextile matting which is much denser than the plastic matting, and it was used on Little Ingleborough and Pen-y-ghent. The major difficulty is that these paths were far too wide (2 m), but the reason was that when the stone chippings were laid, they have to be rolled flat, and therefore the path has to be wide enough for the roller and most of the stone had to be transported by dumper truck. A further variation on the theme was used at Churn Milk Hole where a bituminised geotextile mat was laid, and where the ground was most boggy, chestnut palings were placed underneath. Apart from a tendency to distort to the shape of the palings which produced a slightly corrugated path, this proved successful. There are many different geotextile products, and they serve one of the three primary functions—filtration, separation, and reinforcement—but most geotextiles are only designed to meet one of them. In fact there was an evaluation of a range of geotextiles: initially there was an assessment of spun and needle-punched products such as Terram and Typar. However, the entire range was eliminated as they proved so elastic that they tended to either buckle and then rupture under the load of the machine carrying stone or swell and blister due to excessive groundwater pressure. Then six varieties of woven geotextiles were tested between 1987 and 1991 and high tensile strength geotextiles and geogrids in 1988–1989. The geogrids laid directly onto the ground and then covered with at least 200 mm of stone chippings were most effective, although there was some stone migration from the two sloping edges of the path. Over 2.85 km of path was constructed on Whernside from Bruntscar and on Simon Fell Breast and proved very effective in canalising use.

2.20.3 Temporary Boardwalks

These were used well before the Three Peaks project began such as in the mid-1980s across High Lot Nature Reserve on the flanks of Ingleborough built of old railway sleepers and with pile-driven posts deep into the peat, giving more of a bridge over the muddy peat. However, this type of temporary boardwalk is costly to lay, and therefore a decking system called Flexboard was evaluated on Whernside at Force Gill (1989) and on High Lot (1990). Here flexible boards constructed of treated timber plank, bound by iron bands and joined together by iron hinges and pins and held in place with anchoring pins were airlifted in by helicopter. They were surfaced in two different ways with chicken wire on High Lot and with tarmac and chippings at Force Gill. The latter is much harder-wearing and effective except in severe frosts when ice fills in voids in the tarmac and therefore becomes slippery and the chicken wire tends to rip as the walkers slip on wet boards. However, they are highly effective in canalising use, and 200 m/day could be laid and so was very efficient. Due to the seed bank in the soil, the areas which were up to 50 m wide of black mud became recolonised right up to the edges of the flexboard. They were effective and could be laid on slope up to 15°, but one problem was that the toe of the walker's boot was not able to gain sufficient purchase, especially when ascending.

2.20.4 Stone Pitching

This was used on severely sloping sites up to 40^0 where aggregate paths were unsuitable, especially in the upper reaches of paths where they meet the Millstone Grit summits of the mountains. Stones are placed vertically in the ground so that two thirds of each stone is buried, and the technique has been widely used successfully in the Lake District and Snowdonia (Fig. 2.8). However, in the Three Peaks unfortunately a year later, the stones had moved as subsurface channels had been breached during path construction, so the path had to be re-laid and had to incorporate a base layer of concrete which proved effective.

Fig. 2.8 Stone pitching in Snowdonia, North Wales. Photo by David Huddart

2.20.5 Mechanised Path Construction Using Subsoil

The reorganisation of soil profiles took place on Simon Fell Breast, Little Ingleborough, and Whernside using a 12-ton Hymac digger on loan from the Cairngorm Chairlift Company. The peat was taken off and rearranged on either side of the path to expose the underlying mineral soil. A ditch was dug on either side of the peat to aid drainage, and the path was surfaced using crushed limestone from either side of the path. This technique was relatively cheap and rapid.

2.21 Ecological Trials

These trials focussed on four specific areas: reinforcement of existing vegetation, restoration of severely damaged peat soils, revegetating mineral soils, and revegetating aggregate paths built as part of the engineering programme.

2.21.1 Reinforcement of Existing Vegetation

These trials had actually started in 1986 with fertiliser applications to try and increase the trampling resistance of the existing vegetation, but the results between 1986 and 1989 were inconclusive, and it was decided that fertiliser alone did not increase trampling resistance in a situation of continuing or increased trampling intensity. So synthetics as turf reinforcement were installed to act as reinforcement for the root zone, to allow the plants to grow through the synthetic layer where the roots would become intertwined with the fibres of the synthetic layer and so improve turf strength. Macadamat, a plastic matrix with the voids filled with bitumen, was used at Churn Milk Hole in 1989, and a whole series of products resembling synthetic carpets from the sports turf industry were evaluated at Grain Ings on Whernside. The vegetation was strimmed to provide a level surface, fertilised to promote vegetation growth through the synthetic material and then seeded. Some plots were fenced. However, these trials were not a success as the product costs were high, installation was difficult, and the cover did not achieve 10% on trampled and only 30% on untrampled areas. The Notts Sports Turf VHAF GR700 and the Tensar Mat with Tensar SS35 proved the most effective and were the only ones to allow the growth of the original turf through the synthetic material.

2.21.2 Restoration of Severely Damaged Peat Soils

The first phase involved trying to recreate the native vegetation by using seed mixtures comprising indigenous species, improving species diversity through the introduction of plant litter, germinating the seed bank, and relocating of turves for regenerating the native vegetation. The conclusions were though that large-scale implementation of this policy was not realistic because of the cost and that seed availability was not sufficient. The second phase of this trial was to use a seed mixture whose only function was to rapidly establish and stabilise the bare soil surface. Once this had stabilised, this would generate more favourable environmental conditions for invasion of native plants. Perennial rye grass was applied by hydroseeding, and over 14,000 m^2 of Whernside summit was reseeded by spraying seed, fertiliser, soil, and adhesive out of a spray gun mounted on the back of a vehicle. Within a month of seeding, the grass had begun to seed, and up to an 80% cover was established quickly. The same technique was used to spray the banks and ditches produced by the soil rearrangement and in an attempt to green up the white limestone chip path at Hunt Pot on Pen-y-ghent.

2.21.3 Revegetating Mineral Soils

The same techniques were used as in the previous trials, but tests from Whernside had shown that soil erosion from the high gradients and high rainfall made germination difficult, but pre-seeded erosion control blankets were used where the seed was held on a thin backing layer which was placed directly on the ground. This reduced the erosive potential, but the cost was the problem.

2.21.4 Revegetating Aggregate Path Surfaces

This was to minimise the environmental impact as the roots bind the aggregate and make the path more durable and the surface water and rainfall are intercepted. Hydroseeding was used, although at Ribblehead a material marketed under the name Fibresand was used. This was sand with random plastic fibres, and it did improve the vegetation cover, but as usual the main problem was cost.

2.21.5 Conclusions Related to the Three Peaks Project

The five years of this project proved that tackling footpath erosion and habitat restoration for a major path network is technically possible if the correct intervention strategies are used and it was possible to evaluate the effectiveness of a large range of management options. Specific solutions were found in the following areas:

Aggregate paths can be built over the most severely eroded peat soils which can be cost-effective in construction and maintenance. Subsoil paths are cheaper where the subsoil has a clay fraction, but further work was needed to establish whether this technique was able to guarantee success. Constructed paths can be effectively revegetated through seeding techniques, but the vegetation restoration techniques using native species were not as successful because such seeds are not commercially available from the right provenance and those that were available were not in the quantities needed. Fertiliser application by itself could not increase the trampling resilience of existing turf, but reinforcement through the installation of a synthetic layer appears possible. Hydraulic seeding offers major advantages in vegetation restoration techniques in comparison to more labour-intensive techniques, and erosion control blankets may prove effective at the most difficult sites where hydraulic seeding alone would not. Visitor attitudes towards the need for restoration work and the work itself have proved very positive. The preference of walkers was ranking from most preferred to least preferred: turf reinforcement, soil reorganisation, Flexboard, boardwalk, aggregate path, stepped boardwalk, and steps. However, satisfying walkers' needs can only go so far because, for example, turf reinforcement is only an option where damage is less severe and only on level terrain.

Concluding Remarks

We have established that recreational walking in its various forms has impacts on all types of recreation landscape ranging from wilderness through national parks to country parks in the urban fringes. The vegetation is impacted with any level of walking, but we have illustrated that in certain terrains, severe trampling pressure can result in major vegetation cover loss. At the same time as the vegetation loss occurs, there is parallel soil erosion. In high mountains with severe environmental conditions including high rainfall, low temperatures, and strong winds and on moorlands with blanket bog development and peat soils, there can develop deep gulley erosion. We also know that these recreation impacts will grow because the projected growth figures for recreational walking are high and it has grown consistently to become the most popular active outdoor pursuit in most parts of the world. This means that recreation land managers will continue to need a growth in resources and manpower to manage these ecological impacts, with a need for more research funding and the application of the experience that has built up in the last 30 years or so to counteract the problems and to manage the recreation resource for this activity in a sustainable manner. It seems even more difficult today to preserve the high-quality, natural environments for a high-quality, recreational walking experience, and although the area of land affected by such impacts remains relatively low on a world scale, some individual mountains and natural landscapes seem to have been lost as "honeypots," such as Snowdon in North Wales, the Yosemite valley in California, and some of the national parks, like Arches and Zion National Parks in the southwest of the USA where at certain times of the year the carrying capacity of the recreation land has been far exceeded. What is needed here is more effective education and regulation of the recreation walkers so that all areas are managed sustainably. There is no doubt that there is plenty of recreation resource available for walkers, even in such a small island as the UK, so on a world scale, there should be no problem in providing the demand for recreational walking whilst maintaining, conserving, and sustaining the natural landscape and its wildlife.

References

Agate, E. (1983). *Footpaths—A Practical Handbook*. British Trust for Conservation Volunteers.

Alwis, N. S., Peers, P., & Dayawansa, N. R. (2016). Response of tropical avifauna to visitor-recreational disturbances: A case study from the Sinharaja World Heritage Forest, Sri Lanka. *Avian Research 7*. https://doi.org/10.1186/s40657-016-0050-5.

Amar, A., Henson, C. M., Thewlis, R. M., Smith, K. W., Fuller, R. J., Lindsell, J. A., et al. (2006). *What's Happening to Our Woodland Birds? Long-term Changes in the Populations of Woodland Birds*. BTO Research Report 169, Royal Society for the Protection of Birds Research Report 19, BTO, Thetford and RSPB, Sandy, Bedfordshire.

Andersen, R., Linnell, J. D. C., & Langvatn, R. (1996). Short-term behavioural and physiological response of moose *Alces alces* to military disturbance in Norway. *Biological Conservation, 77*, 169–176.

Anderson, D. W. (1988). Dose-response relationship between human disturbance and brown pelican breeding success. *Wildlife Society Bulletin, 16*, 339–345.

Backshall, J., Manley, J., & Rebane, M. (2001). *The Upland Management Handbook*. Peterborough: English Nature.

Baines, D., & Richardson, M. (2007). An experimental assessment of the potential effects of human disturbance on black grouse *Tetrao tetrix* in the north Pennines, England. *Ibis, 149*, 56–64.

Banks, P. B., & Bryant, J. V. (2007). Four-legged friend or foe? Dog walking displaces native birds from natural areas. *Biological Letters, 3*, 611–613.

Barlow, J., & Thomas, M. (1998). *Mending Our Ways—The Quality Approach to Managing Upland Paths*. British Upland Path Trust.

Bayfield, N. G. (1971). Thin-wire tramplometers—A simple method for detecting variations in walker pressure across paths. *Journal of Applied Ecology, 8*, 533–536.

Bayfield, N. G. (1985). The effect of extended use on mountain footpaths in Britain. In N. G. Bayfield & G. C. Barrow (Eds.), *The Ecological Impacts of Outdoor Recreation in Mountain Areas in Europe and North America* (pp. 100–111). Recreation Ecology Research Group Report 9, London: RERG, Wye.

Bayfield, N. G. (1987). Approaches to reinstatement of damaged footpaths in the three peaks area of the Yorkshire Dales National Park. In M. Bell & R. G. H. Bunce (Eds.), *Agriculture and Conservation in the Hills and Uplands* (pp. 78–87). Natural Environment Research Council and Institute of Terrestrial Ecology, ITE Symposium 23.

Bayfield, N. G., et al. (1990). *Vegetation Reinforcement Trials, 1986–1989*. Banchory, Scotland: Institute of Terrestrial Ecology.

Bayfield, N. G., et al. (1991a). *Restoring Native Plant Cover by Seeding and Live Mulching: Trials in the Three Peaks Area on Peat and Mineral Soils, 1986–1989*. Banchory, Scotland: Institute of Terrestrial Ecology.

Bayfield, N. G. et al. (1991b). *Monitoring of Seeding Trials in the Three Peaks, 1991*. ITE Report No. 12, Institute of Terrestrial Ecology, Banchory, Scotland.

Bayfield, N. G., & Aitken, R. (1992). *Managing the Impacts of Recreation on Vegetation and Soils: A Review of Techniques*. Institute of Terrestrial Ecology, Banchory Research Station, Scotland, 100pp.

Bayfield, N. G., & McGowan, G. M. (1986). *Footpath Survey 1986*. ITE Report No. 1, Institute of Terrestrial Ecology, Banchory, Scotland.

Bayfield, N., & McGowan, G. M. (1990). *Re-establishment of Mountain Vegetation: Laboratory Screening Trials, 1988–9*. Banchory, Scotland: Institute of Terrestrial Ecology.

Bayfield, N. G., & Miller, G. R. (1988). *1987 Progress Report: Revegetation Trials*. ITE Report No. 4, Institute of Terrestrial Ecology, Banchory, Scotland.

Bayfield, N. G., Watson, A., & Miller, G. R. (1988). Assessing and managing the effects of recreational use on British hills. In M. B. Usher & D. B. A. Thompson (Eds.), *Ecological Change in the Uplands*. Special Publication Number 7, British Ecological Society, Blackwell, Oxford.

Beedie, P., & Hudson, S. (2003). Emergence of mountain-based adventure tourism. *Annals of Tourism Research, 30*, 625–643.

Bell, S. L., Tyrvächen, L., Sievänen, T., Pröbstl, U., & Simpson, M. (2007). Outdoor recreation and nature tourism: A European perspective. *Landscape Research Living Reviews* Retrieved from www.Livingreviews.org/lrlr-2007-2.

Bergmann, R. (1995). *Soil Stabilizer for Use on Universally Accessible Trails*. USDA Forest Service Publication 9523-1804-MTDC-P, Technology and Development Program, Beltsville, MD, 11pp.

Birchard, W., & Proudman, R. D. (2000). *Appalachian Trail Design, Construction and Maintenance* (2nd ed.). Harpers Ferry, WV: Appalachian Trail Conference.

Bolduc, F., & Guillemette, M. (2003). Human disturbance and nesting access of common eiders, interaction between visitors and gulls. *Biological Conservation, 110*, 77–83.

Borcard, D., & Matthey, W. (1995). Effect of a controlled trampling of sphagnum mosses on their Oribatid mite assemblages. *Pedobiologia, 39*, 219.

Bowker, J. M., Askew, A. E., Cordell, H. K., Betz, C. J., Zarnock, S. J., & Seymour, L. (2012). *Outdoor Recreation Participation in the United States—Projections to 2060: A Technical Document Supporting the Forest Service 2010 RPA Assessment*. General Technical Report SRS-160, US Department of Agriculture Forest Service, Southern Research Station, Asheville, NC, 34pp.

Bratton, S. P. (1985). Effect of disturbance by visitors on two woodland orchid species in Great Smoky Mountains National Park, USA. *Biological Conservation, 31*, 211–227.

Bratton, S. P., Hickler, M. G., & Graves, J. H. (1979). Trail erosion patterns in great Smoky Mountain National Park. *Environmental Management, 3*, 431–445.

Buchanan, K. (1976). *Some Effects of Trampling on the Flora and Invertebrate Fauna of Sand Dunes*. Discussion Papers in Conservation No. 13, University College, London, 42pp.

Buckley, R. (2004). Impacts of ecotourism on birds. In R. Buckley (Ed.), *Environmental Impacts of Ecotourism* (pp. 187–210). Wallingford and Cambridge, MA: CABI.

Burger, J., Gochfield, M., & Niles, T. H. B. (1995). Ecotourism and birds in coastal New Jersey: Contrasting responses of birds, tourists and managers. *Environmental Conservation, 22*, 56–65.

Burger, J., Jeitnert, C., Clark, K., & Niles, L. J. (2004). The effect of human activities on migrant shorebirds: Successful adaptive management. *Environmental Conservation, 31*, 283–288.

Cardoni, D. A., Favers, M., & Isacch, J. P. (2008). Recreational activities affecting the habitat use by birds in Pampa's wetlands, Argentina: Implications for waterbird conservation. *Biological Conservation, 141*, 797–806.

Carney, K. M., & Sydeman, W. J. (1999). A review of human disturbance effects on nesting colonial waterbirds. *Waterbirds, 22*, 68–79.

Clark, A., & Leung, Y.-F. (2007). *Research on Off-Trail Walking Behaviour, Impacts and Education Efficacy* (pp. 147–168). Department of Park, Recreation and Tourism, North Carolina State University.

Clow, D. W., Forrester, H., Miller, B., Ropp, H., Sickman, J. O., Ryu, H., et al. (2013). Effects of stock use and backpackers on water quality in wilderness in Sequoia and kings canyon National Parks, USA. *Environmental Management, 52*, 1400–1414.

Cole, D. N. (1991). *Changes on Trails in the Selway-Bitterroot Wilderness, Montana, 1978–1989*. USDA Forest Service, Research Paper INT-450, Ogden, UT, 5pp.

Cole, D. N. (1995a). Experimental trampling of vegetation. I relationship between trampling intensity and vegetation response. *Journal of Applied Ecology, 32*, 203–214.

Cole, D. N. (1995b). *Recreational Trampling Experiments: Effects of Trampler Weight and Shoe Type*. USDA Forest Service Research Note INT-RN-425, Intermountain Research Station, Ogden, UT, 4pp.

Cole, D. N. (2004). Impacts of hiking and camping on soils and vegetation: A review. In R. Buckley (Ed.), *Environmental Impacts of Ecotourism* (pp. 41–60). Ecotourism Series No. 2. Wallingford and Cambridge, MA: CABI.

Cole, D. N., & Bayfield, N. G. (1993). Recreational trampling of vegetation: Standard experimental procedures. *Biological Conservation, 63*, 209–215.

Cole, D. N., & Monz, C. A. (2002). *Trampling Disturbance of High-Elevation Vegetation, Wind River Mountains, Wyoming, USA* (pp. 365–370). Arctic, Antarctic and Alpine Research 34.

Cole, D. N., Petersen, M. E., & Lucas, R. C. (1987). *Managing Wilderness Recreation Use: Common Problems and Potential Solutions*. USDA Forest Service General Technical Report INT-GTR-230, Inter Mountain Research Station, Ogden, UT, 30pp.

Cordell, H. K. (2012). *Outdoor Recreation Trends and Futures: A Technical Document Supporting the Forest Service 2010 RPA Assessment*. General Technical Report SRS-150. Asheville, NC: USDA Forest Service, Southern Research Station, 167pp.

Dahlgren, R. B., & Korschgen, C. E. (1992). *Human Disturbances of Waterfowl: An Annotated Bibliography*. Washington, DC: US Department of the Interior, Fish and Wildlife Service.

Davies, P., & Loxham, J. (1996). *Repairing Upland Path Erosion*. Lake District National Park Authority, The National Trust and English Nature.

Davis, A. K. (2007). Walking trails in a nature preserve alter terrestrial salamander distributions. *Natural Areas Journal, 27*, 385–389.

de Boer, H. Y., Van Breukelen, L., & Hootsmans, M. J. M. (2004). Flight distance in roe deer *Capreolus capreolus* and fallow deer *Dama dama* as related to hunting and other factors. *Wildlife Biology, 10*, 35–41.

Derlet, W., & Carlson, J. R. (2006). Coliform bacteria in Sierra Nevada wilderness lakes and streams: What is the impact of backpackers, pack animals and cattle? *Wilderness Environmental Medicine, 17*, 15–20.

Edwards, T. (2007). *National Trails: Results of the National Trail User Survey 2007*. Natural England and the Countryside Council for Wales.

References

Fernández-Juricic, E. (2000). Local and regional effects of pedestrians on forest birds in a fragmented landscape. *Condor, 102*, 247–255.

Fernández-Juricic, E., Jiminiz, M. D., & Lucas, E. (2001a). Alert distance as an alternative measure of bird tolerance to human disturbance: Implications for park design. *Environmental Conservation, 28*, 263–269.

Fernández-Juricic, E., Jiminiz, M. D., & Lucas, E. (2001b). Bird tolerance to human disturbance in urban parks of Madrid (Spain): Management implication. In J. M. Marziuff, R. Bownan, & R. Donnelly (Eds.), *Avian Ecology and Conservation in an Urbanizing World* (pp. 259–273). Berlin: Springer-Verlag.

Fernández-Juricic, E., & Tellería, J. L. (2000). Effects of human disturbance on spatial and temporal feeding patterns of blackbird *Turdus merula* in urban parks in Madrid, Spain. *Bird Study, 47*, 13–21.

Finney, S. K., Pearce-Higgins, J. W., & Yalden, D. W. (2005). The effect of recreational disturbance on an upland breeding bird, the golden plover *Pluvialis apricaria*. *Biological Conservation, 121*, 53–63.

Fox, A. D., & Madsen, J. (1997). Behavioural and distributional effects of hunting disturbance on waterbirds in Europe: Implications for refuge design. *Journal of Applied Ecology, 34*, 1–13.

Gallet, S., Lemauviel, S., & Roz'e, F. (2004). Responses of three heathland shrubs to single or repeated experimental trampling. *Environmental Management, 33*, 821–829.

Gallet, S., & Roz'e, F. (2001). Resistance of Atlantic heathlands to trampling in Brittany (France): Influence of vegetation type, season and weather conditions. *Biological Conservation, 97*, 189–198.

Gallet, S., & Roz'e, F. (2002). Long-term effects of trampling on Atlantic heathland in Brittany (France): Resilience and tolerance in relation to season and meteorological conditions. *Biological Conservation, 103*, 267–275.

Gill, J. A. (2007). Approaches to measuring the effects of human disturbance on birds. *Ibis, 149*, 9–14.

Gill, J. A., Norris, K., & Sutherland, W. J. (2001). The effects of disturbance on habitat use by black-tailed godwits. *Limosa limosa*. *Journal of Applied Ecology, 38*, 846–856.

Gill, J. A., Sutherland, W. J., & Watkinson, A. R. (1996). A method to quantify the effects of human disturbance on animal populations. *Journal of Applied Ecology, 33*, 786–792.

Goodrich, J. M., & Berger, J. (1994). Winter recreation and hibernating black bears *Ursus americanus*. *Biological Conservation, 67*, 105–110.

Grieve, I. C. (2001). Human impacts on soil properties and their implications for the sensitivity of soil systems in Scotland. *Catena, 42*, 361–374.

Grieve, I. C., Davidson, D. A., & Gordon, J. E. (1995). Nature, extent and severity of soil erosion in upland Scotland. *Land Degradation and Development, 6*, 41–55.

Groenier, J. S., Monlux, S., & Vachowski, B. (2008). *Geosynthetics for Trails in Wet Areas* (2nd ed.). USDA Forest Service Publication 0023-2838, MTDC Technology and Development Center, Missoula, MT, 26pp.

Gutzwiller, K. J., Clements, K. L., Marcum, H. A., Wilkins, C. A., & Anderson, S. H. (1998). Vertical distribution of breeding-season birds: Is human intrusion influential? *Wilson Bulletin, 110*, 497–503.

Gutzwiller, K. J., & Riffell, S. K. (2008). Does repeated human intrusion alter use of wildland sites by red squirrels? Multiyear experimental evidence. *Journal of Mammalogy, 89*, 374–380.

Hammitt, W. E., Cole, D. N., & Monz, C. A. (2015). *Wildland Recreation. Ecology and Management* (3rd ed.). Chichester: Wiley-Blackwell, 313pp.

Hayes, P. (1997, February). In defence of erosion. *The Great Outdoors Magazine*.

Hecnar, S. J., & M'Closkey, R. T. (1998). Effects of human disturbance on five-lined skink *Eumeces fasciatus* abundance and distribution. *Biological Conservation, 85*, 213–222.

Hellmund, P. C. (1998). *Planning Trails with Wildlife in Mind: A Handbook for Trail Planners*. Denver, CO: Colorado Dept. of Natural Resources, Colorado State Parks, Colorado State Trails Program.

Hesselbarth, W., Vachowski, B., & Davies, M. A. (2007). *UFS Trail Construction and Maintenance Notebook*. Forest Service Technology and Development, Recreational Trails Program, Missoula, MT, 166pp.

Hill, D., Hochin, D., Price, D., Tucker, G., Morris, R., & Treweek, J. (1997). Bird disturbance: Improving the quality and utility of disturbance research. *Journal of Applied Ecology, 34*, 275–288.

Hingston, S. G. (1982). *Revegetation of Subalpine Backcountry Campaigns: Principle and Guidelines*. Resource Management Report Series KR-3, Alberta Recreation and Parks Division, Kananashis region.

Hockett, K., Clark, A., Leung, Y.-F., Marion, J. L., & Park, L. (2010). *Deterring Off-trail Hiking in Protected Natural Areas: Evaluating Options with Surveys and Unobtrusive Observation*. Final Research Report, Virginia Polytechnic Institute and State University and Canal National Historical Park, Hagerstone, MD, 178pp.

Hockett, K. S., Marion, J. L., & Leung, Y.-F. (2017). The efficacy of combined educational and site management actions in reducing off-trail hiking in an urban-proximate protected area. *Journal of Environmental Management, 203*, 17–28.

Holm, T. E., & Laursen, K. (2009). Experimental disturbance by walkers affects behaviour and territory density nesting black-tailed godwit *Limosa limosa*. *Ibis, 151*, 77–87.

Huxley, T. (1970). *Footpaths in the Countryside*. Countryside Commission for Scotland, Perth.

Ibanez-Alamo, J. D., & Soler, M. (2010). Investigator activities reduce nest predation in blackbirds *Turdus merula*. *Journal of Avian Biology, 41*, 208–212.

Jones, E. (2013). *Croagh Patrick Condition Report*. Retrieved from http://www.mountaineering.ie/_files/Elfyn%20Jones%20Croagh%20Patrick.

Kasworm, W. F., & Manley, T. L. (1990). Road and trail influences on grizzly bears and black bears in Northwest Montana. In *Bears: Their Biology and Management, Volume 8. 8th International Conference on Bear Research and Management, Victoria, BC, Canada, February 1989* (pp. 79–84).

Kellogg, D. S., Rosenbaum, P. F., Kiska, D. L., Riddell, S. W., Welch, T. R., & Shaw, J. (2012). High fecal hand contamination among wilderness hikers. *American Journal of Infection Control, 40*, 893–895.

Kent, M. (2012). *Vegetation Description and Data Analysis: A Practical Approach* (2nd ed.). Chichester: Wiley-Blackwell, 428pp.

Ketchledge, E. H., Leonard, R. E., & Richards, N. A. (1985). *Rehabilitation of Alpine Vegetation in the Adirondack Mountains of New York State*. USDA Forest Service NE Forest Experiment Research Station, Paper NE 533, Broomall, PA.

Knight, R. L., & Cole, D. N. (1991). Effects of recreational activity on wildlife in wildlands. *Transactions of the North American Wildlife and National Resources Conference, 56*, 238–247.

Kuss, F. R., & Graefe, A. R. (1985). Effects of recreation trampling on natural area vegetation. *Journal of Leisure Research, 17*, 165–183.

Langbein, J., & Putnam, R. J. (1998). Behavioral responses of park red deer and fallow deer to disturbance and effects on population performance. *Animal Welfare, 1*, 19–38.

Langston, R. H. W., Liley, D., Murison, G., Woodfield, E., & Clarke, R. T. (2007). What effects do walkers and dogs have on the distribution and productivity of breeding European nightjar *Caprimulgus europaeus*? *Ibis, 149*, 27–36.

Leseberg, A., Hockey, P. A. R., & Loewenthal, D. (2000). Human disturbance and the chick rearing ability of African black oystercatchers (*Haematopus moquini*): A geographical perspective. *Biological Conservation, 96*, 379–385.

Leung, Y.-F., & Marion, J. L. (1996). Trail degradation as influenced by environmental factors. A state-of-knowledge review. *Journal of Soil and Water Conservation, 51*, 130–136.

Leung, Y.-F., & Marion, J. L. (2000). *Recreation Impacts and Management in Wilderness: A State-of-Knowledge Review* (pp. 23–47). USDA Forest Service Proceedings RMRS-P-15-VOL-5.

Leung, Y.-F., Newburger, T., Jones, M., Kuhn, B., & Woiderski, B. (2011). Developing a monitoring protocol for visitor-created informal trails in Yosemite National Park, USA. *Environmental Management, 47*, 93–106.

Liddle, M. J. (1975). A selective review of the ecological effects of human trampling on natural ecosystems. *Biological Conservation, 7*, 17–36.

Liddle, M. J. (1989). *Recreation and the Environment: The Ecology of Recreation Impacts, Section 2: Vegetation and Wear*. AES Working Paper 1/89, Australian Environmental Studies, Griffith University, Brisbane, Australia.

Liddle, M. J. (1997). *Recreation Ecology*. London: Chapman and Hall, 639pp.

Lord, A., Waas, J. R., & Innes, J. (1997). Effects of human activity on the behaviour of northern New Zealand dotterel *Charadrius obscurus aquilonius* chicks. *Biological Conservation, 82*, 15–20.

Lord, A., Waas, J. R., Innes, J., & Whittingham, M. J. (2001). Effects of human approaches to nests of northern New Zealand dotterels. *Biological Conservation, 98*, 233–240.

Luka, C. G., & Hrsak, V. (2005). Influence of visitor numbers on breeding birds in the Paklenica National Park. Croatia. *Ekologia (Bratislava), 24*, 186–199.

Madej, M. A., Weaver, W. E., & Hogans, D. K. (1994). Analysis of bank erosion on the Merced River, Yosemite Valley, Yosemite National Park., California, USA. *Environmental Management, 18*, 235–250.

Manning, R. E., & Anderson, L. E. (2012). *Managing Outdoor Recreation. Case Studies in the National Parks*. Cambridge, MA: CABI, 264pp.

Marcum, H. A. (2005). *The Effects of Human Disturbance on Birds in Bastrop State Park*. PhD thesis, Texas A&M University, USA.

Marini, J. L., Dvorak, R. G., & Manning, R. E. (2008). Wildlife feeding in parks: Methods for monitoring the effectiveness of educational interventions and wildlife food attractions behaviors. *Human Dimensions of Wildlife, 13*, 429–442.

Marion, J. L. (2016). A review and synthesis of recreation ecology research supporting carrying capacity and visitor use management decisionmaking. *Journal of Forestry, 114*, 339–351.

Marion, J. L., & Cole, D. N. (1996). Spatial and temporal variations in soil and vegetation impacts on campsites: Delaware Gap National Recreation Area. *Ecological Applications, 6*, 520–530.

Marion, J. L., & Leung, Y.-F. (2004). Environmentally sustainable trail management. In R. Buckley (Ed.), *Environmental Impacts of Ecotourism* (Chap. 13, pp. 229–243), Ecotourism Series No. 2. Wallingford and Cambridge, MA: CABI.

Marion, J. L., Leung, Y.-F., Eagleston, H., & Burroughs, K. (2016). A review and synthesis of recreation ecology research findings on visitor impacts to wilderness and protected natural areas. *Journal of Forestry, 114*, 352–362.

Marion, J. L., Martinez, T., & Proudman, D. (2000). Trekking poles: Can save your knees and the environment? Retrieved from http://appalachiantrail.org/docs/default_source/recreation-ecology-resource.

Marion, J. L., & Olive, N. (2006). *Addressing and Understanding Trail Degradation: Results from Big South Fork National River and Recreation Area National Park Service*. Final Research Report, State University Department of Forestry, Blacksburg, VA, 80pp.

Marion, J. L., Roggenbuck, J. W., & Manning, R. E. (1993). *Problems and Practices in Backcountry Recreation Management: A Survey of National Park Service Managers*. Natural Resources Report NPS/NRV/NRR 93/12.

References

Marion, J. L., & Sober, T. (1987). Environmental impact management in the boundary waters canoe area wilderness. *Northern Journal of Applied Forestry, 4*, 7–10.

Marion, J. L., Wimpey, J. F., & Park, L. O. (2011). The science of trail surveys: Recreation ecology provides new tools for managing wilderness trails. *Park Science, 28*, 60–65.

McGowan, C. P., & Simons, T. R. (2006). Effects of human recreation on the incubation behaviour of American oystercatchers. *Wilson Journal of Ornithology, 118*, 485–493.

Meek, E. R. (1988). The breeding ecology and decline of the merlin *Falco columbarius* in Orkney. *Bird Study, 35*, 209–218.

Meyer, K. G. (2002). *Managing Degraded Off-highway Vehicle Trails in Wet, Unstable and Sensitive Environments*. USDA Forest Service Publication 0223-2821-MTDC, Missoula Technology and Development Center, Missoula, MT, 48pp.

Miller, S. G., Knight, R. L., & Miller, C. K. (2001). Wildlife responses to pedestrians and dogs. *Wildlife Society Bulletin, 29*, 124–132.

Monlux, S., & Vachowski, B. (2000). *Geosynthetics for Trails in Wet Areas*. USDA Forest Service, Publication 0023-2838-MTDC, Technology and Development Center, Missoula, MT, 23pp.

Monz, C. A., Marion, J. L., Croonan, K. A., Manning, R. E., Wimpey, J., & Carr, C. (2010). Assessment and monitoring of recreation impacts and resource conditions on mountain summits: Examples from the northern forest, USA. *Mountain Research and Development, 30*, 332–343.

Morgan, R. P. C. (1995). *Soil Erosion and Conservation* (2nd ed.). Harlow, Essex: Longman, 198pp.

Müller, C., Jenni-Eirmann, S., Blondel, J., Perret, P., Caro, S. P., Lambrechts, M., et al. (2006). Effect of human presence and handling on circulating corticosterone levels in breeding blue tits (*Parus coeruleus*). *General and Comparative Endrocrinology, 148*, 163–171.

National Trust for Scotland. (2003). *Upland Footpath Repair Techniques in the Cairngorm Mountains: A Review and Recommendations*. Scottish Natural Heritage, Report No. 008 (ROAME NoF99LF02).

Newton, I., Davis, P. E., & Moss, D. (1981). Distribution and breeding of red kites *Milvus milvus* in relation to land use in Wales, UK. *Journal of Applied Ecology, 18*, 173–186.

Nisbet, I. C. T. (2000). Disturbance, habituation and management of waterbird colonies. *Waterbirds, 23*, 312–332.

Nudds, R. L., & Bryant, D. M. (2000). The energetic cost of short flights in birds. *Journal of Experimental Biology, 203*, 1561–1572.

Patthey, P. S., Wirthuer, S., Signorell, N., & Arlettae, R. (2008). Impact of outdoor winter sports on the abundance of key indicator species of alpine ecosystems. *Journal of Applied Ecology, 45*, 1704–1711.

Pease, M. L., Rose, R. K., & Butler, M. J. (2005). Effects of human recreation on the behavior of wintering ducks. *Wildlife Society Bulletin, 33*, 103–112.

Pescott, O. L., & Stewart, G. B. (2014). Assessing the impact of human trampling on vegetation: A systematic review and meta-analysis of experimental evidence. *Peer Journal, 2*, e360. https://doi.org/10.7717/peerj.360.

Picozzi, N. (1971). Breeding performance and shooting bags of red grouse in relation to public access in the Peak District National Park, England. *Biological Conservation, 3*, 211–215.

Porter, M. (1990). *Pennine Way Management Project, Final Report*. Unpublished report to the Peak District National Park.

Price, E. (1987). *Footpath Erosion and Vegetation Changes, Beer Head (Devon)*. Unpublished BEd thesis, Liverpool Polytechnic.

Pryor, P. (1986). The effects of disturbance on open *Juncus trifidus* heath in the Cairngorm mountains, Scotland. In N. G. Bayfield & G. C. Barrow (Eds.), *Recreational Ecology Research Report 9*. London: RERG, Wye.

Rasmussen, H., & Simpson, S. (2010). Disturbance of waterfowl by boaters on pool 4 of the upper Mississippi River National Wildlife and fish refuge. *Society and Natural Resources, 23*, 322–331.

Ramos-Scharrón, C. E., Reale-Munroe, K., & Atkinson, S. C. (2014). Quantification and modelling of foot trail surface erosion in a dry sub-tropical setting. *Earth Surface Processes and Landforms, 39*, 1764–1777.

Reby, D., Cargnelutti, B., & Hewison, A. J. M. (1999). Contexts and possible functions of barking in roe deer. *Animal Behavior, 57*, 1121–1128.

Recarte, J. M., Vincent, J. P., & Hewison, A. J. M. (1998). Flight responses of park fallow deer to the human observer. *Behavioral Processes, 44*, 65–72.

Reed, B. C., & Rasnake, M. S. (2016). An assessment of coliform bacteria in water sources near Appalachian Trail shelters within the Great Smoky Mountains National Park. *Wilderness Environmental Medicine, 27*, 107–110.

Roovers, R., Verkeyen, K., Hermy, M., & Gulinck, H. (2004). Experimental trampling and vegetation recovery in some forest and heathland communities. *Applied Vegetation Science, 7*, 111–118.

Ruddock, M., & Whitfield, D. P. (2007). *A Review of Disturbance Distances in Selected Bird Species*. Inverness, UK: Scottish Natural Heritage.

Schlacher, T. A., Neilsen, T., & Weston, M. A. (2013). Human recreation alters behaviour profiles of non-breeding birds on open-coast sandy shores. *Estuarine and Coastal Shelf Science, 118*, 31–42.

Showler, D. A., Stewart, G. B., Sutherland, W. J., & Pullin, A. S. (2010). *What is the Impact of Public Access on the Breeding Success of Ground-nesting and Cliff-nesting Birds?* Systematic Review CE 005-010, (Sr16) Collaboration for Environmental Evidence, 74pp. Norwich. Retrieved from www.environmentalevidence.org/SR16.html

Sidaway, R. (1990). *Birds and Walkers: A Review of Existing Research on Access to the Countryside and Disturbance to Birds*. London: Ramblers Association.

Smith, R. (1987, April). The three peaks project. *The Great Outdoors Magazine*, pp. 53–57.

Smith-Castro, J. R., & Rodewald, A. D. (2010). Behavioral responses of nesting birds to human disturbance along recreational trails. *Journal of Field Ornithology, 81*, 130–138.

South China Morning Post. (2018, April 6). Could a hiking pole ban protect Hong Kong's country park trails, some of them rapidly eroding? Retrieved from www.scmp.com/lifestyle/health-wellness/2140427/could-hikin.

Speight, M. C. D. (1973). *Outdoor Recreation and Its Ecological Effects: A Bibliography and Review*. Discussion Paper in Conservation No. 4, London.

Steven, R., Pickering, C., & Castley, J. G. (2011). A review of the impacts of nature based recreation on birds. *Journal of Environmental Management, 92*, 2287–2294.

Stolen, E. D. (2003). The effect of vehicle passage on foraging behavior of wading birds. *Waterbirds, 26*, 429–436.

Summers, R. W., McFarlane, J., & Pearce-Higgins, J. W. (2007). Measuring avoidance by capercaillies *Tetrao urogallus* of woodland close to tracks. *Wildlife Biology, 13*, 19–27.

Summers, R. W., Proctor, R., Thornton, M., & Avey, G. (2004). Habitat selection and diet of the Capercaillie *Tetrao urogallus* in abernethy Forest, Strathspey, Scotland. *Bird Study, 51*, 58–68.

Sun, D., & Liddle, M. J. (1993). Trampling resistance, stem flexibility and leaf strength in nine Australian grasses and herbs. *Biological Conservation, 65*, 35–41.

Talbot, L. M., Turton, S. M., & Graheul, A. L. (2003). Trampling resistance of tropical rainforest soils and vegetation in the wet tropics of north east Australia. *Journal of Environmental Management, 69*, 63–69.

Taylor, K., Anderson, P., Taylor, R., Longden, K., & Fisher, P. (2005). *Dogs, Access and Nature Conservation*. English Nature Research Report No. 649, English Nature, Peterborough.

Theil, D., Jenni-Eirmann, S., Palme, R., & Jenni, L. (2011). Winter tourism increases stress hormone levels in the Capercaillie *Tetrao urogallus*. *Ibis, 153*, 122–133.

TNS Travel and Tourism. (2008). *Scottish Recreational Survey*. Annual Summary Report, No. 295, Scottish Natural Heritage, Edinburgh.

Trew, M. (1973). The effects and management of trampling in coastal sand dunes. *Journal of Environmental Planning and Pollution Control, 1*, 38–49.

Tyser, R. W., & Worley, C. A. (1992). Alien flora in grasslands adjacent to road and trail corridors in glacier National Park, Montana (USA). *Conservation Biology, 6*, 253–262.

Ursem, C., Evans, S., Ger, K. A., Richards, J. R., & Derley, R. W. (2009). Surface water quality along the central John Muir Trail in the Sierra Nevada Mountains: Coliforms and algae. *High Altitude Medicine and Biology, 10*, 249–255.

USDA Forest Service. (1985). *Trails Management Handbook*. FSH 2307-18, USDA, Washington, 144pp.

Verboven, N., Ens, B. J., & Decherns, S. (2001). Effects of investigator disturbance on nest attendance and egg predation in Eurasian oystercatchers. *The Auk, 118*, 503–508.

Wagar, J. A. (1964). *The Carrying Capacity of Wildlands for Recreation*. Forest Service Monograph, No. 77, 23pp.

Ward, A. I., White, P. C. L., & Critchley, C. H. (2004). Roe deer *Capreolus capreolus* behaviour affects density estimates from distance sampling surveys. *Mammal Review, 34*, 315–319.

Watson, A. (1984). Paths and people in the Cairngorms. *Scottish Geographical Magazine 100*, pp. 151–160.

Weimerskirch, H., Schaffer, S. A., Mabille, G., Martin, J., Boutard, O., & Rouanet, J. L. (2002). Heart rate and energy expenditure of incubating wandering albatrosses: Basal levels, natural variation and the effects of human disturbance. *Journal of Experimental Biology, 205*, 475–483.

Whitfield, D. P., & Rae, R. (2014). Human disturbance of breeding wood sandpipers *Tringa glareola*: Implications for "alert distances" in prescribing buffer zones. *Ornis Fennica, 91*, 57–66.

Yalden, D. W. (1992). The influence of recreational disturbance on common sandpipers *Actitis hypoleucos* breeding by an upland reservoir, in England. *Biological Conservation, 61*, 41–49.

Yalden, D. W., & Yalden, P. E. (1989). The sensitivity of breeding golden plovers *Pluvialis apricaria* to human intruders. *Bird Study, 36*, 49–55.

Mountain Marathons, Adventure Racing, and Mountain Tours

3

Chapter Summary

This chapter first defines mountain marathons, adventure racing, and mountain tours and gives examples of a range of such activities and events. It then briefly discusses the history and diversity of mountain marathon, adventure racing, and high mountain tours and safety/legal issues before presenting recent data on user numbers. The final part of the chapter focuses on specific environmental impacts associated with particular events such as the UK's National Three Peaks Challenge and the Yorkshire Three Peaks Challenge and highlights the need for more research. The final section considers the management of these activities and gives some examples of education initiatives that have been used in management attempts.

3.1 Definitions

3.1.1 Mountain Marathon

Mountain marathon is an extended form of fell running. Races usually take place over two days and often have a strong element of orienteering (i.e. competitors must plan their own route and navigate using map and compass). Competitors usually participate in teams of two and have to carry their own food and tent. There are various classes of event (e.g. for the Original Mountain Marathon—Elite, A, B, C, D and Long, Medium, and Short Score).

Some of the more well-known events include:

- The Original Mountain Marathon (OMM—formerly the Karrimor International Mountain Marathon/KIMM) held in a UK hill or mountain area in the last weekend in October (www.theomm.com).
- The Saunders Lakeland Mountain Marathon (SLMM) held in or near the Lake District in early July.
- The Swiss International Mountain Marathon (since 1976: formerly the Karrimor International Mountain Marathon/KIMM Switzerland/Mammut International Mountain Marathon/MIMM Switzerland/R'adys Mountain Marathon) held in Switzerland in mid-August.
- Marmot Dark Mountains held on the last weekend of January each year.
- The Lowe Alpine Mountain Marathon (LAMM) held in the Scottish Highlands in June.
- The Mourne Mountain Marathon held in Mourne Mountains, County Down, Northern Ireland, in September.
- The ROC Mountain Marathon held on the last weekend of September each year.
- The SCOTT Snowdonia Trail Marathon, which is a challenge in every sense of the

word. Ascending 1685 metres over 26 miles of iconic and spectacular trails, this epic race circumnavigates and eventually climbs Wales' highest peak—Snowdon.

- The Longmynd Hike—a 50-mile competition hike open to anyone aged 18 or over—which takes place over the first weekend of October every year.
- Starting and finishing at Church Stretton, the hike follows a set figure-of-eight route over the rugged countryside of South Shropshire and the Welsh Marches, with about 8000 ft of climbing covering eight summits.

There are also newer events springing up and attracting increasing numbers, including the Highlander Mountain Marathon which began in 2007 and is held in April at a Scottish location within a two-hour drive of Inverness.

3.1.2 Adventure or Expedition Racing

Adventure racing has been characterised as a new "lifestyle sport": "a non-stop, self-sufficient, multi-day, multidiscipline, mixed gender endurance competition that takes place in the wilderness over a designated but unmarked course" (Kay and Laberge 2002, p. 25).

Adventure racing (also called expedition racing) is typically a multidisciplinary team sport involving navigation over an unmarked wilderness course with races extending anywhere from two hours up to two weeks in length. Some races offer solo competition as well. The principle disciplines in adventure racing include trekking, mountain biking (Fig. 3.1), and paddling although races can incorporate a multitude of other disciplines including climbing, abseiling, horse riding, skiing, and white-water rafting. Teams generally vary in gender mix and in size from two to five competitors. The most popular format is generally a mixed-gender team of four racers. There is typically no suspension of the clock during races, irrespective of length; elapsed competition time runs concurrently with real time, and competitors must choose if or when to rest.

3.1.3 High Mountain Tours

A high mountain tour (German: *Hochtour*) is a mountain tour that takes place in the zone that is covered by ice all year round, the nival zone,

Fig. 3.1 Competitor on a cycling leg of an adventure race in the English Lake District. Photo by Tim Stott

3.1 Definitions

Fig. 3.2 The start of the high mountain tour at the end of the eighteenth century: contemporary portrait of Horace-Bénédict de Saussure on Mont Blanc in 1787. Source: By Marquard Wocher—www.unil.ch/webdav/site/viaticalpes/users/dvaj/public/colloqueprojet/Vaj_Viaticalpes.pdf

above a height of about 3000 metres (High Alps) where many mountains are at least partly glaciated (Fig. 3.2). Important historic milestones in the development of high mountain touring in the Alps were the first ascents of the Ankogel (3262 m) in 1762, Mont Blanc (4810 m) in 1786, the Großglockner (3798 m) in 1800, and the Ortler (3905 m) in 1804 as well as the conquest of many high western Alpine summits during the golden age of Alpinism around the middle of the nineteenth century. In other parts of the world, the term may be misleading. For example, in many non-Alpine areas, such as the polar regions, much lower mountains are glaciated. On the other hand, the summits of much higher peaks in the tropics are not always in the nival zone. As a result, their ascent cannot automatically be described as a high mountain tour using the Alpine definition, even if they share some of the features of Alpinism, such as requiring acclimatisation. Mountaineering expeditions in which elevation plays a particularly important role, especially those from about 7000 m, are no longer referred to as high mountain tours but tend to be described by the term high-altitude mountaineering.

Other forms of mountain tours might be known by the term "peak bagging." Examples include:

- *The Seven Summits*: the Seven Summits are the highest mountains of each of the seven continents. Summiting all of them is regarded as a mountaineering challenge, first achieved on 30 April 1985 by Richard Bass who summited Everest, Aconcagua, Denali,

Kilimanjaro, Elbrus, Kosciuszko, and Vinson. However, there are other versions of the list which depend on how one defines the continents, so sometimes Puncak Jaya (also known as "Carstensz Pyramid") and Mont Blanc are included. The Seven Summits achievement has become noted as an exploration and mountaineering accomplishment.

- *The eight-thousanders*: these are the 14 independent mountains on Earth that are more than 8000 metres (26,247 ft) high above sea level. All eight-thousanders are located in the Himalayan and Karakoram mountain ranges in Asia. Their summits are in the death zone. The first person to climb all 14 eight-thousanders was the Italian Reinhold Messner, who completed this feat on 16 October 1986.

The mountains and hills of Great Britain, and to a lesser extent Ireland, are the subject of a considerable number of lists that categorise them by height, topographic prominence, or other criteria. They are commonly used as a basis for peak bagging, whereby hillwalkers attempt to reach all the summits on a given list. The oldest and best known of these lists is that of the Munros, mountains in Scotland over 3000 ft (914.4 m); other well-known lists include, for example, the Corbetts, Wainwrights, and Marilyns.

- *The Scottish Munros*: the Munro is a mountain in Scotland with a height over 3000 ft (914 m) named after Sir Hugh Munro, who produced the first list of such hills, known as Munro's Tables, in 1891. The publication of the original list is usually considered to be the epoch event of modern peak bagging. The list has been the subject of subsequent variation. The 2012 revision, published by the Scottish Mountaineering Club, has 282 Munros and 227 subsidiary tops. "Munro bagging" is the activity of climbing all the listed Munros. They present challenging conditions to walkers, particularly in winter. As of 2017, more than 6000 people had reported completing a round. The first continuous round was completed by Hamish Brown in 1974, whilst the record for the fastest continuous round is currently held by Stephen Pyke, who completed a round in just under 40 days in 2010.
- *The Corbetts*: these are peaks in Scotland that are between 2500 and 3000 ft (762.0 and 914.4 m) high with a prominence of at least 500 ft (152.4 m). There are currently 222 Corbetts.
- *The Grahams*: these are mountains in Scotland between 2000 and 2499 ft (610 and 762 metres) high, with a drop of at least 150 metres (490 ft) all round. There are currently 221 hills in this list.
- *The Donalds*: these are mountains in the Scottish Lowlands over 2000 ft (610 m). A mountain with a prominence of at least 30 metres (98 ft) is automatically a Donald, but one with a relative height of 15 metres (49 ft) may be one if it is of sufficient topographic interest. There are 140 Donalds, comprising 89 mountains and 51 tops.
- *The Furths*: these are those mountains in Great Britain and Ireland Furth of (i.e. "outside") Scotland that would otherwise qualify as Munros or Munro Tops. They are sometimes referred to as the Irish, English, or Welsh Munros. There are 34 Furths: 15 in Wales, 13 in Ireland, and 6 in England. The highest is Snowdon.
- *The Hewitts*: these are hills in England, Wales, and Ireland over 2000 feet (609.6 m), with a relative height of at least 30 metres (98 ft). There are 528 Hewitts in total: 179 in England, 138 in Wales, and 211 in Ireland. The current TACit booklets contain 525 mountains, with Black Mountain being counted in both England and Wales.
- *The Nuttalls*: these are mountains in England and Wales over 2000 ft (610 m) with a relative height of at least 15 metres (49 ft). There are 444 Nuttalls in total (254 in England and 190 in Wales).
- *The Wainwrights*: these are mountains or hills (locally known as fells) in the English Lake District National Park that have a chapter in one of Alfred Wainwright's *Pictorial Guides to the Lakeland Fells*. There are 214 in the

seven guides. There are no qualifications for inclusion other than an implied requirement of being at least 1000 ft (300 m) high, to which Castle Crag in Borrowdale is the sole exception. A further 116 summits were included in the supplementary guide, *The Outlying Fells of Lakeland*.

- *The Birketts*: these are all the fell tops over 1000 ft high (about 305 m) within the boundaries of the Lake District National Park. Height and location, but not prominence, are the criteria. There are 541 of these tops.
- *The Marilyns*: these are mountains and hills in the British Isles that have a prominence of at least 150 metres (490 ft), regardless of absolute height or other merits. There are currently 1556 Marilyns in Great Britain: 1217 in Scotland, 176 in England, 158 in Wales, and 5 on the Isle of Man (Black Mountain, on the England-Wales border, is counted as being in Wales). There are a further 454 Marilyns in Ireland.

In the English Lake District especially, there is a tradition of finding the maximum number of tops, including all the major summits, which can be visited in a 24-hour period. This usually requires fell running and a support team. The pre-war record, set by Bob Graham, of 42 tops, has become a standard round, which has been repeated by over 1000 people. In 1975 Joss Naylor, the famous English fell runner, and a sheep farmer, born in the English Lake District, ran over 72 peaks, claimed to involve over 100 miles and about 38,000 ft of ascent in 23h20m, a record which stood unbroken for 13 years.

In Wales, Joss Naylor also completed the Welsh 3000s—the 14 peaks of Snowdonia in 1973 in another record-breaking time of 4 h 46 m, which stood until 1988 when Colin Donnelly set his, still-standing, record for the traverse of the Welsh 3000s with a time of 4 h 19 m.

These "lists" and "rounds" or tours can be done continuously (as in these past examples), or they may be completed over a lifetime. However, in terms of environmental damage, they do not see the huge numbers of participants at one time which modern events (discussed hereafter) bring. It is the huge influx or masses of participants which create the biggest impacts on the environment, and so the rest of this chapter tends to focus on such events.

3.2 History, Diversity, and Participation Numbers

3.2.1 Mountain Marathons

The OMM, formerly known as the Karrimor International Mountain Marathon (or KIMM), and initially simply The Karrimor, is a two-day mountain event, held in a different region across the UK every year. It was first held in 1968 and continues today. The full-length KIMM course is a double marathon in length. The team must carry all their gear, including equipment for an overnight camp. The course is not disclosed until the race begins, so each team must have good navigation skills since it is not possible to practice running the course beforehand. Some have called the KIMM the forerunner of modern adventure racing. For its first eight years, the event was known as "The Karrimor." In addition to the "Elite category" double marathon, other course lengths have been added over the years to suit a greater variety of competitors. In 2004 the event became known as the OMM after Karrimor's sponsorship was withdrawn. In 2013, the organisers of the OMM revealed plans for a summer version of the event, along with a mountain biking marathon.

The SLMM is a two-day mountain marathon held in the English Lake District ("or its environs," such as the adjoining Howgill Fells) in early July. It was founded by Robert Saunders, a long-time UK manufacturer of lightweight tents. The SLMM has been held annually since 1978, apart from 2001, when the Lakeland Fells were closed because of the foot and mouth crisis. 2018 will therefore be the 40th event. It is usually considered to be less tough than the slightly older OMM, since the weather is often mild, the courses are slightly shorter, and the overnight camp is often found to be within walking distance of a pub. The event comprises eight courses of which six are solely for pairs of runners, one is exclusively for solo competitors, and one course is open for both pairs and solo entrants. The organisers encourage young competitors, with lower entry fees for under 25s, and there is a specific, handicapped class for parent and child (age 14+). Because of the popularity of the Lakes, courses are usually set to run "across the grain" of the country, away from popular paths, so as to minimise erosion due to the race.

3.2.2 Adventure Racing

The roots of adventure racing are deep, and people debate the origin of the modern adventure race. Some point to the two-day KIMM, first held in 1968 as the birth of modern adventure racing. The Karrimor Marathon required two-person teams to traverse mountainous terrain while carrying all the supplies required to support themselves through the double-length marathon run.

The Adventure Racing World Series (ARWS) is a number of expedition-length adventure races that push the world's best endurance athletes to their limits in a season of competition that tests their skills in a range of disciplines including navigation, trekking, mountain biking, paddling, and climbing. Mixed-gender teams of four competitors compete in a series of up to a dozen races held in locations spread across the globe. These races culminate in the staging of the Adventure Racing World Championships, the winners of which earn the title of World Champions. The competition's format ensures that each of the individual events of the World Series functions as a qualifier for the World Championships. The actual World Championship race rotates each year. One of the qualifying events is singled out and designated as the World Championship event, and this event provides a dramatic conclusion to the end of the World Series racing season.

The first World Series event was held in Switzerland in 2001; there was a gap of two years, and it has been held every year since 2004.

In 1980, the Alpine Ironman was held in New Zealand. Individual competitors ran, paddled, and skied to a distant finish line. Later that year, the Alpine Ironman's creator launched the better-known Coast to Coast race, which involved most of the elements of modern adventure racing: trail running, cycling, and paddling. Independently, a North American race, the Alaska Mountain Wilderness Classic started in 1982 and involved six days of unsupported wilderness racing (carry all food and equipment, no roads, no support) over a 150-mile course. It continues today, changing courses every three years.

In 1989, the modern era of adventure racing began with the launch of the Raid Gauloises in New Zealand. This is an expanded expedition-style race in which competitors rely on their own strengths and abilities to traverse big and challenging terrain. The race incorporates all the modern elements of adventure racing, including mixed-gender teams competing in a multi-day 400+ mile race. The United States Adventure Racing Association (USARA) was formed in 1998 and was the first national governing body for the sport of adventure racing which arose from the need for safety standards, insurance, and to promote the growth of adventure racing in the USA. The USARA has added national rankings, a national championship, and ecological standards to the list of benefits provided for the sport of adventure racing. The USARA National Championship has been held on the first weekend in October since 2000 and is considered the premier adventure race in the USA. The USARA Adventure Racing National Championship has continued each year drawing the best US teams for a chance at earning the title of national champion.

In 2001, the inaugural World Championships were held in Switzerland with Team Nokia Adventure crossing the finishing line first. The concept of a world championship lay dormant until it was revived in 2004, with Canada's Raid the North Extreme serving as the AR World Championship event in Newfoundland and Labrador. The Adventure Racing World Series and its penultimate event, the AR World Championships, have been held every year since. In 2002, the first major expedition length race to be held exclusively in the USA was launched. Primal Quest has become the premier US expedition race, being held each year since its launch. In 2004, the death of veteran racer Nigel Aylott overshadowed the race and raised debates about the nature of Primal Quest and adventure racing.

In 2004, a professional geologist Stjepan Pavicic organised the first Patagonian Expedition Race at the bottom tip of the American continent, in the Chilean Tierra del Fuego. Truly demanding routes through rough terrain of often more than 600 km soon made it be known as "the last wild race." In 2010, the German Adventure Race Series was held for the first time in three different locations all over Germany. Since then the popularity of the sport in Germany has grown every year. More races and venues have joined the series, and the number of competitors is still growing from year to year. In 2012, Commander Forer of the Royal Navy organised the first sea-land navigation discipline race The Solent Amphibious Challenge. The race demanded the competitors to split up between sailing, running, and cycling in parts of the race and rendezvous at the end and sail the yacht to the finish line.

In the USA, during the 2016 calendar year, a total of 24,134 online interviews were carried out with a nationwide sample of individuals and households from the US Online Panel of over one million people operated by Synovate/IPSOS (Outdoor Foundation 2017). A total of 11,453 individual and 12,681 household surveys were completed. The total panel is maintained to be representative of the US population for people aged six and older. Oversampling of ethnic groups took place to boost response from typically under responding groups. The 2016 participation survey sample size of 24,134 completed interviews provides a high degree of statistical accuracy.

Table 3.1 shows that participation numbers in adventure racing rose from 725,000 in 2006 to 2,999,000 (almost 3 million) in 2016, showing a three-year change of 35.5%. Of all the activities surveyed by the Outdoor Foundation (2017) shown in Table 3.1, only BMX biking, cross-country skiing, and stand-up paddleboarding showed higher three-year changes.

3.2.3 High Mountain Tours

The classic high mountain tours require sure-footedness, a head for heights, and the ability to handle greater technical difficulty in rock and ice climbing as well as mixed climbing in combined rock and ice terrain. In glaciated terrain the risk of crevasses means that even technically easy walks require the use of rope, crampons, and ice axes as well as knowledge of safety and rescue techniques. Walking with a rope requires a roped team to be formed and makes trekking alone dangerous. In addition, a certain level of fitness and height acclimatisation is usually necessary. For mountain tours in high mountains such as the Himalayas, the Karakorum, or the Andes, which reach elevations of over 6000 m above sea level, one or two weeks should be allowed for acclimatisation. Low temperatures may also be an important factor. The dangers and problems presented by high mountain touring, as in sports climbing, are caused less by the actual technical difficulty of climbing than by the (often rapidly changing) external conditions. The description of the requirements of a tour with the aid of climbing grade scales is therefore problematic. As a result, such scales attempt to take into account to a greater extent as the severity of a route or its fitness requirements. An example of an established rating system for Alpinism is the SAC Mountain and High Mountain Tour Scale (Table 3.2).

Map reading and the ability to read the weather may also be important in high mountain touring. When snow falls, a knowledge of

Table 3.1 Outdoor participation by activity in the USA, 2006–2016 (The Outdoor Foundation, 2017, p. 8)

	2006	2007	2008	2009	2010	2011	2012	2013	2014	2015	2016	3-year change (%)
Adventure racing	**725**	**698**	**920**	**1089**	**1339**	**1065**	**2170**	**2213**	**2368**	**2864**	**2999**	**35.5**
Backpacking overnight >¼ mile from vehicle/home	7076	6637	7867	7647	8349	7095	8771	9069	10,101	10,100	10,151	11.9
Bicycling (BMX)	1655	1887	1904	1811	2369	1547	2175	2168	2350	2690	3104	43.2
Bicycling (mountain/non-paved surface)	6751	6892	7592	7142	7161	6816	7714	8542	8044	8316	8615	0.9
Bicycling (roads/paved surface)	38,457	38,940	38,114	40,140	39,320	40,349	39,232	40,888	39,725	38,280	38,365	−6.2
Birdwatching (more and ¼ mile from home/vehicle)	11,070	13,476	14,399	13,294	13,339	12,794	14,275	14,152	13,179	13,093	11,589	−18.1
Boardsailing/windsurfing	938	1118	1307	1128	1617	1151	1593	1324	1562	1766	1737	31.2
Camping (RV)	16,946	16,168	16,517	17,436	15,865	16,698	15,108	14,556	14,663	14,699	15,855	8.9
Camping (with ¼ mile of home/vehicle)	35,618	31,375	33,686	34,338	30,996	32,925	29,982	29,269	28,660	27,742	26,467	−9.6
Canoeing	9154	9797	9935	10,058	10,553	9787	9839	10,153	10,044	10,236	10,046	−1.1
Climbing (sports/indoor/boulder)	4728	4514	4769	4313	4770	4119	4592	4745	4536	4684	4905	3.4
Climbing (traditional/ice/mountaineering)	1586	2062	2288	1835	2198	1609	2189	2319	2457	2571	2790	20.3
Fishing (fly)	6071	5756	5941	5568	5478	5683	6012	5878	5842	6089	6456	9.8
Fishing (freshwater/other)	43,100	43,859	40,331	40,961	38,860	38,868	39,135	37,796	37,821	37,682	38,121	0.9
Fishing (saltwater)	12,466	14,437	13,804	12,303	11,809	11,983	12,017	11,790	11,817	11,975	12,266	4.0
Hiking (day)	29,863	29,965	32,511	32,572	32,496	34,491	34,545	34,378	36,222	37,232	42,128	22.5
Hunting (bow)	3875	3818	3722	4226	3908	4633	4075	4079	4411	4564	4427	8.5
Hunting (handgun)	2525	2595	2873	2276	2709	2671	3553	3198	3091	3400	3512	9.8
Hunting (rifle)	11,242	10,635	10,344	11,114	10,150	10,807	10,164	9792	10,081	10,778	10,797	10.3
Hunting (shotgun)	8987	8545	8731	8490	8062	8678	8174	7894	8220	8438	8271	4.8
Kayak fishing	n/a	n/a	n/a	n/a	1044	1201	1409	1798	2074	2265	2371	31.8
Kayaking (recreational)	4134	5070	6240	6212	6465	8229	8144	8716	8855	9499	10,017	14.9
Kayaking (sea/touring)	1136	1485	1780	1771	2144	2029	2446	2694	2912	3079	3124	16.0
Kayaking (white water)	828	1207	1242	1369	1842	1546	1878	2146	2351	2518	2552	18.9

(continued)

Table 3.1 (continued)

	2006	2007	2008	2009	2010	2011	2012	2013	2014	2015	2016	3-year change (%)
Rafting	3609	3786	4226	4342	3869	3725	3958	3915	3924	4099	4095	−10.6
Running/jogging	38,559	41,064	41,130	43,892	49,408	50,713	52,187	54,188	51,127	48,496	47,384	−12.6
Sailing	3390	3786	4226	4342	3869	3725	3958	3915	3924	4099	4095	4.6
Scuba diving	2965	2965	3216	2723	3153	2579	2982	3174	3145	3274	3111	−2.0
Skateboarding	10,130	8429	7807	7352	6808	5827	6627	6350	6582	6436	6442	1.5
Skiing (alpine/downhill)	n/a	10,362	10,346	10,919	11,504	10,201	8243	8044	8649	9378	9267	12.4
Skiing (cross-country)	n/a	3530	3848	4157	4530	3641	3307	3377	3820	4146	4640	40.3
Skiing (freestyle)	n/a	2817	2711	2950	3647	4318	5357	4007	4564	4465	4640	2.7
Snorkelling	8395	9294	10,296	9358	9305	9318	8011	8700	8752	8874	8717	0.2
Snowboarding	n/a	6841	7159	7421	8196	7579	7351	6418	6785	7676	7602	3.4
Snowshoeing	n/a	2400	2922	3431	3823	4111	4029	3012	3501	3885	3533	−12.3
Stand up paddling	n/a	n/a	n/a	n/a	1050	1242	1542	1993	2751	3020	3220	61.6
Surfing	2170	2206	2607	2403	2767	2195	2895	2658	2721	2701	2793	3.0
Telemarking (downhill)	n/a	1173	1435	1482	1821	2099	2766	1732	2188	2569	2848	3.0
Trail running	4558	4216	4857	4833	5136	5610	6003	6792	7531	8139	8582	26.4

Note: All participation numbers are in thousands (000)

Table 3.2 The EXCEDO hiking difficulty scale based on the classification of the Swiss Alpine Club

Scale	Trail/terrain	Requirements
T1 = hiking	Hiking trail well cleared. Flat or slightly sloped terrain. If present, exposed areas are well equipped and secured. No risk of falling given normal conduct and regular circumstances	None. Accessible even with sports shoes. Easy orientation, in general even without a map
T2 = mountain hiking	A continuous hiking trail with balanced ascent. Terrain partially steep, possible risk of falling	Some sure-footedness. Trekking shoes and basic orientation skills recommended
T3 = demanding mountain hiking	Hiking path not necessarily visible along the entire trail. Exposed passages may be secured with ropes and chains. Possible need to use hands for balance. Single exposed passages with risk of falling, scree, pathless grassy slopes, and jagged rocks	Sure-footedness is required. Good trekking shoes and advanced orientation skills recommended. Basic alpine experience
T4 = alpine hiking	Hiking trail not necessarily marked and/or visible. The use of hands might be required for advancing in certain passages. Terrain quite exposed, precarious grassy acclivities, pathless steep scree and jagged rock sections, easy firn fields	Familiarity with exposed terrain. Solid trekking shoes. Some experience in terrain assessment and good orientation skills. Alpine experience
T5 = demanding alpine hiking	Hiking often without trail. Single easy climbing sections. Exposed, demanding terrain, steep grassy acclivities, and jagged rocks. Firn fields with risk of slipping	Climbing boots. Reliable terrain assessment and very good orientation skills. Profound alpine experience. Basic skills in the use of ice axe and rope
T6 = difficult alpine hiking	Generally, hiking without a trail to follow. Climbing sections up to second grade. Terrain often very exposed, very precarious grassy and rocky slopes, glaciers with high risk of slipping and falling. Most often unmarked	Excellent orientation skills. Advanced alpine experience and familiarity with the use of alpine equipment

Source: http://www.excedotravel.com/en/hiking-difficulty-scale/

avalanche behaviour is necessary, even in the summer months. High Alpine terrain is currently subject to a particularly high degree of change in terms of glacier retreat and climate change, which can both increase or decrease the difficulty and dangers of high mountain touring.

NB. A serious misunderstanding, which can lead to tricky situations, is the belief that hiking stops where the Alpine Climb Scale begins. In reality, an alpine hike in the upper range of the T5 and T6 difficulty is usually significantly more demanding than, for example, an "F" rated Alpine Climb. A major difference, as compared to an easy Alpine Climb, for example, is that in case of a T5 and T6 hike, one can rarely or almost never use protective gear such as a rope or other equipment, meaning that the terrain must be perfectly mastered. Often this requires high technical as well as psychological skills. Typical examples are extremely steep grassy slopes, scree, pathless steep slopes with jagged rocks, or very exposed ridges. Due to their different characteristics, a typical Alpine Climb and a typical extreme hike can hardly be compared, but one can assume that a T6 hiking route requires a similar set of skills and experience as an Alpine Climb up to F.

3.3 Safety and Legal Regulation

The 2008 OMM was abandoned, for the first time in the race's history, due to ill-informed media coverage which suggested that the very challenging weather conditions (100 mph winds and

Table 3.3 Example of adventure race event rules from Marmot Dark Mountains

General event rules for all Ourea Ltd events:

1. The Participant must abide by the Event rules as laid out below by the Organiser. Ignorance of these rules by the Participant is no excuse and failure to comply with these rules will result in disqualification from the Event. In the event of disqualification the Participant may be required to leave the Event and travel back to the start at their own expense. In these circumstances no refund of the Participant's entry Fee will be given
2. The Golden Rule. Once registered, each Participant must download their SI data at the Event Centre before departing regardless if they have retired or not (or even not started). This is our check to account for everyone being safely off the hill
3. All Participants are expected to enter into the spirit of this mountain running race and not seek to gain any unfair advantage
4. Participants must comply with our basic safety rules and obey any reasonable instruction given by an event official
5. On open hills and mountains, which are generally defined as Access Land, Participants may cross walls or fences but are encouraged to use gates and stiles where available
6. On agricultural and farmland, Participants must follow rights of way, established footpaths and tracks and must NOT cross walls and fences except at designated crossing points, gates and stiles
7. Any Participant seen dropping litter will be disqualified
8. Participants must comply with the 'Equipment List' and carry all mandatory items as specified. Any breach of the mandatory kit list will result in disqualification
9. Any Participant who acts in a manner that brings the Event into disrepute or endangers another competitor, marshal or member of the public will receive a life ban from Ourea Ltd events

Specific Event Rules for Marmot Dark Mountains:

10. The Event is a team event and each pair must maintain both voice and visual contact with each other for the duration of the Event. Both team members must visit each checkpoint together
11. If one member of the team must retire, then both team members must retire. It is not possible for an individual to continue alone or join another team
12. The competition map may have Out of Bounds Areas, Uncrossable Boundaries and Crossing Points marked on it and these must be respected. An Uncrossable Boundary doesn't necessary mean it is physically uncrossable, but crossing it would be deemed a breach of the rules
13. The competition area is embargoed. If a competitor or team becomes aware of the competition area they are not allowed to reconnoitre or inspect it in advance of the event
14. Absolutely no GPS / Satellite navigation devices (including watches, phones, etc.) are allowed. This includes GPS watches that can display distance travelled or speed even if they cannot display location data. Altimeters that work via barometric pressure are allowed
15. We encourage teams to carry a mobile phone with them but it must be turned off and sealed in bag at registration. Unless required for a genuine emergency the mobile phone must remain sealed in a bag for the duration of the event and this will be checked at Kit Check

Source: http://www.marmot-dark-mountains.com/information/#displayEventRules, accessed 10/01/18

extremely heavy rain) placed competitors and potential rescuers in danger. Reference was made to "1700 people unaccounted for in the hills" though in fact all of these were still competing and unaware that anyone was concerned for them; as usual a significant number of competitors were current or former mountain rescue team members.

The USARA was formed in 1998 and was the first national governing body for the sport of adventure racing which arose from the need for safety standards, insurance, and to promote the growth of adventure racing in the USA. Race organisers have developed event rules (see Table 3.2 for an example) which are there to ensure the safety of the competitors, spectators, and, to some degree, the environment (see Table 3.3, points 5, 6, 7).

3.4 Environmental Impact, Management, and Education

3.4.1 Research Needs

There appear to be no complete systematic scientific studies of the full impact of particular mountain marathons/tours or adventure races on the

Table 3.4 General reviews, recent Australian research, and activity-specific issues/impacts associated with activities that are often part of adventure races

Potential activities	General reviews of impacts	Activity-specific issues/impacts
Walking/running	Liddle (1997), Buckley et al. (2004), Cole (2004), Pickering et al. (2010b), Pickering and Mount (2010), Stevens et al. (2011)	Boots, socks, and other clothing items can spread large numbers of weed seeds from a wide variety of species. It is also likely that shoes spread fungal pathogens, including root rot fungus (*Phytophthora cinnamomi*) in Australia. Spread of weeds and pathogens on boots and clothing. Spread of pathogens in human waste, increased nitrification from human waste
Mountain biking	Liddle (1997), Marion and Wimpey (2007), Pickering et al. (2010b), Stevens et al. (2011)	Likely that spreads weeds and pathogens on tyres, but limited actual research. Damage from construction and use of trail technical features. Some impacts of mountain biking are similar in intensity per km as hiking but likely go further so have more impact per unit of time
Horse riding	Liddle (1997), Newsome et al. (2004, 2008), Pickering and Mount (2010), Pickering et al. (2010b)	Additional nutrients and spread of weeds and pathogens in dung and on hair. Has higher impact per user than mountain biking, walking, and running due to weight per unit area
Abseiling, climbing	Cater and Hales (2008)[a]	Damage to fragile vegetation and lichens growing on cliffs. May also damage nesting birds, depending on location
Camping	Liddle (1997), Smith and Newsome (2002), Smith (2003), Cole (2004)	Longer time periods, may involve deliberate alteration to the site such as creation of campfires
Canyoning, white-water rafting, swimming	Liddle (1997), Stevens et al. (2011)	Often in remote "pristine" water bodies where few other impacts. Damage includes to aquatic system but also to vegetation and soils at access points. Introduction of pollutants including from human waste but also sunscreens, and so on

Source: Newsome et al. (2011, p. 409)
[a]Research by Vogler and Reisch (2011) in Europe indicated that rock climbing reduces the abundance of, and affects the population structure of, cliff vegetation

environment. However, there is a growing body of research associated with the various impacts of walking, running, mountain biking, horse riding, camping, abseiling-climbing, canyoning, white-water rafting, and swimming on the environment (Liddle 1997; Newsome et al. 2002; Buckley 2004; Turton 2005; Pickering and Hill 2007; Monz et al. 2010; Pickering et al. 2010a) which are all commonly undertaken in adventure racing (see Table 3.4).

Newsome (2014) raises awareness about the potential environmental impacts of such activities and sporting events taking place in protected areas (such as national parks and Areas of Outstanding Natural Beauty). Adventure racing participants are most likely focused on risky, thrill-seeking activities where the overall goal is to complete the event in as fast a time as possible. Newsome argues that such a philosophical standpoint and competitive attitude towards the environment is therefore likely to be suboptimal in terms of such visitors appreciating the natural values and conservation function of a protected area.

The rapid increase of adventure racing and its possible impacts on the environment as well as social aspects are thus in need of further research and policy development. Newsome's analysis demonstrated that there was a lack of data concerning the impacts of adventure racing on conservation values, environmental resilience, wildlife disturbance, and ecotourism importance where sporting activities take place in protected areas. Because protected areas, such as national parks, play an important role for conservation and other (more passive) kinds of recreation, the issue of appropriate use of such lands is a cause for concern. Newsome calls for a research agenda that explores the approval process for these events so that park managers can assess the capacity along with existing recreational impacts. There is an urgent need for policy guidelines that can assist managers in making the best environmental decisions.

Next we look at some examples from the UK which seem to generate quite a bit of controversy and debate. One which triggers emotion widely is the National Three Peaks Challenge which takes place in the UK, normally during the summer months and often centred around the third weekend in June which is closest to the longest day of the year (21 June) so maximising the amount of daylight in which to complete the route.

3.4.2 The National Three Peaks Challenge

The National Three Peaks Challenge involves climbing the three highest peaks of Scotland, England, and Wales, often within 24 hours. The total walking distance is 23 miles (37 km), and the total ascent is 3064 metres (10,052 ft). The total driving distance is 462 miles. People can take part in the challenge in two ways—as a self-organised group or a professionally organised event. Self-organised events are the cheapest way to take part, but many groups will hire professional mountain guides. The three mountains are Snowdon, in Wales (1085 m); Scafell Pike, in England (978 m); and Ben Nevis, in Scotland (1345 m). A popular misconception is that the three mountains that form the challenge are the three tallest on the British mainland. Rather, they are the tallest mountains within each representative country: Scafell Pike is the tallest in England, Snowdon the tallest in Wales, and Ben Nevis the tallest in Scotland—over 100 peaks in Scotland are higher than Scafell Pike and 56 higher than Snowdon.

James Keen's article "The Big Debate" in *The Great Outdoors* magazine, January 2009, which can be viewed at http://www.mountainadventures.co.uk/documents/TGO041028_002.pdf illustrates how there are a number of direct criticisms concerning the National Three Peaks Challenge, many of which can be prevented with a bit of consideration and planning.

- *Lack of support for local businesses.* While many participants will spend time in Fort William and Llanberis before and after their challenge, Wasdale Head can be seen to miss out somewhat as groups rush through. The growing popularity of the Three Peaks Challenge over three days helps this matter somewhat.
- *Large groups taking over mountain paths.* While the recommendation is that challenge groups should be kept to an appropriate size, to be considerate to others using the mountains, this is not always adhered to.
- *Walkers don't always stay on the mountain paths* (which on these popular routes have been largely paved to manage erosion caused in the past). Walking off the paths on scree can cause loose rock to be displaced (or even fall on others), and trampled vegetation can quickly be destroyed resulting in additional damage the mountain environment.
- *Reliance on mountain rescue teams.* Ill-prepared and inexperienced groups which are not proficient in mountain navigation and safety can result in unnecessary call-outs for the voluntary mountain rescue teams.
- *Littering.* Walkers or runners who are competitive can be thoughtless and inadvertently drop litter which gets blown around the mountains or left on roadsides to be cleared up by locals after the event.
- Driving over the speed limit to complete the challenge within the 24-hour time has been witnessed; driving through the night is also necessary and needs to be planned in advance.
- *Noise.* Groups passing through isolated farms, hamlets, or small settlements in the dead of night can be disruptive to local residents.
- *Pollution.* Groups from Southern England attempting the challenge will travel nearly 1500 miles in total. This has a carbon cost as it is not possible to complete the challenge in the 24-hour time by public transport.

Table 3.5 The National Three Peaks Challenge: Ben Nevis, Scafell Pike, and Snowdon—for or against?

For the challenge	Against the challenge
• A great test of stamina and mental strength and an excellent personal development tool • Good team-building platform and a chance for people with minimal experience to enjoy the outdoors • Raises substantial funds for the chosen charities (which participants/groups choose) • A challenging objective requiring commitment and discipline, factors often the catalyst for positive change for the individual or organisation	• Attracts thousands of participants from across the world. This has both macro and micro effects on the environment, resources, and local communities • Most people do it in high summer to make the most of the better weather and longer hours of daylight • Most participants have day jobs so do it at weekends which focuses large numbers of people into the certain pressure spots like Wasdale valley over a few weekends • Wasdale, for example, has just one public lavatory for all those people • All those minibuses and one narrow road with limited parking facilities • Large groups pass through farms, hamlets, and small settlements, often at unsocial hours • All those participants and just one stretched mountain rescue team for each peak

Table 3.5 summarises the arguments for and against the UK's National Three Peaks Challenge.

The Institute of Fundraising's Outdoor Events Code of Practice includes some specific guidelines on the Three Peaks Challenge. The code of practice does not apply to privately organised challenges, so does not affect most groups. Applicable only to challenges organised by charities directly, the Code of Practice sets out guidelines to ensure that the potential negative effects of the Three Peaks Challenge are minimised (Table 3.6).

Many protagonists of the National Three Peaks Challenge do admit that it's time for a radical rethink to ensure that its environmental impact is kept to a tolerable level. To ensure a long-term future, future revisions to the Code of Practice for Outdoor Fundraising in the UK (Table 3.6) should reflect the event's popularity and incorporate a registration system so that organisers can submit applications in accordance with a predetermined standard. This should look at scaling down from the current recommended maximum of 200 participants. If a registration system were adopted which required some kind of prior approval by, for example, a national park authority or alternative designated body, it would mean some kind of control over numbers. Then, organisations who have hitherto subscribed to such mass challenge events would have to address their inherent problems and adhere to codes of conduct which would be kinder to the environment.

Table 3.6 Three Peaks Challenge Code of Practice

Specific Three Peaks Challenge guidelines

- Limit the number of walkers to no more than 200 per event
- Avoid the peak holiday times, for example, bank holidays and summer solstice. Events should not cause overcrowding on the mountains and the respective valleys' infrastructure
- Be aware toilet provisions are very limited; plan accordingly when obtaining local permissions
- Strongly discourage racing between teams on and between mountains
- Agree designated rest stops and driving times beforehand that respect speed limits, road safety, and other road users
- Include a policy to remove the time pressure element categorically excluding the driving time between mountains as part of the challenge by allocating a minimum driving time of ten hours for all participants which is added to the walking time, regardless of the actual duration of the drive
- Provide information to participants on the environmental and land management sensitivities of the areas they will be visiting and give participants guidance on how to mitigate their impact as far as is possible
- Individual mountain and site specific codes of conduct should be followed
- To minimise disturbance and adverse impact, organisers ought to consider the timing of the event for the least disturbance
- In settlement areas, arrival or departure ought not to be between the hours of 23:00 and 06:00
- Coaches block narrow roads so should not be used
- Local facilities are inadequate for large events. Organisers ought to identify and use motorway services and other facilities en route especially to top-up water supplies and use the toilets
- If using Pen-y-Pass (Snowdon), parking is usually difficult and waiting not possible so disembark only. Use local bus services when you can

Source: https://www.threepeakschallenge.uk/national-three-peaks-challenge/code-of-practice, accessed 10/01/18

Other practical measures which can help might be if organisers offered alternative and/or mid-week and off-peak events, sensible routing, and planned start times to minimise impact on local communities and a responsible attitude to litter, ensuring that all waste remained in vehicles and was disposed of outside the national parks. Perhaps the best we can hope for is responsible self-regulation on the part of organisers.

In addition to the National Three Peaks Challenge, there are two other "Three Peak Challenges" in the UK: (1) the Welsh Three Peaks Challenge, lesser known than the National or Yorkshire Challenges, takes in the three peaks of Wales—Snowdon (Yr Wyddfa) in the North, Cadair Idris in mid-Wales, and Pen y Fan in the South. The Welsh Three Peaks Challenge includes a total walking distance of 17 miles (27.4 km) and an ascent of 2334 m (7657 ft), usually in less than 24 hours; and (2) the Yorkshire Three Peaks is 24 miles (39 km) and includes the summits of Whernside, Ingleborough and Pen-y-Gent.

3.4.3 The Yorkshire Three Peaks

The Yorkshire Three Peaks Route is about 24 miles (39 km) in length and involves 5000 ft (1600 m) of vertical ascent taking in the summits of Pen-y-ghent (694 m), Whernside (736 m), and Ingleborough (723 m). The terrain underfoot is varied and includes mountain paths, grassy slopes, farm tracks, short sections of steep rocky scrambling, and a bit of tarmac. The organisers have a section of the website about environmental concerns (Table 3.7).

3.4.4 Management Approaches to Minimise Damage

As we have seen, approaches to managing the environmental damage resulting from mountain marathons/tours and adventure racing are still evolving. Event organisers are taking some responsibility through publicising codes of conduct (such as seen in Table 3.4) to participants via entry information and their websites. However, the Institute of Fundraising's Outdoor Events

Table 3.7 The Annual Yorkshire Three Peaks Challenge: environmental concerns

Environmental concerns

Some parts of the route can receive a battering due to the sheer weight of numbers. With the large numbers of people attempting this challenge, it is more important than ever to make sure that we 'tread lightly' and help to maintain this landscape for the people who visit or work on it and the wildlife that lives in it.

When on your challenge
- Stick to the path, even in mud. This helps to minimise erosion.
- Consider doing the challenge mid-week in order to 'spread the load'.
- Cross walls and fences only where there is a stile or a gate.
- Leave no litter (not even a banana skin!!). If you see litter pick it up.
- Leave no food waste. Some otherwise conscientious people leave fruit peel and the like, not realising the problems it can cause for marginal upland species.
- Do not allow dogs to chase sheep or wildlife. The law requires you to keep them on a short lead in areas of Access Land between March 1st and July 31st.
- Keep noise to a minimum especially when near houses, late at night and early in the morning.
- Close gates carefully behind you, avoiding slamming them. Use the latch if present and do not just push it 'to'. If you believe that the gate has been left open deliberately by the landowner then leave it open, but if in doubt, close it.
- Be considerate about where you go to the toilet. Use the public toilets at Horton and avoid going to the toilet where it could offend people. If going to the toilet on the hill then make sure you are nowhere near any path or stream bed (at least 30 m away). Treat all waste as litter; burn it, bury it or bag it up and remove it as appropriate.
- Become a Friend of the 3 Peaks.

Source: http://yorkshire3peaks.org.uk/environmental-concerns.html, accessed 10/01/18

Code of Practice does not apply to privately organised challenges, so does not affect most groups. These codes are still voluntary, and there are no penalties (as far as we are aware) for breaching them. There is some discussion about some kind of registration system so that organisers would be required to submit applications in accordance with a predetermined standard. This system might look at negotiating (and limiting) the number of participants who could take part in a particular event. Event organisations which have previously organised mass challenge events would, hopefully, have to address their inherent problems and adhere to codes of conduct which would be kinder to the environment.

Other practical measures which might help alleviate pressure might be for event organisers to offer alternative and/or mid-week and off-peak events, sensible routing, and planned start times. For example, the Yorkshire Three Peaks Challenge website on environmental concerns (Table 3.7) suggests that in settlement areas, arrival or departure ought not to be between the hours of 23:00 and 06:00 to minimise impact on local communities.

Perhaps the best we can hope for is responsible self-regulation on the part of the organisers.

Conclusions

1. Mountain marathon is an extended form of fell running. Races usually take place over two days and often have a strong element of orienteering. Well-known events include the OMM, the SLMM, the Swiss International Mountain Marathon, Marmot Dark Mountains, the LAMM, the Mourne Mountain Marathon, the ROC Mountain Marathon, the SCOTT Snowdonia Trail Marathon, and the Longmynd Hike. The first OMM was held in 1968, and the size and number of events annually have grown since then.
2. Adventure racing (or expedition racing) is typically a multidisciplinary team sport involving navigation over an unmarked wilderness course with races extending anywhere from two hours up to two weeks in length and often include trekking, mountain biking, and paddling although races can incorporate a multitude of other disciplines including climbing, abseiling, horse riding, skiing, and white-water rafting. The first Adventure Racing World Series event was held in Switzerland in 2001.
3. According to the Outdoor Foundation (2017) survey, in the USA participation numbers in adventure racing rose from 725,000 in 2006 to 2,999,000 (almost 3 million) in 2016, showing a three-year change of 35.5%, with only BMX biking, cross-country skiing, and stand-up paddleboarding showing higher three-year changes in a list of over 40 outdoor activities.
4. Mountain tours are more difficult to define. They began in the European Alps in the 1700s and take place in the zone that is covered by ice all year round (above a height of about 3000 metres) where mountains are at least partly glaciated. Modern mountain tours may include the Seven Summits; the eight-thousanders; the Scottish Munros; Corbetts, Grahams, and Donalds; and the Furths, Hewitts, Nuttalls, Wainwrights, Birketts, and Marylins. Challenges like the Bob Graham Round and Welsh 3000s are still popular races for some or personal challenges undertaken over a lifetime for others.
5. The rapid growth of mountain marathons (since the 1960s) and adventure racing (since 2001) means that there is still a lack of research and evolving policy development on the environmental and social impacts of such events.
6. Most large events today have codes of conduct and environmental guidelines for participants, but these codes are still voluntary. There appears to be a need to undertake more research and develop policies further.

References

Buckley, R. (Ed.). (2004). *Environmental Impacts of Ecotourism*. New York: CABI.

Buckley, R., King, N., & Zubrinich, T. (2004). The role of tourism in spreading dieback disease in Australian vegetation. In R. Buckley (Ed.), *Environmental Impacts of Ecotourism* (pp. 317–324). New York: CABI.

Cater, C., & Hales, R. (2008). Impacts and management of rock climbing in protected areas. In C. Cater, R. Buckley, R. Hales, D. Newsome, C. Pickering,

& A. Smith (Eds.), *High Impact Activities in Parks: Best Management Practice and Future Research* (pp. 24–36). Gold Coast: Sustainable Tourism Cooperative Research Centre, Griffith University.

Cole, D. N. (2004). Impacts of hiking and camping on soils and vegetation: A review. In R. Buckley (Ed.), *Environmental Impacts of Ecotourism* (pp. 41–60). New York: CABI.

Kay, J., & Laberge, S. (2002). Mapping the field of 'AR': Adventure racing and Bourdieu's concept of field. *Sociology of Sport Journal, 19*, 25–46.

Liddle, M. J. (1997). *Recreation Ecology*. London: Chapman & Hall.

Marion, J. L., & Wimpey, J. (2007). Environmental impacts of mountain biking: Science review and best practices. In P. Webber (Ed.), *Managing Mountain Biking: IMBA's Guide to Providing Great Riding* (pp. 94–111). Boulder, CO: International Mountain Bicycling Association (IMBA).

Monz, C. A., Cole, D. N., Leung, Y.-F., & Marion, J. L. (2010). Sustaining visitor use in protected areas: Future opportunities in recreation ecology research based on the USA experience. *Environmental Management, 45*, 551–562.

Newsome, D. (2014). Appropriate policy development and research needs in response to adventure racing in protected areas. *Biological Conservation, 171*, 259–269.

Newsome, D., Cole, D. N., & Marion, J. (2004). Environmental impacts associated with recreational horse-riding. In R. Buckley (Ed.), *Environmental Impacts of Ecotourism* (pp. 61–82). New York: CABI.

Newsome, D., Lacroix, C., & Pickering, C. (2011). Adventure racing events in Australia: Context, assessment and implications for protected area management. *Australian Geographer, 42*(4), 403–418.

Newsome, D., Moore, S., & Dowling, R. (2002). *Natural Area Tourism: Ecology, Impacts and Management*. London: Channel View.

Newsome, D., Smith, A., & Moore, S. (2008). Horse riding in protected areas: A critical review and implications for research and management. *Current Issues in Tourism, 11*, 144–166.

Pickering, C. M., Castley, C., Newsome, D., & Hill, W. (2010b). Environmental, safety and management issues of unauthorised trail technical features for mountain bicycling. *Landscape and Urban Planning, 97*, 58–67.

Pickering, C. M., & Hill, W. (2007). Impacts of recreation and tourism on plant biodiversity and vegetation in protected areas in Australia. *Journal of Environmental Management, 85*, 791–800.

Pickering, C. M., Hill, W., Newsome, D., & Leung, Y.-F. (2010a). Comparing hiking, mountain biking and horse riding impacts on vegetation and soils in Australia and the United States of America. *Journal of Environmental Management, 91*, 551–562.

Pickering, C. M., & Mount, A. (2010). Do tourists disperse weed seed? A global review of unintentional human-mediated terrestrial seed dispersal on clothing, vehicles and horses. *Journal of Sustainable Tourism, 18*, 239–256.

Smith, A. J. (2003). *Campsite Impact Monitoring in the Temperate Eucalypt Forests of Western Australia: An Integrated Approach*. Unpublished PhD thesis, Murdoch University, Perth, Australia.

Smith, A. J., & Newsome, D. (2002). An integrated approach to assessing, managing and monitoring campsite impacts in Warren National Park, Western Australia. *Journal of Sustainable Tourism, 10*, 343–359.

Stevens, R., Pickering, C. M., & Castley, G. (2011). A review of the impacts of nature based recreation on birds. *Journal of Environmental Management, 92*, 2287–2294.

The Outdoor Foundation. (2017). *Outdoor Participation Topline Report 2017*. Washington: The Outdoor Foundation. Retrieved from www.outdoorfoundation.org.

Turton, S. M. (2005). Managing environmental impacts of recreation and tourism in rainforests at the wet tropics of Queensland world heritage area. *Geographical Research, 43*, 140–151.

Vogler, F., & Reisch, C. (2011). Genetic variation on the rocks—The impact of cliff climbing on the population ecology of a typical cliff plant. *Journal of Applied Ecology, 48*, 899–905.

Recreational Climbing and Scrambling

4

Chapter Summary

In this chapter, the types of climbing are defined and numbers involved are estimated. The effects of traditional summer climbing on cliff vegetation and other biota, like gastropods, are evaluated. This includes "gardening," footpath erosion up to the crags, at the tops, and a decrease in rare and endangered species and tree damage from abseiling and belaying. The effects of cliff micro-topography and climbers' preferences are discussed with regard to the vegetation distribution and the fact that not all effects are negative. The impacts on bird populations are evaluated, and the damage to the rock by ropes, chalk, protection, and the creation of rock polish is discussed. Other environmental effects from bouldering, winter, and mixed climbing are evaluated. The management to counteract these impacts such as management plans, memorandums of understanding, liaison groups, closures, seasonal restrictions, star systems in guides, permits, and outreach and education, including codes of conduct, are discussed and evaluated

4.1 Introduction

Climbers cannot deny that their sport causes some ecological damage. It can disturb birds and particularly their breeding, it clears vegetation from the cliff face and specific routes, and it erodes tracks to and from the route. However, climbers often argue that the damage caused is insignificant to that of other outdoor activities, and it is true when one considers other major land-use activities. As Hearn (1994) stated when considering the effects of nature conservation and cliff climbing in Cornwall (UK), "Nationally the effect of climbing, or any other recreational pursuit, is minor when compared with agriculture, pollution, development and other major land-use activities." Conservationists could hold the view that the path climbers leave behind is a scar on the landscape, for example, the paths leading up to Dinas Cromlech in Snowdonia. Would they argue so strongly if the damage had been caused by rockfall? Would attempts be made to repair the damage, and would it even be an issue if there was not a road so ideally placed from which to view it from? A climbers' attitude might be to argue that what they create is a part of the natural environment. However, this is far from the truth, and climbers can damage a specialised environmental niche. Cliffs can be scarce compared with other ecosystems, and the species that have adapted to live there are often rare. So, for example, Raven and Walters (1956) suggested that whole communities of plants, containing many characteristic mountain plants, are in Britain represented by the merest fragments clinging to relatively inaccessible ledges where the most ubiquitous of sheep cannot penetrate. Wyatt (1988) stated that "A crag is not just a rock.

It is a living community of plants and animals. Who decides to climb it becomes part of it. He or she must judge how much of that community of which they are a part must die to satisfy their Adventure." Is the answer none of it, some of it, or all of it? Before considering some of these issues, we have to realise that there are several activities under the umbrella term "climbing."

4.2 Types of Climbing

Traditional climbing or summer climbing takes place on inland cliffs or on coastal cliffs (Fig. 4.1).

This can include instructional cliffs where low technical ability is required, but there may be high numbers involved on some of these cliffs. The climbers ascend rock walls and cliffs that are protected with gear; they have to insert artificial safety equipment, such as wedges or cams which are jammed into cracks in the cliffs and which can often erode the rock. This gear is both placed and removed by the climbers and is considered to have least impact on the rock as a climbing team leaves no trace. The protection that developed differed from country to country and ranged from slings, pegs, and pitons to chocks and camming devices. On the soft Bohemian sandstone, removable jammed knots prevented damage to the rock. It has been suggested that this type of climbing should be called adventure climbing (UIAA 2014). It is largely environmentally friendly. Bouldering is a shorter, power-driven ascent of difficult problems where there are no ropes involved and the routes are usually under 5 m high on boulders or small cliffs (Fig. 4.2). As a discipline it dates from 1993 when Black Diamond sold the first commercially available crash pad which is needed at the cliff base to cushion any falls and make a safer landing if there are falls. Winter climbing might involve snow gullies, ice climbing on waterfalls or glaciers, or mixed climbing on a combination of snow, ice, and rock. Dry tooling is a form of climbing taking place on outdoor crags and indoor climbing walls using ice axes and crampons. Dry tooling on outdoor crags in "summer conditions" typically occurs on overhanging quarried rock or on rock faces which are generally unsuitable for

Fig. 4.1 Big wall climbing, El Capitan southeast face, Yosemite. Photo by D. Huddart

Fig. 4.2 Erosion at base of crags, Harrison's Rocks, South East England. Photo by D. Huddart

conventional rock climbing, and it usually involves the use of fixed equipment and drilled/manufactured axe and crampon placements. The British Mountaineering Council (BMC 2014) has acknowledged it has a place in British climbing, but it is not considered an acceptable practice on established rock climbs. The BMC states that the suitability of individual sites for dry tooling should be considered on a case-by-case basis by the relevant BMC area meeting. Unfortunately it has already caused noticeable, irreparable damage to high-quality rock climbs in the Lake District (UK) with scratch marks and damage to crucial gear placements or handholds. Free soloing requires no safety or aid equipment allowing the climber to ascend a route using no safety equipment and so is the least invasive form of climbing. Sport climbing involves participants setting their own route, using a rope, and can take place either on cliffs or indoor climbing walls. The routes are pre-bolted by the first ascent party who drill the rock and place a bolt, and therefore this eliminates the need for safety equipment. A bolt hanger is used for clipping a carabiner to the bolt. Speed climbing is where the route and the handholds are standardised, and the participants are timed over the route. The latter two types have increased in popularity as we will see from the next section and are important if it can be shown that they lead to the increased popularity of outdoor climbing. Top rope climbing is where a safety rope is always anchored above the climber either in scaling cliffs or artificial walls. A top rope route is set up by a climber either leading a sport or traditional route from the base to a set of anchors, where the rope is attached with locking carabiners. Aid climbing is where climbers ascend steep rock faces with the use of specialised climbing equipment that allows mechanical upward progress rather than free climbing. Climbers can place pitons, cams, and nuts to attach the rope and themselves to as they climb upwards. Scrambling is where easy rock faces and ridges are climbed, usually in mountains, either with or without a rope. It requires basic climbing skills using hands and feet and belaying, descending, and abseiling. So we can see that there are several types of climbing, and logically they can have different environmental and ecological impacts.

4.3 Numbers Involved in Climbing

The numbers involved in the activities may illustrate the relative impact potential either in total number of climbers or the relative importance of the types, although we will see that the impacts vary between types. However, because climbers often just carry out their activities in small independent groups or a pair, are not members of clubs, and may not figure in the surveys that have been carried out, it is extremely difficult to estimate realistically the numbers involved in the activities.

This is especially the case when the types of climbing are grouped in different categories of activity. However, in general there has been a growth worldwide in climbing as a series of recreation pursuits, although climbing did not appear in the top twelve recreation activities in six European countries, Canada, and the USA in the 1990s and 2000s (Bell et al. 2007) and it did not appear to be measured or in the forecast changes up to 2050 in the USA (Cordell et al. 1997; NSRE 2003; Cordell 2004). Cordell (2008, 2012) suggested climbing was the fastest-growing nature-based activity in the USA in the period 2000–2007, with a growth from 7.5 million in 1994–1995, to 9.0 million in 1999–2001, 8.7 million in 2007 to 9.8 million between 2005 and 2009. This was a percentage increase of 9.5% from 1994–1995 to 2009, 8% growth in the period 1980–1984, and a 12% growth in the period 1985–1989. Total number of days of participation increased by 23.8% in the period 2000–2007, up to 0.1 B. However, the Outdoor Foundation, Topline Survey (2017) suggested in the USA a growth of 6% for traditional and ice climbing between 2013 and 2016, with a 20.3% growth between 2006 and 2016 from 1.586 million to 2.79 million. However, in the same period, there was a growth of 3.4% in sport/indoor and bouldering from 4.728 million to 4.905 million. Unfortunately in Bowker et al. (2012), in a survey of potential changes up to 2060, rock climbing was included in backcountry activities, which also included mountain climbing and caving. The total for 2008 was 25 million, and the range for 2060 was estimated to be 38–48 million, with an increase of participation days from 121 million days in 2008 to be between an estimated figure of 178 and 219 million days by 2060. The figures quoted in a report for Colorado (2013) stated that climbing (which included sport, indoor, and bouldering) was the 23rd most popular outdoor recreation activity in the USA with numbers of 4.592 million, 1.5% of the US population, whilst in Colorado it was ranked 21st, carried out by 12.5% of the population, with 3.912 million activity days in total. This illustrates the regional variation within the USA as might be expected, and the numbers had increased from an estimated half a million to over 1 million from 2006 (Climbing Management Guide 2008).

In the UK the BMC membership rose from 25,000 in 1990 to 82,000 currently, and individual membership rose from 25,000 in 2000 to 55,000 in 2015. National Trust figures for rock climbers estimated 1.35 million casual climbers and 150,000 active climbers which illustrates the great difference. In the late 1990s, it was thought that there had been an increase of 140% of active summer climbers in the previous 15 years. The estimated increase in winter climbers had been 20% of the summer climbers in the same time period. Technology had brought about the pushing of climbing limits which has caused new routes to be put up on existing crags, the continued search for new routes, and the development of winter routes that follow summer rock lines and have been climbed using winter tools under winter conditions. Until recently, with the more obvious impact of global warming, the popularity of winter climbing was seen from the number of new routes that have been put up. For example, from October 1996 to the New Year of 1997, 60 new routes were reported in *High* magazine (no.170), and in the same magazine (No.172), another reason for expanded climbing in the mixed scene was restricted resources in the UK and the need to find new ground and to diversify new ascents. In the UK Gordon et al. (2015) suggested that 14% of 8.96 million people who were active in the outdoors participated in mountaineering, abseiling, and climbing on a weekly basis. This figure was 93,000 which was an increase in 16,000 compared to 2014. According to Adventure Participatory Survey Adventure Sports (APS) in 2013–2014, once a week participation in rock climbing age over 14 was 61,300, whilst 28,900 climbed at an indoor wall.

Sport climbing (which includes sport, bouldering, and indoor) has grown tremendously recently. Between 2007 and 2014 the number of people participating grew by 18,732 in the USA, while from 2014 to 2015 the number grew by 148,287 to 4.6 million, which was 70th from a list of the top 111 outdoor activities. In the same country, the Outdoor Industry Association puts the total participation in bouldering at between 4.7 and 6.9 million people. This is also shown by Eastern Mountains Sports who state that its sales of crash pads for bouldering had grown by 15% in the last year (2017), sales of rock shoes had grown by 70% over the last five years, and ropes and other gear by 40%. The International Federation of Sport Climbing echoes this boom and now has over 87 member federations over five continents and claims over 25 million people climb regularly. In England, for example, the number of young people taking part in the BMC Youth Climbing Series has risen by 50% in the last five years, whilst in Austria the number of members and member clubs in 2008 was 23,170 in 141 clubs, whilst in 2016 the figures were 61,140 members in 176 clubs. In 2010 the BMC Climbing Wall Survey to over 3000 members had over 5 million climbing wall user visits each year at 350 walls listed in the BMC Wall Directory. One of the crucial statistics was that over 75% of the respondents said they climbed at both indoor walls and outdoor crags. This boom in sport climbing can only escalate much further as the activity has been added to the 2020 Summer Olympics in Tokyo.

Despite the difficulties of obtaining accurate numbers and the variation in figures put forward, we can establish that the numbers of climbers of all types have increased, as have the number of climbing participation days, and so it is no surprise that there is an increase in noticeable and measurable ecological effects on the crag ecosystems.

4.4 How Do Climbers Affect Crag Ecosystems?

We need to itemise what climbers need from their crags and then look at what plants need and see if the two are compatible. Climbers need good aspect, usually south-facing crags; the rock needs to be clean and free from vegetation; the rock needs to be dry and steep; the rock needs to be good and solid; the faces need to be of a reasonable height, and there needs to be interesting route diversity. Plants need to be protected from grazing and competition. They need available water with nutrient leaching from above a useful bonus; the slope angle should not be too steep or the soil will not build up and support plant life; a southerly aspect is useful; the rock type can be important as base-rich rock tends to support a wider species range than acidic rock. Looking at the needs, it might appear that the rock that is climbed on the most often is the type not suitable for plants to live on and suggests no great conflict of interests. However, this is not so because the cliffs that are climbed on now have been extensively stripped of their vegetation, cleaned, and "gardened" in the past. As an example in the UK climbers tend to affect the arctic-alpine group of plants which are linked to past cold climates and which are increasingly rare and likely to be on the verge of extinction under global warming; there are woodland species which can survive because of high humidity and damp hollows, meadow and flush species which can survive on ledges, and specialised plants where water pours down a rock face or gully, such as the range of mosses and liverworts and the rock pioneer species, the lichens. There are also moorland species characteristic of peaty, acidic soils and the upland grasses and heath species which would be expected in the general area of the higher crags but are selectively grazed by sheep on the UK crags.

4.5 Effects of Climbing on Cliff Vegetation and Other Biota

4.5.1 Traditional Summer Climber Impacts on the Vegetated Crag Environment

Climbing can impact in the following ways (for an excellent overview on Cliff Resource Impacts, see Marion et al. (2011), pp. 105–112):

- Development of new routes where the vegetation and loose rock are gardened, either accidently or on purpose. This gardening sanitises the intended route line, leaving barren rock. Soil can be mechanically scraped from the rock face by using a wire brush, toothbrush, or trowel as the soil and/or loose rock can hinder a climber's ability to ascend a cliff safely. They need a clean surface so that their hands and feet are able to grip the micro-topographical features of the rock. Trundling can take place where the loose, unstable rock and soil can be removed from a climb. This extracted loosened rock means that future climbers will not injure themselves or the person on belay will not be hurt by falling rock. In some areas a few climbers have resorted to chiselling the rock to create handholds. So cliffs are often modified so that the climber may have better traction with the rock.
- Climbing established routes which might need placement of protection in cracks and clean routes and belays is often taken on ledges which are suitable for plant colonisation, and so standing on these ledges will result in erosion. In this way colonisation by plants is prevented, scraping away mosses and lichen which escaped the initial gardening, exposed rock becomes progressively more polished, and rockfall becomes more likely. However, the National Trust illustrated that a recent survey at Avon Gorge (Bristol area, UK) showed that climbers actually had a beneficial effect by keeping paths and belay ledges open and trodden which is exactly the kind of habitat preferred by such plant rarities as the Bristol rock cress, (*Arabis scabra*) which needs open ground for germination.

Access to the base of established routes at the base of the cliff can cause damage. Ascending steep terrain to the crag has sometimes ill-defined, access routes which cause climbers to unknowingly take several paths to the same destination. For example, the 22 climbing sites in the Pinnacles National Monument, California (USA), were reported to have access leading to them, but in a short period of time, steep secondary trails appeared which began to divert water, causing soil loss, trenching, the development of shortcuts, and vegetation loss. The result is increased footpath erosion as at Harrison's Rocks (South East England) and Stanage (Peak District) and damage to areas such as delicate scree slopes with accelerated erosion (e.g. World's End screes at Llangollen, North Wales). If a wider view is taken, then there could be a greater number of cars, car parking problems, sometimes on vegetation, damage to gates and fences, and perhaps disruption to livestock. Better access often leads to more climbers, and a vicious circle can occur. There is also damage to cliff-top vegetation by trampling around abseil points or from footpaths along the crag top. At Harrison's Rocks, Baker (2005) found that ground levels had been lowered by more than 1 m in some places due to intense foot traffic at the crag base (Fig. 4.2).

- At Harrison's Rocks photos of the outcrops in the early 1950s show dense vegetation cover around the rocks, and 50 years later, the position had greatly deteriorated. At Stanage where there are over 8500 graded climbs, with many regional classics and a few of national classic status, Smith (1997) found a variation in crag foot erosion depending on the crag use. At buttresses, like crags 1–3, where there was no climber use, there was little, if any, bare ground at the immediate crag foot, and there was a moorland ecosystem composed of heather, bilberry, bracken, and wavy hair grass. On the other hand, at the buttresses of high popularity, like Tippler Buttress, Flying Buttress, and Mississippi Buttress, the bare ground width averaged between 7 and 10 m, and around 50% of the surveyed area was bare ground. Heath species like heather and bilberry were all but eradicated from the crag base, and those plants that remain lie at 15–20 m from the crag. There are now large areas of bare ground and grassland in an area of heather moorland. The popular buttresses have long since exceeded their carrying capacity in this area.
- Carr (2007) also measured the environmental impact of climbs in the Red River Geological Area of the Daniel Boone National Forest, a world-class climbing destination in Kentucky (USA). He used bare ground as a surrogate for environmental impact, and climbing impacts the ground in a concentrated use area at the base of climbs, an area called a climbsite (with analogy to a campsite). The climbers also impact the ground on the access trails which may either be along the cliff base between climbsites or the access trail from the parking area to the cliff base. The most important factor in determining the amount of impact was found to be the type of climb, whether a sport climb or a traditional climb. Sport climbs were found to have three times as much impact. As the difficulty of a traditional climb increased, the impact reduced. Similarly as the access to the climb gets more difficult either in trail quality (sport climbs) or length (traditional climb), the impact is reduced. Better climbs and more quality stars lead to more impact with traditional climbs. The ground impacted due to climbing was estimated to be 2.8 acres out of a total area of 29,000 acres, which is only 0.01%. The length of cliffline base impacted was estimated to be 0.9 miles out of a total of 254 miles which is only 0.04%. So using these measures of environmental impact, climbing in this area has a limited impact. Moreover Walendziak (2015) measured the environmental area of impact at eight climbing areas in the same area over a six-year period from 2007 to 2013 to determine the impact trends. He found that the total mean area of impact did not change significantly and climbsites remained stable and seem to have reached an area of impact maturity rather like what Cole (2013) found for campsites.
- Retreat from the top of routes can cause erosion on descent footpaths.
- There can be tree damage from abseiling points and belaying from trees. This can cause damage to the bark by friction and abrasion, with the most severe damage caused by friction from ropes being pulled down after abseiling. This friction wears the tree bark and with repeated use can lead to ring barking of the tree. Where there is the removal of a complete ring of bark around the tree trunk, this can lead to disease caused by fresh-wound parasites which can cause cankers, decay, and other diseases and lead to loss of tree strength. Damage to trees is evident in two stages: stress which is a reversible condition where energy or other survival factors become limiting and the tree or one of its parts or processes begins to operate near its limits through root or bark damage. Stress can be reversed. However, if nothing is done about the problem causing stress, a tree can die, or the affected part is severely affected. Once damaged a tree will attempt to self-repair. An example is where Scots pine responded to the removal of 50% of their bark at the base of the trunk by increasing the production of resin acids. Two and a half times the normal concentration of acids was found in the bark adjacent to the wounds and includes dehydroabietic acid, which may be a deterrent to larval feeding and

provide increased decay resistance against white-rot and brown-rot fungi (Liddle 1997). For example, Richards (1997) reported damage from abseiling at Craig Bwlch y Moch, Tremadog (North Wales), and Webber (2002) showed the damage caused by abseil anchors from trees at Clogwyn Cyrau, near Betws-y-Coed (Snowdonia, North Wales). Webber (2002) conducted simulated abseils from three species of tree known to be used by climbers using two 50 m, 9-mm thick ropes, tied together and tied round the tree base. The simulation was carried out 100 times and the depth of any rope groove recorded. The results are illustrated in Table 4.1. The trees were on private land and were already marked for felling. Although the overall final rope groove depth was hardly different, the smoother, thin barked holly was damaged quicker than those species with thicker, rougher, flaky bark like the oak and the Scots pine. This suggests that other smooth-barked trees located at other climbing venues in North Wales, like the rowan and silver birch, will be susceptible to rope damage.

However, a final point worth making is that at the top of Craig Bwlch y Moch (Tremadog), there are young oak trees stripped of bark by sheep, and there are no trees in the agricultural land where sheep ensure no woodland regeneration takes place. So it is worth remembering that it is not only climbers that cause tree damage.

- An example from North America looked at the Eastern white cedar (*Thuja occidentalis*), tree density and age structure that have been shown to be affected by climbing (Kelly and Larson 1997). The authors compared four climbed and three unclimbed sites on the Niagara Escarpment, near Milton (Ontario). The results showed that living tree density on the cliff face was lower in the climbed areas. The age structure of these forests show that the numbers of older and younger age classes have been reduced on the climbed cliff faces. A high percentage of the trees on the climbed cliff faces showed evidence of damage.
- There are positive results too as the National Trust in the UK has shown that climbers have assisted with vegetation clearance in some areas which have difficult access, for example, the clearing of rhododendron on Lundy Island, their assistance with botanical cliff survey, the provision of advice on cliff stability and dangers, and the undertaking of bird-ringing projects, such as the peregrine falcon in the Lake District and choughs in Pembrokeshire (southwest Wales). The presence of climbers can help deter egg thieves as has been demonstrated from the sandstone cliffs at Helsby (Cheshire).
- Rock climbing significantly reduces the vascular and non-vascular vegetation cover on cliffs according to Ruby (2015), although as we will see, not all researchers agree. There must be damage to lichens on crag faces by removal and wear as it might be expected that the pioneer species on rock faces, the lichens, must be affected by climbing. Adams and Zaniewski (2012) evaluated the lichen community composition for both cover and richness on a cliff face commonly used for climbing on a sandstone outcrop on the Sibley Peninsula on the north shore of Lake Superior. They found that both the richness and cover were significantly lower on the climbing cliff

Table 4.1 Simulated abseils (all trees were of maturity class 4) (after Webber 2002)

No. of abseils	Depth of any rope groove in mm: sessile oak (*Quercus petraea*)	Depth of any rope groove in mm: Holly (*Ilex aquifolium*)	Depth of any rope groove in mm: Scots pine (*Pinus sylvestris*)
10	No groove, moss removed	0.5	No groove, some bark smoothing
20	No groove but beginning to smooth	0.8	No groove, more smoothing
30	0.5	1.2	0.5
40	1.0	1.4	0.6
50	1.2	1.5	0.8
60	1.5	1.8	1.1
70	1.6	1.9	1.5
80	1.8	1.9	1.7
90	2.0	2.0	2.0
100	2.1	2.2	2.4

sections compared with the unclimbed sections. Regression models indicated significant relationships with cover and richness to environmental response variables and the climbing treatment. There was major lichen community group separation between climbed and unclimbed areas, and they thought that this was linked to both aspect of the measured plots and the climbing treatment. Baur et al. (2007) assessed the lichen species diversity and cover on climbed and unclimbed areas of ten isolated cliffs in the northern Swiss Jura Mountains. A total of 38 calcicolous lichen species, three bryophyte species, and one alga were found on the rock faces. Climbed and unclimbed rock areas did not differ in the total number of lichen species, species density (the number of species/100 cm^2), or total lichen cover. However, the frequency of occurrence of epilithic lichens (those that grow on top of rock without penetrating the rock substrate) was lower along climbing routes than in unclimbed areas. The lichen community composition of climbed areas differed from that of unclimbed areas, and the dissimilarity increased with increasing climbing intensity on the focal route in climbed areas but not with the age of the climbing route. Nuzzo (1996) on exposed dolomitic cliffs in Illinois (USA) had suggested that climbing had apparently significantly reduced the lichen cover and lichen species density by 50% from 13.7% cover and 2.4 species/0.25^2 on unclimbed areas to 6.7% cover and 1.2 species/0.25 m^2 on climbed cliffs. However, although the climbed cliffs had lower lichen cover, the distribution of community groups was similar on both sets of cliffs, which indicated to the author that environmental and physical variables were the primary determinants of cliff flora on the vertical exposed cliffs. This was because climbing too did not have an apparent effect on the vascular plants, which ranged from 2.74% to 10.62% cover on individual cliffs. Studlar et al. (2015) used a simple force meter to measure the adherence of bryophytes and lichens to try and understand the vulnerability to removal by both climbing route preparation (cleaning) and accidental dislodging. They did this on the Pottsville conglomerate caprock at Coopers Rock and at the New River Gorge National River in West Virginia (USA). They found that lantern moss (*Andreaea rothii*) is held relatively tightly, probably benefitting the more weakly attached species such as the liverwort (*Diplophyllum apiculatum*) which commonly grows epiphytically on *A. rothii*. The umbilicate lichens, the smooth rock tripe (*Umbilicaria mammulata*), and common toadskin (*Lasallia papulosa*) were more tenacious than the lantern moss, but the loss of extensive smooth rock tripe colonies was found to be one of the most visible climbing consequences. However, they recommended long-term studies with different experimental disturbance regimes to more fully evaluate climbing impacts on bryophyte-lichen communities.

- Camp and Knight (1998) showed that in the Joshua Tree National Park (California), six cliffs with no climbing, moderate climbing, and intensive climbing were sampled for plant diversity and community structure. Results showed that the number of individual plants decreased with increased climbing use both on and off the cliff face. Similarly Farris (1998) showed that on three Minnesota cliff systems, the frequencies of most plant taxa were lower in the climbed areas (Table 4.2).

Table 4.2 Mean number of taxa per plot in three climbed and unclimbed areas

From Farris (1998) from Minnesota State Parks					
Blue mounds		Interstate		Shovel point	
Climbed	Unclimbed	Climbed	Unclimbed	Climbed	Unclimbed
2.88 ± 0.24	3.98 ± 0.35	0.88 ± 0.17	4.35 ± 0.29	6.97 ± 0.39	10.18 ± 0.44
$n = 48$	$n = 45$	$n = 48$	$n = 48$	$n = 33$	$n = 27$

There are significant differences between climbed and unclimbed plots at all three locations

However, the total plant cover was significantly lower in the climbed plots. Fragile forms such as the umbilicate and fruticose lichens were especially sensitive. It was found that the microtopography of the rock had a significant impact on both the amount of vegetation present and the use of a cliff area by climbers. It was suggested that the identification of causal links between climbers and vegetation structure must include a careful assessment of the geological and environmental factors that strongly influence both the climbing use and the vegetation dynamics. We will return to this later.

- McMillan and Larson (2002) evaluated the effects of climbing on the heavily climbed limestone cliffs of the Niagara Escarpment, which have the most ancient forest east of the Rocky Mountains, with Eastern white cedar that are over 1000 years old. They compared the vegetation on three parts: the top edge (plateau), the cliff face, and base (talus) of both climbed and unclimbed cliffs. The climbed faces had only 4% as many vascular plant species as those that were unclimbed. The diversity of bryophytes and lichens in climbed areas was 30–40%, respectively, of the unclimbed areas. There was a decreased vegetation cover on the cliffs, and for vascular plants the cover on climbed plateau and talus was around 60% of the unclimbed areas. For bryophytes the cover on the climbed areas was about 20% of unclimbed areas. While climbing did not affect the extent of lichen cover, it did change the types of species that grew on the cliffs. The delicate lichens were replaced by tough species. In the climbed areas, the proportion of non-native plants in the climbed areas was three times higher (81% compared to 27%).
- Climbing reduces the plant density which increases the number of sites where non-native plants can grow. Climbers can introduce seeds and living non-native plants via their boots, clothing, and equipment, but it is not only non-native species or alien species which can colonise but weed species and disturbance-indicator species as well.
- Baur (2004) and Rusterholz et al. (2004) have shown that the exposed cliffs of the Swiss Jura also harbour a highly diverse flora with many rare and endangered species. They examined the impact of rock climbing on the vascular plants in the lower part of four cliffs of the Gerstelflue (NW Switzerland) by comparing the vegetation of climbed and unclimbed areas. In the climbed areas, the plant cover and species diversity was reduced, and the density of forbs (any herbaceous plants but not grasses) and shrubs decreased, whilst the density of ferns tended to increase. The climbing also caused a significant shift in plant species composition and altered the proportions of different plant life-forms. These authors considered that rock climbing can be a threat to sensitive plants of this limestone cliff community. Müller et al. (2004) in the Northern Swiss Jura studied the ecological effects of climbing on the vascular plants at the cliff base and on the cliff face by assessing the plant cover and species density at various distances from the frequently used climbing routes. The plant cover was found to be significantly reduced at the base of the climbing routes, as was the species density too. The plant cover and species density at the cliff face tended to increase with distance from the route. When the vegetation of five frequently climbed cliffs was compared with that of seven unclimbed cliffs, the climbing significantly altered the plant composition. Specialised rock species occurred less frequently on the climbed compared with the unclimbed cliffs.
- It has also been claimed that climbing leads to a decrease in rare and endangered species because some of these are arctic-alpine species that live on high mountain and cliff ecosystems. They are relicts from the last cold Pleistocene phase, and others are endemic to certain restricted areas. We will discuss two examples here. The cliff goldenrod (*Solidago sciaphila*) is an endemic species restricted to dolomitic or sandstone cliffs in or near the driftless region of the upper Midwest USA. A demographic study of 544 genets (a colony of plants that come from a single genetic source)

on currently climbed, previously climbed, and unclimbed dolomite cliffs in north-west Illinois indicates that the position on the cliff face was the most significant factor affecting growth. 70% of all plants grew within 3 m of the cliff top, and this was only 18% of the cliff face (Nuzzo 1995). Within the upper cliff zone, climbing significantly reduced Solidago density. The cliffs that were actively climbed had few genets in the upper 3 m, averaging $3.2/m^2$, whilst unclimbed cliffs and cliffs not climbed for two years supported $14.2/m^2$ and $12.0/m^2$, respectively. The basal area/m^2 and the plant production/m^2 were significantly lower and the inflorescence length non-significantly lower on currently climbed cliffs due to the lower genet density. In the lower cliff face (over 3 m from the top), genet density was low (0.2–2.1 m^2) on all cliffs regardless of the climbing intensity. Two years after cessation of climbing, Solidago in the upper zone of the previously climbed cliffs had similarly high density, basal area, plant production, and inflorescence production as on the unclimbed cliffs. So this suggests a quick recovery of this species.

- A second example is that of the effect of climbing on the yellow whitlow grass (*Draba aizoides*) and the alpine saxifrage (*Saxifraga paniculata*) which occurs on the limestone cliffs of the Swabian Jura in Southern Germany. Both are rare and endangered species, particularly where climbing is allowed. Wezel (2005) measured the following: the number of cushions/cliff, the number of rosettes per cushion, and the aspect and the inclination of the sites. *Saxifraga paniculata* was found on 75% of the cliffs, whilst *D. aizoides* only on 30%. In cliff areas where climbing took place, *S. paniculata* was only found in relatively uninfluenced parts of the cliffs. In contrast *D. aizoides* was found with a significantly higher frequency on climbed cliffs and at the foot of the cliffs which means that climbing even increased the numbers of this plant. This was explained by climbers spreading the seeds, and when cleaning the climbs, cushions and rosettes fell to the talus slope and were re-established because they can put down adventitious roots. This was backed up by the work of Vogler and Reisch (2011) in the Franconian Jura and the Swabian Alb where they compared the number and distribution of *D. aizoides* plants on eight cliffs that had been climbed for the last 50 years with eight pristine, unclimbed cliffs of similar size and aspect. They also collected plants for DNA analysis. This plant was known to be highly sensitive to climbing disturbance (Herter 1996) and is a cushion plant of low competitive ability, highly adapted to the limestone cliff habitat, and very sensitive to trampling or sideways shearing forces. The climbed cliff faces had smaller and fewer plants when compared with the pristine, unclimbed cliffs. On plateau sites species occurred unaffected by climbing. They were significantly less frequent on the cliff faces but more frequent on the talus of climbed in comparison with unclimbed cliffs. The reasons were those suggested first by Wezel (2005). The DNA analysis showed that compared with the climbed cliffs, there were greater genetic differences between plants living at different heights on the unclimbed cliffs. This was attributed to the displacement of plants by climbers who move rosettes down the cliff, displacing genes in the process. This shift in turn reduces the genetic differentiation between the upper and lower subpopulations. This may affect the long-term fitness of this plant species to survive in an environment to which it has become adapted since the beginning of the postglacial period after thousands of years of natural selection.

- Clark and Hessl (2015) from the New River Gorge National River in West Virginia (USA) compared species richness, abundance, and composition of vascular plants, bryophytes, and lichens on 79 pre-established climbing sites and 32 unclimbed control sites across the potential climbing-use intensity and the cliff structure. Differences in species richness and abundance associated with potential climbing-use intensity and cliff structure were variable across the taxonomic groups. Linear models indicated that cliff angle was the strongest explanatory variable of species richness and abundance for all three taxonomic groups. Potential climbing-use intensity had a small but negative effect on species richness and abundance of vascular plants, no effect on bryophytes, and a substantial effect on lichens. Cliff angle and canopy height and aspect were the primary drivers of species composition. They observed no change in community composition due to climbing, and the fundamental control on cliff vegetation was cliff angle.

- Many researchers as we have seen report that climbing has significant negative effects on the cliff biota, but most of the previous work has not been controlled for variation in micro-site characteristics when comparing sites with and without climbing. Most researchers did not identify either the style or difficulty level of the climbing routes. However, Kuntz and Larson (2006) tried to solve these problems by sampling climbing areas used by advanced sports climbers and quantified differences in micro-topography between climbed and control cliffs. They determined whether differences in vegetation existed between pristine and sport-climbed cliffs when micro-site factors were not controlled. Then they determined the relative influence of the climbing, cliff-face micro-topography, local physical factors, and regional geography on the richness, abundance, and community composition of cliff-face vascular plants, bryophytes, and lichens. When they did not control for micro-site differences among cliffs, the results were consistent with the majority of prior work on impacts of climbing: that is, the sport-climbed cliffs supported a lower mean richness of vascular plants and bryophytes and significant different frequencies of individual species when compared with pristine cliff faces. When the relative influences of micro-topography and climbing disturbance were taken into account, however, the differences in vegetation were not related to climbing disturbance but rather to the selection by sport climbers of cliff faces with micro-site characteristics that support less vegetation. Climbed sites had not diverged towards a separate vegetation community, but they supported a subset of species found on the pristine cliff faces. Marion et al. (2011) also suggest that it is clear that the correlation between fewer plants and climbing impacts does not automatically mean recreation has caused the reduction in plants. They suggest that micro-topographical differences and the climbers' preferences must also be factors.
- In the Baetic mountain range (SE Spain) on limestone, Lorite et al. (2017) studied three climbing sites to look at the influence of a range of qualitative categories of climbing frequency impacting on the vegetation. They looked at low frequency (intermittent climbing), medium frequency (high frequency without overcrowding), and high frequency (high frequency with overcrowding). Within each site climbing routes and adjacent areas free of climbing were selected and sampled by photo-plots obtained by abseiling. The images were then analysed to calculate species cover, richness, and total cover. It was found that climbing negatively affected the cliff vegetation community at all sites. There was a significant decrease in plant cover, species richness, and a shift in community composition recorded, with the cover the variable most sensitive to climbing. The impact observed related to the frequency: low-frequency sites with usually more specialised climbers underwent only mild damage, whereas at the high-frequency sites, the impact was severe, and they suggested that the conservation of the species, especially the rare ones, was jeopardised.

4.5.2 Impacts on Gastropods

When we look at other biota on cliffs affected by climbing, we find impacts too. Baur et al. (2007) in the Northern Swiss Jura found that five out of the eleven snail species were specialised lichen-feeders, and plots along climbing routes had fewer snail species than plots in unclimbed areas. Total snail abundance was positively correlated with lichen species richness, but there was no correlation between snail species richness found. The results show that frequent climbing can change the lichen community and reduce the snail community on limestone cliffs. A climbing-related reduction of snail abundance may also alter the lichen-herbivore interaction and indirectly change the competitive interaction among lichen species. Ten years later Baur et al. (2017) reported a diverse gastropod community with some rare species in the similar mountain area. They examined the effects of sports climbing and micro-topographical features of the rock faces on the terrestrial gastropods by assessing species diversity and abundance on climbing routes and in unclimbed areas of seven isolated cliffs. They considered exclusively living individuals attached to rock faces. A total of 19 gastropod species were recorded. Six of them were specialised rock-dwelling species whose individuals spend their entire lives on rock faces, feeding on algae and lichens. Plots along climbing routes showed fewer species of rock-dwelling snails as well as other gastropod species than plots in unclimbed control areas. Similarly both the density of individuals and the frequency of occurrence in the climbing route plots were reduced in both groups of snails. This work indicates that terrestrial snail diversity and abundance are suitable indicators for impact assessment.

4.5.3 Climbing Effects on Bird Populations, Particularly Raptors

Cliffs provide nesting sites for some bird species, and disturbance from climbers who are on routes close to nests on both inland and sea cliffs can affect breeding success. However, climbers sometimes cannot avoid birds on cliffs, although one view is that the cliffs are the birds' home, and climbers could be looked upon as trespassers on the birds' territory simply to achieve their quest for adventure. Generally though climbers can co-exist without detriment to a bird's breeding success, and it is often stated that it is egg collectors, unscrupulous gamekeepers, overzealous bird-watchers, and photographers who can cause the greatest harm. Climbers are therefore not the only cause of bird disturbance or failure to breed, and sometimes the issues are complex. Hearn (1994) suggested that "climbing probably contributes to the disturbance of sea birds," but we know that there are effects of marine oil pollution and climate change, which might be affecting food supplies. Recreation however can have many documented effects on bird populations, and these are illustrated, for example, in reviews by Buckley (2004) and Steven et al. (2011). These can range from the physiology to immediate behaviour that can be affected, including changes in temperature, heart rate, or stress hormone secretion; changes in foraging, vigilance, and evasion; and changes in reproductive success and/or the number or density of birds where there can be reduced number of nests built and eggs laid—a prospecting bird will not establish a nesting territory at a site which is noisy and disturbed. If continually frustrated, it may abandon the cliff resulting in a total loss of that bird. Climbers can also disturb nesting birds on the way to crags as not all endangered species are cliff nesting but nest on moorland or breed around small water bodies (see Table 4.3).

There have been conflicts concerning birds and climbers, and undoubtedly climbers have affected some birds; there have been documented incidents but relatively few, and climbing restrictions have been placed on many crags during breeding seasons (see the BMC Regional Access Database, the Green Guide to the Uplands, and the excellent traffic light system on the Mountaineering Scotland website which informs climbers which crags had nesting raptors and which crags were safe to climb, and all that climbers need to do is check websites before they

Table 4.3 Birds at risk from climbing disturbance in the UK

Sea cliffs	Cliff nesting	On the way to
Auk species, for example:	Chough	crags
	Peregrine	Dotterel
Puffin	falcon	Fieldfare
Razorbill[a]	Golden eagle	Harriers
Guillemot	White-tailed	Hobby
Kittiwake	eagle	Divers
	Raven	Grebes
		Greenshank
		Curlew

[a]80% of the West European razorbill population

go climbing and then change their plans if they spot a nesting raptor whilst climbing; Advice on Disturbance of Birds and Nesting). Nevertheless the climbing restriction system which is in place in the UK and a similar system in the USA seem to work well. Examples will now be discussed using the relationship between the peregrine falcon (*Falco peregrinus*) and climbers in the UK and the USA.

In the UK during the Second World War, the British Air Ministry ordered the destruction of peregrine falcons to protect against losing domestic carrier pigeons. Before 1939 the populations stood at approximately 1050 pairs, and wartime killings resulted in a 13% decrease, but by 1955 the population was restored to 95% of the pre-war figure. However, due to an increase in deaths of homing pigeons, a Peregrine Enquiry was set up in 1960. Instead of finding an abundance of peregrines, the enquiry found a serious decline in numbers. This was due to the use of organochlorines, particularly DDT, and then cyclodiene compounds such as dieldrin and because the peregrine was one of the most seriously affected as it was at the top of the food chain. By 1967 the population had declined to 44%, but there were voluntary and then legally enforced bans on these chemicals by the 1980s. The peregrine is classed as a Schedule 1 species which means it is a rare or endangered species protected by special penalties, and it is a Red Data species, along with the razorbill, chough, and guillemot. Birds can be included in the Red Data Book if the population is internationally significant, a scarce breeder in the UK, restricted in distribution, and vulnerable or of special concern (BMC 1995). By this time the peregrine population was at least 1300 pairs, with the greatest concentration in the Lake District, and currently that area is thought to have the highest density of breeding peregrines in the world. A report by King (1991) was produced for the BMC because it became obvious the birds were nesting in areas that were being climbed and there could be conflicts. In this report 56 occupied crag sites used by climbers were identified in England and Wales which represented between 5 and 10% of the UK peregrine population. At least half of these sites had access restrictions or wardens which minimised disturbance. Only five disturbance incidents had been traced over a number of years, and therefore probably under 1% of the UK peregrine population had been disturbed. Various guidelines though were recommended which we will refer to later. Some of the nesting sites were in the Yorkshire Dales, and this area is popular for climbing on the limestone. There are seasonal climbing bans during the breeding season on several crags, and the use of crag names in guidebooks where peregrines are nesting are only used once the birds have established and restrictions are in place, but certain peregrine sites are not advertised. Guides and BMC restrictions of course give information for egg collectors. Smith (1995) looked at the relationship between climbers and the peregrine falcon in the Yorkshire Dales by means of a questionnaire to 78 climbers. There were some disturbing conclusions outlined: there are restrictions at Malham, Blue Scar, and Great Close Scar, but only 46% of the climbers knew of the Malham restrictions, 38% at Blue Scar, and only 10% at Great Close Scar; some crags that do not have restrictions were thought to have them, so there was confusion related to the knowledge of climbing restrictions; 53% of climbers could not identify the peregrine falcon from bird silhouettes, and the climbers' source of restriction information came from magazines and guidebooks. A good indicator though was 78% of the climbers thought that the restrictions are generally adhered to, but how accurate this figure is in the light of the answers to some of the questions is debateable. Also there were six observations of climbers on restricted routes, and three climbers stated that they had disturbed peregrines in the Dales. Generally though a Yorkshire Naturalists Union representative thought that the climbing restrictions were working well and that climbers have a good attitude towards the restrictions and hinted that some of the birdwatchers were more of a problem. Disturbing a peregrine from the nest can have varying possible effects: eggs and their young are particularly prone to chilling and dehydration if the parents are kept away from the nest while feathered young, which are not yet ready to fly, can be scared out of the nest prematurely. Time off the nest can be the most dangerous as egg cooling can have a disastrous effect and the distressed birds may not return to the nests at all. 91% of the climbers had a good idea of the harm that disturbances can cause. However, the general conclusion was that more education is needed for climbers with regard to their potential effect on bird populations.

In Southern Europe, Brambilla et al. (2004) noted that breeding success and productivity were lower for peregrine pairs that co-existed alternatively with ravens or climbers, compared to undisturbed pairs. Pairs settled at cliffs simultaneously occupied by ravens and frequented by climbers did not fledge any young which suggested that raven predation on peregrine eggs or chicks may be predisposed by human disturbance. Clearly what was needed was a seasonal ban on climbing.

In the USA the peregrine falcon has been protected as an endangered species under state and federal laws. In New York State, certain climbing routes have been closed by the Department of Environmental Conservation (DEC) during the breeding season in the Adirondacks. It is illegal to climb these routes as falcons are very territorial and will use their razor-sharp talons in defence of their territory, including attacks on humans. Just like in the UK with the banning of DDT, in New York in 1971, efforts began to re-establish a breeding population within the state. The Peregrine Fund initiated this reintroduction programme in 1974 which ended in 1988 with 159 peregrines released. The actual areas of cliffs quarantined have to represent a balance between the recreational climbing interests and the need to protect the nesting and breeding activities, although the priority is to protect an endangered species whilst maximising climbing opportunities at the same time. So individual climbing routes rather than whole cliffs are closed. Climbers have helped by providing Department of Environmental Conservation (DEC) with peregrine sightings which are hopefully accurate and by helping to identify climbing routes which should be closed to protect the birds. The closure of routes is based on a number of factors, and a specific distance from the nest site cannot be used to make a closure determination. At the beginning of the breeding season, DEC closes whole sections, or large portions, of the cliff where peregrines have regularly nested. This allows them to choose a nesting site without being affected by climbers. They often chose a site quickly and begin nesting earlier than when the cliffs were subject to climbing. Climbers benefit because an earlier nesting start results in an earlier fledging of the young, and the closed routes are reopened sooner. At the Devil's Tower in Nevada, 45 climbing routes are affected by closures, and there is also the prairie falcon (*Falco mexicanus*) which occupies a similar niche. At Zion National Park, 13 different routes are closed to protect the peregrine, but the cliffs that do not have nest sites are opened sometimes in late April to early May, and cliffs with nests remain closed until the chicks fledge usually in late July. At the Pinnacles National Park in California, there are peregrines, prairie falcons, and golden eagles. In 2016 there were thirteen pairs of peregrines in the region, eight pairs actively nested, and six nests fledged twenty-two chicks. The failures are not usually due to climbers but are caused by egg predation, often by ravens. In Yosemite there are 14 pairs of peregrines. The peregrine was delisted from the Federal Endangered Species List in 1998, and part of the recovery has been due to the respect that climbers have shown to the nesting sites and the nesting restrictions (Fig. 4.3).

4.5.4 Unusual Potential Impacts on Mammals

The pine marten, which is a rare mammal species in the UK, built a den at the top of a crag on May Rock, near Inverness in Scotland. In April 2015 this resulted in the closure of the *Ephemeral Artery* and *Venus Return* routes on that crag. This was because it is an offence to harm pine marten or their dens. This was an extremely unusual closure for a mammal species.

Fig. 4.3 El Capitan (Yosemite) SE face closures due to peregrine nesting areas. Photo by D. Huddart

4.5.5 Overview of the Climbing Impacts on Cliff Biotic Diversity

In a critical review of whether climbing affects and threatens cliff biotic diversity, Holzschuh (2016) indicated that the majority of the published results into these issues may be thwarted by systematic abiotic differences between climbed and unclimbed cliffs. The lack of proper controls may have led to an overestimation of the negative effects of climbing on biodiversity. She suggests that the evidence for the impact of climbing for most taxa is inconclusive. Studies on lichens and vascular plants have described evidence for negative, positive, and no effects as we have illustrated. Gastropod biodiversity seems negatively affected, whilst the evidence for the impact on birds is still lacking. Bryophytes are generally unaffected. She thinks that further research is urgently needed because of the mixed results from previous studies that do not allow final conclusions. Future studies need to select comparable controls for biodiversity comparisons, widen the focus to further cliff-associated taxa, such as micro-invertebrates, and investigate how any climbing effects vary with climbing intensity. This is needed to facilitate improved management of climbing areas that are rich in biodiversity and which contain rare and threatened species. However, it seems extremely difficult to study climbing's potential impacts on the biodiversity as the biggest challenge is to find appropriate unclimbed cliffs to compare with the climbed ones. Cliffs that share characteristics such as slope angle, the amount of sunlight a face gets, the minor geological and geomorphological differences, and so forth are difficult to find. Often all cliffs in a region that are attractive for climbers are climbed, and only cliffs that do not resemble the climbed cliffs in all abiotic characteristics remained unclimbed. If this is the case, then no reliable study can be carried out. The other obvious issue is that there are a limited number of scientists who have the climbing skills necessary to conduct this type of research.

4.5.6 Damage to the Rock

This can occur by chipping and other mechanical wears such as polishing; the impact of chocks, pitons, and bolts; increased rockfall rate; and mainly aesthetic damage caused by chalk and rope wear on relatively soft rocks, like the Wealden sandstones in South East England

4.5 Effects of Climbing on Cliff Vegetation and Other Biota

Fig. 4.4 Rope damage to Sandstone, Zion National Park, Utah. Photo by D. Huddart

(Fig. 4.4). There are other types of damage which may be considered as acts of vandalism, where graffiti, or what the perpetrators would consider artwork, is painted on crags. This can be removed but with considerable effort and potential damage to the rock.

4.5.6.1 Chalk

Most climbers use white chalk ($MgCO_3$) to dry their hands whilst climbing, but its use is controversial, and it is banned in some climbing areas like the Garden of the Gods and Red Rock Canyon Open Space in Colorado Springs and at the Arches National Park in Utah. This is because its long-term use is thought to damage the rock surface, particularly on porous sandstone cliffs. It is aesthetically unsightly too as it can cause white blobs, blemishes, and stains on darker rock if porous, such as sandstone or some types of limestone. The effects of chalk on plants, lichens, and liverworts still needs further study. It is thought that, although it may change the pH, chalk use generally does not harm the cliff environment. The aesthetic impact can be lessened by either using coloured chalk to match the rock colour or no chalk at all whenever possible. Liquid chalk or other types of drying agent that do not leave a colour because they leave no trace on the rock could be used. One type is Metolius Eco Ball which is a non-marking substitute for climbing chalk as it leaves no visible marks on the rock and has a chalk feel. Many climbers ignore the rules in some areas, and it is difficult to clean white chalk from the sandstone surface, especially since cleaners and solvents may damage the rock and should not be used, and any brushes used should have soft bristles. Some local climber's organisations like the Pikes Peak Climber's Alliance schedule chalk cleaning days every year at Garden of the Gods to scrub the white blotches off the sandstone.

4.5.6.2 Effects of Protection

In traditional and aid climbing, it has been suggested that the placement of protection like nuts causes no damage to the rock (Long 2004), but it seems logical to suggest that this is unlikely to be the case and that climbing does have a weathering effect on the cracks where the protection is placed (Fig. 4.5).

A general rule is to use the largest nut that will fit the crack as it is more secure than a smaller one. Using the largest size that will fit appropriately will exert less pressure on the rock surface. The smaller pieces of gear will concentrate the force on a tiny area making the rock more susceptible to splintering and weathering. The smaller

Fig. 4.5 Nut placement. Source: Iwona Erskine-Kellie, Vancouver, British Columbia, Canada

nut will affect a lesser surface area but should put more pressure on it thus eroding the rock surface more, although the larger nut, whilst exerting less pressure, will affect a larger surface area. To try to measure the impact of such protection, Hamilton (2004) adopted an experimental approach but in the field. These experiments occurred on the Fahan Grit on the climbing area known as Rubonid Point on Dunmore Head and on Thur Sandstone on Thur Mountain in County Cavan, both in Ireland. It was found that after 100 repetitions of placing "Wild Country" size 1 and 10 nuts into cracks, weathering did occur. How much weathering occurred depended on several factors that included the type of rock, the rock strength, grain size, and the size and shape of the crack used. Nut movement within the cracks varied between distances of 4 and 127 mm, and the amount of debris displaced from the mechanical action of the nut placement was between 1.66 and 2.19 gm. Additional findings were that there was significant wear to the nuts used and that additional weathering was caused by the swaged cable that had replaced the spectra cord which was originally strung to the chocks. The advantage of this stiff cable is that it allows for easier placement of the nuts.

Aid climbing has resulted in the actual cracks on routes from the placement of pitons becoming eroded, and the climbs have become easier in nature. The original pitons were made of soft iron, and because removal could ruin them, they were usually left fixed in the crack. However, since then the almost indestructible chrome-molybdenum steel pitons have been used. Owing to their repeated placement and removal, the cracks have become brutalised and have ruined more than one climb, leaving large and unsightly scars. Every classic climb in Yosemite was destined to become a string of piton scars, for example, Serenity Crack, until nuts replaced pitons as the common means of protection. The same occurred in the UK and, for example, on Idwal Slabs in Snowdonia where the easy-angled, easy-grade multi-pitch climbs have numerous scars where climbers have used the same placements for running belays and main belays while protecting themselves on classic routes, such as "Faith," "Hope," and "Tennis Shoe." Repeated use of the same placements has in some cases widened the crack and changed the size of the protection that had originally fitted at that location. There were also distinct changes in colour of the rock at the placements to white from beige colour of the rest of the crag. At Idwal Slabs they are very prominent at the bottom of the crag where instructors and belayers have practiced placing nut protection into the convenient vertical cracks (Fig. 4.6A). Here natural weathering morphology can also be seen which seems to have a bigger effect on the dolerite rock than the effects of the climbers (Fig. 4.6B).

There has been a major ethical debate over many years particularly in the USA and the UK over the use of bolts and fixed anchors and how best to create a compromise between the adventure climbers on the one hand and the sport climbers on the other. In some parts of Europe, it is difficult to find natural rock and routes which are unbolted, like in Hungary and Poland. This debate is summarised well in Richards (1997), Achey (2013), and UIAA (2014). It has been argued that bolts prevent cliff-top erosion and tree damage at popular belay stances as we have seen from Tremadog in North Wales, although if it is recommended that descent paths are used, it will create more footpath erosion, and if these are not maintained to a high standard, the climbers will abseil from trees or bolts. It has also been suggested that by following lines of bolts, this can steer climbers away from more vulnerable areas, although this has not been proved. It has been shown too that bolting has increased the attraction of an area like in the Red River Gorge (Kentucky) which has world-class sport climbs.

Within the UK there are policies which govern the installation of bolts and other fixed protections on cliffs. The BMC's position on bolting was restated in 2014. There are crags in the UK which are traditionally bolt-free like the Pembrokeshire and Cornish sea cliffs, the mountain crags of Snowdonia, and the gritstone edges of the Peak District. However, there are many crags across the country where sport climbing is agreed and accepted and even some crags where

Fig. 4.6 (A) Idwal Slabs (Snowdonia): erosion at base of crag, oblique and vertical jointing, and weathering holes. Photo by D. Huddart. (B) Weathering pits in dolerite, Idwal Slabs, Snowdonia. Photo by D. Huddart

both climbing approaches co-exist. Site-specific regional drilled equipment agreements and proposed change to those agreements must be debated and agreed by climbers at open meetings arranged by the BMC. Proposed revisions to drilled equipment policies and/or retro-bolting proposals should be widely publicised prior to discussion and agreed on a consensus basis. In the case of substantive and potentially controversial proposals to use drilled equipment, wider consultation should be carried out through the National Council, the BMC area structure, and the BMC's media outlets prior to agreement. In the 1990s the BMC supported some projects on the grounds of "exceptional access and conservation circumstances," such as on Pen Trwyn, Marine Drive, the Great Orme (North Wales 1991), where to reduce the risk to public safety from loose rock on the exits of climbs, over 300 lower-off points were installed as an alternative to a climbing ban from the council. At Wintour's Leap on GO Wall in the Wye Valley (1994), abseil stations were installed as the landowner specified that unless an alternative means of exit from the cliff could be found, instead of the owner's back garden, there would be access restrictions from April to September when family members required full privacy. A final example comes from Stone Farm Rocks, Sussex, where fixed anchor points to be used as belay points were installed because on the very soft sandstone, there was a lack of suitable belay anchors and the rock showed severe grooves caused by rope cutting into the rock in badly positioned belay systems.

4.5.6.3 Rock Damage by Ropes

Where the rock is extremely soft sandstone like the Ardingly Sandstone of the Weald in South East England and the demand for climbing on natural rock is high as the area is close to London, there are bound to be problems related to erosion of the rock caused by climbing. The Southern Sandstone as it is called is probably the most fragile climbing area in the UK and possibly the busiest area, and the ongoing erosion means that holds are constantly changing, routes are in flux, and repairs have to take place on a regular basis to protect the area. The main problem is that the hardened outer surface crust of the sandstone, which is 2–3 mm thick, gets easily eroded by any weighted rope moving over the sandstone which causes a sawing effect, rock groove formation, and permanent damage. At the crag top, this is most obvious as moving top ropes belayed to trees and tree roots well back from the crag edge. Once this crust has been eroded, the underlying sand grains are further eroded by climbing, and some climbers have chipped handholds which are easy to do in this soft rock where the grains are easily detached by minimal pressure. The rock is far too friable and unsuitable for conventional leader-placed protection, so climbs are carried out using a top rope or by soloing, and careful positioning of ropes is necessary. Belay anchors have been installed at the top of most climbs at Bowles, Harrisons' Rocks, Stone Farm, Bulls Hollow, and the Hut Boulder at High Rocks. An interesting ethical conundrum is that at these sandstone crags, natural weathering has produced honeycomb weathering, and there is differential weathering along bedding planes and vertical joints in the rock, and as we will see later in a section discussing management issues, the methods to try and rectify these anthropogenic erosional problems may well be unsuccessful.

4.5.6.4 Rock Polish

This is a rock feature which is widely recognised by climbers and is often quoted in guidebooks. For example, Jones (1993) in the Ogwen Valley in North Wales suggested that on p. 47 "An obvious line which only just manages to attain the grade due to the polished nature of the rock," and the BMC (1996) highlights the polishing of holds on popular routes as a problem at climbing sites. Although climbers tend to favour the more solid rock types which resist erosion, Bunce (1985) believed that geomorphological damage on stronger rock types may occur if crag use intensified which it has in subsequent years, and he quoted the very polished and dangerous rock on Broad Stand (Scafell). However, limited research has been carried out as to its nature. Curry (1997) looked at the impact of climbing on dolerite at Little Tryfan (Tryfan Fach), a crag 50 m high and 200 m long. Generally it is thought that rock polish is characterised by smoothness and glossiness of the rock surface in relation to the surrounding rock surface and edges may also be rounded. It leads to a loss of friction due to the surface becoming smooth. However, Curry (1997) found that there is smoothness but the surface does not necessarily become glossy, but it can be characterised by discolouration of the rock. However, there is no obvious consistency in this colour change, and it must be affected by the original rock character, but there is rounding off of the polished feature. There also seems to be a gradual reduction in the size of the grains on the polished rock surface. As rock surface weathering usually leads to an increase in surface roughness, it is clear that climbers cause this phenomenon.

4.6 Bouldering and Its Environmental Impacts

There seems to have been a lack of research on any impact of bouldering, but this is strange because this type of climbing involves many of the same types of potential impacts as traditional and sport climbing, such as removal of vegetation and soil, crag base trampling and therefore lowered vegetation diversity, possible effects on gastropods, creation of rock polish, and chipping of the rock. The sport has been growing in popularity, and it is no longer just used as a training regime for climbers, but it is a mainstream recreation activity. However, Tessler and Clark (2016) sampled sites in the Shawangunks (New York). They implemented a paired climbed-unclimbed research design that successfully removed any potentially confounding environmental variation, and so bouldering appeared to have caused the observed differences in vegetation between pairs. Climbed boulders have a lower species richness and cover, with the greatest reduction found on mid-height boulder faces where most climbing seems to occur. Community composition and species frequency did not differ between pairs. This impact is weaker than that reported in most cliff-climbing studies, but the number of climbers and usage at the studied sites is lower.

In some bouldering areas, climbers have modified the rock to create finger or footholds (chipping) or have used epoxy adhesive to prevent friable flakes or lips of rock from breaking off (gluing) and making a problem easier, more difficult, or even impossible. These practices are very uncommon but should be universally condemned by all climbers as it degrades the climbing resource and eliminates challenges for future generations.

The Access Fund (2004) outlines further environmental impacts created by bouldering: paths gradually appear to the individual boulders, and if boulder use is heavy in a popular boulder area, then a weblike trail may link boulders as the boulderers move from one to another. There can be obvious impacts to the base of the boulders, and the soils and vegetation can quickly become compressed; if flat, the vegetation can be eroded, and if on a slope, erosion may occur. Crash pads, whilst they may reduce erosion by distributing and absorbing the force of falls, may damage the vegetation. Bigger issues may occur where there are archaeological remains and particularly rock art on faces. Some federal lands in the USA maintain a 50-foot buffer zone between rock art and climbing/bouldering, but at Hueco Tanks State Historical Park in Texas, there is no minimum linear distance established between rock art panels and climbing routes. Climbing here is simply prohibited directly on, above, and adjacent to these panels.

4.7 Winter Climbing and Its Environmental Impacts

First there needs to be clarification of some of the terms and techniques used in the broad definition of winter climbing. As the name suggests, it involves the use of ice in order to climb inclined, steep terrain, although it is not always the case that winter climbing relies purely on the use of ice for upward progress. It is common that during an ascent of a winter climb, areas that are totally free of ice are climbed; such areas might include snow fields, snow gullies, rock buttresses, frozen vegetation, or a combination of all of these. Generally though, it refers to roped or protected climbing of icefalls and glaciers, frozen waterfalls, and cliffs and rock slabs covered with ice refrozen from water flows. There are two types: alpine ice is frozen precipitation formed in a mountain environment and that usually requires an approach to reach, and water ice is formed from frozen liquid water usually found on a cliff or slab beneath water flows. Mixed climbing is an ascent involving both ice and rock climbing. Progress is made easier by the use of specialist equipment: boots that take crampons, usually plastic, ice axes for each hand, and crampons and protection equipment like pitons and ice screws. It has the potential to be far more environmentally damaging than other types of climbing because the tools and crampons have the potential to damage rock and vegetation far more than rock boots and chalk.

The climbing techniques used vary according to the type of climbing. Where climbing occurs on snow, climbers step kick combined with the use of the ice axe for support or arrest if the feet slip. Front pointing commonly occurs whilst climbing ice, but in the case of mixed climbing, the techniques are more complex, and there is a combination of techniques, including those from summer climbing. However, there is group of techniques which have developed which are unique to mixed climbing and where most of the environmental damage can occur: (a) torquing is where parts of the ice axe are wedged into cracks by a turning force being applied to them after they have been inserted in the crack. This can cause scratching, widening, and the pulling out of any soil and vegetation from inside the crack. Climbers can chip the rock to make the hold that much more secure. (b) Hooking is where the pick of the ice axe is hooked over an edge or spike of rock which can cause scratching and vegetation damage. (c) Turf placement is where the ice axe pick is driven into frozen soil or vegetation in order to gain some purchase. This frozen vegetation is considered the perfect placement. This has been called tufting where low temperatures cause the soil moisture to freeze and so provides an ideal medium for ice axe placements. Many routes can only be climbed in winter because of the amount of turf on the route line.

Other techniques for protecting climbs and setting up belays are slightly different to those used in summer, for example, many icefalls in the Alps use bolts in the rock at the side of the icefall which can be unsightly especially with tapes and slings attached.

The difficulty and insecurity of winter climbing enhance the potential for the climber to commit acts of environmental damage that they would not dream of in summer: the placement of pegs, pitons, and warthogs (all forms of protection) can cause environmental damage as they have to be hammered in and the effect on the crack is to widen it; the removal of a peg also can cause further widening or extra impact on the rock surface because it requires multidirectional hits with a hammer; they can also be left in situ because they will not come out and are unsightly; crampon scratches on rock and stances are more likely to be potential plant habitats which due to the amount of traffic, even in good conditions, are liable to heavy erosion. Damage potential of front pointing is less in good conditions but can be high, for example, an 80 kg person will apply around 100 kg/mm^2 on each of their front points, and this can cause smearing and the production of deep scratches that look unnatural. For example, an account of a new route records the damage done on a single ascent of a new line by Garthwaite and Clarke in December 1993 when they ascended Punster's Crack on the Cobbler, a recognised summer route in Scotland. The ascent left much damage on the soft mica schist. Further damage can be seen on hillwalking paths which regularly become iced or covered in snow in the winter months. An example is the massive scratching and erosion through the Devil's Kitchen in Cwm Idwal (Snowdonia). There are also deep and long scratches on winter ascents of rock climbing classics such as First Pinnacle Rib and Grooved Arête on the east face of Tryfan (Snowdonia). Yet it must be stated that not many kilometres away in Llyn Llydaw, there are many striations and chatter marks caused by ice erosion, and the visual impact is surely not that different. The climbs tend to be on less steep rock which are more likely to have vegetation and in wetter areas like gullies and even on crags that would be ignored in summer, for example, in the Devil's Kitchen. The problems can occur because of the condition of the routes. Where there is plenty of snow and ice, there are no real problems for the flora as all the plants are frozen under a thick snow cover. However, marginal conditions can prove disastrous to sensitive plants, often arctic-alpine rarities which tend to grow best on north-facing slopes where snow lies longest. Also mixed climbing can be undertaken when ice climbs in the same area are out of condition. This is because it often takes a long period of freeze-thaw for ice climbs to come into condition and ice formation can be unpredictable. All that is needed for mixed climbing is that the turf is frozen, there is snow on the ledges, and there is enough hoar frost (rime) on the walls to give a wintery atmosphere and prevent climbing without crampons. From this it can be seen that mixed climbing may cause considerable damage to vegetation and the soil, considerably more so than snow and ice climbing which theoretically should not be damaging. The route up on ice climbs should cause minimal environmental impact, but the route down can be very different and abseiling down can cause crampon scratching on the rock.

Even when there is good ice, it has been suggested that on busier routes, pick marks can spoil the appearance of the ice which is obvious but cannot be permanent and is just aesthetic in nature. However, it has been suggested that some climbers are upset at the appearance of the route that is made more obvious, and they know that they are not the first up the route as they had perhaps hoped.

As an example of the effects of winter climbing, Thompson (1992) conducted a short experiment on a frozen watercourse on the flanks of Clogwyn Du'r Arddu (Snowdonia) on 9 December 1991. The site chosen was not of any great botanical interest, there was no obvious previous disturbance, and the site is of a similar grade to that climbed by most people and of such difficulty and character that it may be disturbed by the techniques described before in mixed climbing (Grade IV); the climb was only ascended, and the climb was in moderate to poor condition at the start of the experiment and deteriorated to poor after six tramples/climbs. The observations during the experiment were as follows: little disturbance during the initial trample, damage to vegetation/soil significantly increased with successive tramples (the climb's condition was becoming poorer), removal of ice tools caused significantly more disturbance than placement, vegetation incorporated in the ice was disturbed, and pieces of this vegetation and soil were removed. One of the significant factors that will affect the degree to which damage will occur will be the amount of pressure that the climber puts upon the climb, and it is likely that a confident and skilful climber will put less pressure on a climb than a climber of lesser ability. When nervous, there is a tendency to drive the picks in deeper and to kick the front points in harder to create more security, but this results in more pressure, and the climb comes into poorer condition faster and with greater disturbance. Climbers who are pushing in terms of difficulty may cause considerably more damage than climbers who are climbing well within their ability. Three control sites were surveyed immediately adjacent to the trample site and were of a similar nature. The changes that were observed after trampling were that chunks of vegetation/soil were found on ledges, some vegetation was detached from the rock but had not fallen off completely, some areas of vegetation had been disturbed in that the soil was extremely loose and more susceptible to any further shearing in comparison to undisturbed areas, small scratch marks were visible on the rock, and soil/vegetation had been removed from cracks. The lycopodium and sphagnum moss species seemed particularly affected. These were species that grew in drier areas and hence might have been less protected by the formation of ice, and their resistance to shear was less. Future potential damage might be far greater as it is possible that upon subsequent climbing, the effect will be greater than if the area had not been disturbed at all. In addition due to the weakened soil and vegetation, it is more likely that natural erosion will be greater during rapid thaw and high rainfall. If the vegetation could recover is uncertain, but the time scale would be considerable. This crag too has rare arctic-alpine species like the Snowdon lily (Fig. 4.7), which may be in danger due to effects of winter climbing in poor conditions.

A questionnaire survey of 100 winter climbers (Thompson 1992) suggested some interesting results: 81% of the climbers would attempt a route in poor conditions provided that there was no excessive danger suggesting that the climbers were unaware of the possibility of environmental damage or were not concerned by it; 52% suggested

Fig. 4.7 Snowdon lily (*Lloydia serotina*), Clogwyn Du'r Arddu, Snowdonia. Photo by D. Huddart

that they would only climb in good conditions; the 41% who indicated that they would continue as normal but with extra care is a little unrealistic, so despite their good intentions, it may be of little help; 7% would climb regardless. If damage is shown to be occurring as a result of winter climbing, most climbers would modify their activities which may be environmentally good news.

4.8 Management of Climbing

In many parts of the world, there are well-developed voluntary systems to manage climbing activities so that the recreation activity can be maximised whilst the environmental impacts are minimised. For example, the Access Fund in the USA, formed in 1991, exists on a mission to protect climbing access and the integrity of America's climbing areas and represents climber's interests and values. They suggest that 60% of climbing areas are on public land and that these lands are currently being covertly dismantled by Congress and the President and that one in five climbing areas in the USA is threatened by an access issue. The organisation maintains a broad network of partnerships within the climbing community and the outdoors industry and other like-minded advocacy groups. It exists to keep climbing areas open and has bought 52 acquisitions through the Access Fund Climbing Preservation Grant Program and the Access Fund Land Conservation Campaign which has helped preserve over 15,500 acres of land for climbing. In the UK the BMC formed in 1944 is the national representative body for England and Wales existing to protect the freedoms and promote the interests of climbers, hillwalkers, and mountaineers. Mountaineering Scotland is the national representative body and membership organisation for those who live in Scotland or enjoy Scotland's mountains. It encourages participation and progression in the mountain activities we have discussed and promotes safety and skills, campaigns to safeguard access rights and responsibilities, and seeks to protect Scotland's mountain landscapes from insensitive development. As can be seen, there are political dimensions too and many government organisations manage much of the climbing areas in the USA, so there has to be liaison between all these organisations and the climbers.

4.8.1 Management Plans

In the USA this has resulted in the management of climbing through management plans, like the New River Gorge National River in West Virginia (NPS 2005). This centres on over 1600 climbing routes along extensive cliffs and is one of the largest climbing locations in the Eastern USA. This plan was published in 2005 because of increasing popularity of climbing in this area, the more obvious impacts, and the conflicts between commercial use groups, non-profit groups like scouts, and individual climbers. The purpose was to present a strategy that protected the natural and cultural resources whilst continuing to provide opportunities for a high-quality climbing experience. The plan was to meet the following objectives:

1. To create a management tool to adequately address resource protection and visitor use issues related to climbing.
2. To build partnerships with climbers, climbing groups, and commercial organisations in managing climbing.

3. To provide guidance on managing commercial groups.
4. To maximise input from the public and the climbing community throughout the planning process.
5. To provide a forum for public involvement and collaboration—this was a crucial aim. The National Park Service (NPS) considers the long-term partnership with climbers and other interested parties to be a crucial component of an effective climbing management plan. The plan gave three alternatives for climbing management: the alternative (A) or the no-action alternative would continue with the existing management; the alternative (B) which is the preferred NPS option would involve climbing management, including education and outreach efforts, improvements to facilities, and the use of new and existing trails; the alternative (C) is similar to (B), but concession contracts would be required for commercial use, and there would be a more restrictive, pre-emptive closure for peregrine falcon nesting on Endless Wall. A detailed inventory of the environmental consequences and their intensity with the three alternatives is discussed. There are several appendices and one on ethics and education which is excellent and another on various aspects of research. There were opportunities for individuals and organisations to respond to the plan.

Early example of the way in which the NPS established the management direction and actions was the Devils Tower National Monument Final Climbing Management Plan (1995, updated 2006) which was a precedent-setting management plan between the NPS, climbers, and Native Americans. At least six Northern Plains Indian tribes had long considered the site to be a sacred place for meditation, offerings, sweat lodges, and ceremonies, particularly important in June, and they would have liked a complete climbing ban. The 1995 plan aimed to preserve cultural values and climbing opportunities by establishing a "voluntary closure" to climbing during June each year which was a compromise with climbers (see the detailed background to the case in Cross and Brennenan 1997). Although climbing has been reduced by about 85% in June since the adoption of the plan, there has been a recent increase in June, and if climbers do not respect this partial closure, it is feared that there well may be a total ban eventually. The current plan also provides direction for management of resource impacts associated with climbing, including fixed anchors, trails, chalk, slings/webbing, and restrictions to protect wildlife.

The Wichita Mountains Wildlife Refuge produced a Final Environmental Assessment for Technical Rock Climbing (1995) which gave a general appraisal of environmental effects in a climbing area and evaluation in relation to impacts from other recreation types. It provided a model for dealing with fixed anchors, permissions for new routes, and fixed anchor replacements in a USFW refuge. The Chickamauga and Chattanooga National Military Park Final Climbing Management Plan (1998) was an early example which closed some routes, authorised placement of bolt top anchors below the rim of mountains to protect summit vegetation and soils, set group size limits, directed park staff to analyse the potential for use fees, urged voluntary trailhead registration by climbers for monitoring purposes, and prohibited new fixed anchors and the use of power drills without NPS approval. Other examples of climbing management within planning documents are the Joshua Tree National Park (2000), Rocky Mountain National Park (2001), Denali National Park and Preserve (2006), and Zion National Park plans (2007). In the Arches National Park, there was a combined Canyoning and Climbing Plan (2013b). Within the Shenandoah National Park plan (NPS 2010), there was a set of climbing guidelines, but for all the national parks in the USA, guidance is given in Managing Climbing Activities in Wilderness (NPS Reference Manual # 41, 2013a). Over the years there has been much discussion over the use of bolts, anchors (see the United States Forest Service policy for anchors and the heated discussion that this generated), drilling, dry tooling, and the impact of sport climbing in both the USA and UK, and the ethics of climbing seems to have gradually changed from the original traditional climbing ethics to a more liberal and environmentally insensitive set of ethics based around sport climbing. In the light of increasing bans of fixed climbing anchors in American parks, Jones (2004) examined the visual impacts of rock climbing in the Rock Canyon Park, Provo (Utah). Using photos he examined the views of both climbers and non-climbers and found there were no significant differences between anchors only, anchors and chalk excluded factors, the chalk only factor, or between climbers and non-climbers. This seemed consistent with the results of Schuster et al. (2001) who found that the traditional sport climbers perceived that managers had little understanding of climbing as an activity and that resources were often micro-managed as a result.

4.8.2 Memorandums of Understanding or Agreements

In the USA there have been several examples of memorandums of understanding or agreements which are written frameworks that establish cooperative relationships between two or more parties that define the common interests that parties share and define the way they will work together to reach common goals. A good example is the MOU between the Wichita Mountains Climbers' Coalition and the Department of the Interior US Fish and Wildlife Service, Wichita Mountains Wildlife Refuge. Here the Climbers' Coalition will provide volunteer assistance through an advisory bolting committee; educate and inform the climbing community about conservation issues, resource protection, leave-no-trace ethics, and stewardship philosophy; assist refuge managers in monitoring the effectiveness of the FWS rock climbing management plan; maintain all climbing areas free of rubbish or abandoned equipment; remove excess chalk from the rock as needed; and work in partnership with the FWS.

4.8.3 Liaison Groups

In the UK there have been established liaison groups for various climbing areas with representatives of all the interested parties, for example, landowners, local clubs, the BMC, guidebook writers, and Natural England so that management can be integrated. There are regular meetings so that there can be a discussion of all relevant issues.

4.8.4 Closures

There have been closures of some climbing cliffs and routes, for example, on the famous arch landforms in the Arches National Park (Fig. 4.8), in Monument Valley, and Canyon de Chelly in Arizona. The latter two have been for cultural reasons, and there has been a complete ban on climbing on Cave Rock in Nevada from 2007 which is a sacred site for the Washoe tribe, and all the bolts were removed by 2009 (see a detailed discussion of this case in Makley and Makley 2010).

4.8.5 Seasonal Restrictions

There have been many seasonal restrictions in the USA, mainly for raptor breeding which are well publicised and generally well respected by climbers. In the UK the BMC has excellent relationships with many conservation and countryside organisations, and it supports reasonable restrictions and recognises the great importance of certain cliff and mountain areas for bird breeding. In the UK, there are well over 160 individual sites in England and Wales that have restrictions to protect sensitive species. Access agreements depend on the species, numbers, and distribution of birds present, but most apply between February and mid-August, but there are variable restrictions at some sites, and some restrictions are lifted if birds do not nest on certain crags. These types of restriction are best because they offer protection for birds but a greater freedom and flexibility for climbers. An example of the restrictions can be given from the island of Lundy where restrictions are in place between 1

4.8 Management of Climbing

Fig. 4.8 Landscape Arch, Arches National Park, Utah. Closure of all climbing routes in this national park on arches. Photo by D. Huddart

April and 31 July, but of the over 100 climbing areas listed in the guide, only 71 have climbing restrictions in this time period and many cliffs are unrestricted. A number of classic climbs have been left open even though they are near to bird colonies, and it is important that there is no encroaching onto nearby restricted climbs or doing anything that causes disturbance to breeding birds. Sometimes because of late fledging, climbs can be closed after 31 July, and checks have to be made with the warden. Care has to be taken that access routes to open climbs do not go through restricted areas. In the UK too, climbers need to be aware of the Wildlife and Countryside Act (1981) and the Nature Conservation Act (Scotland) (2004); the Conservation Regulations (1994) implements the Council Directive 79/409/EEC on the conservation of wild birds (the "birds directive") and the Council Directive 92/43/EEC on the conservation of natural habitats and of wild flora and fauna (the "habitats directive"). Climbers need to know that it is an offence to interfere with the nest of any wild bird or obstruct a bird from using it and that certain birds which are rare or endangered have increased levels of protection and penalties for abusing them. This Schedule 1 status means it is an offence to disturb these birds whilst they are building a nest, are near a nest containing young or dependent young, and even when they are away from the nest. It is also an offence to uproot any wild plant, and on National Trust, land digging, cutting, injuring, or taking any soil or plant is against by-laws. In some areas like in National Trust in Cornwall, seasonal restrictions operate in the vicinity of all known seal pupping caves in September and October.

4.8.6 "Remote Areas"

Some areas of cliff could be designated or scheduled as "remote areas" as in the National Trust policy for Pembrokeshire and in Cornwall. This means minimal recreation use, no facilities like car parks, and no promotion, although climbers will be welcome to find and use these areas by their own initiatives, but with the extensive use of the internet and social media, it is doubtful if this lack of promotion is very successful in deterring climbing. No promotion means no publication in guidebooks, so close liaison is required with editors and writers.

4.8.7 Rerouting

There can be management of the descent routes from crags down the gullies which can be temporarily closed for recovery; the routes can be rationalised and rerouted as appropriate or feasible, for example, erosion control was carried out by the BMC at Harrison's Rocks, Stone Farm and Cadshaw Rocks in 1996, and at Stanage more recently.

4.8.8 Use of a Star System in Guidebooks

It is possible that climbs which have been badly damaged could be closed so that recovery if it occurs could be studied and new climbs might not be publicised without reference to the liaison group, but this type of policy has not really taken place. Use of certain climbs though could be discouraged by some agreed means like the star system could be taken out of guides, climbs could be taken out of the next guidebook edition, or some explanatory text could be included in certain areas which have been identified as vulnerable, like Gurnard's Head, Cornwall, or there is a need for a strong conservation message. The Climbers Club took the controversial decision at the time to omit the star-rating system from the new Tremadog guide, arguing that the three-star routes were eroding too quickly. An argument from James and Wightman (2003) from the Rockfax guides was that the principle of attaching a single star to any climb that is worthwhile was a useful principle. This is because it in turn provided 75–85% of the routes at a crag with stars and therefore attracted climbers to that percentage of routes and so may spread the rock weathering load. These arguments related to stars in guidebooks, and whether they are useful in controlling rock damage and erosion seems a little irrelevant because it is likely that there are many reasons why climbers chose to climb where they do, not just guidebook information.

4.8.9 Booking or Permit System

There may well be limits imposed where there is overcrowding or damage is occurring. This could be a booking or permit system for both individuals and groups. For commercial use there could be a charge for group use and an annual licence system in place. Car park management can play an important role in limiting numbers. Certain crags might be regarded as sacrificial crags, used especially for beginners, group use, like with the scouts and the army, and for commercial groups, for example, Idwal Slabs in Snowdonia.

4.8.10 Outreach and Education

Generally as part of a management plan and policy, there has been an important role for outreach and education, not only to the general public but in this case to the individual climbers, climbing clubs, and other organisations. This may be through close liaison where climbers have been involved in close partnerships with land managers, often to the mutual benefit of both parties. In the Potomac Gorge final report, Marion et al. (2011) suggested that in some instances, it may be possible to alter climbing routes to avoid impact to rare plants. The removal or addition of anchor bolts and/or alteration of climbing guides, which include descriptions and photos or diagrams of each climbing route, can accomplish this. While most climbers own and consult climbing guides, on-site signage may also be necessary to inform all climbers of climbing route changes. Attarian and Keith (2008) note that the strategic placement of fixed anchors on the cliff face can also be used to protect trees or vegetation communities by diverting use away from them. For example, this was done in North Carolina state parks and at Sunset Rocks, Tennessee, in the NPS Chickamauga and Chattanooga National Military Park. They also suggest targeted outreach on species recognition and avoidance practices and individual climbing route restrictions. Collaborations with climbers and the authors of climbing guides are critical to the successful application of this strategy.

Within the Potomac Gorge, climbers have generally used natural anchors rather than fixed protection. However, Marion et al.'s (2011) survey did reveal the presence of older pitons, top and face bolts, and other mostly historic artefacts. Cliff-top trees are numerous and provide the most commonly used anchor. While abrasions to tree bark and small limb cutting are common, their use as anchors rarely causes damage that is more deleterious (but see earlier in this chapter). Nevertheless, if trees are used, one recommended low-impact practice is to install webbing slings around trees to avoid trampling damage and impacts from the repeated wrapping and unwrapping of ropes around the base of trees. If used, periodic inspections are necessary to replace webbing that has deteriorated from exposure to the sun.

The trampling and loss of ground vegetation and organic litter cover and soil around trees used as anchors are perhaps more significant forms of visitor impact. Such impacts could affect rare plants, and the removal of vegetation and organic litter substantially increases the rate of soil drying and loss and may increase tree mortality during severe droughts. The installation of bolt anchors just below the cliff top is an effective management practice that avoids impacts to trees and adjacent vegetation and soils. According to Attarian and Keith (2008), strategic bolt placement is increasingly being used by land managers to protect sensitive resources such as cliff-edge vegetation, soils, and specimen cliff trees.

Another low-impact practice is the placement of permanent bolt anchors at carefully selected, impact-resistant abseiling stations, allowing climbers to descend the cliff without using steep descent trails (which often have vegetation and easily eroded substrates). Substituting an abseil station in a location with naturally barren rock at the top, bottom, and along the intervening cliff face will result in less environmental damage than the same traffic on a descent trail, which generally has soils and vegetation. One abseil station can generally service a cluster of climbing routes. Such stations have been successfully implemented at the New River Gorge National River, WV; Shiprock, NC; and other areas (Attarian and Keith 2008). Alternately, descent trails can be stabilised by rockwork, which can also channel traffic around vegetation.

4.8.10.1 Provision of Information

There have been many attempts to produce leaflets, booklets, guides, codes of conduct, advice through websites, signage (Fig. 4.9), and recently from 2015 the Rock Project Tours to help reduce environmental impacts related to climbing and to promote climbers who are stewards for the climbing resources and responsible users of those resources. The biggest problem is the lack of research to indicate which methods of education and outreach can be most successful in these attempts.

BMC's Tread Lightly Booklet

An early attempt in the UK was the BMC's Tread Lightly booklet *Conserving Britain's Mountains and Crags* (1988), published with the financial support from the Nature Conservancy Council. It was thought that much of the damage and disturbance can be reduced by encouraging a minimum-impact approach to recreation: to climb the rocks and wander the hills yet leave no trace of our passing; that should be the aim. With regard to climbing, it was suggested that climbers should find out about and respect any restrictions placed on climbing for the protection of birds, animals, and plants; always keep gardening to an absolute minimum, especially when opening up new crags; be particularly careful not to damage vegetation when climbing in gills and gullies and keep to the gully floor where possible; try to minimise impact on the rock with no paint or scratched routes and leave no graffiti; not chip handholds; avoid placing pegs or bolts whenever possible (do not remove in situ belay pegs or bolts); avoid leaving any slings hanging from the crag; try to keep chalk use to the minimum; and, when climbing in winter, keep to well-frozen routes and avoid climbing at the start of a general thaw.

Fig. 4.9 Signage at Harrisons Rocks, South East England: information related to climbing etiquette. Photo by D. Huddart

Green Guides

The Green Guide was compiled by the BMC, the National Trust, English Nature, and the Land's End Climbing Club. These groups represent a wide range of interests with different issues that can affect the cliff environment, and possibly this means a more workable and balanced strategy for conservation. The following quote is taken from the Chair Ladder guide to Cornwall (Hannigan 1992), and this type of guidance needs to be more widespread if our natural environments are to be preserved. The attitudes promoted and specific information regarding the protection of particular species are publicised in the Green Guide, and it needs to be written into all guides concerning outdoor recreation: "Individual climbers, instructors and climbing schools are asked to react positively and helpfully to the aims of the National Trust in these matters. Response should include making personal judgements about whether or not to climb established routes in obviously eroded areas, on in fragile plant and bird habitats. Climbers wishing to establish new climbs should also think seriously about the impact of gardening, excessive rock removal and intrusion on ecologically vulnerable sites." Such responses should extend to all cliff areas whether or not they are owned by the National Trust (Hannigan 1992). The BMC has published Green Guides to the Lake District (Green Climbing guide) and the Yorkshire Dales, the Green Guide to the Uplands (2009), the Green Guide to Groups of Climbers (2015), and the Crag and Habitat Management Green Guide, all of which can be downloaded from the BMC website. The BMC have also established the Cwm Idwal Winter Climbing Information Project (2013) which gives live information on winter conditions on the BMC website from the Devil's Kitchen and Clogwyn Du'r Arddu. A remote sensing station was installed at a rock bluff below the cliffs of the Devil's Kitchen, and this generates live data and records the air temperature and the temperature of the ground at 5, 15, and 30 cm in the turf. This complements the BMC's North Wales White Guide (2011) which can be downloaded from the BMC website. Further live information on Lake District winter conditions can be obtained at Great End that is monitored at 750 m in a similar way. This information complements the Lake District winter conditions guide (2015).

Leave No Trace

The Leave No Trace Centre for Outdoor Ethics strives to educate all those who enjoy the outdoors about the nature of their recreation impacts as well as techniques to minimise such impacts. It is best understood as an ethical and educational programme not as a set of rules and regulations. It was created in the 1960s by the US Forest Service and became increasingly necessary as land managers witnessed the biophysical effects of increasing use. By the mid-1980s, the Forest Service had a formal "no-trace" programme emphasising wilderness ethics and sustainable travel and camping practices. The success of this programme led to cooperation between the Forest Service, NPS, and Bureau of Land Management (BLM)'s authorship of a pamphlet called "Leave No Trace Land Ethics." In the early 1990s, the Forest Service worked with the National Outdoor Leadership School (NOLS) to develop hands-on, science-based minimum-impact education training for non-motorised recreation activities. An outdoor recreation summit in 1993 created an independent, non-profit organisation called Leave No Trace Inc. (now the Centre). In 1994 the Centre entered into a series of MOUs with four federal land management agencies, and in 2007 the National Association of State Parks Directors developed a formal affiliate partnership to expand the possible use of the Leave No Trace programme on state park lands. Major programme development has taken place focusing on providing quality Leave No Trace education whilst broadening the programme's reach by including a Travelling Trainer Program consisting of mobile educators that travel throughout the USA teaching Leave No Trace and providing support at the local level. In 2007 a mobile summer educational programme called the "e-tour" was added to the educational offering, and in 2011 a seasonal team for Colorado was added; added also are a youth programme called PEAK (Promoting Environmental Awareness in Kids) that reaches over 150,000 annually with direct programming, frontcountry and urban-based Leave No Trace programmes and training for state and city natural areas and parks, community-based initiatives, including local and regional educational events, and volunteer opportunities coordinated by a State Advocate Network. Lots of educational materials have been produced, like the skills and ethics booklets *101 Ways to Teach Leave No Trace* and Marion's (2014) book *Leave No Trace in the Outdoor*, and courses like the Leave No Trace Youth Program Accreditation and five-day master educator course and workshops that run from one hour to two days.

Access Fund Projects

The Rock Project Tour is part of the Access Fund's mission to keep outdoor rock climbing areas open and protected. They do this by working to engage the climbing community and activate positive social norms, backed by consistent educational content, messaging, and programming that are specific to regional access issues and environmental concerns. It is ultimately climbers that are stewards and responsible users of the outdoor climbing resources, and they have to promote a positive identity for themselves and climbers as a whole group of recreationalists, mitigating the threats to access. The Rock Project is at the centre of this movement. To achieve this, climbers will be presented with the Climbers Pact which is a promise or covenant between climbers to practice a set of responsible outdoor ethics that protect climbing access. The pact outlines ten responsible outdoor behaviours in order to protect climbing access: commit to your fellow climbers to be considerate to other users; park and camp in designated areas; dispose of human waste properly; stay on trails wherever possible; place gear and pads on durable surfaces; respect wildlife, sensitive plants, soils, and cultural resources; clean up chalk and tick marks; minimise group size and noise; pack out all trash, crash pads, and gear; learn the local ethics for the places you climb; respect regulations and closures; and use, install, and replace bolts and fixed anchors responsibly. This is a national programme to encourage and inspire responsible outdoor climbing knowledge to protect the places they climb through responsible, low-impact climbing behaviours; be an upstander, not a bystander. The Access Fund will work closely with Black Diamond and other leaders in the climbing community to build and distribute educational content through a web-based tool box that aggregates existing and newly developed educational materials, programming, templates, exemplar case studies, and community contacts.

Codes of Conduct

There have also been educational resources published by climbing organisations to act as codes of conduct in special circumstances like the Southern Sandstone Code of Practice (BMC) to try and prevent erosion of the soft sandstone (Table 4.4).

Bouldering and Climbing Conservation Codes

There are several bouldering and climbing conservation codes published by the BMC, such as A Code for Winter Climbers (n.d.-a), the Lake District Winter Climbing code (n.d.-b), the Sandstone Bouldering Ten Commandments (2017) which provide excellent advice to help preserve the rock and vegetation and are taken from Panton (2017). Other codes include the BMC's Crag Code (2007) which included "do not disturb livestock, wildlife or cliff vegetation, respect seasonal bird nesting restrictions"; the Mountaineers' Climbing Code (2014) included "behave at all times in a manner favour-

Table 4.4 Southern Sandstone Code of Practice (after the BMC)

Setting up a belay
At Bowles, Harrison's, Stone Farm, and Bulls Hollow, belay anchors have been installed at the top of most climbs. Do not thread the climbing rope through these, but set up a non-stretch belay. Use a static belay rope of 11 mm diameter. A convenient length would be 5 m, and it is also useful to have a longer belay rope of 12 m when there is no anchor and you are using a tree some distance back from the top of the crag. Putting a permanent sleeve around the knot just above the karabiner helps to protect both the knot and the rock. Alternatively, tape slings can be used. Note: the bolts are only to be used for top-roping
When setting up a belay, adjust the height of the karabiner to hang far enough over the edge of the crag so that the climbing rope will not touch the rock. Moving or stretching ropes should never come into contact with the rock. The sawing action destroys the weathered crust and cuts deep grooves in the top of the crag
On an isolated buttress, the first member of a party will usually have to solo to the top. Do not throw a rope over the buttress from the ground
Footwear
Wear light soft-soled footwear. If you do not have specialist rock shoes, lightweight gym shoes are the best. Clean your shoes before starting each climb
Descending
Walk off after completing a climb, do not lower off or abseil. On isolated buttresses all members of a party except the last should down-climb on a slack rope, and the last member should solo down
Climbing style
Top-roping and soloing are the only acceptable methods. A non-stretch belay should be used. Do not use pitons, bolts, nuts, camming devices, or any sort of leader protection
Choose a climb of the right standard. Please do not spend a long time "working" a climb that is too hard
The preferred ethic is not to use chalk. Please keep the use of chalk to a minimum. Do not use resin powder. Avoid cleaning holds if possible, but if cleaning is essential, do it gently with a very soft brush
Sandstone is softer when wet, and climbers must exercise caution by avoiding sharp or fragile holds and ensuring good footwork. If it is wet, why not embrace the opportunity to explore the wonderful cracks and chimneys on offer
Never climb with axes and crampons, even on the rare occasions when ice forms on the rock
Camping
Do not camp, bivouac, light fires, barbecues, or stoves near the crags. There is a campsite next to the car park at Harrison's Rocks. There is a fire area and in situ BBQ at the car park
Additional notes for leaders of organised groups
Please avoid the most popular climbs at weekends, particularly Sundays. Limit the time your group occupies a climb or a section of the crag. Remove ropes when you have stopped using a climb. Choose climbs to suit the ability of the group. If members are having trouble on a climb, their feet will slip repeatedly causing rock erosion, and additionally a weighted rope is likely to cause damage to the rock. Try something easier

(continued)

4.8 Management of Climbing

Table 4.4 (continued)

Abseiling

Please do not abseil at any Southern Sandstone crag. The top of the crag fragile holds on the face will be damaged. Abseiling is not permitted by the owners of the major outcrops

Malicious damage

Unfortunately this still happens from time to time, for example, graffiti, chipping new holds, or enlarging existing holds. If you see anyone doing this, please stop them

Ground erosion

Use established descent paths. Step on rock rather than earth, and avoid treading on vegetation if possible

Trees and vegetation

If there is no belay anchor, it is usually necessary to belay to a tree. Tie the belay rope round the tree as low as possible to minimise leverage. Please do not cut down or prune trees. Do not use herbicides for clearing vegetation from the rock. Regrowth will occur, and it will be the commonest species that recolonise, the rare species being eliminated

ably upon mountaineering, including adherence to Leave No Trace principles and the erosion; respect our wild neighbours and follow our rock climbing ethics." If climbing clubs adopt a crag, then this helps the environment by cleaning up litter, graffiti, and chalk and imbues the correct crag stewardship that needs to be fostered in all climbers and models behaviour by taking the time to educate others (see Organising a Successful Adopt a Crag p5 Vertical Times volume 97, Access Fund 2013).

Concluding Remarks

One obvious conclusion is that instead of encouraging climbing in the mountain environment, we should continue to foster the recent trend where there has been an explosion of indoor climbing, artificial climbing competitions, and even artificial ice climbing. The latter is especially the case with global warming inevitably meaning a decline in resources outdoors for ice climbing and a search to higher and more northern regions for decent climbs in condition. For artificial ice climbing, there are currently indoor resources at, for example, Winona Ice Climbing Park at Sandstone, Minnesota; Lake City Ice Park, Colorado; Stone Gardens, Seattle; Spire Climbing Centre, Bozeman (Montana); Hukawai Glacier Centre in Franz Josef (New Zealand); O2 World in Seoul (South Korea); Ice Factor (Kinlochleven) and Snow Factor (Glasgow) in Scotland; and even Vertical Chill in the basement of Ellis Brigham's shop in Covent Garden, London, and Kong Adventure in Keswick (Lake District). It is even possible to use Nicros FoamIce panels which can allow ice climbing indoors on simulated ice. However, to get the best of both natural environments and guaranteed ice climbing, there is nothing to beat the Ouray Ice Park in the San Juan Mountains (Colorado) which opened in 1995, and each November until spring, ice farmers spray water down the canyon walls of the Uncompahgre Gorge and create over 150 man-made ice and mixed climbs in 11 distinct climbing areas. All the climbing is free and is very close to the town of Ouray. Of course there literally are now hundreds of artificial climbing walls, for example, Basecamp in Reno (Nevada) and Holdistic in Edmond (Oklahoma), to cater for the needs of sport and climbing wall enthusiasts.

Education and appropriate stewardship are the keys to conserve the climbing natural environment. We need much more further research and much more education for the public and the climbing community about the impact of this recreational activity and a restriction of access to the more pristine cliff communities which means better management of the activity to the benefit of both conservationists and climbers. The more liaison there is between the climbing community and the land managers, the better will be the end product for both parties. However, Carr (2007) suggested that climbing areas have no avoidable impacts, rather the impacts are unavoidable, such as trampling and erosion, and that improvement efforts at climbing sites may have to focus more on site hardening and actions to spatially concentrate climbing activities. He thought there should be less emphasis on climber education. This generally is not an accepted view, and a combination of outreach, education, and site management is the way forward. Baur et al. (2017) suggested that any management plan should contain a comprehensive information campaign to show the potential impact of intensive climbing on the specialised flora and fauna with the aim of educating the climbers so as to increase their compliance with any management measures. There are doubts too as to how much damage climbers really cause the natural rock, and compared with natural processes such as slope processes like rockfall or glacial erosion, the impacts appear trivial. There have also been some doubts expressed related to the impact of climbers on natural vegetation, and more research is needed to establish the truth. Seasonal closures of climbs appear to have been successful for the breeding bird populations who use the cliffs, and climbers and other factors may well be having bigger impacts on seabird populations. It seems likely that the educational materials and outreach programmes that have been evident in the last ten years have helped enormously in producing a more environmentally aware climbing population who has a greater concern for the stewardship of the mountain and crag environment.

References

Access Fund. (2004). *Bouldering: Understanding and Managing Climbing on Small Rock Formations*. The Access Fund, 10pp.

Access Fund. (2013). Organising a successful adopt a crag. Local Climbing Organisation 101. *Vertical Times*, 97, p. 5.

Achey, J. (2013, April 8). Summary of fixed anchor controversy in the Wilderness in the USA. *Climbing*.

Adams, M. D., & Zaniewski, K. (2012). Effects of recreational rock climbing and environmental variation on a sandstone cliff face lichen community. *Botany, 90*, 253–259.

Attarian, A., & Keith, J. (2008). *Climbing Management: A Guide to Climbing Issues and the Production of a Climbing Management Plan*. Boulder, CO: The Access Fund.

Baker, G. (2005). *Harrison's Rocks: A Study to Investigate the Erosion and Weathering Processes that are Occurring on the Footpath and Crag Face as well as the Management Techniques that have been Adopted in the Area*. Undergraduate dissertation, Outdoor and Science Education, Liverpool John Moores University.

Baur, B. (2004). Rock climbing alters the vegetation of limestone cliffs in the northern Swiss Jura Mountains. *Canadian Journal of Botany, 82*, 862–870.

Baur, B., Baur, A., & Schmera, D. (2017). Impact assessment of intense sport climbing on limestone cliffs: Response of rock-dwelling land snails. *Ecological Indicators, 72*, 260–267.

Baur, B., Fröberg, L., & Müller, S. W. (2007). Effects of rock climbing on the calcicolous lichen community of limestone cliffs in the northern Swiss Jura Mountains. *Nova Hedwigia, 85*, 429–444.

Bell, S., Tyrvainen, L., Sievanen, T., Probstl-Haider, U., & Simpson, M. (2007). Outdoor recreation and nature tourism: A European perspective. *Living Reviews in landscape Research* Retrieved from http://www.livingreviews.org/lrlr-2007-2.

BMC. (1995). *Bird Nesting Agreement 1995*. Quoted in High Magazine No. 147.

BMC. (1996). *Erosion Problems at Climbing Sites- Guidelines on Reducing Impact at Climbing Sites- Guidelines on Reducing Impact*. Manchester: BMC Information Services.

BMC. (2007). *Crag Code*. Retrieved from http://www.thebmc.co.uk/cragcode

BMC. (2011). *North Wales White Guide*. Manchester: BMC.

BMC. (2014). *BMC Position Statement*. Report on Drilled Equipment and Dry Tooling, Manchester, 2pp.

BMC. (2015). *Lake District Winter Conditions Guide*. Manchester: BMC.

BMC. (2017). *Southern Sandstone Code of Practice*. Manchester: BMC, 8pp.

BMC. (n.d.-a). *A Code for Winter Climbers*. Manchester: BMC.

References

BMC. (n.d.-b). *The Lake District Winter Climbing and Avoidance of Damage*. Manchester: BMC.

BMC Green Guides: Lake District Green Climbing Guide; Yorkshire Dales Green Guide, Green Guide to the Uplands (2009); Crag and Habitat Management Green Guide; Green Guide to Groups of Climbers (2015).

Bowker, J. M., Askew, A. E., Cordell, H. K., Betz, C. J., Zarnock, S. J., & Seymour, L. (2012). *Outdoor Recreation Participation in the United States—Projections to 2060: A Technical Document Supporting the Forest Service 2010 RPA Assessment*. General Technical Report SRS-160, US Department of Agriculture Forest Service, Southern Research Station, Asheville, NC, 34pp.

Brambilla, M., Rudoline, D., & Guidali, F. (2004). Rock climbing and raven *Corvus corax* occurrence depress breeding success of cliff-nesting peregrines *Falco peregrinus* Ardeola. *International Journal of Ornithology, 51*, 425–430.

Buckley, R. (2004). Impacts of ecotourism on birds. In R. Buckley (Ed.), *Environmental Impacts of Ecotourism* (pp. 187–200). Cambridge: CAB International.

Bunce, R. (1985). *Impact Assessment of Cliff Vegetation in the Lake District*. Institute of Terrestrial Ecology, Grange-over-Sands, 8pp.

Camp, R. J., & Knight, R. L. (1998). Effects of rock climbing on Cliff Plant Communities at Joshua Tree National Park, California. *Conservation Biology, 12*, 1302–1306.

Carr, C. (2007). *Variation in Environmental Impact at Rock Climb Areas in RED RIVER GORGE Geological Area and Adjacent Clifty Wilderness, Daniel Boone National Forest, Kentucky*. MA thesis, University of Cincinnati, 258pp.

Clark, P., & Hessl, A. (2015). The effects of climbing on cliff face vegetation. *Applied Vegetation Science, 18*, 705–715.

Climbing Management Guide. (2008). *A Guide to Climbing Issues and the Production of a Climbing Management Plan*. Compiled by Aran Attarian and Jason Keith, Access Fund, Boulder, CO, 76pp.

Cole, D. N. (2013). *Changing Conditions on Wilderness Campsites: Seven Case Studies of Trends over 13 to 32 Years*. USFS Rocky Mountains Research Station Report, Fort Collins, CO.

Colorado Parks and Wildlife. (2013). *Outdoor Recreation Participation Public Survey Summary Report*. Research, Planning and Policy Unit.

Cordell, H. K. (2004). *Outdoor Recreation for the 21st Century: A Report to the Nation. The National Survey on Recreation and the Environment*. State College, PA: Venture Publishing.

Cordell, H. K. (2008, Spring). The latest on trends in nature-based outdoor recreation. *Forest History Today*, pp. 4–10.

Cordell, H. K. (2012). *Outdoor Recreation Trends and Futures: A Technical Document Supporting the Forest Service 2010 RPA Assessment*. General Technical Report SRS-150, US Department of Agriculture Forest Service, Southern Research Station, Asheville, NC, 167pp.

Cordell, H. K., Teasley, R. J., Super, G., Bergstrom, J. C., & McDonald, R. (1997). *Outdoor Recreation in the United States: Results from the National Survey on Recreation and the Environment*. Athens, GA: USDA Forest Service.

Cross, R., & Brennenan, E. (1997). Devils tower at the Crossroads. The National Park Service and the preservation of native American cultural resources in the 21st century. *Public Land and Resources Law Review, 18*, 5–45.

Curry, E. (1997). *A Study of the Effects of Rock Climbing on Rock Surfaces*. Undergraduate thesis, Liverpool John Moores University, 64pp plus appendices.

Farris, M. A. (1998). The effects of rock climbing on the vegetation of three Minnesota cliff systems. *Canadian Journal of Botany, 76*, 1981–1990.

Gordon, K., Chester, M., Corp, N., Edhouse, K., Hayes, G., & Denton, A. (2015). *Getting Active Outdoors: A Study of Demography, Motivation, Participation and Provision in Outdoor Sport and Recreation in England*. Sport England and the Outdoor Industries Association, 64pp.

Hamilton, S. (2004). *Crag Erosion Study: Does Placing Passive Chock Protection (Wild Country size 1 and 10 nut) into Cracks Cause Weathering of Rock?* Undergraduate dissertation, Outdoor and Science Education, Liverpool John Moores University.

Hannigan, D. (1992). *Chair Ladder and the South Coast*. Climbers Club.

Hearn, K. (1994). *Nature Conservation and Cliff Climbing in West Penwith and the Lizard, Cornwall*. Report for the National Trust.

Herter, W. (1996). Die Xerothermvegetation des Oberen Donau tals- Gefahrdung durch Mensch und Wild Sowie Schutz-und Erhaltungs vor schlage Veröffentlichungen des Projektes für Angewandte. *Ökologie, 10*, 1–274.

Holzschuh, A. (2016). Does rock climbing threaten cliff diversity? A critical review. *Biological Conservation, 204*, 153–162.

James, A., & Wightman, B. (2003). *Stars in Guidebooks: Yes or No?* Retrieved from www.ukclimbing.com/articles/page.php?id=17.htp

Jones, C. D. (2004). Evaluating visual impacts of near-view rock climbing scenes. *Journal of Park and Recreation Administration, 22*, 39–49.

Jones, I. A. (1993). *Ogwen and Carneddau*. Climbers Club of Wales, Cordee, Leicester.

Kelly, P. E., & Larson, D. W. (1997). Effects of rock climbing on populations of pre-settlement eastern White Cedar (*Thuja occidentalis*) on cliffs of the Niagara Escarpment, Canada. *Conservation Biology, 11*, 1125–1132.

Kuntz, K. L., & Larson, D. W. (2006). Influences of microhabitat constraints and rock climbing disturbance on cliff-face vegetation communities. *Conservation Biology, 20*, 821–832.

Liddle, M. 1997. *Recreation Ecology*. London: Chapman and Hall, 639pp.

Long, J. (2004). *How to Rock Climb*. Guildford, CT: Falcon Publishing.

Lorite, J., Serrano, F., Lorenzo, A., Cañadas, M., Ballesteros, M., & Peñas, J. (2017). Rock climbing alters plant species composition, cover, and richness in Mediterranean limestone cliffs. *PLOS.One*. https://doi.org/10.1371/journal.p.one.0182414.

Makley, S., & Makley, M. J. (2010). *Cave Rock: Climbers, Courts and a Washoe Indian Sacred Place*. Reno, NV: University of Nevada Press.

Marion, J. (2014). *Leave No Trace in the Outdoor*. Mechanicsburg, PA: Stackpole Books, 118pp.

Marion, J. L., Carr, C., & Davis, C. A. (2011). *Recreation Impacts to Cliff Resources in the Potomac Gorge*. Final Report for the USDI, National Park Service, Chesapeake and Ohio National Historical Park and the George Washington Memorial Parkway. USDI and United States Geological Survey, 127pp.

McMillan, M. A., & Larson, D. W. (2002). Effects of rock climbing on the vegetation of the Niagara Escarpment, Southern Ontario, Canada. *Conservation Biology, 16*, 389–398.

Müller, S. W., Rusterholz, H.-P., & Baur, B. (2004). Rock climbing alters the vegetation of limestone cliffs in the northern Swiss Jura mountains. *Canadian Journal of Botany, 82*, 862–870.

National Park Service. (1995). *Devils Tower: Final Climbing Management Plan and Finding of No Significant Impact*. US Department of the Interior, Devils Tower, Wyoming, 112pp.

National Park Service. (1998). *Chickamauga and Chattanooga National Military Park Final Climbing Management Plan*. Unpublished manuscript, 25pp, US Department of the Interior, Fort Oglethorpe, GA.

National Park Service. (2000). *Joshua Tree National Park Backcountry and Wilderness Management Plan*. National Park Service, US Department of the Interior.

National Park Service. (2001). *Rocky Mountains National Park Backcountry/Wilderness Management Plan*. National Park Service, US Department of the Interior.

National Park Service. (2005). *Climbing Management Plan. Environmental Assessment*. New River Gorge National River, WV. National Park Service, US Department of the Interior, Philadelphia, PA, 196pp.

National Park Service. (2006). *Denali National Park and Preserve Management Plan*. National Park Service, US Department of the Interior.

National Park Service. (2007). *Zion National Park Management Plan*. US Department of the Interior.

National Park Service. (2010). *Climbing Guidelines, Shenandoah National Park*. National Park Service, US Department of the Interior.

National Park Service. (2013a). *Managing Climbing Activities in Wilderness*. NPS Reference Manual #41.

National Park Service. (2013b). *Climbing and Canyoneering Management Plan*. Arches National Park, 21pp. National Park Service, US Department of the Interior, Moab, UT.

NSRE. (2003). *National Survey on Recreation and the Environment 2000–2002. America's Participation in Outdoor Recreation: Results from NSRE*. Athens, GA: USDA Forest Service; University of Tennessee, Knoxville, TN.

Nuzzo, V. A. (1995). Effects of rock climbing on Cliff Goldenrod (*Solidago sciaphila* Steele) in Northwest Illinois. *The American Midland Naturalist, 133*, 229–241.

Nuzzo, V. A. (1996). Structure of cliff vegetation on exposed cliffs and the effect of rock climbing. *Canadian Journal of Botany, 74*, 607–617.

Panton, S. (2017). *North Wales Bouldering*. Groundup, 650pp.

Raven, J., & Walters, M. (1956). *Mountain Flowers*. Collins: The New Naturalist, 240pp.

Richards, M. (1997). *Climbers Attitudes and Opinions towards Abseiling at Craig Bwlch y Moch*. Undergraduate dissertation, Outdoor and Science Education, Liverpool John Moores University.

Ruby, C. (2015). *The Effects of Recreational Rock Climbing on Vascular and Non-vascular Plant Communities in Southeastern Tennessee*. Undergraduate dissertation, University of Southern Mississippi.

Rusterholz, H.-P., Müller, S. W., & Baur, B. (2004). Effects of rock climbing on plant communities on exposed limestone cliffs in the Swiss Jura mountains. *Applied Vegetation Science, 7*, 35–40.

Schuster, R. M., Thompson, J. G., & Hammitt, W. F. (2001). Rock climbers' attitudes toward management of climbing and the use of bolts. *Environmental Management, 28*, 403–412.

Smith, A. (1997). *The Ecological Impact of Rock Climbing on the Crag Foot of Stanage Edge*. Unpublished undergraduate dissertation, Outdoor and Science Education, Liverpool John Moores University.

Smith, J. (1995). *Rock Climbing and the Peregrine Falcon in the Yorkshire Dales: A Study of the Relationship*. Unpublished undergraduate dissertation, Outdoor and Science Education, Liverpool John Moores University.

Steven, R., Pickering, C., & Guy Castley, J. (2011). A review of the impacts of nature based recreation on birds. *Journal of Environmental Management, 92*, 2287–2294.

Studlar, S. M., Fuselier, L., & Clark, P. (2015). Tenacity of bryophytes and lichens on sandstone cliffs in West Virginia and relevance to recreational climbing impacts. *Evansia, 32*, 121–135.

Tessler, M., & Clark, T. A. (2016). The impact of bouldering on rock-associated vegetation. *Biological Conservation, 204*, 426–433.

Thompson, M. (1992). *A Study into the Effects Caused by Winter Climbing on the Natural Vegetation, Soil and Rock Appearance*. Unpublished undergraduate dissertation, B.Ed. (Hons), Secondary Liverpool Polytechnic.

Topline. (2017). *Outdoor Recreation Participation*. Topline Report.

UIAA. (2014). *Recommendations on the Preservation of Natural Rock for Adventure Climbing*. UIAA, 19pp.

References

Vogler, F., & Reisch, C. (2011). Genetic variation on the rocks- the impact of climbing on the population of a typical cliff plant. *Journal of Applied Ecology, 48*, 899–905.

Walendziak, N. (2015). *Longitudinal Variation in Environmental Impact at Rock Climbing Areas in the Red River Gorge Limits of Acceptable Change Study Area, Daniel Boone National Forest, Kentucky.* Online theses and dissertations paper 239. Master of Science, Eastern Kentucky University.

Webber, D. (2002). *Mechanical Rope Damage to Trees by Climbers.* Undergraduate dissertation, Outdoor and Science Education, Liverpool John Moores University.

Wezel, A. (2005). Changes between 1927 and 2004 and effect of rock climbing on occurrence of *Saxifraga paniculata* and *Draba aizoides* two glacial relicts on limestone cliffs of the Swabian Jura, southern Germany. *Journal of Nature Conservation, 15*, 84–93.

Wyatt, J. (1988). *Rock Climbing and Environmental Awareness.* Adventure and Environment Awareness Group conference report, The Lake District.

Gorge Walking, Canyoneering, or Canyoning

5

Chapter Summary

Gorge walking, canyoneering, and canyoning use similar techniques and are undertaken by a relatively small number of participants. The impacts and management approaches from three case studies are illustrated. In the UK there is preserved a rare, specialised but diverse flora where the ecology can suffer types of impact and floral loss. Controlling impacts are suggested, like sacrificial gorges, gorge rotation, Adopt-a-Gorge schemes, and educational methods.

In the Blue Mountains (Australia), the assumption by managers was that canyons were fragile, at risk from degradation, leading to unsustainable biological impact. It has been shown that the participants were lower than thought, concentrated in a few locations, and that the impacts on stream macroinvertebrates and water quality negligible.

In the classic canyons of the USA, impacts by bolting, rock damage by ropes, and tree damage from anchors and slings have been documented. Management plans for the Arches and the Grand Canyon National Parks include booking systems and group-size regulations. General management issues include banning, restoration, and clean-up projects and conduct codes and ethics and education through skills and leadership training.

5.1 Introduction

This refers to a series of specialised outdoor activities that although similar are referred to under slightly different terminology in different parts of the world but undertaken by a relatively small but growing number of participants. In the UK it is referred to as gorge walking or scrambling in Wales or in the Lake District as Gill or Ghyll (Victorian spelling) scrambling. In Europe and many parts of the world, it is known as canyoning and in the USA as canyoneering. The canyon or gorge is a deep, narrow, steep-sided valley, with usually vertical walls. It often has a flowing river and is formed by erosion of the surrounding rock by this river which has created pools, waterfalls, and rock bars because of the geological variability throughout its course, eroded sometimes over long periods of geological time (Fig. 5.1).

The environment can vary tremendously and can include narrow slots, pools, and moving and cold water, but the major difference is the climate. Canyoneering takes place in the semi-arid, south-west states of Utah and Arizona (Fig. 5.2), whereas canyoning takes place generally in more temperate, wetter climates of Europe or Australasia, but all canyons are eroded by running water. Vegetation is often luxuriant in tropical and temperate regions of the world but not in the South-Western USA. One danger in canyoneering is flash flooding, so sometimes there is little difference between the two activities. Gorge

Fig. 5.1 Gill scrambling often includes climbing or descending waterfalls and rapids. Photo by D. Huddart

Fig. 5.2 Canyoneering, Arches National Park, Utah. Note the sandstone rock and the lack of vegetation. Photo by D. Huddart

walking in the UK can be looked upon as the west coast European temperate climate variety of canyoning, but many of the techniques are the same, or very similar.

Bees (2005) defines canyoneering as descending usually dry or still water canyons normally eroded into sandstone with landforms such as potholes and dry waterfalls and with up and down climbing. It is a multidisciplinary activity that emerged in the 1960s as distinct from mountaineering and rock climbing, although it has similarities to both, particularly with regard to the use of ropes and anchors to facilitate abseiling (rappelling) and to provide protection from falls. It is sometimes called dry canyoneering. It developed primarily in southern Utah in areas near Zion Canyon, the Escalante River, and the San Rafael Swell but has now spread to the Grand Canyon where participants explore, traverse, and descend the park's remote canyon tributaries. Clean canyoneering or leave-no-trace canyoneering would leave only a sling behind and if a flood came through it would erode virtually all trace of the anchors and slings, and there would be no permanent trace. The sling could be carried out when no longer useful, and the canyon would be restored back to its original wilderness state. Ghosting is the term used for the ultimate leave-no-trace method when no protection is used at all. Aid canyoneering would hopefully use natural anchors like trees or boulders but controversially also includes bolts and drilled holes for hooks. This is the most destructive form of canyoneering because there is scarring of the rock which is permanent, and the scar from the bolt is likely to outlast the useful life of the bolt. There are ways to replace the bolts and eradicate some of this scarring which we will discuss later.

Canyoning usually involves flowing water in canyons, often eroded in harder rocks such as limestone, slate, or granite and may be found in Europe, Australia, British Columbia, and the Sierra Nevada. In the USA it is sometimes referred to as swift-water canyoneering. In Japan and Taiwan, the participants usually move upstream, and the activity is called river tracing.

5.2 Numbers of Participants

These are extremely difficult to estimate because, for example, canyoneering does not appear in the top 25 outdoor recreation activities ranked by number of participants in 2012 in the USA (Cordell 2012), or in the 39 outdoor activities in Colorado (Colorado Parks and Wildlife 2013). However, it is thought to be a fast-growing sport in some parts of the world like in the Canyonlands of Utah and Arizona (Canyoneering Report 2015) and generally in America (ACA 1998–2010). In the report of 2015, it was found that over 50% of survey respondents had their first canyon experience in the past five years and only 13% had over 15 years' experience. This has been attributed to newer, safer, and easily available equipment, an increase in college and university outdoor adventure recreation curricula, new instructional texts and videos, the growth of commercial guiding and instructional programmes, and the widespread availability of information on recreation sites through guidebooks and the internet where there are forums on canyoneering. Backcountry patrols in the Grand Canyon National Park have documented increased numbers in side canyons over the last few years (Jenkins 2017), and most of the popular routes are extremely scenic, accessible, moderately difficult, and can be completed within a day with equipment that most canyoneers already own. Moreover they are within a day's drive of population hubs like Phoenix, Flagstaff, and Las Vegas and are located in popular backcountry areas with large campsites. The Zion Wilderness has experienced a 15% increase in demand for Wilderness reservations and permits over the last three years (2017) which again indicates likely growth in canyoneering. There was also growth estimated in continental Europe (IAPCG 2009–2010), Nepal (NCA 2008–2009), New Zealand (NZOAR 2006), and Australia (High n' Wild 2010; Oz Canyons 2010). Nevertheless, in some parts of the world, the interest may have peaked, and despite the Blue Mountains World Heritage management authorities suggesting that the level of canyoning along accessible sections of the canyons is "high" and continuing to increase, Hardiman and Burgin

(2010b) were able to demonstrate that overall visitation was lower than perceived in part of Australia and probably declining. So actual numbers participating in these activities are thought to be relatively low and in different parts of the world either increasing or even decreasing, and it has been suggested that these activities are one of those undertaken as "one-off" adventure activities by many participants which they never go back to. As the sport is restricted to wilderness areas, the real quantification is extremely difficult.

5.3 Case Studies of Three Types of Canyoning

To outline some of the environmental impacts, the possible management of such activities, and the education of the participants, three case studies from the UK, Australia, and the South-Western USA will now be discussed.

5.3.1 Gorge Walking or Gill Scrambling in the UK

A gorge is a steep-sided variant of a V-shaped valley which results from a river flowing through a particularly resistant rock band. In the UK they occur on mountainsides as the result of a river eroding a line of weakness, usually along a fault line after thousands of years of post-glacial downcutting to become the gorges or ravines we see today. They often have high levels of minerals present due to fluid percolation from underlying igneous intrusions and have steep gradients, and where there is some geological variability, there are often waterfalls, cascade sections, and pools. In the UK because they occur in mountain areas, on the western side of the country, the rainfall is high and run-off very high, especially over impervious rocks, and there are impressive waterfalls along the ravines (Fig. 5.3). They can also be eroded along other types of geological weakness such as where there are softer dyke intrusions. Limestone gorges are a different category and can result from subglacial fluvial ero-

Fig. 5.3 Abseiling down a waterfall close to the heavily vegetated rock walls. Photo by D. Huddart

sion, like Gordale or Trow Gill in the Yorkshire Dales, normal fluvial erosion with extensive downcutting like the Samaria Gorge in Crete, or occasionally through cave collapse.

In the UK after the last cold phase of the Pleistocene which finished about 10,000 years ago, the climate became warmer during the current interglacial (Holocene), and a rich and complex vegetation clothed the hillsides. Since then there has been an anthropogenic impact with the deforestation of the uplands and the introduction of sheep farming which has cleared much of this varied flora. The gorges have become "slices" of ancient ecosystems and examples of what the mountainsides used to look like. This is because they are often inaccessible to sheep and were of no use to early settlers. The result is that the gorges are the home to some relict and rare plant species. The association of plants too is relatively uncommon in the UK today. So the vegetation has been protected by the physical nature of the gorge, and they are the only places where some plant species survive due to the sheep grazing pressure outside the gorges, and the gorges have a rich flora due to the complex ecosystem based on high mineral content, high water level, rock

Fig. 5.4 (A) Sea plantain (*Plantago maritimus*). Photo by D. Huddart. (B) Scarce Turf Moss (*Rhytidiadelphus subpinnatus*). Photo by Hermann Schachner. (C) Starry Saxifrage (*Saxifraga stellaris*). Photo by D. Huddart

exposure and rock types, and differing aspects and dryness. There are great variations in local conditions. The gorge environment is one of the least disturbed areas in the Lake District and represents possibly the last remaining vegetation in the conditions which would have existed before the arrival of man. This environment is rare, for example, the ecologically interesting gorge vegetation covers only 10 hectares in gorges and cliffs which represent 100 hectares, whilst the Lake District mountains cover approximately 160,000 hectares.

Gorges therefore provide a suitable set of environmental conditions for a range of plants: the steep walls protect plants from grazing; they can grow on the gorge walls, particularly on crumbling rock; they do so because of a combination of cool, wet climate and variable soils. The plants that are present are from many habitats which thrive in a complex environment. As Bunce (1983, 1985) suggested, many species are found growing elsewhere in Europe today but are growing in the UK in gorges at the western end of their distribution, and they occur in unique combinations. This is their special significance. The proximity of the British uplands to the coast has resulted in some maritime species living high up in the sheltered gorge environment, like the sea plantain (*Plantago maritima*, Fig. 5.4A).

As the gorge environment is very wet, damp, and humid, it has become the home of very rare examples of bryophyte flora, and there are certain mosses and liverworts of particular conservation concern in North and South Wales. For example, the bryophyte *Rhytidiadelphus subpinnatus* (Fig. 5.4B) has not been seen since 1912 in the Nantcol Gorge and is only found in the Mellte valley in South Wales, the Torrent Walk near Dolgellau, and isolated places in Carmarthenshire.

Some of the species are illustrated in Table 5.1 but include arctic-alpine species growing at low levels in the Arctic (Fig. 5.4C); woodland species

Table 5.1 Species growing in Lake District Gorges (based on over 200 plant species living in gorges) (after Piggot 1986)

Streams and streamsides: *Dicranum majus*, *Hookeria lucens*, *Ptilidium pulcherrimum*, *Rhytidiadelphus loreus*. Rare mosses and liverworts which live in, or close to, the water in sheltered areas but not in the main flow
Woodland (calcicole): *Geranium robertianum* (Herb Robert), *Fraxinus excelsior* (common ash), *Prunella vulgaris* (selfheal), *Mycelis muralis* (wall lettuce)
Woodland (neutral): *Athyrium filix-femina* (lady fern), *Hyacinthoides non-scriptus* (bluebell), *Oxalis acetosella* (wood sorrel), *Luzula sylvatica* (greater woodrush)
Woodland (calcifuge): *Betula* spp. (birch), *Calluna vulgaris* (ling heather), *Sorbus aucuparia* (rowan or mountain ash), *Vaccinium myrtillus* (bilberry). The most striking feature of gorge vegetation is the dominance of woodland species (c.38%), and this dominance shows that the gorges are relict woodland fragments, although the dominance is surprising (Bunce 1984)
Meadow: *Alchemilla glabra* (lady's mantle), *Festuca rubra* (red fescue), *Angelica sylvestris* (wild angelica), *Pimpinella saxifraga* (burnet saxifrage)
Maritime: *Armeria maritima* (sea thrift), *Cochlearia officinalis* (common scurvy grass), *Plantago maritima* (sea plantain), *Silene maritima* (sea campion)
Weed: *Cerastium fontanum* (mouse-ear), *Cirsium vulgare* (spear thistle), *Taraxacum officinale* (dandelion), *Urtica dioica* (common nettle)
Scrub: *Crataegus monogyna* (hawthorn), *Erica cinerea* (bell heather), *Juniperus communis* (common juniper), *Rhododendron ponticum* (rhododendron)
Grassland: *Agrostis tenuis* (common bent), *Digitalis purpurea* (foxglove), *Festuca ovina* (sheep's fescue), *Potentilla erecta* (tormentil)
Moorland: *Juncus squarrosus* (heath rush), *Nardus stricta* (mat grass), *Carex binervis* (green-ribbed sedge), *Polygala vulgaris* (milkwort)
Flush (acid): *Cirsium palustre* (marsh thistle), *Juncus effusus* (soft rush), *Juncus bulbosus* (bulbous rush), *Pinguicula vulgaris* (common butterwort)
Flush (enriched): *Cardamine hirsuta* (hairy bitter grass), *Carex demissa* (club sedge), *Ranunculus repens* (creeping buttercup), *Epilobium montanum* (broad-leaved willowherb)
Crevice: *Thelypteris dryopteris* (oak fern), *Asplenium viride* (green spleenwort), *Cystopteris fragilis* (brittle bladder fern), *Thymus drucei* (wild thyme)
Arctic-alpine: *Alchemilla alpina* (alpine lady's mantle), *Saxifraga stellaris* (starry saxifrage), *Oxyria digyna* (mountain sorrel), *Saxifraga aizoides* (yellow mountain saxifrage)
Northern: *Festuca vivipara* (viviparous fescue), *Lycopodium selago* (fir clubmoss), *Rubus saxatilis* (stone bramble), *Vaccinium vitis-idaea* (cowberry). These are plants which only grow at high levels in the Lake District but are not arctic-alpines
Any one gorge will contain different assemblages depending on the habitats they provide

mainly growing at low levels in a valley but due to sheltered areas and similar soils which can grow in gorges, especially on rocky ledges on the gorge side; meadow species which have declined in lowland areas due to land use changes but due to rich soils and a moist habitat can survive in gorges; and moorland species which favour the peaty, poor upland soils and flush and crevice environments.

Gorges are used as a valuable resource for outdoor activity providers and by individual participants because they are free and therefore cheap; the activities are relatively easy to run because gorges generally have easy access, by vehicles initially and then by foot. Gorge walking or gill scrambling is a popular activity as it presents an exciting and challenging experience which takes place in a beautiful and aesthetic location. It is about travelling along the gorge, usually from the bottom to the top, and it can involve scrambling or climbing, and occasionally some rope work may be necessary. The concept of climbing a gorge can be conveyed as a kind of "journey into the unknown." It takes place in an interesting, constantly changing environment in terms of vegetation, geomorphology, and rock structures. The activity is good for mixed-ability teaching and is easily differentiated to suit all abilities, including special needs pupils. There is also minimal equipment needed, just personal gear, possibly wet suits, and a rope. A questionnaire survey by Palmer (1994) revealed that there was a great range of interest in, and use of, gorges by outdoor centres for the following reasons: general adventurous activities particularly by school groups, navigation, mountain leadership courses for rope work, adventure walks and access to higher ground and activities like "Jungle Abseils," ice climbing, bird-watching and nature appreciation, management training courses, and geography field trips from field study centres and the general public where walkers are looking for a less crowded, more adventurous route to the higher mountains. The gorge can provide for an extremely versatile activity both in terms of objectives that can be achieved and the variety of abilities that can be catered for. Beasley (1997) found that the most frequently given reason for using gorges in the Lake District was that it is an adventurous, challenging and fun activity, it provides many opportunities for personal and group development, it can be used effectively for educational purposes, and it is a very flexible activity.

However, is there a danger to the ecology of these gorges? The answer has to be yes. Surveys show that there are more people involved in gorge walking each year, but the figures presented here represent usage by data from outdoor centres, and they do not account for visits by the general public and so represent minimum numbers. Nevertheless, it is difficult to know if the figures are accurate, but the figures and the evidence for recreation damage are important as we need to convince some well-respected mountain writers that there is not "indiscriminate conservation" of British gorges (e.g. see Perrin 1990). In 1986 Piggot suggested that for the Lake District National Park, there were 15,000 participants per year in 46 different gorges, but the majority were concentrated in only a few. In North Wales Palmer (1994) suggested 12,000 per year in 21 gorges, whilst in the Brecon Beacons National Park there were 2000 per year in 3 gorges. In 1997 there were conducted vegetation and erosion surveys by the Snowdonia National Park Authority in three gorges that were identified as frequently used. Erosion damage was found to be significant in two of these, and the intended management strategies resulted in the prevention of outdoor activity centres using one gorge and the erection of fences around the more important and vulnerable areas within the others. We have seen that gorges are a valuable resource to outdoor activity providers; they have relatively high usage pressure, especially in what are considered to be "ideal" sites; some gorges are environmentally sensitive due to the wide range of flora which live there and only there, and occasional very rare species are found in gorges. The users though have a limited knowledge about the gorge environment and its ecology, and so an increase in use results in an increase in pressure. There is a risk that unique gorges could be innocently destroyed by uneducated users, especially with the publication of guidebooks and magazine articles that draw attention to these environments like Evans' (1982) book *Scrambles in the Lake District*, which lists 36 gorges and a star rating for gorges, Allen's (1987) *On High Lakeland Fells, the 100 Best Walks and Scrambles* and Ashton's (1988 and 11 subsequent editions) book on Snowdonia. In "The Slaying of Stock Ghyll," there is an obvious disregard for the sensitivity of this environment, and Greenbank (1986) advertises gorges as an area for adventurous conquest. Also there appeared to be some changes to the ways in which some outdoor activity providers were using gorges, for example, abseiling down some botanically important areas which prior to the early 1980s had remained unaffected. Hence the Lake District National Parks Authority set up a gorge monitoring project to provide factual evidence of any changes taking place in the distribution and intensity of erosion in the period 1990–1993.

5.3.1.1 What Are the Needs of the Gorge Walker?

The activity needs clean rock on which to move because the aim of the gorge walker is to travel along the rock in the gorge, and so they avoid turf if possible. They stay on rock until it becomes impassable, and it may then be necessary to move to the vegetated gorge sides and create a new route around an obstacle. If this is required, then the gorge walkers will follow each other's footsteps when an alternative route is obvious. This route will continue to be used even when the path has been eroded to the bedrock beneath, and the walkers will create a new path to the side of the eroded one and widespread, rather than localised, erosion will result. However, Bunce (1983) suggested "with the walls of the gorge covered in carpets of mosses and liverworts, the rocky nature of the floor of the gorge has meant that there has been virtually no damage." So the preliminary early conclusion was that there was minimal conflict between gorge scrambling and the conservation of the vegetation provided the participants do not stray onto the sensitive walls. In the early 1980s, it was thought that although the most popular gorges had sustained damage, any subsequent damage should be kept to a minimum if the simple guidelines that were drawn up were followed and the more sensitive and ecologically important gorges were avoided by outdoor activity centres. If the participants stayed on the rocky floor, this might be the case, but there is a range of pressures unfortunately in gorges. As the main drainage channel, the gorges are the wettest places on the mountainside, with rich, although thin soils. These often waterlogged soils are easily removed by hand or foot and once removed may take thousands of years to reform. Of course once a breach in the vegetation occurs, the rate of erosion accelerates at that point, and the vegetation may never regrow. According to surveys in the Lake District, the damage type varies from gorge to gorge, but the damage could be categorised as follows with the six main forms of erosion: creation of paths and escape routes where the gorge becomes impassable, kicking of steps in soft, wet turf, removal of moss cover when foot and hand holds are covered and become slippery, damage to trees such as branches as aids; excavation of hand pockets in turf or moss, and trampling of vegetation on the rock surface or on less steep slopes. These were ranked according to impact (Pigott 1986). He also assigned a weighting factor to each factor: paths 4; steps 3; moss removal 2; tree damage 2; pockets 1; and trampling 1. Other impacts included litter, water contamination, and the scarring of trees and rock by the use of rope. The main areas of erosion seem to be the access, egress, and escape routes as these areas are the ones that carry more vegetation and are not as resistant to erosion as bare rock. As the gorge walkers follow marked routes, the erosion will be localised and more intense because of the increased pressure.

The damage is caused by people entering a fragile part of the gorge where, through a lack of knowledge and care, they can innocently destroy rare flora as a result of their passage. In a vegetation survey by Grant (1992), the dominance index for a used gorge was one and a half times that of a control gorge. In the Nantcol Gorge using Bunce's Erosion Index, the Nantcol Gorge scored 290, whilst the control gorge (Coed Caerwych) only scored 20. This Erosion Index is produced by noting the occurrences of the six types of erosional damage at four levels of intensity and by multiplying the number of occurrences by their intensity and a weighting related to the seriousness of the type of erosion. The sum of these figures produces the Erosion Index. For each type of erosion, there are four possible scores for each occurrence. The score is multiplied by the number of occurrences and the weighting for the erosion type to produce an index for each intensity of each erosion type. The sum of these indices produces the Erosion Index for the whole gorge ((1) is slight level of damage, (2) moderate, (3) severe, and (4) very severe). In the Lake District, plants occur on the steep rock walls and rocky crevices (e.g. Alpine species) and close to the water's edge (e.g. fern species). In North Wales there is a similar but higher concentration of rare plants at the water's edge, for example, rare ferns and bryophytes, especially in the waterfall splash zones, and in South Wales there is considerable concern for the rare bryophyte flora in or very close to the water.

How can we be certain there is a problem with floral loss in gorges? Grant (1992) looked at the difference between the Nantcol Gorge (Fig. 5.5) (about 4 km inland from Llanbedr, near Harlech) and another control gorge of the same altitude, bedrock, and average water flow.

His results showed that the control gorge had a total of 32 species, whilst the Nantcol Gorge had 19 species. This means the Nantcol Gorge has only 60% of the total number of species of the control gorge. Only two species found in the Nantcol are not present in the control gorge

Fig. 5.5 Nantcol Gorge (North Wales). Photo by D. Huddart

(*Cladonia floerekana*, a lichen; and *Betula pubescens*, the downy birch). He also discovered that in the Nantcol Gorge, two species were dominant over the rest. Grant (1992) suggested that from his vegetational analyses, the ecology of the Nantcol Gorge had been destabilised by the pressure of recreation use. Which species did Grant (1992) find have survived best in the Nantcol Gorge, and what are the reasons for this? The most important species are *Polytrichum commune* (a moss), *Lycopodium selago* (the fir clubmoss), the Festuca (a family of upland grasses), *Thuidium tamariscinum* (a moss), *Baeomyces rufus* (a grey-green lichen), and *Calluna vulgaris* (ling heather). The three mosses all lie very close to the ground, and they can grow in small crevices and cracks in the rock. This means that in areas where they are not being directly affected by trampling, or removal by hand and foot, they are well placed to prosper. The grey-green lichen is virtually flush with the surface on which it grows. It does not protrude upwards from its growth surface at all, except for its fruiting period. It can grow on vertical surfaces and can survive for long periods in conditions of extreme water stress and can survive on areas of rock which have been cleared of the moss cover which could provide the humus and water retention needed for many other plants. The Festuca grass family is a hardy upland family with strong stalks and long rhizomes so is able to withstand high levels of abuse such as pulling and crushing as when used for handholds. They can reproduce vegetatively and so suffer less from any trampling of seed and flower heads. The ling heather is also a hardy and physically tough species which means it can survive a high level of pulling and trampling from people passing through the gorge. It can also propagate itself by spreading along the ground and putting down roots in new places. None of the species which are successful in Nantcol rely entirely on flowers and seed for reproduction, since flowering is the most delicate stage of any plant's reproductive cycle and the flowers appear at one of the periods of the year which sees the heaviest traffic in the gorge. It is clear from the species in the control gorge that species which rely on flowers for the species' propagation are in a far stronger position when there is not the pressure of people passing through the gorge. There are several species which appear in small quantities

Fig. 5.6 Padley Gorge: note the extensive moss and bryophyte cover close to the river. Photo by D. Huddart

in the control gorge that do not appear at all in the Nantcol Gorge. Hence the advantageous characteristics in such an environment are strong roots and attachment to the ground, being low to the ground in order to avoid being used as a handhold, and finally being able to reproduce without the use of flowers which are at their most delicate and vulnerable during the time periods when there is the highest use.

Other major surveys have been carried out using the methodology developed by Bunce, but the techniques depend on the researcher's breadth of experience in able to score accurately the erosion at a particular site compared to others. This makes the Erosion Index a somewhat subjective technique, and although it seems to provide a quantitative and therefore accurate figure, it all depends on a degree of subjectivity in the original data collection. Nevertheless it does provide a relative scale of erosion damage in gorges, although it does not link the intensity of erosion in gorges to factors like the usage on a regular basis by an outdoor centre or centres, the inclusion and star rating in a guidebook, and the proximity of a car park. All must have an important role to play in the rate of erosion.

5.3.1.2 How Can We Control Any Erosion in Gorges?

Possible conflicts arising from the increasing recreation use are centred around how to manage gorges in order to conserve their unique ecology whilst at the same time maintaining access for the various user groups. Banning gorge walking is impractical and impossible to justify since one of the responsibilities of the national parks is to encourage leisure and recreation within the parks. So the question is how we can preserve these habitats (Fig. 5.6).

One method would be to do nothing because the argument is that the activity is only accelerating nature, and if there is a sudden flood, any impact that man has had is washed away anyhow, and this natural erosion is often far greater. However, gradually one by one, the popular gorges would become visually scarred and their ecology irreparably damaged. Users would move on and damage the next gorge, so this is not really a viable approach to the problem.

We could accept the sacrifice of some gorges, use these as "honeypots," and use them for recreation but place bans in others. This keeps the erosion limited and concentrates most in some areas. Grant (1992) came to the conclusion that a number of gorges that have already suffered damage due to gorge walking and scrambling should be sacrificed further to avoid spreading recreation pressure to other gorges. This could serve as a suitable management strategy, and Beasley (1997) considered that the idea should be considered further by concentrating use to those gorges that are able to withstand heavy recreation use. For example, use could be steered towards those gorges with a broad, rocky base and with vegetation that is not particularly well developed, like Stickle Gill, in favour of easing pressure on gorges of more botanical importance and where more difficult terrain inevitably leads to wandering over more vulnerable vegetation areas, such as Dungeon Ghyll (Beasley 1997).

An alternative might be to use gorges in rotation which might allow vegetation regrowth and regeneration. However, it would be difficult to implement this policy, and there is no guarantee that vegetation will regrow. In fact it is unlikely. How long could a particular gorge be used before it needs to be rested, and how long would a rest period be? Experimental research to try and answer such questions would not be justified, and the management of such a scheme would be impractical. It would require constant monitoring of the ecosystems in the gorges both to check on degradation and, in the recovering gorges, to assess the levels of recovery. This would require large staffing levels, and such monitoring would represent a constant level of pressure on the very environments being investigated. As a minor way to help the environment, the Adopt-a-Gorge scheme for outdoor centres has been useful to deal with problems such as litter, and if the clients help in this process, it serves as a valuable educational tool.

However, the best method is to attempt to educate gorge users by other approaches, both individuals and centre instructors. It all comes down to better education about the environment and its problems in relation to the activity and to incorporate a conservation message into the activity. For example, when using gorges, be aware that this is the habitat of rare and sensitive species, and use them as a learning point for the activity not just as a footpath. There needs to be good practice, such as following the Gorge Code of Conduct with guidelines drawn up which gorge walkers and walk organisers are obliged to follow: keep to the rocky bed of the gorge; groups should keep in a line; leave plants for others to enjoy; avoid crumbling rock; and follow only established routes. Environmentally friendly gorge walking and scrambling are possible. Sport and recreation can go hand in hand with gorge ecology but only if people are aware of the damage they can cause. However, a conservation message and the code have not appeared in the obvious gorge walking-related outdoor journals, and centres do not always know of its existence. The code of conduct moreover seems to have disappeared since it first appeared during the 1990s. Gorge walking could become only a wet activity; individuals must avoid scrambling up vegetated rock, but always the impact of the group through mismanagement is a potential problem, such as abseiling from a specific location which could result in abrasion from ropes, or the impact by trampling by people waiting around at the base of a waterfall. Discussions between centre staff and the national park ecologist in relation to the plant rarities can result in avoidance of certain locations or gorges. This will aid an individual's own knowledge about the environments they are using and enhance the centre's relationship with the national park authorities and helps preserve the unique nature of the gorge. This is assuming that both parties share the philosophies of respecting, preserving, using, and enjoying the environment. Instructors should be aware of the important plant species within a gorge they are using, but they should also be aware of the specific location of the important species so that they can avoid the more sensitive areas. The promotion of an environmental discovery approach to the activity in order to provide all users with a knowledge and respect for the gorge environment is an approach with a chance to promote environmental enthusiasm. The general public is more difficult to educate than the outdoor centre staff and their clients, although the national parks must have a role to play here. The centres are usually staffed by people who are sympathetic to an environmentally responsible use of the outdoors, and they see the gorges regularly over extended periods of time.

In the Lake District, it would appear that organised groups of interested people such as those that came together to form the "Gill Group" have done much to raise the awareness of the Cumbrian-based, outdoor activity providers to the unique and special environment of Lakeland gorges and the impact that use can have upon these environments. One such provider with a large client capacity made the decision to stop gorge walking with groups after becoming informed about the effects on the gill environment of scrambling. They also considered that they were not able to provide sufficient education regarding gorge ecology to enable their clients to make informed decisions about using gorges. They used to take groups of people three to four days a week in four to five gorges involving around 200 clients, so this decision could have a big impact on these locations. It appears that education can play a big role in the conservation of gorge vegetation and through events (October 1996) like an Association of Mountain Instructors Gill Scrambling workshop where there was an environmental input to help the education of the instructors.

The promotion of the activity has sometimes been portrayed in a very poor manner in terms of conquering the environment. In magazines and guidebooks, there has been little effort to raise environmental awareness related to the sensitivity of the gorge environment (see Evans' books 1994) where there has been a missed opportunity as 97 gorges are described and 30,000 copies plus had been sold before 1997, yet there was minimal ecological input). Sometimes articles offer potentially damaging advice in terms of gorge vegetation to its readers, for example: "this is an excellent pitch, in tremendous surroundings which appear quite formidable to novice scramblers who might prefer the vegetated wall further left" (Rawson 1996, p. 66). There have also been threats to the gorge environment from the development of hydroelectric power schemes, especially in North Wales. The threat is in the form of power stations diverting all available water on a mountain to the main collection lakes to be used to increase the water volume. This action involves the construction of dams at gorge heads which cuts off their water supply. Fortunately this has not yet damaged the recreation use of gorges and the important ecological gorges.

5.3.2 Canyoning in the Greater Blue Mountains World Heritage Area, Australia

The closest part of this area is located approximately 50 km west of Sydney in New South Wales, and therefore a large population has easy access to this outstanding region of natural scenic beauty. There are at least 400 explored canyons and new ones that continue to be discovered. The Park Management assumed a growth of canyoning in the area. However, the PhD research of Hardiman (2003) suggested that both the numbers of canyoners were not as great as assumed and that the impact of the canyoners was not as great as feared (Hardiman and Burgin 2010a, b, 2011a, b). The Park Management considered that the canyon ecosystems were fragile and at risk from degradation, although the level of such visitation and the degree of the biological fragility were not defined or backed up by any scientific evidence. This has meant that the management policy was made on the assumption that any activity was harmful unless shown to be otherwise. This precautionary policy lacked objective data and relied on anecdotal data, believing that the canyoning levels along accessible sections were "high" and were continuing to increase, that the majority of participants were with commercial tour operators, and that canyoning was spread over many locations. This then was thought to be leading to unsustainable biological impact. The result was that the adopted policy was a limit to both the size of parties and number of trips made by commercial companies to a small number of designated sites and physical closure of, or the making of, access to other sites increasingly difficult.

The land managers also assumed that canyoning was having an environmental impact. This impact could be on the aquatic system with increased turbidity from rive-bed trampling, pollution as a result of poor toilet practice and personal hygiene, littering, and reduced abundance and diversity of aquatic flora and fauna; or it could be terrestrial impact, such as riverbank erosion, trail erosion, flora trampling, rock damage, and littering. Concern over such potential impacts is apparent by a comparison of management plans for the World Heritage Area of 1988, 1998, and 2001 (NPWS 1988, 1998, 2001). While the 1988 Plan of Management (NPWS 1988) makes negligible mention of adventure recreation, a decade later major sections of the management plans (NPWS 1998, 2001) were devoted to addressing canyoning and other forms of adventure recreation, and there were specific policies proposed to manage the potential ecological impacts of these sports. In contrast, most experienced canyoners perceived that the levels of crowding and aesthetics (e.g. track erosion, abseiling slings/bolts) were within acceptable limits. They also typically did not comment on environmental damage and did not support additional management restrictions unless overuse could be demonstrated by park management sometime in the future (Hardiman and Burgin 2010a). In order to provide scientific data to underpin management of canyons, Hardiman (2003) undertook studies of the impact of canyoners in the years from 2000 to 2002 and in subsequent years.

Hardiman and Burgin (2010b) demonstrated that the participant numbers were lower than perceived and probably declining, that the majority of these participants were independent and non-commercial groups or friends, and that these participants were highly concentrated in a very small number of canyons and not widely dispersed as assumed. They also looked at the impacts caused by canyoners in terms of the benthic animal life in the rivers. Hardiman and Burgin (2011a) compared macro-invertebrate assemblages and water quality in those rivers that had, and had no, canyoner traffic. Data was collected over two canyoning seasons from early spring to late autumn. There was no significant relationship between the macro-invertebrate assemblage composition and the water quality observed due to visitor numbers within the canyons, although there were inherent differences between canyons. Pristine water quality was found in all locations. It was therefore concluded that at current canyon participant levels, there was no measureable impact on the macro-invertebrate populations of the canyon's rivers. This conclusion was contrary to the expectations of the land managers of the area who traditionally had perceived the canyons as fragile ecosystems and the passage of people through them would stir up the sediment in the rivers, disturb the macro-invertebrates, and lead to changes in numbers and composition. This was not the case. Hardiman and Burgin (2010a) had showed that there was no statistical difference in the macro-invertebrate communities between the canyon locations subject to high (20–100 visits/week) and low (0–5 visits/week) visitation over many years. They also showed that the ecosystem health in these canyons can show adaptation to this physical disturbance and that this was sustainable.

Since the macro-invertebrate data collection and analysis were time-consuming, Hardiman and Burgin (2010c) collected preliminary data on an assessment of the use of freshwater crayfish as indicators of environmental impact in the canyons because the data collected was far quicker. The crayfish could be used as a rapid bioindicator in canyons used at a relatively high level and those with low use. They found no indicator of human impacts. They recorded a single species, *Euastacus spinifer*, with no significant difference in crayfish abundance or size between the two visitation levels, but there was a difference in abundance between individual canyons. They thought that with a better baseline dataset, crayfish could potentially provide a rapid assessment method for use by canyoners and other non-specialists to underpin management decisions. Canyoners perceived that there was no reason to manage usage at the current visitation levels but that in the future intervention should occur if overuse occurred. The most popular way to minimise impacts was thought to be education by awareness programmes and information related to vegetation regeneration of impacted sites. The least popular way was a management system that limited the number of groups/canyon/day in all the canyons (Hardiman and Burgin 2010a).

5.3.3 Canyoneering in Utah and Arizona

In the Canyoneering Report (CAC 2015), the % of respondents who used the various areas for canyoneering was reported as Zion National Park 66% (with an average of 4.5 canyons descended), North Wash (Utah) 28%, Grand

Staircase-Escalante 24%, San Rafael Swell (Utah) 22%, Grand Canyon 20.5%, Capitol Reef 18%, and Arches National Park 16%. Most of these canyons are in sandstone and in an ecological zone that is completely different in terms of climate and vegetation than the previously discussed case studies. In Grand Canyon National Park, Jenkins (2017) described the side canyons of Garden Creek, Pipe Creek, Phantom Creek, and Ribbon Falls which are all hidden from three of the most popular hiking trails in the national park system (Fig. 5.7). All require technical canyoneering skills to reach, and they remained relatively untouched until recently, but as we saw earlier, there are several reasons why canyoneering in these areas has grown in popularity.

In the Grand Canyon NP, one reason was certainly the publication of Martin's (2011) guidebook which was the first one to feature technical canyoneering routes. Here he lists 105 canyoneering routes of which 68 need abseiling skills, 63 require swimming or wading, and 78 usually take over one day to complete. Many of the routes require the use of packrafts which are small lightweight inflatable boats to navigate the Colorado River in order to allow a return back to one of the canyon rims. Jenkins estimated too that there were about 200 more routes that could exist in the park, so numbers of participants are likely to increase in the future. Due to the fact that many of the canyoneering routes attract those who have not learned low-impact ethics related to technical canyoneering, there is a priority for park management to look carefully at this sport. This is especially so because inexperienced canyoneers may go on to change the character of other routes in the future. This has resulted in a canyon monitoring programme in 2012 which was to serve as a basis for a new strategy for revised management to address the rising popularity of canyoneering. Data were collected on route characteristics such as the levels of use, the impacts on the route approach and exit (erosion on trails and multiple trails); and the impact on natural resources, such as soils, springs, and wildlife, and cultural resources, such as petroglyphs. Some of the questions that the data gather-

Fig. 5.7 Grand Canyon with many tributary canyons hidden from the main canyon rim. Photo by D. Huddart

ing was to answer were how canyoneering might be affecting the wildlife habitat, particularly the bighorn sheep, how to avoid conflicts between user groups, and how to minimise the unnecessary placement of fixed anchors. Other impacts were the number of abandoned ropes, excluding handlines on routes, the number of abandoned handlines on routes, and unnecessary anchors on routes, rope grooves, cairns, graffiti, litter, both inorganic and organic, and human waste.

A case study of one route proves interesting as it illustrates many of the problems of canyoneering. This route was initially descended in the 1990s but remained relatively unused because of the high discharge waterfalls, challenging downclimbs, and the lack of publicly available information. Sometime in 2010 someone installed a single abseil anchor bolt midway down the largest waterfall. Before that at least 213 m of rope was required to descend because of a 107–122 m abseil halfway through the canyon. The publication of Martin's (2011) book also helped to make the route one of the most popular and sought-after canyoneering routes in the park. The issue of fixed anchors is most instructive and controversial and applies to all canyon and canyoneering routes. The situation in the Grand Canyon National Park with regard to anchors per route and clean versus fixed anchors has been much discussed, and in various other parts of the National Park System in the USA, there is a variable policy with regard to bolting, although generally no motorised drills are allowed. At the Grand Canyon, the guidelines in Director's Order 41, Section 7.2, are followed with respect to fixed anchors (USDI 2013). An alternative to the fixed anchor is a loop of removable nylon webbing slung around a natural feature, like rock or a tree, which is the most common type of anchor in the park. The bolts and pitons are generally not used because special equipment is needed and time and the expenditure of physical effort too are needed for their placement. However, in 2012 a second bolt was installed next to the pre-existing bolt in Garden Creek, and then in 2013 three canyoneers reported numerous additional bolts, and a follow-up monitoring trip documented 12 new bolts, including some clean anchors which had been replaced by bolts. This was clearly felt by Jenkins (2017) to be a degradation of the park's natural resources. It resulted in the American Canyoneering Association being asked by the National Park Service (NPS) to remove the illegal entry expansion bolts at Garden Creek. Epoxy resins were used to fill the holes, and some sand from the surrounding area was used to blend the colour to the rock, and the result appeared as a good solution.

Another example of the bolting issues can be taken from the Capitol Reef National Park when the Coalition of American Canyoneers was asked to consult on anchors in Cassidy Arch Canyon. This canyon is located near Torrey (Utah). It became very popular after 2007 with groups who had limited expertise and experience and is close to a dramatic Wingate sandstone arch at the top of the escarpment above the western end of Grand Wash. It is accessed by a constructed trail off the scenic drive, not far from the visitor centre, and this trail is very popular with hikers. The traditional first anchor was off a tree above the arch on the canyon left, but it was showing signs of wear and tear, and numerous rope grooves were present on the rock, either from abseiling or pulling the rope. To try and mitigate this damage, canyoneers were asked to use a retrievable anchor off the tree. However, there was far from total compliance, there was continuing rock damage, and a sling was left on the tree much of the time. In consultation the national park requested a permanent bolt anchor be installed. Several problematic anchors were identified down-canyon, either causing rope groove damage or relying on stressed vegetation. The park has approved replacement fixed anchors to address these issues. However, bolts should really be reserved for locations where no natural anchors can be found and in canyons with a lot of traffic where they can be used to minimise traffic damage, if they are to be used at all.

When the objectives of the national park are considered—to provide for visitor safety within the expectations of reasonably equipped and experienced canyoneers; to protect the resource, including preserving vegetation, minimally altering the rock; to minimise any visual impact for the hiking visitor; and to preserve the original flavour and ambiance of the canyon—it is really debateable whether bolts should have been used at all. Certainly when we consider the ethical guidelines for canyoneers, bolting should not be allowed, and when there is a differing fixed bolt policy in different national parks, this sends mixed messages to the canyoneers. The "Technical Canyoneering" written by William Bees (2005) to encourage people to lessen their impact on canyons suggests that bolts are not needed on the Colorado Plateau (http://climb-utah.com/Misc/natural.htm). However, when the canyoneers were asked whether canyoneering caused environmental impact (CAC 2015), the response was that they caused minimal damage (78%). This clearly is not the case, yet this canyoneer population was opposed to a permit system—51% opposed, or strongly opposed, such a system. Such a system however is in place in some national parks, like the Grand Canyon and particularly the extremely popular Zion National Park and especially the Subway and Mystery Canyon. Here there is an online lottery for reservations. The Advance lottery does not run from November to March because of lower demand, but the rest of the year, applications have to be made three months before the trip, and if awarded a trip, the permit has to be picked up from the park visitor centre before the trip. There is a last minute drawing system to deal with any permits which are left, but only a total of 20 per day are granted.

In the Arches National Park (2013), a Climbing and Canyoneering Management Plan was published in December 2013 to protect the natural environment and the park's resources and visitors' experience. There are certain regulations to follow, and each group must register. Canyoneers must complete a free self-registration permit. Group size is limited in the Fiery Furnace or Lost Spring Canyon to six persons, but in all other canyons, groups are limited to ten. If groups are larger, they must split up and use different routes or use the same route at different times of the day to avoid queuing at rappel sites and to minimise impact on resources and on other visitors. It is prohibited to climb, scramble, walk upon, wrap webbing, or rope around or rappel off any named or unnamed arch with an opening greater than 3 ft. The physical alteration of rock from its natural state such as chiselling, breaking rock to reinforce crevices and pockets as anchors, glue reinforcement of existing holds, and gluing new holds is prohibited. The intentional removal or "gardening" of lichen or plants from rock is prohibited. Use of deadman anchors is prohibited, and any new installations of fixed gear requires a permit. If an existing item or fixed anchor is judged unsafe, it may be replaced in kind without a permit. Software left in place is required to match the rock surface in colour, and bolts, hangars, and chains must be painted the colour of the rock surface before installation. The installation of pitons and the use of motorised drills within proposed wilderness areas are prohibited. Drilling outside these wilderness areas requires a special permit. Slacklining, or highlining, where walking or balancing along a suspended length of webbing tensioned between two anchors, is prohibited, sometimes at elevations above water or the ground, as is guided canyoneering services.

All canyoneers must register by obtaining a free permit, and there are no daily limits on routes (except for the Fiery Furnace), so a permit can be obtained on the day. Registration is free, it increases safety, and it helps maintain the desired conditions of the backcountry zone. There are two ways to obtain a permit: via the online reservation system and self-registration at the kiosk outside the visitor centre. For the Fiery Furnace, the entire party must go to the visitor centre front desk to get a permit at a cost of $6 per person. It is limited to 75 per day and often sells out during the busy season. New routes can be established, but there must be a special use permit obtained before establishing any new route requiring the installation of new fixed gear. Travel to and from these new routes must only be within sandy wash systems, on rock, or on delineated trails.

In the Grand Canyon, there was a draft Backcountry Management Plan published in November 2015 (NPS 2015) which plans to use an adaptive management process to address the increasing demands for recreation access and the uncertainty of how different recreation uses would impact the park resources. When this new version of the BCMP is likely to be adopted in 2018, the following changes for canyoneering would apply:

- A day use permit which would identify the canyoneering route
- Use limits for specific locations
- Maximum group size for both day and overnight (group sizes of 11)
- Seasonal or permanent restrictions for natural or cultural resource protection
- A decision framework for a new fixed anchor placement policy
- The overnight backcountry permit which would identify the activity
- Backcountry permitting process and field surveys which would monitor use and resource impacts
- Ban on use of power drills in the wilderness
- A requirement for human waste to be packed out of narrow canyons
- Implementation of minimum impact canyoneering education
- Development of a canyoneering management plan

These changes should limit the environmental impacts in this area and allow for the further education of canyoneers.

5.4 General Management Issues for Canyoning and Canyoneering

5.4.1 Banning the Activity Completely

Whilst this extreme measure would stop impacts, it is very rarely put into practice. However, the official Triglav National Park Management Plan in Slovenia forbids tourist sports or recreation activities on watercourses, or in standing waters, which do not include fishing, in what is called the first and second preservational zones. Most of the canyons are located in these zones and there are also community bans for canyoning around Kobarid, Tolmin, and Idrija so that only 12 out of 37 canyons can be visited. The alternative is a large fine if caught in a canyon. From this area an example of the minor impact of canyoning compared with natural geomorphological processes can be illustrated. Six years ago a rockfall in the canyon Mlinarica changed from an almost 100 m deep and 1 m wide crack into a 30 m wide canyon. In 2011 Mlinarica collapsed into itself resulting in a major impact, so there is no way that canyoning could possibly have had such an impact.

5.4.2 Restoration and Clean-Up Projects

The Coalition of American Canyoneers, a non-profit organisation, have organised several clean-up and restoration projects in the past few years, and in general this is good environmental practice by the canyoneers and serves as education for the participants. For example, they organised graffiti removal at Monkey Face Falls in San Bernardino National Forest (California) on 1 August 2015, and a clean-up and graffiti removal event in Frye Canyon in Southern Arizona on 21 May 2016 using a pressure washer. At Eaton Canyon (California), there was a clean-up organised on 23 August 2014 where rubbish was picked up and graffiti cleaned from the canyon walls. Unfortunately in this case hammers and chisels were used as the graffiti was painstakingly chipped from the rock surface. It was then that the rock was painted with a thin coating of mud and a dusting of fine sand from the nearby stream bed. Subsequently it was found that rough surfaces are not as appealing as smooth ones, and so by painting on cement and throwing on sand, the canvas was not as attractive for graffiti artists. Hence the graffiti has been covered with a mixture of Portland cement and sand which is not ideal. However, there is no way to deal with graffiti which is great, or 100% successful, although the pressure-washer method seems best and has been used in caves too. There is no doubt though that that the offenders were canyoneers in the locations where the graffiti was found.

5.4.3 Code of Conduct Issues and Ethics Guidelines for Canyoneers and Canyoners

In France there are canyoning guidelines produced by Mountain Wilderness France (www.petzel.com/fondation/projets/zone-de-tranquilite-montagne?language=en.), and in Australia there is an excellent Canyoners Code of Ethics at http://www.fatcanyoners.org/canyoning/canyoners-code-of-ethics. As part of the wilderness management in the Grand Canyon, it has been suggested that canyoneering specific leave-no-trace information should be distributed to all visitors who identify themselves as canyoneers during the backcountry permitting process. There is also a useful article on Ethics: Code of Conduct that all canyoners should follow published in Canyon Magazine on the Environment, Land Management, Canyon other Users and Canyoning Companies (http://canyon-mag.net). The Minimal Environmental Impact Plan, LeaveNoTraceCanyoneering (http://www.canyoneeringusa.com/utah/introduction/minimum-impact), has principles for travelling in technical canyons to minimise impacts:

- *Stay in the watercourse*: this tends to have zero impact as this is either rock or sand and gravel. Taking side paths to avoid small drops has an enormous impact and should be avoided.
- *Do not bolt*: if you come to a drop that you think needs a bolt anchor, look around and work out another way as many people have passed this way before.
- *Do not leave gear such as slings, fixed ropes, and lines*: those left in place are litter and should be removed and carried out.
- *Watch your step*: stay on established trails, or walk on zero impact surfaces like rock or sand gravel. Stay on the main trail to avoid proliferation, and walk single file rather than side by side. When having the choice, follow the edge of the stream rather than taking a side-trail climbing over a hill.
- *Social impacts count*: respect others' right to solitude by giving them space; invite other parties to pass; and create separation between parties by speeding up or slowing down.
- *Travel in small groups*: large groups tend to have significantly more physical impacts per person than small groups and have a greater impact on other people's feeling of solitude.
- This code is similar to the Gorge Code of Conduct discussed earlier in this chapter, and if they are discussed by groups leaders, they can help in the minimising of environmental impacts overall.

5.5 Education

There are plenty of courses for canyoners so that they can improve their skills or leadership. This can only benefit them personally and lessen their impact on the environment. Some are provided by private adventure companies like the Zion Adventure Company which offers Canyon Leader Certificate (CLC) (www.zionadventures.co.courses/courses/canyon-leader-certificate.html) which would allow participants to plan and lead safe, low-impact trips in beginner and intermediate canyons. Others are organised by canyoning organisations like the International Canyoning Organisation for Professionals (ICOPRO) which offers training courses. Then there are some organised by government training organisations, such as the Single-Pitch Canyon Guide trip run by the Australian Government Training (http://www.training.gov.au/Training/Details/SISOCAY304A/html)—the list of skills includes courses to Minimize Environmental Impact (SIS00PS201A) and Plan for Environmental Impact (SIS00PS 304A) and the Department of Sport and Recreation (2013) in Western Australia where there is a canyoning version. There are also more detailed monographs provided by government organisations such as the *Outdoor Activities: Guidelines for Leaders* published by Sport New Zealand (2009).

Problems related to some of these types of course though can be shown by those run by the American Canyoneering Association (ACA) which since 1999 provided training and certification for canyon leaders, professional canyon guides, and rescue team members. Their vision was to provide renowned, world-class training and certification, and they provided ACA certificate level training for recreation leaders at three levels and professional canyon guides training at three levels. However, the organisation was dissolved in 2011. The American Canyon Guides Association was formed following this to become America's first professional association for the canyoneering industry, but in 2016 it too was dissolved because its voluntary directors and limited membership were unable to host the guide assessments for which it was created. The Association for Canyoneering Education independently filled this void in guide certification whilst also establishing a standard for continuity in education and supporting a vibrant community of recreation canyoneers (www.acecanyoneering.com.html), but this industry does seem to have been short of stability in the last few years. Nevertheless the courses offered seem good and can only benefit the participants and help minimise environmental damage.

In the Greater Blue Mountains in Australia, the assumption by managers was that canyon ecosystems were fragile and at risk from degradation, leading to unsustainable biological impact. In fact it has been shown that the numbers of participants were lower than thought and were concentrated in a few locations and that the impacts on stream macro-invertebrates and water quality were negligible.

In Utah and Arizona in the classic canyon country, impacts on the natural landscape by bolting and rock damage by ropes causing grooves and tree damage from anchors and slings have been documented. Management plans for the Arches National Park and the Grand Canyon include booking systems and regulations to group size and on the use of equipment. General management issues include banning the activity, restoration, and clean-up projects and codes of conduct and ethics and education through skills and leadership training.

Concluding Remarks

The activities referred to as gorge walking, canyoneering, and canyoning are similar in the techniques used and are undertaken by a relatively small number of participants worldwide. The impacts and management approaches from three very different case studies are illustrated. In the UK gorges, there is preserved a rare, specialised but diverse flora which potentially can suffer from the gorge walking activity where the gorges are used by outdoor activity providers and individual participants. The gorge ecology can suffer from various types of impact, and there has been floral loss. Various ways of controlling erosion are suggested, such as the use of sacrificial gorges, gorge rotation, Adopt-a-Gorge schemes, and educational methods.

References

Allen, B. (1987). *On High Lakeland Fells: The 100 Best Walks and Scrambles*. Pic Publications, 224pp.

Arches National Park. (2013). *Climbing and Canyoneering Management Plan*. National Park Service, US Department of the Interior, Moab, UT, 21pp.

Ashton, S. (1988) and eleven editions up to 2013. *Scrambles in Snowdonia*. Milnthorpe: Cicerone Press.

Beasley, F. (1997). *A Study of the Recreational Impact of Gill Use within the Lake District National Park*. Unpublished undergraduate dissertation, Liverpool John Moores University, 127pp.

Bees, W. (2005). Technical canyoneering. Retrieved from http://climb-utah.com/Misc/natural.htm.

Bunce, R. G. H. (1983). Is climbing killing off Lakeland's Plant Life? *Lakescene Magazine*.

Bunce, R. G. H. (1984). *Gills. Adventure and Environmental Awareness* (pp. 36–38). Report of a Conference, National Park Visitor Centre, Windermere.

Bunce, R. G. H. (1985). *Impact Assessment of Cliff Vegetation in the Lake District*. Merlewood Research Station, Institute of Terrestrial Ecology, Grange-over-Sands.

Coalition of American Canyoneers. (2015). *Canyoneering in the United States 2015 Final Project Report*. Southern Utah University, 32pp.

Colorado Parks and Wildlife. (2013). *Outdoor Recreation Participation Public Survey.* Summary Report, Research, Planning and Policy Unit.

Cordell, H.K. 2012. *Outdoor Recreation Trends and Futures: A Technical Document Supporting the Forest Service 2010 RPA Assessment.* General Technical Paper SRS-150. Asheville, NC: USDA Forest Services Southern Research Appalachian Trail Design, Construction and Maintenance, second edition. Harpers Ferry, WV: Appalachian Trail Conference.

Department of Sport and Recreation. (2013). Western Australian adventure activity standard. *Canyoning version 2.0.*, 17pp.

Evans, R. B. (1982), 1994 (second edition). *Scrambles in the Lake District.* Milnthorpe: Cicerone Press.

Evans, R. B. (1991), 1994 (second edition). *More Scrambles in the Lake District.* Milnthorpe: Cicerone Press.

Grant, A. (1992). *An Investigation of the Environmental and Ecological Impact of the Recreational Use of Gorges, with Particular Reference to Gorge Walking and Scrambling.* Unpublished undergraduate dissertation, Liverpool John Moores University, 40pp.

Greenbank, T. (1986, March). The Slaying of Stock Ghyll. *Climber and Rambler*, pp. 51–53.

Hardiman, N. (2003). *Visitor Impact Management in Canyons of the Blue Mountains, New South Wales.* PhD thesis, University of Western Sydney, South Penrith.

Hardiman, N., & Burgin, S. (2010a). Canyoners' perceptions, their evaluation of visit impacts and acceptable policies for canyon management in the Blue Mountains (Australia). *Managing Leisure, 15*, 264–278.

Hardiman, N., & Burgin, S. (2010b). Adventure recreation in Australia: A case study that investigated the profile of recreational canyoners, their impact attitudes, and response to potential management options. *Journal of Ecotourism, 9*, 36–44.

Hardiman, N., & Burgin, S. (2010c). Preliminary assessment of freshwater crayfish as environmental indicators of human impacts in canyons of the Blue Mountains, Australia. *Journal of Crustacean Biology, 30*, 771–778.

Hardiman, N., & Burgin, S. (2011a). Comparison of stream macroinvertebrate assemblages in canyon ecosystems of the Blue Mountains (Australia) with and without recreational traffic. *Australian Zoologist, 35*, 757–769.

Hardiman, N., & Burgin, S. (2011b). Effects of trampling on in-stream macroinvertebrate communities from canyoning activity in the Greater Blue Mountains World Heritage Area. *Wetlands Ecology and Management, 19*, 61–71.

High n Wild Mountain Adventures. (2010). Welcome to high n wild mountain adventures. Retrieved from http://www.high-n-wild.com.au/.

IAPCG. (2009–2010). Die CIC ist der. International Association of Professional Canyon Guides (IAPC). Retrieved from http://www.cic-canyoning.org/.

Jenkins, M. (2017). Canyoneering at Grand Canyon National Park: Monitoring pockets of wilderness in the canyon corridor. *Park Science, 33*, 91–98.

Martin, T. (2011). *Grand Canyoneering: Exploring the Rugged and Secret Clots of the Grand Canyon.* Phoenix, AZ: Todd's Desert Hiking Guide.

National Park Service. (2015). *Backcountry Management Plan, Draft Environmental Impact Statement.* US Department of Interior, Grand Canyon National Park.

NCA. (2008–2009). NCA introduction. Nepal Canyoning Association. Retrieved from http://www.nepalcanyoning.org.np/.

NPWS. (1988). *Blue Mountains National Park: Draft Plan of Management.* Hurstville: National Parks and Wildlife Service.

NPWS. (1998). *Blue Mountains National Park: Plan of Management.* Hurstville: National Parks and Wildlife Service.

NPWS. (2001). *Blue Mountains National Park: Plan of Management.* Hurstville: National Parks and Wildlife Service.

NZOAR. (2006). Caving and canyoning. New Zealand Outdoor Adventure and Recreation (NZOAR). Retrieved from http://adventure.thedefinitiveguide.co.nz/caving.html. 618pp.

Oz Canyons.com. (2010). Canyoning near Sydney. Retrieved from http://OzCanyons.com/canyoning/.

Palmer, C. (1994). *A Study of the Gorge Environment and the Impact of Recreational Use.* Unpublished undergraduate dissertation, BSc Outdoor and Science Education, Liverpool John Moores University, 120pp.

Perrin, J. (1990, December). Out of bounds. *Climber and Hillwalker.*

Piggot, C. R. (1986). *Lake District Gills.* Undergraduate dissertation, Faculty of Geography and Landscape Studies, University of Southampton.

Rawson, J. (1996). Fair weather on fairfield. *Great Outdoors Magazine*, p. 66.

Sport New Zealand. (2009). *Outdoor Activities: Guidelines for Leaders.* Sport New Zealand, 197pp.

United States Department of the Interior (USDI). (2013). *Director's Order 41: Wilderness Stewardship.* Washington, DC: National Park Service.

Off-Road and All-Terrain Vehicles, Including Snowmobiling

6

Chapter Summary

This chapter first defines off-road and all-terrain vehicles, including snowmobile, and gives illustrated examples of a range of such vehicles. It then briefly discusses the history and diversity of use of such vehicles before presenting recent data on user numbers. The final part of the chapter focuses on specific environmental impacts associated with off-road, ATV, and snowmobiles which include damage to soil and vegetation, water/air pollution and noise, and disturbance to wildlife. The final section considers the management of these activities and gives some examples of education initiatives that have been used in management attempts.

6.1 Definitions

6.1.1 Off-Road Vehicles (ORVs)

An ORV is considered to be any type of vehicle that is capable of driving on and off paved or gravel roads. These vehicles generally have large tyres with deep, open treads and flexible suspension (Fig. 6.1) or sometimes caterpillar tracks. Other vehicles that do not travel on public highways are generally termed off-highway vehicles and may include tractors, forklifts, cranes, bulldozers, and golf carts.

Due to their many uses and versatility, ORVs are popular, and several types of motorsports involve racing ORVs. The three largest "four-wheel vehicle" off-road types of competitions are rally, desert racing, and rock crawling.

Rallying is a form of motorsport that takes place on public or private roads (sometimes in forests) with modified production or specially built road-legal cars. It is characterised by driving in a point-to-point format (rather than on a circuit) in which participants and their co-drivers drive between set control points (special stages), departing at regular intervals from one or more start points. Rallies may be won by pure speed within the stages or alternatively by driving to a predetermined ideal journey time within the stages.

Desert racing, in its most organised form, began in Southern California in the 1920s. Desert racing takes place, as the name would suggest, in desert in two- or four-wheeled ORVs. Races usually consist of two or more laps around a course covering up to 40 miles. Races can take the form of hare and hound or hare scramble-style event and are often laid out over a long and difficult track through relatively barren terrain. Point-to-point races such as the famous Baja 1000 attract nationally ranked and celebrity drivers. The endurance and capabilities of driver and machine are tested. Sometimes organised clubs or teams may field multiple sponsored riders for particular events, but usually desert racing in its purest form is largely an individual endeavour. Winning

Fig. 6.1 A Ford Bronco dune bashing. Photo by Mtxchevy

drivers accrue points to improve their rank and placement in future competitions. Desert racing vehicles can include rugged enduro-style motorcycles, four-wheeled all-terrain vehicles (ATVs), pickup trucks, and dune buggies, all of which have specialised suspension with increased wheel travel.

Rock crawling is an extreme form of off-road driving using vehicles ranging from stock to highly modified types designed to overcome obstacles. In rock crawling, drivers drive highly modified four-wheel-drive (4WD) vehicles such as trucks, jeeps, and "buggies" over very harsh terrain. Driving locations include boulders, mountain foothills, rock piles, and even mountain trails. Rock crawling is about slow-speed, careful and precise driving and high torque generated through large gear reductions in the vehicles' drivetrain. Rock crawlers often drive up, down, and across obstacles that would appear impassable. The vehicles used to rock climb are primarily 4x4s. Rock crawling competitions range from local events to national series. A rock crawling competition consists of obstacle courses that are 100–200 m long with each obstacle set up with gates, similar to a ski or canoe slalom course.

Dune bashing (Fig. 6.1) is a form of off-roading on sand dunes where a large sport utility vehicle such as the Toyota Land Cruiser is used (however, lightweight vehicles often fare better in the extremely soft sand found on sand dunes). Vehicles driven on dunes may be equipped with a roll cage in case of an overturn. Experience and skill are required to manoeuver the car and prevent accidents. Before entering the dune system, tyre pressure is reduced to gain more traction by increasing the footprint of the tyre and, therefore, reducing the ground pressure of the vehicle on the sand as there is a greater surface area (much like a person wearing snowshoes can walk on a soft surface without sinking, but a person without them cannot). For example, tyres with a recommended pressure of 35 psi would be reduced to approximately 12–14 psi.

6.1.2 All-Terrain Vehicles (ATVs)

An ATV, also known as a quad, quad bike, three-wheeler, four-wheeler, or quadricycle as defined by the American National Standards Institute (ANSI), is a vehicle that travels on low-pressure tyres, with a seat that is straddled by the operator and handlebars for steering control. As the name implies, ATVs are designed to handle a wider variety of terrain than most other vehicles. Although street-legal in some countries, ATVs are not street-legal within most states and provinces of Australia, the USA, or Canada. By the current ANSI definition, ATVs are intended for use by a single operator, although some companies have developed ATVs intended for use by the operator and one passenger (referred to as tandem ATVs).

ATV riders sit on and operate these vehicles like a motorcycle, but the extra wheels give more stability at slower speeds. Although equipped with three or four wheels, six-wheel models exist for specialised applications. Engine sizes of ATVs generally range from 49 to 1000 cc. The three largest types of ATV/motorcycle competitions are motocross, enduro, and also desert racing (see above). The most common use of ATVs is for sightseeing in areas distant from roads. The use of higher-clearance and higher-traction vehicles enables access on trails and forest roads that have rough and low-traction surfaces. Other uses include border patrol, construction, emergency medical services, land management, law enforcement, military, mineral and oil exploration, pipeline transport, search and rescue, small-scale forestry, surveying, and wild land fire control.

6.1.3 Off-Road Motorcycles: Motocross and Enduro Motorcycling

Motocross is a form of off-road motorcycle racing held on enclosed off-road circuits. The sport evolved from motorcycle trial competitions held in the UK (Fig. 6.2). Motocross first evolved in the UK from motorcycle trial competitions, such as the Auto-Cycle Club's first quarterly trial in 1909 and the Scottish Six Days Trial that began in 1912. When later it became a race to become the fastest rider to the finish, the activity became known as "scrambling." The sport grew in popularity, and the competitions became known internationally as "motocross racing," by combining the French word for motorcycle, *motocyclette*, or moto for short, into a portmanteau with "cross-country." The sport has since evolved with sub-disciplines such as stadium events known as supercross and arenacross held in indoor arenas. Classes were also formed for ATVs. Freestyle motocross (FMX) events where riders are judged on their jumping and aerial acrobatic skills have gained popularity, as well as supermoto, where motocross machines race both on tarmac and off-road. Vintage motocross (VMX) events take place usually for motorcycles predating 1975.

Fig. 6.2 A motocross rider coming off a jump. Photo by Adriskala

An enduro motorcycle is an off-road racing motorcycle used in enduros, which are long-distance cross-country time-trial competitions.

Enduro motorcycles closely resemble motocross or "MX" bikes (upon which they are often based). They sometimes have special features such as oversized gas tanks, engines tuned for reliability and longevity, sump protectors, and more durable (and heavier) components. Enduro bikes combine the long-travel suspension of an off-road motocross bike with engines that are reliable and durable over long distances. Some enduro bikes have street-legal features such as headlights and quiet mufflers to enable them to be used in public roadways. The engine of an enduro bike is usually a single-cylinder two-stroke between 125 cc and 360 cc or four-stroke between 195 and 650 cc.

6.1.4 Snowmobiles

The challenges of cross-country transportation in winter led to the invention of the snowmobile, sometimes called ski-doo or snow scooter, an ATV specifically designed for travel across deep snow where other vehicles floundered. Snowmobiles, also known as snow machines, are vehicles designed for travel and recreation on snow. As they can be operated on snow and ice, they do not require a road or trail and so can be driven over frozen lakes or tundra. Snowmobiling is a sport that many people have taken on as a serious hobby. Older snowmobiles could accommodate two people, but most modern ones are for a single rider. Snowmobiles which can carry two riders are referred to as "two-up" snowmobiles or "touring" models and account for a very small share of the market. Snowmobile engines drive a continuous track at the rear and skis at the front provide directional control. Drivers are not enclosed, and there is normally just a windshield for protection (Fig. 6.3A). While early snowmobiles used rubber tracks, modern tracks are typically made of a Kevlar composite. Originally powered by two-stroke gasoline internal combustion engines, snowmobiles powered by four-stroke engines have more recently entered the market.

Recreational snowmobiling has become popular since the second half of the twentieth century. Riders are called snowmobilers or sledders, and recreational riding can take various forms such as snowcross/racing, trail riding, freestyle, mountain climbing, boondocking, carving, ditchbanging, and grass drags. In the summertime snowmobilers can drag race on grass, asphalt strips, or even across water. Snowmobiles are sometimes modified to compete in long-distance off-road races such as Trevor Erickson's #901 entry in the 2014 Vegas to Reno race.

6.2 History, Diversity, and Participation Numbers

6.2.1 ORVs

One of the first modified ORVs developed in Russia between 1906 and 1916 used an unusual caterpillar track which had a flexible belt instead

Fig. 6.3 (A) A snowmobile. Photo by Greg Gjerdingen. (B) Snowmobile being used by reindeer herders. Photo courtesy of altapulken.no

of interlocking metal segments. It could be fitted to a conventional car or truck to turn it into a half-track suitable for use over rough or soft ground. The system was later used in France on Citroën cars between 1921 and 1937 for off-road and military vehicles. There was a surplus of light ORVs like the Jeep and heavier lorries after Second World War. The Jeeps were popular as utility vehicles and "off-roading" as a hobby was born. When the wartime Jeeps wore out, the Jeep Company began to produce vehicles for the civilian market, when British Land Rover and Japanese Toyota, Datsun/Nissan, Suzuki, and Mitsubishi joined in the manufacture of these ORVs. Popular models included the US Jeep Wagoneer and the Ford Bronco, the British Range Rover, the station wagon-bodied Japanese Toyota Land Cruiser, the Nissan Patrol, and Suzuki Lj's series. Later, during the 1990s, manufacturers started to add more luxuries to bring these ORVs on par with regular cars, and what is now known as the SUV (Sports Utility Vehicle) evolved.

In order to be able to successfully drive off paved surfaces, ORVs need several characteristics: (1) a low ground pressure, so as not to sink into soft ground, (2) ground clearance so that they don't get hung up on obstacles, and (3) to keep their wheels or tracks on the ground so they don't lose traction. In wheeled vehicles this is accomplished by having a suitable balance of large or additional tyres combined with tall and flexible suspension. Tracked vehicles achieve this by having wide tracks and a flexible suspension on the road wheels. Most ORVs have special low gearing which allows the operator to utilise the engine's available power while moving slowly through challenging terrain. Power must be provided to all wheels in order to keep traction on slippery surfaces, so for a typical four-wheel vehicle, this is known as four-wheel drive. Vehicles designed for use both on- and off-road may be designed to be switched between two-wheel drive and four-wheel drive so that the vehicle uses fewer driven wheels when driven on the road.

Table 6.1 shows that the number of people (aged 16 and over) participating in off-highway driving in the USA rose by 16.9 million between 1982 and 2001, while snowmobiling rose by 6.1 million in the same period.

This rising trend in off-highway vehicle driving continued into this century (Table 6.2), with 20.6 million participants between 2005 and 2009, a percentage change of 34.5% (1999–2009).

Table 6.3, however, indicates that in the USA, snowmobiling participation in the 2005–2009 period was 5.5% lower than in the 1999–2001 period (a drop of 0.6 million).

Off-highway driving also made it into Cordell's (2012) list of activities which added more than 100 million participation days between 1999–2001 and 2005–2009 (Table 6.4).

6.2.2 ATVs

The first powered quadricycle was built and sold by Royal Enfield 1893 and resembled a modern ATV-style quad bike. The term "ATV" was originally coined to refer to non-straddle ridden six-wheeled amphibious ATVs used in military contexts. The first three-wheeled ATV was designed in 1967, and larger motorcycle companies like Honda entered the market in 1969. Suspension and lower-profile tyres were introduced in the early 1980s, and in 1982 Honda produced a model which featured both suspension and racks, making it the first utility three-wheeled ATV. The ability to go anywhere on terrain that most other vehicles could not cross soon made them popular with US and Canadian hunters and those just looking for a good trail ride. Soon other manufacturers introduced their own models. Sales of utility machines skyrocketed as high-performance three-wheeler sport models were developed featuring full suspension, 248 cc air-cooled two-stroke engines, five-speed transmission with manual clutch, and a front disc brake.

Due to safety concerns of three-wheelers ceased, there was a ten-year voluntary cease of production between 1987 and 1997: three-wheelers were seen as being more unstable than four-wheelers (although accidents were equally severe in both classes). This led to widespread thought that the three-wheel machines were unregistrable, uninsurable, dangerous, and even illegal, but this was false. A Consumer Product Safety Commission (CPSC) study actually found that three-wheelers were no more dangerous than other

Table 6.1 Trends in number of people ages 16 and older participating in recreation activities by historic period in the USA, 1982–2001 (Source: Cordell 2012, p. 33)

Activity	Total participants			Change
	1982–1983	1994–1995	1999–2001	1982–1983 to 1999–2001
Walk for pleasure	91.9	138.5	175.6	83.7
View/photograph birds	20.8	54.3	68.5	47.7
Day hiking	24.3	53.6	69.1	44.8
Picnicking	83.3	112.2	118.3	35.0
Visit outdoor nature centre/zoo	86.7	110.9	121.0	34.3
Swimming in lakes/streams etc.	55.5	87.4	85.5	30.0
Sightseeing	79.8	117.5	109.0	29.2
Boating	48.6	76.2	75.0	26.4
Bicycling	55.5	77.8	81.9	26.4
Developed camping	29.5	46.5	55.3	25.8
Driving for pleasure	83.3	–	107.9	24.6
Motorboating	33.0	59.5	50.7	17.7
Off-highway vehicle driving	**19.1**	**35.9**	**36.0**	**16.9**
Primitive camping	17.3	31.4	33.1	15.8
Sledding	17.3	27.7	30.8	13.5
Backpacking	8.7	17.0	21.5	12.8
Fishing	59.0	70.4	71.6	12.6
Swimming in outdoor pool	74.6	99.1	85.0	10.4
Canoeing or kayaking	13.9	19.2	23.0	9.1
Downhill skiing	10.4	22.8	17.4	7.0
Snowmobiling	5.2	9.6	11.3	6.1
Horseback riding	15.6	20.7	19.8	4.2
Ice skating outdoors	10.4	14.2	13.6	3.2
Hunting	20.8	25.3	23.6	2.8
Cross-country skiing	5.2	8.8	7.8	2.6
Waterskiing	15.6	22.7	16.0	0.4
Sailing	10.4	12.1	10.4	0.0

Missing data are denoted with "–" and indicate that participation data for that activity were not collected during that time period
Source: NRS 1982–1983 (n = 5757), USDA Forest Service (1995) (n = 17,217), and USDA Forest Service (2001) (n = 52,607)
Note: The numbers in this table are *annual* participant estimates on data collected during the three time periods
1982–1983 participants based on 173.5 million people ages 16+ (U.S. Department of the Interior 1986)
1994–1995 participants based on 201.3 million people ages 16+ (Woods & Poole Economics, Inc. 2007)
1999–2001 participants based on 214.0 million people ages 16+ (U.S. Department of Commerce 2000)

Table 6.2 Trends in number of people ages 16 and older participating in recreation activities in the USA, 1999–2001 and 2005–2009 for activities with between 25 and 49 million participants from 2005 through 2009 (Source: Cordell 2012, p. 37)

	Total participants (*millions*)			Percent participating	Percent change
	1994–1995	1999–2001	2005–2009	2005–2009	1999–2001 to 2005–2009
Visit archaeological sites	36.1	44.0	48.8	20.8	11.1
Off-highway vehicle driving	**35.9**	**36.0**	**48.4**	**20.6**	**34.5**
Boat tours or excursions	–	40.8	46.1	19.6	13.1
Bicycling on mountain/hybrid bike	–	44.0	42.7	18.1	−3.0
Primitive camping	31.4	33.1	34.2	14.5	3.2
Sledding	27.7	30.8	32.0	13.6	3.9
Coldwater fishing	25.1	28.4	30.9	13.1	8.7
Saltwater fishing	22.9	21.4	25.1	10.7	17.2

Missing data are denoted with "–" and indicate that participation data for that activity were not collected during that time period. Percent change was calculated before rounding
Source: USDA Forest Service (1995) (n = 17,217), USDA Forest Service (2001) (n = 52,607), and USDA Forest Service (2009) (n = 30,398)
Note: The numbers in this table are *annual* participant estimates on data collected during the three time periods
1994–1995 participants based on 201.3 million people ages 16+ (Woods & Poole Economics, Inc. 2007)
1999–2001 participants based on 214.0 million people ages 16+ (U.S. Department of Commerce 2000)
2005–2009 participants based on 235.3 million people ages 16+ (U.S. Department of Commerce 2008)

ATVs. Nevertheless, from the mid-1980s, Suzuki took the lead in the development of four-wheeled ATVs, with models from Honda and Yamaha following. Suzuki manufactured a 500 cc liquid-cooled two-stroke engine model which reached a top speed of over 79 mph (127 km/h) in the late 1980s. At the same time, development of utility ATVs was rapidly escalating with 4x4 models being the most popular type with hunters, farmers, ranchers, and workers at construction sites.

Models today are divided between the sport and utility markets. Sport models are generally small, light, two-wheel drive vehicles that accelerate quickly, have a manual transmission, and run at speeds up to approximately 80 mph (130 km/h). Utility models are generally bigger 4WD vehicles with a maximum speed of up to approximately 70 mph (110 km/h). Utility models can haul small loads on attached racks or trailers.

6.2.3 Snowmobiles

The first motor sleigh or "traineau automobile" was developed in the USA in 1915, a snow vehicle using the traditional format of rear track(s) and front skis. Later Ford Model Ts were modified to have their undercarriage replaced by tracks and skis and were called Snowflyers. Snowmobiles are today widely used for travel in arctic regions though the small Arctic population means a correspondingly small market. Most snowmobiles are sold for recreational use in places where snow cover is stable during winter. The number of snowmobiles in Europe and other parts of the world is small but growing. Today there are four large North American snowmobile manufacturers: Bombardier Recreational Products (BRP), Arctic Cat, Yamaha, and Polaris. Modern higher-powered snowmobiles, with engine sizes up to

Table 6.3 Trends in number of people ages 16 and older participating in recreation activities in the USA, 1999–2001 and 2005–2009 for activities with fewer than 15 million participants from 2005 through 2009 (Source: Cordell 2012, p. 40)

	Total participants (*millions*)			Percent participating	Percent change
	1994–1995	1999–2001	2005–2009	2005–2009	1999–2001 to 2005–2009
Kayaking	3.4	7.0	14.2	6.0	103.8
Mountain climbing	9.0	13.2	12.4	5.3	−5.9
Snowboarding	6.1	9.1	12.2	5.2	33.7
Ice skating outdoors	14.2	13.6	12.0	5.1	−11.5
Snowmobiling	**9.6**	**11.3**	**10.7**	**4.5**	**−5.5**
Anadromous fishing	11.0	8.6	10.7	4.5	24.1
Sailing	12.1	10.4	10.4	4.4	−0.4
Caving	9.5	8.8	10.4	4.4	18.4
Rock climbing	7.5	9.0	9.8	4.2	9.5
Rowing	10.7	8.6	9.4	4.0	8.9
Orienteering	4.8	3.7	6.2	2.6	−21.7
Cross-country skiing	8.8	7.8	6.1	2.6	−21.7
Migratory bird hunting	5.7	4.9	4.9	2.1	−1.1
Ice fishing	4.8	5.7	4.8	2.1	−15.5
Surfing	2.9	3.2	4.7	2.0	46.3
Snowshoeing	–	4.5	4.1	1.7	−9.4
Scuba diving	–	3.8	3.6	1.5	−5.6
Windsurfing	2.8	1.5	1.4	0.6	−10.1

Missing data are denoted with "–" and indicate that participation data for that activity were not collected during that time period. Percent change was calculated before rounding.
Source: USDA Forest Service (1995) ($n = 17,217$), USDA Forest Service (2001) ($n = 52,607$), and USDA Forest Service (2009) ($n = 30,398$)
Note: The numbers in this table are *annual* participant estimates on data collected during the three time periods 1994–1995 participants based on 201.3 million people ages 16+ (Woods & Poole Economics, Inc. 2007)
1999–2001 participants based on 214.0 million people ages 16+ (U.S. Department of Commerce 2000)
2005–2009 participants based on 235.3 million people ages 16+ (U.S. Department of Commerce 2008)

1200 cc, can reach speeds of 150 mph (240 km/h), while drag racing snowmobiles can reach speeds in excess of 200 mph (320 km/h).

Recreational snowmobilers look for jumps for aerial manoeuvers. Riders often search for non-tracked, virgin terrain and are known to "trailblaze" or "boondock" deep into remote territory where there is absolutely no visible path to follow. This type of trailblazing is known to be dangerous as contact with buried rocks, logs, and frozen ground can cause extensive damage and injuries. Riders tend to look for large open fields of fresh snow where they can carve. Some riders use extensively modified snowmobiles, customised with aftermarket accessories like handlebar risers, handguards, custom/lightweight hoods, windshields and seats, running board supports, studs, and numerous other modifications that increase power and manoeuvrability.

6.3 Safety and Legal Regulation

6.3.1 ORVs

ORVs are built with higher-ground clearance to enable off-road use and thus have a higher centre of gravity which increases the risk of rollover. When an ORV turns, the vehicle's mass resists the turn and carries the weight forward, thus allowing the traction from the tyres to create a lateral centripetal force as the vehicle continues through the turn. ORVs are more likely to be in rollover accidents than passenger cars. According to a study conducted in the USA, ORVs have twice the fatality rate of cars and have nearly triple the fatality rate in rollover accidents. Of vehicles in the USA, light trucks (including ORVs and SUVs) represent 36% of all registered vehicles. They are involved in about half of the fatal two-vehicle

Table 6.4 Mean and total annual days for activities adding more than 100 million participation days between 1999–2001 and 2005–2009 (Source: Cordell 2012, p. 42)

	1999–2001		2005–2009		Percent change in total days	Change in total days (*millions*)
	Mean annual days (*millions*)	Total annual days (*millions*)	Mean annual days (*millions*)	Total annual days (*millions*)	1999–2001 to 2005–2009	1999–2001 to 2005–2009
View wildflowers/trees	61.2	5739.9	86.8	10,532.2	83.5	4792.3
View natural scenery	56.2	7141.5	77.5	11,608.6	62.6	4467.1
Walk for pleasure	103.2	18,109.3	104.6	20,927.8	36.7	2205.7
View/photograph birds	87.8	6009.3	97.7	8215.0	36.7	2205.7
Visit farm/agricultural setting	29.9	1750.4	48.5	3655.3	108.8	1904.9
View wildlife besides birds/fish	38.5	3630.6	46.7	5509.5	51.8	1878.9
Swimming outdoor pool	23.2	1971.1	25.7	2621.1	33.0	650.0
Off-highway driving	**19.7**	**710.4**	**21.6**	**1048.2**	**47.6**	**337.8**
Visit a beach	10.9	924.0	11.6	1184.2	28.2	260.2
Sightseeing	14.8	1616.5	14.9	1842.5	14.0	226.0
Gathering of family/friends	6.2	970.4	6.8	1179.3	21.5	208.9
Gather mushrooms/berries	10.2	614.3	10.3	733.0	30.1	184.7
Visit a wilderness	8.3	558.7	9.3	783.4	28.1	172.0
Visit a waterside besides beach	11.5	611.4	13.9	783.4	28.1	172.0
Swimming in lakes, streams etc.	12.4	1062.4	12.6	1232.4	16.0	170.0
Visit outdoor nature centre/zoo	5.1	620.9	5.5	736.4	18.6	115.5

Source: USDA Forest (2001) ($n = 52,607$) and USDA Forest Service (2009) ($n = 30,398$). Change in total days may not exactly equal the difference between the two time periods due to rounding

Note: The numbers are annual activity days estimates based on data collected during the two time periods. Mean days are the average annual number of days in which participants engage in an activity. Total annual days (in millions) is the product of the estimated number of participants and the mean annual days

crashes with passenger cars, and 80% of these fatalities are to occupants of the passenger cars.

6.3.2 ATVs

Safety has been a major issue with ATVs due to the high number of deaths and injuries associated with them and the negligible protection offered by the machine. The modern breed of ATVs was introduced in the early 1970s and was almost immediately followed by alarming injury rates for children and adolescents. Based on analysis of the National Trauma Data Bank, ATVs are more dangerous than dirt bikes, possibly due to crush injuries and failure to wear safety gear such as helmets. They are as dangerous as motorcycles, based on mortality and injury scores. More children and women are injured on ATVs, who also present a lower rate of helmet usage (US Consumer Product Safety Commission 2011). Jennissen et al. (2014) claimed that more youth are killed every year in the USA in ATV crashes than on bicycles, and since 2001, one-fifth of all ATV fatalities have involved victims aged 15 years or younger. They administered a cross-sectional survey to 4684 youths aged 11 to 16 years at 30 schools across Iowa from November 2010 to April 2013 (Table 6.5) and found that regardless of rurality, at least 75% of students reported having been on an ATV, with 38% of those riding daily or weekly. Among ATV riders, 57% had been in a crash. Most riders engaged in risky behaviours, including riding with passengers (92%), on public roads (81%), or without a helmet (64%). Almost 60% reported engaging in all three behaviours; only 2% engaged in none.

Fatal accidents typically occur when the ATV rolls over. The correct use of protective equipment can prevent many common injuries. ATV manufacturers recommend at least a suitable approved helmet, protective eyewear, gloves, and suitable riding boots for all riding conditions. Sport or aggressive riders, or riders on challenging terrain (such as those rock crawling or hill-climbing), may additionally opt for a motocross-style chest protector and knee/shin guards for further protection. Proper tyres (suited to a particular terrain) can also play a vital role in preventing injuries.

In the USA, statistics released by the CPSC showed that in 2005, there were an estimated 136,700 injuries associated with ATVs treated in US hospital emergency rooms. In 2004, the latest year for which estimates are available, 767 people died in ATV-associated incidents. According to statistics released by CPSC, the risk of injury in 2005 was 171.5 injuries per 10,000 four-wheel ATVs in use or 1.7%. The risk of death in 2004 was 1.1 deaths per 10,000 four-wheelers in use or 0.01%. ATVs must bear a label from the manufacturer stating that the use of machines greater than 90 cc by riders under the age of 12 is prohibited. However, this is a "manufacturer/CPSC recommendation" and not necessarily state law. The American Academy of Pediatrics and the CPSC recommended that no children under the age of 16 should ride ATVs. A Canadian study stated that "associated injury patterns, severity, and costs to the healthcare system" of paediatric injuries associated to ATVs resemble those caused by motor vehicles and that public policies should reflect this fact. The United States government maintains a website about the safety of ATVs (United States Consumer Product Safety Commission 2018) where safety tips are provided, such as not driving ATVs with a passenger (passengers make it difficult or impossible for the driver to shift their weight, as required to drive an ATV) or not driving ATVs on paved roads (ATVs usually have a solid rear axle with no differential).

In 1988, the ATV Safety Institute (ASI) was formed to provide training and education for ATV riders. The cost of attending the training is low and is free for purchasers of new machines that fall within the correct age and size guidelines. Successful completion of a safety training class is, in many states, a minimum requirement for minor-age children to be granted permission to ride on state land. Some states have had to implement their own safety training programmes, as the ASI programme cannot include those riders with ATVs outside of the age and size guidelines, which may still

Table 6.5 ATV exposure and riding behaviours of pupils in Iowa schools from 2010 to 2013 ($n = 4320$)

Exposure/behaviour	All, no. (%)	Male, no. (%)	Female, no. (%)	P value[a]	Isolated rural, no. (%)	Rural, no. (%)	Urban, no. (%)	P value[a]
Exposed to ATVs								
Yes	3344 (77)	1620 (79)	1626 (76)	0.095	1196 (82)	1267 (75)	881 (76)	<0.001
No	976 (23)	442 (21)	502 (24)		269 (18)	429 (25)	278 (24)	
How often do you drive/ride on an ATV?								
Almost daily	666 (20)	378 (23)	263 (16)	<0.001	274 (23)	221 (17)	171 (20)	0.005
About once a week	606 (18)	301 (19)	289 (18)		226 (19)	231 (18)	149 (18)	
About once a month	636 (19)	276 (17)	341 (21)		221 (18)	231 (18)	184 (22)	
Only a few times a year	1436 (43)	665 (41)	733 (45)		475 (40)	584 (46)	337 (40)	
Have you ever ridden or driven an ATV with more than one person?								
Yes	2948 (92)	1409 (90)	1455 (94)	<0.001	1086 (92)	1065 (91)	797 (93)	0.19
No	261 (8)	155 (10)	100 (6)		89 (8)	109 (9)	63 (7)	
Have you ever ridden or driven an ATV on a public road?								
Yes	2534 (81)	1237 (81)	1226 (80)	0.75	889 (82)	952 (79)	693 (81)	0.096
No	604 (19)	292 (19)	298 (20)		191 (18)	255 (21)	158 (19)	
How often do you wear a helmet when riding an ATV?								
Always, almost always	530 (17)	288 (19)	233 (15)	<0.001	185 (14)	163 (22)	182 (22)	<0.001
More than half the time	251 (8)	139 (9)	104 (7)		99 (6)	75 (9)	77 (9)	
Less than half the time	366 (12)	200 (13)	157 (10)		142 (10)	119 (13)	105 (13)	
Never, almost never	2032 (64)	921 (59)	1052 (68)		731 (70)	833 (56)	468 (56)	

Data are provided as no. (column percentages). Column totals may not equal overall population totals because of missing indeterminate responses
Source: Jennissen et al. (2014, p. 312)
ATV = All-terrain vehicle
[a]Chi-square analysis for comparison of proportions

fall within the states' laws. In industry, agriculture workers are disproportionately at risk for ATV accidents. Most fatalities occur in white men over the age of 55.

In the UK a "quad bike" is recognised by law as a vehicle with four wheels and a mass of less than 550 kg. To drive a quad bike legally on a public road, in the UK, requires a B1 licence as well as tax, insurance, and registration. After consultation with stakeholders including farmers and quad bike manufacturers, Australia's Heads of Workplace Safety Authorities (HWSA) in 2011 released a strategy intended to reduce the number of deaths and serious injuries associated with quad-bike use. The strategy encouraged standard safety measures such as helmet wearing, recommended the development of a national training curriculum and point-of-sale material for purchasers, and, controversially, recommended that owners consider fitting of an aftermarket anti-crush device which may offer added protection in the event of a rollover. However, the industry argued that such devices had not been properly tested and that past studies of tractor-style full-frame "cages" around the operator were not only ineffective but could add to the risk to injury or death.

6.3.3 Snowmobiles

Snowmobiles are highly manoeuvrable and can accelerate quickly and reach high speeds. Skill and physical strength are required to operate snowmobiles, and snowmobile injuries and fatalities are high in comparison to road motor vehicle traffic. Losing control of a snowmobile can easily cause extensive damage, injury, or death. One such cause of snowmobile accidents is loss of control from a loose grip. If the rider falls off, the loss of control can easily result in the snowmobile colliding with a nearby object, such as a rock, tree, or other vehicles. Most snowmobiles are fitted with a cord connected to a kill switch, which would stop the snowmobile if the rider falls off; however, not all riders use this device every time they operate a snowmobile. Swerving off of the trail may result in rolling the snowmobile or crashing into an obstacle. Riders unfamiliar with their route have been known to crash into suspended barbed wire or fences (which may be concealed by fresh or blown snow) at high speeds. Each year a number of serious/fatal accidents have been caused by these factors.

Deaths of snowmobile riders occur every year from them colliding with other snowmobiles, automobiles, pedestrians, rocks, trees, or fences, or falling through thin ice. On average, ten people a year have died in such crashes in Minnesota alone, with alcohol a contributing factor in many cases. In Saskatchewan, 16 out of 21 deaths in snowmobile collisions between 1996 and 2000 were caused by the effects of alcohol. In the USA fatal collisions with trains have also occurred when snowmobile operators engage in the illegal practice of "rail riding," riding between railroad track rails over snow-covered sleepers. The inability to hear the sound of an oncoming train over the engine noise of a snowmobile makes this activity extremely dangerous. Collision with large animals such as moose and deer, which may venture onto a snowmobile trail, is another major cause of snowmobile accidents. Most often such encounters occur at night or in low-visibility conditions when the animal could not be seen in time to prevent a collision. A sudden manoeuver to miss hitting an animal crossing a trail can result in the operator losing control of the snowmobile. Many snowmobile deaths in Alaska are caused by drowning. Rivers and lakes are generally frozen over in winter. Riders who operate early or late in the season run the risk of falling through weak ice, and heavy winter clothing can make it extremely difficult to escape the frozen water. While a snowmobile is heavy, it also distributes its weight over a larger area than a standing person, so a driver who has stopped his vehicle out on the ice of a frozen lake can go through the ice just by stepping off the snowmobile. The next leading cause of injury and death in snowmobiling is avalanches, which can result from the practice of "highmarking," or driving a snowmobile as far up a hill as it can go. Risks from avalanches can be reduced through education, proper training, appropriate equipment (such as the wearing of avalanche airbags or transceivers as used by skiers and mountaineers), and attention to published avalanche warnings. It is recommended that snowmobile riders wear a helmet and a snowmobile suit.

6.4 Environmental Impact

In this section we see the environmental impacts of ORVs, ATVs, and motorcycles to be broadly similar, and so we discuss these in one section, whereas those of snowmobiles are in many ways different and, we feel, warrant a separate section.

6.4.1 ORVs, ATVs, and Motorcycles

6.4.1.1 Damage to Soil and Vegetation

In the USA the number of ORV users climbed sevenfold from 5 million in 1972 to 36 million in 2000. Government policies that protect wilderness but also allow recreational ORV use have caused debate across many countries. All trail and off-trail activities impact natural vegetation and wildlife, which can lead to erosion, introduce invasive species on tyres, and cause habitat loss and ultimately species loss decreasing an ecosystem's ability to maintain equilibrium. ORVs cause greater stress to the environment than foot traffic alone, and ORV operators who attempt to test their vehicles against natural obstacles can do significantly more damage than those who simply follow legal trails. Illegal use of ORVs has been identified as a serious land management problem ranked with dumping garbage and other forms of vandalism. Many user organisations, such as Tread Lightly! and the Sierra Club, publish and encourage appropriate trail ethics.

Since ORVs can cover large distances more rapidly than walkers or horse riders, their potential to cause substantial impact on the environment, even on one trip, is greater (Webb and Wilshire 1983; Parikesit et al. 1995). They have the potential to reach remote wilderness areas quickly. The forces created by spinning wheels in association with deep tread tyres can dislodge soil and vegetation rapidly (Cambi et al. 2015a, b). This damage is compounded by the desire of many ORV users to seek out steep, sometimes unstable slopes (Fig. 6.4A, B) where erosion is more easily started (Hammitt and Cole 1998), and once the stable vegetated surface is broken, an erosion cycle begins which can be one of the most significant impacts because of its irreversibility and its tendency to get progressively worse even without continued use. Motorised recreational vehicles can also cause damage to water quality (Marion et al. 2014) particularly where vehicles cross stream banks and enter watercourses (Fig. 6.4C). Damage to stream banks results in eroded soil being deposited in streams which increases sediment loads and turbidity. This extra sediment can clog up fish breeding grounds (e.g. trout and salmon) causing the eggs laid in gravels to be deprived of oxygen, resulting in reduced breeding success. ORVs driving along stream beds (Fig. 6.4C) can also dislodge fish eggs laid in gravels, again resulting in reduced breeding success.

In 2010 Roy Hattersley wrote in the *Guardian* newspaper on how Peak District parkland was under threat from the rise of the 4x4s, stating that off-road motoring was literally wearing parts of the Peak District away. The link to the article is here:

https://www.theguardian.com/uk/2010/mar/21/off-roading-national-park-countryside.

Weaver and Dale (1978) carried out an experimental study in Montana where they examined the effects of horses, hikers, and a lightweight, slowly driven motorcycle on trail erosion. Trails produced by 1000 horse passes were two to three times as wide and 1.5 to 7 times as deep as trails produced by 1000 hiker passes, whereas the effects of the motorcycle were in between these. Soil bulk density increased 1.5 to 2 times as rapidly on horse trails as on hiker trails. The motorcycle was, again, intermediate in severity. However, vegetation loss occurred much more rapidly on horse and motorcycle trails than on hiker trails. Motorcycle damage was greatest when going uphill (Fig. 6.4A), whereas horse and hiker damage was greatest when going downhill.

Anders and Leatherman (1987) examined the effects of ORVs on coastal foredunes at Fire Island, New York, USA. Using sequential quadrat surveys at adjacent control and impact sites over a two-year field study, they found significant loss of vegetation resulting from ORV impacts and concluded that this loss of vegetation resulted in an alteration of the natural foredune profile, which could increase dune erosion during storm wave attack.

Slaughter et al. (1990) reviewed the use of ORVs in permafrost-affected terrain of Alaska which had increased sharply. Until the early 1960s, most ORV use was by industry or government, which employed heavy vehicles such as industrial tractors and tracked carriers. Smaller,

6.4 Environmental Impact

Fig. 6.4 (A) Negative environmental effects caused by a motorcycle to a portion of the Los Padres National Forest, California. Photo by BeenAroundAWhile. (B) Land Rover Series III mud plugging. Photo by AndrewH. (C) A Jeep Grand Cherokee, in action, drives through a watercourse. Photo by DarkSaturos90

commercial ORVs became available in the 1960s, with the variety and number in use rapidly increasing. Wheeled and tracked ORVs, many used exclusively for recreation or subsistence harvesting by individuals, are now ubiquitous in Alaska. This increased use had led to concern over the cumulative effects of such vehicles on vegetation, soils, and environmental variables including offsite values. Factors affecting impact and subsequent restoration include specific environmental setting; vegetation; presence and ice content of permafrost; microtopography; vehicle design, weight, and ground pressure; traffic frequency; season of traffic; and individual operator practices. Approaches for mitigating adverse effects of ORVs include regulation and zoning, terrain analysis and sensitivity mapping, route selection, surface protection, and operator training.

During the 1960s and 1970s, concern about the long-term effect of off-road driving on tundra surfaces increased due to observed damage in northern Alaska and in the Canadian Northwest Territory (Råheim 1992). It was realised that the thermal instability associated with the removal of vegetation could lead to permanent surface modification. Commonly described effects were the deepening of the summer-thawed layer, the creation of permanent depressions due to differential thaw settlement, sediment instability and possible outwash, and scar development on slopes. Stott and Short (1996) studied erosion attributed to mining track in the Gipsdalen Valley in Svalbard (Fig. 6.5A, B). A mining exploration track was established by a Finnish company, which was prospecting for coal and oil in 1985. The track ran for some 18.5 km on the east side

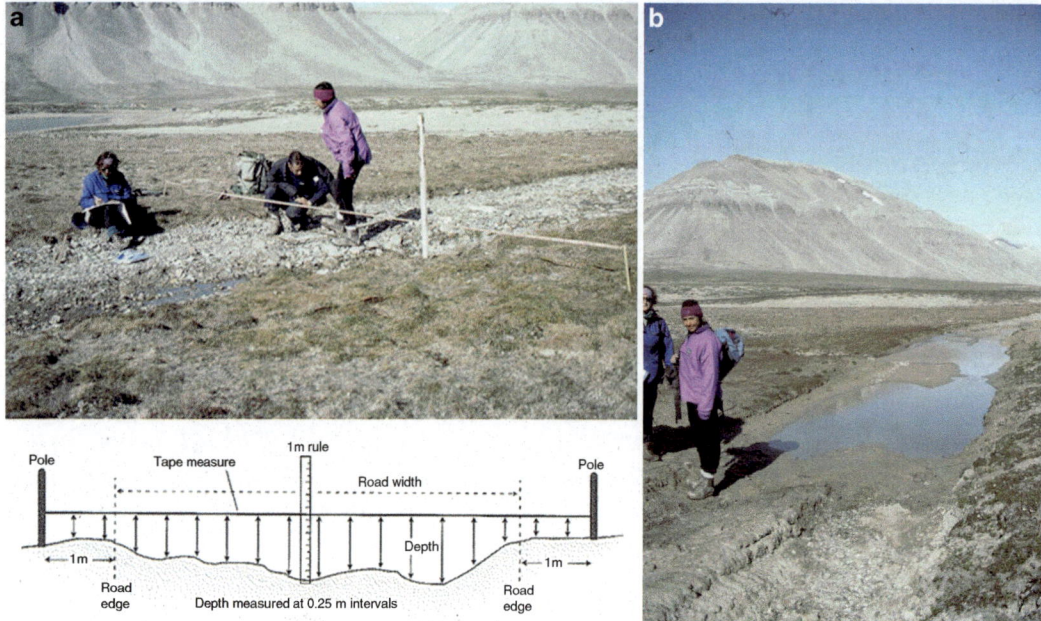

Fig. 6.5 (A) Surveying an eight-year-old vehicular track in Gipsdalen, Svalbard, to measure the eroded cross-section area (Stott and Short 1996). (B) An eight-year-old vehicular track in Gipsdalen, Svalbard, shows how erosion and compaction by ORVs in Arctic tundra can lead to deepening of the summer-thawed layer, the creation of permanent depressions due to differential thaw settlement, waterlogging, and removal of vegetation (Stott and Short 1996). Photos taken by Tim Stott

of Gipsdalen. An erosion survey of 10.25 km of this track was undertaken in summer 1993. It was estimated that approximately 6093 tonnes of soil had been removed due to the presence of this track. Assuming the track had been eroding for eight years, it had produced an average of 762 t yr.$^{-1}$. In order to compare this with the approximate drainage basin erosion rate estimated from fluvially transported sediments, this total was divided by the drainage basin area to give an erosion rate equivalent of 6 t km^{-2} yr.$^{-1}$. In other words, over eight years this track seems to have contributed more than half of the estimated natural background erosion in Gipsdalen. By 1987 the Norwegian government had banned use of the track by vehicles unless it was snow covered and frozen.

Kevan et al. (1995) reported that their examination of the effects of vehicle and pedestrian tracks of known age (13 or more years) and intensity of use (single to multiple passages) on high Arctic tundra vegetation, soil chemistry, soil arthropods, soil thaw characteristics, and small-scale hydrological changes showed clear and interrelated patterns. In general, all tracks, regardless of age, showed small increases in the depth of thaw beneath them (c. 2.8 cm). Tracks were generally depleted of carbon and, to a lesser but significant extent, of potassium and phosphorus. Slight increases in NO_3, NH_4, and calcium were noted. Magnesium and total nitrogen seemed unaffected. On all tracks which had suffered multiple passages, vegetation cover was significantly reduced. In a few sites where single passages were recorded, cover increased through proliferation of the sedge, *Kobresia myosuroides*. Abundance of soil arthropods was significantly reduced on tracks, but the diversity was not. In most sites, soil moisture and overground water flow did not seem affected. Only in sedge meadows where compression from a single passage resulted in channelling of water, and where multiple passages removed vegetation and initiated gulley erosion, were effects serious.

6.4.1.2 Pollution and Noise

ORVs, ATVs, and motorcycles have also been criticised for producing more pollution in areas that might normally have none. In addition to noise pollution that can cause hearing impairment and stress in wildlife, Brattstrom and Bondello (1983) reported on the effects of off-road vehicle noise on desert vertebrates. According to the US Forest Service, old-style two-stroke engines (no longer a component of new ORVs, although some are still in use) emit about 20–33% of the consumed fuel through the exhaust. In 2002, the United States Environmental Protection Agency adopted emissions standards for ATVs that "when fully implemented in 2012… were expected to prevent the release of more than two million tons of air pollution each year—the equivalent of removing the pollution from more than 32 million cars every year." Uberti et al. (2017) proposed the application of an integrated eco-innovation and technical contaminated approach for the design and development of a new concept of off-road motorcycle to meet the requirements of low environmental impact and light weight of the vehicle while maintaining the pleasure of riding in the nature.

The environmental impact of burning fossil fuels, in particular diesel, by ORVs in wilderness areas has led to investigation of alternative fuels such as biofuels. Popovicheva et al. (2015) evaluated the chemical composition of combustion aerosols emitted by off-road engines fuelled by diesel and biofuels. Particles produced by burning diesel, heated rapeseed oil (RO), RO with ethylhexylnitrate, and heated palm oil were sampled from exhausts of representative in-use diesel engines. Their findings provided functional markers of organic surface structure of off-road diesel emission, allowing for a better evaluation of relation between engine, fuel, operation condition, and particle composition, thus improving the quantification of environmental impacts of alternative energy source emissions.

6.4.1.3 Wildlife

The increasing popularity of recreational activities in the wild has led to concerns about their potential impacts on wildlife. ATVs often bring people into wildlife habitats, where they may disturb animal populations. There has been a plethora of studies on the impact of ORVs on wildlife, a few examples of which are summarised here.

ORV use is arguably one of the most environmentally damaging human activities undertaken on sandy beaches worldwide. Studies which focused on areas of high traffic volumes have demonstrated significantly lower abundance, diversity, and species richness of fauna in zones where traffic is concentrated. For example, in the 1970s Busack and Bury (1974) reported that the use of ORVs vehicles on the Mojave Desert eliminated vegetation and adversely affected lizard populations. Wolcott and Wolcott (1984) examined the potential and actual impacts of ORV use on beach macroinvertebrates on the Cape Lookout National Seashore (North Carolina). Mole crabs *Emerita talpoida* and coquinas *Donax variabilis* were not damaged. Ghost crabs *Ocypode quadrata* were completely protected by burrows as shallow as 5 cm and therefore were not subject to injury during the day, but they could be killed in large numbers by vehicles while feeding on the foreshore at night. Ghost crab populations on the seashore were large (10,000 km^{-1} of beach), and a small proportion of the population would be killed by a single vehicle pass. Nevertheless, predicted population mortalities calculated from observed kills of ghost crabs per vehicle-km ranged from 14% to 98% for 100 vehicle passes. Vehicle use on the beach was light, and essentially none occurred on the foreshore after dark. Little impact on beach macroinvertebrates would be expected from this usage pattern. Actual impact on ghost crab populations, assessed by burrow censuses, was negligible. No differences were detected between heavy-use and light-use sites in total population size, average crab size, or population change through the heaviest traffic season. However, increases in traffic to levels seen on other beaches, especially night driving, would probably have devastating effects on ghost crab populations. In heavily used areas, they suggested that banning of ORVs from the foreshore between dusk and dawn may be required to protect this species.

Davies et al. (2016) investigated the impacts of relatively low-level vehicle traffic on sandy beach fauna by sampling invertebrate communities at eight beaches located in south-western Australia. They found that even low-level vehicle traffic negatively impacts the physical beach environment and, consequently, the ability of many species to survive in this habitat in the face of this disturbance. Compaction, rutting, and displacement of the sand matrix were observed over a large area, resulting in significant decreases in species diversity and density and measurable shifts in community structure on beaches that experienced ORV traffic. Communities at impact sites did not display seasonal recovery as traffic was not significantly different between seasons. Given a choice between either reducing traffic volumes or excluding ORV traffic from beaches, their results suggested that the latter would be more appropriate when the retention of ecological integrity is the objective.

Borneman et al. (2016) investigated how ORVs affected nesting behaviour and reproductive success of American oystercatchers. Felton et al. (2017) designed a field experiment to study the responses of nesting American oystercatchers (*Haematopus palliatus*) to off-road passenger vehicles (ORVs) at Cape Hatteras and Cape Lookout National Seashores in North Carolina, USA. Using continuous video and heart rate recordings to assess changes in the behaviour and physiology of incubating oystercatchers, they conducted driving experiments affecting 7 nesting pairs in 2014 and 19 nesting pairs in 2015, between April and July of each year. They concluded that beach-nesting birds may benefit from reduced vehicle traffic at their nesting sites, allowing parents to spend more time attending the nest and less time on defensive behaviours.

In South Africa, ORVs are banned from most coastal areas, while some areas are designated for restricted ORV use, providing an opportunity to assess whether ORV traffic restrictions translate into biological returns. Working in Sodwana Bay, Lucrezi et al. (2014) investigated the impact of ORVs on ghost crab populations. During Easter 2012, ghost crab burrows were counted on beach sections opened and closed to traffic. Burrow density in the impact section was less than a third that of the reference section, and by the end of the study, burrow size in the impact section was half that of the reference section. ORV traffic caused a shift in burrow distribution to the lower beach. However, differences in burrow densities between sections were 14 times smaller than differences obtained at a time when ORV use in Sodwana Bay was not controlled. While confirming the well-established detrimental effects of ORV use on sandy beach ecosystems, these results demonstrated that traffic restrictions on beaches measurably minimise impacts to the fauna, thus translating into worthwhile biological returns.

St-Louis et al. (2013) assessed the influence of ATVs on the behaviour of mountain goats (*Oreamnos americanus*) in a long-term study population at Caw Ridge, Alberta, Canada. They used multinomial models containing environment-, disturbance-, and group-related factors, to evaluate the response of mountain goats to the approach of ATVs. Goats were moderately to strongly disturbed by ATVs 44% of the time, and disturbance levels were mainly influenced by the direction and speed of the approaching vehicles. Environment- or group-related factors (e.g. time of year, distance to escape terrain, group size or type) did not affect mountain goat responses to ATVs. Because goat reactions were influenced by disturbance-level factors, they proposed mitigating measures regarding the use of ATVs in the wild to minimise the disturbance to mountain goats and potentially other alpine ungulates.

6.4.2 Snowmobiles

6.4.2.1 Damage to Soil and Vegetation

Greller et al. (1974) described the effects of 1020 passages of snowmobiles, made over two winters, on three regularly winter snow-free alpine tundra plant communities. A cushion-plant community on a seven-degree slope showed a 31% reduction in total living plant coverage due to snowmobile impact. Destruction was greatest to soil lichens, rock lichens, and the cushion plants *Arenaria obtusiloba*, *Arenaria fendleri*, *Paronychia sessiliflora var. pulvinata*, *Silene acaulis*, *Eritrichium aretioides*, and *Phlox pulvinata*. Graminoids generally survived to increase in importance. On a flat site, a cushion-plant community with *Kobresia myosuroides* as its most important species showed the greatest loss of living plant coverage, namely 46%. This was due primarily to the destruction of *Kobresia*, although *Selaginella densa*, *Arenaria obtusiloba*, *Hymenoxys acaulis*, and *Eritrichium aretioides* also showed heavy losses. In a *Kobresia* turf community, destruction was decidedly less severe than in the cushion-plant communities, reduction in total living plant coverage being only 19%. Greller et al. suggested that the closed nature of the *Kobresia* turf, with its stiff tussocks, enabled it to absorb impact well. They recommended that snowmobile travel be confined to *Kobresia* or similar turfs, when such travel is necessary under snow-free conditions.

Keddy et al. (1979) carried out a study in Nova Scotia, Canada, to experimentally assess the effect of snowmobiles on old field and marsh vegetation. Snowmobile treatments ranging from a single pass to 25 passes (five passes on five separate days) were administered. The first pass by a snowmobile caused the greatest increase in snow compaction—roughly 75% of that observed after five sequential passes. Snowmobile treatment resulted in highly significant increases in snow retention in spring. Frequency was more important than intensity in this regard. Standing crop and species composition were measured the following summer. Standing crop in the field showed a significant reduction with increasing snowmobile use; frequency of treatment ($p < 0.01$) was more important than intensity ($p = 0.125$). *Stellaria graminea*, *Aster cordifolius*, *Ranunculus repens*, and *Equisetum arvense* all showed significant ($p < 0.05$) differences in percent cover resulting from the treatment. Marginally significant changes were observed in *Agrostis tenuis* and *Phleum pratense*. Marsh vegetation showed no significant effects of snowmobile treatment. This may have been because of solid ice cover during the winter. Overall it was concluded that snowmobile use could have a highly significant effect upon natural vegetation.

6.4.2.2 Pollution and Noise

There has been interest in snowmobile emissions for several decades. For example, Hare et al. (1974) described a research programme on exhaust emissions from snowmobile engines, including both emissions characterisation and estimation of national emissions impact. Tests were conducted on three popular two-stroke twins and on one rotary (Wankel) engine. Emissions that were measured included total hydrocarbons, (paraffinic) hydrocarbons by NDIR, CO, CO_2, NO (by two methods), NO_x, O_2, aldehydes, light hydrocarbons, particulate, and smoke. Emissions of SO_x were estimated on the basis of fuel consumed, and evaporative hydrocarbons were projected to be negligible for actual snowmobile operation. Based on test results and the best snowmobile population and usage data available, impact of snowmobile emissions on a national scale was computed to be minimal.

However, during a two-month monitoring campaign in 2007 in the Arctic town of Longyearbyen (Spitsbergen, Svalbard), Reimann et al. (2009) measured the aromatic hydrocarbons benzene, toluene, and C_2-benzenes (ethyl benzene and m-,p-,o-xylene) (BTEX). Reflecting the remoteness of the location, very low mixing ratios were observed during night and in windy conditions. In late spring (April–May), however, the high frequency of guided snowmobile tours resulted in "rush-hour" maximum values of more than 10 ppb of BTEX. These concentration levels are comparable to those in European towns and are caused predominately by the outdated two-stroke engines, which are still used by approximately 30% of the snowmobiles in Longyearbyen. During summer, peak events were about a factor of 100 lower compared to those during the snowmobile season. Emissions in summer were mainly caused by diesel-fuelled heavy-duty vehicles (HDVs), permanently used for coal transport from the adjacent coal mines. Reimann et al. concluded that these documented high BTEX mixing ratios from snowmobiles in the Arctic provide an obvious incentive to change the regulation practice to a cleaner engine technology.

Discharge of unburnt fuel from two-stroke snowmobile engines can lead to indirect pollutant deposition into the top layer of snow and subsequently into the associated surface and groundwater. As early as the 1970s, Adams (1975) reported that lead and hydrocarbons from snowmobile exhaust were found in the water at high levels during the week following ice-out in a Maine pond. Fingerling brook trout (*Salvelinus fontinalis*) held in fish cages in the pond showed lead and hydrocarbon uptake. These contaminants accumulated during the previous winter when snowmobile operation on the pond was equivalent to one snowmobile burning 250 litres of fuel per season on a 0.405 hectare pond with average depth of 1 m. Lead content of the water rose from 4.1 ppb before snowmobiling to 135 ppb at ice-out, with exposed trout contained 9 to 16 times more lead than controls. Hydrocarbon levels undetectable prior to snowmobiling reached 10 ppm in the water and 1 ppm in exposed fish. Trout held in aquaria for three weeks in melted snow containing three different concentrations of snowmobile exhaust also showed lead and hydrocarbon uptake. Their digestive tract tissue contained the most lead (2 ppm) and gills the least (0.2 ppm). Stamina, as measured by the ability to swim against a current, was significantly less in trout exposed to snowmobile exhaust than in control fish.

Maximum noise restrictions have been enacted by law for both production of snowmobiles and aftermarket components. For instance, in Quebec (Canada) noise levels must be 78 decibels or less at 20 m from a snowmobile path. As of 2009, snowmobiles produce 90% less noise than in the 1960s, but there are still numerous complaints. Efforts to reduce noise focus on suppressing mechanical noise of the suspension components and tracks. Arctic Cat in 2005 introduced "Silent Track technology" on some touring snowmobile models.

According to Mullet et al. (2017), snowmobiling in congressionally designated wilderness (CW) in Alaska is a contentious issue in the arena of appropriate use of public lands. The 1980 Alaska National Interests Lands Conservation Act permits snowmobiling in CW for traditional activities. Conversely, the 1964 Wilderness Act prohibits motor vehicles in CW to preserve its naturalness and opportunities for solitude. These conflicting mandates challenge the ability of managers to preserve CW character. The Kenai National Wildlife Refuge (KENWR) manages 534,300 ha of CW, where 253,200 ha are open to snowmobiling. Snowmobile noise degrades CW character, whereas natural quiet is indicative of naturalness and offers opportunities for solitude. In their study Mullet et al. (2017) determined the acoustic footprint of snowmobile noise and areas of natural quiet refugia in CW by recording the soundscape at 27 locations inside, and 37 locations outside, KENWR CW. They calculated soundscape power (normalised watts/kHz) from 59,598 sound recordings and generated spatial models of snowmobile noise and natural quiet using machine learning (TreeNet). They calculated the area of CW with the highest and lowest soundscape power for snowmobile noise and natural quiet, respectively. Snowmobile noise occurred during daylight hours, while natural quiet was predominant at night. Snowmobile noise was higher in February and March, while January was quieter. Snowmobile noise affected 39% of CW open to snowmobiling, while natural quiet made up 36%. Natural quiet occurred in 51% of all KENWR CW of which 39% was prohibited by management or inaccessible by snowmobiles. These models identified areas where conservation of winter soundscapes in CW should be focused (Fig. 6.6).

6.4.2.3 Wildlife

In an observational experiment, Fuglei et al. (2017) examined the impact of snowmobile traffic on the diurnal activity of Arctic fox in high Arctic Svalbard. They conducted the study in two areas in Svalbard, one control area with low snowmobile traffic and one experimental area with high snowmobile traffic. In each area ten camera traps, baited with reindeer carcasses, were positioned and programmed to take photographs every five minutes. The proportion of photographs with foxes was higher during the night than during the day, and the difference

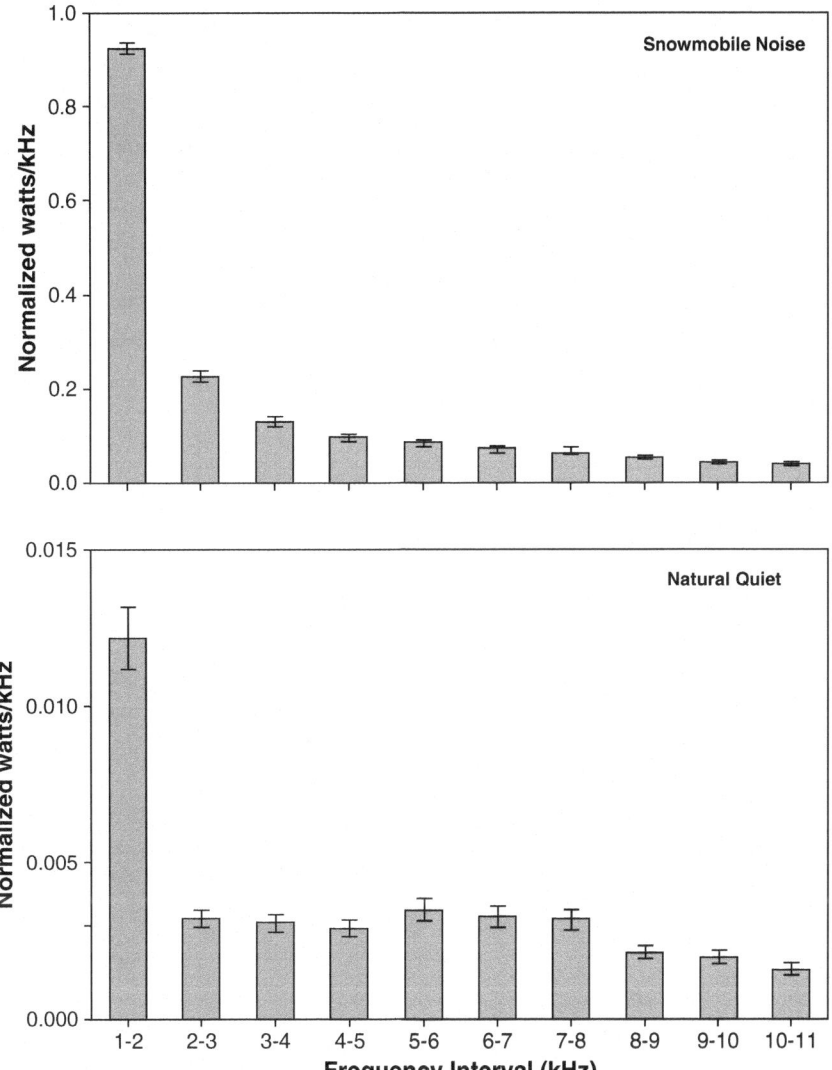

Fig. 6.6 Average soundscape power (normalised watts/kHz) and 95% confidence intervals within ten 1-kHz frequency intervals summarised for snowmobile noise and natural quiet identified from 59,598 sound recordings acquired over winter (December 2011–April 2012) in Kenai National Wildlife Refuge, Alaska. Source: Mullet et al. (2017)

between night and day was larger in the area with more snowmobile traffic. By using data obtained according to a similar study design in two Arctic Russian sites, Yamal and Nenetsky, with little human activity and low snowmobile traffic, they were able to compare Arctic fox activity patterns in Svalbard on a larger scale. Their results indicated that snowmobile traffic had an impact on the diurnal activity of the Arctic fox in Svalbard, while there were no obvious diurnal activity patterns among Russian foxes. Even the area with low snowmobile traffic in Svalbard showed increased use of the reindeer carcasses during the night compared to one of the Russian sites, where foxes used carcasses equally during day and night. Such knowledge is of importance in designing cautious management practices.

Andersen and Aars (2008) measured the distance at which polar bears detected and actively responded to approaching snowmobiles on Svalbard, an arctic island where human traffic had increased substantially in recent years. Fieldwork was conducted in April and/or May during the years 2003–2005. Polar bears were observed on ice with telescopes and binoculars. Undisturbed polar bears were observed continuously and their behaviours recorded, during the time when two snowmobiles moved towards the bear(s). Distances between the bear, the observer, and the approaching snowmobiles were measured using GPS positions taken on the track towards the bear. The responses of the polar bears to the snowmobiles were categorised according to intensity and persistence of reactions. Females with cubs and single medium-sized bears tended to show more intense responses than adult males and lone adult females. Wind direction affects sound and odour transmission, and although an effect on response distance was not found, the response intensity was affected by wind direction. They concluded that female polar bears with small cubs in particular may have a greater risk to being disturbed, since they reacted at greater distances with amplified reactions; thus, users of snowmobiles should take particular care in areas where females with cubs are present.

6.5 Management and Education

6.5.1 Introduction

In 1987 a report was published by the Royal Society for Nature Conservation in the UK entitled "Damage to Wildlife Sites by Off-Road Motor Vehicles" (RSNC 1987) which reported on a survey which had identified at least 130 sites of wildlife importance had been damaged by ORVs. Over half of the sites were Sites of Special Scientific Interest (SSSIs), and many had other conservation designations. The most affected habitats were grasslands and woodlands, followed by heathlands, saltmarsh, and moorland. It concluded that the following measures were needed to alleviate this damage to wildlife interests:

1. Education of the participants, so that they understood the damage that they could cause
2. Provision of alternative sites where ORV could be used without causing problems
3. A requirement for normal planning permission requirements to apply if a site was to be used for any off-road events
4. Stronger measures to prevent third-party damage to SSSIs

Measures to control ORVs used by conservation bodies included the erection of physical barriers like fences and gates the digging of ditches or erection of signs. However, it was recognised that physical barriers were not usually effective as determined riders could usually find their way onto a site. Many sites used a warden which was more effective as they were able to educate the riders rather than just imposing a prohibition without explanation. However, provision of wardens was seen to be expensive. Only in a minority of cases were conservation bodies able to reach agreements with a club or to suggest an alternative site. For some riders/drivers, the notion of using an alternative site can be a problem if the site is some distance away. The police had become involved at around half of the 130 damaged sites noted in the RSNC report, and the apprehension of just one or two persistent offenders by the police soon becomes news and can discourage others. However, police resources are stretched, and it is not always possible to access police support wherever and whenever it is needed. ORV damage is quite low on the list of police priorities.

6.5.2 Legal Controls

In the UK informal access to land by ORVs is controlled, in theory, by the Road Traffic Act 1972 (section 36) which specifies that it is a criminal offence to drive any motor vehicle off-road without the landowner's permission (except within 15 m of the centre of a road for parking, for the saving of life, fire, or other emergencies). However, in practice the temporary use of land for organised scrambling or ORV events is permitted for short periods, with only activities lasting more than 14 days per year requiring planning permission. Problems can arise where there is informal use of a site by individuals without consent of the owner, who in some cases may turn a blind eye in the hope that the damage to the site will eventually lead to the removal of an SSSI designation.

In the USA the Bureau of Land Management (BLM) supervised several large ORV areas in California's Mojave Desert. In what has become known as the Mojave Desert controversy in 2009, US District Judge Susan Illston ruled against the BLM's proposed designation of additional off-road use on designated open routes on public land. According to the ruling, the BLM violated its own regulations when it designated approximately 5000 miles of ORV routes in 2006. According to Judge Illston, the BLM's designation was "flawed because it did not contain a reasonable range of alternatives" to limit damage to sensitive habitat, as required under the National Environmental Policy Act. Illston found that the bureau had inadequately analysed the route's impact on air quality, soils, plant communities, and sensitive species such as the endangered Mojave fringe-toed lizard, pointing out that the United States Congress has declared that the California desert and its resources are "extremely fragile, easily scarred, and slowly healed." The court also found that the BLM failed to follow route restrictions established in the agency's own conservation plan, resulting in the establishment of hundreds of illegal ORV routes during the previous three decades. The plan violated the BLM's own regulations, specifically the Federal Land Policy and Management Act of 1976 (FLPMA) and the National Environmental Policy Act of 1969 (NEPA). The ruling was considered a success for a coalition of conservation groups including the Friends of Juniper Flats, Community Off-Road Vehicle Watch, California Native Plant Society, the Centre for Biological Diversity, the Sierra Club, and The Wilderness Society who initiated the legal challenge in late 2006.

Many US national parks have discussed or enacted roadless rules and partial or total bans on ORVs. To accommodate enthusiasts, some parks like Big Cypress National Preserve in Florida were created specifically for ORVs and related purposes. However, such designations have not prevented damage or abuse of the policy. In 2004, several environmental organisations sent a letter to Dale Bosworth, Chief of the United States Forest Service, and described the extent of damage caused by ORV use, including health threats to other people. It was articulated that the proliferation of ORV and snowmobile use placed soil, vegetation, air and water quality, and wildlife at risk through pollution, erosion, sedimentation of streams, habitat fragmentation and disturbance, and other adverse impacts to resources. These impacts were causing severe and lasting damage to the natural environment on which human-powered and equestrian recreation depended and altered the remote and wild character of the backcountry. Motorised recreation sometimes monopolised forest areas by denying other users the quiet, pristine, backcountry experience they seek. It also presented safety and health threats to other recreationists. In 2004 the Supreme Court Justice Antonin Scalia listed several problems that result from ORV use in natural areas. From the Environment News Service article, Scalia noted that ORV use on federal land has "negative environmental consequences including soil disruption and compaction, harassment of animals, and annoyance of wilderness lovers."

A number of environmental organisations, including the Rangers for Responsible Recreation, continue to campaign to draw attention to a growing threat posed by ORV misuse and to assist land managers in addressing ORV use impacts. These campaigns in part have prompted congressional hearings, and key to the discussions was the "travel planning process," a complex analysis and decision-making procedure with the aim of designating appropriate roads and trails. Both the Forest Service and the BLM have identified unmanaged recreation—including ORV use—as one of the top four threats to the management and health of the National Forest System.

Adams and McCool (2009) outlined how the Forest Service and the BLM were revising their local travel management plans. These plans govern much of the allocation of recreation experience opportunities, including the balance between ORV and non-motorised opportunities. Their paper explored the historic management of ORVs by the Forest Service and the BLM, as well as laws and regulations governing ORV management, in order to (1) explain how Forest Service and BLM travel management works, (2) evaluate the ORV and non-motorised allocations for multiple-use lands, and (3) provide suggestions for improved agency management of ORVs. Ultimately, concerns regarding appropriate allocations, the escalating conflicts between recreationists, increasing demand for outdoor recreation, the rising stakes associated with allocation decisions, and the plainly political nature of allocation decisions all point to a better, long-term solution: a new statutory recreation policy for multiple-use lands.

6.5.3 Managing ORV Use and Experiences

While most research on ORV use has focused on its potential environmental impacts and conflicts with other types of recreation, Hallo et al. (2009) examined the ORV experience itself. The concept of indicators and standards of quality has emerged in the literature as a conceptual framework for understanding and managing outdoor recreation. Their study applied the concept of indicators of quality to ORV use at Cape Cod National Seashore (Cape Cod). Qualitative interviews were conducted during the 2004 use season with 61 ORV users at Cape Cod to gather information on indicators of quality for the ORV experience. A content analysis was performed on the transcripts, and results suggested that crowding, the portion of the ORV route open, ease of obtaining a permit, amount of litter, availability of support facilities, and the behaviour and actions of ORV users are potential indicators of quality. The portion of the ORV route open may be a less useful indicator because it is not easily adaptable due to current legislative and regulatory guidance for the management of threatened shorebirds. The other variables are better potential indicators because they may be more readily measured and managed.

New research methods such as the use of airborne remote sensing (e.g. Dewidar et al. 2016) and GIS (Geographical Information Systems) (e.g. Rybansky 2014; Westcott and Andrew 2015; Kobryn et al. 2017) offer managers concerned with ORV impacts new tools for gathering data and informing planning and decision-making.

The tremendous growth of recreational ATV use in Canada and the USA has led to rapidly increasing pressure on local authorities and provincial/state governments to either sanction increased access to ATVs or restrict their use on community trails and local roadways. Given this increased pressure, there is a growing need for a policy development tool to assist decision-makers in making prudent policy decisions that carefully balance special interest lobbying with the broader public interest, whether that be at the local or community level or in the broader context of provincial or state legislative policy. Bissix (2015) presented a decision support framework that guided policy decision-makers to consider a broad range of health and safety factors along with environmental, social, and economic considerations. He hoped that by using this multidimensional assessment framework in an open and transparent way, policy actors would be encouraged to rely on defensible scientific evidence and best practices rather than react to the vociferous advocacy of policy champions.

SAE International, the Society of Automotive Engineers, is a US-based, globally active professional association and standards developing organisation for engineering professionals in various industries. Each year it opens a competition called the SAE Clean Snowmobile Challenge. The University of Idaho's entry into the 2016 SAE Clean Snowmobile Challenge (Savage et al. 2016) was a 2013 Ski-Doo MXZ-TNT chassis with a reduced-speed 797 cc direct injected two-stroke engine, modified for flex fuel use on blended ethanol fuel. A battery-less direct injection system was used to improve fuel economy and decrease emissions while maintaining a high power-to-weight ratio. A new tuned exhaust was designed to accommodate the lowered operating speed of the engine while improving the peak power output from 65.6 kW (88 hp) to 77.6 kW (104 hp). Noise was reduced through the implementation of a mechanically active quarter-wave resonator, Helmholtz resonator, and strategically placed sound-absorbing/deadening materials and by operating the engine at a lower speed. A muffler was modified to incorporate a three-way catalyst, which reduces engine emissions while not significantly reducing power output or increasing sound output. Such developments are promising and show that the manufacturing industry is making efforts to minimise the environmental impacts of snowmobiles.

6.5.4 Education and ORV Users

Like naturalists, motor sports enthusiasts are legitimate users of the countryside, but they need to recognise that their activities can cause damage and conflicts with other interests. Such recognition can be brought about by education of the participants by interested parties. Once ORV users begin to realise the problems they can cause on inappropriate sites, they may be more willing to move to alternative areas. However, there may still need to be strong protective measures for the most important nature conservation sites.

It appears that many users of ORVs regard areas of wildlife habitat, even nature reserves, simply as areas of derelict land, and this ignorance in many cases is genuine. The majority of ORV users cannot be accused of mindless vandalism; they are often genuinely concerned about the problems they may have caused and may not be aware of the damage done until it is pointed out to them. The more responsible and organised motorsport participants will belong to an organisation or club. For example, in the UK, Offroad Motorsport UK is a professional organisation which understands the requirements for holding events whether they are motocross, trials, enduro, practice events, or other off-road activities and can offer their services and insurance at competitive costs. Their aim is to promote good family events, and to do this they issue their members with an exemption permit under the Motor Vehicle (off-road events) Regulations 1995. Most genuine off-road motorsports clubs in the UK are affiliated to Offroad Motorsport UK. Organisations like this ensure that their affiliated clubs and their members are well aware of the damage that can be done to wildlife by their activities and, where possible, encourage members to enter organised events where measures are taken to control their impact and operate legally.

It is more difficult, however, to get the nature conservation message across to the large body of off-road motor sport participants who do not belong to recognised clubs or organisations. Articles in motorsport magazines can reach some as can websites and social media or even reports published by pressure groups (see, e.g., Fig. 6.7). Another widely effective measure is for manufacturers or retailers of machines intended to be used off-road to provide point-of-sale information about specific rights and responsibilities of those riding or driving off-road in the countryside (including nature conservation information). This, of course, does not necessarily reach people buying ORVs in the second-hand market, however.

In response to issues and challenges facing the operators of nature-based tours, Lyngnes and Prebensen (2014) examined snowmobile tours in Svalbard, Norway. They explored the awareness of the tour guides of the fragile nature of the environment in which they were operating and how this awareness was implemented in their offerings. The results showed that the guides were aware of the fragile nature and did strive to promote sustainable behaviour during tours. In particular, they focused on informing and teaching the tourists about environmental aspects of the tour through storytelling and staging during the tour. By empowering the tourists through education and involvement, they aimed to make the tourists change their focus from riding the snowmobile to learning about the fragility of the nature and wildlife. Further, the guides stated that the tourists may even become spokespersons for sustainable tourism due to the tour which they had experienced.

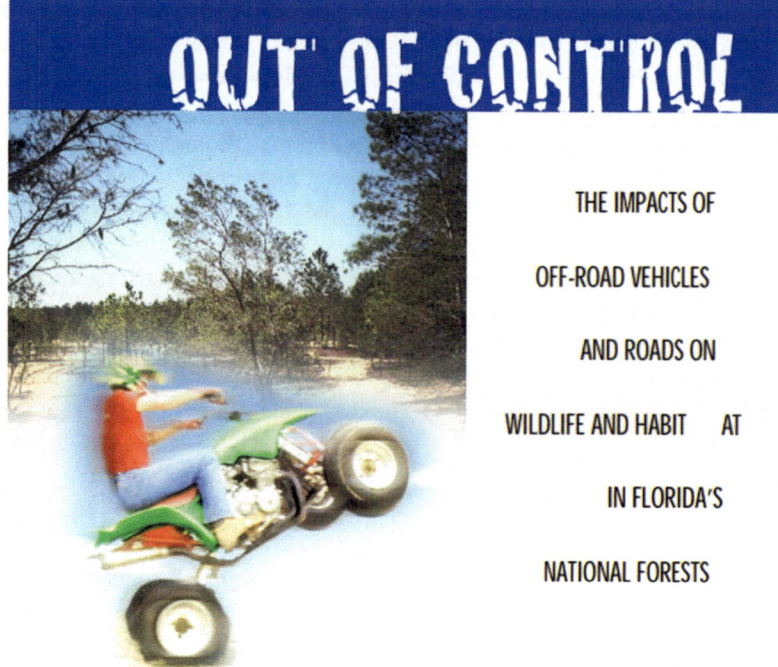

Fig. 6.7 Out of control: the impact of ORVs and roads on wildlife and habitat in Florida's national forests: an educational publication by Defenders of Wildlife. Source: https://defenders.org/sites/default/files/publications/out_of_control_orvs_and_floridas_forests.pdf

6.6 Future Trends

Table 6.6 shows the 2008 number of participants, participation rate, days, and days per participant for motorised activities (motorised off-road use and motorised snow use) as compared with other outdoor recreation activities in the USA.

Bowker et al. (2012) used sophisticated modelling procedures to projected motorised off-road participation and use (off-road driving) by adult US residents, 2008–2060 (Table 6.7).

According to Bowker et al. (2012) participation in off-road driving averaged about 20% of the adult US population, or about 48 million adults annually between 2005 and 2009 (Table 6.7). Based on the projections made by Bowker et al. (2012), future participation rates in off-road driving are expected to decline under two of three Resources Planning Act (RPA) scenarios, RPA A2 (16–18%) and RPA B2 (7–8%), while the percent of adult participants under RPA A1B will be about the same in 2060 as today. The relatively larger decline in participation rate under RPA A2 can be attributed to smaller income growth than under RPA A1B and a larger decline in federal and private forest and rangeland than under either RPA B2 or A1B. Despite the static or declining rate of growth in per capita participation, the number of participants in off-road driving will increase between 28% and 58% under the assessment scenarios to somewhere between 62 and 75 million people, because the rate of population growth will outstrip any decline in per capita participation through 2060. Alternative climate futures do not appear to have an appreciable effect on participation percentages or actual numbers. Annual days per off-road driving participant are projected to decline by 3–7%, or about 1.4 participant days, annually by 2060 (Table 6.7). The decline, consistent across the

Table 6.6 Outdoor recreation activities for 2008 by participants, participation rate, days, and days per participant

Activity[a]	Participants (millions)[b]	Percent participating	Days (millions)[b]	Days per participant
Visiting developed sites				
Developed site use—family gatherings, picnicking, developed camping	194	82	2246	11.7
Visiting interpretive sites—nature centres, zoos, historic sites, prehistoric sites	158	67	1249	7.8
Viewing and photographing nature				
Birding—viewing/photographing birds	82	35	8255	97.7
Nature viewing—viewing, photography, study, or nature gathering related to fauna, flora, or natural settings	190	81	32461	169.6
Backcountry activities				
Challenge activities—caving, mountain biking, mountain climbing, rock climbing	25	11	121	4.8
Equestrian	17	7	263	16.3
Hiking—day hiking	79	33	1835	22.9
Visiting primitive areas—backpacking, primitive camping, wilderness	91	38	1239	13.2
Motorised activities				
Motorised off-road use	48	20	1053	21.6
Motorised snow use	10	4	69	7.3
Motorised water use	62	26	958	15.3
Hunting and fishing				
Hunting—small game, big game, migratory bird, others	28	12	538	19.1
Fishing—anadromous, cold water, saltwater, warm water	73	31	1369	18.5
Non-motorised winter activities				
Downhill skiing, snowboarding	24	11	178	7.2
Undeveloped skiing—cross-country, snow-shoeing	8	3	52	6.6
Non-motorised water activities				
Swimming, snorkelling, surfing, diving	144	61	3476	24.0
Floating—canoeing, kayaking, rafting	40	17	262	6.5

Source: Bowker et al. (2012, p. 3); National Survey of Recreation and the Environment (NSRE) 2005–2009, Versions 1 to 4 (Jan 2005 to Apr 2009) ($n = 24,073$) (USDA Forest Service 2009)
[a]Activities are individual or activity composites derived from the NSRE. Participants are determined by the product of the average weighted frequency of participation by activity for NSRE data from 2005 to 2009 and the adult (>16) population in the USA during 2008 (235.4 million)
[b]Because of small population and income difference, initial participant and day values for 2008 differ across Resources Planning Act scenarios; thus an average is used for a starting value

RPA scenarios, is invariant to climate alternatives. The declines in participation rate and average annual days per participant imply that, under all scenarios, the total number of days of off-road driving will increase by less than the respective population growth rates. Nevertheless, RPA A1B yields a potential increase of about 500 million days of off-roading per year by 2060, while RPA B2 implies an increase of a little more than 200 million days.

Bowker et al. (2012) also included motorised snow use (snowmobiling) in their data and projections. Snowmobiling was noted to be a geographically limited activity undertaken by 4% of the adult population, or nine to ten million people in 2008. Per capita participation in snowmobiling is expected to decline between 13% and 72% under all assessment and climate scenarios (Table 6.8).

Table 6.9 summarises the changes in total outdoor recreation participants in the USA between

Table 6.7 Projected motorised off-road participation and use (off-road driving) by adult US residents, 2008–2060, by Resources Planning Act (RPA) scenario and related climate futures

RPA scenario	2008	2060 No CC	2060 No CC[a]	2060 Climate 1[b]	2060 Climate 2[c]	2060 Climate 3[d]
	Per capita participation		**Percent increase (decrease) from 2008**			
A1B	0.204	0.203	0	(1)[e]	1	1
A2	0.204	0.169	(18)	(18)	(18)	(16)
B2	0.204	0.189	(8)	(7)	(7)	(8)
	Adult participation (millions)		**Percent increase (decrease) from 2008**			
A1B	47.9	75.0	56	55	57	58
A2	48.8	70.2	44	42	42	45
B2	47.8	61.7	29	29	29	28
	Days per participant		**Percent increase (decrease) from 2008**			
A1B	21.6	20.2	(6)	(6)	(3)	(3)
A2	21.6	20.2	(7)	(5)	(4)	(4)
B2	21.6	20.3	(6)	(5)	(5)	(5)
	Total days (millions)		**Percent increase (decrease) from 2008**			
A1B	1048	1532	46	46	53	53
A2	1066	1433	34	36	36	39
B2	1045	1264	21	23	22	21

Source: Bowker et al. (2012, p. 20)
[a]Climate variable omitted from model and projection
[b]Climate 1 uses forecast data from CGCM3.1 for scenarios A1B and A2; CGCM2 for scenario B2
[c]Climate 2 uses forecast data from CSIRO-MK3.5 for scenarios A1B and A2; CSIRO-MK2 for scenario B2
[d]Climate 3 uses forecast data from MIROC3.2 for scenarios A1B and A2; UKMO-HADCM3 for scenario B2
[e]Parentheses denote a decrease

Table 6.8 Projected motorised snow activity participation and use (snowmobiling) by adult US residents, 2008–2060, by Resources Planning Act (RPA) scenario and related climate futures

RPA scenario	2008	2060 No CC	2060 No CC[a]	2060 Climate 1[b]	2060 Climate 2[c]	2060 Climate 3[d]
	Per capita participation		**Percent increase (decrease) from 2008**			
A1B	0.04	0.035	(13)[e]	(32)	(49)	(72)
A2	0.04	0.031	(23)	(60)	(43)	(69)
B2	0.04	0.032	(21)	(49)	(46)	(51)
	Adult participation (millions)		**Percent increase (decrease) from 2008**			
A1B	9.44	12.99	37	6	(20)	(56)
A2	9.60	12.94	35	(31)	1	(46)
B2	9.42	10.39	10	(29)	(25)	(32)
	Days per participant		**Percent increase (decrease) from 2008**			
A1B	7.25	7.04	(3)	(10)	(24)	(24)
A2	7.25	6.95	(4)	(9)	(18)	(22)
B2	7.25	7.12	(2)	(13)	(14)	(13)
	Total days (millions)		**Percent increase (decrease) from 2008**			
A1B	68.4	91.0	33	(6)	(40)	(67)
A2	69.6	89.8	29	(44)	(17)	(58)
B2	68.3	73.8	8	(38)	(36)	(41)

Source: Bowker et al. (2012, p. 21)
[a]Climate variable omitted from model and projection
[b]Climate 1 uses forecast data from CGCM3.1 for scenarios A1B and A2; CGCM2 for scenario B2
[c]Climate 2 uses forecast data from CSIRO-MK3.5 for scenarios A1B and A2; CSIRO-MK2 for scenario B2
[d]Climate 3 uses forecast data from MIROC3.2 for scenarios A1B and A2; UKMO-HADCM3 for scenario B2
[e]Parentheses denote a decrease

Table 6.9 Changes in total outdoor recreation participants between 2008 and 2060 across all activities and scenarios

Activity[a]	2008 Participants[b] (millions)	2060 Participant range[c] (millions/[percent])	2060 Average participant change (millions)	2060 Participant range[d] (millions/[percent])	2060 Average participant change[d] (millions)
Visiting developed sites					
Developed site use—family gatherings, picnicking, developed camping	194	273–346 [42–77]	116	271–339 [40–75]	112
Visiting interpretive sites—nature centres, zoos, historic sites, prehistoric sites	158	231–294 [48–84]	106	231–289 [46–83]	104
Viewing and photographic nature					
Birding—viewing/photographing birds	82	118–149 [42–76]	53	115–144 [40–76]	47
Nature viewing—viewing, photography, study, or nature gathering related to fauna, flora, or natural settings	190	267–338 [42–76]	114	268–333 [41–75]	112
Backcountry activities					
Challenge activities—caving, mountain biking, mountain climbing, rock climbing	25	38–48 [50–86]	19	37–48 [47–90]	18
Equestrian	17	24–31 [44–87]	11	25–35 [50–110]	13
Hiking—day hiking	79	117–150 [50–88]	55	114–143 [45–82]	50
Visiting primitive areas—backpacking, primitive camping, wilderness	91	120–152 [34–65]	47	119–145 [31–60]	42
Motorised activities					
Motorised off-road use	48	62–75 [29–56]	21	62–76 [28–58]	21
Motorised snow use (snowmobiles)	10	10–13 [10–37]	3	4–10 [(56)–6]	(2.5)[e]
Motorised water use	62	87–112 [41–81]	40	84–111 [35–78]	35
Consumptive					
Hunting—all types of legal hunting	28	30–34 [8–23]	5	29–34 [5–21]	4
Fishing—anadromous, cold water, saltwater, warm water	73	92–115 [28–56]	33	89–115 [22–58]	30
Non-motorised winter activities					
Downhill skiing, snowboarding	24	38–54 [58–127]	23	36–54 [50–126]	21
Undeveloped skiing—cross-country, snow-shoeing	8	10–13 [32–67]	4	5–10 [(42)–28]	(1)
Non-motorised water					
Swimming, snorkelling, surfing, diving	144	210–268 [47–85]	99	212–266 [47–85]	99
Floating—canoeing, kayaking, rafting	40	52–65 [30–62]	20	47–62 [18–56]	13

(continued)

Table 6.9 (continued)

Source: Bowker et al. (2012, p. 28); National Survey of Recreation and the Environment (NSRE) 2005–09, Versions 1 to 4 (Jan 2005 to Apr 2009) (n = 24,073) (USDA Forest Service 2009)
[a]Activities are individual or activity composites derived from the NSRE. Participants are determined by the product of the average weighted frequency of participation by activity for NSRE data from 2005 to 2009 and the adult (>16) population in the USA during 2008 (235.4 million)
[b]Because of small population and income differences, initial values for 2008 differ across PRA scenarios; thus an average is used for a starting value
[c]Participant range across Resources Planning Act (RPA) scenarios A1B, A2, and B2, without climate considerations
[d]Participant range across RPA scenarios A1B, A2, and B2, each with three selected climate futures
[e]Parentheses denote negative number

Table 6.10 Changes in total outdoor recreation days between 2008 and 2060 across all activities and scenarios

Activity[a]	2008 Days[b] (millions)	2060 Days range[c] (millions/[percent])	2060 Average days change (millions)	2060 Days range[d] (millions/[percent])	2060 Average days change[d] (millions)
Visiting developed sites					
Developed site use—family gatherings, picnicking, developed camping	2246	3121–3949 [40–74]	1294	3055–3796 [36–69]	1185
Visiting interpretive sites—nature centres, zoos, historic sites, prehistoric sites	1249	1899–2417 [53–91]	952	1935–2435 [55–95]	988
Viewing and photographic nature					
Birding—viewing/photographing birds	8255	11680–14322 [40–74]	4859	10050–13313 [36–69]	3764
Nature viewing—viewing, photography, study, or nature gathering related to fauna, flora, or natural settings	32461	41805–52835 [31–61]	14635	41550–51288 [28–58]	13597
Backcountry activities					
Challenge activities—caving, mountain biking, mountain climbing, rock climbing	121	178–219 [49–83]	4859	179–232 [48–92]	89
Equestrian	263	388–503 [49–92]	196	369–482 [40–83]	166
Hiking—day hiking	1835	2901–3682 [59–98]	1470	2825–3541 [54–93]	1366
Visiting primitive areas—backpacking, primitive camping, wilderness	1239	2046	622	1562–1946 [26–57]	519

(continued)

2008 and 2060 across all activities and scenarios. Off-road driving, according to the projections made by Bowker et al. (2012), will see an additional 21 million participants in 2060 as compared with 2008. Snowmobiling in the USA will see an additional three million participants in 2060 as compared to 2008.

Table 6.10 summarises the changes in total outdoor recreation days in the USA between 2008 and 2060 across all activities and scenarios. Off-road driving, according to the projections made by Bowker et al. (2012), will see an additional 357 million days in 2060 as compared with 2008. Snowmobiling in the USA will see an additional 16 million days in 2060 as compared to 2008. This is the smallest increase of all the activities included in the projections made by Bowker et al. (2012); seen in Table 6.10.

Table 6.10 (continued)

Activity[a]	2008 Days[b] (millions)	2060 Days range[c] (millions/[percent])	2060 Average days change (millions)	2060 Days range[d] (millions/[percent])	2060 Average days change[d] (millions)
Motorised activities					
Motorised off-road use	1053	1264–1532 [21–46]	357	1274–1611 [21–53]	385
Motorised snow use (snowmobiles)	69	74–91 [8–33]	16	23–65 [(6)–(67)]	(27)[e]
Motorised water use	958	1304–1806 [37–90]	596	1245–1763 [30–84]	495
Consumptive					
Hunting—all types of legal hunting	538	506–576 [(5)–8]	14	494–575 [(8)–7]	(8)
Fishing—anadromous, cold water, saltwater, warm water	1369	1665–2020 [23–46]	514	1602–1958 [17–41]	397
Non-motorised winter activities					
Downhill skiing, snowboarding	178	274–437 [61–150]	179	258–422 [50–146]	165
Undeveloped skiing—cross-country, snow-shoeing	52	69–87 [35–70]	29	28–64 [(45)–25]	(5)
Non-motorised water					
Swimming, snorkelling, surfing, diving	3476	5037–6429 [46–83]	2446	4396–6257 [42–80]	2298
Floating—canoeing, kayaking, rafting	262	338–422 [30–62]	128	309–409 [18–56]	83

Source: Bowker et al. (2012, p. 29); National Survey of Recreation and the Environment (NSRE) 2005–2009, Versions 1 to 4 (Jan 2005 to Apr 2009) (n = 24,073) (USDA Forest Service 2009)
[a]Activities are individual or activity composites derived from the NSRE. Participants are determined by the product of the average weighted frequency of participation by activity for NSRE data from 2005 to 2009 and the adult (>16)·population in the USA during 2008 (235.4 million)
[b]Because of small population and income differences, initial values for 2008 differ across PRA scenarios; thus an average is used for a starting value
[c]Participant range across Resources Planning Act (RPA) scenarios A1B, A2, and B2, without climate considerations
[d]Participant range across RPA scenarios A1B, A2, and B2, each with three selected climate futures
[e]Parentheses denote negative number

Conclusions

1. ORVs discussed in this chapter include four-wheel-drive (4WD) vehicles such as Land Rovers and Jeeps (which usually have deep tread tyres and raised suspension) or may be tracked and may, for example, undertake activities such as rallying, desert racing, rock crawling, or dune bashing.
2. ATVs are also known as quad bikes (but can be three-wheeled) that have a seat that is straddled by the operator and handlebars for steering control.
3. Off-road motorcycling has evolved to include various sub-disciplines like motocross, scrambling, supercross and arenacross, supermoto, VMX, and enduro.
4. Snowmobiles are ATVs specifically designed for travel across deep snow where other vehicles flounder.
5. The number of people (aged 16 and over) participating in off-highway driving in the USA rose by 16.9 million between 1982 and 2001, while snowmobiling rose by 6.1 million in the same

period. This rising trend in off-highway vehicle driving continued into this century with 20.6 million participants between 2005 and 2009, a percentage change of 34.5% (1999–2009).
6. According to Bowker et al. (2012), participation in off-road driving averaged about 20% of the adult US population, or about 48 million adults annually between 2005 and 2009 (Table 6.7). Based on the projections made by Bowker et al. (2012), future participation rates in off-road driving are expected to decline under two of three Resources Planning Act (RPA) scenarios, RPA A2 (16–18%) and RPA B2 (7–8%), while the percent of adult participants under RPA A1B will be about the same in 2060 as today.
7. All four groups of ORVs have higher than on-road vehicle accident rates. One particular problem is accident caused by rollovers. The accident rate among children under 16 is a major concern.
8. The environmental impact of these ORVs is a problem in some wildlife and conservation areas, and examples are discussed. The environmental impacts are in three main areas: damage to soil and vegetation (erosion), noise and pollution, and effects on wildlife (damage, disturbance).
9. The management of ORVs is discussed and opportunities for the education of ORV users outlined.

References

Adams, E. S. (1975). Effects of lead and hydrocarbons from snowmobile exhaust on brook trout (*Salvelinus fontinalis*). *Transactions of the American Fisheries Society, 104*(2), 363–373.

Adams, J. C., & McCool, S. F. (2009). Finite recreation opportunities: The forest service, the Bureau of Land Management, and off-road vehicle management. *Natural Resources Journal*, 45–116.

Anders, F. J., & Leatherman, S. P. (1987). Effects of off-road vehicles on coastal foredunes at Fire Island, New York, USA. *Environmental Management, 11*(1), 45–52.

Andersen, M., & Aars, J. (2008). Short-term behavioural response of polar bears (*Ursus maritimus*) to snowmobile disturbance. *Polar Biology, 31*(4), 501.

Bissix, G. (2015). A multidimensional framework for assessing the acceptability of recreational all-terrain vehicle access on community trails and local public highways. *Leisure/Loisir, 39*(3–4), 345–359.

Borneman, T. E., Rose, E. T., & Simons, T. R. (2016). Off-road vehicles affect nesting behaviour and reproductive success of American Oystercatchers *Haematopus palliatus*. *Ibis, 158*(2), 261–278.

Bowker, J. M., Askew, A. E., Cordell, H. K., Betz, C. J., Zarnoch, S. J., & Seymour, L. (2012) *Outdoor Recreation Participation in the United States–Projections to 2060: A Technical Document Supporting the Forest Service 2010 RPA Assessment*. Ashville: Southern Research Station. Retrieved from www.srs.fs.usda.gov.

Brattstrom, B. H., & Bondello, M. C. (1983). Effects of off-road vehicle noise on desert vertebrates. In R. H. Webb & H. G. Wilshire (Eds.), *Environmental Effects of Off-Road Vehicles* (pp. 167–206). New York: Springer.

Busack, S. D., & Bury, R. B. (1974). Some effects of off-road vehicles and sheep grazing on lizard populations in the Mojave Desert. *Biological Conservation, 6*(3), 179–183.

Cambi, M., Certini, G., Fabiano, F., Foderi, C., Laschi, A., & Picchio, R. (2015a). Impact of wheeled and tracked tractors on soil physical properties in a mixed conifer stand. *iForest-Biogeosciences and Forestry, 9*(1), 89.

Cambi, M., Certini, G., Neri, F., & Marchi, E. (2015b). The impact of heavy traffic on forest soils: A review. *Forest Ecology and Management, 338*, 124–138.

Cordell, K. (2012). *Outdoor Recreation Trends and Futures: A Technical Document Supporting the Forest Service 2010 RPA Assessment*. Ashville: Southern Research Station. Retrieved from www.srs.fs.usda.gov.

Davies, R., Speldewinde, P. C., & Stewart, B. A. (2016). Low level off-road vehicle (ORV) traffic negatively impacts macroinvertebrate assemblages at sandy beaches in south-western Australia. *Scientific Reports, 6*, 24899.

Dewidar, K., Thomas, J., & Bayoumi, S. (2016). Detecting the environmental impact of off-road vehicles on Rawdat Al Shams in central Saudi Arabia by remote sensing. *Environmental Monitoring and Assessment, 188*(7), 1–12.

Felton, S. K., Pollock, K. H., & Simons, T. R. (2017). Response of beach-nesting American Oystercatchers to off-road vehicles: An experimental approach reveals physiological nuances and decreased nest attendance. *The Condor, 120*(1), 47–62.

Fuglei, E., Ehrich, D., Killengreen, S. T., Rodnikova, A. Y., Sokolov, A. A., & Pedersen, Å. Ø. (2017). Snowmobile impact on diurnal behaviour in the Arctic fox. *Polar Research, 36*(Suppl. 1), 10.

References

Greller, A. M., Goldstein, M., & Marcus, L. (1974). Snowmobile impact on three alpine tundra plant communities. *Environmental Conservation, 1*(2), 101–110.

Hallo, J. C., Manning, R. E., & Stokowski, P. A. (2009). Understanding and managing the off-road vehicle experience: Indicators of quality. *Managing Leisure, 14*(3), 195–209.

Hammitt, W. E., & Cole, D. N. (1998). *Wildland Recreation: Ecology and Management* (2nd ed.). Chichester: John Wiley & Sons.

Hare, C. T., Springer, K. J., & Huls, T. A. (1974). *Snowmobile Engine Emissions and Their Impact (No. 740735)*. SAE Technical Paper.

Jennissen, C. A., Harland, K. K., Wetjen, K., Peck, J., Hoogerwerf, P., & Denning, G. M. (2014). A school-based study of adolescent all-terrain vehicle exposure, safety behaviors, and crash experience. *The Annals of Family Medicine, 12*(4), 310–316.

Keddy, P. A., Spavold, A. J., & Keddy, C. J. (1979). Snowmobile impact on old field and marsh vegetation in Nova Scotia, Canada: An experimental study. *Environmental Management, 3*(5), 409–415.

Kevan, P. G., Forbes, B. C., Kevan, S. M., & Behan-Pelletier, V. (1995). Vehicle tracks on high Arctic tundra: Their effects on the soil, vegetation, and soil arthropods. *Journal of Applied Ecology, 32*, 655–667.

Kobryn, H. T., Beckley, L. E., Cramer, V., & Newsome, D. (2017). An assessment of coastal land cover and off-road vehicle tracks adjacent to Ningaloo Marine Park, north-western Australia. *Ocean & Coastal Management, 145*, 94–105.

Lucrezi, S., Saayman, M., & Van der Merwe, P. (2014). Impact of off-road vehicles (ORVs) on ghost crabs of sandy beaches with traffic restrictions: A case study of Sodwana Bay, South Africa. *Environmental Management, 53*(3), 520–533.

Lyngnes, S., & Prebensen, N. K. (2014). Sustainable and attractive motorised nature-based experiences: Challenges and opportunities. In J. S. Chen (Ed.), *Advances in Hospitality and Leisure* (pp. 151–171). Bingley, UK: Emerald Group Publishing Limited.

Marion, D. A., Phillips, J. D., Yocum, C., & Mehlhope, S. H. (2014). Stream channel responses and soil loss at off-highway vehicle stream crossings in the Ouachita National Forest. *Geomorphology, 216*, 40–52.

Muller, B. (2016). Mending man's ways: Wickedness, complexity and off-road travel. *Landscape and Urban Planning, 154*, 93–101.

Mullet, T. C., Morton, J. M., Gage, S. H., & Huettmann, F. (2017). Acoustic footprint of snowmobile noise and natural quiet refugia in an Alaskan wilderness. *Natural Areas Journal, 37*(3), 332–349.

Newsome, D. (2014). Appropriate policy development and research needs in response to adventure racing in protected areas. *Biological Conservation, 171*, 259–269.

Parikesit, P., Larson, D. W., & Matthes-Sears, U. (1995). Impacts of trails on cliff-edge forest structure. *Canadian Journal of Botany, 73*(6), 943–953.

Popovicheva, O. B., Kireeva, E. D., Shonija, N. K., Vojtisek-Lom, M., & Schwarz, J. (2015). FTIR analysis of surface functionalities on particulate matter produced by off-road diesel engines operating on diesel and biofuel. *Environmental Science and Pollution Research, 22*(6), 4534–4544.

Råheim, E. (1992). *Registration of Vehicular Tracks on the Svalbard Archipelago*. Oslo, Meddelelser NR. 122.

Reimann, S., Kallenborn, R., & Schmidbauer, N. (2009). Severe aromatic hydrocarbon pollution in the Arctic town of Longyearbyen (Svalbard) caused by snowmobile emissions. *Environmental Science & Technology, 43*(13), 4791–4795.

Royal Society for Nature Conservation. (1987). *Damage to Wildlife Sites by Off-Road Motor Vehicles*. Nettleham, Lincoln: RSNC.

Rybansky, M. (2014). Modelling of the optimal vehicle route in terrain in emergency situations using GIS data. In *IOP Conference Series: Earth and Environmental Science* (Vol. 18, No. 1, p. 012131). IOP Publishing.

Savage, D., Woodland, M., Eliason, A., Lipple, Z., Smith, C., Maas, J., et al. (2016). Design and validation of the 2016 University of Idaho Clean Snowmobile: A reduced speed 797cc Flex-fueled direct-injection two-stroke with active and passive noise cancellation. Retrieved from http://www.mtukrc.org/download/idaho/idaho_ic_design_paper_2016.pdf

Selva, N., Switalski, A., Kreft, S., & Ibisch, P. L. (2015). Why keep areas road-free? The importance of roadless areas. In R. van der Ree, D. J. Smith, & C. Grilo (Eds.), *Handbook of Road Ecology* (pp. 16–26). Chichester: Wiley.

Slaughter, C. W., Racine, C. H., Walker, D. A., Johnson, L. A., & Abele, G. (1990). Use of off-road vehicles and mitigation of effects in Alaska permafrost environments: A review. *Environmental Management, 14*(1), 63–72.

St-Louis, A., Hamel, S., Mainguy, J., & Côté, S. D. (2013). Factors influencing the reaction of mountain goats towards all-terrain vehicles. *The Journal of Wildlife Management, 77*(3), 599–605.

Stott, T. A., & Short, N. (1996). Erosion rates and human impacts in The Arctic Tundra: Findings from a British schools exploring society expedition to Svalbard. *Geography Review, 10*(2), 18–24.

Uberti, S., Copeta, A., Baronio, G., & Motyl, B. (2017). An eco-innovation and technical contaminated approach for designing a low environmental impact off-road motorcycle. *International Journal on Interactive Design and Manufacturing (IJIDeM), 12*, 1–15.

United States Consumer Product Safety Commission. (2018). ATV safety information centre. Retrieved January 2, 2018, from https://www.cpsc.gov/Safety-Education/Safety-Education-Centers/ATV-Safety-Information-Center/.

U.S. Department of Agriculture (USDA) Forest Service. (2009). *National Survey on Recreation and the Environment [Dataset]*. Retrieved September 15, 2010, from www.srs.fs.usda.gov/trends/nsre/nsre2.html

Weaver, T., & Dale, D. (1978). Trampling effects of hikers, motorcycles and horses in meadows and forests. *Journal of Applied Ecology, 15*, 451–457.

Webb, R. H., & Wilshire, H. G. (1983). *Environmental Effects of Off-Road Vehicles: Impacts and Management in Arid Regions.* New York: Springer-Verlag.

Westcott, F., & Andrew, M. E. (2015). Spatial and environmental patterns of off-road vehicle recreation in a semi-arid woodland. *Applied Geography, 62*, 97–106.

Wolcott, T. G., & Wolcott, D. L. (1984). Impact of off-road vehicles on macroinvertebrates of a mid-Atlantic beach. *Biological Conservation, 29*(3), 217–240.

Mountain Biking

Chapter Summary

This chapter first defines mountain biking (MTB), its history and development, and the range of different bikes used. It then examines participation numbers before considering the history, designs, and disciplines with MTB. The final part of the chapter focuses on specific environmental impacts: damage to soil, vegetation, and water and the impacts on wildlife. The final section considers the management of these activities such as trail design and the development of the forest-based MTB centres in the UK and gives some examples of education initiatives such as the International Mountain Biking Association Rules of the Trail which have been used in management attempts.

7.1 Definitions

7.1.1 Mountain Biking (MTB)

The mountain bike was developed in Marin County, California, in the mid-1970s. The conventional bike was strengthened and made more flexible, quickly evolving into the form seen today, with front and rear suspensions available (Fig. 7.1). The first mountain bike was introduced into the UK in 1983, and though no specific records of sales are kept, the overall purchase of cycles is running at a very high level.

Mountain bikes are designed for off-road use, although of course they can be used on paved road, but their heavier weights and more upright sitting position means that they are generally not as fast as modern-day road bikes. Mountain bikes share similarities with other bikes but incorporate features designed to enhance durability and performance in rough terrain. These typically include front or full (front and rear) suspension, large knobby tyres, more durable wheels and spokes, more powerful brakes (usually disc brakes), and lower gear ratios for climbing steep hills. Mountain bikes are typically ridden on mountain trails, which may often be purpose-built single track, fire roads, and other unpaved surfaces. This type of terrain usually has tree roots, loose dirt, rocky surfaces, and steep grades. Many purpose-built trails will also have additional features such as log piles, log rides, rock gardens, gap jumps, and wall rides. Mountain bikes are built to handle these types of terrain and features.

Since the development of the sport in the 1970s, many new subtypes of MTB have developed, such as cross-country (XC), all-day endurance, freeride, downhill, and a variety of track and slalom types. Each of these requires different designs for optimal performance. MTB development has led to an increase in suspension travel, now often up to 8 inches (200 mm) and gearing up to 27 speeds, to facilitate both climbing and

Fig. 7.1 A full-suspension mountain bike. Photo by Tim Stott

rapid descents. Advancements in gearing have also led to a "1x" (pronounced "one-by") trend, simplifying the gearing to one chain ring in the front and a cassette at the rear, typically with 9–12 sprockets. The expression "all terrain bike" and the acronym "ATB" are used as synonyms for "mountain bike."

7.1.2 Fatbikes

A fatbike (also called fat bike or fat-tyre bike) is an off-road bicycle with oversized tyres, typically 3.8 inches (97 mm) or larger and rims 2.6 inches (66 mm) or wider, designed for low ground pressure to allow riding on soft unstable terrain, such as snow, sand, bogs, and mud. Fatbikes are built around frames with wide forks and stays to accommodate the wide rims required to fit these tyres. The wide tyres can be used with inflation pressures as low as 340 hPa (5 psi) to allow for a smooth ride over rough obstacles. A rating of 550–690 hPa (8–10 psi) is suitable for the majority of riders.

7.1.3 BMX Bikes

A BMX bike is an off-road sport bicycle used for racing and stunt riding. BMX means bicycle motocross. Although the term BMX originally meant a bicycle intended for BMX racing, the term "BMX bike" is now used to encompass race bikes, as well as those used for the dirt, vert, park, street, flatland, and BMX freestyle disciplines of BMX.

7.2 Participation Numbers

In the USA, during the 2016 calendar year, a total of 24,134 online interviews were carried out with a nationwide sample of individuals and households from the US Online Panel of over one million people operated by Synovate/IPSOS (Outdoor Foundation 2017). A total of 11,453 individual and 12,681 household surveys were completed. The total panel is maintained to be representative of the US population for people ages six and older. Oversampling of ethnic groups took place to boost response from typically under-responding groups. The 2016 participation survey sample size of 24,134 completed interviews provides a high degree of statistical accuracy.

As can be seen in rows 3–5 of Table 7.1, BMX and mountain (non-paved surface) commanded participation numbers of 1,655,000 and 6,751,000, respectively, in 2006, and these figures rose to 3,104,000 and 8,615,000 in 2016, three-year increases of 43.2% for BMX riding and 0.9% for MTB. These numbers are still a relatively small proportion of the participation numbers for road/paved bicycling which were 38,475,000 in 2006 and 38,365,000 in 2016 with a three-year change of −6.2%. So, the participation in BMX seems to be the biggest growth area, although total numbers are still lower than for MTB.

Although not as up to date at the Outdoor Foundation survey, Cordell's (2012) report lumped together road biking, MTB, and BMX and showed that bicycle participation in outdoor activities in the USA exceeded hiking and was on a par with jogging/trail running and car/back yard/RV camping (Fig. 7.2).

Additionally, in Cordell's (2012) report (Table 7.2), he showed that the number of people in the USA (1999–2001) bicycling on mountain/hybrid bikes was 44 million (18% of the total number aged >16 participating in outdoor activities), although this had declined to 42.7 million by the time of his 2005–2009 survey.

Table 7.1 Outdoor participation by activity (ages 6+) in the USA, 2006–2016 (The Outdoor Foundation 2017, p. 8)

	2006	2007	2008	2009	2010	2011	2012	2013	2014	2015	2016	3-year change (%)
Adventure racing	725	698	920	1089	1339	1065	2170	2213	2368	2864	2999	35.5
Backpacking overnight >¼ mile from vehicle/home	7076	6637	7867	7647	8349	7095	8771	9069	10,101	10,100	10,151	11.9
Bicycling (BMX)	1655	1887	1904	1811	2369	1547	2175	2168	2350	2690	3104	43.2
Bicycling (mountain/non-paved surface)	**6751**	**6892**	**7592**	**7142**	**7161**	**6816**	**7714**	**8542**	**8044**	**8316**	**8615**	**0.9**
Bicycling (roads/paved surface)	38,457	38,940	38,114	40,140	39,320	40,349	39,232	40,888	39,725	38,280	38,365	−6.2
Birdwatching (more and ¼ mile from home/vehicle)	11,070	13,476	14,399	13,294	13,339	12,794	14,275	14,152	13,179	13,093	11,589	−18.1
Boardsailing/windsurfing	938	1118	1307	1128	1617	1151	1593	1324	1562	1766	1737	31.2
Camping (RV)	16,946	16,168	16,517	17,436	15,865	16,698	15,108	14,556	14,663	14,699	15,855	8.9
Camping (with ¼ mile of home/vehicle)	35,618	31,375	33,686	34,338	30,996	32,925	29,982	29,269	28,660	27,742	26,467	−9.6
Canoeing	9154	9797	9935	10,058	10,553	9787	9839	10,153	10,044	10,236	10,046	−1.1
Climbing (sports/indoor/boulder)	4728	4514	4769	4313	4770	4119	4592	4745	4536	4684	4905	3.4
Climbing (traditional/ice/mountaineering)	1586	2062	2288	1835	2198	1609	2189	2319	2457	2571	2790	20.3
Fishing (fly)	6071	5756	5941	5568	5478	5683	6012	5878	5842	6089	6456	9.8
Fishing (freshwater/other)	43,100	43,859	40,331	40,961	38,860	38,868	39,135	37,796	37,821	37,682	38,121	0.9
Fishing (saltwater)	12,466	14,437	13,804	12,303	11,809	11,983	12,017	11,790	11,817	11,975	12,266	4.0
Hiking (day)	29,863	29,965	32,511	32,572	32,496	34,491	34,545	34,378	36,222	37,232	42,128	22.5
Hunting (bow)	3875	3818	3722	4226	3908	4633	4075	4079	4411	4564	4427	8.5
Hunting (handgun)	2525	2595	2873	2276	2709	2671	3553	3198	3091	3400	3512	9.8
Hunting (rifle)	11,242	10,635	10,344	11,114	10,150	10,807	10,164	9792	10,081	10,778	10,797	10.3
Hunting (shotgun)	8987	8545	8731	8490	8062	8678	8174	7894	8220	8438	8271	4.8
Kayak fishing	n/a	n/a	n/a	n/a	1044	1201	1409	1798	2074	2265	2371	31.8
Kayaking (recreational)	4134	5070	6240	6212	6465	8229	8144	8716	8855	9499	10,017	14.9
Kayaking (sea/touring)	1136	1485	1780	1771	2144	2029	2446	2694	2912	3079	3124	16.0
Kayaking (white water)	828	1207	1242	1369	1842	1546	1878	2146	2351	2518	2552	18.9

(continued)

Table 7.1 (continued)

	2006	2007	2008	2009	2010	2011	2012	2013	2014	2015	2016	3-year change (%)
Rafting	3609	3786	4226	4342	3869	3725	3958	3915	3924	4099	4095	-10.6
Running/jogging	38,559	41,064	41,130	43,892	49,408	50,713	52,187	54,188	51,127	48,496	47,384	-12.6
Sailing	3390	3786	4226	4342	3869	3725	3958	3915	3924	4099	4095	4.6
Scuba diving	2965	2965	3216	2723	3153	2579	2982	3174	3145	3274	3111	-2.0
Skateboarding	10,130	8429	7807	7352	6808	5827	6627	6350	6582	6436	6442	1.5
Skiing (alpine/downhill)	n/a	10,362	10,346	10,919	11,504	10,201	8243	8044	8649	9378	9267	12.4
Skiing (cross-country)	n/a	3530	3848	4157	4530	3641	3307	3377	3820	4146	4640	40.3
Skiing (freestyle)	n/a	2817	2711	2950	3647	4318	5357	4007	4564	4465	4640	2.7
Snorkelling	8395	9294	10,296	9358	9305	9318	8011	8700	8752	8874	8717	0.2
Snowboarding	n/a	6841	7159	7421	8196	7579	7351	6418	6785	7676	7602	3.4
Snowshoeing	n/a	2400	2922	3431	3823	4111	4029	3012	3501	3885	3533	-12.3
Stand up paddling	n/a	n/a	n/a	n/a	1050	1242	1542	1993	2751	3020	3220	61.6
Surfing	2170	2206	2607	2403	2767	2195	2895	2658	2721	2701	2793	3.0
Telemarking (downhill)	n/a	1173	1435	1482	1821	2099	2766	1732	2188	2569	2848	3.0
Trail running	4558	4216	4857	4833	5136	5610	6003	6792	7531	8139	8582	26.4

Note: All participation numbers are in thousands (000)

7.2 Participation Numbers

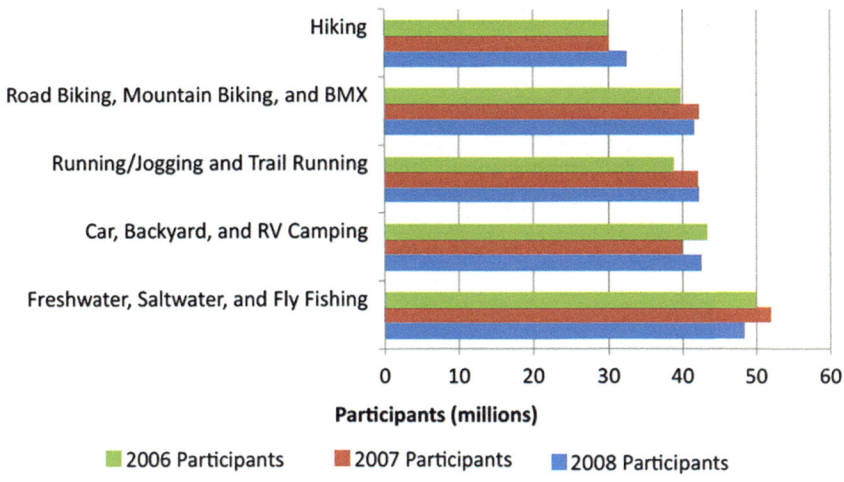

Fig. 7.2 Participation in gateway outdoor activities. Source: Cordell (2012, p. 27)

Table 7.2 Trends in number of people ages 16 and older participating in recreation activities in the USA, 1999–2001 and 2005–2009 for activities with between 25 and 49 million participants from 2005 through 2009 (Source: Cordell 2012, p. 37)

	Total participants (*millions*)			Percent participating	Percent change
	1994–1995	1999–2001	2005–2009	2005–2009	1999–2001 to 2005–2009
Visit archaeological sites	36.1	44.0	48.8	20.8	11.1
Off-highway vehicle driving	35.9	36.0	48.4	20.6	34.5
Boat tours or excursions	–	40.8	46.1	19.6	13.1
Bicycling on mountain/hybrid bike	–	**44.0**	**42.7**	**18.1**	**−3.0**
Primitive camping	31.4	33.1	34.2	14.5	3.2
Sledding	27.7	30.8	32.0	13.6	3.9
Coldwater fishing	25.1	28.4	30.9	13.1	8.7
Saltwater fishing	22.9	21.4	25.1	10.7	17.2

Missing data are denoted with "–" and indicate that participation data for that activity were not collected during that time period. Percent change was calculated before rounding
Source: USDA Forest Service (1995) ($n = 17,217$), USDA Forest Service (2001) ($n = 52,607$), and USDA Forest Service (2009) ($n = 30,398$)
Note: The numbers in this table are *annual* participant estimates on data collected during the three time periods
1994–1995 participants based on 201.3 million people ages 16+ (Woods & Poole Economics, Inc. 2007)
1999–2001 participants based on 214.0 million people ages 16+ (U.S. Department of Commerce 2000)
2005–2009 participants based on 235.3 million people ages 16+ (U.S. Department of Commerce 2008)

Unfortunately, as far as we are aware, there is no such equivalent comprehensive survey of participation numbers in Europe or any other part of the world to date, although Sport England's Active People Survey (Table 7.3) offers some indicative figures for cycling and recreational in general and for BMX, cyclo-cross, and MTB in England for 2012–2013. It would seem that there are in the region of 3.6 million participants over the age of 14 taking part in cycling (8.1% of the adult population) with around 2.1 million classed as recreational cycling. However, it is not clear whether these numbers are in addition to the general cycling figures or part of them. Some 736,900 participated in MTB, 27,300 in cyclo-cross, and 54,000 in BMX. So these figures reflect the trends

Table 7.3 Average monthly participation[a] in sport and recreation in England, October 2012–2013

Activity	Number of people (14+) participating monthly	Percentage of the adult population (14+)
Outdoor recreation group	25,703,100	59.3
Outdoor recreation group (excluding walking)	7,707,500	17.8
Coarse fishing	632,800	1.4
Game fishing	155,800	0.4
Sea fishing	245,900	0.6
Running	2,791,500	6.3
Canoeing	133,300	0.3
Cycling	3,524,400	8.1
BMX	54,000	0.1
Cyclo-cross	27,300	0.1
Mountain biking	736,900	1.7
Recreational cycling	2,159,800	5.0
Pony trekking	35,300	0.1
Other horse riding	301,700	0.7
Outdoor climbing	191,200	0.4
Orienteering	11,800	0.0
Water-based rowing	47,500	0.1
Windsurfing	19,400	0.0
Cruising sailing	47,600	0.1
Alpine skiing	95,900	0.2
Freestyle skiing	22,300	0.1
Nordic skiing	17,400	0.0
Snowboarding	29,100	0.1
Outdoor swimming	826,700	1.9
Recreational walking	23,313,500	53.8

Source: Sport England, 2014, Active People Survey
[a]At least one session of any duration in the last 28 days

seen in the US data provided in the Outdoor Foundation (2017) study (Table 7.1).

Other nations in which MTB is popular include Germany (3.5 million mountain bikers of 7.2 million recreational cyclists) and Switzerland and Austria, with the total number of mountain bikers estimated at 800,000 (Koepke 2005). In Australia, the number of cyclists grew by 15.3% between 2001 and 2004 (Faulks et al. 2008), and of the 753,843 bikes sold in 2004, 70% were mountain bikes (Bradshaw 2006). Although the percentage of such bikes used for off-road riding and their frequency of use are unknown, such data suggest that MTB is growing worldwide (Hardiman and Burgin 2013). International Mountain Biking Association (IMBA) is now represented in 17 countries including the USA, Australia, Canada, Italy, Mexico, Spain, Netherlands, and UK, and their code of conduct is considered universal (International Mountain Biking Association 2013).

7.3 History, Designs, and Disciplines with MTB

7.3.1 History

The original mountain bikes were modified heavy cruiser bicycles used for freewheeling down mountain trails. The sport became popular in the 1970s in Northern California, USA, with riders using older single-speed balloon-tyre bikes to ride down rugged hillsides. By the late 1970s and early 1980s, road bike companies began to manufacture mountain bikes using high-tech lightweight materials, such as M4 aluminium. The first mass production mountain bike was the Specialized Stumpjumper, produced in 1981. Throughout the 1990s and 2000s, MTB moved from a little-known sport to a mainstream activity complete with an international racing circuit and a world championship, in addition to various freeride competitions, such as the FMB World Tour and the Red Bull Rampage.

7.3.2 Designs

Mountain bikes can be divided into three broad categories based on suspension configuration:

- *Rigid*: a bike with neither front nor rear suspension
- *Hardtail*: a bike equipped with a suspension fork (front wheel) but otherwise a rigid frame
- *Full suspension (or dual suspension)*: a bike equipped with both front and rear suspensions. The front suspension is usually a telescopic fork similar to that of a motorcycle, and

the rear by a mechanical linkage with components for absorbing shock.

There are several different styles of MTB, usually defined by the terrain, and therefore bikes employed. Styles of mountain bike riding and mountain bikes have evolved rapidly in recent years leading to terms such as freeride and "Trail bike" being used to categorise mountain bikes.

7.3.3 MTB Disciplines

Cross-country (XC) mountain bikes are designed primarily around the discipline of cross-country racing, placing emphasis on climbing speed and endurance and therefore demanding lightweight, efficient bikes (Fig. 7.4B). In the 1980s and early 1990s, XC mountain bikes typically consisted of a lightweight steel hardtail frame with rigid forks. Throughout the 1990s, XC bikes evolved to incorporate lightweight aluminium frames and short-travel (65–110 mm) front suspension forks. Recently full-suspension designs have become more popular among racers and enthusiasts alike, and the use of advanced carbon fibre composites has allowed bike designers to produce full-suspension designs which weigh under 10 kg. In recent years 29 "wheels have largely replaced the original standard of 26"; the US men's and women's marathon cross-country races were won on 29ers in 2009 and 2010. The geometry of cross-country bikes favours climbing ability and fast responses over descending and stability, and although intended for off-road use, XC bikes are not designed for use on steep or particularly rough terrain. Put in terms of rider emphasis, XC bikes are designed for approximately 80% uphill or flat riding and 20% downhill.

Trail bikes are a development of XC bikes that are generally used by recreational mountain bikers either at purpose-built "trail centres" or on natural off-road trails. They usually have around 120–140 mm (5 inches) of travel, weigh 11–15 kg (24–33 lb), and have geometries situated somewhere between full XC and All-Mountain bikes. With less of an emphasis on weight, Trail bikes are typically built to handle rougher terrain than dedicated XC bikes, and they provide greater stability while descending. Trail bikes are designed for approximately 60–70% uphill and 30–40% downhill riding.

Enduro/All-Mountain (AM) bikes bridge the gap between XC and freeride bikes which typically weigh between 13 and 16 kg (29 to 35 lb). These bikes tend to feature greater suspension travel, frequently as much as 150 mm (6 or 7 inches) of front and rear travel, often adjustable on newer mid- and high-end bikes. Designed to be able to climb and descend well, these bikes are intended to be taken on all-day rides involving both steep climbs and steep descents, hence the term "all-mountain." In terms of aggressiveness, these bikes are intended for anywhere from 50–70% downhill riding to 30–50% uphill riding, bridging the gap between trail and downhill bikes. In recent years, there has been somewhat a split between Enduro and All-Mountain bikes, with the former placing more emphasis on descent due to the increased emphasis on timed downhill runs in enduro racing when compared to more typical All-Mountain riding.

Downhill (DH) bikes typically have eight or more inches (200 mm) of suspension travel and extremely low, slack geometry intended to set the rider in a comfortable position when descending steep trails at high speed (Fig. 7.4C). Due to their often high gear ratios, soft suspension, and aggressive geometry, downhill bikes are ideal only for riding down dedicated trails or race courses. Some mechanical uplift is usually employed which may be using a ski lift or telepherique outside of the ski season, or an agricultural tractor or other vehicle pulls a trailer back up the hill with the bikes and riders on board. Occasionally riders may push or carry their downhill bike uphill as they are too heavy to ride. Downhill frames are often intended for racing and as such are required to be both extremely durable and lightweight. Bicycle designers often make use of similar materials in the construction of downhill and XC frames and components (e.g. carbon fibre), despite their vastly different purposes, as the ultimate goal of a high strength to weight ratio is the same. In recent years, more advanced frame and component designs have produced high-end downhill bikes with similar weights to average Trail and All-Mountain bikes, with an increasing expectation that complete downhill bikes remain below 18 kg (40 lb).

Freeride (FR) mountain bikes are similar to downhill bikes, with less emphasis on weight and more on strength. Freeride bikes have ample suspension and typically have at least 180 mm (7 inches) of travel. Freeride bikes are intended for trail features with large air time, such as jumps and drops, and as such are designed to handle heavy impacts, whether from landings or crashes. Freeride frames and parts are rarely made from carbon fibre due to strength and durability concerns and are instead usually made from aluminium, sacrificing marginal weight gain for more predictable material response under heavy usage. Certain freeride-specific bikes can be ridden uphill more easily than downhill bikes but are nevertheless still inefficient in pedalling and difficult to manoeuver while angled uphill. Originally, freeride bikes sat between All-Mountain and downhill bikes in geometry, with frame angles steeper than those found in downhill bikes with higher rider positioning, enhancing manoeuvrability on technical or low-speed features commonly found on "North Shore"-style trails. Freeride bikes typically range in weight from 14 to 20 kg (31 to 44 lb). Slopestyle and Dirt Jump bikes are included in this category by some, due to similar purposes, but the distinction in bike design is significant between the three.

Dirt jumping, urban, and street mountain bikes lie somewhere in between a BMX bike and a freeride bike. They are rigid or hardtail bikes, with 76–114 mm (3–4.5 inches) of front suspension and rigid, durable frames with low bottom brackets and short chain stays to improve manoeuvrability. Dirt Jump bikes often overlap in design with Four-Cross bikes, though that discipline has dropped in popularity, with many frames including removable derailleur hangers and/or integrated chain tensioners to allow for single-speed and multi-speed arrangements (Four-Cross bikes mostly use derailleurs, while dirt jumpers usually use single-speed setups). Tyres on these bikes are usually 24 or 26″ diameter, fast-rolling slicks (tyres without tread), or semi-slicks. Dirt jumpers usually have low seat posts and oversized handlebars, to make room for tricks. Most dirt jumpers have an extended rear brake cable installed and have no front brake, which allows the rider to spin the handle bars multiple times without tangling the brake cables.

Slopestyle (SS) bikes are a strange blend of Dirt Jump and freeride bikes, having the geometry similar to dirt jumpers but with approximately 100 mm (4 inches) of suspension travel in both the front and rear forks. These bikes are mostly used by professional slopestyle riders, this specific usage being their origin, and as such are designed for the extremely large jumps and high speeds encountered in competition. The frames are either adapted from existing All-Mountain or freeride designs or designed specifically for the purpose, with durable frame designs and sophisticated suspension linkages to make the most of their minimal suspension travel.

Trials bikes are set up very specifically for the purpose of bike trials. Two varieties of trials bike exist, those with 26″ wheels (referred to as "stock") and those with 20″ wheels (referred to as "mod"—because historically they were modified BMX bikes). They typically have no suspension at all, though some still make use of some form of it. Competition rules require stock bikes to have multiple gears for competition, but most riders never use their shifters. Competition rules do not require mod bikes to have any gears. Many non-competitive riders run single-speed, choosing a fairly low-speed, high-torque gear. Most modern trials bikes have no seat at all, as the rider spends all of his time out of the saddle, and trials riding is not conducive to the use of the saddle as a control interface as in normal MTB. These bikes are significantly lighter than almost all other mountain bikes, ranging from 7 to 11 kg (15 to 24 lb) which makes manoeuvring the bike much easier.

Single-speed (SS) mountain bikes have one set gear ratio. The gear ratio chosen depends on the terrain being ridden, the strength and skill of the rider, and the size of the bike (a bike with 29″ wheels often requires a different gearing than a bike with standard 26″ wheels). Often single-speeds are fully rigid, steel-framed bikes.

These are typically ridden by very fit individuals on mild to moderate cross-country terrain.

Mountain cross or Four-Cross (4X) is a type of racing in which four bikers race downhill on a prepared, BMX style track. Four-Cross racing has fallen in popularity recently, with the UCI removing Four-Cross from the World Cup due to excessive erosion and inconvenience caused by the purpose-built race tracks. *Dual slalom (DS)* is similar to Four-Cross, but instead of four competing cyclists during a race, there are only two, racing in parallel lanes. The courses are in general more technical with smaller jumps than Four-Cross courses. Dual slalom races originally took place on grass slopes with gates and minimal jumps but are now held on man-made courses. Dual slalom racers will usually use Slopestyle or Dirt Jump bikes. *Indycross (IX)* is essentially a mountain cross event featuring a wide variety of features run by one competitor per time. *North Shore* bikes are much like freeride bikes in their geometry and downhill bikes in their component makeup. Because North Shore stunts have evolved to not only include simple and complex bridges but also large drops and high-speed descents through a series of stunts, North Shore bikes commonly have as much travel as downhill and freeride bikes, however with much more nimble and manoeuvrable frame designs and often lighter weight.

Circle dirt-track racing is a class of racing in which any kind of bikes are used, most commonly a hardtail mountain bike with front suspension. Many different modifications are made to track racing bikes, such as reducing bike weight, increasing brake power, trying different cambers (so that when the bike leans, the tyre is more level with the track thus creating more grip), and trying different gear ratios.

7.4 Environmental Impact

Infrastructure to support the various forms of MTB such as purpose-built single track trails, uplift facilities for downhill, and bike parks for freeriding/trials is increasing in many countries (Koepke 2005; IMBA 2010). In the USA, locations such as Moab (Utah) and Fruita (Colorado) each offers hundreds of kilometres of single track mountain bike trails in desert ecosystems (MATC 2010). In Canada, alpine resorts such as Whistler Blackcomb offer more than 200 km of trails for MTB, including 34 trails of lift-serviced downhill routes. An indication of how important MTB has become to such resorts is that summer revenue now represents approximately 75% of winter snow recreation revenue (TRC 2005; Whistler Blackcomb n.d.).

Significant economic benefits can be gained from developing and promoting MTB in its various forms. Examples include destination MTB tourism and competitive sporting events, typified by the World Cup Mountain Bike Series, Union Cycliste Internationale Mountain Bike, and Trials Championship. MTB also provides social networking opportunities and supports a substantial industry in both equipment and clothing.

7.4.1 Damage to Soil and Vegetation

The rising popularity of MTB has raised concerns of potential environmental impacts (see Burgin and Hardiman 2012 for review). The IMBA "rules" (see rules 1–3, Table 7.4) include this dimension. Such impacts associated with recreational trails result from their initial design, construction, and subsequent use (e.g. type, user behaviour, frequency, and intensity) (Pickering et al. 2010). Assessing impacts caused by MTB is difficult since mountain bikers often share trails used by others: for hiking, horse riding, and 4WD driving, so the specific effects of MTB often cannot be readily distinguished. Despite this, instances of the creation of unauthorised, informal bike trails and/or construction of bike-specific infrastructure such as concrete-reinforced jumps and wooden board ways used in freeriding/North Shore are becoming more common, even in protected areas (Fig. 7.3).

On flat terrain under dry conditions, recreational MTB impacts on trails, for example, increased water runoff, sediment yield, and/or soil exposure, together with vegetation and/or species loss, have been found to be comparable

Table 7.4 Official IMBA "Mountain Bike Rules of the Trail" in which the IMBA considers that "every mountain biker should know and live by…"

Rule number	Rule	Background
1	Ride on open trails only	Respect trail and road closures—ask if uncertain; avoid trespassing on private land; obtain permits or other authorisation as may be required. Federal and state wilderness areas are closed to cycling. The way you ride will influence trail management decisions and policies
2	Leave no trace	Be sensitive to the dirt beneath you. Recognise different types of soils and trail construction; practice low-impact cycling. Wet and muddy trails are more vulnerable to damage. When the trail bed is soft, consider other riding options. This also means staying on existing trails and not creating new ones. Do not cut switchbacks. Be sure to pack out at least as much as you pack in
3	Control your bicycle	Inattention for even a second can cause problems. Obey all bicycle speed regulations and recommendations
4	Always yield trail	Let your fellow trail users know you are coming. A friendly greeting or a bell is considerate and works well; do not startle others. Show your respect when passing by slowing to a walking pace or even stopping. Anticipate other trail users around corners or in blind spots. Yielding means slow down, establish communication, be prepared to stop if necessary, and pass safely
5	Never scare animals	All animals are startled by an unannounced approach, a sudden movement, or a loud noise. This can be dangerous for you, others, and the animals. Give animals extra room and time to adjust to you. When passing horses use special care and follow directions from horseback riders—ask if uncertain. Running cattle and disturbing wildlife is a serious offence. Leave gates as you find them or as marked
6	Plan ahead	Know your equipment, your ability, and the area in which you are riding—and prepare accordingly. Be self-sufficient at all times, keep your equipment in good repair, and carry necessary supplies for changes in weather or other conditions. A well-executed trip is a satisfaction to you and not a burden to others. Always wear a helmet and appropriate safety gear

Source: International Mountain Biking Association (2013)

with those of walking and less than those from motorised vehicles or horse riding (Chiu and Kriwoken 2003; Thurston and Reader 2001). Figure 7.4 shows the moderate damage to vegetation can easily be caused by the passage of mountain bikes on a moorland in north-east Wales during wet weather in winter. Mountain bikes crossing watercourses (Fig. 7.4B) and passing through waterlogged flushes in upland areas (Fig. 7.4C, D) can release fine sediment which can result in siltation on stream beds. In severe cases, siltation can affect spawning gravels of fish (like salmon) by blocking the flow of water and oxygen to eggs laid within gravels (Wickett 1954; Sear et al. 2017). The severity of impacts depends on climate, slope, and other environmental variables. Steep slopes with sparse vegetation and/or fine homogenous soils are most susceptible to damage from biking (White et al. 2006).

The greatest impacts usually occur early in trail use, on downhill (braking and skidding) and uphill (wheel spinning) slopes (especially when wet), and on curves (braking and skidding) (Goeft and Alder 2001). This damage may increase trail incision and/or widening, soil erosion, and water runoff. The impact of MTB on erosion is, however, cumulative and curvilinear (Chiu and Kriwoken 2003). After rapid initial erosion, the rate of change declines, probably because of increasing soil compaction.

MTB is becoming more and more popular as a competitive sport, and impacts from competitive MTB probably occur faster and are more acute than those from recreational biking. There is little research into the question of use intensity (e.g. under competitive racing conditions) and/or duration, but large organised mountain bike/challenge events such as The Brecon Beast (http://

7.4 Environmental Impact

Fig. 7.3 Example of a North Shore board way at Llandegla mountain bike centre, North Wales. Photo by Tim Stott

www.breconbeast.co.uk/) are likely to have large impacts over a short space of time (one weekend). However, such events are only held once per year—the trails used do have time to recover. The essential thrill element of racing demands technically challenging courses, steep up/downhill slopes, fast, hard braking, more intense use, cutting corners, wet sections, and the inclusion of jumps/drop-offs. Newsome et al. (2011) argued that adventure racing which might include downhill competitive MTB events, for example, probably poses higher risk of environmental impacts than recreational biking since the element of competition means that competitors have less time to consider their impact on the environment and take ameliorating action.

Australian studies of racing events have found that soil loss at sharp corners is greater than on straight sections (Hawes 1997). Under wetter conditions, there are increased off-trail vegetation impacts and trail widening, especially on steep slopes and corners. Racing under such conditions also increases off-trail vegetation impacts and trail widening (Goeft and Alder 2001). Spectator crowds may cause additional impacts (e.g. off-track vegetation trampling). A German study which evaluated the impacts from a World Championship MTB race with 870 participants and 80,000 spectators showed soil compaction that resulted from bikes was less, although deeper, compared to that from the spectators, with recovery taking approximately 19 months (Wöhrstein 1998). Research studies have consistently revealed that most impact occurs with initial or low use, with a diminishing increase in impact associated with increasing levels of traffic (Hammit and Cole 1998; Leung and Marion 1996). Furthermore, once trampling occurs, vegetative recovery is a very slow process. Wilson and Seney (1994) examined the relative impact of hikers, horses, motorcycles, and off-road bicycles in terms of water runoff and sediment yield on existing trails in Montana. They found that horses and hikers (hooves and feet) made more sediment available than wheels (motorcycles and off-road bikes) and that the effect was most pronounced on pre-wetted trails. However, the study was limited to tests of only 50 and 100 passes by the four modes of travel.

Fig. 7.4 (A) Damage to moorland vegetation due to the passage of mountain bikes on a moorland in NE Wales in winter. Photo by Tim Stott. (B) Mountain bikes crossing watercourses can release fine sediment which can result in siltation on stream beds. Photo by Ewan Stott. (C) Mountain bikes passing through waterlogged flushes in upland areas can cause compaction, remove binding vegetation, and release fine sediments. Photo by Ewan Stott. (D) Mountain bikes passing through upland moorland flatten vegetation. Photo by Tim Stott

7.4.2 Impacts of Mountain Biking on Wildlife

The impacts of MTB on wildlife are similar to those of hikers and other non-motorised trail users. In comparison to the off-road vehicles and snowmobiles discussed in an earlier chapter, mountain bikes at least do not have the noise of an engine. Nevertheless, riders can make noise which can disturb wildlife within a certain range of the activity. An investigation of the interactions of wildlife and trail users (hikers and mountain bikers) was carried out at Antelope Island State Park in Utah by Taylor and Knight (2003). In their study a hidden observer used an optical rangefinder to record the response of pronghorn antelope, bison, and mule deer to an assistant who hiked or cycled a section of trail. The observer then measured wildlife reactions, including flight response, flight distance, alert distance, distance fled, and distance from trail. Their results showed that 70% of animals located within 100 m of a trail were likely to flee when a trail user passed and that wildlife exhibited statistically similar responses to MTB and hiking. Wildlife reacted more strongly to off-trail recreationists, suggesting that visitors who stay on trails would reduce wildlife disturbance. While Taylor and Knight found no biological justification for managing MTB any differently than hiking, they note that bikers cover more ground in a given time period than hikers and thus can potentially disturb more wildlife per unit time. Interestingly, in their study Taylor and Knight also surveyed 640 hikers, mountain bikers, and horseback riders on the island to assess their perceptions of the effects of recreation on wildlife. Most respondents felt that they could approach animals far closer than the flight distance suggested by the research, and 50% felt that recreational users did not have a negative effect on wildlife.

Gander and Ingold (1997) conducted an experimental study in Switzerland to evaluate the disturbance associated with MTB, hiking, and jogging on high elevation chamois (goat-like mammals found in the European mountains). They assessed alert distance, flight distance, and distance fled and found that approximately 20% of the animals fled from trailside pastures in response to visitor intrusions. There were no statistically significant differences, however, between the behavioural responses of animals and the three different types of user, and the authors concluded that restrictions on MTB above the timberline could not be recommended from the perspective of chamois disturbance only.

Park staff of Banff National Park noted that hikers were far more numerous than mountain bikers on the Moraine Lake Highline Trail but that the number of encounters between cyclists and grizzly bears was disproportionately high. Benn and Herrero (2002) investigated this and confirmed that three of the four human-grizzly bear encounters that occurred along the trail during 1997–1998 involved mountain bikers. Previous research had shown that grizzly bears are more likely to attack when they first become aware of a human presence at distances of less than 50 m. Herrero and Herrero concluded that mountain bikers travel faster, more quietly, and with closer attention to the tread than hikers, all attributes that limit reaction time for bears and bikers and increase the likelihood of sub-50 m encounters. In addition, most of the bear-cyclist encounters took place on a fast section of trail that went through high-quality bear habitat with abundant berries. To reduce such incidents, they recommended education, seasonal closures of the trail to bikes and/or hikers, construction of an alternate trail, and regulations requiring a minimum group size for bikers.

Papouchis et al. (2001) evaluated the behavioural responses of desert bighorn sheep to disturbance by hikers, mountain bikers, and vehicles in low- and high-use areas of the Canyonlands National Park, USA. A total of 1029 bighorn sheep-human interaction observations were made, and the authors reported that sheep fled 61% of the time from hikers, 17% of the time from vehicles, and 6% of the time from mountain bikers. They attributed the stronger reaction to hikers, particularly in the high-use area, to more off-trail hiking and direct approaches to the sheep. The researchers recommended that park officials restrict recreational uses to trails, particularly during the lambing and rut seasons, in order to minimise disturbance.

Spahr (1990) studied flushing distances of bald eagles along the Boise River in Idaho when they were exposed to actual and simulated walkers, joggers, fishermen, bicyclists, and vehicles. The highest frequency of eagle flushing was associated with walkers (46%), followed by fishermen (34%), bicyclists (15%), joggers (13%), and vehicles (6%). However, cyclists caused eagles to flush at the greatest distances (mean = 148 m), followed by vehicles (107 m), walkers (87 m), fishermen (64 m), and joggers (50 m). Eagles were most likely to flush when recreationists approached slowly or stopped to observe them and were less alarmed when bicyclists or vehicles passed quickly at constant speeds.

Similar findings have been reported in other studies, where the difference in flushing frequency between walkers and cyclists/vehicles was attributed either to the shorter time of disturbance or the additional time an eagle has to "decide" to fly (Van der Zande et al. 1984).

7.5 Management and Education

7.5.1 Impacts to Vegetation: Management Implications

Trail managers can either avoid or minimise impacts to vegetation through careful trail design, construction, maintenance, and management of visitor use. Marion and Wimpey (2007) make some recommendations to reduce vegetation impacts:

- Design trails that provide the experience that trail users seek to reduce their desire to venture off-trail.
- Locate trails away from rare plants and animals and from sensitive or critical habitats of other species. Involve resource professionals in designing and approving new trail alignments.
- Keep trails narrow to reduce the total area of intensive tread disturbance, slow trail users, and minimise vegetation and soil impacts.
- Limit vegetation disturbance outside the corridor when constructing trails. Hand construction is least disruptive; mechanised construction with small equipment is less disruptive than full-sized equipment; skilled operators do less damage than those with limited experience.
- Locate trails on sidehills where possible. Constructing a sidehill trail requires greater initial vegetation and soil disturbance, but sloping topography above and below the trail bench will clearly define the tread and concentrate traffic on it. Trails in flatter terrain or along the fall line may involve less initial disturbance but allow excessive future tread widening and off-tread trampling, which favour non-native plants.
- Use construction techniques that save and redistribute topsoil and excavated plants (Marion and Wimpey 2007, p. 2).

Marion and Wimpey (2007) also go on to suggest important considerations for maintaining and managing trails to avoid unnecessary ongoing impacts to vegetation:

- While it is necessary to keep trail corridors free of obstructing vegetation, such work should seek to avoid "daylighting" the trail corridor when possible. Excessive opening of the overstory vegetation allows greater sunlight penetration that permits greater vegetation compositional change and colonisation by non-native plants.
- An active maintenance programme that removes tree falls and maintains a stable and predictable tread also encourages visitors to remain on the intended narrow tread. A variety of maintenance actions can discourage trail widening, such as only cutting a narrow section out of trees that fall across the trail, limiting the width of vegetation trimming, and defining trail borders with logs, rocks, or other objects that won't impede drainage.
- Use education to discourage off-trail travel, which can quickly lead to the establishment of informal visitor-created trails that unnecessarily remove vegetation cover and spread non-native plants. Such routes often degrade rapidly and are abandoned in favour of adjacent new routes, which unnecessarily magnify the extent and severity of trampling damage.
- Educate visitors to be aware of their ability to carry non-native plant seeds on their bikes or clothing, and encourage them to remove seeds by washing mud from bikes, tyres, shoes, and clothing. Preventing the introduction of non-natives is key, as their subsequent removal is difficult and costly.
- Educate visitors about low-impact riding practices, such as those contained in the IMBA-approved *Leave No Trace Skills & Ethics: Mountain Biking* booklet (www.LNT.org).

7.5.2 Impacts to Soils: Management Implications

Soil loss is among the longest lasting forms of trail impact, and minimising erosion and muddiness is the most important objective for achieving a sustainable trail. Soil cannot easily be replaced on trails, and where soil disappears, it leaves ruts that make travel and water drainage more difficult, prompting further impacts, such as trail widening. Research indicates that MTB and hiking are very similar in their impact on soils. Other factors, particularly trail grade, trail/slope alignment angle, soil type/wetness, and trail maintenance, are more influential determinants of tread erosion or wetness.

Marion and Wimpey (2007) proposed several tactics for avoiding the worst soil-related impacts to trails:

- Discourage or prohibit off-trail travel. Informal trails created by off-trail travel frequently have steep grades and fall-line alignments that quickly erode, particularly in the absence of tread maintenance. Exceptions include areas of solid rock or non-vegetated cobble.
- Design trails with sustainable grades, and avoid fall-line alignments (see p. 112 for more).
- When possible, build trails in dry, cohesive soils that easily compact and contain a larger percentage of coarse material or rocks. These soils better resist erosion by wind and water or displacement by feet, hooves, and tyres.
- Minimise tread muddiness by avoiding flat terrain, wet soils, and drainage-bottom locations.
- Use grade reversals to remove water from trail treads. Grade reversals are permanent and sustainable—when designed into a trail's alignment, they remain 100% effective and rarely require maintenance (Marion and Wimpey 2007, p. 4).

Other more temporary strategies will require periodic maintenance to keep them effective:

- While the use of a substantial slope (e.g. 5%) helps remove water from trail surfaces, it is rarely a long-term solution. Surface cupping and berm development will generally occur within a few years after trail construction. If it is not possible to install additional grade reversals, reshape the trail to re-establish a cambered surface periodically, and install wheel-friendly drainage dips or other drainage structures to help water flow off the trail.
- If it is not possible to install proper drainage on a trail, consider re-routing problematic trail sections or possibly hardening the surface.
- In flatter areas, elevate and crown the surface to prevent muddiness, or add a gravel/soil mixture in low spots.

It is always important to understand that visitor use of any type on trails when soils are wet contributes substantially greater soil impact than the same activities when soils are dry. So, another effective measure would be to discourage or prohibit the use of trails that are prone to muddiness during rainy seasons or snowmelt. Generally such use can be redirected to trails that have design or environmental attributes that allow them to better sustain wet-season uses.

7.5.3 Impacts to Water Resources: Management Implications

Marion and Wimpey (2007) state that the same trail design, construction, and maintenance measures that help minimise vegetation and soil impacts also apply to water, although some additional measures are needed to protect water resources:

- Trails should avoid close proximity to water resources. For example, it is better to build a trail on a sidehill along a lower valley wall than to align it through flat terrain along a stream edge, where trail runoff will drain directly into the stream.
- It is best to minimise the number of stream crossings. Where crossings are necessary, scout the stream carefully to select the most resistant location for the crossing. Look for rocky banks and soils that provide durable surfaces.
- Design water crossings so the trail descends into and climbs out of the steam crossing, pre-

venting stream water from flowing down the trail.
- Armour trails at stream crossings with rock, geotextiles, or gravel to prevent erosion.
- Include grade reversals, regularly maintained cambered trail surface, and/or drainage features to divert water off the trail near stream crossings. This prevents large volumes of water and sediment from flowing down the trail into the stream and allows trailside organic litter, vegetation, and soils to slow and filter water.
- On some heavily used trails, a bridge may be needed to provide a sustainable crossing.
- Where permanent or intermittent stream channels cross trails, use wheel-friendly open rock culverts or properly sized buried drainage culverts to allow water to cross properly, without flowing down the trail (Marion and Wimpey 2007, p. 5).

The environmental impacts of MTB and rider preferences in southwest Western Australia were analysed by Goeft and Alder (2001) to determine appropriate trail design and to ensure that this popular nature-based activity has minimal environmental impact while meeting rider requirements. Environmental impacts such as soil erosion and compaction, trail widening, and changes in vegetation cover on a recreational trail and racing track were monitored for 12 months to determine the short- and long-term effects of riding during winter (rainy) and summer (dry) seasons. Rider preferences were determined through a survey of mountain bike riders in the region. The study found that trail erosion, soil compaction, trail widening, and vegetation damage can occur, but they can be avoided or minimised with appropriate trail siting, design, and management.

The study also found that rider preferences for downhills, steep slopes, curves, and jumps along with water stations and trail markings need to be included in the siting and design of the trails. When multiple-use trails are considered, mountain bikers are willing to share the trail with most other users, the exception being with motorised vehicles.

Owing to the risk of potential environmental impacts (ground and wildlife disturbance) and a relative lack of empirical, comparable data (White et al. 2006), even non-competitive, cross-country recreational MTB remains restricted or banned in many protected areas with a conservation mandate. Examples include parts of the Cairngorm Mountains (Scotland) (e.g. Hanley et al. 2002) and wilderness areas within the Greater Blue Mountains World Heritage Area (Australia) (NPWS 2001). However, lobbying pressure from bikers for greater access to such areas is growing. Management agencies need to provide empirical evidence of environmental impacts when making and/or justifying their decisions of whether or not to permit MTB (e.g. Office of Environment and Heritage 2011).

7.5.4 Managing and Educating Mountain Bikers

Studies carried out in several countries have shown that mountain bike riders' preference is to ride in large, scenic, natural areas on single, unsealed trails with a variety of features that include steep slopes, short and long curves, jumps, rocks, and logs (e.g. Koepke 2005; Goeft and Alder 2001). Traditionally, protected areas like national parks and nature reserves, which have a responsibility for conservation, have provided settings for the specialised activities of adventure recreation such as MTB. Guidebooks have, arguably, encouraged mountain bikers into such areas resulting in environmental damage and sometimes social conflicts.

Social conflicts and/or environmental impacts in such areas resulting from activities such as MTB have usually been handled by land managers, first by establishing standards for the activity and then developing regulations (Ewert et al. 2006). Planning models and management frameworks, such as the "Recreational Opportunity Spectrum" and "Limits of Acceptable Change" (Stankey et al. 1985) and the "International Mountain Biking Association Rules of the Trail" (Table 7.4) have been developed to support such decisions. All depend on agreement among stakeholders on what constitutes acceptable use of public natural areas. If the majority of participants' motivations for using such areas are consumptive, management may find it difficult to apply such tools to MTB, especially in its more extreme derivatives. Issues may be exacerbated when visitors and managers perceive impacts differently (Burgin and Hardiman 2012; Martin et al. 1989).

As a recreational group, the lobbying power of mountain bikers is increasing. Formal groups such as sports associations and clubs, as well as informal groups which may operate through online forums, are becoming more influential (Cleggett 2010). Tourism and retailing industries who have commercial interests also add influence.

Managers of areas with a conservation mandate are sometimes confronted with threats of litigation or appeals against their efforts to restrict what they perceive to be inappropriate recreational activities. Potential for litigation may grow because of the perceived risk of injury (Sarre 1989). The internet enhances the lobbying power of such stakeholders (AARA 2010). For example, in the 1990s in at least three UK national parks, there were campaigns by the MTB lobby that resulted in changes in the decisions of land managers. MTB in Dartmoor was initially a criminal offence, while Exmoor considered it an "unsuitable activity," and Snowdonia attempted to ban mountain bikers from its bridleways. MTB has since become an accepted activity in these parks, although opposition from other users continues (Palmer 2006).

7.5.5 Forest-Based Mountain Biking: The UK Experience

The first purpose-built, forest-based mountain bike venue in the UK was opened in the mid-1990s at Coed-y-Brenin (North Wales), and its success sparked similar developments elsewhere in the country (Table 7.5).

Throughout the UK these venues, together with other cycle ways, provide more than 2600 km of tracks on national forest estate lands. These include "centres" dedicated to single site MTB locations with a visitor centre and support facilities (e.g. café, bike repair shop, showers and toilets, trail guides), offering multiple-way, marked trails of varying difficulty (e.g. Fig. 7.5A). "Bases," on the other hand, host waymarked or mapped trails, together with independently operated support facilities (e.g. accommodation, restaurants/cafés, bike sale, and/or repair shop). Located in sparsely populated, poorer rural areas,

Table 7.5 Mountain biking in England: venues listed on the Forestry Commission website

Whinlatter (NW)
Whinlatter Duathlon Whinlatter puts the mountain back into mountain biking. The Altura Trail is already a classic affording stunning views, crazy descents and leg burning climbs.
Nearest town: Keswick (Sat Nav: CA12 5TW)
Grizedale (NW)
Mountain bikers on The North Face Trail Grizedale. Grizedale plays host to The North Face Trail and is a great base for expeditions along the challenging mountainous routes of the southern Lake District.
Nearest town: Clitheroe (Sat Nav: LA22 0QJ)
Gisburn Forest (NW)
Mountain bikers on The North Face Trail Grizedale. Start your adventure on the Skills Loop where you can learn how to ride or brush up on your trail feature skills from table tops to berms. Or head straight out on the Bottoms Beck or The 8 bike trails.
Nearest town: Ambleside (Sat Nav: BB7 4TS)
Delamere (NW)
Cyclist on a forest trail. Combining miles of XC trails with a great skills area, Delamere's light soils make it an all year round venue. Cheshire's largest woodland area is a stone's throw from the major cities of the North West and an excellent venue for evening rides. Please note that the car parks are locked at 8pm in summer and 5pm in winter.
Nearest town: Northwich (Sat Nav: CW8 2JD)

(continued)

Table 7.5 (continued)

Kielder (NE)
Cyclists riding the Lonesome Pine mountain bike trail in Kielder Forest. Wilderness mountain biking at its best with epic red grade trails and cross boarder links to the Newcastleton Seven Stanes Centre. The lakeside blue grade trails offer spectacular views. Nearest town: Bellingham (Sat Nav: NE48 1ER)
Hamsterley Forest (NE)
Women's mountain bike event, Hamsterley Forest. The North East's hidden gem, combes miles of cross-country routes with the adrenaline fuelled 4X course of Descend Bike Park. A fantastic place for all levels of skill and experience. Nearest town: Bishop Auckland (Sat Nav: DL13 3NL)
Dalby Forest (Yorks)
Mountain biker at Darkgate Dyke in Dalby Forest. With miles of expertly sculpted technical single track you'll want to ride all day. Dixon's Hollow Bike Park ticks all the freeride boxes from North Shore to 4X. Home of the 2010 World Cup XC course that provides a challenge for even the most experienced riders. Nearest town: Pickering (Sat Nav: YO18 7LT)
Cannock Chase (W Midlands)
Mountain biking at Cannock Chase. Great mountain biking for all ages and experience in the heart of the West Midlands. Barrel along the Follow the Dog Trail, or the brand new Monkey Trail, or drop in at Stile Cop Bike Park. Nearest town: Rugeley (Sat Nav: WS15 2UQ)
Sherwood Pines (E Mids)
Cyclist on a single track mountain bike trail. A wide variety of graded trails guarantee that all riders are catered for. to a warp speed ride around the Kitchener Trail. The ever developing bike park and dirt jumps ensure there is plenty of potential for air time. Nearest town: Mansfield (Sat Nav: NG21 9JH)
Thetford Forest (E)
England's largest lowland forest boasts literally hundreds of miles of fast flowing single track. Fast and unrelenting trails make for long and challenging rides but also make it accessible for all level of cyclists. Nearest town: Thetford (Sat Nav: IP27 0AF)
Bedgebury (SE)
Cyclist riding a freeride North Shore course. Not far from London, Bedgebury is a truly stunning location to ride. From family routes to fast red grade single track, there is something for everyone. Nearest town: Goudhurst (Sat Nav: TN17 2SJ)
Forest of Dean (SW)
Downhill cycle trail. Forest of Dean Cannop Cycle Centre is the ideal base to take in the whole of the forest and its myriad of singletrack and trails. The nearby Sallowvallets bike area is home to the Freeminer Trail and some great short downhill runs. Nearest town: Coleford (Sat Nav: GL15 4)
Haldon Forest Park (SW)
Cyclists using the skills area at Haldon Forest Park. Just 15 minutes outside of Exeter, Haldon Forest Park caters for hardened freeriders. The new red grade XC trail gives you a great introduction to the network of trails that wind through the forest. Nearest town: Exeter (Sat Nav: EX6 7XR)

Source: https://www.forestry.gov.uk/forestry/INFD-6QHHV3, accessed 22/2/18

their development also offers substantial economic benefits through employment (TRC 2005).

Although use of the trails is free, supporting facilities are provided on a commercial basis. These initiatives are public-private sector partnerships, led by the respective regional forestry commissions and comprising local governments and national and regional tourism bodies, together with local private enterprises. Although all centres have proved successful, those in Scotland especially have prospered. For example, the Nevis Range and Leanachan Forest venues (Fort William, Scotland) hosted the annual World Cup Mountain Bike Series during 2002–2005 and again in 2010. In 2007, they also hosted the Mountain Bike World Championships with international competition for four mountain bike disciplines: Downhill, Cross-Country, Trials, and 4-Cross. The Scottish town of Dumfries hosted the 2010 World Mountain Bike Conference, and

7.5 Management and Education

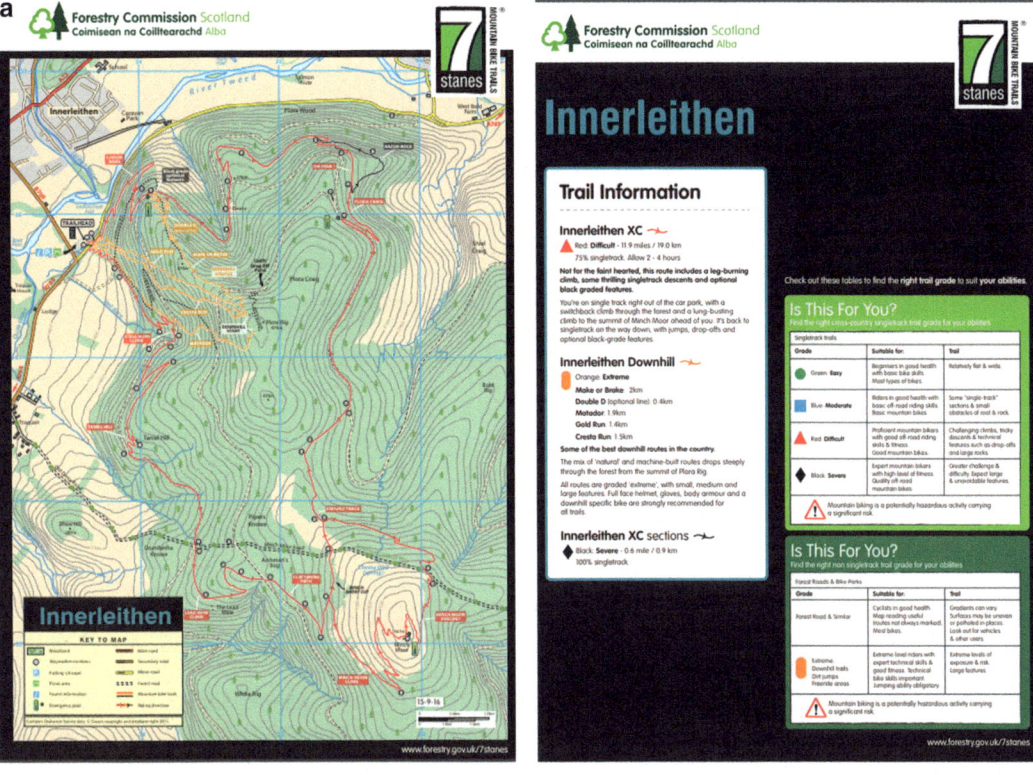

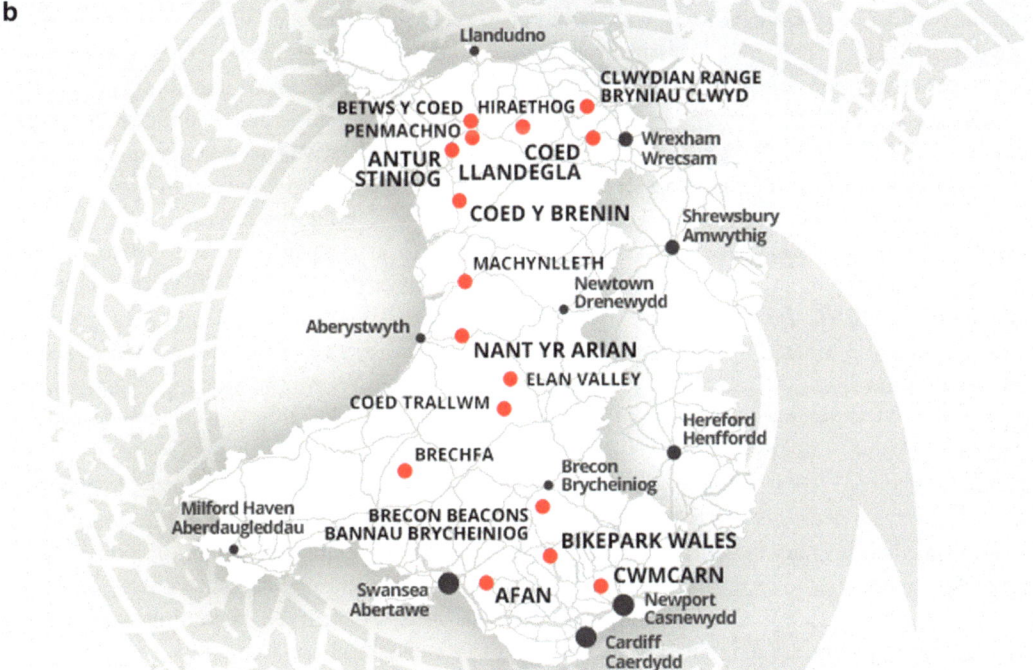

Fig. 7.5 (A) Trail guide for Innerleithen, part of the 7stanes mountain biking suite of trails developed and managed by Forestry Commission Scotland. Source: http://scotland.forestry.gov.uk/images/pdf/rec_pdfs/7stanes-innerleithen.pdf, accessed 22/2/18. (B) Mountain bike centres and bases in Wales which offer a variety of ride experiences. Trails are graded to help riders choose the best to suit their experience. Source: http://www.mbwales.com/, accessed 22/2/18

the 2014 Commonwealth Games was held in Glasgow. The men's and women's cross-country MTB competition was held at the Cathkin Braes Mountain Bike Trails. MTB returned to the Commonwealth Games programme, after last being competed back in 2006. This undoubtedly brought more attention to the sport and the Scottish venues in particular.

The largest of the UK's MTB venues is the 7stanes project in Southern Scotland (Forestry Commission, Scotland, n.d). Opened in 2001, this multi-agency, seven-centre network is a world-class MTB venue that attracts domestic and international visitors. There are nearly 600 km of single track trails of varying levels from "easy" to "severe." The "difficult" trails are most popular. There are also Action Trail Areas for freestyle enthusiasts, and additional non-waymarked and ungraded forest trails. 49% of visitors are "intermediate" riders, 30% "advanced," and 8% "beginners" (TRC/EKOS 2007).

Highly experienced mountain bike riders were targeted as "early adopters," and the focus was on product (e.g. trail building, infrastructure development). The strategy is to widen the user base, attract new users into the sport, and make it more accessible socially, especially to females, families, schools, and older visitors. This equates to the development of a true mass-market tourism/recreation product. There have been substantial economic benefits for a mainly rural region that has traditionally suffered high unemployment (TRC 2005). In 2007, 7stanes attracted an estimated 395,000 visitors (increased from 172,000 in 2004), making it one of the 20 most popular Scottish tourist attractions. Some 43% of visitors came from within Scotland, 32% from elsewhere in the UK, and 5% from overseas. For 78% of visitors, 7stanes was their primary reason for visiting the region and more than one-third stayed at least overnight (up from 25% in 2004). The project's net economic benefits are estimated to be £9.18 million (USA $14.53) in tourism expenditure, the creation of 212 full-time equivalent jobs, and £3.72 million (USA $5.89) gross value added to the regional economy (TRC/EKOS 2007). Other forest-based MTB centres in the UK have produced comparable economic benefits to their respective regions and local communities.

7.5.6 Future Research and Management Implications for Mountain Biking

MTB is very popular in affluent, economically developed countries where governments are keen to promote healthy exercise and whose citizens are expected to enjoy increasing leisure time in the coming decades (Molitor 2000). MTB will probably continue to produce new derivatives undertaken for tourism/recreation and as competitive, formalised sports; for example, night MTB has been popular over the past decade or two in some areas of the UK. With a widening diversity of participants seeking different experiences, there will be more social and/or environmental management challenges for land managers. The main challenge is to innovate and not to react negatively. The UK Forestry Commission case shows that tourism/recreation demand and commercial natural resource production supply have successfully collaborated to produce a, seemingly, "win-win" solution for a range of stakeholders.

The forestry history of the UK has helped. The country became depleted of its own timber resources after WWI, and the government set up the Forestry Commission in 1919 to acquire land and plant conifers. This legacy of commercially harvested forests has become an appropriate and complementary resource to national parks for MTB as they (1) provide the large spaces in natural settings that are required, (2) are less biologically sensitive to anthropogenic impacts, (3) may offer substantial economic benefits to local rural communities, and (4) may offer substantial economic benefits to the owners of the lands.

To assist decision-making by the various public/private stakeholders in the multi-agency partnerships, research is needed to provide a better understanding of (1) the environmental impacts emanating from MTB activities across different ecosystems and (2) the demographics and understanding people's motivation for MTB (Taylor 2010). With such information, and with models of environmentally sustainable operations available, potential conflict over access to and/or inappropriate use of public lands of importance for conservation could be reduced. Land managers could then better manage biodiversity by offering options elsewhere and thus clear the trail for MTB.

Conclusions

1. The mountain bike was developed in Marin County, California, in the mid-1970s since when many new subtypes of MTB have developed, such as cross-country, all-day endurance, freeride, and downhill and a variety of track and slalom types. Fatbikes and BMX are other derivatives.
2. In the USA the Outdoor Foundation's survey (2017) reported that BMX and MTB (non-paved surface) commanded participation numbers of 1,655,000 and 6,751,000, respectively, in 2006, and these figures rose to 3,104,000 and 8,615,000 in 2016, three-year increases of 43.2% for BMX riding and 0.9% for MTB.
3. Cordell (2012) showed that that bicycle participation in outdoor activities in the USA exceeded hiking and was on a par with jogging/trail running and car/backyard/RV camping. He also showed that the number of people in the USA (1999–2001) bicycling on mountain/hybrid bikes was 44 million (18% of the total number aged > 16 participating in outdoor activities), although this had declined to 42.7 million by the time of his 2005–2009 survey.
4. Infrastructure to support the various forms of MTB such as purpose-built single track trails, uplift facilities for downhill, and bike parks for freeriding/trials is increasing in many countries. In the UK there are now over 40 purpose-built MTB/forest cycling centres/bases currently operated by the Forestry Commission of Great Britain. An indication of how important MTB has become to some ski resorts is that summer revenue now represents approximately 75% of winter snow recreation revenue.
5. Environmental damage caused by MTB includes damage to vegetation, soils, and water resources and disturbance to wildlife.
6. Huge progress in managing the impact has been made in the half century since the first mountain bike was developed. Numerous purpose-built centres and bases have specially designed and graded trails for different levels of ability. The IMBA has been educating mountain bikers by commissioning the publication books on managing MTB and its "Mountain Bike Rules of the Trail" which the IMBA considers that "every mountain biker should know and live by...."

References

Americans for Responsible Recreational (ARRA). (2010). About ARRA: Statement of purpose. Retrieved August 5, 2013, from http://www.arra-access.com/site/PageServer?pagename=arra_about.

Benn, B., & Herrero, S. (2002). Grizzly bear mortality and human access in Banff and Yoho National Parks, 1971–98. *Ursus*, 213–221.

Bradshaw, G. (2006). *The Australian Bicycle Industry Report 2006*. Melbourne: Graphyte Media Pty Ltd.

Burgin, S., & Hardiman, N. (2012). Is the evolving sport of mountain biking compatible with fauna conservation in national parks? *Australian Zoologist, 36*, 201–208.

Chiu, L., & Kriwoken, L. (2003). Managing recreational mountain biking in Wellington Park, Tasmania, Australia. *Annals of Leisure Research, 6*, 339–361.

Cleggett, M. (2010, July 7). Wheel turns in favour of downhill riders. *Blue Mountains Gazette*.

Cordell, K. (2012). *Outdoor Recreation Trends and Futures: A Technical Document Supporting the Forest Service 2010 RPA Assessment*. Ashville: Southern Research Station. Retrieved from www.srs.fs.usda.gov.

Ewert, A., Attarian, A., Hollenhorst, S., Russell, K., & Voight, A. (2006). Evolving adventure pursuits on public lands: Emerging challenges for management and public policy. *Journal of Park Recreation Administration, 24*, 125–140.

Faulks, P., Ritchie, B., Brown, G., & Beeton, S. (2008). *Cycle Tourism and South Australia Destination Marketing*. Gold Coast: Sustainable Tourism Cooperative Research Centre.

Forestry Commission, Scotland (FCS). (n.d.). Ride the 7stanes—Scotland's biking heaven. Retrieved August 5, 2013, from http://www.forestry.gov.uk/forestry/achs-5rjeky.

Gander, H., & Ingold, P. (1997). Reactions of male alpine chamois *Rupicapra r. rupicapra* to hikers, joggers and mountain bikers. *Biological Conservation, 79*, 3.

Goeft, U., & Alder, J. (2001). Sustainable mountain biking: A case study from the southwest of western Australia. *Journal of Sustainable Tourism, 9*(3), 193–211.

Hammit, W. E., & Cole, D. N. (1998). *Wildland Recreation: Ecology and Management*. New York: John Wiley and Sons.

Hanley, N., Alvarez-Farizo, B., & Shaw, W. D. (2002). Rationing an open-access resource: Mountaineering in Scotland. *Land Use Policy, 19*, 167–176.

Hardiman, N., & Burgin, S. (2013). Mountain biking: Downhill for the environment or chance to up a gear? *International Journal of Environmental Studies, 70*(6), 976–986.

Hawes, M. (1997). *Environmental Impacts and Management Implications of the 1997 National Mountain Bike Championships*. Canberra: Department of Conservation and Land Management.

International Mountain Biking Association (IMBA). (2013). Official IMBA mountain bike rules of the trail. Retrieved September 12, 2013, from http://mountainbike.about.com/od/tipsandtechniques/a/IMBA_Rules.htm.

International Mountain Biking Association (IMBA), Australia. (2010). Retrieved August 5, 2013, from http://www.imbaau.com/.

Koepke, J. (2005). *Exploring the Market Potential for Yukon Mountain Bike Tourism*. Whitehorse: Cycling Association of Yukon.

Leung, Y. F., & Marion, J. L. (1996). Trail degradation as influenced by environmental factors: A state-of-the-knowledge review. *Journal of Soil and Water Conservation, 51*(2), 130–136.

Marion, J. L., & Wimpey, J. (2007). Environmental impacts of mountain biking: Science review and best practices. In *Managing Mountain Biking, IMBA's Guide to Providing Great Riding* (pp. 94–111). Boulder, CO: International Mountain Bicycling Association (IMBA).

Martin, S. R., McCool, S. F., & Lucas, R. C. (1989). Wilderness campsite impacts: Do managers and visitors see them the same? *Environmental Management, 13*, 623–629.

Moab Area Travel Council (MATC). (2010). Mountain biking in Moab. Retrieved August 5, 2013, from http://www.discovermoab.com/biking.htm.

Molitor, G. T. T. (2000). Five economic activities likely to dominate the new millennium: II the leisure era. *Technological Forecasting and Social Change, 65*, 239–249.

Newsome, D., Lacroix, C., & Pickering, C. (2011). Adventure racing events in Australia: Context, assessment and implications for protected area management. *Australian Geographer, 42*(4), 403–418.

NPWS. (2001). *Blue Mountains National Park: Plan of Management*. Hurstville: New South Wales National Parks and Wildlife Service.

Office of Environment and Heritage. (2011). *Sustainable Mountain Biking Strategy*. Sydney: NSW National Parks and Wildlife Service.

Palmer, C. (2006). Mountain biking: Settling into middle age—or clicking up a gear? *Byway and Bridleway, 3*, 33–34.

Papouchis, C. M., Singer, F. J., & Sloan, W. B. (2001). Responses of desert bighorn sheep to increased human recreation. *The Journal of Wildlife Management, 65*, 573–582.

Pickering, C. M., Hill, W., Newsome, D., & Leung, Y.-F. (2010). Comparing hiking, mountain biking and horse riding impacts on vegetation and soils in Australia and the United States of America. *Journal of Environmental Management, 91*, 551–562.

Sarre, R. (1989). Risk management in recreation: Legal dilemmas and legal solutions. *Australian Parks and Recreation, 25*, 7–14.

Sear, D. A., Pattison, I., Collins, A. L., Smallman, D. J., Jones, J. I., & Naden, P. S. (2017). The magnitude and significance of sediment oxygen demand in gravel spawning beds for the incubation of salmonid embryos. *River Research and Applications, 33*(10), 1642–1654.

Spahr, R. (1990). *Factors Affecting the Distribution of Bald Eagles and Effects of Human Activity on Bald Eagles Wintering along the Boise River, 1990*. Thesis, Boise State University.

Stankey, G. H., Cole, D. N., Lucas, R. C., Petersen, M. E., & Frissell, S. S. (1985). *The Limits of Acceptable Change (LAC) System for Wilderness Planning*. General Technical Report INT-176. Intermountain Forest and Range Experiment Station, United States Department of Agriculture, Forest Service, Ogden, UT.

Taylor, A. R., & Knight, R. L. (2003). Wildlife responses to recreation and associated visitor perceptions. *Ecological Applications, 13*(4), 12.

Taylor, S. (2010). Extending the dream machine': Understanding people's participation in mountain biking. *Annals of Leisure Research, 13*(1–2), 259–281.

The Outdoor Foundation. (2017). *Outdoor Participation Topline Report 2017*. Washington, DC: The Outdoor Foundation. Retrieved from www.outdoorfoundation.org.

References

Thurston, E., & Reader, R. J. (2001). Impacts of experimentally applied mountain biking and hiking on vegetation and soil of a deciduous forest. *Environmental Management, 27*, 397–409.

TRC. (2005). *Forestry Commission Scotland: An Ambition for Forest Cycling and Mountain Biking: Towards a National Strategy, Final Report*. Tourism Resources Company, Glasgow.

TRC/EKOS. (2007). *7stanes Phase 2 Evaluation: Report for Forestry Commission Scotland, October 2007*. Tourism Resources Company and EKOS, Glasgow.

Van der Zande, A. N., Berkhuizen, J. C., van Latesteijn, H. C., ter Keurs, W. J., & Poppelaars, A. J. (1984). Impact of outdoor recreation on the density of a number of breeding bird species in woods adjacent to urban residential areas. *Biological Conservation, 30*, 1–39.

Whistler Blackcomb. (n.d.). Whistler mountain bike park. Retrieved August 5, 2013, from http://www.whistlerbike.com/index.htm.

White, D. D., Waskey, M. T., Brodehl, G. P., & Foti, P. E. (2006). A comparative study of impacts to mountain bike trails in five common ecological regions of the Southwestern US. *Journal of Park Recreation Administration, 24*, 21–41.

Wickett, W. P. (1954). The oxygen supply to salmon eggs in spawning beds. *Journal of the Fisheries Board of Canada, 11*(6), 933–953.

Wilson, J. P., & Seney, J. P. (1994). Erosional impact of hikers, horses, motorcycles, and off-road bicycles on mountain trails in Montana. *Mountain Research and Development*, 77–88.

Wöhrstein, T. (1998). *Mountain Bike and Environment: Ecological Impacts and Use Conflict*. Saarbrücken Dudweiler: Pirrot Verlag und Druck.

Camping, Wild Camping, Snow Holing, and Bothies

8

Chapter Summary

This chapter first defines camping and presents a camping spectrum which ranges from survival camping to trailer tents, caravans, and motorhomes. It then discusses snow caves, quinzhees and igloos, and finally bothies before examining participation numbers. The final part of the chapter focuses on specific environmental impacts: damage to soil and vegetation, impacts on water, and the impacts on wildlife. The final section considers the management of these activities such as trail design and the development of hardened campsites as in the Overland Track in Tasmania. There is discussion on some attempts to manage the impact of human faeces on water resources, with examples from the Cairngorms, UK. Finally, examples of how the impact of camping on wildlife has been managed are presented.

8.1 Definitions

8.1.1 Camping

Camping is an outdoor activity which involves at least one overnight stay away from home in a shelter, such as a tent. Generally participants leave developed areas to spend time outdoors in more natural ones in pursuit of activities which give enjoyment, rejuvenation, and/or recreation. Some research has even indicated that people who camp are happier (Camping and Caravanning Club 2011) and that camping has benefits to families in terms of personal and social development (Jirásek et al. 2017). By our definition "camping" should include a minimum of one night spent outdoors, which distinguishes it from day-tripping, picnicking, and other similarly short-term recreational activities. Camping can be enjoyed through all four seasons. Luxury may be an element, such as in early twentieth-century African safaris or some of the early mountain tours in the Alps before hut accommodation was available (see Chap. 6), but including accommodation in fixed or permanent structures, be they wooden cabins such as may be found in summer camps or sports camps under the banner of "camping" is probably a step too far.

Camping as a recreational activity became popular among the elite middle and upper classes in the early twentieth century. Over time it has become popular with all demographic and economic classes. Modern campers visit publicly owned natural resources such as national and state parks, wilderness areas, Areas of Outstanding Natural Beauty, and commercial campgrounds. Camping is an important ingredient in the programmes of many youth organisations around the world, such as scouting, which is used to teach both self-reliance and teamwork.

Camping encompasses a range of activities and approaches to outdoor accommodation. Table 8.1 shows a spectrum of camping options.

Table 8.1 A proposed camping spectrum

Survival/very close to nature			Luxury camping or "glamping"	
Less insulated from nature			Very insulated from nature	
Survival camping	Wild camping/ expedition camping	Campsite/valley camping	Family/fixed tent camping with electricity	Trailer tent/caravan/ motorhome (recreation vehicle, RV)
Use of a "bivvy" (bivouac) bag or home-made/natural shelter, tarp, or basher (cape used by military personnel). Maybe an emergency shelter or may be a pre-planned camp but emphasis and priority on lightweight and survival *not* comfort	Use of a lightweight backpacking tent which has been carried into a wild area where there are no facilities, perhaps camping by a mountain tarn, remote but some attempt to be comfortable (e.g. sleeping mat)	Any portable tent pitched on a commercial site (farm or holiday park) where basic facilities like toilets, washing-up or shower facilities provided on-site. Usually the camping equipment is carried to the site in a car or van	Usually a large tent, often with two-thirds compartments which may be pre-pitched (e.g. Eurocamp), may have proper beds, raised wooden floor, and some luxury facilities (stove/ fire, electricity) but still canvas roof walls, so insulated from nature to some extent, but not totally	The main difference between this and tent camping is the mobility, raised off the ground. Many caravans or motorhomes have central heating, running water (from an onboard tank), cooker/ oven, permanent beds with comfortable mattress, shower and toilet, and so on. Capability to close blinds, turn on lights, and become largely *isolated from nature*

Survivalist campers (extreme left in Table 8.1) set off with as little as possible to get by and may rely on a waterproof bivouac (bivvy) bag (Fig. 8.1A) or simple tarpaulin (Fig. 8.1B, C) for protection from the weather, or they may use a hammock (Fig. 8.1D). Camping may be combined with hiking, called "backpacking," in which case lightweight backpacking tents such as that in Fig. 8.1E are used and, when pitched in remote regions (as in Fig. 8.1E), is termed "wild camping." However, camping is often enjoyed in conjunction with other outdoor activities such as canoeing, climbing, fishing, hunting, and cycling.

In the middle of the camping spectrum (Table 8.1) is campsite/valley camping where individuals, groups, or families pay a fee (usually per person/tent per night) to pitch on a well-maintained field, next to which there will be, as a minimum, toilets, but often other facilities like showers, washing-up sinks, hot water, and maybe a small shop selling basic provisions like bread and milk. In family/fixed tent camping with electricity (Fig. 8.1 F, G), the campers bring large family tents which sleep up to six or eight people and usually pitch up for a week or more. There will be folding chairs, tables, cookers, and even fridges or TVs when hooked up to electricity points which may be provided on the site (at extra cost).

At the other end of the spectrum (to the right in Table 8.1), trailer tents, touring caravans (Fig. 8.2A, C), motorhomes, and recreational vehicle (RV) travellers arrive equipped with their own electricity, heat, and patio furniture. At this more luxurious end of the spectrum, the term "glamping" is now used, a blend of "glamorous" and "camping" which has evolved from African safaris where demanding European and American travellers slept in luxurious canvas tents, supported by chefs, guides, porters, and butlers.

8.1 Definitions

Fig. 8.1 (A) A hooped bivouac bag used for lightweight survival camping. Photo by Tim Stott. (B) Using a tarpaulin (also called an Army "basher") to create an overnight shelter in Yorkshire, UK. Photo by Tim Stott. (C) A tarpaulin used as a lightweight overnight shelter in a woodland. The campers sleep on the ground. Photo by Tim Stott. (D) A hammock with tarpaulin used as a lightweight overnight shelter in a woodland. The camper sleeps off the ground. Photo by Tim Stott. (E) A mountain tent used for a camp at 2000 m by Castle Creek Glacier in the Cariboo Mountains, British Columbia. Photo by Tim Stott. (F) Typical family camping tent on a campsite in Anglesey, Wales. Photo by Tim Stott. (G) A campsite in Switzerland in summer showing a range of typical tent designs. Photo by Tim Stott

There is no universally held definition of what is and what is not camping. Fundamentally, it reflects a combination of the intent to stay out overnight and the nature of the activities involved. A children's summer camp with dining hall, catered meals, and bunkhouse accommodation may have "camp" in its name but perhaps fails to reflect the spirit and form of "camping" as it is broadly understood. Similarly, a homeless person's lifestyle may involve many common camping activities, such as sleeping out and preparing meals over a fire, but fails to reflect the elective nature and pursuit of spirit rejuvenation that are an integral aspect of camping. Likewise, cultures with itinerant lifestyles or lack of permanent dwellings such as the

Fig. 8.1 (continued)

Bedouin who inhabit desert regions in North Africa cannot truly be said to be "camping", as it is just their way of life. With this in mind, Fig. 8.2C shows typical static caravan park in North Wales. Although technically mobile (with wheels), such mobile homes are raised on blocks and may not be moved for many years. They have full services (water, electricity, gas) and owners typically pay an annual fee for ground rent and services and visit (usually by car) at weekends, for longer periods or even live permanently in them. These are not included in our definition of camping, though perhaps could be in some cases?

8.1.2 Snow Caves, Quinzhees, and Igloos

Snow holing is a general term for the activity of digging or making a shelter in snow as overnight accommodation. A snow cave is a shelter constructed in snow by certain animals (e.g. polar bear) and humans, in particular mountaineers, ski tourers, and anyone who enjoys the challenge of surviving outdoors in winter. Snow caves have thermal properties similar to igloos, both of which are particularly effective at providing protection from wind as well as low temperatures. A well-constructed snow cave can be 0 °C (32 °F) or warmer inside, even when outside temperatures are −40 °C (−40 °F).

Before a snow cave can be constructed, considerable care must be taken to find a suitable location. A bank of deep stable snow is required, preferably with a steep face into which the entrance can be excavated (Fig. 8.3A). The snow cave is then constructed by excavating snow so that the tunnel entrance is below the main space to retain warm air (Fig. 8.3B). By building it on a steep slope and digging slightly upwards and horizontally into the slope, the task is made

Fig. 8.2 (A) A 5-berth UK touring caravan on a campsite in Wales. Note the electricity hook-up point to the right of the picture. Photo by Tim Stott. (B) A typical touring caravan park at Strathyre in central Scotland. Note the spacing between caravans (for fire safety) and communal building which may provide services like washing machines, dryers, and sinks. Photo by Tim Stott. (C) A typical static caravan park in North Wales. Although technically mobile (with wheels), such vans are raised on blocks and may not be moved for many years. They have full services (water, electricity, gas) and owners typically pay an annual fee for ground rent and services, and visit (usually by car) at weekends and for longer periods. Photo by Tim Stott

easier. The roof is domed (Fig. 8.3C) to prevent dripping on the occupants. Adequate snow depth, free of rocks and ice, is needed with a depth of at least 1.5 m (4–5 ft) is required. Normally some kinds of markers (e.g. wands, flags, ski sticks, or avalanche probes, as seen in Fig. 8.3A) are placed around the perimeter to warn other walkers or skiers that the cave is there in case they walk on and collapse the roof. In windy conditions the entrance may be blocked from inside by the occupants using snow blocks or a rucksack to prevent spindrift entering the cave. In such circumstances the occupants need to be sure to maintain airflow in the cave. Where more than one cave is constructed, a climbing rope may be used to connect the occupants of each cave in case a cave collapsed. The rope would make it easier for rescuers to locate and dig out the occupants.

Regardless of construction type, the snow must be consolidated so that it retains its structure. A small pit is often dug deeper into one part of the cave floor (Fig. 8.3D) to provide a place for the coldest air to gather, away from the occupant(s). It is possible to sleep several consecutive nights in a snow cave, but care must be taken since a slight ice surface may develop on the inside of the cave from moisture in the exhaled air of the inhabitants. This is thought to result in reduced air ventilation through the snow cave walls and roof and thus increase risk of suffocation. As a precaution it is common to scrape

Fig. 8.3 (A) Excavating snow caves at Garbh Uist Beag, Cairngorms, Scotland. Photo by Tim Stott. (B) A snow cave is constructed by excavating snow so that the tunnel entrance is below the main space to retain warm air. Photo by Tim Stott. (C) Inside a snow cave in the Cairngorms, Scotland. Photo by Tim Stott. (D) Cooking equipment inside a snow cave. Photo by Tim Stott

off a thin layer from the inside of the cave ceiling each day spent in the cave.

The narrow entrance tunnel, which is just a little wider than the occupants of the cave, leads into a main chamber which consists of a flat area, perhaps with elevated sleeping platform(s), also excavated from snow (Fig. 8.3C). The use of tools such as a shovel and ice axe is vital; digging by hand is for emergencies only. Digging a snow cave can be very physically demanding. In perfect conditions with good snow, digging a snow cave for two or three persons can take three to four hours to complete. Therefore it is usual for a team of four to undertake the task. Two entrances are excavated (approximately 2–3 m apart) with a pair working on each entrance. Each person works in five-minute intervals digging, while the others help remove excess snow outside the cave and prepare food and warm liquids for the group. The aim is for the two tunnels to meet up and at that point the main chamber is excavated.

Another kind of snow cave is the quinzhee (or quinzee) which is a snow shelter that is made from a large pile of loose snow which is shaped then hollowed (Fig. 8.4A). This is in contrast to an igloo, which is built up from blocks of hard snow (Fig. 8.4B–D), and a snow cave (Fig. 8.3A–D), constructed by digging into the snow. The word

Fig. 8.4 (A) A quinzhee (or quinzee) is a snow shelter that is made from a large pile of loose snow which is shaped then hollowed. Photo by Tim Stott. (B) Cutting blocks to make the base of an igloo, Cairngorms, Scotland. Photo by Tim Stott. (C) Construction of an igloo nearing completion. Photo by Tim Stott. (D) A finished igloo. The person standing on top is demonstrating the structural strength of the igloo. Photo by Tim Stott

is of Athabaskan origin. The snow for a quinzhee need not be of the same quality as required for an igloo. Quinzhees are not usually made for permanent shelter, whereas igloos and can be used for seasonal habitation. Quinzhee can be made for winter camping and survival purposes or for fun. The construction of a quinzhee is much easier than the construction of an igloo, although they are somewhat less sturdy and more prone to collapsing in harsh weather conditions. Quinzhees are normally constructed in times of necessity, usually as an instrument of survival, so aesthetic and long-term dwelling considerations are normally exchanged for economy of time and materials. One simple construction technique is to pile up several rucksacks, shovel loose snow on top until it is around 0.5 m (2 ft) thick, then slowly remove the rucksacks from inside. This leaves an inside height after excavation which allows for sitting or crouching but not standing.

An igloo (the name coming from the Inuit language: iglu), also known as a snow house or snow hut, is a type of shelter built of snow, typically built when the snow can be easily compacted. Although igloos are normally associated with all Eskimo peoples, they were traditionally associated with people of Canada's Central Arctic and Greenland's Thule area. Snow is used because

Fig. 8.5 Ryvoan bothy in the Spey Valley, Scotland. Photo by Tim Stott

the air pockets trapped in it makes it an insulator. On the outside, temperatures may be as low as −45 °C (−49 °F), but on the inside the temperature may range from −7 °C (19 °F) to 16 °C (61 °F) when warmed by body heat alone. The snow used to build an igloo must have enough structural strength to be cut and stacked appropriately (Fig. 8.4B). The best snow to use for this purpose is snow which has been blown by wind, which can serve to compact and interlock the ice crystals. The hole left in the snow where the blocks are cut is usually used as the lower half of the shelter. Sometimes, a short tunnel is constructed at the entrance (Fig. 8.4D) to reduce wind and heat loss when the door is opened. Snow's effective insulating properties enable the inside of the igloo to remain relatively warm.

8.1.3 Bothies

A bothy is a basic shelter, usually left unlocked and available for anyone to use free of charge (Fig. 8.5). They are usually remote, with no road access, and may be ruined estate workers' or shepherds' cottages which have been renovated. They may or may not have glass in the windows, but normally the roof will have been maintained to provide a dry shelter. Some have an open fireplace, but visitors need to bring their own fuel and supplies. Most bothies are near a natural source of water. There are usually no toilets but a spade may be provided to bury excrement. Most bothies have designated sleeping areas, which commonly are either an upstairs room or a raised platform, thus allowing visitors to keep clear of cold air and draughts at floor height. No bedding, mattresses, or blankets are provided. Public access to bothies is either on foot, by bicycle, or by boat.

The term bothy was also used for basic accommodation, usually for gardeners or other workers on an estate. Bothies are to be found in remote mountainous areas of Scotland, Northern England, Ireland, and Wales. They are particularly common in the Scottish Highlands, but related buildings can be found around the world (e.g. in the Nordic countries there are wilderness huts).

The aim of the UK's Mountain Bothies Association (MBA) is to maintain simple shelters in remote country for the use and benefit of all who love wild and lonely places (www.mountainbothies.org.uk). It received the Queen's Award for Voluntary Service and celebrated 50 years of its existence in 2015, and its volunteers maintain over 100 bothies.

8.2 Participation Numbers

Estimating the number or people who camp is a difficult task, particularly as camping is so often combined with other activities. In his

Table 8.2 Trends in number of people of ages 16 and older participating in recreation activities by historic period in the USA, 1982–2001 (Source: Cordell 2012, p. 33)

Activity	Total participants			Change
	1982–1983	1994–1995	1999–2001	1982–1983 to 1999–2001
Walk for pleasure	91.9	138.5	175.6	83.7
View/photograph birds	20.8	54.3	68.5	47.7
Day hiking	24.3	53.6	69.1	44.8
Picnicking	83.3	112.2	118.3	35.0
Visit outdoor nature centre/zoo	86.7	110.9	121.0	34.3
Swimming in lakes/streams etc.	55.5	87.4	85.5	30.0
Sightseeing	79.8	117.5	109.0	29.2
Boating	48.6	76.2	75.0	26.4
Bicycling	55.5	77.8	81.9	26.4
Developed camping	**29.5**	**46.5**	**55.3**	**25.8**
Driving for pleasure	83.3	–	107.9	24.6
Motorboating	33.0	59.5	50.7	17.7
Off-highway vehicle driving	19.1	35.9	36.0	16.9
Primitive camping	**17.3**	**31.4**	**33.1**	**15.8**
Sledding	17.3	27.7	30.8	13.5
Backpacking	8.7	17.0	21.5	12.8
Fishing	59.0	70.4	71.6	12.6
Swimming in outdoor pool	74.6	99.1	85.0	10.4
Canoeing or kayaking	13.9	19.2	23.0	9.1
Downhill skiing	10.4	22.8	17.4	7.0
Snowmobiling	5.2	9.6	11.3	6.1
Horseback riding	15.6	20.7	19.8	4.2
Ice skating outdoors	10.4	14.2	13.6	3.2
Hunting	20.8	25.3	23.6	2.8
Cross-country skiing	5.2	8.8	7.8	2.6
Waterskiing	15.6	22.7	16.0	0.4
Sailing	10.4	12.1	10.4	0.0

Missing data are denoted with "–" and indicate that participation data for that activity were not collected during that time period
Source: NRS 1982–1983 ($n = 5757$), USDA Forest Service (1995) ($n = 17,217$), and USDA Forest Service (2001) ($n = 52,607$)
Note: The numbers in this table are *annual* participant estimates on data collected during the three time periods
1982–1983 participants based on 173.5 million people ages 16+ (U.S. Department of the Interior 1986)
1994–1995 participants based on 201.3 million people ages 16+ (Woods & Poole Economics, Inc. 2007)
1999–2001 participants based on 214.0 million people ages 16+ (U.S. Department of Commerce 2000)

survey of trends in number of people ages 16 and older participating in recreation activities by historic period in the USA, 1982 to 2001, Cordell (2012) estimated that there were 25.9 million people taking part in developed camping in 1982–1983, increasing to 55.3 million in 1999–2001, showing an increase of 25.8 million over the two decades (Table 8.3). For what Cordell termed primitive camping, he estimated that there were 17.3 million people taking part in 1982–1983, increasing to 33.1 million in 1999–2001, showing an increase of 15.8 million over the two decades (Table 8.3). In Cordell's table (Table 8.3), he has ranked 27 recreational activities and developed camping is ranked 10th in terms of the 1982–1983 to 1999–2001 increase, and primitive camping was 14th (Table 8.2).

In his more updated survey (Cordell 2012) of trends in number of people ages 16 and older participating in recreation activities in the USA, 1999–2001 and 2005–2009 for activities with between 25 and 49 million participants from 2005 through 2009, he found 34.2 million

Table 8.3 Trends in number of people of ages 16 and older participating in recreation activities in the USA, 1999–2001 and 2005–2009 for activities with between 25 and 49 million participants from 2005 through 2009 (Source: Cordell 2012, p. 37)

	Total participants (*millions*)			Percent participating	Percent change
	1994–1995	1999–2001	2005–2009	2005–2009	1999–2001 to 2005–2009
Visit archaeological sites	36.1	44.0	48.8	20.8	11.1
Off-highway vehicle driving	35.9	36.0	48.4	20.6	34.5
Boat tours or excursions	–	40.8	46.1	19.6	13.1
Bicycling on mountain/hybrid bike	–	44.0	42.7	18.1	−3.0
Primitive camping	**31.4**	**33.1**	**34.2**	**14.5**	**3.2**
Sledding	27.7	30.8	32.0	13.6	3.9
Coldwater fishing	25.1	28.4	30.9	13.1	8.7
Saltwater fishing	22.9	21.4	25.1	10.7	17.2

Missing data are denoted with "–" and indicate that participation data for that activity were not collected during that time period. Percent change was calculated before rounding
Source: USDA Forest Service (1995) ($n = 17,217$), USDA Forest Service (2001) ($n = 52,607$), and USDA Forest Service (2009) ($n = 30,398$)
Note: The numbers in this table are *annual* participant estimates on data collected during the three time periods
1994–1995 participants based on 201.3 million people ages 16+ (Woods & Poole Economics, Inc. 2007)
1999–2001 participants based on 214.0 million people ages 16+ (U.S. Department of Commerce 2000)
2005–2009 participants based on 235.3 million people ages 16+ (U.S. Department of Commerce 2008)

participating in primitive camping in the 2005 to 2009 period (Table 8.3).

In the USA, during the 2016 calendar year, a total of 24,134 online interviews were carried out with a nationwide sample of individuals and households from the US Online Panel of over one million people operated by Synovate/IPSOS (Outdoor Foundation 2017). A total of 11,453 individual and 12,681 household surveys were completed. The total panel is maintained to be representative of the US population for people ages six and older. Over sampling of ethnic groups took place to boost response from typically under responding groups. The 2016 participation survey sample size of 24,134 completed interviews provides a high degree of statistical accuracy.

In this most up-to-date survey that we can find (The Outdoor Foundation 2017), camping (RV) had 15.8 million participants in 2016 (with an 8.9% increase in the previous three years) while camping within one-fourth mile of vehicle/home had 25.5 million participants in 2016 (with an −9.6% decrease in the previous three years)—see Table 8.4.

Although not as up-to-date at the Outdoor Foundation survey, Cordell's (2012) report lumped together car, backyard, and RV camping showed that the number of participants exceeded hiking, cycling, and running, with only fishing having higher numbers participating (Fig. 8.6).

Unfortunately, such comprehensive surveys as those by the Outdoor Foundation and Cordell have not yet been undertaken in other parts of the world. However, Brooker and Joppe (2013) stated that in Europe, one in six of all overnight stays were spent in a campground (Eurostat 2012). The most avid campers are found in Australia and New Zealand where 86% of Australians (Alliance Strategic Research 2011) and 80% of New Zealanders (DOC 2006) have visited a caravan or holiday park at least once in their lifetime.

Whether the numbers for snow holing are included in primitive camping in Cordell's (2012) survey or not is unclear. Obtaining data on the number of snow hole and bothy users is difficult, but in comparison to camping they are likely to be very small, though perhaps comparable in some areas with bothy visits and snow holing.

Table 8.4 Outdoor participation by activity (ages 6+) in the USA, 2006–2016 (The Outdoor Foundation 2017, p. 8)

	2006	2007	2008	2009	2010	2011	2012	2013	2014	2015	2016	3-year change (%)
Adventure racing	725	698	920	1089	1339	1065	2170	2213	2368	2864	2999	35.5
Backpacking overnight >¼ mile from vehicle/home	7076	6637	7867	7647	8349	7095	8771	9069	10,101	10,100	10,151	11.9
Bicycling (BMX)	1655	1887	1904	1811	2369	1547	2175	2168	2350	2690	3104	43.2
Bicycling (mountain/non-paved surface)	6751	6892	7592	7142	7161	6816	7714	8542	8044	8316	8615	0.9
Bicycling (roads/paved surface)	38,457	38,940	38,114	40,140	39,320	40,349	39,232	40,888	39,725	38,280	38,365	−6.2
Birdwatching (more and ¼ mile from home/vehicle)	11,070	13,476	14,399	13,294	13,339	12,794	14,275	14,152	13,179	13,093	11,589	−18.1
Boardsailing/windsurfing	938	1118	1307	1128	1617	1151	1593	1324	1562	1766	1737	31.2
Camping (RV)	**16,946**	**16,168**	**16,517**	**17,436**	**15,865**	**16,698**	**15,108**	**14,556**	**14,663**	**14,699**	**15,855**	**8.9**
Camping (with ¼ mile of home/vehicle)	**35,618**	**31,375**	**33,686**	**34,338**	**30,996**	**32,925**	**29,982**	**29,269**	**28,660**	**27,742**	**26,467**	**−9.6**
Canoeing	9154	9797	9935	10,058	10,553	9787	9839	10,153	10,044	10,236	10,046	−1.1
Climbing (sports/indoor/boulder)	4728	4514	4769	4313	4770	4119	4592	4745	4536	4684	4905	3.4
Climbing (traditional/ice/mountaineering)	1586	2062	2288	1835	2198	1609	2189	2319	2457	2571	2790	20.3
Fishing (fly)	6071	5756	5941	5568	5478	5683	6012	5878	5842	6089	6456	9.8
Fishing (freshwater/other)	43,100	43,859	40,331	40,961	38,860	38,868	39,135	37,796	37,821	37,682	38,121	0.9
Fishing (saltwater)	12,466	14,437	13,804	12,303	11,809	11,983	12,017	11,790	11,817	11,975	12,266	4.0
Hiking (day)	29,863	29,965	32,511	32,572	32,496	34,491	34,545	34,378	36,222	37,232	42,128	22.5
Hunting (bow)	3875	3818	3722	4226	3908	4633	4075	4079	4411	4564	4427	8.5
Hunting (handgun)	2525	2595	2873	2276	2709	2671	3553	3198	3091	3400	3512	9.8
Hunting (rifle)	11,242	10,635	10,344	11,114	10,150	10,807	10,164	9792	10,081	10,778	10,797	10.3
Hunting (shotgun)	8987	8545	8731	8490	8062	8678	8174	7894	8220	8438	8271	4.8
Kayak fishing	n/a	n/a	n/a	n/a	1044	1201	1409	1798	2074	2265	2371	31.8
Kayaking (recreational)	4134	5070	6240	6212	6465	8229	8144	8716	8855	9499	10,017	14.9
Kayaking (sea/touring)	1136	1485	1780	1771	2144	2029	2446	2694	2912	3079	3124	16.0
Kayaking (white water)	828	1207	1242	1369	1842	1546	1878	2146	2351	2518	2552	18.9
Rafting	3609	3786	4226	4342	3869	3725	3958	3915	3924	4099	4095	−10.6
Running/jogging	38,559	41,064	41,130	43,892	49,408	50,713	52,187	54,188	51,127	48,496	47,384	−12.6
Sailing	3390	3786	4226	4342	3869	3725	3958	3915	3924	4099	4095	4.6
Scuba diving	2965	2965	3216	2723	3153	2579	2982	3174	3145	3274	3111	−2.0
Skateboarding	10,130	8429	7807	7352	6808	5827	6627	6350	6582	6436	6442	1.5
Skiing (alpine/downhill)	n/a	10,362	10,346	10,919	11,504	10,201	8243	8044	8649	9378	9267	12.4
Skiing (cross-country)	n/a	3530	3848	4157	4530	3641	3307	3377	3820	4146	4640	40.3
Skiing (freestyle)	n/a	2817	2711	2950	3647	4318	5357	4007	4564	4465	4640	2.7
Snorkelling	8395	9294	10,296	9358	9305	9318	8011	8700	8752	8874	8717	0.2
Snowboarding	n/a	6841	7159	7421	8196	7579	7351	6418	6785	7676	7602	3.4
Snowshoeing	n/a	2400	2922	3431	3823	4111	4029	3012	3501	3885	3533	−12.3
Stand up paddling	n/a	n/a	n/a	n/a	1050	1242	1542	1993	2751	3020	3220	61.6
Surfing	2170	2206	2607	2403	2767	2195	2895	2658	2721	2701	2793	3.0
Telemarking (downhill)	n/a	1173	1435	1482	1821	2099	2766	1732	2188	2569	2848	3.0
Trail running	4558	4216	4857	4833	5136	5610	6003	6792	7531	8139	8582	26.4

Note: All participation numbers are in thousands (000)

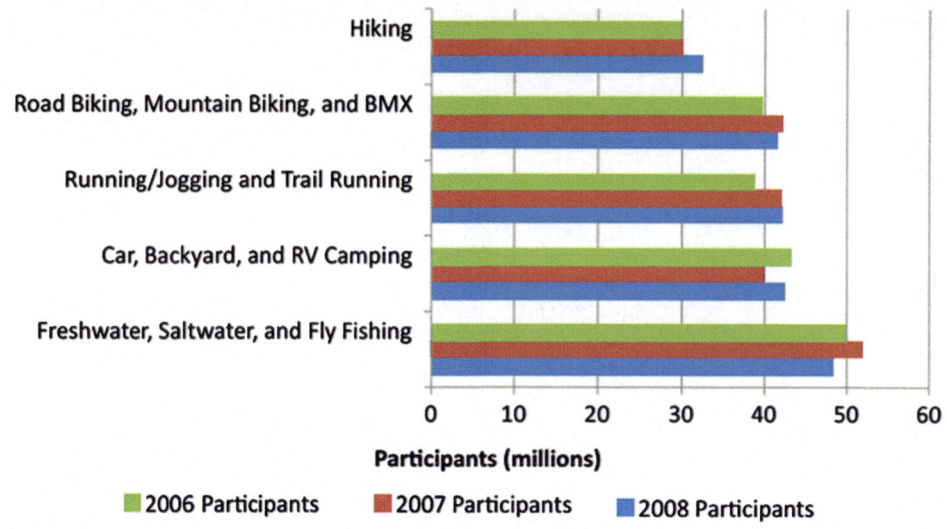

Fig. 8.6 Participation in gateway outdoor activities. Source: Cordell (2012, p. 27)

8.3 Environmental Impact

> Humans have walked and camped for as long as they have existed. Only in recent centuries, particularly in developed countries, has there been little need for large portions of the population to walk from place to place. In the past half century, this trend has reversed. As the proportion of people with substantial leisure time has increased, people are turning to hiking and camping as recreational activities.
> (Cole 2004, p. 2)

Inevitably camping, snow holing, and staying at mountain bothies come with some impact on the environment. This chapter does not seek to comment on off-site or indirect impacts such as travel to/from campsites, snow-hole venues, or bothies. Instead it will focus on the more direct/on-site impacts. These fall into three main categories: damage to soil and vegetation, impacts on water resources, and impacts on wildlife.

Sid Frissell conducted the first study of campsites that received differing levels of use (Frissell and Duncan 1965). His research showed that impact occurs wherever use occurs, leading him to suggest that the decision facing recreation managers is how much impact is acceptable, not whether or not to allow impact. This provided the conceptual foundation for planning processes which followed, such as the Limits of Acceptable Change (LAC) (Stankey et al. 1985). Frissell's data illustrated the curvilinear nature of the relationship between amount of use and amount of impact, although it was another 15 years before this finding and its significance to recreation management was articulated (Cole 1981).

8.3.1 Damage to Soil and Vegetation

Trumbull et al. (1994) compared the woody vegetation and soils on 20–40-year-old military campsites with undisturbed but otherwise similar areas in south-central Missouri. Military camping caused a reduction in the density and species richness of canopy and understory plants. Ground cover on the campsites was found to have less litter and more bare ground, but canopy cover on the campsites was indistinguishable from the control sites. Radial growth of the canopy was unaffected by 40 years of military camping. Soils on the campsite had higher bulk density, less total organic carbon, and a trend towards lower infiltration rates. The percentage rock volume on the surface of the campsites suggests that between 28 and 61 cm of soil has been lost.

Cole and Monz (2003) carried out experiments with controlled levels of recreational camping and compared them with previously undisturbed sites in two different plant communities in the subalpine zone of the Wind River Mountains, Wyoming, USA. The plant communities were coniferous forest with understory dominated by the low shrub *Vaccinium scoparium* and a riparian meadow of intermixed grasses and forbs, of which *Deschampsia cespitosa* was most abundant. Sites were camped on at intensities of either one or four nights per year, for either one (acute disturbance) or three consecutive years (chronic disturbance). Recovery was followed for three years on sites camped on for one year and for one year on sites camped on for three years. Reductions in vegetation cover and vegetation height were much more pronounced on sites in the forest than on sites in the meadow. In both plant communities, increases in vegetation impact were not proportional to increases in either years of camping or nights per year of camping. Close to the centre of campsites, near-maximum levels of impact occurred after the first year of camping on forested sites and after the second year on meadow sites. Meadow sites recovered completely within a year, at the camping intensities employed in the experiments. Forest sites, even those camped on for just one night, did not recover completely within three years. Differences between acute and chronic disturbance were not pronounced.

Pickering and Hill (2007) identified four key effects of camping on soil and vegetation. These were:

1. *Addition of nutrients*: the disposal of human waste (such as urine and faeces) has direct effects such as removal of vegetation in order to dig a hole but also has indirect effects through the addition of nutrients which can result in a change to species composition due to competitive displacement. This can create feedbacks for continuing change and also benefit weed species, leading to changes in vegetation communities. Research in Tasmania found a beneficial effect of low levels of nutrient addition (artificial urine) on vegetation, with increased growth of many taxa, with the only obvious negative effects on moss at one site (Bridle and Kirkpatrick 2003).

2. *Impacts of weeds*: another indirect and potentially self-sustaining impact of camping is the accidental introduction of weed propagules on visitors' shoes, clothing, and equipment. The risk associated with even low numbers of campers visiting remote areas was highlighted by Whinam et al. (2005) who found 981 propagules on the clothing and equipment of just 64 people visiting a remote subantarctic island. High-risk items were equipment cases, daypacks, and the cuffs and Velcro closures on outer clothing. As a result there have been policy changes regarding clothing for people visiting subantarctic islands as part of expeditions from Australia.

   Another important issue is the potential for exotics to spread from areas disturbed by tourism infrastructure into natural vegetation. In protected areas in Australia, for example, the verges of tracks and trails are often characterised by high diversity and cover of exotics, but not all these species spread into undisturbed native vegetation and become important environmental weeds (Godfree et al. 2004; Johnston 2005).

3. *Impact of pathogens*: another important example of an indirect and self-sustaining impact of tourism is the spread of exotic soil-borne pathogen *P. cinnamomi* (Buckley et al. 2004) in protected areas in Australia. This root rot fungus is a threat to vegetation including many plants that are already classified as rare and threatened. This threat has been recognised nationally within Australia, and it is listed as a key threatening process by the Australian Government (Environment Australia 2001) and by the NSW government in the Threatened Species Conservation Act 1995. Tourism contributes to the spread of *P. cinnamomi* by transportation of spores in mud on footwear, tent pegs, trowels, bike tyres, and other types of vehicles.

Marion and Farrell (2002) assessed campsite conditions and the effectiveness of campsite

Fig. 8.7 (A) Damage to grass by trampling around the entrance to a family tent on a commercial campsite in Switzerland. Photo by Tim Stott. (B) Damage to natural vegetation by trampling around tents at a mountain training camp in SW Greenland. Photo by Tim Stott. (C) Building open fires for cooking is a common activity when camping. Photo by Tim Stott

impact management strategies at Isle Royale National Park, USA. Vegetation and soil conditions were assessed at 156 campsites and 88 shelters within 36 backcountry campgrounds. The average site was 68 m² and 83% of sites lost vegetation over areas less than 47 m². Figure 8.7A, B shows examples of where damage has been caused to vegetation around tent sites on both commercial campsite (Fig. 8.7A) and in wilderness areas (Fig. 8.7B).

Building open fires for cooking is a common activity when camping (Fig. 8.7C). Unless carefully managed, this can result in dead wood being collected from the area around the fire. Dead wood provides a habitat for a variety of species, and this habitat is lost when the wood is removed and burned. The wood naturally decomposes over a number of years, returning nutrients to the soil. When wood is burned, this process is interrupted as a large proportion of the nutrients (e.g. carbon) are lost to the atmosphere, though some may be returned to the soil through the burnt wood ash.

Marion and Cole (1996) studied the impacts of camping on soil and vegetation at Delaware Water Gap National Recreation Area. They assessed the magnitude of impact on campsites that varied in amount of use and in topographic position and also evaluated change over a five-year period on long-established, recently opened, and recently closed campsites, as well as on plots subjected to experimental trampling. Campsite impacts were intense and spatially variable. Amount of use and topographic position explained some of this variation. Soil and vegetation conditions changed rapidly when campsites were initially opened to use and when they were closed to use. Changes were less pronounced on the long-established campsites that remained open to use. In the trampling experiments, impact varied greatly with trampling intensity and between vegetation types. An open-canopy grassland vegetation type was much more resistant to trampling than a forb-dominated forest vegetation type. Campsite impacts increased rapidly with initial disturbance, stabilised with ongoing disturbance, and—in contrast to what has been found in most other studies—decreased rapidly once disturbance was terminated. In Table 8.5 (adapted from Marion and Cole's 1996 paper), of all the variables they measured, only soil moisture did not show a statistically significant difference between the campsites and the control.

Provided there is full snow cover, building and staying overnight in snow caves, quinzhees, and igloos are unlikely to have any impact on vegetation and soils. However, the effects around bothies are likely to be similar, if not more intense, than around tents. This is because bothies are permanent structures with fixed entrances which visitors must use each time they enter and exit the bothy. The extent of soil compaction and vegetation damage will be proportional to the intensity of use and other factors like weather/climate, season, soil type/geology, and slope angle.

8.3.2 Impacts of Camping, Snow Holing, and Bothying on Water Resources

Cole and Landres (1996) argued that the effects of recreation on aquatic systems is often more spatially extensive than the effects on soil and vegetation, concluding that most recreational research focused on terrestrial environments. They identified the need for more research into the effects of human activity on individual watercourses. Pringle (1996) reported levels of increased coliform bacteria, up to ten times the normal background levels, at Ryvoan Bothy (Fig. 8.5B) on Mar Lodge Estate in Cairngorm National Park, Scotland, highlighting the correlation between raised coliform levels and areas of human activity.

Table 8.5 Vegetation and soil conditions on 29 campsites and undisturbed control sites at Delaware Water Gap National Recreation Area, 1986

Impact parameter	Campsite Mean	Range	Control Mean	Range	P
Ground vegetation cover (%)	15	0–63	72	1–95	0.001
Floristic dissimilarity (%)	75	23–100	n/a	n/a	
Graminoid cover (%)	58	0–100	26	0–92	0.023
Forb cover (%)	23	0–78	59	5–100	0.001
Mineral soil cover (%)	61	21–94	1	0–15	0.001
Organic horizon thickness (cm)	0.5	0–1.4	1.5	0.2–3.1	0.002
Soil bulk density (g cm^{-3})	1.26	1.0–1.4	1.06	0.7–1.4	0.001
Soil penetration resistance (kPa)[a]	275	137–382	49	0–226	0.001
Soil moisture (g cm^{-3})	18	8–32	17	8–31	0.710
Felled trees (%)	19	0–53	n/a	n/a	
Damaged trees (%)	77	25–100	n/a	n/a	
Tree reproduction (stems ha^{-1})	936	0–6275	10,090	0–56,400	0.001
Non-vegetated area (m^2)	181	0–696	0	0–15	0.001
Campsite area (m^2)	269	51–731	n/a	n/a	
Shoreline disturbance (m)	9	0–20	n/a	n/a	

Source: Marion and Cole (1996, p. 523)
[a] 1 kPa = the pressure corresponding to 1.01971×10^{-2} kg cm^{-2}

Later in the same area, Bryan (2002) reported increased levels of coliform bacteria, up to ten times that of the accepted background levels, around the same bothy though he did not quantify the faecal coliform levels. Although not conclusive evidence of human-derived contamination, these findings highlight a tenuous link between areas of human activity and the potential for raised coliform levels in mountain streams.

In the UK organisations such as the MBA and the Mountaineering Council of Scotland (MCofS) advise recreation seekers to bury their organic waste. Liddle and Scorgie (1980) and Temple et al. (1982) identified the potential persistence and associated effects of the burial of human faeces. The disposal of human waste is a recurring concern among wilderness managers with visitation trends having a potential impact on faecal coliform levels (Cilimburg et al. 2000). The impact of wild camping has been a focus of attention for the various recreational governing bodies, such as the MCofS, MBA, and British Mountaineering Council (BMC). During the winter months concerns generally relate to the activity of snowholing and bothies, since tented camps are less popular in the harsh winter conditions. Anecdotal evidence (Cairngorm Ranger, pers. comm. 16.4.09) suggested that during the 2008/2009 winter season, in the region of 400 snowholing parties had accessed Ciste Mhearad, one of the closest snowholing sites to the Coire Cas ski area in the Cairngorms. Forrester and Stott (2016) investigated the spatial distribution of stream water faecal coliform concentrations in specific winter recreation areas in the northern Corries of the Cairngorm Mountains, Scotland. During two winter seasons, 2007–2008 and 2008–2009, 207 samples were collected from ten sites and analysed for the presence of faecal coliforms, specifically *Escherichia coli* (*E.coli*). E.coli was not detected at the seven above 635 m, but three sites below 635 m (the altitude of the ski area buildings and car park) had positive detection rates for *E.coli*, these being 32%, 35%, and 31% respectively, suggesting that snow holing was not associated with elevated faecal coliform levels (their site 1 was right next to the popular snowholing sites in Ciste Mhearad), but that the ski infrastructure was.

Carter et al. (2015) examined the impact of beach camping on beach freshwater on Fraser Island, a popular tourist destination off the east coast of Australia. Prior to their study the assumption was that the natural assimilative capacity of the fore dune ecosystem was sufficient to dissipate any negative environmental impact. Their study of nutrients, faecal coliforms, and faecal sterols in the water table and beach flows associated with camping and non-camping zones revealed concerning differences between sample sites. The study suggested that nutrient levels in the water table were enriched in camping zones and that, in some areas, faecal coliforms persisted in beach flows. The link to a human cause was supported by the presence of strong faecal sterol signals in soil samples from the water table interface. The risk implications for human health were thought to be significant.

Waters derived from remote "wilderness" locations in the Scottish mountains, unused for agriculture, had long been assumed to be largely free of bacterial contamination. However, McDonald et al. (2008) challenged this assumption after carrying out their bacterial survey of the waters draining several stream catchments on the south side of the Cairngorms (on the Mar Lodge Estate). Over 480 spot samples taken from 59 sites revealed that over 75% of samples tested positive for *Escherichia coli* (*E. coli*) and 85% for total coliforms. Largest values occurred over the summer months and particularly at weekends at sites frequented by visitors, either for "wild" camping or day visits or where water was drawn from the river for drinking. Overall the spatial and temporal variations in bacterial concentrations suggested a relationship with visitor numbers and in particular wild camping.

8.3.3 Impacts of Camping, Snow Holing, and Bothying on Wildlife

Blakesley and Reese (1988) compared use of riparian habitat by 14 bird species during the breeding season on campground ($n = 31$) and non-campground ($n = 80$) sites in northern Utah. Multivariate analysis showed that seven bird species were closely associated with campgrounds, whereas six of seven species associated with non-campgrounds were ground- or shrub-nesting, or ground-foraging. These avian responses may be explained by differences in shrub and sapling density, litter depth, and amount of dead woody vegetation between the campground and non-campground.

Farooquee et al. (2008) studied the environmental and socio-cultural impacts of river rafting and camping on the Ganga in the Uttarakhand Himalaya. They reported that displacement of wildlife had occurred in the region due to bright colours of tents, toilet tents, rafts, and loud music and lights in and around campsites. According to a survey conducted among the rural population of this area, prior to the camping and rafting activities, animals were frequently spotted on river side while drinking water and resting on the sand beach; now they are not visible in the area for months, especially during the camping and rafting season (Table 8.6).

Clevenger (1977) reported on some of the effects of campgrounds on small mammal populations in Canyonlands and Arches National Parks, Utah. Data collection consisted of live-trapping from April to November, 1975 (12,337 trap-nights). The populations of Ordls kangaroo rat (*Dipodomys ordii*), antelope ground squirrels (*Ammospermophilus leucurus*), deer mice (*Peromyscus spp.*), woodrats (*Neotoma spp.*), Colorado chipmunks (*Eutamias guadrivittatus*), and desert cottontails (*Sylvilagus audubonii*) inhabiting campgrounds were compared with non-campground control areas. Clevenger found that Squaw Flat campground in Canyonlands National Park contained significantly higher populations of woodrats and Colorado chipmunks than the control. Devil's Garden campground in Arches National Park exhibited significantly higher populations of deer mice, but a lower population of woodrats than the control. No significant difference was found between campgrounds and control areas for all other species. Occurrence of species in the campground and control areas was identical.

It seems likely that wherever there are concentrations of visitors staying overnight bringing food with them, they are likely to invoke interest from certain groups of animals. One species which receives some attention in the literature is the black bear (*Ursus americanus*). Ayres et al. (1986) working in Sequoia National Park noted that in places where black bears have become pests (in campgrounds and other developments), their visits are frequently at night. However, this is not the case elsewhere and they suggest that human activity, when imposed on black bear habitat, disrupts bear activity patterns. In national parks where hunting is not permitted, the two principal factors affecting the population ecology and behaviour of black bears are the availability of human food and the management practices designed to remove bears from sites with human activity such as campgrounds.

McCutchen (1990) on the other hand, while agreeing that black bears in many US and Canadian national parks become habituated to humans (they are often bold, frequent human use areas, and are generally a nuisance), his study at Rocky Mountain National Park, Colorado found the antithesis of this behaviour. His four-year study of black bears using radio-telemetry and observation indicated that although many bears have home ranges in high human use areas, they are secretive and avoid humans and developed areas.

8.4 Management and Education

8.4.1 Managing the Impacts of Camping, Snow Holing, and Bothying on Vegetation and Soils

Turton (2005) examined environmental impacts of tourism and recreation activities in the world heritage listed rainforests of northeast Australia Visitor use in the World Heritage Area was mostly associated with walking tracks, camping areas, day-use areas, and off-road vehicle use of old forestry roads and tracks. Adverse environmental impacts range from vegetation trampling, soil

Table 8.6 Spotting of animals by villagers of the surrounding areas before and after the start of camping/rafting

Animal	Activity/time/place when spotted	Month spotted											
		J	F	M	A	M	J	J	A	S	O	N	D
Barking deer	Grazing, drinking	-+	-	-	-	-+	-+	-+	-	-	-	-	-
Sambar	Grazing, drinking		-	-	-	-+	-+	-+		-	-	-	
Rabbit	Evening, forest		-	-	-	-+	+		-+	-	-	-	
Monkey	Daytime	-+	-+	+	+	+	+			+	-+	-+	-+
Wild boar	Morning and evening	+	+	-+	-+	-+	+			-+	-+	-+	+
Langur	Forest, drinking	-+	-+	-		-+	-					-+	-+
Fox	Morning and evening	-	-	-		-+	-+		-	-	-		-
Jackal	Barking in evening	-+	-+	-	-	-	-	-+	-	-	-	-	-
Leopard	Night	-+	-	-	-			-+					
Black bear	Late evening	-			-	-	-	-+	-+	-	-	-	-
Goral	Grazing, drinking	-+	-	-			-+	-+		-	-	-+	-
Mongoose	Forest, daytime	-+	-	-	-		-+	-+	-+	-	-	-+	-
Wild cat	Evening and night	-+	-	-	-	-	-+	-		-	-	-	-
Porcupine	Late evening	-+	-	-						-+	-+	-	-+
Common otter	Late evening	-+	-+	-			-	-		-+		-+	-+

Source: Farooquee et al. (2008, p. 592)

Note: −, spotting before start of caming/rafting; +, spotting after the start of camping/rafting compaction, water contamination, and soil erosion at the local scale through to spread of weeds, feral animals, and soil pathogens along extensive networks of old forestry roads and tracks at the regional scale.

In terms of managing these impacts, he concluded that concentration of visitor use is the most desirable management strategy for controlling adverse impacts at most World Heritage Area visitor nodes and sites. This included methods such as site hardening and shielding to contain impacts. For dispersed visitor activities, such as off-road vehicle driving and long-distance walking, the preferred management strategies included procedures like removal of mud and soils from vehicle tyres and hiking boots before entering pathogen-free catchments, seasonal closure of roads and tracks, the retention of canopy cover at camping areas and day-use areas, and along walking tracks and forestry roads. These were simple, yet effective, management strategies for reducing a range of adverse impacts, including dispersal of weeds and feral animals, edge effects, soil erosion and nutrient loss, road kill and linear barrier effects on rainforest fauna.

In order to control invasive species (such as *P. cinnamomiis* as discussed earlier), Pickering and Hill (2007) suggested that quarantine and hygiene were the main strategies that have been implemented by protected area managers to combat this threat. Some parks have permanent or seasonal closures of specific tracks, or sections within a park, or in a few cases whole parks are closed particularly in severely affected areas of Western Australia and South Australia (Newsome 2003; Buckley et al. 2004). Hygiene procedures to minimise the spread of spores were implemented through education programs (signs, leaflets, etc.) which encourage/require visitors to wash down vehicles, boots, tent pegs, and so on when entering and leaving sites and in some cases to visit uninfected sites before infected sites (Buckley et al. 2004). Figure 8.8 shows some examples of management strategies for aquatic invasive species spread by fishing and boat users.

Marion and Farrell (2002) assessed campsite conditions and the effectiveness of campsite impact management strategies at Isle Royale National Park, USA. Vegetation and soil conditions were assessed 156 campsites and 88 shelters within 36 backcountry campgrounds. The average site was 68 m^2 and 83% of sites lost vegetation over areas less than 47 m^2. They concluded that management actions implemented to spatially concentrate camping activities and reduce camping disturbance had been highly successful. Comparisons of disturbed area/overnight stay among other protected areas reinforces this assertion. These reductions in area of camping disturbance are attributed to a designated site camping policy, limitation on site numbers, construction of sites in sloping terrain, use of facilities, and an ongoing programme of campsite maintenance. Such actions are most appropriate in higher use backcountry and wilderness settings.

Dixon (2017) studied the 79 km Overland Track which is Tasmania's premier overnight walking track (trail) and one of Australia's best known and most popular backcountry hikes. Trampling impacts (poor track condition) were recognised in the 1970s and degraded campsites were a concern by the 1980s. Despite three decades of intermittent works, many sections of track remained in poor condition in the early 2000s (Fig. 8.8A), but targeted works since 2006 has addressed many problem areas (Fig. 8.8B). Hardening of campsites at selected overnight nodes (Fig. 8.8C, D) commenced in 2000 and a reduction in overall camping impacts followed, presumed due to a greater concentration of camping use at the hardened sites despite unrestricted camping still being permitted (Fig. 8.9).

Longitudinal monitoring of both track (eight years) and campsite (16–25 years) conditions by Dixon (2017) has successfully described the scale and constrained the location of changes in condition and has provided a useful planning tool for management. In particular, it has contributed to documenting a contemporaneous improvement in track and campsite conditions partly associated with a booking system to regulate walker use of the Overland Track, introduced in 2005. Booking fees have contributed to management successes by providing adequate and consistent resourcing for

Fig. 8.8 (A) Example of a visitor sign used by Maine Lakes Environmental Association, USA, to control the spread of invasive aquatic species. Source: http://www.mainelakes.org/, accessed 10/3/18. (B) Visitor sign in the Eastern USA alerting water users to the spread of the invasive species, Eurasian Watermilfoil. Source: https://amateuranglers.wordpress.com/2016/09/12/the-war-on-milfoil-and-how-it-affects-fish/, accessed 10/3/18

the repair and maintenance of walking track surface infrastructure.

Management implications from Dixon's (2017) study included the following:

- Extensive hardening is an effective way to sustainably manage a moderate to high use of walking track that has not been initially well-designed.
- Adequate and consistent resourcing for the repair and maintenance of walking track surface and infrastructure is necessary to sustainably manage such tracks.
- The provision of inviting facilities, including camping platforms, at selected overnight nodes has resulted in a concentration of visitor camping use on a smaller number of campsites, hence reducing the overall impact of camping along the Overland Track.

8.4.2 Managing the Impacts of Camping, Snow Holing, and Bothying on Water Resources

When heading off to the hills for overnight expeditions (and sometimes on day trips if you get your timings wrong) going to the toilet, where there are none, can result in unsightly and unpleasant piles of human faeces near campsites, bothies or paths. Worse still, if they are near watercourses they can be washed in and cause faecal coliforms in streams to rise, and thereby contaminate the water (Fewtrell 1991: McDonald et al. 2008; Forrester and Stott 2016). It is a matter organisations such as Scottish Natural Heritage (SNH) and the MCofS have tackled. They warn that public and animal health is threatened by irresponsible toileting because the waste could contaminate drinking water which, further downstream, could be someone else's drinking water. People can be put at risk to a cocktail of nasty pathogens, such as *Cryptosporidium*, *Campylobacter*, *Aeromonas*, *E. coli* O157, and giardia. The SNH recommends that if you need to urinate, do so at least 30 m from open water or rivers and streams, and if you need to defecate, do so as far away as possible from buildings, from open water or rivers and streams, and from any farm animals and bury faeces in a shallow hole and replace the turf.

The MCofS offers guidance in its leaflet "Where to 'Go' in the Great Outdoors" (Fig. 8.10). It recommends taking home toilet paper in containers and cleaning hands using

Fig. 8.9 (A) A degraded section of the Tasmanian Overland Track (2005). Photo by Tim Stott. (B) A renovated section of the Tasmanian Overland Track (2005). Photo by Tim Stott. (C) Hardening of campsites at selected overnight nodes (foreground) on the Tasmanian Overland Track commenced in 2000 and a reduction in overall camping impacts followed, presumed due to a greater concentration of camping use at these hardened sites. Photo by Tim Stott, 2005. (D) Camping platform near a overnight stay cabin on the Tasmanian Overland Track. Photo by Tim Stott, 2005

gels. The council asks that people should not "go" near paths, huts, and bothies and never in caves. It suggests carrying a small trowel to make the task of digging a hole to bury waste easier. The leaflet specifies that: "When digging a hole is absolutely impossible and you are in a very remote place, spread excrement thinly or arrange rocks such that air can circulate. Avoid just putting a rock on top as it slows decomposition."

In winter it is a different matter. To be able to dig out enough snow to get to the ground below and then dig a hole may take some serious excavations. If excrement is simply buried in the snow then this is only delaying the time when, after the snow has melted, it appears on the ground surface. Once the temperature rises enough, the various bacteria, microbes, and insects will get to work to break it down. All this takes time and can be very unpleasant in popular spots, specifically around snow holes and bothies.

In order to try to reduce the effect of people staying overnight in snow holes on the Cairngorm plateau, the Cairngorm National Park Ranger Service set up the Snow White facility (formerly known as the Poo Project) to encourage people to bring back all human waste and dispose of it in the disposal facilities at Cairngorm Mountain. This facility is unique in Scotland (Fig. 8.11).

Fig. 8.10 "Where to Go in the Great Outdoors". Mountaineering Council of Scotland Advisory Leaflet. Source: https://www.mountaineering.scot/assets/contentfiles/pdf/where-to-go-leaflet.pdf, accessed 10/3/18

Fig. 8.11 The Cairngorm Poo Project provides visitors with these bottles to bring back their human waste in winter. This one was used by a Winter Mountain Leader Training group. Source: https://www.walkhighlands.co.uk/Forum/viewtopic.php?f=9&t=4082, accessed 10/3/18

Forrester and Stott (2016) attempted to evaluate the success of the Cairngorm Snow White project but, as mentioned earlier, found no evidence of faecal coliform contamination in the stream at their Ciste Mhearad snow-holing site. This, of course, does not confirm that the project works, and they discuss a range of possible reasons as to why they may not have detected faecal coliform contamination.

Due to their permanent nature, perhaps the greatest issues with human excrement arise around bothies and permanent camps which do not have toilet facilities. A number of authors (and the MBA itself) have drawn attention to this problem in Scotland (e.g. Hillbrant 1992; Bryan 2002; McDonald et al. 2008). Bothy users have been encouraged to "do their business" in as considerate and environmentally friendly a way as possible, away from the bothy. This means taking a trowel to a quiet spot well away from streams and paths and digging a little hole in the ground for excrement and the accompanying toilet roll (better still use moss). Most bothies are equipped with a spade for this purpose. Unfortunately some people don't bother to think of others, and when it's cold, wet, and windy, people will often take the easy option of squatting down against the back wall of the bothy, rather than venturing a little further afield. At Corrour bothy at the south end of the Lairig Ghru, Cairngorms, you can see the damage this does—little piles of human waste and toilet paper scattered over the hillside. As well as being unsightly, this presents a very real health problem to walkers and animals and also upsets landowners. The MBA has developed the Bothy Code (Fig. 8.12), and in locations where building and maintaining permanent toilet facilities is not possible, we have to reply on educating walkers and campers through such leaflets, signs, and the internet. Mountain training courses such as those run by the Mountain Training UK (http://www.mountain-training.org/) include environmental responsibility in their syllabi.

Other issues at bothies include the collection of firewood from the surrounding area which can destroy habitats, leaving of rubbish which attracts animals which in turn can end up ingesting or getting tangled in plastic. The Bothy Code attempts to educate visitors against creating these problems.

8.4.3 Managing the Impacts of Camping, Snow Holing, and Bothying on Wildlife

Rogers (2011) studied the effects of what has been termed "diversionary feeding" of black bears (*Ursus americanus*) around campgrounds and residential areas in an attempt to divert nui-

Fig. 8.12 The Mountain Bothies Association's Bothy Code. Source: https://www.mountainbothies.org.uk/wp-content/uploads/2017/07/Responsible-access.pdf, accessed 10/3/18

sance bears away from the public and thereby increase public safety. Rogers studied diversionary feeding, habituation, and food conditioning at a US Forest Service campground and residential complex near Ely, Minnesota. From 1981 to 1983, six bears (two/year) had been removed from this area as nuisances; but during eight years of diversionary feeding (1984–1991), the only removals were two bears that had newly immigrated to the periphery of the study area and had not yet found the diversionary feeding site. The reduction in nuisance activity was significant, despite continued availability of garbage and the fact that the study bears were habituated and food-conditioned. No bear that visited the diversionary-feeding site became a nuisance or jeopardised public safety, even in 1985, the year with the lowest bear food index and the highest number of nuisance complaints ever recorded throughout Minnesota. Diversionary feeding led to greater tolerance of bears by residents.

Hammitt et al. (2015) suggest that bear problems are aggravated by concentrating use on a few sites rather than using the dispersal management technique of spreading visitors over a larger number of campsites. However, small mammals are more likely to be adversely affected by the creation on many moderately impacted sites rather than a few highly impacted sites.

Martin et al. (1989) investigated human-induced impacts from recreational use of wilderness which threaten the integrity of the wilderness resource and the quality of visitor experiences. They noted that campsite impacts are of particular concern to managers. One approach to this problem is the LAC planning system, which focuses attention on the question, "How much change in wilderness conditions is acceptable?" Their study compares and contrasts wilderness manager and visitor perceptions of the acceptability of different levels of campsite impacts, amount of impact, and perceptual zoning of wilderness. The results reinforce previous findings regarding differences between managers and visitors.

Gore et al. (2007) studied negative human-black bear interactions in New York's Adirondack Park campgrounds which pose risk management challenges. They highlighted that communication is one tool available to modify human behaviour and reduce associated risks, but knowledge of constructs influencing risk perception among key stakeholder groups was needed to design effective risk communication approaches. They interviewed managers (n = 14) and users (n = 40) at seven Adirondack Park campgrounds to characterise risk perceptions between groups and identified eight constructs influencing risk perceived by users and/or managers with three constructs on which both groups agreed and five on which they did not agree. They concluded that shared understanding across groups, and explicit recognition by risk communicators of differences between groups, may offer opportunities to maximise successes of risk communication efforts in campgrounds.

Crowe and Reid (1998) examined the future management of mountain bothies in the Scottish Highlands. Their research was undertaken in the context of major changes in the planning and management of countryside recreation opportunities in the UK; not least the increasing pressures to manage such facilities on a more commercial basis. Commercialisation can bring distinct advantages in the management and improvement of facilities, including the manipulation of users and, of course, revenue generation. However, there may be disadvantages in terms of accessibility. Mountain bothies are currently managed by volunteers in the MBA and other organisations and are free of charge to users. However, there are increasing concerns about overuse, vandalism, and pollution, particularly in areas of outstanding wildlife and landscape importance. Crowe and Reid claimed that the system of management must adapt if it is to respond to these growing pressures. In order to propose a way forward, alternative remote accommodation systems in Norway and New Zealand were examined. These systems appear more effective and generally include an element of charging for accommodation. It is suggested that the MBA will need to consider some degree of commercialisation in order to ensure the future protection of the bothy system in Scotland.

Conclusions

1. Camping is an outdoor activity which involves at least one overnight stay away from home in a shelter, such as a tent. The spectrum of types of camping ranges from survival camping (with emphasis and priority on lightweight and survival *not* comfort) through to wild camping using a tent, to camping on a valley campsite/family camping, to camping with electricity and using trailer tents, caravans, and motorhomes (glamping).
2. Snow caves, quinzhees, and igloos are used at alternatives in winter and/or snowy regions.
3. A bothy is a basic shelter, usually left unlocked and available for anyone to use free of charge. They are usually remote, with no road access, and may be ruined estate workers' or shepherds' cottages which have been renovated. The aim of the UK's MBA is to maintain simple shelters in remote country for the use and benefit of all who love wild and lonely places.
4. Cordell's (2012) survey of trends in number of people ages 16 and older participating in recreation activities in the USA, 1999–2001 and 2005–2009 for activities with between 25 and 49 million participants from 2005 through 2009, showed that 34.2 million were participating in primitive camping in the 2005 to 2009 period. The Outdoor Foundation, (2017) reported camping (RV) had 15.8 million participants in 2016 (with an 8.9% increase in the previous three years) while camping within one fourth mile of vehicle/home had 25.5 million participants in 2016 (with an −9.6% decrease in the previous three years). Cordell's (2012) report lumped together car, backyard, and RV camping and showed that the number of participants exceeded hiking, cycling, and running, with only fishing having higher numbers participating.
5. Early research showed that impact occurs wherever use occurs, leading to the suggestion that the decision facing recreation managers is how much impact is acceptable, not whether or not to allow impact. This provided the conceptual foundation for planning processes known as the LAC.
6. Numerous studies have shown that camping impacts soils, vegetation, water resources, and wildlife in a range of environments and over different time scales. Camping effects on soils can result in addition of nutrients, the accidental introduction of weed propagules, the potential for exotics to spread from areas disturbed by tourism infrastructure into natural vegetation, spread of exotic soil-borne pathogens.
7. The effects of recreation on aquatic systems are often more spatially extensive than the effects on soil and vegetation. Camping, bothy use, and snow holing all have the potential to introduce pathogens, such as *Cryptosporidium*, *Campylobacter*, *Aeromonas*, *E. coli* O157 and giardia into watercourses. A range of advice from organisations like the MCofS and MBA is available to backcountry users about how to manage this issue in both summer and winter.
8. The impacts of camping, snow holing, and bothying on wildlife are also well researched, and certain groups of animals such as black bears emerge as nuisances, and a range of measures have been adopted to deal with this problem in certain areas.
9. The management of the impacts of camping, snow holing, and bothying on the environment comes back to the question "How much change in wilderness conditions is acceptable?" or the LAC planning system.

References

Alliance Strategic Research. (2011). *The Caravan and Park Industry*. RV and Accommodation Industry Association of Australia. Brisbane: Caravan.

Ayres, L. A., Chow, L. S., & Graber, D. M. (1986). Black bear activity patterns and human induced modifications in Sequoia National Park. *Bears: Their Biology and Management, 6*, 151–154.

Blakesley, J. A., & Reese, K. P. (1988). Avian use of campground and noncampgound sites in riparian zones. *The Journal of Wildlife Management, 52*, 399–402.

Bridle, K. L., & Kirkpatrick, J. B. (2003). Impacts of nutrient additions and digging for human waste disposal in natural environments, Tasmania, Australia. *Journal of Environmental Management, 69*, 299–306.

Brooker, E., & Joppe, M. (2013). Trends in camping and outdoor hospitality—An international review. *Journal of Outdoor Recreation and Tourism, 3*, 1–6.

Bryan, D. (2002). Joined-up thinking for recreation management? The issue of water pollution by human sanitation on the Mar Lodge Estate Cairngorms. *Countryside Recreation, 10*(1), 18–22.

Buckley, R., King, N., & Zubrinich, T. (2004). The role of tourism in spreading dieback disease in Australian Vegetation. In R. Buckley (Ed.), *Environmental Impacts of Ecotourism* (pp. 317–324). New York: CABI Publishing.

Carter, R. W., Tindale, N., Brooks, P., & Sullivan, D. (2015). Impact of camping on ground and beach flow water quality on the eastern beach of K'gari-Fraser Island: A preliminary study. *Australasian Journal of Environmental Management, 22*(2), 216–232.

Cilimburg, A., Monz, C. A., & Kehoe, S. K. (2000). Wildland recreation and human waste: A review of problems, practices and concerns. *Journal of Environmental Management, 25*(6), 587–598.

Clevenger, G. A. (1977). *The Effects of Campgrounds on Small Mammals in Canyonlands and Arches National Parks, Utah*. All graduate theses and dissertations, 1661. Retrieved from https://digitalcommons.usu.edu/etd/1661.

Cole, D. N. (1981). Managing ecological impacts at wilderness campsites: An evaluation of techniques. *Journal of Forestry, 79*, 86–89.

Cole, D. N. (2004). Impacts of hiking and camping on soils and vegetation: A review. *Environmental impacts of ecotourism, 41*, 60.

Cole, D. N., & Landres, P. B. (1996). Threats to wilderness ecosystems: Impacts and research needs. *Ecological Applications, 6*(1), 168–184.

Cole, D. N., & Monz, C. A. (2003). Impacts of camping on vegetation: Response and recovery following acute and chronic disturbance. *Environmental Management, 32*(6), 693–705.

Cordell, K. (2012). *Outdoor Recreation Trends and Futures: A Technical Document Supporting the Forest Service 2010 RPA Assessment*. Ashville: Southern Research Station. Retrieved from www.srs.fs.usda.gov.

Crowe, L., & Reid, P. (1998). The increasing commercialization of countryside recreation facilities: The case of Scottish mountain bothies. *Managing Leisure, 3*(4), 204–212.

Dixon, G. (2017). A longitudinal study of backcountry track and campsite conditions on the Overland Track, Tasmania, Australia. *Journal of Outdoor Recreation and Tourism, 19*, 25–36.

DOC. (2006). *Review of Camping Opportunities in New Zealand*. Heritage Appreciation Unit Research, Development & Improvement, Department of Conservation.

Environment Australia. (2001). *Threat Abatement Plan for Dieback Caused by the Root-rot Fungus Phytophthora cinnamomi*. Department of the Environment and Heritage, Australian Government, Canberra.

Eurostat. (2012). *Tourism statistics at regional level*. Retrieved December 3, 2012, from http://epp.eurostat.ec.europa.eu/statistics_explained/index.php/Tourism_statistics_at_regional_level#CampingS.

Farooquee, N. A., Budal, T. K., & Maikhuri, R. K. (2008). Environmental and socio-cultural impacts of river rafting and camping on Ganga in Uttarakhand Himalaya. *Current Science, 94*, 587–594.

Fewtrell, L. (1991). Freshwater recreation: A cause for concern? *Applied Geography, II*, 215–226.

Forrester, B. J., & Stott, T. A. (2016). Faecal coliform levels in mountain streams of winter recreation zones in the Cairngorms National Park, Scotland. *Scottish Geographical Journal, 132*(3-4), 246–256.

Frissell, S. S., & Duncan, D. P. (1965). Campsite preference and deterioration in the Quetico-Superior canoe country. *Journal of Forestry, 65*, 256–260.

Godfree, R., Brendan, L., & Mallinson, D. (2004). Ecological filtering of exotic plants in an Australian sub-alpine environment. *Journal of Vegetation Sciences, 15*, 227–236.

Gore, M. L., Knuth, B. A., Curtis, P. D., & Shanahan, J. E. (2007). Campground manager and user perceptions of risk associated with negative human–black bear interactions. *Human Dimensions of Wildlife, 12*(1), 31–43.

Hammitt, W. E., Cole, D. N., & Monz, C. A. (2015). *Wildland Recreation: Ecology and Management*. Hoboken, NJ: John Wiley & Sons.

Hillbrandt, D. (1992). Environmentally friendly dumping. *Mountain Bothies Association Newsletter, 104*, pp. 33–34.

Jirásek, I., Roberson, D. N., & Jirásková, M. (2017). The impact of families camping together: Opportunities for personal and social development. *Leisure Sciences, 39*(1), 79–93.

Johnston, F. (2005). *Exotic Plants in the Australian Alps Including a Case Study of the Ecology of Achillea millefolium in Kosciuszko National Park*. PhD thesis, School of Environmental and Applied Sciences, Griffith University, Gold Coast.

Liddle, M. J., & Scorgie, H. R. A. (1980). The effects of recreation on freshwater plants and animals: A review. *Biological Conservation, 17*(3), 183–206.

Marion, J. L., & Cole, D. N. (1996). Spatial and temporal variation in soil and vegetation impacts on campsites. *Ecological Applications, 6*, 520–530.

Marion, J. L., & Farrell, T. A. (2002). Management practices that concentrate visitor activities: Camping impact management at Isle Royale National Park, USA. *Journal of Environmental Management, 66*(2), 201–212.

Martin, S. R., McCool, S. F., & Lucas, R. C. (1989). Wilderness campsite impacts: Do managers and visitors see them the same? *Environmental Management, 13*, 623–629.

McCutchen, H. E. (1990). Cryptic behaviour of black bears (*Ursus americanus*) in Rocky Mountain National Park, Colorado. *Bears: Their Biology and Management, 8*, 65–72.

McDonald, A. T., Chapman, P. J., & Fukasawa, K. (2008). The microbial status of natural waters in a protected wilderness area. *Journal of Environmental Management, 87*(4), 600–608.

Newsome, D. (2003). The role of an accidentally introduced fungus in degrading the health of the Stirling Range National Park ecosystem in southwestern Australia: Status and prognosis. In D. J. Rapport, W. L. Lasely, D. E. Roslton, N. O. Nielsen, C. O. Qualset, & A. B. Damania (Eds.), *Managing for Healthy Ecosystems* (pp. 375–387). London: Lewis Publishers.

Pickering, C. M., & Hill, W. (2007). Impacts of recreation and tourism on plant biodiversity and vegetation in protected areas in Australia. *Journal of Environmental Management, 85*(4), 791–800.

Pringle, R. (1996). Bothies and bugs. *The Great Outdoors*, pp. 20–22.

Rogers, L. L. (2011) Does diversionary feeding create nuisance bears and jeopardize public safety? *Human–Wildlife Interactions, 5*(2), Article 16. Retrieved from https://digitalcommons.usu.edu/hwi/vol5/iss2/16.

Stankey, G. H., Cole, D. N., Lucas, R. C., Petersen, M. E., & Frissell, S. S. (1985). *The Limits of Acceptable Change (LAC) System for Wilderness Planning*. Research Paper INT-176. U.S. Department of Agriculture, Forest Service, Ogden, UT, 37pp.

Temple, K. L., Camper, A. K., & Lucas, R. C. (1982). Potential health hazard from human wastes in the wilderness. *Journal of Soil Water Conservation, 37*, 357–359.

The Camping and Caravanning Club. (2011). *Are Those Who Camp Richer for It? The Psychological and Social Benefits of the Camping Experience*. Coventry: Camping and Caravanning Club. Retrieved from www.campingandcaravanningclub.co.uk.

The Outdoor Foundation. (2017). *Outdoor Participation Topline Report 2017*. Washington, DC: The Outdoor Foundation. Retrieved from www.outdoorfoundation.org.

Trumbull, V. L., Dubois, P. C., Brozka, R. J., & Guyette, R. (1994). Military camping impacts on vegetation and soils of the Ozark Plateau. *Journal of Environmental Management, 40*(4), 329–339.

Turton, S. M. (2005). Managing environmental impacts of recreation and tourism in rainforests of the wet tropics of Queensland World Heritage Area. *Geographical Research, 43*(2), 140–151.

Whinam, J., Chilcott, N., & Bergstrom, D. M. (2005). Sub-antarctic hitchhikers: Expeditioners as vectors for the introduction of alien organisms. *Biological Conservation, 121*, 207–219.

Horseback Riding

9

Chapter Summary

Due to the scale and varied environmental impact of horseback riding on the environment, its increasing popularity, and the conflict with other recreational users, it needs to be emphasised to horseback riders that for continued access to wilderness areas, national and state parks, and conservation areas, best-practice management is imperative. This includes planning trail location and design; trail construction and maintenance; visitor regulation such as confinement, the amount and timing of use; education to improve user knowledge and change user behaviour using codes of conduct circulated through interpretative information, public lectures, and discussions through the internet, associations, and clubs; and greater policing and enforcement by management of strategies in place. Where unacceptable impacts occur and significant conservation and biodiversity values are threatened, prohibiting horseback riding completely in certain areas might be necessary.

9.1 Introduction

Horse riding is an important recreational pursuit in many countries, especially developed ones, as it tends to be an expensive activity to take part in. It can take place in protected areas such as national and regional parks where land is set aside for both conservation and recreation which may be a problem where there are conflicts of interest. Due to the fact that horse riding has negative environmental connotations, this can cause land-use and user conflicts for management to resolve. This is particularly the case in the USA and Australia, where large areas of public land are set aside as national parks and forests and where there is a strong cultural heritage of horse use. In Australia, many horseback riders wish to emulate the endeavours of early explorers and mountain cattlemen, and the image of the bushman on horseback is an iconic image in that country (Beeton 1999). Horse riding in protected areas usually occurs on designated bridle or multiple-use trails, but there is also cross-country riding where no designated trail exists, and peri-urban protected areas can also experience moderate to high levels of equestrian activity (Landsberg et al. 2001). Today, horses continue to provide transportation and companionship for many wilderness users in the USA and are an integral part of many outdoor recreation activities too in Australia, South Africa, and Europe. Overlaps clearly exist between recreation and commercial horseback riding and tourism based on horse riding operations. Equestrian tourism includes all kinds of leisure activities linked to horses, ponies, and donkeys, practiced by a person outside their normal place of residence for more than 24 hours

© The Author(s) 2019
D. Huddart, T. Stott, *Outdoor Recreation*, https://doi.org/10.1007/978-3-319-97758-4_9

and less than four months (Pickel-Chevalier 2015). It includes any equine-related tourist activity, including guided horse treks, tours, and trail rides, which include overnight accommodation or camping. It also includes activities linked to the horse as an attraction, like shows and events, racing, and museum sites, like National Stud heritage centres. In Europe France is the premier country in terms of equestrian tourism with around 16,000 businesses and over 60,000 km of equestrian trails. Horseback riding holidays in France increased from 4113 in 1990 to over 8000 in 2014. It is also relatively well developed in Iceland where between 15 and 18% of the foreign visitors per year are involved in equestrian tourism, with an emphasis on the Icelandic horse (Helgadóttir and Sigurðardóttir 2008), but under 40% of all riders come from outside that country, and horseback riding is an important type of holiday for Icelandic families (Sigurdsdottir 2011). Horse riding is included in the category of "light" adventure tourism (Adventure Tourism Market Study, Adventure Travel Trade Association, and the George Washington University, 2013). Trail rides commonly take place on established trails, but horse treks of several days or longer often take place off trails, through protected areas and spanning a wide range of environments (Ollenburg 2006). The growth of this horseback tourism, the number of people who own horses, and the large number of horse and pony clubs means that any problems related to environmental impacts are bound to grow in the future, especially when in a later section of this chapter there are figures to suggest there may be a major growth in recreation horseback riding by 2060. This growth in recreation horseback riding also means there will be continuing conflicts with other recreation users with, for example, competition for the same space. This conflict will need careful management by national park and other conservation authorities as we know that horse riders and commercial equestrian tourism operators are often part of strong lobby groups. There is also the issue that in some cases it has been suggested that that horse riders are indifferent to, or unaware of, their effects on the environment (UK CEED 2000; Newsome et al. 2004).

9.1.1 Recreational Horse Riding

The following types of horse riding are included under recreation horse riding:

Trail riding: the use and riding of horses along trails for pleasure and for scenic viewing and enjoyment of the natural environment and cultural heritage of an area (Fig. 9.1). It is generally undertaken at most on a day basis, starting and finishing from a base and may be carried out as either an organised club event, on a casual basis by individuals, or in small groups (usually the latter). Most visitors/tourists/less experienced riders would take part in this activity category, but reasonable horse riding skills are required, although

Fig. 9.1 Pony trekking, Brecon Beacons. Photo by D. Huddart

most horse rental includes some instruction. This kind of riding accounts for the majority of riders. The preferred terrain is on forest roads, bridle paths, and wide, open single tracks but not on roads used regularly by motorised traffic. In the UK this would be called pony trekking. Most of this riding is walking with some trotting, negligible cantering, and no galloping. There is a variation of trail riding, popular in Europe, Australia, and the USA called Le Trec which comprises three phases: trail riding, jumping, and correct basic flatwork. It is an all-day ride across varied terrain, route finding, and negotiating obstacles and natural hazards whilst considering the welfare of the horse, respecting the countryside, and enjoying all it has to offer.

Endurance riding: the use and riding of horses actively involved in competition over relatively long distances. Alternatively, the horses may be engaged in training rides over shorter distances. It is a timed test against the clock which traverses a marked cross-country trail consisting of a distance usually between 80 and 120 km in one day. The popularity of this activity can be illustrated by the fact that the American Endurance Ride Conference (North America's official national governing body) sanctions over 700 rides each year throughout the USA and Canada. In South Africa there is the Simonstown Military Sports Bureau Club which has organised an annual endurance ride since 1990 and is one of ten such events in the Western Cape area. In South East Queensland, there is the Murrumba Downs endurance ride where over 100 horses/riders traverse a trail participating in an event in the Moreton Bay Region.

Horse trekking: here there is the use and riding of horses on a trip that lasts over 24 hours as a means of long-distance travel and carrying all that is required to camp overnight and continue a trip. Often this activity includes packstock which is any authorised animal used to pack or retrieve supplies and equipment and can include horses, mules, donkeys, and llamas. These animals are used to carry people and their supplies to remote, wilderness areas where the packstock are often turned out to forage, drink water, and rest in mountain meadows. In the USA this type of horse riding has been an important recreation activity for over a century, developing from the original pioneers (Fig. 9.2A), and packstock were such a traditional part of wilderness recreation that Leopold (1921) defined wilderness as lands that were large enough to absorb a two-week, packstock trip (Fig. 9.2B).

Hacking: these are usually for more experienced riders who enjoy riding out for fun and exhilaration. These riders may ride out on their own or in small groups as riding pace is very personal. The riding pace will vary but will comprise walking, trotting, cantering, and, occasionally, even galloping when the terrain and circumstances permit.

Recreation horseback riding has also been used in therapy for emotional development.

9.2 Numbers Involved in Horseback Riding

Ten countries in the world have a horse population of over one million, with a total global equine population estimated at between 58 and 60 million. This figure is an unreliable estimate and includes feral and working horses for recreation but not donkeys and mules. The USA has by far the highest total with figures estimated between 9.2 million (American Horse Council) and 9.5 million (Food and Agriculture Organization of the United Nations), with c.3.91 million used for recreation and c.2 million belonging to horse owners, and of these 85% are recreational riders. There could be around 20 million horseback riders across the developed world (Ollenburg 2006; Equus Project 2014), and many own their own horse. Approximate figures for the European Union are 6.4 million, 2% of the population, with the UK at 2.4 million. France has about one million horses and 1.5 million riders, although Pickel-Chevalier (2015) estimates 2.2 million horse riders, of whom one-third are regular riders and two-thirds classed as casual riders, with the number of horses registered as 930,000, and Germany has 760,000. In Australia the estimated horse population is 221,000. In the USA the figures based

Fig. 9.2 (A) Pioneering pack mule and miner on route to mine at Mount Lowe, California, Wikimedia Commons. (B) Packstock on Baja California (Mexico) mountain trail. Author in the saddle. Note the cutting up of the sandy soil by the hooves. Photo by Silvia Gonzalez

Table 9.1 Numbers involved in horseback riding in the USA

1982–1983	1994–1995	1999–2000	2005–2009	% change 1999–2001 to 2005–2003
15.6	20.7	19.8	21.5	+ 8.4
1999–2001: 18.3	290.3	2005–2009: 16.3	262.1	−28.2

Millions of people over 16+ carrying out horseback riding on trails (first line) and mean annual and total annual days (second line) (after Cordell, 2012)

on Cordell (2012) and Bowker et al. (2012) are illustrated in Table 9.1.

Bowker et al. (2012) suggested that the figure for 2008 was 17 million or 7% of the population and 263 million activity days. This was estimated to increase to maximum figures of 30.57 million adult participants by 2060, 16.8 days/participant and 503 million total days, but this was without any climate change variable and was the figure using the higher income/social class in the study. Depending on the figures used, there could be a range in percent growth of between +44.4% and +86.5%. There are regional variations in the USA, but for Colorado horseback riding was surprisingly only the 26th outdoor activity in total numbers with 2.874 million. In the UK *Getting Active Outdoors* (2015) estimated that participants over age 14 riding once a week were 338,000. Again there are regional variations, and Wales Outdoors (2014) estimated that 4% of the population had participated in horseback riding and only 1% had this activity as the main outdoor activity in the last four weeks but that this was only likely to grow in the future to 3%. The figure for females was 2%, and this gender balance is common worldwide. 69% of these activities took over three hours/visit, and so there is the possibility of considerable impact because of this time frame alone. However, as with all figures for recreation, the way the data is collected, the ages of the participants, and the questions asked of the survey participants make the interpretation of the data sometimes difficult.

9.3 Horseback Riding: Biophysical and Social Impacts

In order to understand horseback riding impacts and to arrive at viable solutions regarding their management, it is important to examine and understand the impacts and factors that influence them.

9.3.1 Biophysical Impacts Caused by Horseback Riding

Most of the research addressing horseback riding impacts has concentrated on its physical and ecological impacts and the social impacts on other users, such as hikers. Most of this research has been conducted in the USA and Australia.

Many types of impacts from horses are similar to those from hiking, particularly soil compaction and erosion, loss of organic litter, loss of ground-cover vegetation, loss of species, trail erosion and widening, and potentially the spread of weeds and pathogens into natural vegetation (Table 9.2). What can differ is the severity of the impacts, and horses because of their weight and hooves have the potential to cause considerable damage to vegetation and soils. For example, the greater weight of horses can result in more damage to vegetation and soils than people hiking (Weaver and Dale 1978; Liddle 1997), and quantitatively the impacts are much greater. The results presented in Cole and Spildie (1998) show that horse traffic has more potential to disturb vegetation and ground cover than either llama or hiker traffic. Deluca et al. (1998) experimentally compared the effects of hikers, llamas, and horses on established recreational trails in western Montana. They concluded that horses consistently created more trail sediment than the other two user groups regardless of trail weather conditions or traffic levels. Similarly, Wilson and Seney (1994) found horse trails produce more sediment than bikes or hikers.

There is a difference though in the sense that some impacts such as grazing and confinement are unique to horse riding, and grazing by, and

Table 9.2 Suggested environmental impacts of horse riding on natural ecosystems

Soil	Vegetation	Water	Animals
Soil compaction	Exposed roots	Introduction of exotic species	Increased noise
Loss of organic litter	Tree trunk damage	Increased turbidity and sedimentation	Increased vigilance and potential for displacement of grazing animals
Loss of soil	Loss of vegetation height and vigour	Increased input of nutrients	Damage to burrows
Change in hydrology of soils	Loss of ground cover	Increased levels of pathogens	
Alteration in microbial activity	Loss of fragile species	Degraded water quality	
Nutrient enrichment (manure, urine)	Loss of trees/shrubs	Reduced ecological health of aquatic ecosystems	
Increase in resistant species	Compositional change in aquatic plant and animal communities		
Introduction and spread of exotic plant species (invasive weeds)	Excessive algal growth		
Introduction and spread of fungal pathogens			

Modified from and based on Liddle (1997), Landsberg et al. (2001), Newsome et al. (2002, 2004, 2008), Aust et al. (2005), Beavis (2005), and Pickering et al. (2010)

tethering of, horses can result in considerable damage to grasses and other palatable species. For example, Phillips (2000) noted that grazing is an additional impact associated with horses at tethering areas. Horses temporarily tied to bushes and trees, or man-made facilities, graze and trample the surrounding area of exposed soil, loosening the soil surface and damaging trees by exposing roots. Tethering areas are of particular concern to national park management when the impacts become visually extensive. Horses tied to trees can also result in damage to bark. Ropes or chewing can damage tree bark and may completely girdle and kill trees (Cole 1983). Bark damage weakens trees and opens their inner wood to invasion by insects and diseases. Pawing, scraping, and digging by confined horses erode soils and can expose tree roots.

The growing number of scientific studies demonstrates the high impacts of horse riding, both on the natural environment and on other users (Landsberg et al. 2001; Newsome et al. 2002, 2004, 2008).

Research has documented greater potential for trail degradation from horse use in comparison to other trail users. For example, horse traffic can eliminate vegetation cover more quickly than foot or bike traffic, and their greater ground pressures compact soils to greater densities and depths (Nagy and Scotter 1974; Widner and Marion 1993; Liddle 1997). The resulting hoof prints and rutting retain water and promote muddiness and erosion following rains.

Trail impacts include a wide variety of problems. Even low levels of trampling disturbance reduce ground vegetation height, cover, and biomass and may alter species composition by eliminating fragile species (Cole 1991, 1995a, b; Sun and Liddle 1993). Higher levels of trampling cause more complete ground vegetation loss and compositional change (Cole 1995a, b; Marion and Cole 1996). Changes are apparent away from the trail edges, and not only do some plants common in the forest understory, particularly shrubs, disappear from the trail edges, but other plants, especially introduced and native grassland species, become more common. There are probably multiple reasons for these differences including the direct effects of trampling and changes in light, seed supply, soil water, and nutrient changes following soil disturbance.

In an English heath, Liddle and Chitty (1981) showed that elevated soil nutrients were particularly important along horse tracks. Some came from manure, but much of the increase probably came from soil organic matter breakdown and other soil changes caused by horse trampling. Here concentrated traffic pulverised soil leaf litter and humus layers, which are either lost through erosional processes or intermixed with underlying mineral soils. These soils then become exposed and vulnerable to wind or water erosion and compaction (Cole 1991; Marion and Leung 2001). The compaction of soils decreases soil pore space and water infiltration, which in turn increases water runoff and soil erosion.

The major ecological impacts to trails from horse use are thus vegetation loss, trail widening, erosion, muddiness, and informal trail development. Erosion is considered to be the most severe form of impact because its effects are long lasting, if not permanent. Many studies demonstrate that trampling by a horse is more destructive to vegetation than trampling by foot (Nagy and Scotter 1974; Weaver and Dale 1978). Vegetation on horse trails is churned up and often cut off at the roots, instead of flattened, as on hiking trails. An experimental trampling study by Nagy and Scotter (1974) found vegetation loss to be four to eight times greater from horse trampling than hiker trampling. The greater vegetation loss from horse use tends to widen horse trails, which are often two to three times the width of hiker trails (Weaver and Dale 1978) resulting in greater soil exposure which also contributes to the greater erosion potential of horse trails.

Erosion occurs after vegetation is lost, and vegetation loss exposes soil that can then be eroded by disturbances such as horses' hooves, wind, and water. Horse use can be a significant precursor for increased erosion potential on trails. A horse carries a heavy weight on a small, usually shod, hoof, plus a rider. This weight exerts approximately 1282 g/cm^2 ground pressure for unshod horses to 4380 g/cm^2 for shod horses, compared to 206 g/cm^2 for a hiker in boots (Liddle 1997) and is a function of the gravitational forces associated with the animal and rider's weights and the greatly increased forces created by movement (Liddle 1997). Gravitational force can be converted to pressure by dividing the weight by the area in contact with the ground. The static pressure exerted by a shod horse is over 20 times the pressure exerted by a person wearing boots and over twice the pressure caused by a mountain bike or four-wheel-drive (4WD) vehicle. Horses are also more likely to shear vegetation than hikers (Whinam et al. 1994), and so these processes from horse traffic cause significant compaction to the underlying soil layers, reducing water infiltration and increasing surface runoff. This soil compaction is associated with a decrease in seed germination; it increases the soil strength, reduces the ability of water to soak in, affects plant root growth negatively, and raises soil temperature. In areas with high soil strength and compaction, there is usually low plant diversity along the trail. However, in addition, the action of a horse hoof tends to puncture and dig up the soil surface where the soil is sandy (McQuiad-Cook 1978). This loose, unconsolidated soil is more prone to erosion than compacted soil, and as a result, the potential for erosion increases on some horse trails as compared to hiker trails, depending on the type of soil as horse hoof pressure can either cause compaction (in clay-rich soils) or loosening of soil particles in sandy soils. So it all depends on the soil texture. An evaluation by Deluca et al. (1998) of the mechanisms by which trail traffic leads to accelerated erosion suggested that soil loosening and detachment of soil contributed to the higher erosional rates. Soil compaction and decreased infiltration were not considered as important, a finding supported by the work of Wilson and Seney (1994).

Heavy horse traffic in areas with wet soils though can result in the formation of muddy quagmires and excessive trail widening. Whittaker and Bratton (1978) found loosening of the soil to be a precursor to muddy trail sections. Loose soil is more apt to form mud than compacted soil, and the highly compacted subsurface soils prohibit water infiltration. The resulting impermeable basins retain water and mud long after rainfall. Marion (1994) noted that deep horse hoof prints collect and retain water, providing greater surface contact between water and soil and accelerating the formation of mud. Trail muddiness can be a temporary or seasonal problem, making travel difficult and often results in significant trail widening when trail users try to get round muddy sections.

Other trail problems attributed to horseback riding use include the proliferation of informal trails, manure on trails, tree damage, and the introduction and spread of exotic vegetation. Trail braiding is especially troublesome in meadows, where stock users tend to spread out rather than ride in single file. The creation of side trails to access water, features of interest, or short cuts to other trails are also considered a significant form of trail impact. User-created trails are often poorly routed and not maintained, resulting in an increased potential for degradation. Manure on trails is both an ecological and social problem. Manure can contain the seeds of exotic plants, although seeds may also be introduced from horse feed, equipment, and mud stuck to horse hooves. Large numbers of weed seeds can pass through the gut of horses and germinate in their manure (St John-Sweeting and Morris 1991). However, Whinam et al. (1994) found that weed seeds were limited to the manure, and Whinam and Comfort (1996) revealed no indication of introduced weeds from monitoring. Large amounts of manure may also pose a threat to water quality, but see later for a discussion on these two potential problems (Sects. 9.5–9.7).

Generally though more use results in more impact. Therefore information on how many people use a trail is very important when assessing their impacts. The common model of the relationship between increasing use and damage is curvilinear, where proportionally more damage occurs at lower levels of use, that is, the first foot fall (or hoof fall) causes proportionally more damage than the tenth or the hundredth (Cole 1995a, b; Newsome et al. 2004; Cole et al. 2004). However, recent research indicates that, in more resistant vegetation communities, the relationship is closer to linear, that is, each foot fall/hoof fall causes the same amount of damage. Gillieson et al. (1987) quantified the relationships between soil and vegetation changes on newly developed horse tracks in subalpine grassland in Kosciuszko National Park (Australia) where they found that vegetation damage occurred after only 10 passes by two horses but that soil damage was not apparent until after 30 passes. Nevertheless significant damage from low levels of horse trampling can cause reduction in vegetation height, with fewer plant species on the trampled sites. However, in the mountains of Montana, Weaver and Dale (1978) comparing the effects of hikers, horses, and motorbikes studied the impacts of much greater numbers of passes on natural grassland and shrubby pine forest. After 1000 passes, damage was least on grassy and stony sites and generally greater on slopes. In all environments horses tended to cause most damage and they caused more damage moving downhill. There have generally been relatively high-use thresholds for grasslands in the Rocky Mountains and the Central Plateau of Tasmania (Whinam et al. 1994). However, 20–30 horse passes were sufficient to cause change to shrubland, herbfield, and bolster heath (a cushioned moorland of low-growing, tightly packed, compact plants which are slow growing and therefore fragile) in the latter area but had little effect on dry grassland.

It is important to note that while horse use is often a more impacting type of use, other factors may be more influential determinants of resource degradation. For example, McQuiad-Cook (1978) found trail impact to be more a function of slope and trail location than a result of user type. Nagy and Scotter (1974) concluded that although horse use generally causes more damage than hikers, the degree of difference depends on the soil, vegetation, topographic, and climate characteristics. Summer (1980) identified the most influential landscape factors governing trail deterioration in the Rocky Mountains of Colorado as soil parent material, grade of trail and side slope, soil texture and organic content, rockiness, type of vegetation, and drainage. Measurements of physical changes along trails receiving a constant amount of horse use resulted in a wide spectrum of erosional impacts as influenced by one or more of the landscape factors listed above. Trails were most resistant when crossing rock outcrops, talus slopes, and the tops of moraines. Trails on the level valley floors and terraces with well-drained soils were resistant to erosion but susceptible to trail widening over time, and the trails most vulnerable were those that crossed colluvial slopes, moraine side slopes, wet bogs, and alpine areas. Summer (1980, 1986) concluded that horse traffic was not the most important controlling agent contributing to trail degradation. Gillieson et al. (1987) found similar vulnerability differences along an established horse trail in subalpine woodland with the impacts on plant cover most marked on the wettest trail sections, whilst Whinam and Comfort (1996) showed that there were major differences in the amount of horseback riding trail degradation among vegetation types in Tasmanian subalpine environments. In fact all sites were affected, but those in eucalypt forest and moorland were affected most and those in rainforest least.

9.3.2 Trail Proliferation

McQuiad-Cook (1978) noted that trail proliferation was a major environmental issue, for example, where shortcuts straight up a slope were taken rather than along more gentle switchbacks on a hill. This can occur both from hiking and horse riding, but horses cause more damage on the steep shortcuts. He also suggested that horse trails generally have a less compacted and more incised path than hiking trails, particularly in areas with moderate slopes. This was attributed to the shod hooves tending to loosen and move the soil rather than flatten and compact it. Trail braiding can also be caused by horses moving around treefall and around waterlogged sections of the trail.

9.3.3 Field Experiments on the Impacts

Research into the impacts of horseback riding has been both experimental and quasi-experimental. Experimental approaches have been used for off-trail research, whereas both approaches have been applied to studies of existing trails. In the off-trail studies, the impacts to the experimental plots are taken as being representative of the impacts associated with riding in off-trail environments. The results from these experimental studies clearly demonstrate that the impacts caused by horses in an off-trail situation have the potential to cause considerable damage to soils and vegetation in the studied environments.

Further support for the high-impact potential of cross-country horseback riding or riding on poorly defined trails is afforded by Phillips and Newsome (2002). The study, conducted in a vegetated parabolic dune area in a sub-Mediterranean coastal environment in D'Entrecasteaux National Park (Western Australia), set out to determine the impact of horses by measuring changes in species composition, vegetation cover and height, soil micro-topography and soil penetrometry on previously undisturbed plots. Horse intensities of 20, 100, 200, and 300 passes were systematically applied to each treatment transect and resampled after each level of horse trampling intensity. The horses were of a similar size (400–500 kg), unshod, and included a saddle plus rider. The results showed that horse trampling caused a decrease in vegetation cover and height, a change in species composition, a reduction in the frequency of plant species, and increase in soil depth and amount of bare ground (Fig. 9.3). The most impacted portion of the treatment cross-sectional profile was the central portion (40–60 cm). Field observation showed that horses tended to walk through the centre of the treatment transects following the defined paths made by previous horses. They naturally follow trail lines created by the horses ahead. This study illustrated that there can be significant damage to vegetation if horses are taken cross-country or stray from formed trails.

A weakness and source of criticism in horseback riding impact research is the lack of standardisation in the methodologies employed and the variables studied, which can hinder comparisons between studies. Additionally, variation in environmental conditions further hinders and complicates such comparisons (Leung and Marion 1996). Despite these potential limitations in judging trail degradation from a comparative standpoint, a number of studies conducted in the USA provide firm evidence that horses are

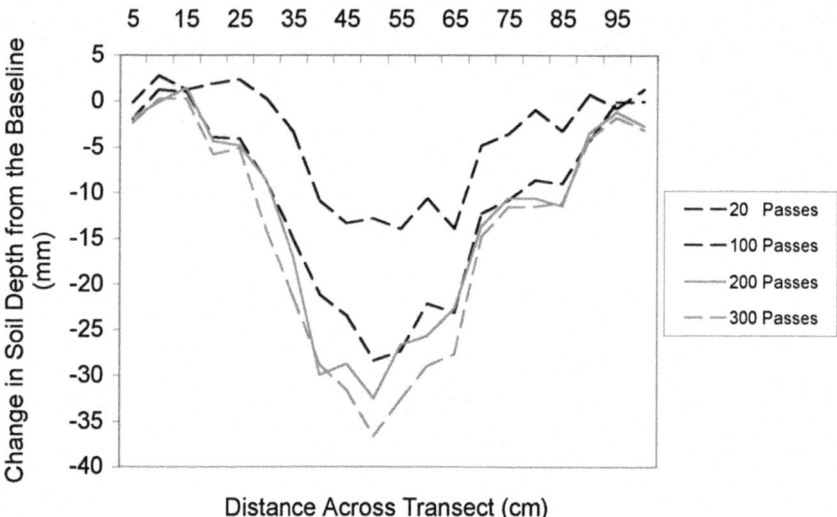

Fig. 9.3 The change in soil depth from the baseline micro-topography across 5–100 cm of the treatment transects cross-sectional profile after various intensities of horse trampling at study site DE3 in D'Entrecasteaux National Park, Western Australia (after Phillips and Newsome 2002)

degrading trails. For example, Weaver and Dale (1978) examined the effects of trampling due to hikers, horses, and motorcycles on multiple-use trails in the northern Rocky Mountains, Montana. Level and sloping (15°) sites were chosen in an alpine forest and meadow. The sites were subjected to 50, 100, and 1000 passes by horses, hikers, and motorcycles. On level ground, horses were the most destructive and hikers the least destructive, but on grassy slopes motorcycles were more destructive than horses. On both level and sloping sites, trail width was greatest for horses and least for hikers. Compaction was greater under horses than hikers or motorcycles because horses tended to exert the greatest downward pressure on the soil.

Wilson and Seney (1994) also examined the relative impact of various users on water runoff and sediment yield on multiple-use trails in the Gallatin National Forest (Montana). The sites were subjected to 50 and 100 passes by horses, hikers, motorcycles, and off-road bicycles. Measures were taken on 108 sample plots on the trails. Multiple comparison tests showed that horses and hikers made more sediment available than wheels (motorcycles and off-road bikes). This effect was most pronounced on pre-wetted trails. Of all users, horses produced the greatest sediment yield on both pre-wetted and dry trails. Cole and Spildie (1998) found that trampling reduced vegetation height more than it reduced vegetation cover, but the differential effects of users were more pronounced for vegetation cover. Only horse use caused mineral soil exposure, and they suggested a six- to tenfold difference in the amount of use the different vegetation types can sustain before a given amount of vegetation cover loss occurs. Two earlier studies had suggested a four- to eightfold difference in the impacts caused by horses and hikers. Nagy and Scotter (1974) studied a prairie grassland substantially more resistant than either the *Equisetum* or *Vaccinium* vegetation types studied by Cole and Spildie (1998). Depending on when trampling occurred, 100–200 passes by a horse resulted in vegetation cover loss equivalent to that caused by 800 hiker passes. Weaver and Dale's (1978) vegetation types were also more resistant (a *Poa pratensis-Festuca idahoensis* grassland and a *Pinus albicaulis-Vaccinium scoparium* forest with an understory very similar to the *Vaccinium* type). They found that in both vegetation types, 400–500 passes by hikers resulted in exposure of bare ground, equivalent to that caused by 100 horse passes. Deluca et al. (1998) conducted a study on trails in the Lubrecht

Experimental Forest (Montana). Horse, llama, and hiker traffic were applied to 56 plots at intensities of 250 and 1000 passes and compared with a control under both pre-wetted and dry trail conditions. Horses consistently liberated more trail sediment, which was then available for erosion, than either llama or hiker traffic. This was especially noticeable following 1000 horse passes. However, pronounced impacts were detectable after 250 passes suggesting that the initial traffic created the bulk of environmental damage. Furthermore, sediment yields were higher on dry trail plots than on pre-wetted plots, indicating that dry trail conditions made the trail more susceptible to sediment detachment. Surface runoff, however, was significantly greater on pre-wetted trail plots compared to dry plots, indicating that traffic moving on wet trails fosters increased runoff, which could result in greater downslope channelling of water and associated sediment transport. It was concluded that horse traffic tends to cause more trail erosion than hiker or llama traffic with the major reason being that horses are heavier and their weight is carried on a shoe with a small bearing surface; moreover, horses' shoes are typically metal and frequently cleated. The observation was also made that horses are less careful and deliberate than llamas or humans about where they place their feet. The differences persisted for at least one year.

9.4 Impact on Water Crossings and Rivers

Trampling damage is often particularly severe near watercourses (creeks, lakes, dams, and wetlands), with a far greater potential for soil erosion and trail incision (Liddle 1997; Newsome et al. 2004). Potential impacts at these sites include introduction of exotic species (aquatic and terrestrial weeds), increased turbidity associated with movement in water and soil erosion of banks, increased input of nutrients and sediments from soil erosion on banks, degraded water quality, compositional change of biota of water bodies and creek banks, potential for excessive algal growth, and reduced ecological health of aquatic ecosystems (Liddle 1997; Newsome et al. 2002).

These environmental effects can also influence the visual amenity of aquatic sites and surrounding landscapes.

Manure can also increase the risk of runoff into local waterways affecting riverbank and aquatic biota. Horseback riding can be considered a "disturbance" activity, and any monitoring programme needs to consider the potential impacts that might occur in such terms (Figs. 9.4).

Lake (2000) has categorised potential disturbance events into three types: pulse (short-termed and sharply delineated), press (may arise sharply, like a pulse, but then reaches a maintained, constant level), or ramp (when the strength of a disturbance increases over time). In the case of horseback riding, it was theorised by Redfearn et al. (2012) that horse riders crossing a stream would create a pulse event, with a potentially large impact that lasts a short amount of time. Trampling by horses causes soil erosion, and therefore it is likely that a high level of sediment will be deposited into aquatic systems during these pulse events (horse crossings). This deposition can be via dirt, seeds, or weeds from their hooves or directly from defecation (Pickering 2008). Mechanisms that increase sediment runoff can be considered a threat in the aquatic environment as it has been shown that increased sedimentation can lead to large changes and in the faunal composition of streams (Wood and Armitage 1997). Redfearn et al. (2012) showed that at horse trail crossings, there were increases in both organic and inorganic sediment, *E. coli*, nitrogen, and phosphorous.

There are, however, several areas that still need assessing. Whilst impacts have been detected, their intensity is determined on the number and regularity of potential disturbance events. This is of particular interest in relation to the management of a trail network, as small numbers of horse riders along a particular trail could be considered pulse events, with little long-term impact, whilst increased traffic and frequency could lead to ramp events and potential degradation of crossings. As part of the Scientific Monitoring Program in South East Queensland (see later in Sect. 9.15.5), the use of some trails is being assessed, and this information could be useful in determining which stream crossings

Fig. 9.4 Horse trekking through river, Cochamó, Chile. Photo by McKay Savage

may need further assessment and management. An assessment of each stream crossing on high-use trails is recommended. Any trail that has organised horse events should certainly have baseline monitoring undertaken. This would determine whether any infrastructure is currently needed and allow for regular reassessment that could detect deterioration of any crossing. Unfortunately, due to the hydrology of streams, most sediment and nutrient runoff will float downstream until it reaches the first pool. This, therefore, is the area that is likely to receive the most impact, and where aquatic health monitoring should be focused, as this deposited sediment can smother macro-invertebrate habitats. Assessment of stream crossings should therefore also include any pools immediately downstream.

Although impacts were found by Redfearn et al. (2012), the runoff during storms and floods, especially in national parks that are downstream of pastures and residential areas, was potentially much greater than anything that occurred during the anthropogenic disturbances captured during the events. Therefore, as long as careful monitoring is maintained, horse riding along these trails should be allowed to continue. Similar low-impact levels were noted by Forrester et al. (2017) from the Yosemite meadows where at packstock stream crossings, *E.coli* concentrations were greater downstream from crossings than upstream with the greatest increases occurring during storms. Differences in water quality downstream v. upstream from meadows grazed by packstock were not detectable for most water quality indicators. Low *E.coli* concentrations decreased downstream suggesting entrapment and die-off of faecal indicator bacteria in meadows. Under current packstock use levels, they are associated with relatively minor effects, although detectable, on water quality which are most pronounced during storms.

9.5 Impacts of Horse Manure

Horses' manure (faeces and urine) contain nitrogen, phosphorous, and various heavy metals. In stables, farms, paddocks, and natural areas, the management of horse waste is an important environmental issue particularly where it may contaminate waterways. The amount of dung produced by an adult horse (400–600 kg body weight) per day is of the order of 17–26 kg, whilst for urine it is around 5–7 litres per day (Mastsui et al. 2003). The addition of nutrients in horse manure is more likely to be an issue where soils are low in nutrients, particularly phosphorous, such as many Australian soils (Newsome et al. 2004, 2008). Horse manure can introduce around 1 g of phosphorous and 2.5 g of nitrogen per horse per day. Along trails and tracks, it can lead to local nutrient hotspots. In tethering areas or other places where horse densities are higher, the amounts of nutrients added can start to affect local vegetation favouring species adapted to higher nutrients (Moiuisse et al. 2005). In fact a potentially important effect of feral horses in Australia and the USA in more arid environments is their role in redistributing seeds, nutrients, and moisture via their faeces. Bienz et al. (1999) quote a figure of 9145 kg of faeces/year for an adult horse, and the extra nutrients and moisture provided by the horse manure probably provide a more favourable environment for both native and non-native plant species. Ostermann-Kelm et al. (2009) have shown that areas near horse manure have significantly increased plant diversity and cover compared to areas further away, yet the diversity consisted mainly of native plants to the study area. This was in habitats used by feral horses, and therefore in some parts of the USA and Australia, these considerable horse populations cause similar and perhaps even greater problems than horseback riding for recreation. For example, in Australia there are estimated to be over 400,000 feral horses, with 100,000 in Queensland alone (Csurhes et al. 2009, 2016), whilst in the Western USA, there are estimated to be over 58,000 (Ward et al. 2016). There has been much discussion as to how to deal with such large numbers of feral horses, especially in Australia, for example, in Nimmo and Miller (2006), but in the USA the US Congress passed the Wild Free-Roaming Horses and Burros Act of 1971 to protect, manage, and control wild horses and burros on public lands. This legislation declared these wild animals to be the living symbols of the historic and pioneer spirit of the West and tasked the United States Bureau of Land Management and the USDA Forest Service for their management for a "thriving ecological balance." This is a difficult task.

9.6 Horses As Agents of Weed Spreading

Manure acts as a favourable growing site and seems an ideal substrate for seedling germination in a controlled laboratory setting (e.g. Weaver and Adams 1996; Törn et al. 2009), and there is no reason to suppose that this should not apply to trails, apart from the factor of constant disturbance. It therefore seems possible that horses can act as shuttles for non-native seeds by consuming them from pasture and transporting them to a new plant community where they could germinate and grow. Seeds from hay could be transported, and it has been shown that horses that consistently consume hay transport more seeds through their manure than by their coats or hooves (Gower 2008). Weaver and Adams (1996) recorded 29 species from three national parks in the USA, and 17 weed species were recorded from track verges and two holding yards. The difference between tracks and verges may be due to the continued churning of the tracks by the horses, whilst Liddle and Chitty (1981) suggested that the lack of water may be the factor inhibiting establishment in drier environments. The sources of potential weed seeds include both local pastures and dried stock feeds which may be rich in weed seeds. After a review of the literature, Pickering et al. (2010) listed 189 seed species which have been found on the coats of horses (epizoochory) or germinated from horse manure (endozoochory). So it is not just horse manure which may be important here, and in Belgium Couvreur et al. (2004) studied 6385 seeds which germinated from 75 species which had been brushed from the coats of 201 horses, donkeys, and Galloway cattle in 27 nature reserves. They concluded that these animals were important dispersers of many plant species by seasonal grazing in different reserves, and that they function as "mobile link organisms" which connect isolated nature reserves through seed dispersal. This possibly influences vegetation development and the long-term survival of plant populations.

Results from Quinn et al. (2008, 2010) and others clearly indicate that seeds are capable of germination in greenhouse conditions after digestion by horses. Van Dyk and Neser (2000) found that the critical time for the dispersal of seeds via faeces was 18–48 hours after ingestion of feed containing seeds, when caution should be exercised when taking horses into sensitive areas. However, immersion of seeds in seawater did not affect the ability of seeds that had passed through the gut of horses to germinate. In fact the ability to germinate after having been exposed to digestive fluids seemed to be enhanced, since swelling of seeds and rupturing of the testa were observed.

More work is required to determine the importance of horses as potential weed dispersal vectors in field conditions, but the fact that weed germination has been documented should caution owners to carefully select feed sources free of weeds before bringing horses onto trails. Yet trails probably provide environments suitable for the establishment of many weed species regardless if they come from horse manure as the primary agent of transport. Many introduced herbs which seed prolifically and opportunistically can occupy the disturbed spaces of horse trails. They also have high relative growth rates and compete vigorously for the available moisture so inhibiting the colonisation of native species. Törn et al. (2009) found that on horse trails there were more forbs and grasses, many of which did not grow naturally in the Finnish forest studied, and it is possible that the species that were limited to horseback riding trails may change the structure of the adjacent plant communities in the long run. Nevertheless, a comprehensive literature review on the role of horses and weed dispersal suggests that in situ germination of invasive species from horse faeces on trails and into adjacent plant communities is uncommon. However, several studies show that in some locations, weeds do become established. When suitable conditions are lacking for germination of weed seeds transported by horse manure, the seeds may remain viable in soil banks for years. Such weeds have the potential to sprout and grow in subsequent years under favourable conditions.

Andrews et al. (1998) suggest that one of the most pernicious exotic species is the spotted knapweed (*Centaurea maculosa* Lam,), which covers vast areas of the American West. This species is thought to be actively invading the Selway-Bitterroot Wilderness Area of Montana and Idaho. Spotted knapweed is native to the steppes of Europe and was introduced to North America in the early twentieth century as a contaminant in Turkestan alfalfa (*Medicago sativa* L) seed. It has, since that time, expanded its range to cover almost three million ha in the Northwest USA (Lacey et al. 1992). Disturbed habitat is a key factor facilitating spotted knapweed invasion. Exposed soil, reduced canopy, irrigation, selective grazing of native species, and contaminated hay all have been cited as causative factors in its spread. Spotted knapweed reproduces only by seeds which are dispersed up to 1 m by a flicking motion when the plant is disturbed. In Montana seed production of spotted knapweed averages 1000 seeds per plant (Chicoine 1984). French and Lacey (1983) found that seeds may remain viable for up to five years, while Davis et al. (1993) continued to find viable seeds into the eighth year of their study. Spotted knapweed rapidly expands along roadways and in fields as plants are caught up in the undercarriage of farm machinery and motor vehicles (Montana Department of Agriculture 1986). In reserves and grasslands, primary roads and vehicles help facilitate seed dispersal into adjacent grasslands and trailheads (Tyser and Worley 1992). Within wilderness packstock camps, where the use of motorised vehicles is prohibited, spotted knapweed can be introduced from seeds in packstock hay (Cole 1983; Marion et al. 1986) or within manure from animals that have consumed weed-infested feed (Dale and Weaver 1974; Marion et al. 1986; Montana Department of Agriculture 1986). Seeds can also adhere to damp tarpaulin or tent bottoms or become attached to humans or packstock as they move along trails. Stock camps are occupied by both humans and animals, but usually packstock and saddle stock are kept separate from the portion of the camp where humans eat, sleep, and socialise. Thus, one might expect more spotted knapweed to be present in stock portions of the camp. It is notable that only 6 of 30 wilderness campsites and very small portions of 5 wilderness trails contained spotted knapweed in an area perceived to be at great risk from infestation. Furthermore, spotted knapweed occurred in only one of four camps infested in 1993 that were revisited the following summer, and seed production was low for specimens collected along the trail during the summer of 1994 (Andrews et al. 1998). If the Bitterroot portion of the Selway-Bitterroot Wilderness is representative of forested wilderness areas in the Northern Rockies, then the perceived threat may substantially exceed the actual danger in many instances. The results from this study suggest general avenues of management responses:

1. Managers should conduct surveys before initiating costly control measures in any wilderness area. Surveys in forested regions similar to the Selway-Bitterroot should initially focus on areas most prone to infestation, that is, areas with open canopy adjacent to trails and in elevations that are optimal for spotted knapweed.
2. Wilderness workers can be trained to remove weeds as part of their normal backcountry duties. Likewise, volunteers can be educated and recruited to remove weeds via existing weed awareness programmes, signing of trails, and information packs given to backcountry users.
3. The association of spotted knapweed at campsites with loss of vegetation, exposed mineral soil, open canopy, and development of facilities emphasises the need for already existing regulations promoting minimum-impact camping in wilderness areas. In particular, backcountry permits should stress packing in camp chairs and using aluminium poles for tent poles and hitches rather than tearing down dead wood and cutting live trees, both of which open up the canopy.

In Table Mountain National Park (South Africa) where horse food seeds have sprouted in the park, it however appears to be on a very limited basis and only in proximity to the paths. Any period of dry and hot weather, or burning of vegetation, seems to destroy the fledgling growth. On the Glencairn Mountain and Noordhoek areas where horses have been ridden over many years, no notable outbreak, or spread of seed as a result of spread from horse manure, has been documented. There are however no known objective studies to test assertions that germinating seed from horse manure does not persist in the park or become invasive.

However, it is thought that each year, noxious weeds cause millions of dollars in losses to the US western rangelands. Since the 1990s, most hunters and recreationists have been confronted with the need to buy "certified weed seed-free forages" to travel with horses in Montana's backcountry areas. Preventing new weed invasions is critical to the resources, and the Montana Noxious Weed Seed Free Forage (MNWSFF) programme was developed for this reason. Each year, the Montana Department of Agriculture coordinates an inspection and verification programme to identify forages and feeds that are packed into federal and state lands. Fields are inspected for the absence of Montana's 23 noxious weeds by qualified inspectors, and "certified" forages or feeds are then marked by a system of tags or colour-coded bale twine. Each year, the Montana Department of Agriculture provides a list of MNWSFF-certified producers, so that horse and livestock owners can buy local weed-free hay or feeds. The MNWSFF programme is overseen by an advisory council comprised of hay producers, backcountry organisations, and state agencies. As a further control in wilderness areas since 2003, all forage brought into trailheads on federally and state-managed properties have had to have weed-free certificates.

The use of MNWSFF forages and feeds is becoming a successful method to restrict new weed introductions on public lands. On many privately owned ranch, farm, and subdivision properties, weeds can invade in many different ways.

It is evident that noxious weeds were likely introduced by feeding weed-infested hay or feeds. Many current weed problems are worsening due to poor grazing practices, lack of weed control, and continual reinfestation and spread by manure and additional contaminated feeds. Currently, the MNWSFF programme is being targeted to horse owners on small acreages to limit the spread of weeds on private lands.

Wildlife and domestic livestock can move weed seeds over long distances on their hair or skin and in their digestive tracts. Depending on weed maturity, weed density, and animal consumption, passage of viable weed seeds in animal manure is a fairly efficient method of weed seed dispersal. Passage of viable weed seeds has been documented in wild birds, deer, mice, swine, cattle, sheep, poultry, and horses. In most cases, mastication and digestion reduces seed viability by over 90%. However, if a large quantity of noxious weed seeds is consumed, some viable seed will successfully pass through the digestive tract. In research at Montana State University (MSU) by Brett Olson and co-workers, sheep and mule deer were fed alfalfa pellets and barley dosed with 5000 spotted knapweed seeds. Manure was collected daily for ten days after feeding the weed seeds, and knapweed seeds were extracted and tested for viability. Over the ten-day period, the mule deer passed 43 and sheep passed 16 viable spotted knapweed seeds (0.9% and 0.3% of the original 5000). Interestingly, seed passage through the mule deer was discontinuous—9 viable seeds passed on day 10, compared to sheep where most viable seeds (14 of 16) had passed by day 3. The implications of this research are that grazing livestock should not be moved from weed-infested sites to noninfested sites for a minimum of four days and that wildlife seed movement is a fairly high risk.

Very little has been published on horse digestion and passage rates relative to weed seed survival, which could be different from that of ruminants such as sheep and cattle. Previous MSU horse research with chemical markers indicate that over 95% turnover occurs in 72 hours. Recently, the MSU group conducted a horse-feeding trial where feed was dosed with known quantities of weed seeds. The weeds included leafy spurge, spotted knapweed, Persian darnel, wild oat, curly dock, and quack grass. Alfalfa seed was also used in the test because of its seed coat and size characteristics and availability. Manure was collected at intervals up to 72 hours, and the manure was subsampled and grown in the greenhouse to detect passage of viable seeds. In this preliminary evaluation, total passage of viable weed seeds through 72 hours ranged from 0% to 2% (weeds) to over 10% for alfalfa. If horses eat high levels of noxious weeds, it appears that a low level of viable seeds will pass and could contaminate lots, pastures, and public lands.

Further, based on these preliminary results, MSU are encouraging people to feed their horses certified hay and grain three to four days before transporting them to the trailheads, to minimise the risk of spreading weed seeds (Cash et al. 2008). However, in Montana, certified MNWSFF hay and feed sources cost 5–20% more than other feeds. For horses that are fed hay on small drylots throughout the year, it is advisable to feed certified hay or feeds year-round. It has been suggested though that if combined with a diligent weed control programme, the 5–20% premium generally pays for itself in future weed control costs on private properties.

The presence of weeds in natural areas has been suggested to pose a significant biodiversity conservation problem and reduces the aesthetic appeal and inherent value of native vegetation. Various studies have also shown that there is potential for horses to act as a vector of weed spread (Campbell and Gibson 2001; Weaver and Adams 1996; Whinam et al. 1994). Nevertheless, Campbell and Gibson (2001) found that of the 23 weed species found in manure samples collected from trails in Southern Illinois, USA, only one species was found in trail plots. Similarly, Whinam et al. (1994) found that four weed species germinated from manure collected in Tasmania (Australia), in the glasshouse but not in field conditions. Field experiments show that weed establishment is highest in areas of previously disturbed ground and where grazing animals had been excluded. However, Barratt (1999) found a lack of viable weed seeds in horse droppings collected on bridle trails in Western Australia, but Weaver and Adams (1996) recorded 29 weed species germinating from horse manure samples collected from bridle trails in Victoria (Australia). The presence of weed seeds in horse manure highlights that horses have the capacity to disperse viable propagules of both woody and herbaceous weeds (Cosyns and Hoffman 2005). This is also borne out by field observations of the rampant weed veldt grass (*Ehrharta calycina*) germinating from horse dung along trails in John Forrest National Park in Western Australia. If invasive weeds are germinating from horse dung as observed, then the use of trails by horses, especially through good quality and mostly undisturbed vegetation with no prior weed invasion, is contributing to degradation by facilitating weed spread through trail corridors (Newsome et al. 2002a, b). Users may find trails degraded by weeds unsightly and not in keeping with the overall concept of protected areas. Weed seeds can also be dispersed by other means such as wind or water; by ingestion or attachment to hair (or clothing) on native, feral, and domestic animals, including humans; and by vehicles through mud encrustations, especially on tyres (Liddle 1997; Weaver and Adams 1996). As such, horses are only one of a number of vectors. They are, however, an important one because of their ability to transport large numbers of seeds and then deposit them, complete with fertiliser, in areas that are otherwise remote from weed sources. Additionally, horseback riding can result in the development of new trails which other vectors, such as hikers and wildlife, can subsequently use and move weed material into previously less inaccessible areas.

Based on 11 documented studies, the review by Pickering et al. (2010) showed that seed from 216 species is known to be viable after passing through the digestive tracks of horses, 45 of which are environmental weeds. For example, in South Australia St. John-Sweeting and Morris (1991) researched what happened to seeds that pass through the horse digestive system. The majority of seeds tested showed little or no loss in viability after transmission. The work showed that horses can disperse weed seed for as long as ten days after ingestion (*Trifolium* and *Medicago*) and they can pass relatively large amounts of seed four days after ingestion. What is less clear is if these species can germinate in situ, become established, and spread in protected areas. There do not appear to be any Australian or USA field studies confirming that weed species germinate in situ from horse manure along trails in protected areas. Nonetheless, the environmental weed *Ehrharta calycina* has been observed germinating from dung deposited by horses on walking trails traversing weed-free natural vegetation in John Forrest National Park (Western Australia). Studies in Europe have confirmed that such seed can germinate from horse dung in a range of environments (Moiuisse et al. 2005; Törn et al. 2009). A field study in subalpine Tasmania found no indication of introduced weeds and that weed seed did not germinate from manure along trails, but they did germinate from horse dung in field plots where soil and vegetation were disturbed (Whinam and Comfort 1996). This was thought to be due to the constant churning of manure on the tracks which inhibited the establishment of weeds. In contrast, weeds did not germinate from manure or hoof debris samples along trails in the Eastern USA (Gower 2008), and there was thought to be no problem because of this poor germination.

The potential for horses to disperse weed seed and facilitate weed establishment and spread along trails and subsequently into natural vegetation in protected areas clearly needs further research, although it does seem a likely possibility, despite the sometimes confusing results from the research.

Horse riding therefore poses a risk for the introduction of weed species and has been noted along the horse trails in South East Queensland. For example, of the 14 significant grass weeds for South East Queensland (Thorp and Wilson 2008), all can be dispersed by seed, all can be found in pastures, and many also occur on road verges. Hence they have the potential to appear in horse feed and dung. Several of them could also be spread on the fur of horses, on riders, or on riding equipment. For example, Mossman River grass (*Cenchrus echinatus*) has burrs that attach to fur, African lovegrass (*Eragrostis curvula*) has seed that can be dispersed on the mud of cars and on animal fur, olive hymenachne (*Hymenachne amplexicaulis*), a weed of national significance, can be dispersed in stock feed, fountain grass (*Pennisetum setaceum*) can be dispersed on clothing, and giant parramatta grass (*Sporobolus fertilis*) and giant rat's tail grass (*Sporobolus pyramidalis*) have seed that can become attached to machinery, and seed that can attach to fur and hair, while grader grass (*Themeda quadrivalvis*) can be dispersed by animals, mud, and on graders (hence the common name) (Thorp and Wilson 2008 and references therein). For South East Queensland horse trails, the potential for long-distance dispersal (introduction into a park) and short-distance dispersal (within the park) of weed species by horses is clearly of concern. Adherence to the horse riding code should reduce the risk of introducing seeds and reduce damage to vegetation that may favour the establishment of weeds. Assessment of the seed load on horses (coat, equipment), in feed, and in manure would help quantify the risk of long- and short-term dispersal of weeds from riding in parks. To assess the relative risk from horses, research on seed introduction via car and bike wheels as well as on walker shoes should also be conducted.

The big problem is that once established, weeds are hard to control and can spread from tracks and tethering areas (Törn 2007; Newsome et al. 2008).

9.7 Horses As Agents of the Spread of Non-native or Alien Species

It has been suggested that exotic, non-native species will colonise forest more frequently by the introduction of seeds through horse manure deposition along trails. However, whether these exotic species have the ability to spread and establish themselves has been much disputed (Johnston and Pickering 2001; Ngugi et al. 2014). Campbell and Gibson (2001) took samples from three areas in Southern Illinois which were open to horse travel recreation. Vegetation data was collected from each of the trails as well as from a trail where horse riding was prohibited. Manure samples were placed in situ along the horse trails at one site to examine seedling germination in natural conditions. Whilst 23 exotic species germinated from samples of manure placed in a greenhouse, only one of these exotic species was also found in the trail plots (*Kummerowia striata*). Similarly while there were empirically more exotic species found along the trails which allowed horseback riding than there were on the trail lacking any horseback riding, the relative importance of these species was negligible along both trails. These results suggested that the migration of exotic species via horse manure does not pose an immediate threat to the plant communities adjacent to the trails in these forest ecosystems. Nevertheless Campbell and Gibson (2001) suggested that the large number of exotic species in horse manure reflects a constant threat to any ecosystem from these species and that care must be taken when allowing horse riding in areas to anticipate an invasion of exotic species from horse manure. In the Oulanka National Park (Finland), Törn (2007) studied the impact of horse riding on the introduction of alien species, vegetation, and soil characteristics by conducting in the period 2002–2005 a full factorial experiment with four factors: (1) disturbance, (2) manure addition, (3) seed addition (*Vaccinium myrtillus*, *Vaccinium vitis-idaea*, and *Empetrum nigrum* seeds), and (4) both manure and seed addition. In the disturbance treatment, organic material was completely removed to expose mineral soil. Manure treatment was applied by the addition of 7.5 litres of horse manure per study plot. Seed addition treatment was carried out by adding seeds of 25 berries per each species, the amount representing the natural seed rain of dwarf shrubs of the area. The number of shoots of each species germinating was recorded in 2002 and 2003. The plant biomass was measured in 2004. The concentrations of phosphorous, calcium, potassium, and magnesium in the soil were measured in 2004, and also the acidity, electrical conductivity, and the content of organic matter and water in the soil were analysed.

Manure addition resulted in the establishment of 15 species of graminoid and forb species on the study plots that were otherwise absent from the boreal forest habitat studied. The total shoot density and the density of shrubs were lower in disturbed plots than undisturbed plots, while the densities of sown shrub seedlings and manure seedlings (seedlings of plants whose seeds are carried in manure) were higher in disturbed plots relative to undisturbed plots. However, the difference between undisturbed and disturbed plots in total shoot density and the density of dwarf shrubs nearly levelled out during the study period. Two years after treatment, total biomass was lower in disturbed plots relative to undisturbed plots. However, the biomass of manure seedlings was 4.4 times greater in disturbed plots relative to undisturbed plots. The concentrations of phosphorous, calcium, magnesium, and potassium were also significantly higher in disturbed plots compared with undisturbed addition. The density of adult shrubs was lower in manure addition treatment compared with plots without manure addition in disturbed plots. In disturbed plots with manure addition, total biomass was higher than plots with no manure addition. The phosphorous concentration in soil was the only measured nutrient that was higher due to manure addition in both disturbed and undisturbed plots. The concentrations of calcium, magnesium, and potassium were also higher in disturbed plots with manure addition. It was generally felt by Törn (2007) and Törn et al. (2010) that the risk of alien species to the biodiversity of natural forest in Finland is likely to be relatively small due to the lack of continuous disturbance in these habitats. Instead they thought that the greatest risk is caused by the possibility of alien species spreading via horse trails to neighbouring sensitive, open habitats.

They also found that horseback riding had the most effects on plant species/genera and trail characteristics compared to hiking and skiing. Horse trails were narrower than hiking trails, but just as deep, even though the number of users on the horse trails was lower (150 times higher on hiking trails). The narrowness of horse trails relative to hiking trails may be because these trails are more recent and because horse tour groups tend to travel in single file. We have seen these greater negative ecological impacts of horses per capita relative to the hikers have also been found in other studies (e.g. Weaver and Dale 1978; Liddle 1997; Cole and Spildie 1998; Bear et al. 2006; Wells and Lauenroth 2007; Wells et al. 2013). It seems that instead of the quantity of users, the quality of activity and the impacts often can have a more pronounced role in the planning and management of tourism.

9.7.1 Summary Related to Impact of Horses on Weed and Alien Plant Spreading

There are many factors affecting the establishment of weed and alien plant species, and it is difficult to tease out which is the major control on the level of trail damage, but track conditions, including soil structure and slope, the level of horse use affecting the amount of churning, the general climatic environmental conditions affecting moisture and dryness, and the relative sensitivity of the vegetation may well combine to affect the ability of weed and alien species to germinate successfully following dispersal by horses and the level of trail damage. Newsome et al. (2004) suggest too that the level of management that is in place, if any, and its effectiveness are important in controlling the impacts of horseback riding. There are many ways in which horses themselves can affect seed germination along trails. They trample the ground opening up bare areas, and they defoliate vegetation which has been shown to assist in the spreading of seeds along trails and to enhance seed germination by creating open gaps in the ground. There are other vector possibilities too on multi-use trails like off-road vehicles, mountain bikers and hikers, and other animals, such as deer, elk, and birds. Ways to help in controlling weed and alien species would be to collect horse manure along trails, but this seems an unlikely solution, and limit the number of horses allowed on a trail, which is much more likely (Törn et al. 2009). Proper disposal of unused or spoiled hay would also decrease the chances of exotic plant growth (Gower 2008).

9.8 Damage to Campsites by Horses

Damage to campsites used by horseback riders also differs from impacts at campsites used by backpackers. The percentage vegetation cover decreased, and there was increased compaction of soil, exposure of tree root, and the introduction and spread of weeds (Cole 1983). In the Bob Marshall Wilderness in Montana, campsites used by horse groups had 11 times as many damaged trees and 25 times more trees with exposed roots than backpacker sites (Cole 1983).

9.9 Impact of Packstock Grazing

Apart from damaging campsites, horses can damage packstock grazing areas. Horses are usually allowed to graze freely, and they need to be confined for long periods. While grazing, they defoliate plants, urinate and produce manure, and trample the soils. The soils of forage- abundant meadows are often moist and therefore prone to trampling. Horses are often tethered with the impacts noted above, but there are less damaging techniques like confinement which include tying stock to a rope tied between trees, known as a high line, and confining stock inside an electric fence. The use of packstock has become a contentious and litigious issue for land management agencies in the Western USA due to concerns over their effects on the environment (McClaran and Cole 1993). The potential environmental effects of packstock on Sierra Nevada meadow ecosystems have been reviewed by Ostoja et al. (2014), and they concluded that the use of packstock has the potential to influence the following: (1) water nutrient dynamics, sedimentation, temperature, and microbial pathogen content; (2) soil chemistry, nutrient cycling, soil compaction, and hydrology; (3) plant individuals, populations and community dynamics, non-native invasive species, and encroachment of woody species; and (4) wildlife individuals, populations, and communities. It is considered from currently available information that management objectives of packstock should include the following: minimise bare ground, maximise plant cover, maintain species composition of native plants, minimise trampling, especially on wet soils and stream banks, and minimise direct urination and defecation by packstock into water. However, incomplete documentation of patterns of packstock use and limited past research limit current understanding of the effects of packstock, especially their effects on water, soils, and wildlife. To improve management of packstock in this region, research is needed on linking measurable monitoring variables (e.g. plant cover) with environmental relevancy (e.g. soil erosion processes, wildlife habitat use) and identifying specific environmental thresholds of degradation along gradients of packstock use, particularly in Sierra Nevada meadows. A case study of the effects of grazing in remote meadows in wilderness areas is illustrated in Cole et al. (2004) and Newsome et al. (2004). Three different meadows were studied: (1) a high-elevation (3100 m) xeric shorthair sedge (*Carex filifolia*) meadow, (2) a mesic shorthair reed grass (*Calamagrostis breweri*) meadow at 2600 m, and (3) a more mesic, tufted hair grass (*Deschampsia cespitosa*) meadow at 2285 m. None had been grazed in the last century. In each of these meadows, horses and mules were allowed to graze at specified intensities each year for four successive years. Four replicate blocks of four grazing intensities (0%, 25%, 50%, 75% forage removal) were achieved by tethering animals to a stake, using a 4 m rope, for as long as was required to remove the target forage level. This produced ~50m^2 grazing plots which were monitored before and after grazing for each of the four years, as well as one year after the final grazing treatment. This caused considerable changes in meadow conditions, and the productivity (vegetation biomass one year after grazing) was significantly reduced after the second grazing season. There were also increased bare soil and small changes in species composition. Grazing reduced the relative cover of grasses in all three meadows.

Lee et al. (2017)) evaluated the influence of horse and mule packstock use on meadow plant communities in Sequoia and Yosemite National

Parks in the Sierra Nevada of California. Meadows were sampled to account for inherent variability across multiple scales by (1) controlling for among-meadow variability by using remotely sensed hydro-climatic and geospatial data to pair stock use meadows with similar non-stock (reference) sites, (2) accounting for within-meadow variation in the local hydrology using in situ soil moisture readings, and (3) incorporating variation in stock use intensity by sampling across the entire available gradient of packstock use. Increased cover of bare ground was detected only within dry meadow areas at the two most heavily used packstock meadows (maximum animals per night per hectare). There was no difference in plant community composition for any level of soil moisture or packstock use. Increased local-scale spatial variability in plant community composition (species dispersion) was detected in wet meadow areas at the two most heavily used meadows. These results suggest that at the meadow scale, plant communities are generally resistant to the contemporary levels of recreation packstock use. However, finer-scale within-meadow responses such as increased bare ground or spatial variability in the plant community can be a function of local-scale hydrological conditions. In reality Sierra Nevada meadows are complex ecosystems that are routinely subjected to natural disturbances, such as seasonal flooding from snowmelt, long-term decadal droughts, and bioturbation from small mammals. Meadow plant communities, along with the environmental variables that help structure those communities, vary greatly within and across individual meadows, watersheds, and elevations. Thus, at the whole-meadow scale, it is not surprising that potential signals of packstock use on meadow vegetation were swamped by these larger-scale environmental factors. It was only by adopting a multi-scale approach that accounted for environmental processes known to influence variation in plant communities, both among and within meadows, that Lee et al. (2017) were able to detect a meaningful ecological signal of packstock use. The above study did not account for past impacts from periods when packstock use levels were much higher than that currently allowed. For example, historical accounts of Sierra Club outings from the early 1900s have descriptions of over 100 pack animals being used on single trips to the Sierra Nevada high country. With historically higher use levels, there is a possibility of legacy effects that are unaccounted for in the meadows that were studied. While legacy effects could include altered hydrologic regimes, other legacies may encompass adaptation to grazing by meadow communities and shifts to a more resilient composition, as well as increased productivity. The number of individual animals allowed in a single group has been limited in Yosemite and Sequoia and Kings Canyon National Parks since 1972, and current numbers are set at 25 animals per group in Yosemite and 20 animals per group in Sequoia and King's Canyon National Parks, with some popular areas restricted to smaller groups still. Total packstock numbers within the national parks have been declining since the 1960s and are currently at historically low levels, so it is possible that there may be some historical legacy.

9.10 Impacts on Animals

Few studies document impacts of horseback riding on animal populations. However, in those that do, horseback riders appear to be on the lower end of the spectrum in causing direct disturbance to wildlife (Wisdom et al. 2004; Taylor and Knight 2003). Concentrations of horses around water can negatively impact habitat quality for aquatic wildlife (Vinson 1998). Horses can attract brown-headed cowbirds and potential predators of some songbirds (Snetsinger and White 2009), particularly where corrals and stables are present (US Fish and Wildlife Service 2002). Klinger et al. (2015) evaluated the potential overlap between packstock and Bighorn Sheep in Sequoia and King's Canyon National Parks in California.

There appears to be greater macro-invertebrate abundance on trails which may reflect superior habitat conditions for the species observed (Osterman-Kelm et al. 2009), but it could be that the visibility is higher in the areas which had less vegetation. However, Holmquist et al. (2010) sampled terrestrial arthropods in paired grazed and ungrazed meadows across Sequoia National Park

(Sierra Nevada), and although they found some negative effects of grazing on vegetation structure, they found few lasting negative or positive effects of packstock grazing on arthropods in these wetlands. It appeared that pack stock do not cause lasting damage to the arthropod assemblages, but they did caution that the extent of the impact at the height of the grazing season remains unknown.

9.10.1 Impacts on Breeding Birds

The Cape Peninsula's beaches in South Africa are home to a variety of waders. The most important resident breeding species is the African black oystercatcher. It breeds on sandy beaches and is very susceptible to disturbance by horses that traverse these areas to access the beach areas below the high-water mark and dried out pans. It has been documented that the oystercatcher breeding colony, which nests above the high-water mark in the southern corners of Chapman's Bay, is under threat from people, horses, and dogs. The other resident wader likely to be impacted is the white-fronted plover which also nests on sandy beaches, and its inconspicuous nest is also vulnerable to disturbance by horses. Horses are likely to disturb breeding birds too in other parts of the world.

9.11 Potential Spread of Pathogens by Horses

The spread of pathogens by visitors into, and within, protected areas has been documented (Newsome et al. 2002, 2008; Pickering and Hill 2007), with the spread of root-rotting fungi (*Phytophthora cinnamomi*) of most concern in Australia. Horse riding, along with other activities such as mountain biking, walking, and 4WD driving, has the potential to introduce pathogens into parks and to enhance their spread within a park. Again sampling along trails and sampling material from different user groups could help quantify this risk in South East Queensland. The severity of the threat has been recognised nationally, and it was listed as a key threatening process by the Australian Government (Environment Australia 2001) and by the NSW government in the *Threatened Species Conservation Act 1995* (*NSW*). There do not appear to be any studies that directly tested horse's hooves as dispersal mechanisms for plant pathogens, such as *Phytophthora* in Australia or the USA. However, horseback riding is considered to be an important risk factor for many protected areas as the pathogens have been definitely been transported on the tyres of vehicles and on human shoes (Newsome et al. 2002, 2008). Therefore, the spread of the pathogens from an infected area to what was a disease-free area may result from horseback riding as an activity in protected areas, even if horses themselves are not the primary vector.

A final comment is worth making regarding the nature and extent of the biophysical damage: damage is highly context-dependent. The Australian continent, for example, is characterised by the widespread occurrence of sandy and nutrient-poor soils. These soils have a low resilience to human-sourced disturbance, and this, combined with arid and semi-arid climates, with long dry seasons and the presence of diverse, complex ecosystems that are susceptible to infection by introduced pathogens, such as the cinnamon fungus (Newsome et al. 2002), makes impacts from horseback riding highly likely. Additionally, these factors mean that recovery from impacts is likely to be a long-, rather than short-, term process. In the USA, extensive damage is also a very real possibility, but for very different reasons. Much of the horseback riding in protected areas takes place in montane wilderness, with the associated issues of waterlogged soils and a short growing season making recovery difficult.

The biophysical damage is summarised in Table 9.3.

There are other reasons too why horseback riding impacts or potential impacts are of concern to Australian ecosystems which is covered in the next section.

9.12 Why Horse Riding Impacts Are of Particular Concern in Australian Ecosystems

Australia's biota is of international significance due, at least in part, to its high levels of endemism (species that are only found in Australia;

Table 9.3 Variables that can be used to monitor horse riding impacts

Biophysical variables	Vegetation-based variables	Visitor data	Social condition variables
Increased amount of bare ground	Reduction in height and vegetation cover on verges	Who visits	User conflicts on multi-use trails
Soil compaction	Presence of weeds on trails and verges	When visitation occurs	Unsafe or difficult travelling conditions
Soil erosion	Increase in width of verge (area of altered vegetation adjacent to trail)	Activities undertaken	Manure on trails
Increased trail width	Increase in resistant species in verge and adjacent areas	Location of activities	Visitor expectations
Wet muddy trails	Evidence of dieback on verges		Visitor satisfaction
Increased surface runoff	Root exposure		

Some types of variables that can be used to form inventory categories or directly measured for monitoring horse riding impacts on existing management trails (modified from Newsome et al. 2008; Hadwen et al. 2008)

Department of the Environment, Sport and Territories 1994). As a result of the long period of geographical isolation of the continent, 85% of vascular plant species and 82% of mammal species are endemic to Australia. Australia is one of the world's 15 mega-diverse countries and contains several critical diversity hotspots, including the south west of Western Australia, the Wet Tropics of Queensland, and the Great Barrier Reef. The Southeast region of Queensland also has high biodiversity, with areas of such significance that are world heritage listed (e.g. the Gondwana Rainforests of Australia World Heritage Area).

One important way in which the Australian flora and fauna differ from other areas is that Australian species have evolved in the absence of large, hard-hoofed herbivores such as horses, antelope, sheep, cattle, or goats (Newsome et al. 2002). Australian grasses, forbs, and shrubs were therefore not subject to the types of grazing and trampling that plants native to the plains of North and South America, Europe, and Africa have experienced. For example, many Australian native grasses are tussock species which can easily be uprooted by grazing animals such as horses, while inter-tussock spaces are important habitats for herbs that are easily damaged by trampling from hard-hoofed animals, such as horses.

A second issue is that Australian soils are often low in nutrients, particularly nitrogen and phosphorous (Newsome et al. 2002). As a result Australian plants show many different adaptations to cope with low nutrient levels, including sclerophyllous form (small leaves, short internodes, hard leaves with waxy layers), proteoid root mats found in many Proteaceae species, nitrogen fixation (leguminous species including many shrubs and wattles and nonlegumes such as she-oaks), and associations with a range of fungi (mycorrhiza) adapted to enhance the uptake of nutrients (Newsome et al. 2002). Damage from horse riding to roots, combined with nutrient supplementation (manure, urine), can reduce the effectiveness of root-based mechanisms for regulating nutrient uptake resulting in increased stress for many Australian plants. In some cases nutrient supplementation can even be toxic (Newsome et al. 2002). The brittle nature of many Australian sclerophyllous shrubs also makes them more susceptible to trampling damage from horses (Whinam et al. 1994; Newsome et al. 2002).

The third major way in which Australian ecosystems are likely to experience greater impacts from horses is by the introduction and dispersal of weeds and pathogens (St John-Sweeting and Morris 1991; Whinam et al. 1994; Weaver and Adams 1996; Newsome et al. 2002). Due to Australia's long period of isolation, exotic weeds pose a significant threat to natural vegetation by competing with native species for light, space, and nutrients and by modifying natural ecosystem functioning (Williams and West 2000). Plant community composition and structure may change, affecting food resources and shelter for native fauna (Adair 1995). Therefore increasing the diversity and abundance of exotic species in protected areas is of concern. This has been acknowledged with environmental weeds (plants that invade natural ecosystems and can cause major modifications to indigenous species and ecosystem function) being recognised as major threats to conservation in Australia (Williams and West 2000; Williams et al. 2001).

9.13 Social Impacts of Horse Riding

Impacts from horse use can be ecological (impacts to the resource) or social (impacts to the experiences of other visitors). For example, many studies have revealed conflicts between hikers and horseback riders. The severity of resource impacts depends on the characteristics and behaviour of the user, environmental attributes, and how visitors and trails are managed. So in addition to the biophysical effects of horseback riding in protected areas, trail impact studies are also concerned with social impacts, such as user conflict, perceptions of users, and depreciative behaviour. Both types of impact serve to bring horse-use concerns to the attention of managers. Conflicts with other users include objection to the presence of horse manure, increased incidence of insects attracted to manure, introduction of smells and the sight of horses and horse-related infrastructure, and general feelings of the inappropriateness of horses in wilderness areas that may conflict or accord with visitors' wilderness values. Information on social impacts has generally been collected using visitor surveys. In most, but not all, cases, surveys include horse riders and other visitors to protected areas. Some of the most cited work is that by Watson et al. (1993, 1994). In their 1993 study, they surveyed hikers and pack stock users in June to November 1990 in three wilderness areas: John Muir Wilderness ($n5501$) in California; the Sierra and Inyo National Forests, Sequoia-Kings Canyon Wilderness ($n5389$) in California; and the Charles C. Deam Wilderness ($n5502$) in Indiana. The objective was to gain a perspective on the interaction between hikers and recreation livestock (primarily horses). Their findings showed that up to 44% of hikers disliked encounters with horseback riders, although not all hikers disliked these encounters. In Deam Wilderness, of the hikers who encountered horses, 20% enjoyed meeting them, and about half reported that they did not mind. Only 4% of horse users disliked their encounters with hikers. A strong predictor of conflict between hikers and horse users were general feelings of inappropriateness of horse use in wilderness. Hikers also rated encounters with horses as somewhat undesirable with almost half indicating the behaviour of horseback riders interfered with their enjoyment of wilderness. The main behaviour of concern in John Muir and Sequoia-Kings Canyon Wilderness was horses defecating in places (mainly along trails) where hikers would have to walk and horse groups making too much noise and being rude. In Deam Wilderness, the main behaviours of concern were horse groups making too much noise and damaging trails. When asked to evaluate the problems they encountered, horse users tended to evaluate problems as less severe than hikers, with litter and human damage to vegetation as the most severe problems. Alternatively, hikers rated impacts to trails by horses, horse manure on trails, and vegetation damaged by horses as most severe. In another wilderness study, 75% of managers reported they received complaints about horses, including excessive trail impacts, manure on trails, and damage to meadows and riparian areas (Shew et al. 1986). A more complete review of the social impacts of horse use can be found in Hammitt et al. 2015, Jacob and Schreyer (1980), McClaran (1989), and Newsome et al. (2002).

9.14 Management of Horseback Riding Impacts

Most management of horseback riding is the responsibility of the national or state parks and other conservation organisations, and these usually have rules and regulations with regard to horseback riding, and there are codes of conduct which should be followed by riders or tour guides. For example, the Brecon Beacons National Park Equine Cluster Group (Wales) follow a horse riding code of conduct. The businesses which form that group agree to their best endeavours to ensure that their staff, volunteers, and customers follow this code of conduct in order to protect the riding resource and wider environment from damage by following the actions:

1. To ride responsibly in a way that has minimum impact on bridleways, restricted byways, unsurfaced roads, and open country.
2. To liaise with the National Park Authority staff about which routes are best used on an intensive, regular basis and how such regular routes are best varied to relieve pressure. The National Park Authority will ensure that any alternative Rights of Way are kept open to minimise overuse of specific routes.
3. To respect land ownership and always gain relevant permissions.
4. To ensure staff, volunteers, and customers understand the importance of protected sites for wildlife and history/archaeology and measures needed to protect their interest. The National Park Authority agree to assist with the identification of those sites and to liaise with regulatory bodies over appropriate usage.
5. To be aware and act sensibly when notified by the National Park Authority of significant and vulnerable ground-nesting bird activity.
6. To organise working parties to assist with the maintenance, repair, and improvement of riding routes in cooperation with the warden's teams.
7. To show respect and consideration for other users of the Rights of Way network.

In Southern England "The Horse Riding Code for the New Forest National Park" states:

1. Keep to the tracks when the ground is soft or muddy.
2. Take an alternative route to avoid soft slopes. Riding straight up and down causes erosion.
3. Avoid widening existing tracks. Keep off reseeded areas and recently reinstated rides.
4. Vary your route in the Forest to spread the wear and tear.
5. Keep to the tracks when birds are nesting on the ground (1 March to 31 July).
6. Slow down and call out a warning when approaching other Forest users. Be courteous and friendly.
7. Do not build jumps or create lunging areas in the Forest.
8. Keep well away from any work going on in the Forest.
9. Never ride more than two abreast. Limit groups to a maximum of eight horses on the road.

In South Africa in the Cape region (Table Mountain National Park), there is an environmental management program for horse riding (South African National Parks and the Cape Peninsula Horse Riding Group (2004), for example, the Sleepy Hollow Horse Riding and Noordhoek Riding Association have drawn up rules and regulations for the commercial beach rides. The rules include:

- May not use the same paths to the beach as beach walkers and their dogs use.
- Paths to the beach that may be used are the pipe path at the bottom of the beach road complexes, the access path (northern boundary or what was previously Mr. van der

Horst's property and now included into the Park), and the path at the southern end of the beach on to the recently acquired Park land.

- May not ride to the north of the pipe path (to ensure that riders stay away from the picnic areas).
- May not ride in the sand dunes (to protect the sand dune flora).
- May not ride along the middle of the beach (between the high-water mark and the lagoon edge), other than to cross over directly to the water edge (this may only be done at the walk) as this is where the oystercatchers and strandloper birds will often nest.
- May not ride to the south of the rocks at the southernmost edge of the beach, where penguins and black mussels are found.

- A maximum of six riders per guide, ensuring small controlled rides.
- Limited number of permits allowed for commercial businesses.
- All guides need to be at least 17 years of age, must have a good knowledge of the history of Noordhoek and its fauna and flora, thereby teaching visiting riders about the environment that they are privileged to be riding in.
- No galloping allowed.
- Schools must abide by the rules and regulations set out by the SPCA and Pony Club in the management of their stable yards and the treatment and condition of their horses.

Similar codes are in place in Queensland, and Landsberg et al. (2001) suggested certain principles to guide management in a peri-urban nature reserve in Australia:

1. Provide for recreation horse riding only.
2. No dogs allowed.
3. Confine horses to specific trails.
4. Locate trails near the perimeter of the reserves and/or in modified zones.
5. Construct and maintain trails to a standard (drained and hardened/stable surface of suitable width).
6. Exclude horse riding from ecologically sensitive areas.
7. Rationalise existing trail networks where horse riding is currently allowed with a view to closing trails and developing alternative routes, and/or construct trails to acceptable standard.
8. Develop a code of conduct that fosters rider compliance to the management system in place.
9. Develop monitoring systems to measure rider compliance and impacts of horse riding.
10. Modify the management programme if unacceptable impacts are detected.

However, experience tells us that self-regulating systems usually fail. Even if most people observe the codes of conduct put in place, a small number will choose to ignore them. Without the appropriate level of management, the total system then often fails, especially with activities such as horseback riding, where we have seen that relatively low levels of inappropriate activity can cause significant impact.

9.15 What Are Other Management Options and How Acceptable Are They?

The management strategies for horse riding in protected areas are summarised in Table 9.4 and are discussed in the following section.

9.15.1 Prohibit Horseback Riding

The simplest and most effective means of minimising horseback riding impacts is to prohibit all use. Accordingly, Royce (1983) concluded that horseback riding had seriously degraded the environmental quality of large areas of John Forrest National Park in Western Australia and recommended that a long-term strategy to phase out horse trails should be proposed in all national parks in the state. However, national park policy in Australia, and in most parts of the world, makes it clear that recreation opportunities are to be provided for the use and enjoyment of the public as well as conserving the natural environment and its ecosystems. In addition to this, there is an increasing lobby for horseback riding opportunities in some natural areas. Prohibiting all horse riding opportunities in national parks would not be a socially or politically acceptable course of action, as there are issues of equity in the limitation of certain recreation opportunities (Landsberg et al. 2001). Furthermore, the issue of changing land use, for example, pastoral land to national park/protected area, and change from traditional grazing and horseback riding, especially in areas with a rich cultural history of wilderness and pioneering activity in both Australia and the USA, has given rise to conflict and pressure from the horse riding lobby. Despite this there is now sufficient evidence (e.g. Landsberg et al. 2001; Phillips 2000; Phillips and Newsome 2002; Newsome et al. 2002, 2004; Pickering 2008) to prohibit any new horse riding activity in national parks and other protected areas. Moreover, the image that equestrian tourism sometimes conveys

9.15 What Are Other Management Options and How Acceptable Are They?

Table 9.4 Management strategies for horse riding in protected areas

Zoning
Use-specific zoning
(1) Set aside areas for horse use only, for example, designated bridle trails.
(2) Locate trails near edges of protected areas or in modified zones.
(3) Exclude horse riding from ecologically sensitive areas.
Site (trail) management
Locating trails
(1) Locate trail on contour and on level ground, trail grade to be below 10%.
(2) Control water, for example, bridges, drainage dips, outsloped treads, water bars, and ditches.
(3) Re-route short sections of trail to stop trail degradation in problem areas.
Managing trails
(1) Apply trail hardening and surfacing techniques, for example, materials such as gravel, earth, crushed stone, and geosynthetics (e.g. geotextiles).
(2) Reinforce soil structure; for example, use chemical binders such as liquid concentrates and latex polymer products.
(3) Clear overhanging vegetation.
Visitor management
Information and education: codes of conduct
(1) Ride and stay on designated trails.
(2) Ride in single file on trails to reduce width.
(3) Use facilities provided.
(4) Pass other people quietly on a track.
(5) Keep horses under control at all times.
(6) Spread out in untracked country.
(7) If possible, do not shoe a horse before a trip as new shoes cut up the ground more.
(8) Feed horses on commercial, processed feeds prior to and during trip to reduce likelihood of introducing weeds. Feed horses using a nosebag while in the protected area.
(9) Hold horses at least 30 m from water sources, for example, lakes and streams, huts, and camping areas. Water downstream and at least 30 m from camping areas on lake foreshores.
(10) Use hitching rails or other holding facilities provided. Keep horses away from tree trunks and roots. Use a low-power electric portable fence or tether where facilities are not provided.
(11) Avoid crossing areas easily damaged by horses such as sphagnum moss beds, swamps, and steep or boggy creek crossings.
(12) Introduce horse users to consequences of use, and encourage them to adopt low-impact practices.
Regulating visitor use
Numbers of horses
(1) Visit in small groups (four to eight people/horses).
(2) Limit use, for example, number of horses that visit per year.
(3) Limit the number of groups with horses and the length of stay.
(4) Limit the length of time horse users can access the area, most often applied to campsites.
Feed type
(5) Encourage use of "permitted feeds" such as good quality, clean chaff, cracked grain, and processed feed. All feeds must be as weed free as possible.

Adapted from Newsome et al. (2008)

does not fit with conservation and a sustainable use of the environment. The image of the tough stockman and a pioneer in the American West and in Australia is one of taming the land and does not fit in with the more sensitive approach of understanding the nature of the environment. It could be argued that there could be a restriction or banning of the activity in the wilderness areas and national parks remote from major centres of population. In the latter areas close to major population centres, it could be argued that this is where horseback riding is more appropriate, especially as this is where the riders and their horses live.

9.15.2 Unrestricted Open Access

In some areas open access may be considered as part of the horseback riding experience because of a desire to add more excitement or variety to a tour. Areas with a strong horse-riding, pioneering heritage often fall into this category, such as the southern coast of Western Australia and the wilderness areas of the Western USA. However, as we have seen, the indications are that there will be an increasing demand for horseback riding opportunities in the next 40 years and that given time, the cumulative impacts of free-range horseback riding, horse trails, tethering facilities, and campsites would cause extensive environmental damage. Horseback riding also has a high potential to reduce the ecological health of a number of Western Australian national parks by providing the conditions for accelerated erosion and the transport of exotic plants and diseases such as *Phytophthora cinnamomi*.

Given that impacts will be inevitable wherever horseback riding occurs, management is required to implement and rigorously enforce a variety of visitor and site-management strategies and techniques capable of controlling impacts. Based on the findings that low levels of horse riding cause a significant degree of vegetation and soil impact, the potential problems of erosion, invasion, and spread of weeds and dieback, combined with limited management resources, open access of protected areas for horseback riding for recreation is not a sensible option.

9.15.3 Managing Horseback Riding Commercial Operators

There are a number of situations, such as areas with a history of horseback riding, where horseback riding will continue to be part of the recreation spectrum of opportunity in a protected area. For example, in Western Australia the current Shannon Park and D'Entrecasteaux National Park Management Plan aims to prohibit open public access but allow a permit system. This management approach is based on the commercial operator applying for a permit having an incentive for the operator to minimise damage. The main advantage of providing horseback riding opportunities through a commercial operator is that, theoretically, they have direct control over the number of users and the areal extent of use. Controlling the areal extent of impact means that potentially severe local impacts only occur in a small part of the national park. This leaves most of the park essentially unvisited by horseback riders. The advantage commercial operators provide is that they are in a position to enforce very low levels of use, over designated areas within the park, and the extent of impact is controlled by confining horse riding to a number of trails within the national park. This also helps prevent the transport of *Phytophthora cinnamomi* and exotic plants and weeds. An informed commercial operator can help to confine horse use to low-risk areas, providing management with a tight control on the transport of dieback whilst still providing for horseback riding opportunities.

Commercial horse riding on a permit basis provides an incentive to minimise damage or face permit suspension or cancellation. This encourages the commercial operators to ensure that all possible precautions are taken to prevent the introduction and transport of exotic plants and weeds, avoid areas of high erosion potential, such as coastal dunes, and use tethering areas, with the aim of minimising both the areal and visual extent of impact. Additionally, it is necessary to implement an ongoing monitoring programme of commercial horse riding operations in all national parks. Such monitoring should include assessments of baseline conditions and changes in resource conditions over time. Some of the suggested parameters for monitoring are the presence of *Phytophthora cinnamomi* and other root pathogens, exotic plants/weeds, permanent quadrats for vegetation parameters, and horse-trail characteristics such as soil and vegetation impact parameters. The approach described here mirrors that recommended by Landsberg et al. (2001) for the Canberra Nature Park. They listed confinement to specified trails; use of trails at the reserve perimeter, or on already modified trails; exclusion from areas of conservation significance; and a high degree of rider compliance and monitoring. Given the high potential for horses to degrade trails, a vital component of trail usage is that they are hardened to control weed invasion and erosion and constructed so that adequate drainage is maintained. Landsberg et al. (2001) suggest that horseback riding activity should be continuously reviewed and modified accordingly.

Permits and fees, which are usually charged in most areas for horseback riding, are a common method to try and minimise the environmental damage. For example, in South Africa a WILD card system is used to identify registered recreational riders in the Table Mountain National Park (TMNP). This Infinity-based "smart card" loyalty programme currently costs R195 per family, R175 per couple, or R95 per individual for a full-year access to the TMNP. All recreation riders are required to acquire a WILD card as a prerequisite for using the park for horseback riding. A short-term WILD card for once-off visitors has been developed, and a commercial WILD card has also been developed that provides a commercial operator access to the park, with a fixed number of clients. For WILD cards bought through the TMNP, the code of conduct will be made available to horseback riders. In time, the capacity to endorse the WILD card for a range of approved recreation activities in the park will be developed. Using the "smart card," enforcement staff in the field will be able to request from any rider their WILD card and, with a portable scanner, obtain all the available information on the rider. It is envisaged that income derived from the sale of the WILD card will be used to offset the park maintenance, security, management, education, and enforcement costs incurred in effectively implementing this scheme. This is part of the problem of permits and fees because there has to be an effective operating system for selling the permits, collecting the fees, and enforcement for non-compliance.

9.15.4 Horseback Riders' Reaction to Management

Mulders (2006) reported that horseback rider groups are prepared to abide by access restrictions, employ a code of conduct, clean horses' hooves, and select appropriate feed before entering a protected area. Both Recreational Riding for Bold Park (2001) and the Australian Horse Alliance (2006) maintain that horseback rider groups will self-manage to a high standard. The former group refer to the use of accredited food sources, restriction to designated bridle trails, manure brochures, and the dissemination of material supplied by park management. The latter group stated that "horse riders have no desire to see bushland destroyed." They promise responsible horse riding on designated trails and the avoidance of sensitive areas.

It is generally the case that horseback riders and associated lobbyists continue to profess enthusiasm for, and laud the effectiveness of, codes of conduct in reducing impacts. Research is urgently needed to test the effectiveness of this management strategy, both against the "do nothing" option and other management possibilities, such as reducing numbers, restricting the season of use, intensive trail design, construction, and maintenance activities. Many protected area agencies have a poor history of reporting on the management effectiveness of their actions so it is difficult to see which management strategies are really effective.

9.15.5 Scientific Monitoring Over an Extended Time Period

The South East Queensland Horse Riding Trail Network (HTN) includes more than 500 km of trails within 29 reserves between Gympie and the State's southern border. The HTN trails mainly occur in forest reserves (FR), but a few are in State forests and timber reserves. The reserves that host the HTN trails are being converted to national parks following the South East Queensland Forests Agreement (SEQFA) signed in 1999. The HTN trails link to a broader trail network that includes about 340 km of trails in Queensland's forest plantations and at least 470 km of trails on other tenures, including several other State forests.

Recent amendments to the Nature Conservation Act 1992 (NCA) specify that the HTN trails must be reviewed by the chief executive and that the review should start as soon as possible and finish before 2026. The NCA also specifies that the review will be informed by an assessment undertaken by a Scientific Advisory Committee (SAC) of the impact of horseback riding use on horseback riding trails and adjacent areas. The SAC's assessment must be based on the results of monitoring and evaluation conducted over a period long enough to assess the likely impacts of horseback riding use and must take account of the cumulative impacts of horseback riding and other activities conducted in the HTN trails and adjacent areas and vegetation communities. The Queensland Government has committed to a detailed Scientific Monitoring Program that will operate over a 20-year period with regular points of review, to monitor any potential impacts that result from horseback riding on SEQFA lands. The monitoring programme will be overseen by an independent SAC.

This Scientific Monitoring Program on the South East Queensland HTN throughout South East Queensland protected areas will identify any impacts of such use and recommend management actions to address such impacts (DERM 2010). It has been set up to study the direct social and biophysical impacts of horse riding on trails within SEQ. While several studies have investigated the potential impacts of horseback riding in conservation areas, there is as yet no substantial research undertaken that is specific to the environs of SEQ, or on such a large scale. Specific issues such as a significant lack of research directly targeting the impact of horseback riding on aquatic ecosystems will be studied. In her review of HRT Network monitoring needs, Pickering (2008) highlighted that "indicators on the impact of horse riding on aquatic ecosystems are required." This type of management approach is excellent as it is based on scientific principles and will provide direct unequivocal evidence of the long-term impacts of horseback riding and will suggest the best management techniques to minimise these impacts. Similar large-scale scientific approaches would be excellent and should be developed in other parts of the world to counteract the problems caused by horseback riding.

9.15.6 Packstock Management Strategies

Cole (2002) provides an overview of the five primary strategies available for managing packstock impacts in wilderness areas in the national parks of the USA:

- The amount of use can be reduced by, for example, prohibiting packstock use or by closing overgrazed meadows.
- Behaviour can be changed, either through restrictions or by education. Critical behaviours include group-size, stock-confinement techniques, carrying feed and steps to insure against the introduction of alien species.
- The timing of use is managed as it is often critical that horses stay off trails and away from meadows shortly after snowmelt when the soils are saturated.
- Hardening trails.
- Impacts can be confined by only allowing packstock use on certain trails and in certain locations.

A good example of the use of concentration and use containment strategies is provided by Newsome et al. (2004) from the Seven Lakes Basin in the Selway-Bitterroot Wilderness, USA.

Concluding Remarks

Due to the high impact of recreational horseback riding on the environment, its likely increasing importance as a recreation activity, and the established conflict with other recreational users, it needs to be emphasised to horseback riders that for continued access to wilderness areas, national and state parks, and conservation areas, best-practice management is imperative. This has to include planning trail location and design; trail construction for drainage and erosion control which probably will include trail hardening using gravel and/or geotextiles; trail maintenance; visitor regulation such as confinement and the amount and timing of use; education to improve user knowledge and change user behaviour using codes of conduct circulated through interpretative information, public lectures, and discussions through associations and clubs and through the internet; and greater policing and enforcement of the management strategies that are in place. The problem with the latter point, and the monitoring of this recreation activity, is that management capacity is often not sufficient and cannot cope adequately with the problems. If it is found that unacceptable impacts are occurring and that significant conservation and biodiversity values are threatened, it may be necessary to prohibit horseback riding completely in certain areas, although this would be an unfortunate outcome.

References

Andrews, W. M., Milner, G., & Maxwell, B. (1998). Spotted knapweed distribution in stock camps and trails of the Selway-bitterroot wilderness. *Great Basin Naturalist, 58*, 156–166.

Aust, M. W., Marion, J. L., & Kyler, K. (2005). *Research for the Development of Best Management Practices to Minimize Horse Trail Impacts on the Hoosier National Forest*. US Forest Service, Virginia Tech., Department of Forestry, Blacksburg, VA.

Australian Horse Alliance. (2006). Horse riding in the Australian Alps.

Barratt, R. (1999). *Effects of Horses in Bold Park, Perth, Western Australia: Ecological Impacts and Management Considerations*. University of Western Australia, Department of Geography, Nedlands, WA, Australia.

Bear, R., Hill, W., & Pickering, C. M. (2006). Distribution and diversity of exotic plant species in montane to alpine areas of Kosciuszko National Park. *Cunninghamia, 9*, 559–570.

Beavis, S. (2005). Biophysical impacts of recreational horse riding in multi-use national parks and reserves. *Australasian Journal of Environmental Management, 12*, 109–116.

Beeton, S. (1999). Visitors to national parks: Attitudes of walkers towards commercial horseback tours. *Pacific Tourism Review, 3*, 49–60.

Bienz, U., Manzi, H., & Frossard, E. (1999). Production and composition of horse feces. *Agraforschung, 6*, 205–208.

Bowker, J. M., Askew, A. E., Cordell, H. K., Betz, C. J., Zarnoch, S. J., & Seymour, L. (2012). *Outdoor Recreation Participation in the United States-Projections to 2060*. A Technical Document Supporting the Forest Service 2010 RPA Assessment. Asheville, NC: Southern Research Station. Retrieved from www.srs.fs.usda.gov.

Campbell, J. E., & Gibson, D. J. (2001). The effect of seeds of exotic species transported via horse dung on vegetation along trail corridors. *Plant Ecology, 157*, 23–35.

Cash, D., Barney, L., & Gagnon, S. (2008). Can Horses spread weeds? Montana State University Extension Service, Bozeman. Retrieved from www.animalrangeextension.montana.edu/Articles/Forage/Fall/horse-weeds.htm.

Chicoine, T. K. (1984). *Spotted Knapweed (Centaurea maculosa L.) Control, Seed Longevity and Migration in Montana*. Unpublished Master's thesis, Department of Agronomy, Montana State University, Bozeman, MT, 83pp.

Cole, D. N. (1983). *Assessing and Monitoring Backcountry Trail Condition (Research Paper INT-303)*. USDA Forest Service, Intermountain Forest and Range Experiment Station, Ogden, UT.

Cole, D. N. (1991). *Changes on Trails in the Selway-Bitterroot Wilderness, Montana, 1978–89 (Research Paper INT-212)*. USDA Forest Service, Intermountain Research Station, Ogden, UT.

Cole, D. N. (1995a). Experimental trampling of vegetation. I. Relationship between trampling intensity and vegetation response. *Journal of Applied Ecology, 32*, 203–214.

Cole, D. N. (1995b). Experimental trampling of vegetation. II. Predictors of resistance and resilience. *Journal of Applied Ecology, 32*, 215–224.

Cole, D. N., & Spildie, D. R. (1998). Hiker, horse and llama trampling effects on native vegetation in Montana, USA. *Journal of Environmental Management, 53*, 61–71.

Cole, D. N., van Wagtendonk, J., McClaran, M., Moore, P., & McDonald, N. (2004). Response of mountain meadows to grazing by recreational pack stock. *Journal of Range Management, 57*, 153–160.

Cordell, H. K. (2012). *Outdoor Participation Trends and Futures*. A Technical Document Supporting the Forest Service 2010 RPA Assessment. Asheville, NC: Southern Research Station. Retrieved from www.srs.fs.usda.gov.

Cosyns, E., & Hoffman, M. (2005). Horse dung germinable seed content in relation to plant species, abundance, diet, composition and seed characteristics. *Basic and Applied Ecology, 6*, 11–24.

Couvreur, M., Christiaen, B., Verheyen, K., & Hermy, M. (2004). Large herbivores as mobile links between isolated nature reserves through adhesive seed dispersal. *Applied Vegetation Science, 7*, 229–236.

Csurhes, S., Paroz, G., & Markula, A. (2009). *Pest Animal Risk Assessment: Feral Horse Equus caballus*. The State of Queensland, Department of Employment, Economic Development and Innovation.

Csurhes, S., Paroz, G., & Markula, A. (2016). *Invasive Animal Risk Assessment: Feral Horse Equus caballus*. Biosecurity, QLD: Queensland Government, Department of Agriculture and Fisheries, 26pp.

Dale, D., & Weaver, T. (1974). Trampling effects on vegetation of the trail corridors of North Rocky Mountain forests. *Journal of Applied Ecology, 11*, 767–772.

Davis, E. S., Fay, P. K., Chicoine, T. K., & Lacey, C. A. (1993). Persistence of spotted knapweed (*Centaurea maculosa*) seed in soil. *Weed Science, 41*, 57–61.

Deluca, T. A., Patterson, W. A., & Freimund, W. A. (1998). Influence of llamas, horse and hikers on soil erosion from established trails in western Montana, USA. *Environmental Management, 22*, 255–262.

DERM. (2010). *Scientific Monitoring Program for the SE Queensland Horse Riding Trail Network*. Department of Environment and Resource Management, State of Queensland, Brisbane.

Equus Project. (2014). *Horse Riding in Europe*. 14pp.

Forrester, H., Clow, D., Rocje, J., Heyvaert, A., & Battaglin, W. (2017). Effects of backpacker use, pack stock trail use, and pack stock grazing on water quality indicators, including nutrients. *E. Coli*, hormones and pharmaceuticals in Yosemite National Park, USA. *Environmental Management, 60*, 526–543.

French, R., & Lacey, J. R. (1983). *Knapweed: Its Cause, Effect and Spreading in Montana*. Montana Coop. Extension Service 307, Montana State University, Bozeman, MT.

Gillieson, D., Davies, J., & Hardy, P. (1987). *Gurragorambla Creek Horse Trail Monitoring, Kosciusko National Park*. Unpublished paper to Royal Australian Institute of Parks and Recreation Conference, Canberra. Quoted in Landsberg et al. 2001.

Gower, S. T. (2008). Are horses responsible for introducing non-native plants along forest trails in the eastern United States? *Forest Ecology and Management, 256*, 997–1003.

Hadwen, W. L., Hill, W., & Pickering, C. (2008). Linking visitor impact research to visitor impact monitoring in protected areas. *Journal of Ecotourism, 7*, 87–93.

Helgadóttir, G., & Sigurðardóttir, I. (2008). Horse-based tourism: Community, quality and disinterest in economic value. *Scandinavian Journal of Hospitality and Tourism, 8*, 1–17.

Holmquist, J. G., Schmidt-Gengenbach, J., & Haultain, S. A. (2010). Does Longterm grazing by pack stock in subalpine wet meadows result in lasting effects on arthropod assemblages? *Wetlands, 30*, 252.

Jacob, G. R., & Schreyer, R. (1980). Conflict in outdoor recreation: A theoretical perspective. *Journal of Leisure Research, 12*, 368–380.

Johnston, F. M., & Pickering, C. M. (2001). Alien plants in the Australian alps. *Mountain Research and Development, 21*, 284–291.

Klinger, R. C., Few, A. P., Knox, K. A., Hatfield, B. E., Clark, J., German, D. W., & et al. (2015). *Evaluating Potential Overlap between Pack Stock and Sierra Nevada Bighorn Sheep (Ovis canadensis sierrae) in Sequoia and Kings Canyon National Parks, California*. United States Geological Survey Open-File Report 2015-1102, 55pp.

Lacey, C. A., Lacey, J. R., Fay, P. K., Story, J. M., and Zamora, D. L. (1992). *Controlling Knapweed on Montana Rangeland*. Montana State University Extension Service Bulletin 2C0311, Bozeman, MT, 6pp.

Lake, P. S. (2000). Disturbance, patchiness, and diversity in streams. *The Journal of the North American Benthological Society, 19*, 573–592.

Landsberg, J., Logan, B., & Shorthouse, D. (2001). Horse riding in urban conservation areas: Reviewing scientific evidence to guide management. *Ecological Management and Restoration, 2*, 36–46.

Lee, R., Berlow, E. L., Ostoja, S. M., Brooks, M. L., Génin, A., Matchett, J. R., et al. (2017). A multi-scale evaluation of pack stock effects on subalpine meadow plant communities in the Sierra Nevada. *PLoS One, 12*(6), e0178536. https://doi.org/10.1371/journal.pone.1078536.

Leopold, A. (1921). The wilderness and its place in recreational policy. *Journal of Forestry, 19*, 718–721.

Leung, Y. F., & Marion, J. L. (1996). Trail degradation as influenced by environmental factors: A state-of-knowledge review. *Journal of Soil and Water Conservation, 51*(2), 130–136.

Liddle, M. J. (1997). *Recreation Ecology*. London: Chapman and Hall.

Liddle, M. J., & Chitty, L. D. (1981). The nutrient-budget of horse tracks on an English lowland heath. *Journal of Applied Ecology, 18*, 841–848.

Marion, J. L. (1994). *An Assessment of Trail Conditions in Great Smoky Mountains National Park*. Research/Resources Management Report. USDI National Park Service, Southeast Region, Atlanta, GA, 155pp.

Marion, J. L., & Cole, D. N. (1996). Spatial and temporal variation in soil and vegetation impacts on campsites. *Ecological Applications, 6*, 520–530.

Marion, J. L., Cole, D. N., & Bratton, S. P. (1986). Exotic vegetation in wilderness areas. In R. C. Lucas (Ed.), *Proceedings of the National Wilderness Research Conference; Current Research*. General Technical Report INT-212 (pp. 114–120). Ogden, UT: Intermountain Research Station, USDA Forest Service.

Marion, J., & Leung, Y. (2001). Trail resource impacts and an examination of alternative assessment techniques. *Journal of Park and Recreation Administration, 19*, 17–37.

Mastsui, A., Inoue, Y., & Asai, Y. (2003). The effects of putting the bag with collecting faeces and urea ("Equine Diaper") to the amount of ammonia gases concentrated in Horse's pen. *Journal of Equine Science, 14*, 75–79.

McClaran, M. P. (1989). Recreational pack stock management in Sequoia and King's Canyon National Parks. *Rangelands, 11*, 3–8.

McClaran, M. P., & Cole, D. N. (1993). *Packstock in Wilderness: Use, Impacts, Monitoring and Management*. General Technical Report INT-301, USDA Forest Service, Fort Ogden, UT, 33pp.

McQuiad-Cook, J. (1978). Effects of hikers and horses on mountain trails. *Journal of Environmental Management, 6*, 209–212.

Moiuisse, A. M., Vos, P., Verhagen, H. M. C., & Bakker, J. P. (2005). Endozoochory by free ranging, large herbivores: Ecological correlates and perspectives for restoration. *Basic and Applied Ecology, 6*, 547–548.

Montana Department of Agriculture. (1986). *Weed Training Manual*. Environmental Management Division, Helena, MT, 3pp.

Mulders, C. G. (2006). *Western Australian Equestrian Tracks and Trails Study*. Unpublished report. Quoted in Newsome et al. 2008.

Nagy, J. A., & Scotter, G. W. (1974). *A Qualitative Assessment of the Effects of Human and Horse Trampling on Natural Areas, Waterton Lakes National Park*. Edmonton, AB, Canada: Canadian Wildlife Service.

Newsome, D., Cole, D. N., & Marion, J. (2004). Environmental impacts associated with recreational horse-riding. In R. Buckley (Ed.), *Environmental Impacts of Ecotourism* (pp. 61–82). New York: CABI.

Newsome, D., Milewski, A., Philips, N., & Annear, R. (2002). Effects of horse-riding on National Parks and other natural ecosystems in Australia: Implications for management. *Journal of Ecotourism, 1*, 52–74.

Newsome, D., Moore, S. A., & Dowling, R. K. (2002a). *Natural Area Tourism: Ecology Impacts and Management*. Sydney: Channel View Publications.

Newsome, D., Phillips, N., Mileswski, A., & Annear, R. (2002b). Effects of horse riding on national parks and other natural ecosystems in Australia: Implications for management. *Journal of Ecotourism, 1*, 52–74.

Newsome, D., Smith, A., & Moore, S. A. (2008). Horse riding in protected areas: A critical review and implications for research and management. *Current Issues in Tourism, 11*, 144–166.

Ngugi, M. R., Neldner, V. J., & Dowling, R. (2014). Non-native plant species richness adjacent to a horse trail network in seven National Parks in south-east Queensland, Australia. *Australasian Journal of Environmental Management, 21*. https://doi.org/10.1080/144865563.2014.952788.

Nimmo, D. G., & Miller, K. K. (2006). Ecological and human dimensions of management of feral horses in Australia: A review. *Wildlife Research, 34*, 408–417.

Ollenburg, C. (2006). Horse riding. In R. C. Buckley (Ed.), *Adventure Tourism* (pp. 305–323). Wallingford, Oxfordshire: CAB International.

Ostermann-Kelm, S. D., Atwill, E. D., Rubin, E. S., Hendrickson, L. E., & Boyce, W. M. (2009). Impacts of feral horses on a desert environment. *BMC Ecology, 9*(22). https://doi.org/10.1186/1472-6785-9-22.

Ostoja, S. M., Brooks, M. L., Moore, P. E., Berlow, E. L., Blank, R., Roche, J., et al. (2014). Potential environmental effects of pack stock on meadow ecosystems of the Sierra Nevada, USA. *The Rangeland Journal, 36*, 411–427.

Phillips, N. (2000). *A Field Experiment to Quantify the Environmental Impacts of Horse Riding in D'Entrecasteaux National Park, Western Australia*. Unpublished Honours thesis, School of Environmental Science, Murdoch University, Perth, WA.

Phillips, N., & Newsome, D. (2002). Understanding the impacts of recreation in Australian protected areas: Quantifying damage caused by horse riding in D'Entrecasteaux National Park, Western Australia. *Pacific Conservation Biology, 7*, 256–273.

Pickel-Chevalier, S. (2015). Can equestrian tourism be a solution for sustainable tourism development in France? *Society and Leisure, 38*, 110–134.

Pickering, C. M. (2008). *Literature Review of Horse Riding Impacts on Protected Areas and a Guide to the Development of an Assessment Program*. Environmental Protection Agency, Brisbane, 50pp.

Pickering, C. M., & Hill, W. (2007). Roadside weeds of the Snowy Mountains, Australia. *Mountain Research and Development, 27*, 359–367.

Pickering, C. M., Hill, W., Newsome, D., & Leung, Y. F. (2010). Comparing hiking, mountain biking and horse riding impacts on vegetation and soils in Australia and the United States of America. *Journal of Environmental Management, 11*, 551–562.

Quinn, L. D., Kolipinski, M., Coelho, V. R., Davis, B., Vianney, J.-M., Batjargal, O., et al. (2008). Germination of invasive plant seeds after digestion by horses in California. *Natural Areas Journal, 28*, 356–362.

Quinn, L. D., Quinn, A., Kolipinski, M., Davis, B., Berto, C., Orcholski, M., et al. (2010). Role of horses as potential vectors of non-native plant invasion: An overview. *Natural Areas Journal, 30*, 408–416.

Recreational Riding for Bold Park. (2001). Submission to the Board of Botanic Gardens and Parks Authority, 22 March 2001.

Redfearn, S., Blessing, J., Marshal, J., Hadwen, W., & Negus, P. (2012). *Detecting Stream Health Impacts of Horse Riding and 4WD Vehicle Water Crossings in South East Queensland: An Event Based Assessment*. Department of Science, Information Technology, Innovation and the Arts, Queensland Government, 21pp.

Royce, P. (1983). *Horse Riding Trails in John Forrest National Park: An Environmental Assessment*. National Park Authority Perth.

Shew, R. L., Saunders, P. R., & Ford, J. D. (1986). Wilderness managers' perceptions of recreational horse use in the Northwestern United States. In *Proceedings of the National Wilderness Conference, Current Research* (pp. 320–325). General Technical Report, INT-212. Fort Collins, CO: USDA Forest Service, Intermountain Research Station.

Sigurdsdottir, I. (2011). *Economical Importance of the Horse Industry in North West Iceland: A Case in Point*. Retrieved from http://holar.academic.edu/IngibjorgSigurdardottir.

Snetsinger, S., & White, K. (2009). *Wildlife Species of Interest in Mount Spokane State Park*. Pacific Biodiversity Institute, Winthrop, WA, 60pp.

South African National Parks and the Cape Peninsula Horse Riding Group. (2004). Environmental management program for horse riding in the Table Mountain National Park.

St. John-Sweeting, R. S., & Morris, K. A. (1991). Seed transmission through the digestive tract of the horse. In *Proceedings of the 9th Australian Weeds Conference*, 6–10 August 1990, Adelaide, SA.

Summer, R. M. (1980). Impacts of horse traffic on trails in Rocky Mountain National Park. *Journal of Soil and Water Conservation, 35*, 85–87.

Summer, R. M. (1986). Geomorphic impacts of horse traffic on montane landforms. *Journal of Soil and Water Conservation, 42*, 126–128.

Sun, D., & Liddle, M. J. (1993). A survey of trampling effects on soils and vegetation in eight tropical and subtropical sites. *Environmental Management, 17*, 497–510.

Taylor, A. R., & Knight, R. L. (2003). Wildlife responses to recreation and associated visitor perceptions. *Ecological Applications, 13*, 951–967.

Thorp, J., & Wilson, M. (2008). Weeds Australia. Retrieved from www.weeds.org.au.

Törn, A. (2007). *Sustainability of Nature-Based Tourism*. Acta Universitatis Ouluensis A Scientiae Rerum Naturalium 498, Oulu University Press, Oulu, 56pp.

Törn, A., Siimaki, P., & Tolvanen, A. (2010). Can horse riding induce the introduction and establishment of alien plants species through endozoochory and gap creation? *Plant Ecology, 208*, 235–244.

Törn, A., Tolvanen, A., Norokorpi, Y., Tervo, R., & Siimaki, P. (2009). Comparing the impacts of hiking, skiing and horse riding on trails in different types of forest. *Journal of Environmental Management, 90*, 1427–1434.

Tyser, R. W., & Worley, C. A. (1992). Alien flora in grasslands adjacent to road and trail corridors in glacier National Park, Montana, U.S.A. *Conservation Biology, 6*, 253–262.

UK CEED. (2000). *A Review of the Effects of Recreational Interactions within European Marine Sites*. UK Marina SACs Project, Countryside Council for Wales, Cambridge, UK.

Van Dyk, E., & Neser, S. (2000). The spread of weeds into sensitive area by seeds in horse faeces. *Journal of the South African Vetinary Association, 71*, 173–174.

Vinson, M. (1998). *Effects of Recreational Activities on Declining Anuran Species in the John Muir Wilderness, CA*. MSc thesis, University of Montana, Missoula, MT, 83pp.

Ward, L., Lindsey, S., Martin, J. M., Nicodemus, M., & Memli, E. (2016). Review, challenges, and opportunities in rising feral horse populations. *The Professional Animal Scientist, 32*, 717–724.

Watson, A. E., Niccolucci, M. J., & Williams, D. R. (1993). *Hikers and Recreational Stock Users: Predicting and Managing Recreation Conflicts in Three Wildernesses*. Research Paper INT-468, USDA Forest Service, Intermountain Forest and Range Experiment Station, Ogden, UT.

Watson, A. E., Niccolucci, M. J., & Williams, D. R. (1994). The nature of conflict between hikers and recreational stock users in the John Muir wilderness. *Journal of Leisure Research, 26*, 372–385.

Weaver, T., & Dale, D. (1978). Effects of hikers, motorcycles and horses in meadows and forests. *Journal of Applied Ecology, 15*, 451–457.

Weaver, V., & Adams, R. (1996). Horse as vectors in the dispersal of weeds into native vegetation. In *Proceedings of the 11th Australian Weeds Conference* (pp. 383–397).

Wells, F. H., & Lauenroth, W. K. (2007). The potential for horses to disperse alien plants along recreation trails. *Rangeland Ecology and Management, 60*, 574–577.

Wells, F. H., Lauenroth, W. K., & Bradford, J. B. (2013). Recreational trails as corridors for alien plants in the Rocky Mountains, USA. *Western North American Naturalist, 72*(4), Article 8. Retrieved from http://scholararchive.byu.edu/whan/vol72/iss4/8.

Whinam, J., Cannell, E. J., Kirkpatrick, J. B., & Comfort, M. (1994). Studies on the potential impact of recreational horse riding on some alpine environments of the central plateau, Tasmania. *Journal of Environmental Management, 40*, 103–117.

Whinam, J., & Comfort, M. (1996). The impact of commercial horse-riding on sub-alpine environments at Cradle Mountain, Tasmania, Australia. *Journal of Environmental Management, 47*, 61–70.

Whittaker, P. L., & Bratton, S. P. (1978). *Comparison of Surface Impact by Hiking and Horseback Riding in the Great Smoky Mountains National Park*. Research/Response Management Report No. 24, Gatlinbug, TN: Great Smoky Mountains National Park, Uplands Field Research Laboratory.

Widner, C., & Marion, J. (1993). *Horse Impacts: Research Findings and Their Implications*. Master Network, a publication of the National Outdoor Leadership School, Lander, Wyoming, Part 1—1993, No 5, pp. 5–14; Part 2—1994, No 6,77, p. 56.

Williams, J., Read, C., Norton, A., Dovers, S., Burgman, M., Proctor, W., et al. (2001). *Biodiversity, Australian State of the Environment Report 2001 (Theme Report)*. CSIRO Publishing on behalf of the Department of the Environment and Heritage, Canberra.

Williams, J., & West, C. J. (2000). Environmental weeds in Australia and New Zealand: Issues and approaches to management. *Austral Ecology, 25*, 425–444.

Wilson, J. P., & Seney, J. R. (1994). Erosional impacts of hikers, horse, motorcycles, and off-road vehicles on mountain trails in Montana. *Mountain Research and Development, 14*, 77–88.

Wisdom, M. J., Ager, A. A., Preisler, H. K., Cimon, N. S., & Johnson, B. K. (2004). Effects of off-road recreation on mule, deer and elk. *Transactions of the North American Wildlife and Natural Resources Conference, 69*, 531–550.

Wood, P. J., & Armitage, P. D. (1997). Biological effects of fine sediment in the lotic environment. *Environmental Management, 21*, 203–217.

Geocaching, Letterboxing, and Orienteering

10

Chapter Summary

One of the opening statements by the New Forest Review Group (1988) that orienteering "by its very nature causes very considerable disturbance to wildlife and damage to flora over wide areas" has been shown to be unsupported. There is trampling of vegetation, but little evidence of disturbance of mammals and birds. It has also been shown that the reason for this is that orienteering, by its very nature, and by procedures of good practice, is a sport of low ecological impact. There also seem to be many wide-ranging educational benefits of orienteering, letterboxing, and geocaching and particularly the development of environmental stewardship, where the benefits outweigh the minor environmental impact. However, the growth in popularity of these activities needs careful monitoring and further research carried out.

10.1 Introduction

Orienteering, geocaching, and letterboxing are linked in that they all depend on following directions using some aspect of technology in the outdoors. Orienteering is a group of competitive sports, originating in Sweden, that involves map-reading skills and cross-country running in which competitors race through an unknown area, usually in wild terrain, to find various checkpoints by only using a compass and a specially prepared topographical map, which has standardised map symbols. The winner is the finisher with the lowest elapsed time. Each competitor has a control card, marked at each control point to show that the competitor has completed the course correctly. Orienteering has developed many variations like canoe, mountain bike, ski, urban orienteering, and rogaining. The latter is an orienteering sport of long distance cross-country navigation, involving route planning and navigation between checkpoints using a variety of map types, originally developed in Australia. In a rogaine, teams of two to five people choose which checkpoints to visit within a time limit, with the intent of maximising their score. The checkpoints are scored differently, depending on their difficulty level to reach them, and therefore a team has to choose a strategy which involves teamwork. The time varies between 2 and 24 hours, and there are now variations, popular in Australia, like metrogaine (predominantly in urban areas), cyclogaine involving mountain biking, snogaine using snowshoes or skis, and paddlegaine using human-powered watercraft in a water-covered area, like a set of interconnected lakes.

Geocaching is a high-technology treasure hunt that involves using handheld global positioning system (GPS) devices and hiking (Fig. 10.1A, B) to find small containers (Fig. 10.2). Geocachers can place caches, record the location's coordinates, and post that information and experiences online at a free account at

Fig. 10.1 (A) Geocache location in Delaware State Park woods. Courtesy of the Delaware Division of Parks & Recreation. (B) Geocaching in Delaware State Park. From www.delawarestatepark.com. Courtesy of the Delaware Division of Parks & Recreation

Fig. 10.2 Small official geocache in Delaware State Park woods. Courtesy of the Delaware Division of Parks & Recreation

the geocaching.com website for others to find and seek out the caches. It is becoming an increasingly popular outdoor activity, drawing a wide range of participants from all age groups. Due to the fact that the activity is largely participant-created and run, there is little oversight for how geocaching is monitored and controlled in environments like open spaces and along hiking trails. Many participant-created caches are placed off-trail and, often enough, in environmentally sensitive areas. The purpose of the activity is not necessarily to obtain the "treasure" within the containers, but rather to take people to places where they would not otherwise find themselves.

The origin of the GPS, formally known as NAVSTAR GPS can be traced back to the US military in the latter half of the twentieth century. The primary purposes of the Department of Defence's development of this technology were twofold: for precise weapon delivery and to provide a unified navigational system that would combine varying technology already being explored within different branches of the military. The US Navy and Air Force had been developing their own navigational systems simultaneously during the late 1960s. By the 1970s, the military began developing a compromise system combining the best elements of both programmes: the signal structure and frequencies from the Air Force's 621B programme and the satellite orbits of the Navy's Timation system. The emerging system that would undergo testing during the next couple of decades would become what is known today as NAVSTAR GPS. The system is composed of 24 satellites orbiting several thousand miles above the earth, completing their orbits around the earth every 12 hours. Radio broadcasts from these satellites are picked up from receivers on the ground to triangulate their location. Reception from four satellites provides information to the receivers in four dimensions: latitude, longitude, altitude, and time, with the latter determine by atomic clocks on board the satellites.

A big trigger that opened up GPS for more general use was actually a tragic event in the skies. President Ronald Reagan opened up GPS to non-military use in 1983, after a passenger plane that had mistakenly flown into Soviet airspace was shot down by Soviet fighter jets, killing all on board. For security reasons, the military intentionally scrambled GPS satellite signals using Selective Availability that could only be removed by receivers that had decryption keys. In 2000, President Bill Clinton announced that practice would stop and civilian demand for GPS products skyrocketed as GPS became nearly ten times more accurate literally overnight. Today, GPS provides two services: one for civilian use referred to as Standard Positioning Service (SPS) and one only available to authorised users referred to as Precise Positioning Service (PPS). SPS, which is used by consumers' GPS receivers, including handheld hiking GPS receivers often used for geocaching, has an accuracy of 95% for horizontal positioning (up to 13 m and even less) and 95% for vertical positioning (22 m and less).

Letterboxing is sometimes called an outdoor hobby that originated on Dartmoor (UK) in 1854 when James Perrott of Chagford set up a cairn at Cranmere Pool on North Dartmoor that contained a glass jar where visitors could leave their visiting cards. From then on it slowly developed, with further letterboxes established at Taw Marsh (1894), Ducks Pool (1938), and Crow Tor (1962) and hikers on the moor began to leave a letter or postcard inside a waterproof box. The next person to discover the site would collect the postcards/letters and post them. Until the 1970s no more than a dozen such sites were placed around the moor, usually in the most inaccessible locations. Increasingly though letterboxes have been located in relatively accessible, scenic sites along trails and now with thousands of letterboxes, the tradition of leaving a letter or postcard has been forgotten, and it has become a popular sport, which has spread worldwide. In the USA it is generally considered to have started with a feature article in the Smithsonian Magazine in April 1998, and by 2001 there were over 1000 letterboxes covering every state. In 2004 Atlas Quest appeared which started with the idea of allowing letterboxers to create a virtual online logbook. So letterboxing combines orienteering with treasure hunting, art, and puzzle solving. The box contains a rubber stamp often of elaborate artistic design and logbook, and a letterboxer uses a series of clues (many published on websites, on published lists, or by word of mouth), a compass, and map to search for the boxes. Once found the finder will stamp the box's logbook with their personal stamp and use the box's stamp to mark their own personal logbook.

Letterboxing on Dartmoor is largely uncontrolled (Parker 2005c). There is a self-appointed "Letterboxing 100 Club" which attempts to apply some structure to the activity by issuing badges for collecting various numbers of stamps, and several members have reported passing the 10,000 mark, new box clues and general advice. The Dartmoor National Park has dialogue with the Club and makes requests for certain areas not to be visited during the bird nesting season. The Club is able to advise members to abide by the request and, where appropriate, to withdraw their boxes during the critical period. However, most letterboxing is outside the sphere of influence of the Club and continues throughout the year, abated only by weather and season. The locating of a letterbox often requires a detailed and possibly extensive search, probing into holes and crevices, many of which are typical of nesting sites of, for example, the wheatear (*Oenanthe oenanthe*). Experienced letterboxes use a stick for this, aware that many crevices on Dartmoor also contain the adder (*Vipera berus*). These potentially disturbing actions are coupled with the large participation numbers who move mostly off-track.

10.2 Numbers of Participants

Orienteering was not included in the Topline 2017 list of Outdoor Activities in the USA (Outdoor Foundation) but in Cordell (2012) the numbers were estimated at 4.8 million in 1994–1995 and 6.2 million in 2005–2009 which is a percentage growth of 67.8%. However, Orienteering USA has only about 1500 family and individual members and 55 member clubs. In the UK the British Orienteering membership was 10,869 in 2012, with 2013 events and 185,009 total event participants. The Active People Survey (2014) documented that 11,800 people aged over 14 participated monthly in the UK. In Norway the Norges Orienteringsforbund has over 400 clubs and over 24,000 members, the Finnish Orienteering Federation has over 400 clubs and about 60,000 members, whilst the Swedish equivalent has 600 clubs, with about 75,000 members. Scandinavia, where orienteering first developed, has relative to the total population by far the greatest number of participants.

The growth of geocaching has been spectacular since its origin around 2000. There are approximately 6 million active geocachers worldwide, with over 830,000 in the USA, over 375,000 in Germany, followed by Canada, and over 160,000 in the UK in 2015. The number of geocaches worldwide was 540,700 in March 2008 and grew to 1.69 million by 2012, to over 2.8 million currently and in over 180 countries (www.geocaching.com). In the USA the current estimate of the percentage of population who were geocachers over 16 was 3.5%, around 8 million people, but the numbers must be higher than this because of the number of under 16 who take part (Schneider and Chavez 2012). There are few figures for letterboxers although we do know that as long ago as 2005 there were 560,000 day visits in that year for Dartmoor for this activity.

10.3 Geocaching

An example of how the official line on geocaching has changed in the USA can be taken from the Minnesota State Parks (Brost 2011; Brost and Quinn 2012). Initially the Minnesota Department of Natural Resources received an increasing number of enquiries regarding the activity on state park lands as the popularity of the activity grew. However, no placement of caches was allowed on Division-administered lands because state park rules prohibited the storage or abandonment of personal property on state park lands without prior approval of the park manager. The Division was also concerned about visitor safety, the potential impacts to natural and cultural resources, liability, and the likely increase of staff workload. In 2005–2006 at the request of the Minnesota Geocaching Association, a number of meetings were held to see if there was a way that geocaching could be allowed on state park lands. As a result the Division determined that it would be allowed in state parks as long as it was managed so that it conformed to statutory direction for outdoor recreation activities allowed in state parks. Minnesota statutes define those as activities "which will not cause material disturbance of the natural features of the park or the introduction of undue artificiality into the natural scene." In 2006 people could apply for permits to place geocaches and letterboxes on Division-administered lands. The approval process intended to minimise the potential to cause impacts to natural and cultural resources and included such things as a review of the proposed location compared to rare natural features, cultural resource sites, high-quality native plant communities, and areas of active management such as those areas that were being burnt as part of active management. There were in fact few applications in the next two years so that by 2008 the park established a geocaching programme called "The History Challenge" to coincide with Minnesota's sesquicentennial. Caches were placed in all 72 state parks and recreational areas for one year. In 2009 a new programme was launched called "The Wildlife Safari" for three years. In view of the perceived success of the programmes in connecting people to the outdoors, it was decided to continue to sponsor such programmes and allow private citizens to establish geocaches.

In Queensland technically a geocache could be considered litter under the relevant legislation, and in the absence of a policy position in the late 2000s, local decisions were made as whether to sanction geocaches in a particular park. This led to the removal of unauthorised caches without notifying the cache owner which potentially caused more environmental disturbance through people trying to locate the cache or replacement of the cache by the cache owner. However, now after this rather chaotic initial situation Queensland has an operational policy on geocaching (2012). These two examples of how policy has changed show that most management authorities now regard geocaching as having a minor impact on the environment and in some ways has a positive role in helping them look after the land's resources.

10.3.1 Types of Geocaches

There are several types of geocache which need to be noted:

1. Standard traditional cache which at least contains a logbook and is located at the posted coordinates.
2. Multi-cache which includes at least one stage in addition to the physical final stage. The posted coordinates are a stage of the multi-cache, and at each stage the geocacher gathers information that leads to the next stage or to the final cache.
3. Mystery cache which can have many different designs, but for most the posted coordinates are not the actual cache location and it is often the case that a puzzle must be solved in advance to determine the next stage or the final coordinates. The cache page must provide information to solve the puzzle, and the information must be publicly available.
4. Offset caches are where published coordinates direct users to a specific location such as an existing historical monument, and from this site the cache hunter must look around to find offset numbers stamped/written in, or on some part of the marker site, or continue based on instructions posted on geocaching.com.
5. Letterbox hybrids require GPS usage for at least part of the search.
6. Wherigo caches require a Wherigo cartridge to find a cache container with a logbook. The cartridge must be hosted at Wherigo.com, and the cache description must include a link to the cartridge. Wherigo posted coordinates must be the same as the "start at" coordinates on Wherigo.com. Wherigo is a GPS location software platform created by Groundspeak and was first released in January 2008 and is a mix between an adventure game and a geocache search. Authors can develop self-enclosed story files called cartridges that are read by the Wherigo player software installed on either a GPS or a smartphone. Completing an adventure can require reaching different locations and solving puzzles.
7. Event caches are where there is a gathering of geocachers focusing on the social aspects of geocaching. Some are bigger and called a mega-event cache attended by over 500 people and those attended by over 5000 people are called giga-event caches.
8. CITO event caches are where there is a Cache-In-Trash-Out gathering to improve

parks and other cache-friendly places. These events are usually for removing rubbish but can be for planting trees, building trails, and removing invasive plants.

9. Virtual caches are just coordinates for existing landmarks or topographical features such as a waterfall or lake. Often a question will be posted about the landmark at the cache website for the finder to answer to act as proof to the cache originator that the finder was really there. EarthCaches are an example of this type of virtual cache where Geocaching HQ partners the Geological Society of America to oversee the EarthCache programme (community.geosociety.org/earthcache/guidelines/guidelines2). An EarthCache provides a learning experience in the geosciences that requires a visit to a unique geological feature, but there are no containers or logbooks but a series of questions have to be answered (Lewis and McLelland 2015). The first EarthCache was created in January 2004 in Australia, and now there are over 20,500 sites in more than 167 countries (March 2016), and over 6.3 million people have logged that they have visited EarthCache sites, and this figure is growing exponentially. A good example of an Earthcache would be the Grand Canyon National Park's EarthCache programme (http://www.nps.gov/grca/planyourvisit/virtual-caching.htm) which follows a trail around the canyon rim for 13.8 km with four sites.

10. Geotrails which are a series of geocaches tied together by a common theme have been set up by many organisations such as national parks, like the Haytor Trail in Dartmoor (UK). The Allegheny GeoTrail (USA) is not an itinerary as in a traditional trail but rather a series of points of interest in a kind of self-guided, goal-orientated plan which encompasses ten counties throughout NW and NE Pennsylvania, and within each county there are between 10 and 20 designated geocaches which make up the trail. It is located in the scenic and rugged Allegheny National Forest. Typically there is an incentive to finding a certain number of caches, for example, find six in a county and earn that county's unique geocoin and find six caches in all ten counties and there is a special geocoin that represents the entire trail. The Wild River State Park (Minnesota, USA) set up their geocaches to take participants past self-guided interpretative signs and on nature trails (Rosier and Yu 2011). Some of the caches require the participants to solve puzzles related to natural resources and history before they can find the cache location.

There seems to be a universal appeal to geocaching which is remarkable but the motivations for taking part seem to be as varied as its participants. Typically the motivations include experiencing nature, discovering and exploring new places, the promotion of physical activity, experiencing competition, and collecting achievements whilst benefitting from the educational aspects (Cord et al. 2015). As O'Hara (2008) states the activity not only consists of "consumption" (the seeking of geocaches) but also "creation" (the hiding of geocaches). It benefitted from the development of relatively cheap GPSs, including smartphones and the widespread access to the internet which allowed a technology-friendly community that was interested in applying these new technologies to the outdoors, although this was not the main reason but a means to an end (Telaar, et al. 2014).

10.4 Geocaching: Environmental Impacts

As with all other recreation activities geocaching and letterboxing must have measurable impacts on the environment in which they take place, and there must be a cumulative negative impact of all those geocachers that have escalated greatly since 2000 as they are following the same coordinates to the same locations. This means that multiple visitations will cause environmental degradation. This was acknowledged by some participants that geocaching could harm a park's natural resources (Rosier and Yu 2011). They could have a negative effect on the vegetation by forging their own trails, compacting soil, digging with spades looking for geocaches, and generally making it more difficult for plants to grow, and animals that consume this vegetation will also suffer. In Minnesota State Parks by 2008 the typical impacts noted were more or less the same as those caused by hillwalking that we have noted in Chap. 2: trampling of vegetation and damage to woody vegetation, exposure of bare soil, soil compaction, and soil erosion. The impacts occurred at the geocache locations, at geocache waypoint places (where a geocacher stops to search for the next set of coordinates on their way to the physical container), and at the travelways (the routes along which geocachers travel to find waypoints and ultimately the geocache) in between cache locations. However, Brost (2011) after a detailed analysis of Itasca State Park (Minnesota) found that the majority of the geocaches were in locations with a low likelihood of negative impact, based on defined vegetation, distance from trails and soil erosion variables, because park staff and natural resource managers who placed the caches were aware of the policy to avoid impacts to natural resources and carefully selected cache locations based on this policy. However, this is where management of geocaches is tightly controlled, and this is not always the case. It was also noted that on Dartmoor (UK), for example, letterboxers had started to pull apart historic rock walls and field boundaries and painted graffiti marking the locations of letterboxes. However, compared with other outdoor activities, most would consider that participants in these activities cause relatively little environmental damage, although there seems to be no published scientific study related to the scale of any environmental damage—because of the likely future growth in this outdoor activity, this seems necessary.

10.5 Geocaching: Educational Benefits and the Promotion of Environmental Stewardship

Despite these apparent inevitable environmental impacts caused by these activities which appear to be relatively minor, there seem to be positive educational benefits across a wide range of areas and for a range of age groups. The use of geocaching has been making its way into education in the last ten years (e.g. Christie 2007) and also into environmental education (e.g. Patubo 2010; Zecha 2012). As an example Zecha (2012) describes a study where each geocaching route deals with a special theme within environmental education. Each cache on the route covers one aspect of the topic. There is learning content in the caches, which is connected to a specific place and which you find with the help of coordinates and a GPS receiver. The consequence is that the children learn how to orientate themselves within nature whilst obtaining useful, interesting information about nature at the same time. This method within the field of environmental education brings children back to nature whilst setting important challenges. High-tech gadgetry provides a bridge to nature, especially for the children, who often previously have shown no interest in nature. Attitudes towards and awareness of the environment can be changed but only when influenced and affected by some of the creative settings in which the caches have been located (Boulaire and Cova 2012). There is no doubt that the best and most effective method with regards to environmental education is when pupils create their own geocaching route which allows the routes to be incorporated into their daily lives. They look for the special locations to install the cache and develop a task for the cache themselves. An extension of this is that the created geocaching route is uploaded onto the internet so that other people can use the route themselves. In an internet forum, they discuss the results of some caches and look for further information which is a very important step in integrating student participation. Inamäki (2012a) reviewed the interactive communication channels around geocaching, and she suggested pupils use GPS technology in a new way, develop it continuously, and create more interactive communication channels and services. They use interactive community instruments such as the Groundspeak discussion board, Groundspeak iPhone application, the Geocache Navigator, Geosetter, and Geocaching chat. Some of the geocachers are innovators, for example, the GPS technology used to create Geopt Geocaching Tools, a set of tools that facilitate the uploading of multiple photos.

Christie (2007) suggested that teachers can, by using GPS receivers and geocaching, create technology-rich, constructivist learning environments that engage students in a student-centred, personally meaningful, authentic, and collaborative learning. This learning is enquiry-based, requires informed decision-making, views mistakes as opportunities for growth, and values information exchange among all learners. She provides a website with many curriculum examples and online GPS and Geocaching Education resources. Lisenbee et al. (2015) suggested that geocaching is an enquiry-based activity that encourages creativity, active learning, and real-world problem solving. A good example of the creativity approach was discussed by Inamäki (2012b) who used a letterboxing, treasure hunt game in outdoor adventure education. Messick (2009) detailed ways of incorporating GPS technology into the geography curriculum, and it has also been suggested by Spencer (2014) that geocaching can be used by libraries as an educational tool.

Robinson and Hardcastle (2015) suggested that geocaching is a feasible activity to promote physical activity and especially increased levels of walking, especially amongst families, whilst Schlatter and Hurd (2005) suggested that geocaching offered physical educators abundant opportunities for multidisciplinary lessons. However, whilst all these benefits that apparently geocaching can bring to education are well worth having, probably the greatest claim is that it can increase environmental stewardship. This was suggested by MacQuaig (2013) who concluded that a large percentage of geocachers take part in other outdoor activities after beginning to participate in geocaching and that they claim to have increased their environmental stewardship. This promotion of environmental stewardship through leisure and educational activities seems to get people and students involved far more in the environment. Many are part of a new generation of technological savvy people who otherwise might chose the WWW or social media rather than a hike in the outdoors, and it certainly seems that it can be the first step in a long-term love of the outdoors (Harmon 2008). These kinds of suggestions seem to bring us to the conclusion that although there are some environmental impacts related to geocaching and letterboxing, the benefits far outweigh the negatives and the growth in numbers involved in a relatively short period of time has been stupendous.

10.6 Management of Geocaching and Letterboxing

These large numbers and the popularity of geocaching have meant that the land managers and geocaching organisations have had to put in place guidelines for geocaching such as those published by Geocaching Ireland or the regulations for Queensland PWS managed areas. Although the policies can be variable, most state parks in the USA, national forest wilderness areas in the USA, national parks in the UK, and the provincial and national parks in Canada have policies with regard to geocaching. There are codes of conduct such as the Code of Conduct for Letterboxing on the Dartmoor National Park (UK). There are leave-no-trace policies for geocachers, tread lightly tips for geocachers, and a set of outdoor ethics for geocachers and letterboxing.

10.7 Orienteering: Environmental Impacts

In response to increasing difficulties in obtaining access to countryside for sport and recreation in the UK, arising from perceptions of the potential for environmental damage (Marzano and Dandy 2012), Sidaway (1991) recommended that sports governing bodies become involved in research. The large numbers that often take part in orienteering events, especially in Scandinavia, were a worry, even if different routes are taken over the same course. Orienteering federations have responded to perceived environmental damage such as trampling of flora and the creation of new paths and faunal disturbance (e.g. by Anderson and Radford 1992; Littlemore and Barlow 2005 and McEvoy et al. 2008) by commissioning, encouraging, and collaborating in research into the ecological impact of their particular sport. Research has been conducted in the three main areas of environmental concern which are the trampling of vegetation, the disturbance of large mammals, and the disturbance of birds.

10.7.1 Disturbance to Vegetation

Parker (2005b) concluded from the general vegetation impact studies that he reviewed (e.g. Breckle et al. 1989; Douglas 1989; Myllyvirta et al. 1998) that orienteering, for events with up to approximately 2500 entrants, has very low impact on vegetation, with rapid recovery. For very large events, there is more significant general vegetation impact with sometimes an additional growing season or part season necessary for full recovery to be achieved at the more heavily used sites within the competition terrain. However, given that long-term damage is defined as that persisting for more than ten years, none is reported from any of the studies, including those events with very large entries of 10,000 or more. Therefore, with respect to general vegetation impact, the perception that orienteering does cause significant long-term damage was rejected.

In addition to the general vegetation impact on the competition terrain, there may be localised areas of more sensitive vegetation sites, such as marshes and lichen-covered rock. The protection of such areas is normally secured by standard planning procedures which route courses away from them or ensure that carrying capacities are not exceeded. The two studies of lichen both indicated low impact. The one result which did raise a question of possible long-term damage is the transect containing moss and lichen on a sandy soil at a New Forest event, which had not fully recovered one year after the event. Whether recovery would have occurred within a decade is unknown. The extent to which this damage is considered significant may be judged by noting that the researcher referred to it only in terms of recommendations for future avoidance and her overall conclusion was that there was very little vegetation damage from the event. With the proviso that there was some uncertainty about this one result, the studies on sensitive vegetation found no significant long-term damage from orienteering.

10.7.2 Disturbance to Mammals

Concerns about the disturbance to non-avian fauna centre on mammals such as deer. Smaller fauna which are nocturnal or, if out in daylight, can retreat to setts, earths, burrows, or other shelter are not considered to be significantly at risk from disturbance, although special protection legislation applies in the UK to badger setts.

A number of studies are reported by Liddle (1997) related to the movement of roe and red deer at orienteering events in Denmark. Observations during an event with 700 participants at Pamhule Forest (Denmark) produced the predictable result that the deer responded to the orienteers and sought shelter or left the area. A more useful related study in Kalo Forest (Denmark) monitored the movements of seven radio-tagged, roe deer. These, when disturbed, sought cover in small, dense plantations or marshes, areas not utilised by the orienteers, within their home ranges and stayed there for two hours until the event was over (Jepperson 1984). A similar study with red deer had two hinds collared with radios. Both hinds left the orienteering area and crossed open land to a neighbouring plantation, returning to their original range within 24 and 48 hours (Jepperson 1987a, b). It is suggested in reports that all of these populations are hunted and this could account for the animals' timidity. However, it has to be noted that the reported behaviour of the deer appears to match those of populations in the UK which are not hunted. The reported studies of deer reaction to disturbance by orienteers show that deer readily flee from disturbance and, if refuge zones are available within, or adjacent to, the competition area, they will take cover in these and then quickly return to a lowered anxiety level and their normal range after the event. The return to normal behaviour after the New Forest orienteering event was prevented by the deer being disturbed by further hunts. However, the researcher considered that the effect of the orienteering event alone would have been short-lived. Whilst disturbance of deer can expose them to risk, there are few reported examples of deer suffering injury or death as a result of orienteering in contrast to the number of deer-related traffic accidents of 20,000–40,000 annually in the UK (Deer Commission 2001) out of a deer population estimated at 500,000–600,000 (Deer Initiative Council 2002). Two additional important studies have been made of the effect of orienteering and other human activities on different deer populations, one in Sweden and the other in the New Forest in southern England, and are discussed in the next section.

10.7.2.1 Disturbance of Elk and Deer in Sweden

Sweden has substantial populations of roe deer *Capreolus capreolus* and elk *Alces alces*. There are approximately one million deer in Sweden. The additional elk population of 250,000 is the highest in any country, and 100,000 of these are shot each year. Clearly, elk and deer hunting is a major activity whose participants take an interest in any other activity which has the potential for disturbing these animals. A comprehensive study of the effect of orienteering competitions on populations of elk and deer was carried out near the Grimsö Game Research Centre (Sweden) by Cederlund et al. (1981). The study was a cooperative venture between the Swedish Orienteering Federation, the Hunters Central Association, the Government Conservation Office, and the Government Real Estate Office. The behaviour of elk and deer was studied in three competitions: in June 1979 with 1800 participants, in October 1979 with 600 participants, and in April 1980 with 2000 participants.

The reactions of elk and deer to disturbance were assessed based on observations of flight length, external signs of stress, physical injury, and mortality. The location and movement of the animals were determined by visual observation of marked animals and by radio location of animals fitted with transmitters. These observations allowed the density and consistency of the population, and its territorial behaviour, to be determined. Some complementary research was carried out to determine the effectiveness of procedures for driving animals from the competition terrain into refuge zones and to test whether tapes used in the competition terrain to delineate out-of-bounds, or dangerous features, affected animals' flight.

In all three competitions, environmental good practice with respect to deer was employed in leaving areas free of the courses to act as deer refuges and in planning the courses so that they progressed in the same general direction to avoid repeated encounters between deer and orienteers appearing from very different directions.

The studies found a significant difference in behaviour between elk and deer. In one test elk took flight at a "flushing distance" of about 200 m from a confronting, walking small group and moved a "flight distance," on average around 1300 m, before stopping. In another test, 20–30 runners approached from a single direction, flushing the elk at a distance of 200–300 m and causing them to take flight for a distance of 1500 m. Deer had a similar flushing distance of around 200 m but with a much shorter flight length before stopping, an average of 600 m. The consequence of this is that, provided that unidirectional course setting is in use, the elk tend to move out of the competition area. Deer, on the other hand, tend to remain in the competition area and seek shelter in small thickets or larger areas set aside as refuges. The observations showed that the elk and deer populations returned to normal within their usual territories within 24 hours. Some elk and deer showed signs of stress, but there was no mortality or injury reported.

In the complementary tests with a line of drivers moving forward to drive the animals from the competition area, similar results were noted, with elk being flushed at 300 m and then taking flight for 1700 m and the deer being flushed at 200–300 m and then moving 700 m. The drive was found to be effective for elk with separations between the drivers of as much as 100 m. Deer, however, were as likely to run back through the driving line as run away from it, even with separations between the drivers as little as 30 m. The deer were also seen to clear 1 m high obstructions (in this case marker tapes) without difficulty.

This work was a comprehensive and sophisticated study of deer and elk behaviour using radio-location devices. There is a suggestion from the figures that both species, not unexpectedly, regard a line of approaching drivers as more threatening than a group of walkers or runners and take flight further, as a consequence. The key conclusions from the studies are that deer and elk behave differently, the deer remaining in the competition terrain and, for whom, sanctuary areas should be designed into the courses, if not naturally present. The need for parallel, unidirectional courses is made clear. This study makes an important contribution to the knowledge base for these two major species.

10.7.2.2 Disturbance of Deer in the New Forest in Southern England

The 1988 orienteering event in the New Forest monitored by Douglas (1989) comprised vegetation trampling observations and extended studies on the sensitivity of the fallow deer *Dama dama* to the orienteering event, a deer hunt, and other human activities. It was intended that the effect of the event on the fallow deer be monitored by regular checking of the nature of the reaction of the deer to the researcher before and after the event. However, interference by the deer hunt required the monitoring programme to be modified and extended, covering in all a period of 16 months and totalling 72 visits. As fallow deer become accustomed to long-term disturbances that remain within set bounds, the researcher varied the direction of the walks determining deer reaction.

The researcher reported that the deer appeared normal on both the morning and evening of the orienteering event. During the day, after the first few competitors had passed, the deer moved to unused areas or hid in thick stands of conifers. One deer was hurt by running into a fence during the competition, but this incident was seen by a research assistant who reported that the deer was one of a group startled by the appearance of some non-orienteers. The detailed deer reaction measurements showed an elevated reading following the orienteering weekend, but this had included the hunt on Saturday before the orienteering event on Sunday. Two further hunt disturbances occurred in the following week but, by three weeks after the event, the deer reaction readings showed full recovery. Since the deer reaction walks showed increased anxiety by deer in the presence and aftermath of the hunt, it was difficult to determine what level of reaction was due to the orienteering event. However, based on her 16 months of observations, Douglas concluded that, had the orienteering event not been compromised by the hunt on the previous day, the deer reaction would have been low, similar to that on a public holiday.

Given the evidence of very low impact on larger mammals from these studies and that if orienteering is conducted in accordance with established environmental procedures for mammals, it is concluded that, for practical purposes, the perception that the sport is damaging to fauna is unlikely.

10.7.3 Disturbance of Birds

The most widespread reason for access restrictions on orienteering and other activities using semi-natural countryside is the potential for disturbing breeding birds and reducing their breeding success. In the UK, evidence to the House of Commons Environment Committee by the Chief Executive of English Nature, the governmental organisation responsible for nature conservation in England, with respect to access to the countryside for recreation, stated that "the most contentious area remains disturbance on breeding birds" (Langslow 1995).

A number of ad hoc observations have been carried out with respect to the disturbance or potential disturbance of birds by orienteering. For example, in the Jukola 1995 relay at Sipoo (Finland), with 10,000 entrants, concern was expressed about nesting capercaillie *Tetrao urogallus*. The organisers cooperated with local ornithologists to mark the nests in the forest before the competition, and the courses were adjusted to avoid the nest sites. One nest was overlooked and, during the competition, some 2500 competitors passed within "a very short distance" (Myllyvirta et al. 1998).

In 1991 the British Orienteering Federation, concerned at the increasing restriction of its activities in the stated interests of protecting nesting birds and at the lack of meaningful data on the effect of orienteering on nesting birds (Parker 2005c), commissioned an environmental consultancy to carry out an independent comprehensive survey at an event (Goodall and Gregory 1991). The event chosen was in Brandon Park, a section of Thetford Forest in Suffolk in May 1991, with 480 competitors. The aim of the study was to determine whether the orienteering activity affected the breeding bird community. Given that there was an effect, further questions that arise concern its significance, how long it lasts, whether the different species are similarly affected, and whether it relates to the density and duration of the disturbance. The survey took place over a 20-day period symmetrically spanning the event. A total of 54 species were monitored, comprising canopy-, hole-, scrub- and ground-nesting guilds in roughly equal proportions. About two thirds of the species had individuals within high activity (high potential disturbance) zones. The results of the study were that after intensive survey, none of the methods produced sufficient evidence to reject the original hypothesis that the orienteering event would have no significant effect on the breeding bird population (Goodall and Gregory 1991). It was suggested that because this area was widely used by the general public that birds would be tolerant to some disturbance which might explain some of the findings. An important finding in the body of the report but not drawn out in the conclusions or summary concerned the nesting in the area of a highly protected bird, a species listed in Schedule 1 of the Wildlife and Countryside Act 1981. The disturbance of such a bird whilst it is nesting is an offence. The listed bird in Brandon Park was the woodlark (*Lullula arborea*) of which at least four pairs were nesting in the clear felled areas. Following a close study of the courses and the likely routes of the competitors, some changes to the courses were applied in order to avoid the woodlark breeding areas. By chance the prediction of the likely routes chosen by competitors was partly in error, and two woodlark territories received medium levels of activity, with up to 50 competitors crossing each. The post-event survey found no change in activity or number or territories of this species. The particular importance of the Brandon Park finding, which two nesting woodlarks were tolerant of medium orienteering disturbance across their territories, is that it represents a study that cannot be set up by design, because any disturbance that might result to a scheduled bird constitutes an offence under the 1981 Act.

10.7.3.1 Disturbance of Nesting Birds in Drumore Wood, Aberfoyle, Scotland

In June 1987 the Scottish Orienteering Championships were held in Drumore Wood, Aberfoyle. This wood is a deciduous enclave, mostly of oak *Quercus robur* and *Q. petraea*, in the predominantly coniferous Loch Ard Forest, part of the Queen Elizabeth Forest Park. Concerns were expressed by the Scottish Ornithologists' Club that the orienteering event, by virtue of its date and the expected number of participants (around 1000), could result in potential problems of damage to the vegetation and particularly of disturbance at a critical period for certain ground-nesting birds, including capercaillie and migratory songbirds (Brackenridge 1988). Discussion with the Scottish Orienteering Association resulted in the ornithologists carrying out a Common Bird Census in a section of the competition area from late April to early July, spanning the day of the event. In this they were assisted by the orienteers who volunteered maps and information about the courses. Brackenridge reported that torrential rain coincided with the event on what was one of the wettest days of the summer of 1987 in that part of Scotland. The analysis of the nine visits which made up the bird census showed that some birds did abandon territories about the beginning of June. In particular, out of ten redstart *Phoenicurus phoenicurus* territories located in May, only two or three showed birds to be present after the event. The ground-nesting wood warbler *Phylloscopus sibilatrix*, found singing in two or three sites during May, was found only once after the event. Out of five or six tree pipit *Anthus trivialis* territories, only three were certainly maintained.

In his conclusions Brackenridge did not review the results but did observe that the coincidence of two significant stress factors (prolonged human disturbance in an area unused to this, coupled with the heavy rain) obviously does lessen the ability to lay the blame for the local desertion of territory solely on the orienteering event, without a comparative study in an undisturbed plot nearby (Brackenridge 1988).

10.7.3.2 Discussion of Research into the Disturbance of Birds by Orienteers

The study of the impact of orienteering on birds is a complex issue, but the Brandon Park research showed no impact on 54 species from a medium-sized orienteering event and it was unable to reject the null hypothesis that the orienteering event would have no significant effect on the breeding bird population. To what extent this result is likely to be replicated at an orienteering event with over 1000 competitors was tested at Titterstone Clee in the West Midlands (UK) on 31 May 1999 (Parker 2005a). The area supports a population of about 40 breeding pairs of migrant wheatear (*Oenanthe oenanthe*) (Fig. 10.3) which were monitored by local ornithologists. The event had no observable effect on the breeding success of the nests within the competition area, and the wheatear was tolerant of disturbance levels of more than 200 competitors per hour passing within 25 m of the nests. However, four nests were abandoned in the derelict quarry used for the car parking and the competition facilities. This area had been selected for this purpose

Fig. 10.3 Wheatear (*Oenanthe oenanthe*). Photo by Philippe Kurapski

according to established successful practice, to minimise visual intrusion and ecological disturbance, but the propensity for wheatear to nest in such man-made terrain was unknown to the event organisers. Measures have been put in place since to prevent a recurrence. Retrospective analysis of data from the orienteers and the ornithologists has quantified the levels of potentially disturbing orienteering activity near nests. It is concluded that breeding wheatears are very tolerant of transient disturbance. Similar tolerance is expected for stonechat (*Saxicola torquata*), also present on Titterstone Clee.

Given the above evidence, some land managers might take the view that the restrictions they place on orienteering during bird nesting seasons could be eased. Others might consider that, despite the clear results of the Brandon Park and Titterstone Clee studies, too much uncertainty remains and they would follow the same procedure as at Drumore Wood and, as a precautionary measure, apply the restrictions in full.

However, reviewing the idea that orienteering causes significant long-term damage, in this case to birds, there is no evidence to support it. As to whether there is sufficient evidence to reject the hypothesis is a matter of subjective judgement. Given that the limited evidence so far available indicates that 56 breeding species were, at a particular time and place, unaffected by orienteers, the starting point for any discussion should be that the sport, conducted in accordance with its own environmental procedures, appears to be non-damaging to breeding birds.

It is apparent from this analysis of research into the environmental impact of orienteering that this is a subject with a great many variables. Therefore, the total of reported research in this field appears sparse. Those managing recreation opportunities in semi-natural countryside may have concerns about the extent to which the lack of objective data on environmental impact is affecting their decision-making on access. Orienteers who consider that their activities are significantly and adversely affected by lack of sound knowledge of their environmental impact regard further research as a prime objective. The fear of environmental damage by recreationists is still pretty much a presumption, and there has been surprisingly little research undertaken in this area. Until such work has been carried out, it is hard to develop a very strong case for exclusion of the public. Of the three areas of environmental concern, the disturbance of large mammals, the trampling of vegetation, and the disturbance of birds, it is not likely that further research into the effect of orienteering on large mammals, such as deer, is a prime requirement. The reported studies of the reactions of three internationally widespread species to orienteering disturbance appear to provide sufficient information for current management purposes about the effect of orienteering and similar off-track activities on general deer populations. With regard to the trampling of vegetation, the conclusions from the studies reviewed point to a generally low impact and rapid regeneration, but there may be reluctance to apply such general observations to specific sites on the grounds that different vegetation communities may have different sensitivities to trampling. The diversity of such communities is extensive. With respect to bird disturbance from orienteering activity, further research appears to be a clear priority. Formal recreation activities are often severely curtailed in the interests of preventing potential damaging disturbance to bird populations. There have been general bans on orienteering during extended bird breeding seasons in many locations. In the UK, for instance, in the important orienteering area of Cannock Chase, the bird breeding season is deemed to run from April to August (with localised embargoes from the beginning of February) and is likely to extend forward if the trend of milder winters and earlier springs continues to encourage earlier nesting. A similar embargo places the whole of Dartmoor, in South West England, out of bounds to orienteers from the end of February to mid-July. It may be that further research would demonstrate very low disturbance, as at Brandon Park, and that access constraints could be reduced or removed. However, although it might be expected that there should be substantial gains for recreationists from improved clarity on the environmental impact of their activities on birds, there could be a major problem in the unwillingness of conservation bodies to accept studies of the disturbance of birds by ephemeral, short-term activities. This arises from their focus of avian research being on the long-term viability of bird populations, which does not readily accommodate short-term, one-off disturbances, like an orienteering event.

In order to make balanced judgements, land managers need to relate any impact that is reported for orienteering to the ecological pressure as a whole on their sites. Also important is that research which shows no or negligible impact are reported and not abandoned, as is often the case as nil results have value. The general lack of research may well reflect the idea that orienteering has a minimal impact on the environment and that the research that has been published, much sponsored by vested interests like the national orienteering federations, should not be viewed with suspicion as biased.

10.8 A Comparison Between Orienteering and Letterboxing Impacts

Instructive in the impact debate is a comparison between orienteering and letterboxing made by Parker (2005c) on Dartmoor. Both activities involve convergence of dispersed participants onto particular points, a letterbox or a control flag. It is this convergence which gives rise to ecological concerns. In terms of the average time off-track that the exponents of the two activities spend on a single day (one hour and four hours), it might be assumed that one letterboxer is equivalent to four orienteers. However, this does not take note of the important differences in the way the activities are conducted in the terrain and which affect their potential ecological impact.

Letterboxers make their way at leisure to the target letterbox site using written descriptions and then have to search around for the hidden box. This will take ten minutes, perhaps less, perhaps more. Having found the box, letterboxers will open it, take out the stamp, ink it with their own ink pad, stamp their logbook, ink and stamp their own mark in the book in the box, read any instructions or clues within the box, put it all back together, ensuring it is waterproof, and restore it to its cavity. This will take five minutes or more. In comparison, orienteers make their way at speed to the target feature by map reading; the marker flag is not hidden and is visible from 15 m distance and usually from very much further. Having found the marker flag, orienteers will record their visit by using a punch on a card they carry or by swiping an electronic sensor. This will take five seconds or less. All in all, on average, a letterboxer will spend 15 minutes in the vicinity of a letterbox, and an orienteer will spend 15 seconds in the vicinity of a control. In terms of total disturbance, comparing time at one site, one letterboxer is equivalent to 60 orienteers. This is the typical total number of orienteers that pass through a busy control site during the three hour period of a club orienteering event on Dartmoor.

In considering the disturbance of ground-nesting birds, the impact of one prolonged 15-minute disturbance is different from 60 ephemeral 15-second disturbances spread over a 3-hour period. The prolonged disturbance is potentially more damaging. Although there may be differences between species, observations reported earlier by Parker (2005b) are consistent with repeated short disturbances being non-damaging.

This comparison refers to a single site. Of course, an orienteering event has more than one control site and a letterboxer will visit more than one letterbox in a day's outing. An average tally of letterboxes for one day's outing may be taken as being in the range 10–20 boxes and the number of reasonably busy control sites in a club orienteering event is much the same. Therefore, one club orienteering event of around 100 competitors is roughly equivalent to two letterboxers in terms of potential ecological impact around the control points where the competitors converge.

This equivalence of two letterboxers and a club-sized orienteering event may be scaled, with caution, to more letterboxers in a single day and larger orienteering events. The most important difference, however, between letterboxing and orienteering is that, in a particular area, the letterboxing will occur day-on-day (subject to season and weather) whereas the orienteering will only take place once in a year or perhaps at longer intervals. The point being made by Parker (2005c) is that if letterboxing is considered not to be environmentally damaging then orienteering should be considered in the same light.

Concluding Remarks

One of the opening statements by the New Forest Review Group (1988) that orienteering "by its very nature causes very considerable disturbance to wildlife and damage to flora over wide areas" has been shown to be unsupported by Parker (2005a, b, c, 2010) through firm evidence. It has also been shown that the reason for this is that orienteering, by its very nature and by additional procedures of good practice, is a sport of low ecological impact. There also seem to be many wide-ranging educational benefits of orienteering and geocaching and particularly the development of environmental stewardship, where the benefits outweigh the minor environmental impact.

References

Anderson, P., & Radford, E. (1992). *A Review of the Effect of Recreation on Woodland Soils, Vegetation and Flora*. English Nature Research Report No. 27, English Nature, Peterborough.

Boulaire, C., & Cova, B. (2012). The dynamics and trajectory of creative consumption practices as revealed by the postmodern game of geocaching. *Journal of Consumption, Markets and Culture, 16*. https://doi.org/10.1080/10253866.2012.659434.

Brackenridge, W. R. (1988, March/April). Effects of a major orienteering event in Drumore Wood, Aberfoyle, in June 1987. *Compass Sport*, pp. 20–22.

Breckle, S.-W., Breckle, H., & Breckle, U. (1989). Vegetation impact by orienteering? *Scientific Journal of Orienteering, 5*, 25–39.

Brost, J. T. (2011). *Assessing Effects of Geocaching as a Recreational Activity on Natural Resources within Minnesota State Parks*. Papers in Resource Analysis v.13, 20pp. Saint Mary's University of Minnesota. Retrieved from http://www.gis.smumn.edu.

Brost, J. T., & Quinn, E. M. (2012). *An Assessment of the Impact of Geocaching on Natural Resources*. Minnesota State Parks, MNDNR, Division of Parks and Trails.

Cederlund, G., Larsson, K., & Lemnell, P. A. (1981). *Orienteringstävlingars inverkan på älg och rådjur, (Impact of orienteering on elk and red deer)*. Sweden: Naturvårdverkets.

Christie, A. (2007). *Using GPS and Geocaching Engages, Empowers and Enlightens Middle School Teachers and Students*. Online GPS and Geocaching in Education. Retrieved from http://www.alicechristie.org/pubs/E6/print.html.

Cord, A., Roessiger, F., & Schwarz, N. (2015). Geocaching data as an indicator for recreational ecosystem services in urban areas: Exploring spatial gradients, preferences and motivations. *Landscape and Urban Planning, 144*, 151–162.

Cordell, H. K. (2012). *Outdoor Recreation Trends and Futures: A Technical Document Supporting the Forest Service 2010 RPA Assessment*. General Technical Report SRS-150. Asheville, NC: US Forest Service Southern Research Station, 167pp.

Deer Commission. (2001). The Deer Commission for Scotland. Retrieved from http://www.dcs.gov.uk/.

Deer Initiative Council. (2002). Retrieved from http://www.thedeerinitiative.co.uk.

Douglas, E. A. (1989). *Assessment of the impact of the November Classic Badge Event 1988 on the New Forest*. Matlock: British Orienteering Federation.

Goodall, A., & Gregory, C. (1991). *The Effect of the May 1991 Orienteering Event on the Breeding Bird Community in Brandon Park. Ecosurveys, Spilsby, Lincs*. Report IOF/Env/003, International Orienteering Federation, Finland.

Harmon, L. K. (2008). Get out and stay out: Technology can be the first step in a long-term love of the outdoors. *Parks and Recreation, 43*, 50–56.

Inamäki, P. J. (2012a). Geocaching: Interactive communication instruments around the game. *Eludamos Computer Game Culture, 6*, 133–152.

Inamäki, P. J. (2012b). Fare tale orienteering: Developing art word by letterboxing event. *Tourismos: An International Multidisciplinary Journal of Tourism, 7*, 253–268.

Jepperson, J. L. (1984). Human disturbance of roe deer and red deer: Preliminary results. In O. Saastamoiren, S. G. Hultman, N. Elerstioch & L. Mattison (Eds.), *Multiple-use Forestry in the Scandinavian countries* (pp. 113–118). Communication Institute Foresta.

Jepperson, J. L. (1987a). Impact of human disturbance on home range movements and activity of red deer *Cervus elaphus* in a Danish environment. *Danish Review of Game Biology, 13*, 1–38.

Jepperson, J. L. (1987b). The disturbing effects of orienteering and hunting on roe deer (*Capreolus capreolus*). *Danish Review of Game Biology, 13*, 24pp.

Langslow, D. (1995). *Evidence by English Nature*. House of Commons Environment Committee 1995, vol. II, Minute 334.

Lewis, G. B., & McLelland, C. V. (2015). *Earthcaching: An Educator's Guide*. Geological Society of America free guide.

Liddle, M. J. (1997). *Recreational Ecology*. London: Chapman and Hall.

Lisenbee, P., Hallman, C., & Landry, D. (2015). Geocaching is catching Students' Attention in the classroom. *The Geography Teacher, 12*, 7–16.

Littlemore, J., & Barlow, C. (2005). Managing public access for wildlife in woodlands- ecological principles and guidelines for best practice. *Quarterly Journal of Forestry, 99*, 271–286.

MacQuaig, J. K. (2013). *Does Participating in Geocaching Activities lead to Participation in Other Types of Outdoor Activities and Increased Environmental Stewardship?* MAEd. thesis, Hamline University, St. Paul, MN.

Marzano, M., & Dandy, N. (2012). *Recreational Use of Forests and Disturbance of Wildlife. A Literature Review*. Forestry Commission Research Report. Forestry Commission, Edinburgh, 40pp.

McEvoy, D., Cavan, G., Handley, J., McMorrow, J., & Lindley, S. (2008). Changes to climate and visitor behaviour: Implications for vulnerable landscapes in the north west region of England. *Journal of Sustainable Tourism, 16*, 101–121.

Messick, B. E. (2009). *Incorporating GPS Technology in Geography Curriculum*. MAEd. thesis, Hamline University, St. Paul, MN.

Myllyvirta, T., Henriksson, M., & Aalto, V. (1998). Sipoon Jukolan viestin 1995 kasvillisuusvaikutustutkimus (A study of the Sipoo Jukola relay 1995 impact on vegetation). Itä-Uudenmaan ja Porvoonjoen vesien-ja ilmansuojeluyhdistys r.y. [Summary in English].

New Forest Review Group. (1988). *New Forest Review*. Forestry Commission, Lyndhurst.

O'Hara, K. (2008). Understanding geocaching practices and motivations. In *Proceedings of the SIGCHI Conference on Human Factors in Computing Systems* (pp. 1177–1186), Florence, Italy, April 5–10, 2008. https://doi.org/10.1145/1357054.1357239.

Parker, B. H. (2005a). *The Effect of an Orienteering Event on Breeding Wheatear Oenanthe oenanthe at Titterstone Clee, Shropshire, UK*. Report IOF/Env/001. International Orienteering Federation, Finland, 18pp.

Parker, B. H. (2005b). *Review of Research into the Ecological Impact of Orienteering*. Report IOF/ENV/02. International Orienteering Federation, Finland, 43pp.

Parker, B. H. (2005c). *Environmental Impact of Orienteering and Other Off-track Recreations in the Dartmoor National Park, UK*. Report IOF/ENV/004. International Orienteering Federation, Finland, 30pp.

References

Parker, B. H. (2010). *Orienteering. A Nature Sport with Low Ecological Impact*. Report IOF/ENV/007. International Orienteering Federation, Finland, 21pp.

Patubo, B. G. (2010). *Environmental Impacts of Human Activity Associated with Geocaching*. Social Sciences Department, California Polytechnic State University, San Luis Obispo, 14pp. Retrieved from http://www.digitalcommons.calpoly.edu/erscsp/12.html.

Robinson, S., & Hardcastle, S. J. (2015). Exploring the attitudes towards and experiences of geocaching amongst families in the community. *International Journal of Environmental Health Research, 26*. https://doi.org/10.1080/09603123.2015.1061116.

Rosier, J. L., & Yu, A.H-C. (2011). Exploring the effects of geocaching on understanding Natural Resources and History. In *Proceedings of the Northeastern Recreation Research Symposium*, 5pp.

Schlatter, B. E., & Hurd, A. (2005). Geocaching 21stC hide and seek. *Journal of Physical Education, Recreation and Dance, 76*, 28–32.

Schneider, I. E., & Chavez, D. (2012). Geocaching: Form, function and opportunity. In K. Cordell (Ed.), *Outdoor Trends and Futures* (pp. 55–58). USDA Forest Service.

Sidaway, R. (1991). *Good Conservation Practice for Sport and Recreation*. Study 37, Sports Council, together with Countryside Commission, Nature Conservancy Council, World Wide Fund for Nature, London.

Spencer, A. (2014). Hide-and-seek in Macquarie University library: Geocaching as an educational and outreach tool. *The Australian Library Journal, 64*, 35–39.

Telaar, D., Krüger, A., & Schöning, J. (2014). A large-scale quantitative survey of the german geocaching community in 2007. *Advances in Human-Computer Interaction* v.2014, ID.101038, 13pp.

Zecha, S. (2012). Geocaching, a tool to support environmental education !?—An explorative study. *Educational Research 1*, eJournal. https://doi.org/10.5838/erej.2012.12.06.

Skiing, Snowboarding, and Snowshoeing

Chapter Summary

This chapter first defines some of the different disciplines of skiing (alpine, Nordic, telemark, ski mountaineering), snowboarding, and snowshoeing. It then examines snow sport competition before examining participation numbers. The final part of the chapter focuses on specific environmental impacts of snow sports: damage to soil, vegetation, water, and the impacts on wildlife. There is discussion about artificial snowmaking and its environmental impacts. The final section considers the management of these activities and the impacts of climate change on snow sport.

11.1 Definitions

Skiing can be a means of transport, a recreational activity, or a competitive winter sport in which skis are used to glide on snow. Many types of competitive skiing events are recognised by the International Olympic Committee (IOC) and the International Ski Federation (FIS). The term snow sport encompasses alpine, freestyle, snowboard, and Nordic skiing (UK Snow Sports 2018).

11.1.1 Alpine Skiing

Alpine skiing, also called downhill skiing, typically takes place on a piste at a ski resort. It is characterised by fixed-heel bindings that attach at both the toe and the heel of the skier's boot (Fig. 11.1A, B). Because the alpine equipment is somewhat difficult to walk in, ski lifts (Fig. 11.1C), including chairlifts (Fig. 11.1D), bring skiers up the slope. Clearly these have impacts on the environment which are discussed later in this chapter. Backcountry skiing can be accessed by helicopter, snowcat, hiking, and snowmobile. Facilities at resorts can include night skiing, après-ski (which can include bars with outdoor music and dancing on the ski slopes), and snowmaking facilities (discussed later). Alpine skiing branched off from the older Nordic skiing around the 1920s when the advent of ski lifts meant that it was not necessary to walk any longer.

11.1.2 Nordic Skiing

The Nordic disciplines include cross-country skiing and ski jumping, which both use bindings that attach at the toes of the skier's boots but not at the heels (Fig. 11.2A). Cross-country skiing may be practised on groomed trails or in undeveloped backcountry areas. Ski jumping skiing is practised at certain areas that are deemed for ski jumping only.

Fig. 11.1 (A) An Alpine ski boot and bindings. The boot is held at the toe and the heel. The bindings will release under force to minimise the risk of injury in even of a fall. Photo by Tim Stott. (B) An Alpine ski is shaped to enhance its turning capability. Photo by Tim Stott. (C) Alpine ski technique with skis parallel. Photo by Tim Stott. (D) Drag lift (also called a poma tow) at Chamrousse 1650, France, used to take Alpine skiers up the mountain. Photo by Tim Stott. (E) Chairlift and ski resort infrastructure at Les Deux Alpes, France, for transporting Alpine skiers up the mountain. Photo by Tim Stott

11.1.3 Telemark

Telemark skiing is a ski turning technique and FIS-sanctioned discipline. It is named after the Telemark region of Norway. The equipment similar to Nordic skiing, with the ski bindings having the ski boot attached only at the toe (Fig. 11.3A) and the ski shaped (Fig. 11.3B) more like an Alpine ski, which facilitates turning. The "free-heel" binding allows the skier to raise his/her

11.1 Definitions

Fig. 11.2 (A) Nordic ski boot and binding. This type of binding, known as a 3-pin binding, attached to the toe of the boot only. The ski and boot are much lighter and more flexible than the Alpine set up so permitting faster travel. Photo by Tim Stott. (B) A Nordic ski is longer and narrower than an Alpine ski. It is designed to move faster and has more directional stability (but is more difficult to turn). Photo by Tim Stott. (C) Nordic skiing at the Plateau D'Alsace, Chamrousse, France. The technique in this photograph is called "skating" which allows the skiers to go uphill. At other times the skiers follow pre-cut parallel tracks, cut by a special machine. Photo by Tim Stott

heel throughout the turn (Fig. 11.3C). Hence, telemark skiing is known as "free-heel" skiing in North America.

11.1.4 Ski Mountaineering

Ski mountaineering is a skiing discipline that involves climbing mountains either on skis (skins which grip the snow are fixed to the base of the ski for uphill travel) or carrying them (Fig. 11.4G), depending on the steepness of the ascent, and then descending on skis. There are two major categories of equipment used, free-heel telemark skis and skis based on Alpine skis (Figs. 11.4A, B), where the heel is free for ascent but is fixed during descent. The discipline may be practised recreationally or as a competitive sport.

Competitive ski mountaineering is typically a timed racing event that follows an established trail through challenging winter alpine terrain while passing through a series of checkpoints. Racers climb and descend under their own power using backcountry skiing equipment and techniques. More generally, ski mountaineering is an activity that variously combines ski touring, telemark, backcountry skiing, and mountaineering.

11.1.5 Snowboarding

Snowboarding is a recreational activity and Olympic and Paralympic sport that involves

Fig. 11.3 (A) Telemark boot and binding. The boot is made from plastic but a rubber section above the toe makes it flexible. The boot is only fixed to the binding at the toe. There is a cable around the heel. In the event of a fall, the boot and metal part of the binding and cable come away from the ski. Photo by Tim Stott. (B) A telemark ski is shaped like an Alpine ski. Photo by Tim Stott. (C) Telemark ski technique. Photo by Tim Stott

descending a snow-covered slope while standing on a snowboard attached to a rider's feet (Fig. 11.5A, B).

The development of snowboarding was inspired by skateboarding, sledding, surfing, and skiing. It was developed in the USA in the 1960s, became a Winter Olympic Sport at Nagano in 1998, and was first featured in the Winter Paralympics at Sochi in 2014. Its popularity (as measured by equipment sales) in the USA peaked in 2007 and has been in a decline since. Snowboarding styles now include jibbing, freeriding, freestyle, Alpine snowboarding (Fig. 11.5B), slopestyle, big air; half-pipe; boardercross, and snowboard racing (Fig. 11.5C).

11.1.6 Snowshoeing

A snowshoe is footwear for walking over snow. Snowshoes work by distributing the weight of the person over a larger area so that the person's foot does not sink completely into the snow, a quality called "flotation." Snowshoeing is a form of hiking. The origin and age of snowshoes is not precisely known, although historians believe they were invented from 4000 to 6000 years ago, probably starting in Central Asia. Two groups of snowshoe pioneers diverged early on, setting patterns that can still be seen today. One group abandoned the snowshoe as it migrated north to what is now Scandinavia, eventually turning the design into the forerunners of the Nordic ski. The other went northeast, eventually crossing the Bering Strait into North America. In 2016, Italian scientists reported "the oldest snowshoe in the world" discovered in the Dolomites and dated to between 3800 and 3700 BC. As snowshoes resemble a tennis racquet, the French use the term is *raquette de neige*.

Traditional snowshoes have a hardwood frame with rawhide lacings. Most modern snowshoes

11.1 Definitions

Fig. 11.4 (A) Ski mountaineering boot and binding. These bindings are fixed at the toe with two pins so the boot articulates. The heel can be free (for uphill travel) with three risers (for increasingly steep slopes) and are then clipped down for the ski descent. Photo by Tim Stott. (B) Ski mountaineering boots, bindings, and skis are lighter than Alpine equivalents. The skis are usually wider and longer to support the skier in soft snow "off-piste." Photo by Tim Stott. (C) A ski mountaineering party in northern Norway skinning uphill. Photo by Tim Stott. (D) Ski mountaineers make zigzag tracks up steep slopes. Photo by Tim Stott. (E) Ski mountaineering ski technique for the descent is basically the same as for Alpine skiing, though the snow conditions can be variable. Photo by Tim Stott. (F) Ski mountaineering descent in Kvaloya, near Tromso, Norway. Photo by Tim Stott. (G) Ski mountaineers sometimes have to walk in carrying their skis (on approach to the Komna plateau, Slovenia). Photo by Tim Stott. (H) Ski mountaineers reaching a summit in Norway. Sometimes they take off their skis and replace them with crampons and an ice axe for the final part of the ascent. Photo by Tim Stott

Fig. 11.4 (continued)

Fig. 11.5 (A) Snowboarders use soft boots which they strap into bindings. Photo by Tim Stott. (B) Snowboarding on a piste. Photo by Tim Stott. (C) Snowboarders in a play park. Photo by Tim Stott

Fig. 11.6 Modern snowshoes made from plastic have bindings, heel lift, and small spikes on the underside for grip on hard snow. Photo by Tim Stott

are made of materials such as lightweight metal, plastic, and synthetic fabric (Fig. 11.6). In addition to distributing the weight, snowshoes are generally raised at the toe for manoeuvrability. They must not accumulate snow, hence the latticework, and require bindings to attach them to the feet.

In the past, snowshoes were essential tools for fur traders, trappers, and anyone whose life or living depended on the ability to get around in areas of deep and frequent snowfall, and they remain necessary equipment for forest rangers and others who must be able to get around areas inaccessible to motorised vehicles when the snow is deep. However, snowshoes are mainly used today for recreation, primarily by hikers and runners who like to continue their hobby in wintertime. Snowshoeing is easy to learn and in appropriate conditions is a relatively safe and inexpensive recreational activity.

- *Cross-country*: the sport encompasses a variety of formats for cross-country skiing races over courses of varying lengths. Such races occur over homologated, groomed courses designed to support classic (in-track) and freestyle events, where the skiers may employ skate skiing. It also encompasses cross-country ski marathon events, sanctioned by the Worldloppet Ski Federation, and cross-country ski orienteering events, sanctioned by the International Orienteering Federation, and biathlon, a combination of cross-country and shooting.
- *Ski jumping*: contested at the Olympics, the FIS Ski Jumping World Cup, the summer FIS Grand Prix Ski Jumping, and the FIS Ski-Flying World Championships.
- *Nordic combined*: contested at the Olympics and at the FIS Nordic Combined World Cup, it is a combination of cross-country skiing and ski jumping.
- *Alpine skiing* disciplines include combined, downhill, slalom, giant slalom, super-G, and para-alpine. There are also combined events that include two events, one run of each event like one run of super-G and one run of slalom skiing, called a super combined.
- *Speed skiing* dates from 1898 with official records being set as of 1932 with an 89-mile-per-hour (143 km/h) run by Leo Gasperi. It became an FIS sport in the 1960s and a demonstration Olympic sport at the 1992 Winter Olympics in Albertville.
- *Freestyle skiing*: includes mogul skiing, aerials, ski cross, half-pipe, and slopestyle.
- *Snowboard* competition includes slopestyle, cross, half-pipe, alpine, parallel slalom, and parallel giant slalom.
- Other competition includes grass skiing and *telemark*.
- *Skiboarding*: consists of a combination of skiing and snowboarding. It uses ski boots with a snowboard.

11.2 Snow Sport Competition

The following disciplines are sanctioned by the FIS. Many have their own world cups and are in the Winter Olympic Games.

11.3 Participation Numbers

In the USA, during the 2016 calendar year, a total of 24,134 online interviews were carried out with a nationwide sample of individuals and households from the US Online Panel of over one million people operated by Synovate/IPSOS (Outdoor Foundation 2017). A total of 11,453 individual and 12,681 household surveys were completed. The total panel is maintained to be representative of the US population for people ages six and older. Over sampling of ethnic groups took place to boost response from typically under responding groups. The 2016 participation survey sample size of 24,134 completed interviews provides a high degree of statistical accuracy.

As can be seen in Table 11.1, The Outdoor Foundation (2017) survey data for the USA show that, of the six snow sports disciplines measured in the survey:

- Alpine/downhill skiing had the greatest number of participants in 2016 (9,267,000) and a 12.4% increase over the previous three years (2014–2016),
- Snowboarding had the second greatest number of participants in 2016 (7,602,000), showing a 3.4% increase over the previous three years.
- Cross-country skiing had 4,640,000 participants but the greatest three-year increase of 40.3%.
- Freestyle skiing with 4,640,000 participants in 2016 had the same number participating as cross-country skiing but with only a 2.7% three-year increase.
- Telemark (downhill) skiing has the smallest participation numbers in the six disciplines surveyed (2,848,000) and a 3.0% three-year increase.
- Snowshoeing had 3,533,000 participants but showed a −12.3% decrease in participation in the 2014–2016 period.

Cordell's (2012) survey showed snowboarding (Table 11.2) as having 12,200,000 participants in the 2005–2009 period with a 33.7% increase between the 1999–2001 and 2005–2009 periods. Comparing that with the Outdoor Foundation's (2017) data, it appears that snowboarding had declined after it peaked in 2010 at 8,196,000.

Table 11.2 also shows that both cross-country skiing and snowshoeing showed declines of −21.7% and −9.4%, respectively, in the numbers participating between the 1999–2001 and 2005–2009 periods

Cordell (2012) stated that "across the demography of Americans generally, just over 11% participated in some form of snow skiing or boarding in 2005–2009. Participation rates are high relative to the general population for males, non-Hispanic Whites, people ages 16–34 (especially those under age 25), people with college to postgraduate education, people earning more than $75,000 annually, and urban residents (Table 11.3). Less likely than the population to participate in snow skiing or boarding activities are females, Blacks, Native Americans, people over 55 years of age, those lacking college degrees, people with low incomes, and rural residents" (Cordell 2012, p. 65).

Bowker et al. (2012) projected changes in total outdoor recreation participants between 2008 and 2060 (Table 11.4). There was an estimated 24 million participants in 2008 taking part in developed skiing (downhill skiing and snowboarding), and this was predicted to become 21–23 million by 2060. For undeveloped skiing (cross-country skiing and showshoeing), there was an estimated eight million participants in 2008 taking part), and this was predicted to decline to one to four million by 2060.

Table 11.5 shows the changes over the same period for the total number of outdoor recreation days. There was an estimated 178 million days in 2008 where people took part in developed skiing (downhill skiing and snowboarding), and this was predicted to become 165–179 million days by 2060. For undeveloped skiing (cross-country skiing and showshoeing), there was an estimated 52 million days in 2008 where people took part, and this was predicted to decline to 5–29 million days by 2060 (Bowker et al. 2012).

In England, data are available from the Sport England Active People Survey (APS). For this survey Sport England defined snow sport to

Table 11.1 Outdoor participation by activity (ages 6+) in the USA, 2006–2016 (The Outdoor Foundation 2017, p. 8)

	2006	2007	2008	2009	2010	2011	2012	2013	2014	2015	2016	3-year change (%)
Adventure racing	725	698	920	1089	1339	1065	2170	2213	2368	2864	2999	35.5
Backpacking overnight >¼ mile from vehicle/home	7076	6637	7867	7647	8349	7095	8771	9069	10,101	10,100	10,151	11.9
Bicycling (BMX)	1655	1887	1904	1811	2369	1547	2175	2168	2350	2690	3104	43.2
Bicycling (mountain/non-paved surface)	6751	6892	7592	7142	7161	6816	7714	8542	8044	8316	8615	0.9
Bicycling (roads/paved surface)	38,457	38,940	38,114	40,140	39,320	40,349	39,232	40,888	39,725	38,280	38,365	−6.2
Birdwatching (more and ¼ mile from home/vehicle)	11,070	13,476	14,399	13,294	13,339	12,794	14,275	14,152	13,179	13,093	11,589	−18.1
Boardsailing/windsurfing	938	1118	1307	1128	1617	1151	1593	1324	1562	1766	1737	31.2
Camping (RV)	16,946	16,168	16,517	17,436	15,865	16,698	15,108	14,556	14,663	14,699	15,855	8.9
Camping (with ¼ mile of home/vehicle)	35,618	31,375	33,686	34,338	30,996	32,925	29,982	29,269	28,660	27,742	26,467	−9.6
Canoeing	9154	9797	9935	10,058	10,553	9787	9839	10,153	10,044	10,236	10,046	−1.1
Climbing (sports/indoor/boulder)	4728	4514	4769	4313	4770	4119	4592	4745	4536	4684	4905	3.4
Climbing (traditional/ice/mountaineering)	1586	2062	2288	1835	2198	1609	2189	2319	2457	2571	2790	20.3
Fishing (fly)	6071	5756	5941	5568	5478	5683	6012	5878	5842	6089	6456	9.8
Fishing (freshwater/other)	43,100	43,859	40,331	40,961	38,860	38,868	39,135	37,796	37,821	37,682	38,121	0.9
Fishing (saltwater)	12,466	14,437	13,804	12,303	11,809	11,983	12,017	11,790	11,817	11,975	12,266	4.0
Hiking (day)	29,863	29,965	32,511	32,572	32,496	34,491	34,545	34,378	36,222	37,232	42,128	22.5
Hunting (bow)	3875	3818	3722	4226	3908	4633	4075	4079	4411	4564	4427	8.5
Hunting (handgun)	2525	2595	2873	2276	2709	2671	3553	3198	3091	3400	3512	9.8
Hunting (rifle)	11,242	10,635	10,344	11,114	10,150	10,807	10,164	9792	10,081	10,778	10,797	10.3
Hunting (shotgun)	8987	8545	8731	8490	8062	8678	8174	7894	8220	8438	8271	4.8
Kayak fishing	n/a	n/a	n/a	n/a	1044	1201	1409	1798	2074	2265	2371	31.8

(continued)

Table 11.1 (continued)

	2006	2007	2008	2009	2010	2011	2012	2013	2014	2015	2016	3-year change (%)
Kayaking (recreational)	4134	5070	6240	6212	6465	8229	8144	8716	8855	9499	10,017	14.9
Kayaking (sea/touring)	1136	1485	1780	1771	2144	2029	2446	2694	2912	3079	3124	16.0
Kayaking (white water)	828	1207	1242	1369	1842	1546	1878	2146	2351	2518	2552	18.9
Rafting	3609	3786	4226	4342	3869	3725	3958	3915	3924	4099	4095	−10.6
Running/jogging	38,559	41,064	41,130	43,892	49,408	50,713	52,187	54,188	51,127	48,496	47,384	−12.6
Sailing	3390	3786	4226	4342	3869	3725	3958	3915	3924	4099	4095	4.6
Scuba diving	2965	2965	3216	2723	3153	2579	2982	3174	3145	3274	3111	−2.0
Skateboarding	10,130	8429	7807	7352	6808	5827	6627	6350	6582	6436	6442	1.5
Skiing (alpine/downhill)	**n/a**	**10,362**	**10,346**	**10,919**	**11,504**	**10,201**	**8243**	**8044**	**8649**	**9378**	**9267**	**12.4**
Skiing (cross-country)	**n/a**	**3530**	**3848**	**4157**	**4530**	**3641**	**3307**	**3377**	**3820**	**4146**	**4640**	**40.3**
Skiing (freestyle)	**n/a**	**2817**	**2711**	**2950**	**3647**	**4318**	**5357**	**4007**	**4564**	**4465**	**4640**	**2.7**
Snorkelling	8395	9294	10,296	9358	9305	9318	8011	8700	8752	8874	8717	0.2
Snowboarding	**n/a**	**6841**	**7159**	**7421**	**8196**	**7579**	**7351**	**6418**	**6785**	**7676**	**7602**	**3.4**
Snowshoeing	**n/a**	**2400**	**2922**	**3431**	**3823**	**4111**	**4029**	**3012**	**3501**	**3885**	**3533**	**−12.3**
Stand up paddling	n/a	n/a	n/a	n/a	1050	1242	1542	1993	2751	3020	3220	61.6
Surfing	2170	2206	2607	2403	2767	2195	2895	2658	2721	2701	2793	3.0
Telemarking (downhill)	**n/a**	**1173**	**1435**	**1482**	**1821**	**2099**	**2766**	**1732**	**2188**	**2569**	**2848**	**3.0**
Trail running	4558	4216	4857	4833	5136	5610	6003	6792	7531	8139	8582	26.4

Note: All participation numbers are in thousands (000)

11.3 Participation Numbers

Table 11.2 Trends in number of people of ages 16 and older participating in recreation activities in the USA, 1999–2001 and 2005–2009 for activities with fewer than 15 million participants from 2005 through 2009 (Source: Cordell 2012, p.40)

	Total participants (*millions*)			Percent participating	Percent change
	1994–1995	1999–2001	2005–2009	2005–2009	1999–2001 to 2005–2009
Kayaking	3.4	7.0	14.2	6.0	103.8
Mountain climbing	9.0	13.2	12.4	5.3	−5.9
Snowboarding	**6.1**	**9.1**	**12.2**	**5.2**	**33.7**
Ice skating outdoors	14.2	13.6	12.0	5.1	−11.5
Snowmobiling	9.6	11.3	10.7	4.5	−5.5
Anadromous fishing	11.0	8.6	10.7	4.5	24.1
Sailing	12.1	10.4	10.4	4.4	−0.4
Caving	9.5	8.8	10.4	4.4	18.4
Rock climbing	7.5	9.0	9.8	4.2	9.5
Rowing	10.7	8.6	9.4	4.0	8.9
Orienteering	4.8	3.7	6.2	2.6	−21.7
Cross-country skiing	**8.8**	**7.8**	**6.1**	**2.6**	**−21.7**
Migratory bird hunting	5.7	4.9	4.9	2.1	−1.1
Ice fishing	4.8	5.7	4.8	2.1	−15.5
Surfing	2.9	3.2	4.7	2.0	46.3
Snowshoeing	–	**4.5**	**4.1**	**1.7**	**−9.4**
Scuba diving	–	3.8	3.6	1.5	−5.6
Windsurfing	2.8	1.5	1.4	0.6	−10.1

Missing data are denoted with "–" and indicate that participation data for that activity were not collected during that time period. Percent change was calculated before rounding
Source: USDA Forest Service (1995) ($n = 17,217$), USDA Forest Service (2001) ($n = 52,607$), and USDA Forest Service (2009) ($n = 30,398$)
Note: The numbers in this table are *annual* participant estimates on data collected during the three time periods
1994–1995 participants based on 201.3 million people of ages 16+ (Woods & Poole Economics, Inc. 2007)
1999–2001 participants based on 214.0 million people of ages 16+ (U.S. Department of Commerce 2000)
2005–2009 participants based on 235.3 million people of ages 16+ (U.S. Department of Commerce 2008)

include Alpine skiing, freestyle skiing, Nordic skiing, and snowboarding. Table 11.6 shows that there were an estimated 127,400 participants (0.31% of the population of England) in the October 2005–October 2006 survey period. This had declined to 99,800 (0.23% of the population) by the October 2016–October 2066 survey period.

Figure 11.7 displays the number of people who ski per country in Europe as of 2016. In 2016, there were approximately 1.1 million people in the Ukraine who skied. Germany had the highest number of ski participants with a total of approximately 14.6 million, followed by France with 8.6 million and the UK with 6.3 million. This estimate is clearly significantly higher than the Sport England APS (Table 11.6). When considering the total population, the share of people who ski in European countries was the highest in Switzerland with 37% and in Austria with 36%.

Clearly there are some discrepancies in the various estimates for snow sport, but from Table 11.1 (The Outdoor Foundation 2017) we can see that in the US snow sport activities fall behind running/jogging (47,384,000), hiking (42,128,000), road cycling (38,365,000), fishing (38,121,000), camping (26,467,000), hunting/canoeing/kayaking (all around the 10 million mark), then follows Alpine skiing (9,267,000).

Table 11.3 Percentage of participants and population, rations of percentages, and statistical test results for the activity group snow skiing or boarding

Demographic	Stratum	Percent of participants	Percent of nation	Ratio (1)/(2)	Percent participating
All groups	All people age 16 and older	100.0	100.0	1.00	11.2
Gender*	Male*	63.0	48.2	1.31	14.5
	Female*	37.0	51.8	0.71	8.1
Race/ethnicity*	White, non-Hispanic*	75.9	67.3	1.13	12.7
	Black, non-Hispanic*	5.5	13.9	0.40	4.2
	American Indian, non-Hispanic	0.4	0.8	0.50	6.5
	Asian or Pacific Islander, non-Hispanic	3.6	3.6	1.00	11.3
	Hispanic	14.6	14.4	1.01	11.5
Age*	16–24*	38.8	15.8	2.46	27.6
	25–34*	18.6	16.2	1.15	13.0
	35–44	17.7	16.9	1.05	12.1
	45–54*	16.5	17.6	0.94	10.4
	55–64*	5.7	13.6	0.42	4.6
	65+*	2.7	20.0	0.14	1.5
Education*	Less than high school**	21.3	24.0	0.89	9.9
	High school graduate*	19.4	26.9	0.72	8.1
	Some college***	24.6	26.8	0.92	10.3
	College degree*	21.5	14.4	1.49	16.7
	Postgraduate degree*	13.2	7.9	1.67	18.7
Annual family income*	<$15,000*	9.3	16.5	0.56	6.6
	$15,000–$24,999*	5.7	11.4	0.50	5.8
	$25,000–$49,999*	18.7	27.4	0.68	7.9
	$50,000–$74,999***	19.0	18.3	1.04	12.5
	$75,000–$99,999*	14.5	11.1	1.31	14.5
	$100,000–$149,999*	18.0	9.4	1.91	21.5
	$150,000+*	14.8	6.0	2.47	27.6
Place of residence*	Nonmetro resident*	13.0	17.5	0.74	8.4
	Metro area resident**	87.0	82.5	1.05	11.8
Residence status	Native born or US citizen	96.3	96.7	1.00	11.2
	Foreign born	3.7	3.3	1.12	11.7

Source: Cordell (2012, p. 69); USDA Forest Service (2009), Versions 1–4 (N = 14,070). Interview dates: 1/05 to 4/09
Note: Test statistic in the "Demographic" column is chi-square goodness of fit which tests independent of the observed proportions in the categories of each demographic group. Test statistic in the "Stratum" column are binomial tests of significance between the stratum participation rate ("Percent participating") and the participation rate for all people age 16 and older shown in line 1. Significance levels indicated by: *$p < 0.01$, ** $p < 0.05$, *** $p < 0.10$
Percentages sum down to 100 within each demographic group in the first two columns—may not sum to 100% exactly due to rounding. In fourth column, compare stratum percent to the percent participating for all respondents in line 1. Sample sizes may vary by activity because not all activities were asked in every NSRE version

11.4 Environmental Impact

As in previous chapters, it is reasonable to examine the impacts of snow sport on the environment in three categories: impacts on soils and vegetation, impacts on wildlife, impacts on water resources.

11.4.1 Impacts of Snow Sport on Soil and Vegetation

The rising popularity of Alpine skiing and snowboarding has raised concerns of potential environmental impacts. A considerable amount of research has been carried out to assess the impact of the development of a relatively small (by world

11.4 Environmental Impact

Table 11.4 Changes in total outdoor recreation participants between 2008 and 2060 across all activities and scenarios (Source: Bowker et al. 2012, p. 28)

Activity[a]	2008 Participants[b] (millions)	2060 Participant range[c] (millions/[percent])	2060 Average participant change[c] (millions)	2060 Participant range[d] (millions/[percent])	2060 Average participant change[d] (millions)
Visiting developed sites					
Developed site use—family gatherings, picnicking, developed camping	194	273–346 [42–77]	116	271–339 [40–75]	112
Visiting interpretive sites—nature centres, zoos, historic sites, prehistoric sites	158	231–294 [48–84]	106	231–289 [46–83]	104
Viewing and photographic nature					
Birding—viewing/photographing birds	82	118–149 [42–76]	53	115–144 [40–76]	47
Nature viewing—viewing, photography, study, or nature gathering related to fauna, flora, or natural settings	190	267–338 [42–76]	114	268–333 [41–75]	112
Backcountry activities					
Challenge activities—caving, mountain biking, mountain climbing, rock climbing	25	38–48 [50–86]	19	37–48 [47–90]	18
Equestrian	17	24–31 [44–87]	11	25–35 [50–110]	13
Hiking—day hiking	79	117–150 [50–88]	55	114–143 [45–82]	50
Visiting primitive areas—backpacking, primitive camping, wilderness	91	120–152 [34–65]	47	119–145 [31–60]	42
Motorised activities					
Motorised off-road use	48	62–75 [29–56]	21	62–76 [28–58]	21
Motorised snow use (snowmobiles)	10	10–13 [10–37]	3	4–10 [(56)–6]	(2.5)[e]
Motorised water use	62	87–112 [41–81]	40	84–111 [35–78]	35
Consumptive					
Hunting—all types of legal hunting	28	30–34 [8–23]	5	29–34 [5–21]	4
Fishing—anadromous, cold-water, saltwater, warm water	73	92–115 [28–56]	33	89–115 [22–58]	30
Non-motorised winter activities					
Downhill skiing, snowboarding	**24**	**38–54 [58–127]**	**23**	**36–54 [50–126]**	**21**
Undeveloped skiing—cross-country, snow-shoeing	**8**	**10–13 [32–67]**	**4**	**5–10 [(42)–28]**	**(1)**
Non-motorised water					
Swimming, snorkelling, surfing, diving	144	210–268 [47–85]	99	212–266 [47–85]	99
Floating—canoeing, kayaking, rafting	40	52–65 [30–62]	20	47–62 [18–56]	13

Source: National Survey of Recreation and the Environment (NSRE) 2005–09, Versions 1 to 4 (January 2005 to April 2009). $n = 24{,}073$ (USDA Forest Service 2009)

[a]Activities are individual or activity composites derived from the NSRE. Participants are determined by the product of the average weighted frequency of participation by activity for NSRE data from 2005 to 2009 and the adult (>16) population in the USA during 2008 (235.4 million)

[b]Because of small population and income differences, initial values for 2008 differ across PRA scenarios, thus an average is used for a starting value

[c]Participant range across Resources Planning Act (RPA) scenarios A1B, A2, and B2, without climate considerations

[d]Participant range across RPA scenarios A1B, A2, and B2, each with three selected climate futures

[e]Parentheses denote negative number

Table 11.5 Changes in total outdoor recreation days between 2008 and 2060 across all activities and scenarios (Source: Bowker et al. 2012, p. 29)

Activity[a]	2008 Days[b] (millions)	2060 Days range[c] (millions/[percent])	2060 Average days change[c] (millions)	2060 Days range[d] (millions/[percent])	2060 Average days change[d] (millions)
Visiting developed sites					
Developed site use—family gatherings, picnicking, developed camping	2246	3121–3949 [40–74]	1294	3055–3796 [36–69]	1185
Visiting interpretive sites—nature centres, zoos, historic sites, prehistoric sites	1249	1899–2417 [53–91]	952	1935–2435 [55–95]	988
Viewing and photographic nature					
Birding—viewing/photographing birds	8255	11,680–14,322 [40–74]	4859	10,050–13,313 [36–69]	3764
Nature viewing—viewing, photography, study, or nature gathering related to fauna, flora, or natural settings	32,461	41,805–52,835 [31–61]	14,635	41,550–51,288 [28–58]	13,597
Backcountry activities					
Challenge activities—caving, mountain biking, mountain climbing, rock climbing	121	178–219 [49–83]	4859	179–232 [48–92]	89
Equestrian	263	388–503 [49–92]	196	369–482 [40–83]	166
Hiking—day hiking	1835	2901–3682 [59–98]	1470	2825–3541 [54–93]	1366
Visiting primitive areas—backpacking, primitive camping, wilderness	1239	2046	622	1562–1946 [26–57]	519
Motorised activities					
Motorised off-road use	1053	1264–1532 [21–46]	357	1274–1611 [21–53]	385
Motorised snow use (snowmobiles)	69	74–91 [8–33]	16	23–65 [(6)–(67)]	(27)[e]
Motorised water use	958	1304–1806 [37–90]	596	1245–1763 [30–84]	495
Consumptive					
Hunting—all types of legal hunting	538	506–576 [(5)–8]	14	494–575 [(8)–7]	(8)
Fishing—anadromous, cold-water, saltwater, warm water	1369	1665–2020 [23–46]	514	1602–1958 [17–41]	397
Non-motorised winter activities					
Downhill skiing, snowboarding	**178**	**274–437 [61–150]**	**179**	**258–422 [50–146]**	**165**
Undeveloped skiing—cross-country, snow-shoeing	**52**	**69–87 [35–70]**	**29**	**28–64 [(45)–25]**	**(5)**
Non-motorised water					
Swimming, snorkelling, surfing, diving	3476	5037–6429 [46–83]	2446	4396–6257 [42–80]	2298
Floating—canoeing, kayaking, rafting	262	338–422 [30–62]	128	309–409 [18–56]	83

Source: National Survey of Recreation and the Environment (NSRE) 2005–2009, Versions 1 to 4 (January 2005 to April 2009). $n = 24{,}073$ (USDA Forest Service 2009)

[a]Activities are individual or activity composites derived from the NSRE. Participants are determined by the product of the average weighted frequency of participation by activity for NSRE data from 2005 to 2009 and the adult (>16) population in the USA during 2008 (235.4 million)
[b]Because of small population and income differences, initial values for 2008 differ across PRA scenarios, thus an average is used for a starting value
[c]Participant range across Resources Planning Act (RPA) scenarios A1B, A2, and B2, without climate considerations
[d]Participant range across RPA scenarios A1B, A2, and B2, each with three selected climate futures
[e]Parentheses denote negative number

11.4 Environmental Impact

Table 11.6 Once a week participation in funded sports (16 years and over)—Sport England: Active People Survey 10 (October 2015–September 2016) (Source: https://www.sportengland.org/media/11746/1x30_sport_16plus-factsheet_aps10.pdf)

Sport England NGB 13-17 Funded sports	Active People Survey 10 (October 2015–September 2016)	
	%	n
Swimming	5.67	2,516,700
Athletics	5.01	2,217,800
Cycling	4.40	1,950,300
Football	4.21	1,844,900
Golf	1.64	729,300
Exercise, movement, & dance	0.98	437,200
Badminton	0.97	425,800
Tennis	0.90	398,100
Equestrian	0.64	282,400
Bowls	1.33	211,900
Squash & racquetball	0.45	199,500
Rugby union	0.46	199,000
Netball	0.42	180,200
Boxing	0.36	159,000
Cricket	0.36	158,500
Basketball	0.35	150,800
Mountaineering	0.25	110,200
Table tennis	0.24	107,100
Angling	0.24	106,200
Snowsport	**0.23**	**99,800**
Hockey	0.22	92,700
Weightlifting	0.20	88,100
Rowing	0.19	83,400
Gymnastics	0.15	65,100
Shooting	0.13	56,600
Sailing	0.10	45,600
Rugby league	0.10	44,900
Canoeing	0.09	41,900
Volleyball	0.08	33,800
Archery	0.07	32,400
Taekwondo	0.06	23,900
Judo	0.04	18,900
Rounders	0.03	12,800

Source: Sport England's Active People Survey

standards) ski resort in the northern Cairngorm mountains in Eastern Scotland. Since 1960, the construction of chairlifts, ski tows, and buildings on Cairngorm has resulted in the exposure of areas of mineral soil. Watson et al. (1970) reported that the building of new roads and ski lifts, and a consequent increase of human traffic in summer and winter, had damaged vegetation and soils on mountain tundra in Scotland. This made an eyesore in a tourist area of high environmental quality and led to erosion which is a serious potential threat to the roads and ski lifts themselves. Watson et al. (1970) carried out field surveys and experimental studies of erosion, compaction, vegetation damage, and the use of paths, as well as assessments of the success of various methods of rehabilitation on different kinds of substrate. In places accelerated erosion (since 1964) had occurred and surface material had been deposited in fans downslope, often onto otherwise undamaged vegetation. Later, Bayfield (1974) found that the extent of accelerated ero-

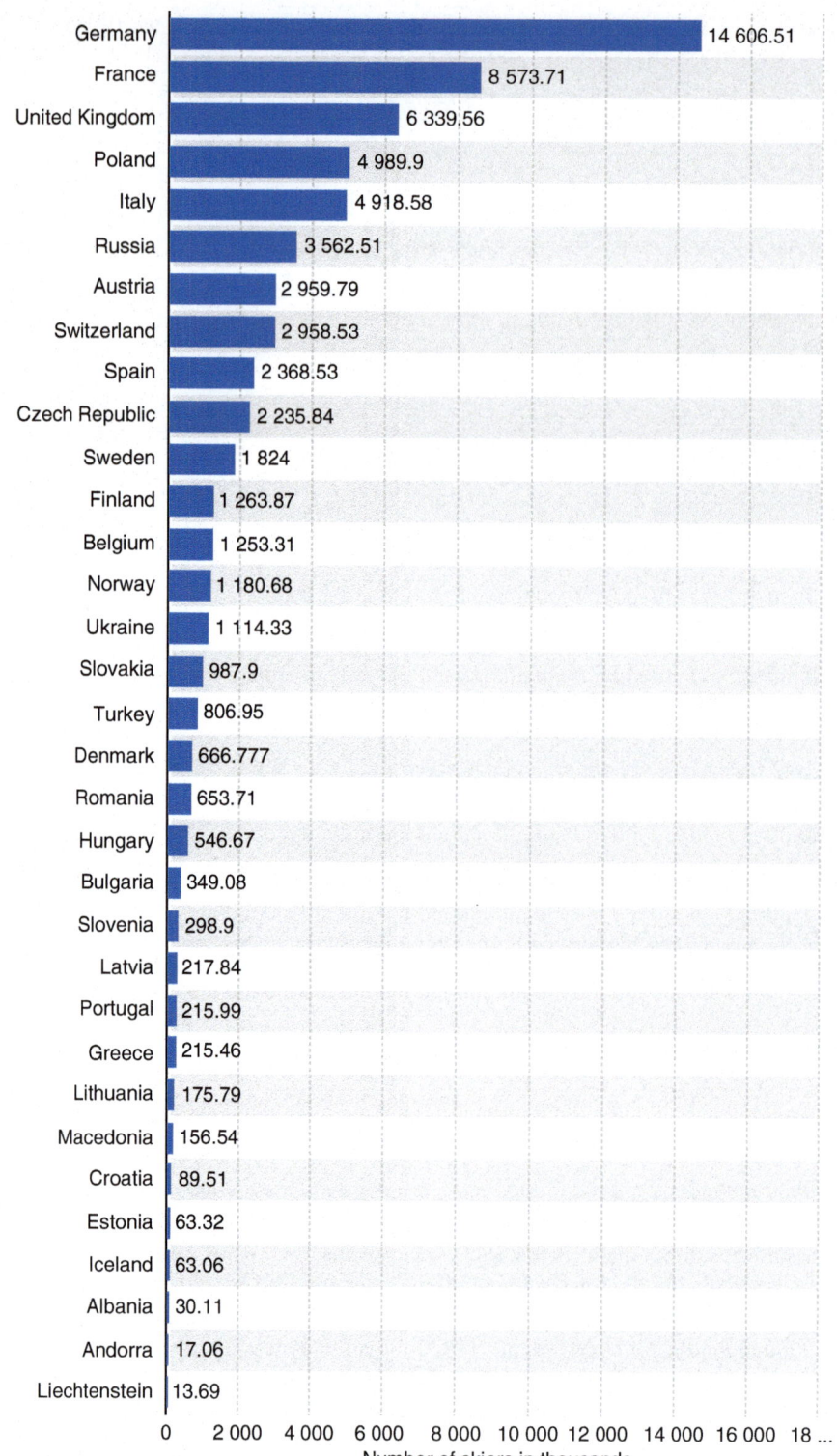

Fig. 11.7 Number of people who ski in Europe as of 2016, by country (in 1000). Source: https://www.statista.com/statistics/660546/europe-number-of-people-skiing-by-country/, accessed 20 March 2018

sion from ground damaged near ski lifts reached a peak in about 1969, with a marked decline since then. This decline was attributable to reseeding of damaged ground, provision of drains, and grading of dirt roads. Burial experiments showed that where erosion debris had covered vegetation, recovery at best took several years and with depths above about 7 cm, was almost negligible.

Bayfield (1980) reported that areas of disturbed ground up to 1100 m in altitude, used for skiing in the Cairngorm Mountains, Scotland, and the verges of a new road to the ski area, had been seeded by the chairlift operators and the roads authorities, between 1966 and 1968. The establishment of the seed mixtures, and the subsequent invasion by self-sown species, was followed from 1969 to 1976. On the lower ground (up to 850 m) invasion by heather (*Calluna vulgaris*) was very successful, but on higher ground colonisation by indigenous species was poor. Bryophytes, however, were successful at all altitudes, producing about 20% cover after one year and 50% or more after eight years. Most colonising vascular plants were also present in the surrounding undisturbed vegetation, but there were also a number of opportunist species. These survived better than the local species in places such as road margins, where disturbance continued. Lowland (60–210 m) turves transplanted satisfactorily to altitudes up to 1200 m but did not increase much in size except at the lowest sites (650–690 m).

Watson (1985) conducted a survey at Cairngorm during 1981 which showed severe damage extending on to the adjacent plateau well inside the Cairngorms National Nature Reserve. It was distinguished from natural damage by diagnostic features associated with human footprints. Areas visited by many people showed more plant damage and soil erosion than areas seldom visited. Disturbed land covered 403 ha, 17% of it in the Reserve. Disturbed land had a higher proportion of grit lying on vegetation than undisturbed land, a lower proportion of ground covered by vegetation, a higher proportion of damaged vegetation, and a higher frequency of plant burial, rill erosion, and dislodged stones and soil. Disturbed land had less bilberry, least willow, ground lichens and mosses, and other species besides grasses, sedges, and rushes. On slopes of 15–29°, foot-slipping increased with slope gradient on disturbed but not undisturbed land. Disturbed soil had less water, fine particles, and organic matter.

Bayfield (1996) examined the long-term effects of grass seeding at the Cairngorm ski area on the colonisation of bulldozed ground by native species. He monitored the colonisation of three bulldozed pistes (ski runs) on Cairngorm over 25 years. Two pistes were seeded and fertilised at the time of construction and the third was left unsown. By the end of the study, the seeded ground blended well with the surrounding ground, but the unsown piste remained visually conspicuous because of the high proportion of bare ground (>60%). Cover on seeded ground was mainly sown grasses and mosses for the first nine years, but after that the cover of sown grasses declined whereas moss cover peaked after 18 years. Cover of local vascular plant species gradually increased, and after 25 years it exceeded that of the sown grasses. Vascular species made up 21% of the total at 1180 m and 32% at 1000 m. On unseeded ground, vegetation cover was much lower than on seeded ground on every occasion. Mosses, grasses, and forbs tended to be more prevalent at seeded sites than on intact ground. Some characteristic species of intact ground, such as *Empetrum nigrum* and *Carex bigelowii*, were uncommon on seeded ground. Most local vascular species were more effective colonists of seeded than of untreated ground. An exception was *Juncus trifidus*, which was more successful on unsown ground. Some sown species had persisted for 25 years and might take another 10–15 years or longer to disappear. It seems likely that the vegetation of disturbed ground will remain botanically distinct from that of the surroundings because of ineffective colonisation by certain key species and because of the influence of late snowlie. Bayfield concluded that grass seeding substantially enhanced colonisation by native species.

Snow grooming is the process of manipulating snow for recreational uses with a tractor, snowmobile, piste caterpillar, truck, or snowcat towing specialised equipment. The process is used to maintain ski hills, cross-country ski trails, and snowmobile trails by grooming (moving, flattening, or compacting) the snow on them. A snow groomer is usually employed to pack snow and improve skiing and snowboarding and snowmobile trail conditions. The resulting pattern on the snow is known as corduroy and is widely regarded as a good surface on which to ski or ride. Snow groomers can also move accumulated snow made by snow machines as part of a process, called "snow farming." Skiing at the end of season (Fig. 11.8A) or on thin snow cover (Fig. 11.8B) can result in damage to soil and vegetation from the passing of the sharp edges of skis.

Fahey et al. (1999) investigated the effect of snow grooming on snow properties at Treble Cone Ski Field north of Queenstown in South Island, New Zealand. Snow depth, density, equivalent water content, and hardness were monitored in the NZ winter along transects at five non-groomed and four groomed slopes in late July, late August, and late September 1997. Average densities measured for transects on groomed slopes were 36% higher than those on non-groomed slopes. There was 45% more water available on average from the snowpack on groomed slopes than on non-groomed slopes. Snow hardness was 400% higher across groomed transects in late July but only 40% higher in late September. Increases in snow density and hardness attributed to snow grooming are similar to those observed overseas but lie at the low end of the range. They are probably sufficient to inhibit or delay soil bacterial activity and subsequent litter decomposition.

The production of artificial snow and the use of snow additives in ski resorts have increased considerably during the last two to three decades. The impact on the environment provokes concern. Rixen et al. (2003) compiled a review of studies about the ecological implications of preparing and grooming ski-pistes in general and of artificial snow production in particular. In common with Fahey et al. (1999), they found that the main direct impacts of ski-piste preparation on the vegetation were related to the compaction of the snow cover, namely, the induction of soil frost, the formation of ice layers, mechanical damage, and a delay in plant development. The vegetation reacts with changes in species composition and a decrease in biodiversity. Artificial snowmaking modifies some of these impacts: the soil frost is mitigated due to better insulation of the snowpack, whereas the formation of ice layers is not considerably changed.

The mechanical impacts of snow-grooming vehicles are reduced after artificial snowmaking due to the deeper snow cover. The delay of the vegetation development is enhanced by a considerably postponed snowmelt. Furthermore, artificial snowmaking induces new impacts to the alpine environment which include the input of water and ions to ski-pistes, which can have a fertilising effect and hence change the plant species

Fig. 11.8 (A) End-of-season skiing at Val Thorens, France, shows some damage to the soil and vegetation on runs and how artificial snowmaking can be used to extend the ski season at lower altitudes. Photo by Tim Stott. (B) Skiing on thin snow cover at Lecht Ski Resort in Scotland can result in damage to soil and vegetation around the edges of runs. Photo by Tim Stott

composition. Increasingly, snow additives, made of potentially phytopathogenic bacteria (Lagriffoul et al. 2010), are used for snow production. They enhance ice crystal formation due to their ice nucleation activity. Although sterilised, additives are reported to have affected the growth of some alpine plant species in laboratory experiments. Salts are applied not only but preferably on snowed pistes to improve the snow quality for ski races. The environmental impacts of most salts have not yet been investigated, but a commonly used nitrate salt has intense fertilising properties. Although snowmaking reduces some of the negative impacts of ski-piste preparation in general, new impacts induced by snowmaking might be non-beneficial to the vegetation.

Following their review, Rixen et al. (2004) measured snow depth and density from groomed ski-pistes with compacted snow and their effects on ground temperatures and timing of snowmelt. They analysed groomed pistes with and without artificial snow (ten each) as well as adjacent ungroomed off-piste control plots beside the piste. On pistes with natural snow, the thin and compacted snow cover led to severe and long-lasting seasonal soil frost. On pistes with artificial snow, soil frost occurred less frequently because of increased insulation due to the greater snow depth. However, due to the greater snow mass, the beginning of the snow-free season was delayed by more than two weeks. Average winter ground temperatures under a continuous snow cover were decreased by approximately 1 °C on both piste types compared with off-piste control plots. The results suggest that the heat balance of alpine soils is changed by both piste types, either by an extensive heat loss on pistes with natural snow or by prolonged snow cover on pistes with artificial snow.

Wipf et al. (2005) investigated the effects of ski-piste management on vegetation structure and composition in 12 Swiss alpine ski resorts using a pairwise design of 38 plots on ski-pistes and 38 adjacent plots off-piste. Plots on ski-pistes had lower species richness and productivity, and lower abundance and cover of woody plants and early flowering species, than reference plots. Plots on machine-graded pistes had higher indicator values for nutrients and light and lower vegetation cover, productivity, species diversity, and abundance of early flowering and woody plants. Time since machine-grading did not reduce the impacts of machine-grading, even for those plots where revegetation had been attempted by sowing. The longer artificial snow had been used on ski-pistes (2–15 years), the higher the moisture and nutrient indicator values. Longer use also affected species composition by increasing the abundance of woody plants, snowbed species, and late-flowering species and decreasing wind-edge species. Wipf et al. (2005) concluded that all types of ski-piste management cause deviations from the natural structure and composition of alpine vegetation and lead to lower plant species diversity. Machine-grading causes particularly severe and lasting impacts on alpine vegetation, which do not improve with time or by revegetation measures. The impacts of artificial snow increase with the period of time since it was first applied to ski-piste vegetation. Extensive machine-grading and snow production should be avoided, especially in areas where nutrient and water input are a concern. They recommended that ski-pistes should not be established in areas where the alpine vegetation has a high conservation value.

Rolando et al. (2007) stated that treeless mountainous areas at high altitudes have increased in value as wildlife habitat but have been increasingly threatened by ski-resort developments, in particular by the construction and enlargement of ski-pistes. They compared bird diversity and community composition in circular plots centred on (1) ski runs of recent construction, (2) grassland habitats adjacent to ski runs, and (3) natural grassland habitats far from the ski runs. Not surprisingly, plots located in natural grasslands supported the greatest bird species richness and diversity and the greatest grassland species density, whereas those set in ski-pistes presented the lowest values. Plots located beside ski runs did not support smaller numbers of bird species and diversity than plots of natural areas, but they supported a significantly lower bird density. This suggests that ski-pistes, besides exerting a negative direct effect on the structure of local bird communities, may also exert an indirect, detrimental effect on bird density in nearby patches. Generalised linear models showed that species richness and diversity, and abundance of grassland species were best modelled by combinations of factors, including habitat type (the three categories defined above) and altitude. The category ski run, in particular, was negatively correlated with species richness, diversity, and abundance, and altitude was negatively associated with richness and diversity. Richness and abundance of arthropods were significantly lower in ski-pistes than in the other plot types. Given that many invertebrates were preyed upon by birds, low food availability on ski runs may be one of the factors reducing the attractiveness of these patches to birds. Retaining the avifauna around ski resorts is likely to involve developing new, environmentally friendly ways of constructing pistes, such as only removing rocks and/or levelling the roughest ground surfaces, to preserve as much soil and natural vegetation as possible. They recommended that restoration of ski-pistes should promote the recovery and maintenance of local vegetation to enhance invertebrate and bird assemblages. However, in order to not compromise the safety of the ski runs, they conceded that it may be necessary to control encroaching shrubs through pruning and/or cattle grazing.

Barni et al. (2007) conducted their study at the Monterosa Ski Resort (Val d'Ayas, Aosta, Italy) and aimed to evaluate (1) how disturbance related to ski-run construction at high altitude (2200–2600 m a.s.l.) had affected vegetation and soil properties compared to undisturbed sites and (2) how vegetation and soil properties changed in machine-graded ski runs with increasing time after hydroseeding (a planting process that uses a slurry of seed and mulch). Herbaceous cover and specific composition, root density, physico-chemical soil properties, and aggregate stability were evaluated to determine the vegetation and soil dynamics of four runs constructed above timberline and hydroseeded 4, 6, 10, and 12 years ago, respectively, and of the adjacent undisturbed alpine pasture as control. The seeded species had quickly formed a cover that was still high even after ten years. However, cover values were always extremely low for wild species, and this could be related to their strategies and to altered soil properties (higher pH, organic matter impoverishment, and loss of both fine particles and aggregates). The study indicated that more has to be done to conserve or restore physico-chemical soil properties as a decisive factor in establishing a self-sustaining native plant community.

Burt and Rice (2009) pointed out that ski-run creation always results in some level of disturbance, but disturbance intensity varies greatly with the construction method. Ski runs may be established either by clearing (cutting and removing tall vegetation) or by clearing and then machine-grading (levelling the soil surface with heavy equipment). In order to quantify how these different intensities of initial disturbance affected ecosystem properties, they extensively surveyed vegetation, soils, and environmental characteristics on cleared ski runs, graded ski runs, and adjacent reference forests across seven large downhill ski resorts in the northern Sierra Nevada, USA. They found that the greater disturbance intensity associated with grading resulted in greater impacts on all ecosystem properties considered, including plant community composition and diversity, soil characteristics relating to processes of nutrient cycling and retention, and measures of erosion potential. They also found that cleared ski runs retained many ecological similarities to reference forests and may even offer some added benefits by possessing greater plant species and functional diversity than either forests or graded runs. Because grading is more damaging to multiple indicators of ecosystem function, they recommended that clearing rather than grading should be used to create ski slopes wherever practical.

Using the latest survey technology, Fidelus-Orzechowska et al. (2018) examined the impact of new ski developments in Poland. The purpose of their study was to quantify ongoing change patterns via: (1) a determination of spatial and quantitative changes in catchment covered by new ski runs, (2) a determination of the effect of new ski runs on the rejuvenation of relief in valleys adjacent to ski runs, and (3) an identification of changes in the surface runoff pattern before and after the construction of ski runs. The research was carried out on two ski runs in the Remiaszów catchment (southern Poland). Airborne Laser Scanning (ALS) data from 2013 and 2016 were used in the study along with Terrestrial Laser Scanning (TLS) data from 2015. LiDAR (Light Detection and Ranging) point clouds were interpolated to create multi-temporal digital elevation models (DEMs), and then by subtraction, these DEMs were used to identify erosion and accumulation zones. The largest changes in relief were observed in areas with ski runs, with one ski run lowering an average of 0.07 m (±0.03 m) and the other an average of 0.12 m (±0.03 m). By comparison the entire area lowered about 0.02 m. The construction of new ski runs therefore resulted in a rejuvenation of denudation valleys located in the vicinity of existing ski runs. Valley incisions reaching 1.5 m (±0.15 m) were observed. Both the convergence and divergence zones for surface runoff were identified, which made it possible to show changes in the geometry of flow direction. The identification of these sites may help forecast erosion and deposition zones.

de Jong et al. (2014) investigated the soil properties, compaction, and infiltration characteristics on ski slopes and compared to natural sites for three different ski resorts (Les Menuires, La Rosière, and Foppolo) in the French and Italian Alps. The results showed that soil properties differed substantially, with lower nitrogen and carbon content and higher pH on ski runs. Soil compaction was up to three times higher, and infiltration took up to four times longer on ski slopes compared to natural sites. Some new ski slopes were even 100% impermeable. This explains why ski slopes are more prone to landslides and sheet, rill, and gully erosion and have a distinct vegetation cover.

There is less research into the environmental impacts of "off-piste" skiing which includes cross-country (or Nordic) skiing, ski mountaineering, and snowshoeing. Törn et al. (2009) examined the impacts of hiking, cross-country skiing, and horse riding on trail characteristics and vegetation in Northern Finland. Widths and depths of existing trails and vegetation on trails and in the neighbouring forests were monitored in two research sites during 2001 and 2002. Trail characteristics and vegetation were clearly related to the recreational activity, research site, and forest type. Horse trails were as deep as hiking trails, even though the annual number of users was 150-fold higher on the hiking trails. Simultaneously, cross-country skiing had the least effect on trails due to the protective snow cover during winter.

Steinbauer et al. (2017) proposed that space requirements by winter sports and accelerating global warming are usually perceived as stressors for mountain meadow plant communities. Cross-country ski track preparation (i.e. grooming), however, might retard the effects of climate change and, being limited in space requirements, might increase abiotic heterogeneity. The effect of cross-country ski tracks on meadow vegetation was quantified along a representative ski track that had been operated for 30 years in the Fichtelgebirge, a low mountain range in central Europe. Paired sampling was implemented to assess the effect of skiing operations on snow and soil properties, plant phenology, biomass production, and species composition. Additionally, boosted regression tree analyses were used to quantify the relative importance of the cross-country ski track compared to other environmental conditions. In common with the aforementioned research on snow grooming on pistes, they found that the cross-country ski track strongly increased snow density, enhanced soil frost, and retarded snowmelt, thereby delaying flower phenology (by 2.1 days) and the early development stages of plant species on the track. However, biomass, species richness, and species composition were unaffected by skiing operations except for one species (*Leontodon autumnalis*) which showed exclusive occurrence on the track, while four others showed reduced relative occurrence on the track.

While snow and soil properties were influenced by cross-country ski track preparation, natural environmental variability was more influential for species composition and biomass production than the ski track. They therefore concluded that the ski track—without artificial snow—did not negatively affect species composition. By delaying flower phenology, the effects of the ski track even counteracted global warming to some degree. Due to their small spatial extent in the landscape, these ski tracks may add to environmental heterogeneity and thus support sustaining diverse species compositions during environmental changes.

11.4.2 Impacts of Snow Sport on Wildlife

Since 1960, the construction of chairlifts, ski tows, and buildings on Cairngorm has resulted in the exposure of areas of mineral soil. Watson et al. (1970) noted that these developments posed a threat to populations of animals, due to direct disturbance or indirectly to habitat change. In their 1970 paper, Watson et al. compared conditions in the Scottish mountains and in arctic North America and described a research programme by a small team studying human impact near the Scottish ski lifts. Research on animal populations showed that there had been no effect at that point on rock ptarmigan or red grouse (willow ptarmigan) populations or on dotterel and other species; however there had been less daylight use of the developed areas by deer.

In further work, Watson (1979) made counts of animals on Scottish skiing areas to find whether human impact was affecting their numbers. More people, dogs, crows, and snow buntings were seen on these disturbed areas after than before the ski developments. After the developments, however, more people and dogs were on disturbed areas than on undisturbed areas visited by very few people. He found that spring densities and breeding success of the native ptarmigan and red grouse did not differ between disturbed and undisturbed areas, and likewise neither did spring densities of meadow pipits and wheatears. Although ski-lift wires killed some ptarmigan and red grouse, this had no detectable influence on their breeding populations. At Cairngorm, more sheep, reindeer, and the native mountain hares occurred on disturbed areas; they concentrated on small patches that had been treated with grass seeds and fertiliser to reduce soil erosion. More pied wagtails, crows, rooks, gulls, and snow buntings, which fed frequently on waste human food, were seen on disturbed than on undisturbed areas, especially around car parks. Watson concluded by saying that the influx and increase of these scavenging bird species had occurred on ground adjacent to two national nature reserves on fairly natural, rare arctic-alpine habitats.

Continuing these studies at the same ski area in the Cairngorms, Watson and Moss (2004) reported adverse impacts on numbers and breeding success of ptarmigan (*Lagopus mutus*) between 1967 and 1996, where ptarmigan normally shows ten-year population cycles. An influx of carrion crows (*Corvus corone*), generalist predators, followed the ski area development. On the most developed area near the main car park, ptarmigan occurred at high density but then lost nests to frequent crows, reared abnormally few broods, died flying into ski-lift wires, and declined until none bred for many summers. On a nearby higher area with fewer wires, ptarmigan lost nests to frequent crows and reared abnormally few broods, but seldom died on wires. Adult numbers declined and then became unusually steady for over two decades, with no significant cycle. On a third area further from the car park, ptarmigan lost fewer nests to the less frequent crows but bred more poorly than in the massif's centre and showed cycles of lower amplitude. On a fourth area yet further away, with few or no crows, ptarmigan bred as well as in the massif's centre and showed cycles of the same amplitude.

Negro et al. (2009) assessed the effect of forest clearing for winter sport activities on ground-dwelling arthropods (viz. ground beetles and spiders) and small mammals (shrews and voles) at two ski resorts in north-western Italian Alps by pitfall trapping. Measures of diversity (mean abundance, species richness, and Shannon index) of spiders and macropterous carabids increased from forest interior to open habitats (i.e. ski-piste or pasture), whereas parameters of brachypterous carabids significantly decreased from forest interior to open habitats. Diversity measures of macropterous ground beetles were higher on pastures than on ski-pistes. Small mammals were virtually absent from ski-pistes. Observed frequencies in the three adjacent habitats were significantly different from expected ones for the bank vole *Myodes glareolus* and the pygmy shrew *Sorex minutus*. Generalised linear models showed that abundance, species richness, and diversity of spiders and macropterous carabids of ski-pistes were best modelled by a combination of factors, including grass cover and width of the ski-piste. Indicator species analysis showed that species that significantly preferred ski-pistes were less than those preferring pastures, and species which were exclusive of ski-pistes were very few. They recommended that in order to retain the arthropod ground-dwelling fauna of open habitats, environmentally friendly ways of constructing pistes should be developed. After tree clearing, only the roughest ground surfaces should be levelled, in order to preserve as much natural vegetation as possible. Where necessary, ski-pistes should be restored through the recovery of local vegetation. In a second study, Negro et al. (2010) analysed the effect of ski-pistes on the abundance and species richness of arthropods (viz. carabids, spiders, opilionids, and grasshoppers) trapped in grasslands adjacent to the ski run, on ski-pistes, and at the edge between these two habitat types. Their results showed that diversity of brachypterous carabids, spiders, and grasshoppers decreased significantly from natural grasslands to ski-pistes. This was not true for the macropterous carabid guild, which included species with contrasting ecological requirements. Analysis of indicator species showed that most of the species (some of them restricted areas in the north-western Alps) had clear preferences for natural grassland and few taxa were limited to ski-pistes. Generalised linear models suggested that the local extent of grass and rock cover can significantly affect assemblages: the low grass cover of ski-pistes, in particular, was a serious hindrance to colonisation by spider, grasshopper, brachypterous, and some macropterous carabid species. The results obtained, support concerns over the possible disruption of local ecosystem functionality, and over the conservation of arthropod species which are endemic to restricted alpine areas. In order to retain arthropod ground-dwelling fauna they suggested that: (i) new, environmentally friendly ways of constructing pistes should be developed to preserve as much soil and grass cover as possible and (ii) existing ski-pistes should be restored through management to promote the recovery of local vegetation.

Finally, Caprio et al. (2016) found that restoration of grasslands on ski-pistes caused a recovery in the bird community, but not to the extent that it was equivalent to a natural Alpine grassland community. The bird communities in two ski resorts in the Italian Maritime Alps were surveyed using a standardised area count method in three different plot types: non-restored ski-pistes (newly constructed), restored ski-pistes, and control plots in grassland far from ski-pistes. In 49 independent plots, 32 species were recorded. Species richness and abundance of birds were significantly higher on restored than on non-restored ski-pistes, independently of the species group considered and the analyses carried out. Bird community parameters of restored ski-pistes were still lower than those of natural grassland, as shown by results of typical grassland species. Their results suggest that an apparently successful restoration of ski-pistes may be not enough to promote a complete recovery of bird communities. The complete recovery of local bird communities may be promoted only if an integral recovery of the original vegetal communities is achieved. They suggested that the best conservation option is to adopt techniques to maintain as far as possible original grassland if construction of new ski-pistes is unavoidable.

11.4.3 Impacts of Snow Sport on Water Resources

Artificial snowmaking discussed earlier presents challenges to the water resource requirements of many ski resorts (de Jong and Barth 2007; de Jong and Biedler 2012) because it usually draws water directly from local lakes which are also the main water supply of the resort. The Alpine Convention (2011), in its report "Water in the alps; striking the balance" stated that:

> …. water requirements for snowmaking can be substantial at a local level, using a considerable share of the annual water abstraction and can lead to water conflicts especially in the winter season in areas where snowmaking stations are connected to the drinking supply network; this can cause temporary water shortages. (p. 92)

In an earlier report, the same organisation (Alpine Convention 2011) criticises that: "The EU Framework Directive does not include problems relating to artificial snow installations such as the impact of chemical snow-making additives on water quality or the tapping of water to create artificial snow in times of water shortage." However, the 2010 report no longer addresses this issue. Such shortcomings were mentioned by de Jong (2014) proposing a new EU tourist directive, that is, to restrict increasing water use for tourist resorts and carefully monitor environmental impacts from an interdisciplinary point of view to avoid opposing the EU Water and EU Soils directive.

Kangas et al. (2012) investigated the impact of ski resorts on water quality of lakes near two popular ski resorts in Finland. They examined how water quality problems induced by ski resorts relate to effects of agriculture and forestry on similar lake types. Human impact significantly increased nutrient concentrations, although the differences observed between impact and control lakes were generally small. Water quality of the ski resort lakes and lakes polluted by agriculture and forestry appeared to be quite similar, with the exception of a small, humic ski-resort lake with extremely high nutrient concentrations. Two ski-resort lakes and one agricultural lake failed the total phosphorus criteria set for reaching good ecological status. Their results indicated that water protection measures should be considered more carefully in management of ski resorts.

Forrester and Stott (2016) investigated the spatial distribution of stream water faecal coliform concentrations in specific winter recreation areas in the northern Corries of the Cairngorm Mountains, Scotland. A total of 207 water samples were collected from ten sites during two winter seasons, 2007–2008 and 2008–2009, and analysed for the presence of faecal coliforms, specifically *Escherichia coli* (*E.coli*). *E.coli* was not detected at the seven sites above 635 m, but three sites below 635 m (the altitude of the ski area buildings and car park) had positive detection rates for *E.coli*, these being 32%, 35%, and 31% respectively, suggesting that snow holing was not associated with elevated faecal coliform levels (their site 1 was right next to the popular snow-holing sites in Ciste Mhearad) but that the ski infrastructure was somehow contributing *E. coli* to the streams running through it.

Apollo (2017) pointed out how millions of mountaineers (which presumably includes ski mountaineers) visiting high-mountain areas generate tonnes of faeces and cubic hectometres of urine annually. The proper disposal of human waste is important for the conservation and appropriate management of high-mountain areas. The management can address the issue in three ways: the good (complete/non-invasive), the bad (partial/superficial), and the ugly (invasive). With use of those categories, 20 selected summits from different parts of the world were evaluated, separately in respect to faeces and urine. It was expected that correct or incorrect disposal of human waste would depend on the changing altitude and/or development level. Disappointingly, the correlation between selection criteria (better or worse solution) and the increase of altitude does not exist. Similarly, the increase of the development level does not play a significant role, especially when urine is taken under consideration. The problem is more global than was thought. The paper makes recommendations which could lead to reduction of this problem.

In Chap. 5 we discussed how lead and hydrocarbons from snowmobile exhaust were found in the water at high levels during the week following ice-out in a Maine pond (Adams 1975), and that fingerling brook trout (*Salvelinus fontinalis*) held in fish cages in the pond showed lead and hydrocarbon uptake. Snowmobiles, and a range of other snow-grooming machines, are used at most ski resorts and so it is quite possible that hydrocarbons from both exhaust fumes and petrol or oil spills or leakages can contaminate snow which, when it melts, reaches soil water, groundwater, and streams/lakes.

During a three-month monitoring campaign by Kallenborn et al. (2011) in very high atmospheric levels of benzene-toluene-xylene (BTX)-related emissions were found during daytime along the main snowmobile routes in Longyearbyen, Svalbard, during the late winter season (main tourist season) in 2007. Total emissions of about 81 t/year were estimated for 2007 solely for snowmobile activities. Two-stroke engine-driven vehicles were estimated to contribute around 92% (74 t/a) of the hydrocarbon emissions. Such data are likely to apply to snowmobiles used for transport (and sometimes tourist tours) in ski resorts.

11.5 Management and Education

Future management of snow sport must surely take account of the predicted future changes in climate.

11.5.1 Effects of Climate Change on Snow Sport

Changes in the spatial extent and duration of winter snow cover, both in Scotland and in a wider global context, have a number of socio-economic and environmental implications. Evidence from Scottish climatological presented by Harrison et al. (2001) stations appeared to suggest that the most marked decrease in the number of days with snow lying has occurred since the late 1970s. Information on the effects of these changes was gathered using a questionnaire which was sent to key stakeholders. Responses suggest deleterious effects on winter recreation and sports, upland habitats, and flood regimes in Scottish rivers. An extended snow-free season has affected access to, and management of, Scottish land.

Moen and Fredman (2007) reported that annual snow cover extent in the northern hemisphere had decreased by about 10% since 1966, and in Sweden, the last decade was wetter and warmer than the preceding 30-year period. These changes will affect many aspects of utilisation patterns that are dependent on the physical environment, such as alpine winter tourism. In their paper they discuss the future development of the downhill skiing industry in Sweden, (1) reviewing trends in alpine winter tourism in relation to climate change, (2) examining trends in climate parameters relevant to alpine winter tourism in Sweden during the last 30 years, and (3) using these parameters, together with regional projections of climate change, predicting the effects on the number of skiing days in order to estimate the monetary loss for the skiing industry in Sweden. Their analyses showed predicted losses that were larger than current ski-ticket sales. Adaptation strategies such as the development of year-round tourist activities that should be developed as soon as possible were recommended.

While the ski industry has become one of the main economic activities for many mountain regions worldwide, Gilaberte-Búrdalo et al. (2014) examined the economic viability of this activity and stated how it is highly dependent of the interannual variability of the snow and climatic conditions and that it is jeopardised by climate warming. In their study, they reviewed the main scientific literature on the relationship between climate change and the ski feasibility under different climate change scenarios. In spite of the different methodologies and climate change scenarios used in the reviewed studies, their findings generally point to a significant impact of climate change on the ski industry caused by a reduction in the natural availability of snow as well as a contraction in the duration of seasonal conditions suitable for skiing. It emphasises that the problem was real and should not be ignored in the study and management of tourism in mountain regions. However, there were significant differences in the impacts between different areas. These differences were mainly associated with the elevation of the ski resorts, their infrastructures for snowmaking, and the various climate models, emission scenarios, time horizons, and scales of analysis used. Their review high-

Fig. 11.9 Reduction in the number of ski days and the percentage closure of ski resorts in various regions as a function of temperature increase (adapted from Gilaberte-Búrdalo et al. 2014)

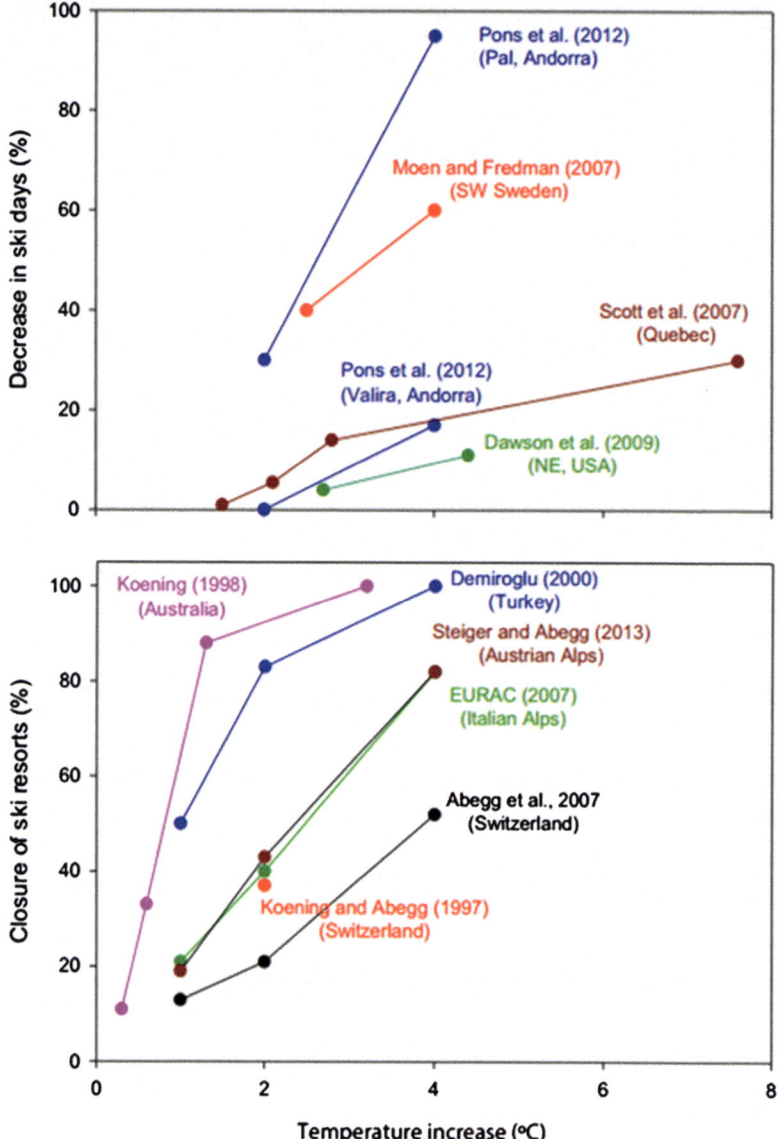

lighted the necessity for scientists to harmonise indicators and methodologies to allow a better comparison of the results from different studies and to increase the clarity of the conclusions transmitted to land managers and policy makers. Moreover, a better integration of the uncertainty in the model's outputs, as well as the treatment applied to the snowpack in ski slopes is necessary to provide more accurate indications on how this sector will respond to climate change.

Figure 11.9 shows the decrease in the number of ski days and the percentage of ski resorts within a particular region that should be closed as a function of various scenarios of climate warming. The studies from which Fig. 11.9 was derived, and the other studies reviewed by Gilaberte-Búrdalo et al. (2014), highlight how sensitive the ski industry is to increasing temperature but also the large variability in skiability in response to similar warming rates. For example, changes in the number of ski days in response to climate warming at two resorts located in Andorra differed markedly (Pons-Pons et al. 2012). These differences can be largely explained by the average altitude of the resorts, or the latitude of the mountain regions involved. Thus, colder areas (because of high altitude or latitudinal location) are less affected by climate change or may even benefit through spillover from lower-altitude resorts that are more vulnerable to the effects of climate change.

Wobus et al. (2017) used a physically based water and energy balance model to simulate natural snow accumulation at 247 winter recreation locations across the continental USA. They combined this model with projections of snowmaking conditions to determine downhill skiing, cross-country skiing, and snowmobiling season lengths under baseline and future climates, using data from five climate models and two emissions scenarios. Projected season lengths were combined with baseline estimates of winter recreation activity, entrance fee information, and potential changes in population to estimate the financial cost of impacts to the selected winter recreation activity categories for the years 2050 and 2090. Their results identified changes in winter recreation season lengths across the USA that vary by location, recreational activity type, and climate scenario. However, virtually all locations were projected to see reductions in winter recreation season lengths, exceeding 50% by 2050 and 80% in 2090 for some downhill skiing locations. They estimated that these season length changes could result in millions to tens of millions of foregone recreational visits annually by 2050, with an annual financial cost of hundreds of millions of dollars. Comparing results from the alternative emission scenarios shows that limiting global greenhouse gas emissions could both delay and substantially reduce adverse impacts to the winter recreation industry.

11.5.2 Do Ski Resorts Need to Become "Greener" for Tourism to Become Sustainable?

For a growing number of skiers, the sport represents one of the great dilemmas and conflicts between recreational enjoyment of the countryside and the conservation of the fragile Alpine and mountain areas where skiing inevitably takes place. Hudson (1996) pointed out how the situation was particularly acute in the European Alps where, in the 1990s, the issue had been given a high profile by many concerned environmentalists. Hudson's paper looks at the impact of skiing on the environment both in North America and in Europe and at the emerging concept of sustainability. The author looks at the marketing opportunities for destinations seeking to green their operations and uses Verbier in Switzerland as a case study.

The development of tourism in mountain areas can have profound influences on both the local economy and physical environment. Holden (1999) claimed that as the concept of sustainability became more important in policy making and that the future of downhill skiing in mountain areas would become more uncertain. However, the extent to which policy might shift from an anthropocentric bias towards a more eco-centric approach is uncertain. One mountainous area that has recently developed a sustainable management strategy is the Cairngorms area of the Scottish Highlands. The development of downhill skiing in this area is highly contentious owing to the uniqueness of the physical environment. Using the case study and different perspectives on sustainability, Holden evaluated the role of downhill skiing as part of a sustainable policy for mountain areas.

One could argue that in terms of agricultural production, the high altitude mountain areas used for skiing are not taking up land which much economic value in terms of agriculture production. Instead, they are turning this marginal (almost waste?) land into a valuable economic resource. The actual area of the world's mountains that is taken up by ski infrastructure is tiny in comparison to the total. So, there are still plenty of mountains left for others to enjoy.

The United Nations (UN) claimed that "whenever a person engages in sport there is an impact on the environment" (UN 2010). Spector et al. (2012) examined the safeguarding of the natural environment, or environmental sustainability (ES), in sport by studying the level of environmentally responsible actions for ski resorts in the USA. They focused specifically on the USA ski industry and examined Ski Resort's

Environmental Communications (SRECs) stated on each of 82 resort websites. The methods included rating these communications for their prominence, breadth, and depth based on the environmental categories in the US Sustainable Slopes Program (SSP) Charter. Based on both these SREC ratings and the grades assigned to each resort by the Ski Area Citizens Coalition (SACC), the resorts were classified as inactive, exploitive, reactive, or proactive using an adaption of Hudson and Miller's (2005) model. The results provided an assessment of the level of environmentally responsible actions taken by the ski resorts. Future research still needs to examine the motivations behind ski resort publications on environmental communications and the likelihood of skiers selecting resorts based on the environmental communications posted on websites.

11.5.3 Concluding Comments

Clearly there are a number examples of research which indicate that the development of ski resorts, in particular bulldozing the ground to make it suitable for ski-pistes, results in soil erosion; modification of the natural landscape shape, gullies, and fans; removal of vegetation; and changes in species composition years after the runs were constructed. Reseeding experiments have had some degree of success but still do not return the flora back to its natural state. There are reported changes in soil structure, compaction, and the addition of chemicals resulting from artificial snowmaking. The research also showed reductions in bird species like ptarmigan and increases in crows. All these studies demonstrate some impact, but how to mitigate this is more difficult. It is likely that with the future climate scenarios (or warming) that ski resorts are likely to become even more dependent on artificial snowmaking, bringing greater pressures on the water resources of ski resorts and surrounding mountain communities.

Off-piste skiing (ski mountaineering, cross-country/Nordic, and showshoeing) is not associated with these same issues. The main concerns with these pursuits are those discussed in Chap. 8 about waste disposal in wilderness environments in winter.

Conclusions

1. Skiing can be a means of transport, a recreational activity, or a competitive winter sport in which skis are used to glide on snow. Many types of competitive skiing events are recognised by the IOC and the FIS. The term snow sport encompasses alpine, freestyle, snowboard, and Nordic skiing.
2. Off-piste snow sport includes ski mountaineering, cross-country, and snowshoeing.
3. The Outdoor Foundation (2017) survey data for the USA show that, of the six snow sports disciplines measured in the survey:
   - Alpine/downhill skiing had the greatest number of participants in 2016 (9,267,000) and a 12.4% increase over the previous three years (2014–2016).
   - Snowboarding had the second greatest number of participants in 2016 (7,602,000), showing a 3.4% increase over the previous three years.
   - Cross-country skiing had 4,640,000 participants but the greatest three-year increase of 40.3%.
   - Freestyle skiing with 4,640,000 participants in 2016 had the same number participating as cross-country skiing but with only a 2.7% three-year increase.
   - Telemark (downhill) skiing has the smallest participation numbers in the six disciplines surveyed (2,848,000) and a 3.0% three-year increase.
   - Snowshoeing had 3,533,000 participants but showed a −12.3% decrease in participation in the 2014–2016 period.

4. Some of the first research into the impacts of ski resort development on soils and vegetation from the Cairngorms, Scotland, began in the late 1960s and involved field surveys and experimental studies of erosion, compaction, vegetation damage, and the use of paths, as well as assessments of the success of various methods of rehabilitation on different kinds of substrate. The construction of chairlifts, ski tows, and buildings on Cairngorm has resulted in the exposure of areas of mineral soil followed by erosion.
5. Plant species diversity was changed following ski-run construction and bird species and arthropod populations declined.
6. Snow grooming (moving, flattening, or compacting the snow) increases snow density and hardness which in turn inhibits or delays soil bacterial activity and subsequent litter decomposition.
7. Artificial snowmaking induces new impacts to the alpine environment which include the input of water and ions to ski-pistes, which can have a fertilising effect and hence change the plant species composition.
8. While snow and soil properties were influenced by cross-country ski track preparation, natural environmental variability was more influential for species composition and biomass production than the ski track. The cross-country ski track—without artificial snow—did not negatively affect species composition. By delaying flower phenology, the effects of the ski track even counteracted global warming to some degree.
9. Water consumption for snowmaking can be substantial at a local level, using a considerable share of the annual water abstraction and can lead to water conflicts especially in the winter season in areas where snowmaking stations are connected to the drinking supply network, which can cause temporary water shortages.
10. Several studies have examined the effect that future predictions for climate change will have on the winter tourism industry. The relationship between temperature increase and the percent of resorts predicted to close is generally linear with between 50% and 90% of ski resorts closing with a temperature increase of 4 °C, but the relationship varies from region to region. High altitude resorts would be least affected.

References

Adams, E. S. (1975). Effects of lead and hydrocarbons from snowmobile exhaust on brook trout (*Salvelinus fontinalis*). *Transactions of the American Fisheries Society, 104*(2), 363–373.

Alpine Convention. (2011). Water in the alps; striking the balance. In *3rd International Conference and Preparatory Workshops 2010*, Rome, p. 104.

Apollo, M. (2017). The good, the bad and the ugly—Three approaches to management of human waste in a high-mountain environment. *International Journal of Environmental Studies, 74*(1), 129–158.

Barni, E., Freppaz, M., & Siniscalco, C. (2007). Interactions between vegetation, roots, and soil stability in restored high-altitude ski runs in the Alps. *Arctic, Antarctic, and Alpine Research, 39*(1), 25–33.

Bayfield, N. G. (1974). Burial of vegetation by erosion debris near ski lifts on Cairngorm, Scotland. *Biological Conservation, 6*(4), 246–251.

Bayfield, N. G. (1980). Replacement of vegetation on disturbed ground near ski lifts in the Cairngorm Mountains, Scotland. *Journal of Biogeography, 7*, 249–260.

Bayfield, N. G. (1996). Long-term changes in colonization of bulldozed ski pistes at Cairn Gorm, Scotland. *Journal of Applied Ecology, 33*, 1359–1365.

Bowker, J. M., Askew, A. E., Cordell, H. K., Betz, C. J., Zarnoch, S. J., & Seymour, L. (2012). *Outdoor Recreation Participation in the United States—Projections to 2060: A Technical Document Supporting the Forest Service 2010 RPA Assessment.*

Ashville: Southern Research Station. Retrieved from www.srs.fs.usda.gov.

Burt, J. W., & Rice, K. J. (2009). Not all ski slopes are created equal: Disturbance intensity affects ecosystem properties. *Ecological Applications, 19*(8), 2242–2253.

Caprio, E., Chamberlain, D., & Rolando, A. (2016). Ski-piste revegetation promotes partial bird community recovery in the European Alps. *Bird Study, 63*(4), 470–478.

Cordell, K. (2012). *Outdoor Recreation Trends and Futures: A Technical Document Supporting the Forest Service 2010 RPA Assessment.* Ashville: Southern Research Station. Retrieved from www.srs.fs.usda.gov.

de Jong, C. (2014). A white decay of winter tourism. Climate change adaptation manual. In A. Prutsch, S. McCallum, T. Grothmann, R. Swart, I. Schauser, & R. Swart (Eds.), *Lessons Learned from European and Other Industrialized Countries* (pp. 226–233). London and New York, NY: Routledge.

de Jong, C., & Barth, T. (2007). Challenges in hydrology of mountain ski resorts under changing climate and human pressures. In *ESA, Proceedings of the 2nd Space for Hydrology Workshop, "Water Storage and Runoff: Modeling, In-Situ Data and Remote Sensing"*, Geneva.

de Jong, C., & Biedler, M. (2012). Shadow of a drought. *European Science and Technology, 14*, 208–209.

de Jong, C., Carletti, G., & Previtali, F. (2014). Assessing impacts of climate change, ski slope, snow and hydraulic engineering on slope stability in ski resorts (French and Italian Alps). In G. Lollino, A. Manconi, J. Clague, W. Shan, & M. Chiarle (Eds.), *Engineering Geology for Society and Territory—Climate Change and Engineering Geology* (pp. 51–55). Cham: Springer.

Fahey, B. D., Wardle, K., & Weir, P. (1999). *Environmental Effects Associated with Snow Grooming and Skiing at Treble Cone Ski Field*. Department of Conservation.

Fidelus-Orzechowska, J., Wrońska-Wałach, D., Cebulski, J., & Żelazny, M. (2018). Effect of the construction of ski runs on changes in relief in a mountain catchment (Inner Carpathians, Southern Poland). *Science of the Total Environment, 630*, 1298–1308.

Forrester, B. J., & Stott, T. A. (2016). Faecal coliform levels in mountain streams of winter recreation zones in the Cairngorms National Park, Scotland. *Scottish Geographical Journal, 132*(3–4), 246–256.

Gilaberte-Búrdalo, M., López-Martín, F., Pino-Otín, M. R., & López-Moreno, J. I. (2014). Impacts of climate change on ski industry. *Environmental Science and Policy, 44*, 51–61.

Harrison, S. J., Winterbottom, S. J., & Johnson, R. C. (2001). A preliminary assessment of the socio-economic and environmental impacts of recent changes in winter snow cover in Scotland. *Scottish Geographical Journal, 117*(4), 297–312.

Holden, A. (1999). High impact tourism: A suitable component of sustainable policy? The case of downhill skiing development at Cairngorm, Scotland. *Journal of Sustainable Tourism, 7*(2), 97–107.

Hudson, S. (1996). The 'greening' of ski resorts: A necessity for sustainable tourism, or a marketing opportunity for skiing communities? *Journal of Vacation Marketing, 2*(2), 176–185.

Hudson, S., & Miller, G. (2005). The responsible marketing of tourism: The case of Canadian Mountain Holidays. *Tourism Management, 26*(2), 133–142.

Kallenborn, R., Schmidbauer, N., Reinmann, S., Trümper, M., & Tessmann, M. (2011). Local contaminant sources in the Arctic: Volatile and non-volatile residues from combustion engines in surface soils from snow mobile tracks in the vicinity of Longyearbyen (Svalbard Norway). Retrieved from https://www.researchgate.net/profile/Roland_Kallenborn/publication/282317733_Local_contaminant_sources_in_the_Arctic_Volatile and nonvolatile_residues_from_combustion_engines_in_surface_soils_from_snow_mobile_tracks_in_the_vicinity_of_Longyearbyen_Svalbard_Norway/links/560b8cda08ae576ce6411cf2/Local-contaminant-sources-in-the-Arctic-Volatile-and- nonvolatile-residues-from-combustionengines-in-surface-soils-from-snow-mobile-tracks-in-the-vicinity-of-Longyearbyen-Svalbard-Norway.pdf

Kangas, K., Vuori, K. M., Määttä-Juntunen, H., & Siikamäki, P. (2012). Impacts of ski resorts on water quality of boreal lakes: A case study in northern Finland. *Boreal Environment Research, 17*, 313–325.

Lagriffoul, A., Boudene, J. L., Absi, R., Ballet, J. J., Berjeaud, J. M., Chevalier, S., et al. (2010). Bacterial-based additives for the production of artificial snow: What are the risks to human health? *Science of the Total Environment, 408*, 1659–1666.

Moen, J., & Fredman, P. (2007). Effects of climate change on alpine skiing in Sweden. *Journal of Sustainable Tourism, 15*(4), 418–437.

Negro, M., Isaia, M., Palestrini, C., & Rolando, A. (2009). The impact of forest ski-pistes on diversity of ground-dwelling arthropods and small mammals in the Alps. *Biodiversity and Conservation, 18*(11), 2799–2821.

Negro, M., Isaia, M., Palestrini, C., Schoenhofer, A., & Rolando, A. (2010). The impact of high-altitude ski pistes on ground-dwelling arthropods in the Alps. *Biodiversity and Conservation, 19*(7), 1853–1870.

Pons-Pons, M., Johnson, P. A., Rosas-Casals, M., Sureda, B., & Jover, E. (2012). Modelling climate change effects on winter ski tourism in Andorra. *Climate Research, 54*, 197–207.

Rixen, C., Haeberli, W., & Stoeckli, V. (2004). Ground temperatures under ski pistes with artificial and natural snow. *Arctic, Antarctic, and Alpine Research, 36*(4), 419–427.

Rixen, C., Stoeckli, V., & Ammann, W. (2003). Does artificial snow production affect soil and vegetation of ski pistes? A review. *Perspectives in Plant Ecology, Evolution and Systematics, 5*(4), 219–230.

Rolando, A., Caprio, E., Rinaldi, E., & Ellena, I. (2007). The impact of high-altitude ski-runs on alpine grassland bird communities. *Journal of Applied Ecology, 44*(1), 210–219.

Spector, S., Chard, C., Mallen, C., & Hyatt, C. (2012). Socially constructed environmental issues and sport: A content analysis of ski resort environmental communications. *Sport Management Review, 15*(4), 416–433.

Steinbauer, M. J., Kreyling, J., Stöhr, C., & Audorff, V. (2017). Positive sport—Biosphere interactions?—Cross-country skiing delays spring phenology of meadow vegetation. *Basic and Applied Ecology., 20*(11), 1405–1413.

The Outdoor Foundation. (2017). *Outdoor Participation Topline Report 2017*. Washington, DC: The Outdoor Foundation. Retrieved from www.outdoorfoundation.org.

Törn, A., Tolvanen, A., Norokorpi, Y., Tervo, R., & Siikamäki, P. (2009). Comparing the impacts of hiking, skiing and horse riding on trail and vegetation in different types of forest. *Journal of Environmental Management, 90*(3), 1427–1434.

UK Snow Sports. (2018). Retrieved March 20, 2018, from http://www.uksnowsports.co.uk/.

United Nations. (2010). United Nations Environmental Programme: Sport and environment. Retrieved from http://www.unep.org/sport_env/.

Watson, A. (1979). Bird and mammal numbers in relation to human impact at ski lifts on Scottish hills. *Journal of Applied Ecology, 16*, 753–764.

Watson, A. (1985). Soil erosion and vegetation damage near ski lifts at Cairn Gorm, Scotland. *Biological Conservation, 33*(4), 363–381.

Watson, A., Bayfield, N., & Moyes, S. M. (1970). Research on human pressures on Scottish mountain tundra, soils and animals. In W. A. Fuller & P. G. Kevan International Union for Conservation of Nature and Natural Resources (Eds.), *Productivity and Conservation in Northern Circumpolar Lands* (Vol. 16, pp. 256–266). Switzerland: IUCN Publications Morges. Retrieved from https://portals.iucn.org/library/sites/library/files/documents/NS-016.pdf#page=258

Watson, A., & Moss, R. (2004). Impacts of ski-development on ptarmigan (*Lagopus mutus*) at Cairn Gorm, Scotland. *Biological Conservation, 116*(2), 267–275.

Wipf, S., Rixen, C., Fischer, M., Schmid, B., & Stoeckli, V. (2005). Effects of ski piste preparation on alpine vegetation. *Journal of Applied Ecology, 42*(2), 306–316.

Wobus, C., Small, E. E., Hosterman, H., Mills, D., Stein, J., Rissing, M., et al. (2017). Projected climate change impacts on skiing and snowmobiling: A case study of the United States. *Global Environmental Change, 45*, 1–14.

Caving

12

Chapter Summary

Caving damages a specialised, rare, and delicate environment which can never recover. The aim is to minimise this mostly unintentional damage. The number of cavers is small but they can damage caves in many ways: the geological environment, cave fauna and flora which is specialised and often endemic. Bats are important and have suffered in North America from white nose syndrome, partly spread by cavers. Many potential management strategies for caves exist including cave plans, conservation codes, and National Conservation policies. Controlling access can be important and there are access agreements, zero, restricted or periodic access, booking systems, gating, sacrificial caves, zoning in caves, cave exploration policies, and cave fauna management, including building artificial bat caves. Education for cavers includes minimal impact codes, websites, leader and instructor schemes, involvement in cave conservation planning, and cave adoption schemes and alternatives to caving, such as the use of mines and artificial caves.

12.1 Introduction and Numbers Involved in Caving

Caving as an activity is a specialist outdoor pursuit that is practised by a relatively small number of practitioners. Although it is possible to subdivide the activity into several types: speleology (the science related to caves), spelunking (US term), and caving (the recreation activity related to caves) and potholing (for caving where there are vertical pitches), we will consider all three as one in this chapter and call the activities caving. There is a further category to be considered which is the tourist show cave. The numbers involved in the activity are difficult to ascertain but the following are some minimum estimates: In Britain the numbers caving in 1971–1972 were estimated at a conservative figure of 16,000–17,000 with 400 caving clubs (Wilmut 1972). By 1990 the numbers caving each year had risen to 30,000 based on the total members of caving clubs (Ford 1990). As an indication of the importance of caving as an outdoor recreation activity though in the report by Gordon et al. (2015) caving is not considered in the section on Outdoor Activity participation. In the USA in 2012 caving did not appear in the top 25 outdoor activities, and out of 43 activities in terms of outdoor participation it does not appear (Outdoor Foundation 2016). In the USA the figures are surprisingly low with the National Speleological Society (NSS) having around 2000 new members in 1950, 2100 in 1980, 4900 in 2000, but with the figure total down to 9256 in February 2016. In the USA each member belongs to local chapters known as Grottos. The National Survey on Recreation for the period 2005–2008 shows that the total US participants in caving were 9871, an 8.9% positive change as compared to the period 1999–2001. During 2005–2009 these figures were 10,400, a 4.4% increase, whilst the participation days were 19.5 million, a 2.4% positive change (Bowker et al. (2012); according to Cordell et al. (2008) and Cordell (2012) the total annual participation days were 21.6 million in 2005–2008, a 3.3% positive change. The projected backcountry outdoor recreation participation rates from 2008 to 2060 and under the category of challenge (mountain climbing, rock climbing, and caving) the participants in 2008 were 25 million, but by 2060 this figure was predicted to change to anywhere between 38 and 48 million, with an average change of plus 19 million. The number of participation days in 2008 was 121 million for these activities, but by 2060 this figure is estimated to increase to between 178 and 219 million, with an average change of 86 million. The figures vary because of the estimates of population growth used, income growth, and the various effects of climate change. However, based on the figures for caving noted earlier, cavers make up a small percentage of participants in these three challenge activities.

In response to the question "How many cavers are there in Europe?," the Association Hommes des Cavernes (2015) estimated a total figure of 40,300 based on figures from national federations who are members of the European Speleological Federation, with France having the largest number.

The cavers in New Zealand belonging to the New Zealand Speleological Society currently stands at around 300 members, whilst in Australia the Australian Speleological Federation represents the interest of 28 caving clubs and has over 700 members. Whilst the membership might only be a conservative estimate of the numbers actively caving it gives a good indication of the small total numbers involved. The bigger problem might be, however, that the small number of cavers not members of clubs may well cause a disproportionately greater amount of damage to caves and because of the high turnover of participants and the fact that at any given time period the numbers caving consist of a very high proportion of relative beginners. This is especially so when one considers the numbers of these beginners who are taken into caves from outdoor centres for adventure experiences with perhaps greater numbers than we would ideally recommend in a group and hence the greater difficulties of controlling that group. Against this view would be the fact that the leaders of these groups are generally well qualified and probably use easy, popular caves considered as sacrificial.

Recreational cavers can be subdivided into individuals and small groups using caves for outdoor pursuits and exploration; outdoor centres using caving as an activity or for educational purposes; and caving clubs using caves for exploration. Speleologists include geologists, mineralogists, geomorphologists, hydrologists, biologists, and archaeologists simply because caves are repositories for a wealth of scientific knowledge about landscape development, specialised forms of mammals, fish, and insects and how man and animals have evolved and lived at various time periods. The include Minchin Hole on Gower, Wookey Hole and Gough's Cave (Cheddar), the Buckfastleigh caves (Devon), Pontnewydd and Clwyd caves in North Wales in Britain; the important South African hominid and animal bone caves (Fig. 12.1) at the Cradle of Humankind World Heritage site in Gauteng Province (South Africa), which have famous cave sites for Australopithecus, *Homo erectus*, *Homo naledi*, and *Homo sapiens sapiens* fossils; Natural Trap Cave (Wyoming), Boodie Cave, Barrow Island off the west Australian coast which was first occupied by man 51,100–46,200 years ago and the important European cave sites like Altamira in Spain and in the Dordogne in France.

There are many commercial show caves too which can have detrimental effects as we will see later. Occasionally caves are used for industry and with specialised forms of agriculture such as mushroom farming, fish breeding, and

12.1 Introduction and Numbers Involved in Caving

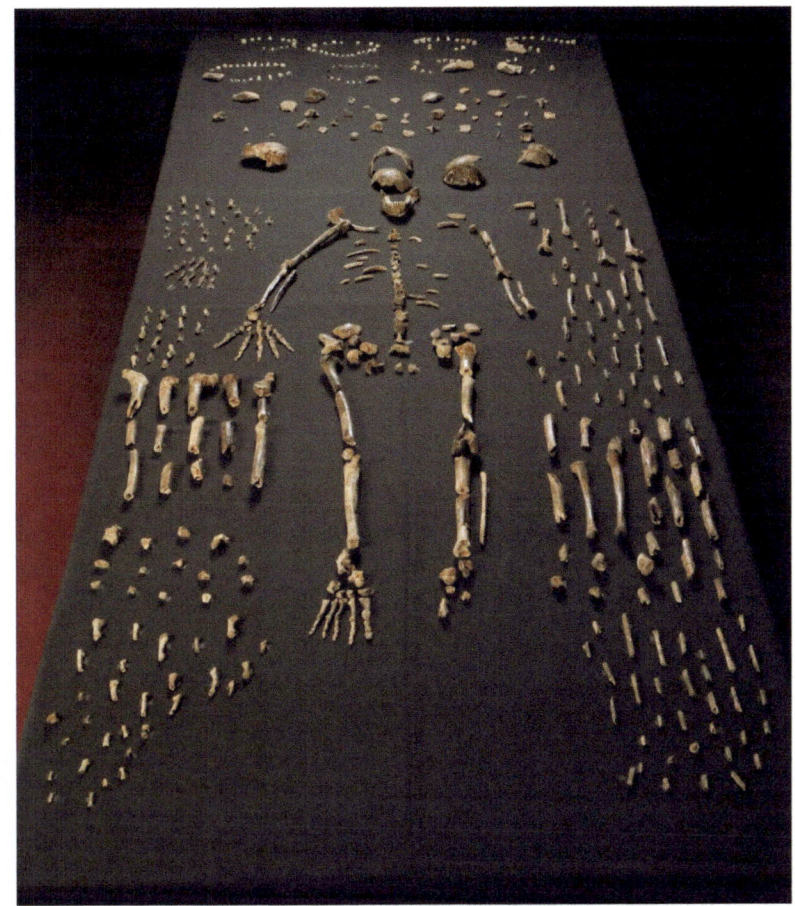

Fig. 12.1 *Homo naledi* bones from Rising Star Cave, part of the Cradle of Humankind World Heritage site about 50 km north-west of Johannesburg. Found by recreation cavers in 2013 and first described in 2015 as probably an offshoot of modern man, although there is still discussion as to its exact position in human evolution. Source: http://elifesciences.org/content/4/e09560. Author Lee Roger Berger research team

cheese production, but these are on a very small scale. It is not just in the USA, Britain, Europe, and Australasia that there is documented cave damage. There seems to be a growing problem in the spectacular Asian karst with fragile caves facing growing development risks and the biodiversity living in these landscapes facing major problems (Clements et al. 2006). As tourism expands in response to a growing Asian middle class, caves are being developed as scenic attractions. There is also a thriving construction industry which needs limestone and logging and land clearance which can affect karst systems. Hence there are threats to cave biodiversity, and there is a lack of management, with a low priority for cave conservation. It is not just the fact that there is a high level of endemic species and the caves act as biodiversity reservoirs but these caves have the potential for future archaeological and palaeontological discoveries. There are many invertebrates, bats, and fishes living in these caves too. Of 143 species from karst regions of the world which are globally threatened (IUCN), 31 occur in South East Asia, and these figures are thought to be conservative. There have been many species lost and, for example, at least 18 species of karst plants in Peninsular Malaysia have become extinct.

There are several types of caves in terms of their geology and geomorphology, but a predominant number are in limestone, dolomite, and gypsum and are formed because these rock types are soluble (Fig. 12.2A). Other types of cave are formed as basalt lava tubes (Fig. 12.2B) and in glaciers by meltwater (Fig. 12.2C), but these are much less important in total number, especially the latter as they are so ephemeral because of glacier movement and meltwater stream changes.

Fig. 12.2 (A) Long Churn Cave in soluble limestone: one of the entrances with Ingleborough in the background. Photo by D. Huddart. (B) Lava tube and lake, Jameos del Agua, Lanzarote: formed by lava flow from an eruption of La Corona volcano, which then partially collapsed to give several underground access points. Photo by D. Huddart. (C) Glacier cave in Iceland. Photo by D. Huddart

12.2 How Can Caves Be Damaged?

We start by looking at how caves can be damaged, what can be done to prevent this damage, and then go on to evaluate the management techniques that are currently being used and what is our responsibility as outdoor and environmental educators to preserve the cave environment. Generally most damage is probably accidental but it is insidious, and although each incident like a straw stalactite being broken is hardly noticeable, cumulatively the result is irreparable wear and tear. Within the British Isles, for example, new cave systems are discovered only rarely and a cave destroyed is a cave destroyed forever. Some cave passageways are active streamways, with no deposits and few formations. These are robust environments where little or no damage to the cave by recreation is likely to occur but many caves are small and constricted with fragile formations and vulnerable sediments which are easily damaged. High level, inactive caves and those that are well decorated with speleothems are much more prone to damage. However, it is not practical to generalise, and each cave should be seen as unique.

A newly discovered cave system is the only true natural ecosystem available for study in a country such as Britain, but although caves provide a unique environment, they also provide a unique challenge to the conservationist. An example of this type of problem was documented by Hewitt (1992) who described the effects of a new discovery in Lathkiller Hall in 1990 in Lathkill Head, the wet weather resurgence of the river Lathkill in the Peak District (Derbyshire). It was well decorated with all kinds of flowstone decorations, and with excellent cave sediment sequences. Up to January 1991 damage was minimal, with less than eight people visiting and under 20 hours of caving time spent in the cave. Baseline observations were made, and there was a system to record the numbers visiting. In 14 months 95 people visited, there was damage to two stalagmite curtains, and there was erosion of mud in a crawl. The author realised that he was documenting at least partly his own effect on the cave and two others who were digging to extend the cave. These three cavers accounted for 34 out of the total visits, and the author realised he was significantly eroding the mud at the exit of the crawl while trying to determine the potential for change.

Caves are vulnerable to human pressure because this pressure can be so concentrated. It is also true that there has been relatively little scientific work examining the ecological effects of caving. Williams (1966) found reduced algal and bacterial populations when investigating specific sites in two well-used Welsh caves (Ogof Ffynnon Dhu and Porth yr Ogof), and threats to bat roosts have been well documented, but Sidaway (1988) suggested that the changes occurring are only apparent over long cycles which makes population trends difficult to assess. In 1972 a National Cave Association report on the state of conservation and access in caves and mines in England and Wales made one obvious conclusion: the extent of usage a cave receives reflects the extent and nature of damage occurring, but as we have already seen it can depend on the cave as each cave is unique and can present unique problems to solve. Trampling pressure can occur on the access routes to the caves and around cave entrances and exits. Inside the cave because usually cavers follow well-defined routes, trampling can cause unintentional wear and tear by cavers and their equipment. The passage floors can be eroded, especially where soft cave sediment floors are lowered, or compacted which might mean loss of habitat as the fauna cannot penetrate the compacted sediment. The cave can be made more accessible by increasing the size of the passages, for example, the crawl in Cwmdwr Quarry Cave (South Wales), and in some caves there can be erosion of the rock surfaces by ladder and rope grooves. This trampling can disturb cave biota, fungi can be scraped away, and the flora around the cave entrances/exits can be damaged. It is particularly important not to disturb roosting bats, and in Britain it is illegal to do so under the Wildlife and Countryside Act (1981).

There has been deliberate and unintentional vandalism to some caves, and Wilmut (1972) showed that even 45 years ago 13% of features in all known British caves and 27% of major caves had been damaged. All it takes is a muddy hand to destroy a white calcite formation or the accidental emplacement of grease from fingertips onto calcite to inhibit formation growth. Rather more obvious are the effects of Victorian collectors who removed flowstone formations, or Alexander Pope who had most of the stalactites shot down in one cave for his collection, or the deliberate removal of formations to make a squeeze passable. Graffiti is obvious on some cave walls, for example, the spectacular spray painting at Buckner Cave (Richard Blenz Nature Preserve, Indiana), and the building of waymarking cairns and scratched signs and arrows can be viewed as deliberate vandalism, a classic example being the route from Bar Pot to the main chamber of Gaping Gill in the Yorkshire Dales.

All kinds of rubbish can be dropped or deliberately left underground by cavers, like food waste, wrappers, items of broken equipment, cigarette ends, and rubber from decomposing clothing. Some of this enriches the food supply which can result in, for example, an increase in one springtail species around Camp One in Otter Hole (South Wales) and a loss of three other species at a popular eating/lunch stop in that cave. Spent carbide used to be a major problem and could find its way into the food chain and also leave small piles and soot on the passage walls and roof. This spent carbide eventually reacts to produce acetylene gas and lime, sulphides, and metals which may kill cave fauna, especially in low-energy cave systems. Skin cells and fine lint introduced by humans bring bacteria which is an extra source of energy for the cave biota. Tobacco smoke contains a powerful insecticide which challenges, if not kills, many invertebrates in the relatively enclosed cave. To illustrate the amount of litter that can be found in caves, the Red Rose Cave and Pothole Club removes hundreds of kilos from the Easegill system (Yorkshire Dales) during its annual clean-up weekend.

To extend cave systems or to find new ones, some cavers and caving clubs deliberately dig in caves or at potential entrances. The result can be piles of debris and worn out or abandoned equip-

Fig. 12.3 Great Douk dig, vertically from the collapsed cave entrance section to try and connect the active streamway from the cave to the River Greta to the west. Note the shoring, scaffolding, and ladder. Photo by D. Huddart

ment and some of these digs can last for tens of years as at Great Douk (Fig. 12.3).

Passage shapes can be altered, or the character can be changed due to digging or accumulation of waste material originating from a dig. On a bigger scale, blasting, the drilling of shot holes, and the use of explosives has in the past changed the morphology of some caves. This can also lead to the subsequent temporary abandonment of a cave due to toxic fumes as in Sell Gill Holes (Yorkshire Dales) in October 1993 which may also have an effect on cave fauna. Deliberate sump drainage and other hydrological alterations in the name of exploration allow inaccessible passages to be used. Usually they fill up again fairly soon but the duck in Valley Entrance (West Kingdale, Yorkshire Dales) was lowered. As it was near the entrance, it formerly used to dissuade many leaders of novice groups from using the cave since most of the trip would have to be made subsequently in wet gear. Now this cave is subject to much greater use as the lowering of the duck allows a much drier and more comfortable trip. Enlarging of entrances during digging can lead to changes in air flow which may lead to desiccation of rock and sediment surfaces but this damage must be minor. Where cave diving is necessary for exploration, the damage caused by airbottles carried to the first dive site can be a big problem and formation damage in constricted sumps is inevitable. The potential downsides to digging include:

- Alteration of the natural appearance of the entrance or landscape.
- Changes to the patterns of air circulation within the cave and the accompanying impacts to the ecosystem and mineral growth.
- An increase in drying, especially during the winter, due to cave microclimate disruption.
- Alteration in drainage characteristics and patterns of sediment transport within the cave due to changes around the entrance zone.
- Possible creation of an unstable passage with an increased danger of rockfall.
- Potential increase in the number of visitors to the cave.

There used to be much in-situ equipment such as bolts, pegs, permanent ladders and handlines, and sump rope, especially at pitch heads, much of it old and abandoned, but with the use of ecobolts this has considerably reduced. Dye testing resulting in the temporary coloration of the water in an effort to trace the hydrology of a cave is not really a problem but the loss and abandonment of dye detectors can be. Camping underground is relatively uncommon but where it does occur it poses much the same kind of environmental problems as on the surface: littering potential, disposal of human waste, and large amounts of gear. Occasionally deaths occur underground and in Britain at least one caver has been concreted into a squeeze where he became stuck and more bodies lie lost in sumps where they drowned.

An unusual attempt was made to document the environmental impact of caving at Pridhamsleigh in Devon by Sargent (1998) as he had noted that mud was disappearing or had totally disappeared from parts of the cave, for example, there was a 2 m trench in Bishop's Chamber. He measured the amount of mud from different types of caving suits and estimated that 0.771 kg was the average amount of mud extracted from the suits/trip, which if the estimated 9000 cavers per year was correct was a total of 6939 kg. However, different caving suits absorb different amounts of moisture and the amount of mud absorbed. The technique was based on 30 measured suits with three different types: a cotton boiler suit, a Daleswear lightweight Cordura suit, and a Warmbac heavy duty Cordura suit, and weights were measured before caving, after caving, and after drying. There was a lack of data for mud on other caving gear and on hands and faces, but after one trip it was found that the total mud that could be added to the suit figures was 0.224 kg for the belt, 0.346 kg for the helmet, 1.218 kg for the boots, 1.137 kg for the battery, and 0.187 kg for headset. This totalled 3.112 kg for the gear. This cave is very muddy but the figures of mud eroded from the cave and extracted are large, even if the figures are only best estimates and the total number of cavers is again an estimate. There could be a major loss of scientific information from some caves because of this type of erosion where there are large muddy sections or even mixing of the sediment stratigraphy as at Black Chasm Cave (California).

Light pollution in many popular caves and show caves can cause unnatural algal growth on cave walls as the atmosphere and temperature can be changed. For example, a single person releases heat equivalent to that of a light bulb, and a single party of 87 tourists raised the temperature of the cave air by 1.5 °C during a five-minute visit at White Scar Caves (Yorkshire Dales). This can affect the water vapour capacity as a 1 °C temperature rise results in an eightfold increase.

In the artificially illuminated parts of caves, the development of heterotrophic biofilms and phototrophic communities serving as primary producers is common. This community, generally known as lampenflora, is usually composed of different microbes, eukaryotic algae, cyanobacteria, bryophytes, mosses, and ferns, for example, in Reed Flute Cave, Guilin, Guangxi (China), where there is multi-coloured lighting and extensive lampenflora in the show cave and in Korean show caves (Byoung-woo 2002). The lampenflora adheres strongly to the substratum and deteriorates speleothems. Nutrients and moisture levels are often sufficient to support its growth. Rock surfaces, sediments, and artificial materials around lamps often become colonised by these phototrophs.

Biomass fixed due to light energy, together with other organic matter brought by tourists on clothing and skin, becomes available to cave organisms. The lighting system can alter the microclimate, favouring the growth of photosynthetic organisms, as happened in the Lascaux caves, France, where algal colonisation damaged the cave paintings. Lampenflora is completely dependent on light, as the light saturation point of these species is quickly reached at the cave temperature. These phototrophic communities are inappropriate from an aesthetic point of view, cause degradation of colonised substrata, and produce weak organic acids that can slowly corrode the speleothems.

Direct impacts that are particularly relevant to cave microclimate include construction of access routes through caves and entrance modifications that alter cave airflow and elevated air temperatures from the accumulated body heat from large numbers of visitors. The build-up of carbon dioxide in the cave from human breath can combine with moisture to corrode speleothems and bedrock. Dust accumulation in the cave can also be a problem. Cave dust is composed of lint from clothes, hair, and flakes of dry skin that provide additional food sources for carbon dioxide-producing bacteria and from microbial activity in general. Similarly, abandoned wooden walkways and railings provide food sources for micro-organisms, resulting in decomposition and increased carbon dioxide emissions into the cave air (Russell and MacLean 2008). Cave lighting may heat up and dry the ambient air, inhibiting speleothem growth. Although broad spectrum emission lighting commonly leads to the growth of lampenflora (algae and mosses) on clastic sediments, speleothems, and cave walls, narrow spectrum and relatively cool LED lights reduce lampenflora growth and heat output. Many of these impacts are cumulative and often lead to irreversible degradation to the cave ecosystem.

Sources of carbon dioxide in show caves, such as the Waitomo Glowworm Cave in Tasmania are:

1. respiration of people in the cave
2. outgassing from water flowing through the cave and from vadose waters
3. oxidation of organic material and respiration by micro-organisms
4. diffusion of soil gas through soil and rock into the cave.

In the absence of air exchange with the outside environment, the concentration of CO_2 in the cave air is a function solely of the rate of CO_2 input from sources 1 to 4 above.

Sinks of carbon dioxide in caves are:

1. airflow and air exchange with the outside (ventilation)
2. solution in undersaturated cave water
3. diffusion through (porous) cave walls.

CO_2 concentration in the cave air is normally greater than that outside, so ventilation is the major control on the concentration of CO_2 in cave air. In show caves, humans are clearly the major cause of elevated concentrations of CO_2, directly through respiration and, to a lesser extent, indirectly by promoting the activity of bacteria and other micro-organisms that feed on organic matter, including skin and hair shed from the human body. People exhale air that is slightly depleted in oxygen and enriched in CO_2 (approximately 4% CO_2). Concentrations depend on visitor numbers and ventilation rates through the cave. A single person exhales CO_2 at approximately 17 l hr^{-1}, and thus a tour group of 200 visitors expels about 3360 l hr^{-1}. Concentrations of carbon dioxide of up to 5000 ppm have been recorded in the Waitomo Glowworm Cave, but the allowable level that should be specified in cave management guidelines is open to debate (Cigni and Burri 2000; Dragovitch and Grose 1990). Added to this is the concern that when carbon dioxide concentrations exceed about 2400 ppm in the Waitomo caves, water can combine with CO_2, forming a weak acid, which can lead to corrosion of limestone features of the cave.

As Gillieson (1996) estimated the number of visitors globally to show caves at the end of the twentieth century at over 20 million and Aley (1976) suggested at least 5 million/yr visit show caves in the USA alone, the light pollution problems outlined above are important and have been discussed by Dragovitch and Grose (1990), Russell and McLean (2008), de Freitas (2010), and D'Agostino et al. (2015).

External impacts to cave sites include all the usual problems associated with the use of an outdoor recreation site such as parking area problems; litter; congestion; damage to flora, walls, and fences; and footpath erosion and damage to walls and fences resulting from the access routes to caves. There is also the dumping of waste into caves by farmers.

Most threats to caves can be attributed to unintentional damage and general wear and tear by cavers. To reduce this threat, leaders of groups should encourage the careful movement of individuals and their equipment underground. Much of the threat which remains is specialised in nature and to reduce it requires debate and decision-making by individuals with a similarly specialist knowledge, and usually this will be by those who are the cause of the threat originally. As Britton (1975) suggested though "only unpopular actions are effective in preserving caves and unpopular actions succeed only when the prevailing climate of opinion renders them acceptable."

12.3 Cave Fauna and Flora

Cave fauna is often highly specialised, adapted to live in such an environment and often endemic and very rare and includes snails, spiders, beetles, shrimps, pseudoscorpions, and cave fish (examples in Fig. 12.4A, B).

Fig. 12.4 (A) Blind albino crab (*Munidopsis polymorpha*), Jameos del Agua lake, Malpais de la Corona, Lanzarote. Photo by D. Huddart. (B) *Zospeum tholussum*, a microscopic cave snail completely blind with a translucent shell, from the Lukina Jama-Trojama cave system (Velebit Mountains, Croatia). Ospeum species (Gastropoda, Ellobiidea, Carychiides) from 980 m depth in the Lukina Jama-Trojama cave system (Velebit Mtns, Croatia) [*Subterranean Biology*, 45–53. Authors: J. Bedekand and Alexander M. Weigand]. Photo by J. Bedek. (C) Devil's Hole (*Cyprinodon diabolis*). Source: National Digital Library of the United States Fish and Wildlife Service. Photo by Olin Feuerbacher (USFWS)

Examples include the Bone Cave harvestman in Travis and Williamson Counties (Texas) which was added to the United States Fish and Wildlife Service (USFWS) Endangered Species list in 1988, along with others such as the Kretschmarr Cave mold beetle and the Tooth Cave pseudoscorpion. Warton's cave spider in 1992 was described from a single individual in Pichule Pit, a shallow cave, in Travis County and was last seen in 2000. There have been lawsuits from the landowner, who gated the cave with a lock which is now rusted shut, and this has been a test case for conservation against the rights of landowners. Another example is the blind cave beetle (*Leptodirus hochenwartii*) has been found in a cave in the Inner Carniola region, Slovenia, but the exact location has not been given because of conservation concerns. The Devil's Hole Pupfish exists with a population of under 200 individuals in a single groundwater pool in a Nevada cavern (Fig. 12.4C).

This fauna plays an essential role in underground ecosystems by decomposing organic matter and recycling nutrients through the food web. Many of them are very rare and include ancient, primitive forms no longer found on the surface. They provide important information for studies of evolution and ecology. However, although cave animals are adapted to living underground, it is vital we recognise that their ecosystem is linked to the surface above and any changes we make here can affect their subterranean habitat. Many cave creatures live in the water and feed on debris washed into the cave. Others feed on creatures that live in the water. For example, the glow-worm builds its silken nest above streams and uses its light to attract caddisflies and other insects (caddisflies have an aquatic larval stage). For all these creatures, maintaining an unpolluted water supply is vital.

Cave environments are strongly buffered against the daily, seasonal, and longer-term surface climatic changes. They provide stable, sheltered, and moist refuges for animals which might otherwise not survive on the surface. Green plants cannot grow in the complete darkness of caves, so the food supply for cave creatures must ultimately come from the surface. Plant material falls or is carried in by streams, while animals wander into caves, fall, or become swept underground.

Hence, cave ecosystems directly depend upon the surrounding surface environment. This means it is essential that we maintain the natural soil, vegetation, and water quality around caves. The special nature of karst makes it particularly vulnerable to degradation and such areas should be treated with special care.

12.3.1 Zones in Caves

The cave environment can be divided into four distinct zones:

- *Entrance zone*: here the surface and underground environments meet.
- *Twilight zone*: here light progressively diminishes to zero. Plants such as ferns, mosses, liverworts and algae cannot grow beyond the limit of light penetration.
- *Transition zone*: light is absent here although surface environmental fluctuations such as temperature and moisture are still felt. Cave crickets often congregate here, and on suitable nights venture outside the cave to forage for food.
- *Deep zone*: remote from entrances, the deep zone is completely dark. Here the relative humidity is high and evaporation rate is low. Temperature is nearly constant all year around.

Creatures living in this latter zone have become adapted for life in the dark, no longer needing vision. They are called troglobites, they may have reduced body pigment and eyes and longer legs and antennae to help them find food in the darkness. Only small amounts of food ever reach the deep zone so troglobites have to survive long periods without food.

There is also a long list of specialised marine cave species, many again extremely rare and endemic to a single cave or area. The world register of marine cave species can be located at www.marinespecies.org/docs/activities/2015/WORCS_report.pdf. This is now an opportune place to briefly discuss the possible cave diving effects and suggest that preventing cave damage is every cave diver's responsibility. Whether diving in marine or terrestrial caves, divers should try not to disturb silt, they should avoid pull-and-glide propulsion where they grasp the rock and pull themselves forward to advance movement because this can cause cave damage, especially in low-flow caves. Ceiling push-off should be avoided due to potential damage by feet, and they have to be careful with back-mounted tanks to avoid damage.

12.3.2 Bats in Caves

The largest aggregates of living vertebrates are found in caves (bats), and in the 1960s the midsummer colonies of adult Mexican free-tailed bats (*Tadarida brasiliensis*) in 17 caves in SW USA were estimated at 150 million individuals. However, the survival of many bat species depends on natural caves. For example, of 39 bat species in temperate America, 18 rely substantially on caves, including 13 species that dwell in them all year whilst the remaining 5 depend on caves for hibernation sites. The figure in China is much greater as 77% of the known bat fauna (101 out of 131 species) roosts in caves there (Luo et al. 2013). The bats provide many benefits for ecology outlined by Furey and Racey (2015) and Medellin et al. (2017), including the guano as a source of food for many invertebrates.

Recreation users of caves can disturb both hibernating and nursing bat colonies, and disturbance is likely to be more severe if there is a large party in a system occupied by bats. Thomas (1997) showed that non-tactile disturbance from seemingly innocent cave visits during hibernation periods can cause bats to arouse and maintain significantly greater flight activity for up to eight hours afterwards. Such arousals are highly detrimental to their overwinter survival, and non-tactile disturbance during other critical periods such as reproduction may lead to: (1) death of young that lose their roost-hold and fall to the cave floor, (2) females abandoning the roost for less ideal sites where prospects for reproductive success may be reduced, (3) greater energy expenditure among females and less efficient energy transfer to young (translating into slower growth of young and increased foraging demands on females), and (4) reductions in the thermoregulatory benefits of a roost as a result of decreased numbers of bats frequenting the site.

Bats have a fat reserve which they use during hibernation and if a bat is awoken from hibernation then the fat reserve is consumed far faster than that consumed during hibernation. Responses to tactile stimulation showed a significant increase in energy expenditure when handled (Speakman et al. 1991) which shows that disturbance can also be caused by conservationists monitoring bat populations and in some circumstances may be of more significance than the more general disturbance exerted by cavers or tourists. Speakman et al. (1991) predicted that each non-tactile disturbance decreased fat stores by 0.01 g whilst each tactile disturbance decreased the stores by 0.05 g.

As a result, uncontrolled human disturbance often leads to decreases in numbers of bats roosting in caves and mines (Tuttle 1977). For instance, disturbance in caves in West Virginia, USA, occupied by the Indiana myotis (*Myotis sodalis*) and Townsend's big bat (*Corynorhinus townsendii*) resulted in a decline from 1137 bats to 286 in one cave and from 560 to 168 in another (Stihler and Hall 1993).

The increase in cave tourism has caused problems, especially in South East Asia, and the commercialisation of Fourth Chute Cave in Quebec resulted in the abandonment of the largest hibernacula of the eastern small-footed Myotis known at the time in Eastern North America (Mohr 1972).

The biggest problem facing bat populations in North America since 2006 has been the accelerated loss of bats due to the white nose syndrome (WNS) (Fig. 12.5A for a healthy bat and Fig. 12.5B for an affected bat).

Populations in some caves have dropped by 90–100% caused by the spread of the fungus *Pseudogymnoascus destructans*. It has killed over 6 million hibernating bats in North America since the winter of 2006–2007 when it was detected on a bat in a cave near Albany (New York). It has spread to 29 states as well as 5 Canadian prov-

Fig. 12.5 (A) Healthy little brown bats, Aeolus Cave or Dorset Bat Cave in the Taconic Mountains in East Dorset, Vermont. Before white nose syndrome reduced the bat population, it was known as the largest bat hibernaculum in the north-east USA. Source: US Fish and Wildlife Service Headquarters. Photo by Dolovis. (B) Bat roosting in cave with white nose syndrome, Greeley Mine. Vermont. Source: National Digital Library of the United States Fish and Wildlife Service. Photo by Marvin Moriarty/USFWS

inces and is spread from bat to bat, although it is partly spread by cavers. The USFWS recommend anyone in caves follow the agency's white-nose decontamination protocol. This elaborate procedure involves the washing of bodies, clothing, vehicles, and equipment, and there is restricted access to caves where detected.

12.4 Management Strategies to Conserve Caves

12.4.1 Potential Strategies

There are many potential management strategies, and each one depends on the individual cave concerned, and each cave should have one or find one of its own. More popular caves will need a more in-depth strategy than caves that are rarely used. However, whatever the status of the cave, it has to be appreciated that once damage has been done there usually can be no rectification in our lifetime as caves have developed over thousands of years and sometimes much longer. However an example of the removal of graffiti from Buckner Cave in Indiana on the Richard Blenz Nature Preserve which was donated to the NSS which was heavily graffitied by spray painting can be seen where sandblasting removed the worst damage.

As Aley (1976) suggested "The carrying capacity of a cave is zero," and because most damage and environmental change are irreversible, there needs to be determined the environmental management techniques that are appropriate for a given cave. The cave manager should be concerned with defining the desired, or optimal level, or range of environmental conditions that should occur and then maintain them in that cave. The cave system is the only true natural ecosystem available for study in a country such as the UK but only lasts a very short time period after the cave discovery. Caves provide a unique environment but also a unique challenge to the conservationist. They are vulnerable yet are also subjected to concentrated human pressure. However, throughout the world there are various organisations responsible for cave management. For example, in the USA the Bureau of Land Management (BLM) manages nearly 800 caves in the 11 western states, whilst the National Park Service (NPS) manages caves and karst scenery in 120 parks (81 contain caves and over 3900 caves are known throughout the park system). The United States Forest Service (USFS), under the Department of Agriculture (DOA), manages 193 million acres in the form of 155 National Forests and 20 National Grasslands. The mission of the United States Department of Agriculture Forest Service is to sustain the health, diversity, and productivity of the Nation's forests and grasslands to meet the needs of present and future generations. National Forests provide sustainable forest products, mining leases, and recreation opportunities, including in caves across the country. Good examples of cave and karst management can be taken from the Arizona National Forests where the following cave and karst management guides have been published for Kaibab National Forest (2014), Apache-Sitgreaves National Forest (2014), Coconino National Forest (2014), Tonto National Forest (2014), and Coronado National Forest (2012). Along with items such as cave classification, monitoring, and inventory procedures, there are caving ethics for both Forest personnel and the general public. There is also a general Arizona National Forest Cave and Karst Management plan (Keeler and Bohman 2013) which illustrates clear and acceptable guidelines and policies that can be implemented in a uniform way. Karst and Cave Areas are designated as a separate land use designation in the Forest Plan, and the latter can be updated without having to go through the extremely long Forest Plan amendment process. The Plan draws from and highlights the relevant sections of federal laws and statutes, including the United States Code (USC), the Code of Federal Regulations (CFR), and Forest Service manuals (FSM, US Forest Service 2009). The USFWS manages approximately 96.4 million acres of land in the form of roughly 545 national wildlife refuges and approximately another 90 districts and areas. The National Wildlife Refuge System Administration Act of 1966 identified lands under which the USFWS was to manage for the protection of wildlife and wildlife habitat. Their mission is working with others to conserve, protect, and enhance fish, wildlife, and plants and their habitats for the continuing benefit of the American people. The NPS manages approximately 84.6 million acres in the form of 391 units, 58 of which have national park designation. Over 4000 caves have been identified from 85 NPS units. The NPS mission is: "to promote and regulate the use of the national parks…which purpose is to conserve the scenery and the natural and historic objects and the wildlife therein and to provide for the enjoyment of the same in such manner and by such means as will leave them unimpaired for the enjoyment of future generations." NPS policy also states that all caves within their management are significant and thus will be managed to their fullest protection. Nolfi (2011) illustrates that this always has not always been the situation with a case study approach from the Great Smoky Mountains National Park, Harley et al. (2011) established a cave inventory for West Central Florida caves to stimulate the development of management strategies, whilst Donato et al. (2014) described a conservation status index for the management of cave environments.

At 66 million acres the Bureau of Indian Affairs (BIA) manages a substantial amount of land, but not all the land is managed for public use and within the lands, management varies significantly based on resources and needs. Although cave protection is provided through several acts of congress, the Federal Cave Resources Protection Act (FCRPA) does not apply to BIA lands and thus provides no protection to caves they manage. It is important to note that additional tracts of federal land are managed by agencies that do not fall under the jurisdiction of the Department of the Interior (DOI) or DOA and therefore are not bound by FCRPA. That does not imply that cave resources are not considered in land management. For example, the Department of Defense (DoD) manages over 25 million acres and the Department of Energy 2.4 million acres. Significant cave resources fall under management of each of these agencies. The DoD's Legacy Program has assisted in the identification of 18 new cave species from 2 Army bases in Texas. Close to 1 million dollars was spent over 12 years to find and research caves and cave fauna at those two bases (Elliott 2005). Tennessee Valley Authority manages over 293,000 acres and has known cave resources. When the FCRPA does not apply, cave protection is often afforded under the Endangered Species Act, 1973.

12.4.2 Federal Cave Management and the National Park Service

With the development of the FCRPA, its federal land management agencies under both the DOI and the DOA are required to inventory and list significant caves on federal lands and to provide management and dissemination of information about caves. In 1998 Congress passed the National Cave and Karst Research Institute Act of 1998 in order to further promote cave and karst research. In addition to these broad federal regulations regarding cave and karst management, the NPS is also guided by more specific legislation, such as the Lechuguilla Cave Protection Act of 1993 which protects land above and around the cave. In order to fulfil these obligations, federal land management agencies are continually devoting increased resources to karst management, as concepts and practices in cave and karst management continue to evolve. The NPS' cave management falls under the advisory of the Cave and Karst Program. One-hundred and twenty park lands have identified cave and karst features, with 85 containing caves. Under the FCRPA and CFR Title 43—Public Lands: Interior, Part 37 - Cave Management, the NPS designates all caves as significant caves and manages accordingly. NPS resource managers are guided in managing, protecting, and conserving all natural resources in their unit by the NPS' Director's Orders guidance; Natural Resources Management Reference Manual (RM#77). The guidelines under RM#77 specify the policy and programme directives, the authoritative legislation, methods of protection and fulfilment of legislation, as well as an explanation of the roles and responsibilities of those who are in position to manage caves and karst. Within the NPS' RM#77, the Cave and Karst Management section provides guidelines for the management of caves, encompassing the many disciplines necessary to protect and perpetuate natural cave systems. Guidance is oriented towards the needs of anthropogenic challenges within caves ranging from resource planning for karst protection to direct management of developed caves (as in "cave parks," such as Mammoth Cave or Carlsbad Caverns National Park). It is stated that parks with small, undeveloped caves should adapt and apply relevant management as they see fit for their conditions. Management of caves includes protection of soils, surface landforms, natural drainage patterns and hydrologic systems, and cave microclimate and ecosystems (RM#77). Although NPS units with cave resources are mandated by RM#77 to develop and implement a cave management plan, many currently do not employ cave management plans.

However, several NPS units employ cave and/or karst specific management plans for optimal management. Several of these provide developers of management plans with an understanding of the concerns and the needs to make plans effective and efficient. Plans from Carlsbad Caverns NP (2006), Grand Canyon NP (GRCA) (2007), Sequoia and Kings Canyon NP (1998), Timpanogos Cave NM (TICA) (1993), Cumberland Gap NHP (1998), Wind Cave NP (2007), and Jewel Cave NM (2007) are good sources of information applicable to most managers in developing specific cave and/or karst specific plans.

12.4.3 NPS Cooperative Relationships

The NPS has a memorandum of understanding (MOU) with the NSS for the purpose of support and encouragement of the NSS' involvement in the inventory, scientific study, management, planning, and protection of cave resources on agency-administered lands. In accordance with this MOU, the NPS will provide access to caves under their management, advise opportunities for cave-related studies and projects, advise of NPS research and cave management policy, assist to develop and implement safety programmes and search and rescue plans for the cave- and karst-related projects/studies, and acknowledge the work products and data gathered by the NSS. There is also a specific MOU with the Cave Research Foundation (CRF) to facilitate project development where they are the primary collaborator for in-cave scientific research. The American Cave Conservation Association has an MOU with the NPS to foster stewardship relationships with commercial cave interests in national parks. They also have worked to define guidelines and assistance for cave gating projects. In addition, Bat Conservation International (BCI) works within an MOU with the NPS to provide guidance, support, and protection of bats in the USA. These MOUs, as well as others, all foster protection to cave- and karst-related resources within the NPS.

12.4 Management Strategies to Conserve Caves

As mentioned earlier, there is also an Interagency Agreement for the Collaboration in Cave and Karst Resources Management between the NPS, USFWS, BLM, United States Geological Survey, and USFS which addresses a need for collaboration to achieve efficient management. The purpose is to achieve a more effective and efficient management of caves through their cooperation in understanding mutual concerns and avenues for better management. This need for cooperation fulfils the FCRPA and the National Cave and Karst Research Institute (NCKRI) requirements for exchange of information and cooperation.

Then there are various privately owned caves, especially show caves, and in the USA the National Caves Association was founded in 1965 by a small group of over 80 show cave owners. In the USA the National Cave and Karst Management Symposium has been an important forum for promoting, advancing, and sharing concepts of effective management of cave and karst resources for over 30 years.

12.4.4 Canadian National Parks

At least 12 of 41 national parks in Canada have caves. A group of six parks in western Canada are adopting cave management guidelines using a three-tier classification system to manage access (Horne 2005). Class 1 caves are access by application: highest resource value, not for recreation, each visit must add knowledge, or give net benefit to the cave.

Class 2 caves are access by permit where recreation use is allowed, there are some management concerns and education/orientation is possible during permit process. Class 3 caves have unrestricted public access with few or no management concerns and no permit is required.

In order to determine which class each known cave is in, three sets of factors are considered: (a) cave resources, (b) surface resources, and (c) accident and rescue potential. Cave exploration in the western Canadian mountain national parks only began in the 1960s, and this current access policy has been influenced by the remote rugged nature of the landscape and the need to work with speleological groups to explore and document park features. A change in park staff awareness of the resource has contributed greater exchange of information and opportunities for cavers to gain access and the park to know more about its resources.

12.4.5 Access Agreements and Physical Barriers

The most obvious way of controlling damage is controlling the access to caves and/or imposing a physical barrier to the entrances. For example, in the UK, access agreements negotiated by the Council for Northern Caving Clubs (CNCC) are given on their website, and examples are given below to illustrate some of the restrictions. For Birks Fell, Redmire Farm, Buckden, near Skipton, the agreement is with Messrs. Dacre Son and Hartley, for, and on behalf of, W.A.G. Watson:

- Access to the cave is by track from Redmire Farm only.
- Agreed access is through Birks Fell Cave entrance only unless written consent is obtained from the agent and tenant of Redmire Farm.
- No cars to be taken to Redmire Farm. Buckden car park to be used.
- All gates on the access track must be closed.
- No camping permitted.
- Access to be granted to member clubs, one per day.
- No access from 1 November to 15 April the following year.
- CNCC to be responsible for making good any damage resulting from the access to the cave and any claim arising from the damage.
- CNCC will indemnify Mr. Watson, his agents, and tenants against any claims for accidents or damage. All persons will visit the cave at their own risk.
- Agents and tenant Mr. Horner to be notified monthly in advance of all bookings.
- Member clubs must call at Redmire Farm, when going to and when returning from the cave, but, the tenant does not accept responsibility for notifying the authorities in an emergency.

- The tenant may deny access on any day by giving reasonable notice.
- The owner retains the right to terminate the agreement at any time by giving written notice.

For Casterton Fell the access agreement is administered on behalf of the Whelprigg Estate by the CNCC. Club access is only for CNCC and British Caving Association (BCA) member clubs. It is a condition of the access agreement with the Whelprigg Estate that novice cavers are not permitted into the Easegill system and that the system is not used for training cavers in caving techniques (other than the techniques used by experienced cavers, for example, photography, surveying, and conservation).

- Five permits per day at weekends and two permits per day on weekdays with a maximum of eight cavers per permit and a maximum of two cars per permit.
- Written application on club letter headed paper (with stamped addressed envelope) or via email if you have applied in the past.
- Subject to availability, permits can be issued at short notice, however the fell is often booked up several weeks in advance so as much notice as possible is best.
- CNCC must provide a list of authorised clubs. Access to all the caves must be on the agreed routes; these routes are displayed on the reverse of the permit.
- Cavers must abide by the countryside code and the cave conservation code. Particular attending must be applied during the breeding and nesting season for birds and also at lambing time.
- Breaches of the access conditions can result in the withdrawal of future permits and can in certain circumstances cause the Fell to be closed.
- No digging or explosives are allowed on the Whelprigg Estate's land.

12.4.6 Secret Conservation

This is where a cave discovery is not publicised so few cavers visit. It is frequently adopted at the start of a find but it is only suitable in the short term as it is elitist, divisive, controversial, and often counter-productive as a conservation technique. We have seen from an earlier example information related to cave discoveries spreads quickly so secret conservation does not work.

12.4.7 Zero Access

The most radical and revolutionary form of conservation is to have zero access to a cave and the thinking behind this is simple: people damage caves, caves are delicate environments susceptible to damage, and therefore no people, no damage. However, this raises many arguments regarding the educational, censorship, and freedom of movement aspects and has moral issues too but it cannot be neglected as a conservation tool. In Britain an example of this approach are the Stump Cross Caverns which is really a show cave, but this does not cover the whole cave and there are other sections which could be explored. The book *Northern Caving* suggests that permission to enter these sections of the cave is unlikely to be given but it is possible to write to the show cave to try and gain access.

12.4.8 Restricted Access

This management tool only allows certain groups to enter the cave system, for example, educational groups, research groups, exploration groups, caving club groups affiliated to a national park committee and general public groups in show caves. There is a set system of entry organised. An example would be the leader system in the Mendips for St. Cuthberts, Shatter, Withyhill, and Reservoir caves where parties are limited to not more than five people as greater numbers are believed to increase carelessness and damage. However, Stanton (1982) suggests that in this system "they demand a good deal of determination and dedication on the part of the leaders, deterioration still occurs but at a much slower rate." Some caves require a recognised leader for the trip as in Dan yr Ogof (South Wales) where the leader is generally someone from the South Wales Caving Club who has visited the cave at least three times and has proven that they are aware of conservation and safety issues. Again groups are generally limited to five people so that each person can talk to each other without having to overtake another person and risk straying from the path and disturbing formations. It may be looked upon as elitist, and if combined with gating then it could be visually offensive to the natural environment.

12.4.9 Periodical Access

This refers to where certain groups have access only at a certain time of year, in other words a caving season. This would intensify the use over a single time period and the idea is that the cave could recover over the closed season. However there is no real evidence that this is the case. In the Yorkshire Dales, this closed season is different for different systems, for example on Leck Fell and the Pipperkin-Nipperkin system, it is between 1 April and 30 June whilst in the Mongo Gill-Shockle Shaft, it is during May and July. The major difficulty here though is the enforcement of access.

12.4.10 Booking

This can be best achieved through a management strategy, but this lack of spontaneity and rigidity by having to book to take part in one's chosen pastime can put people off either taking part or booking. Again the Leck Fell system requires written application to the CNCC one month in advance, and the problem is again how to enforce this system. In the USA, for example, the Great Basin National Park in Nevada has over 40 caves, and to cave there has to be an application for a cave permit at least 2 weeks before the trip. They are approved for those who can demonstrate experience with both horizontal and vertical techniques, cave conservation ethics, and expertise with the required equipment and can certify that their equipment is clean and disinfected. This permit must be in possession whilst caving, and the group is limited to between three and six people.

12.4.11 Gating

This creates a barrier to the cave user. The most common form is a padlocked gate to the system where the caver first must obtain the key before descending (Fig. 12.6A, B). This form of restriction is effective as the caver must belong to a recognised caving club and it is thought that such club members have a greater respect for the cave in question and use it with greater sensitivity, but it does not always work as damage is still caused by club cavers. It is not always practical either as some caves have too many entrances like Porth yr Ogof. An example is Craig y Ciliau, National Nature Reserve, Agen Allwedd, where permission to use the cave must first be received from the Agen Allwedd Cave Management Committee. This involves applying and booking at least two weeks in advance of the proposed trip and involves such information as the name of the caving club or organisation, the leader's name, the number in the group, the date of the proposed visit, a deposit, and a stamped addressed envelope. On receipt of this information, a decision will be made to allow access, and if access is allowed a key is sent to the leader. This only applied to a "normal" caving trip, and extra permission is required for underground camping, surveying, exploration, or diving. Permission may be refused for any reason which the permit secretary considers is valid.

Occasionally gating might be considered actually in the cave, for example, in the White River Series in Peak Cavern (Peak District) where the discovery team (13 May 1991) felt that the formations were too delicate. This they felt justified and the gating was established from a conservation viewpoint, but they also considered that they had worked very hard in opening up the system and they deserved the satisfaction of completing the survey (Hewitt 1992). This is an example of selective access which stopped not just inexperienced but experienced cavers too, although it did not last long.

Fig. 12.6 (A) Gating at Agen Allwedd, Llangattock, South Wales. Photo by D. Huddart. (B) Bat Gate at the entrance of Skeleton Cave, near Bend, Deschutes County, Oregon; lava tube on the northern flank of Newberry volcano. Photo by United States Forest Service

12.4.12 Sacrificial Caves

In Britain the BCA encourages novice groups to avoid sensitive caves and focus activities on those caves that are capable of sustaining the pressure. This is a honeypot management approach and that by agreeing on specific caves as sacrificial where conservation interests are no longer the prime consideration, it will reduce pressure on other caves. The best examples in the Yorkshire Dales are the Long Churn system or Great Douk and in South Wales, the Porth yr Ogof system, where a car park was built to encourage use (Fig. 12.7A–C). These caves must have a high educational value, and there have to be examples of the need for conservation, otherwise the danger is that too low a priority is placed on the education in the activity and a new generation of cavers may be created oblivious to the need for conservation.

12.4 Management Strategies to Conserve Caves

Fig. 12.7 (A) Great Douk Cave entrance and Great Douk Pot (collapse section of the cave). Entrance on ledge to the right of the photo or up the waterfall in the middle. The cave dig illustrated in Fig. 12.3 is up against the left wall of the collapsed section. Photo by Tim Stott. (B) Long Churn passage (Yorkshire Dales), active streamway, phreatic upper passage with vadose trench incised showing several water levels. Sacrificial cave used by many centres and schools. Photo by D. Huddart. (C) Great Douk sacrificial cave. The author in the upper part of the cave, note the flat bedding plane, the vadose trench, and the flowstone curtains on the left hand wall. Photo by Tim Stott

12.4.13 Endurance Conservation

Endurance conservation is where the caving is hard, awkward, and tight or the inner parts of the cave are at a distance from the entrance and therefore experience fewer cavers. There can be natural barriers to cavers that can protect passages such as sumps, ducks, canals, climbs, bolt routes, and big pitches, such as Titan, in Peak Cavern (Peak District). However, damage can still occur and examples quoted in the Cave Conservation Handbook (from National Caving Association—NCA) reinforce the view that "competent cavers are not necessarily good conservationists" and that education is an essential requirement for any long-term strategy.

12.4.14 Artificial Obstacles

Where there are vertical pitches, each party should have to place and retrieve their own bolts so that it will reduce access to those groups with the correct gear. It may be possible to create artificial sumps or block popular routes through popular caves leaving only the harder routes so as to discourage the numbers. For example, it had been suggested that in the Swildon's Hole, (Mendips) there should be the creation of an artificial sump at the bottom of this cave which would conserve formations in Barnes' Loop where there had been considerable damage, but this seems too late for this cave once damage has occurred.

12.4.15 Zoning Off

The process of zoning off certain sections or formations (spot taping) in order to make clear to all cavers not to proceed beyond a certain point has been tried, but plastic tape can move on uneven cave floors and not all cavers will follow the tape (Fig. 12.8). However, tape can mark a path to be followed, or cavers can walk as close to the tape as possible or on the far side of any undisturbed sediments or formations. Raised taping about 20 cm above the floor is maybe a better system to highlight the areas to be avoided. Route finding is often used as a justification for taping, but it does not do anything for conservation and should be unnecessary with good education. Taping is impractical too for some of the bigger and well-decorated caves. A solid version is where a boulder wall is erected to mark a formation, sometimes with the addition of tape. However, although

Fig. 12.8 Zoning off of stalagmites, Matienzo Caves, Spain. With kind permission from Matienzo Caves Project

12.4 Management Strategies to Conserve Caves

successful in places as in Midwinter Chambers in Ogof Draenen (South Wales), it is again not a natural system to be recommended.

12.4.16 Formation Repair Work

It may be possible to rebuild broken formations using resin for small areas or pins for bigger repairs, but usually the pieces cannot be located and repair is generally not a realistic proposition. However, this has been attempted reasonably successfully at Matienzo Caves in central Spain (matienzocaves.org.uk/miscpics/conserve/intro.htm) and at many caves in the USA as at Carlsbad Caverns (New Mexico), Oregon Caves National Monument, and Kartchner Caverns (Arizona), and detailed methodology can be found in Hildreth-Werker and Werker (2006).

12.4.17 Exploration Policy

The potential benefits of any dig should be weighed against the disadvantages; the dig should be organised by cavers, either under the auspices of the appropriate Regional Council in Britain or through a caving club which has access agreements. In the case of open-access caves, a liaison group of interested parties should be established. Where digs occur care should be taken to minimise any damage done, and excavation should be kept to the absolute minimum. Speleothems and any cave sediments of archaeological value should be left undisturbed, but if disturbance is unavoidable, everything must be recorded and made available for research. Sections cut in sediments should be sampled, recorded, and made available for research, but of course the big problem here is that often the explorer does not have the experience or skills to record accurately the cave deposits. Dig care should adopt a commonsense attitude such as building a wall for protection from blasting so reducing potential damage from flying debris, hiding debris in dead-end passages, although creating a path with the debris produced is controversial as it gives the cave an unnatural appearance. The CNCC issues digging guidelines for sites of scientific interest which involve initially obtaining the landowner's permission. Then the landowner must submit a "Notification to Undertake Works" form to Natural England whose Conservation Adviser will make a site visit and a consent form will be issued for a specific time. There are also guidelines for digging published for cavers and resource managers by Jones et al. (2005) and an online journal and website Cave-diggers.com edited by Passerby (2002) up to the present.

12.4.18 Cave Adoption Schemes

Where clubs and cavers take responsibility for a particular cave, monitor its condition and undertake regular clean-ups. For example, the Red Rose Caving Club and the Easegill area (Pennines) and the Buttered Badger Potholing Club were cleaning up Oxlow Caverns East Chamber and North Rift in Giant's Hole (Peak District) with the latter club collecting 11 bags of rubbish in two trips. In the USA there have been many cave clean-up schemes, such as the graffiti clean-up at Bloomington Cave just outside St. George (Utah) when 48 volunteers took part over 7 weekends in 2005, which involved over 1000 hours of volunteer time. Not only did the cavers contribute time but they contributed over 90% of the project's total cost. The sandblasting technique used was discussed by Jasper and Voyles (2005), but great care must be taken as damage can occur to the rock, formations, and cultural artefacts, and some of the chemicals used in graffiti removal can be dangerous to the users. There was a case in South West France in 1992 at Maynieries Cave, near Braniavel (Tarn et Garonne) where prehistoric cave art (15,000 years old bison) was partially cleaned off the walls with steel brushes by 70 scouts before it was realised what they were doing.

12.4.19 Cave Fauna Management

The first priority in developing a strategy to manage cave fauna is to monitor the fauna by surveying to identify rare and sensitive species and

Fig. 12.9 Waitomo Glowworm Caves, New Zealand. Glowworms require careful cave management along with other specialised fauna. Photo by Shaun Jeffers with kind permission from Waitomo Glowworm Caves

habitats. This can result in the development of management options for the fauna's protection especially in relation to visitor use. Sometimes this visitor use is very high and, for example, the Waitomo Glowworm Cave is the most visited cave in Australia (Fig. 12.9), with the average visitor use just below half a million/year, with a daily average of 2296 at its peak in the 1990s.

A research programme was developed to monitor the impacts of cave users on the fauna in Ida Bay, Tasmania (Eberhard 1999), including glow-worms, cave crickets, spiders and beetles which gives baseline data for monitoring comparisons. It was noted that there were over 100 invertebrate species, with some endemic like the very rare and highly adapted blind cave beetle (*Goedetrechus mendumae*) and the glow-worm colonies are the best developed in Australia and are amongst the best in the world. This work developed from the detailed pioneering survey by Clarke (1997). Here a number of Tasmanian cave species have ancient lineages and are considered Gondwana or Pangean relicts. He documented 643 invertebrate species in Tasmanian caves and 159 of these were considered possibly rare or rare, 6 were rare or vulnerable and three were endangered. Two species had not been reported since 1910 and are considered likely to be extinct. Sixty four of the species were considered to be rare or threatened. Although trampling by visitors occurs and there has been collection for scientific research, the managers walk a difficult line between enforcement of conservation ideals and their need to maintain public access (Fig. 12.10).

Many of the threats were outside the caves associated with land-use changes and the effects of forest practice. In the Tasmanian Wilderness World Heritage Area at Ida Bay, Eberhard (1999) suggested that education of cave users was critical to the fauna's protection. Specific Minimal Impact Caving guidelines to protect the fauna were developed and promoted and the vulnerability to visitor impacts of habitat types was assessed. This included illustrated factsheets on cave fauna and Minimum Impact techniques which were made available at the Parks and Wildlife Service shop fronts and distributed to cave users from regional offices and cave sites, from cave permit applications as well as the Parks and Wildlife Service website.

There was an article on Tasmanian cave fauna and Minimum Impact caving published in the Australian Caver journal (Eberhard 1998) and public lectures were given to the local caving club, scientists, cave mangers and cave guides. The restoration of sites was carried out such as the breaking up of compacted sediment floors and the restoration of cave climatic conditions for glow-worm colonies as in Waitomo (de Freitas and Pugsley 1997). Seven Faunal Sanctuaries have been created, such as Keller's Squeeze and the Ball Room stream passage in Exit Cave for a blind cave beetle habitat. These are sites worthy of special protection because of their vulnerability because they have conservation value as examples of optimum representative, or rare habitats and/or animal communities, or because of their value for public interpretation. These

12.4 Management Strategies to Conserve Caves

Fig. 12.10 Green Glow Caves, New Zealand. With kind permission from photographer Donnie Ray Jones

sanctuaries are not open to general access for cavers. Each Sanctuary is delineated by a string line across the passage clearly indicating to cavers that further access is barred and there is an explanatory sign. Route markers were installed to protect sensitive habitats, and there was an installation of a gate in Arthur Folly's Cave and the provision of a permit system.

The use of external cave gates to restrict public access is a common management technique for bat conservation, and an internal cave gating system was used in eastern Oklahoma (Martin et al. 2000). However, cave gating can also impede bat access to caves, and early attempts from the 1950s to 1970s often resulted in abandonment (Tuttle 1977). Gates that are more "bat friendly" have since been designed (see Fig. 12.6B). Berthinussen et al. (2017) suggested that cave gates should be used to restrict public access, although there is some evidence that no increase in bat populations always occurred.

The US NPS has allowed the viewing of the dusk departure and dawn return of a large colony of *Tadarida brasiliensis* bats from an amphitheatre at the entrance of Carlsbad Caverns (New Mexico) but has banned the use of flash photography because of concerns that it disturbs the bats.

12.4.20 Artificial Bat Caves

These have been built to try and fight WNS. The Nature Conservancy in Tennessee have embarked on a radical scheme by building an artificial cave next to an existing natural bat cave (Bellamy Cave). It began construction in August 2012 but by 2017 bats have not hibernated in it in large numbers, although researchers suggest that it can take years for bats to choose a new hibernation location. In Bellamy Cave 40,000–50,000 bats used the cave but after the Nature Conservancy bought the site in 2006 and fenced it, the numbers rocketed to 60,000 in 2010 and 265,000 in 2017, but WNS has recently been found here. Recent research projects which may well prove an answer to the syndrome are the testing of chlorine dioxide, an environmental cleaning agent which may be used to clean man-made hibernating sites such as mines; testing the effectiveness of a natural biopolymer, chitosan, to cure the threatened bats and to test the safety and effectiveness of two antimicrobial and enzyme inhibitors on the affected bats. A further artificial cave has been built at Selah, Bamberger Ranch Preserve (Bronco County, Texas), which has been a success, and over 156,000 Mexican free-tailed bats were counted in May 2011.

The mission of the WCS (Wildlife Conservation Society, Canada) BatCaver programme is to identify and study bat hibernation sites in Western Canada using the resources of cavers and the public to expand knowledge. This information is crucial to conserving bat populations from threats to their survival, such as WNS. In 2017 the organisation produced brochures aimed at people who are visiting caves which explains the risks of inadvertently transporting WNS spores from one region to another. It also contains conservation messaging, decontamination protocols for WNS, and contacts for further information. The brochures have been sent to tourist caves in Western Canada and caving organisations. They have also produced signage regarding bat conservation messaging intended for posting at bat cave entrances and a sign, after consultation with British Columbia Parks, for posting at trailheads to provincial parks.

12.4.21 Management of Lampenflora in Show Caves

Lampenflora is a problem in show caves and particularly when it becomes covered in $CaCO_3$, and such an amorphous mix of dead phototrophs and $CaCO_3$ irreversibly destroys the speleothem's natural heritage and other cultural artefacts, like prehistoric cave paintings. The simplest solution to the problem would be the complete removal of existing phototrophic communities, getting rid of the lights, and the abolition of tourist visits, but this would not be acceptable to cave management. The methods to try and counteract the problems caused by lampenflora are reviewed by Mulec and Kosi (2009) and Cigna (2011), but there is no easy solution. The physical methods include the cleaning of the speleothems overgrown by algae with brushes and water, but this is not recommended because the infestation can be more easily spread throughout the cave, there can be damage to the speleothems, and small fragile flowstones are easily destroyed. Lighting should be shut down when not needed by automatic switches as it has been estimated that the lampenflora cannot develop to any great extent where the illumination does not exceed 100/h/year. Switching off lamps for a prolonged period, for example, one month, counteracts the proliferation of phototrophic organisms. However, this may favour the diffusion of species which are especially resilient like *Phormidium autumnale* and generally cyanobacteria by reducing competition (Montechiano and Giordano 2006). The reduction in light intensity and the use of special lamps that emit light at wavelengths which do not support maximum absorption of the main photosynthetic pigments should help. UV lamps should be switched on when visitors are absent. In Mammoth Cave (USA), light-emitting diodes have controlled the lampenflora, using yellow-light LEDs at an intensity of 49.5 lux, preventing growth for 1.5 years after complete lampenflora removal (Olson 2002).

Any chemical methods for removal must have minimal side effects on the cave environment and its organisms, and biocides used should have long-lasting effects without any negative influence on the rock, the speleothems, and the electrical installations. No herbicides should be used as they are toxic to the cave environment. The possible chemical methods that could be used have been evaluated by Mulec and Kosi (2009). There seems to be no ideal solution but hydrogen peroxide, 15% by volume, was thought to be best. Meyer et al. (2017) from research at Crystal Cave, Sequoia National Park (California), evaluated various treatments of sodium hypochlorite and decided that 0.5% by volume achieved management goals for eradication of lampenflora, with limited impacts to the presence or diet of a common cave-adapted indicator species, cave springtail (*Tomocerus celsus*).

12.5 Education

12.5.1 Cave Conservation and Responsible Caving Practices

There are several important educational provisions that have been developed here by most of the national caving organisations. These include Conservation Codes and there have also been

12.5 Education

developed general Cave Conservation Codes and National Cave Conservation Policies which have a role to play in educating cavers.

(a) *Cave Conservation Code, Cave Conservation Handbook,* and *Protect Our Caves* from the NCA, although the BCA took over its functions in 2006.
(b) *Minimal Impact Caving Code* (1995) from the Australian Speleological Federation where the message is cave softly and Cave SAFE and Low Impact Caving from the Tasmanian Parks and Wildlife Service, the *Speleological Union of Ireland Conservation and Access Policy,* the *Caving Care Code* published by the Department of Conservation, Te Papa Atawhai (New Zealand), and the *Minimal Impact Guidelines* from the BCA (2017).
(c) *A Guide to Responsible Caving* published by the National Speleological Society (2009).

The NCA, the British Cave Research Association (BCRA), and the Speleological Union of Ireland are constantly trying to enforce the following objectives to make the cave environment a better place: a more detailed documentation of features of particular importance and vulnerability within cave systems; the establishment of special designation for particular features, for example, voluntary special conservation areas; and joint management groups for caves requiring special conservation, with the objective of setting up a management plan.

(d) The BCRA holds a conference each year about the ecological impact of caving called the "Hidden Earth" which brings the most recent developments to the caving community. *Descent,* the British national caver's magazine also covers environmental, ecological, and research aspects of the speleological world.
(e) The US NPS has a comprehensive website that provides information to the general public, teachers, and scientific readers. The role of the NPS Cave and Karst Program in the management of caves and karst is explained with the emphasis on stewardship, responsibility, science, cooperation, coordination, and education. The importance of threats to, and management of, NPS cave and karst resources are described within the broader framework of other federal agencies' cave and karst management and programmes. Various NPS units have successful outreach programmes, as well, describing the importance and protection of cave and karst areas in their park and managed within the NPS. In 1998, the NPS developed a newsletter as an avenue for NPS cave and karst managers to share ideas about the management of cave and karst resources which is called the *NPS Cave and Karst Outreach Inside Earth Newsletter,* where topics discussed range from wilderness cave management to major construction within tourist caves (NPS Cave and Karst Program Website 2008).
(f) *Cave Leaders and Cave Instructors Certificate Schemes.* In Britain education in all aspects of cave conservation is covered within the NCA Local Cave Leader Assessment scheme and the Cave Instructor's Certificate. The theory and practice of cave conservation should be covered with ideas presented on the formulation of cave conservation plans. Certification is important because it should introduce newcomers to the activity in a safe manner via individuals or clubs where cave conservation can be enhanced. It is also important to introduce the non-caving public, especially children, to the importance of caves and their susceptibility to damage. Show caves have an important role to play here as well as the existing educational system, scouting, and other youth organisations. In New Zealand there is a similar graded system with currently five commonly used caving-specific qualifications. They are administered by the New Zealand Outdoor Instructors Association (NZOIA) and Skills Active Aotearoa Industry Training Organisation (Skills Active). The five qualifications are:
- *Skills Active Cave Streamway award*: this qualification is for people who guide clients in streamway caves where single rope technique (SRT) is not required.

- *NZOIA Cave 1*: this qualification is for people who deal with clients in easier caves with short pitches that can be negotiated using ladders.
- *Skills Active SRT Cave Guide certificate*: this qualification is for people who guide clients in caves where SRT is required.
- *Skills Active Caving SRT Instruction certificate*: this qualification is for people who guide clients in the caving environment and has particular emphasis on SRT cave guiding instruction.
- *NZOIA Cave 2*: this qualification is for people who deal with clients in all aspects of caving, including SRT, and for those who organise and supervise caving programmes.

There is also a comprehensive set of guidelines for caving published aimed at instructor education (ASG Activity Safety Guidelines-Caving 2013).

(g) *National Caving Association Cave Conservation Policy 1990*:

This was developed by a joint English Nature and the NCA initiative for cave conservation. It suggested the formulation of cave conservation plans for individual sites designated as Cave Sites of Special Scientific Interest (SSSI), with cavers to devise and implement these plans; there should be area specific conservation committees to assist in the formulation and implementation and to encourage education and training around issues related to cave conservation and to commission research by setting up scientific databases and developing management and conservation techniques. In Britain there are 813 caves notified as sites of scientific interest for biological or geological reasons. This however gives limited powers as noted by Chapman (1993).

(h) *Cave Conservation Plans*: The development of Cave Conservation Plans involves a fourfold process (Glasser and Barber 1995). The plans for each Cave SSSI will be to integrate management of these caves for geology with management for wildlife. Within the plans the scientific interest should be documented, including the type of interest, the location, and the current condition. The pressures and threats to the cave should be described and recommended actions to counter these threats. The practical conservation measures to be realistically implemented should be suggested, and over time the effectiveness of the conservation should be monitored, and the deficiencies should be identified and addressed. An example of the recommendations for Knock Fell Caverns (Northern Pennines) suggested permit-only access, with control and monitoring of use; access should be restricted to experienced cavers, and the use of the cave should be stopped for novices and outdoor activity groups; access documentation should give the status of the cave, access restrictions, and the dangers; documentation, including a survey showing the normal visitor routes and the dangerous and sensitive areas, should be given; there should be some taped-off areas, baseline data collected, and continuous monitoring should take place. The scheme has taken a long time to develop, but there are now several examples of cave conservation plans like the Witches Cave Conservation Plan (2012), in the Leck Fell area of South Cumbria; the caves underlying "Gruffy Field" in Charterhouse-on-Mendip, including GB and Charterhouse Cave (2015); and Stoney Middleton and Waterways Swallet in the Peak District (2012, 2013).

12.5.2 Alternatives to Caving to Take Pressure Off the Caves

Alternatives can be suggested, such the use of mines which are safe and well regulated by the leaders (Fig. 12.11), such as the slate mines in the Machno Valley and Tanygrisiau in Snowdonia operated by GoBelow and Artificial Caves such as the one in the Belfast Adventure Centre (Northern Ireland).

The latter, designed in conjunction with an experienced caver to ensure as much authenticity as possible, features one of the biggest man-made caves ever seen and the world's largest artificial

Fig. 12.11 Swan Mine, Mendips. Photo by D. Huddart

cavern and tallest (8 m) waterfall as part of the Adventure Learning Park. It has over 200 m of tunnels and passageways and a cavern which is 9.5 m long, 4.5 m wide, and 8 m high. The three caving passages are pumped with water and terminate in a sump or egress pond which users must swim through to leave the tunnels. It uses sprayed concrete on pre-bent reinforcement cages. Despite the fact that the firm that has developed artificial caves (Entre-Prises) can produce stalagmites and stalactites, fossils and cave paintings to add interest using a polyester resin system, modular speleo-systems and sprayed concrete, with linear sections, chambers, arches, and squeezes, and the fact that the BCA has introduced many children to the activity at roadshow events using a portable artificial cave, there is no doubt that, despite some advantages, this is not real caving. However an alternative with an educational message seems a good alternative. This has been developed at The Cave at CityROCK Climbing Gym in Colorado Springs where there is realistic passage, over 50 formations, each with electronic sensors to detect when a caver bumps into the formations. These sensors beep and LEDs light up around them to teach the participants about their soft caving skills, a recorded voice tells the user to be more careful next time, and a computer tracks their score. This theoretically allows the users to learn about caving and cave conservation but whether the skills learned by the use of this modern technology can be transferred to real caves or will stop more cavers using real caves is debatable.

12.6 Cave Art Teaching and Experimental Archaeology

The most famous cave painting sites have been closed for many years as it became apparent that the visitor numbers changed the delicate natural environment and had degraded the images. Hence it was decided by Liverpool University that to teach archaeology students representational art it was best to build an artificial cave in the University's Central Teaching Laboratories. Of course students could get access to the representational art by visiting some of the purpose-built replica, tourist show caves (Fig. 12.12) in France and Spain such as the Caverne du Pont-d'Arc close to Chauvet created by national heritage agencies to cater for the desire of many thousands of tourists who want to see the images in their natural environment. However, these caves, apart from the cost of getting to them, are too well lit to convey a proper experience that

Fig. 12.12 Replica of lionesses painting from Chauvet Cave, Ardeche, in the Moravian Museum, Brno. Photo by HTO

students should have and they cannot be physically handled. Hence in 2014 in conjunction with Hangfast, a climbing wall manufacturer, an artificial cave was constructed. The wall replicates some of the better known painted caves like Lascaux, Altamira, and Gargas. The light used mimics the original lamps and students can use original materials to replicate images, like different types of animal fats, and can experience the difficulties of making brushes with animal hair and original glues and experience how these images might have been seen at the time. It allows experimental research (see Nelson et al. 2017) and is a novel approach to teaching that students respond to.

12.7 Minimum Impact Caving Techniques for Fauna Developed in Tasmania

This Minimum Impact Caving Techniques policy for fauna suggests:

- Keep to a single path throughout the cave and follow marked routes. Do not wander about the place.
- Move slowly and carefully at all times, taking care where you place your hands, feet and body, whilst looking out for small animals.
- Where possible, use routes which avoid interfering with fauna and sensitive habitats (Avoid trampling on wood and leaf litter, tree roots, or other organic material).
- Avoid trampling on riparian sediment banks—step on solid rock surfaces where possible.
- Avoid walking in pools and small water courses.
- In medium-energy and high-energy stream passages, walk in the stream bed in preference to riparian sediment banks or other fossil substrates. In low-energy streamways, try to avoid walking in the stream bed, but not if this causes greater degradation to riparian or other fossil substrates alongside.
- Avoid making loud noises or shining lights directly onto animals.
- Avoid breaking spider webs or entangling glow-worm snares.
- Do not leave any foreign material in the cave, including food scraps, human waste, or spent carbide.
- Cave softly!

Cave softly is as follows: *Cave S.A.F.E.*:

S—tread *slowly* and *softly* at all times. Take care where you place your hands and feet.
A—be *aware* of sensitive features, including fauna and their habitats. Walk carefully around waterways, tree roots, sediment banks and organic deposits (leaf litter, wood, dead animals). Look at, but do not disturb, spider webs and glow-worm threads.
F—be *fit*. Fitness enables you to move through the cave efficiently, so you can better appreciate the environment and experience. Tiredness and lack of fitness can contribute to cave degradation.
E—*experience*. Join a caving club—you can learn a lot this way.

Living on Karst Awareness has been created to educate landowners living in limestone regions of the potential sensitivities and environmental dangers associated with karst geology. It was published originally by the Cave Conservancy of the Virginias in the USA in 1997 by their editor Carol Zokaites.

Concluding Remarks

Caving inevitably causes damage to a specialised, rare, and delicate environment which once damaged will never recover. The aim is to minimise this damage, although most is unintentional. The number of cavers is relatively small but they can damage caves in many ways, and it is not just the geological environment which can suffer change by caving as an activity but cave fauna and flora is highly specialised, often endemic to restricted areas and they live in different cave zones. Bats are an important cave species, and they have been suffering in North America from WNS, partly spread by cavers. There are many potential management strategies for caves which include the compilation of cave plans, conservation codes, and National Conservation policies. Controlling access can be important, and there are access agreements, zero, restricted or periodic access, booking systems, gating, sacrificial caves, zoning in caves, cave exploration policies, and cave fauna management, including the building of artificial bat caves. Education for cavers includes provisions including minimal impact codes, websites, leader and instructor schemes, involvement in cave conservation planning and cave adoption schemes, and the provision of alternatives to caving, such as the use of mines and artificial caves.

References

Aley, T. (1976). Caves, cows and carrying capacity. In *National Cave Management Symposium Proceedings 1975* (pp. 70–71). Albuquerque, NM: Speleobooks.

Berthinussen, A., Richardson, O. C., Smith, R. K., Altringham, J. D., & Sutherland, W. J. (2017). Bat conservation. In W. J. Sutherland, L. V. Dicks, N. Ochenden, & R. K. Smith (Eds.), *What Works in Conservation* (pp. 67–93). Cambridge, UK: Open Book Publishers.

Bowker, J. M., Askew, A., Cordell, H. K., & et al. (2012). *Outdoor Recreation Participation in the United States Projections to 2060: A Technical Document Supporting the Forest Service 2010 RPA Assessment.* General Technical Report SRS-GRT-160. US Department of Agriculture, Forest Service, Southern Research Station, Asheville, NC, 34pp.

British Caving Association. (2017). *Minimal Impact Guidelines.* BCA.

Britton, K. (1975). British caving experience relating to conservation and cave management practices. In *Proceedings of the National Cave Management Symposium* (pp. 104–106). Albuquerque, NM: Speleobooks.

Byoung-woo, K. (2002). Ecological study for the control of green contamination in Korean show caves. In *Proceedings of the 4th Samcheok International Cave Symposium* (pp. 74–76), Kangwon Development Research Institute.

Chapman, P. (1993). *Caves and Cave Life.* Collins, 224pp.

Cigna, A. A. (2011). Problem of Lampenflora in show caves. In P. Bella & P. Grazik (Eds.), *Proceedings of the 6th International Show Cave Association Conference* (pp. 201–205), Slovakia.

Cigni, A. A., & Burri, E. (2000). Development, management and ecology of show caves. *International Journal of Speleology, 29*, 1–27.

Clarke, A. (1997). *Management Prescriptions for Tasmania's Cave Fauna.* Report to Tasmanian Regional Forest Agreement Environment and Heritage Technical Committee, 185pp.

Clements, R., Sodhi, N. S., Schilthuizen, M., & Ng, P. K. L. (2006). Limestone karsts of Southeast Asia: Imperilled arks of biodiversity. *Bioscience, 56*, 733–742.

Cordell, H. K. (2012). *Outdoor Recreation Trends and Futures: A Technical Document Supporting the Forest Service 1010 RPA Assessment.* General Technical Report, US Department of Agriculture Forest Service, Southern Research Station, Asheville, NC, 167pp.

Cordell, H. K., Betz, C. J., Green, G. T., & Mou, S. H. (2008). *Outdoor Recreation Activity Trends: What's Growing, What's Showing.* IRIS Internet Research Information Series, USDA, Forest Service Department of Agriculture, 5pp.

D'Agostino, D., Beccarisi, L., Canassa, M., & Febbroriello, P. (2015). Microclimate and microbial characterization in the Zinzulusa show cave (South Italy) after switching to LED lighting. *Journal of Cave and Karst Studies, 77*, 133–144.

De Freitas, C. R. (2010). The role and importance of cave microclimate in the sustainable use and management of show caves. *Acta Carsologica, 39*, 437–489.

de Freitas, C. & Pugsley, C. (1997). *Monitoring and Management of the Glowworm Cave, Waitomo.* Commissioned report to general manager. Tourist Holdings Ltd., 22pp.

Donato, C. R., Ribeiro de Sousa, A., & de Sousa, L. (2014). A conservation status index as an auxiliary tool for the management of cave environments. *International Journal of Speleology, 43*, 315–322.

Dragovitch, D., & Grose, J. (1990). Impact of tourists on carbon dioxide levels at Jenolan caves, Australia: An examination of microclimate constraints on tourist cave management. *Geoforum, 21*, 111–120.

Eberhard, S. M. (1998). Cava Fauna. *Australian Caver, 144*, 15–20.

Eberhard, S. M. (1999). *Cave Fauna Management and Monitoring at Ida Bay, Tasmania*. Nature Conservation Report 99/1, Parks and Wildlife Service, Tasmania, 42pp.

Elliott, W. R. (2005). Critical issues in cave biology. In *2005 National Cave and Karst Management Symposium*, pp. 35–39.

Ford, T. (1990). Caves and conservation. *Earth Science Conservation, 28*, 6–8.

Furey, N. M., & Racey, P. A. (2015). Conservation ecology of cave bats. In C. C. Voight & T. Kingston (Eds.), *Bats in the Anthropocene: Conservation of Bats in a Changing World* (pp. 463–500). Cham: Springer International Publishing.

Gillieson, D. (1996). *Caves: Processes, Development and Management*. Wiley-Blackwell, 340pp.

Glasser, N. F., & Barber, G. (1995). Cave conservation plans: The role of English nature. *Cave and Karst Science, 21*, 33–36.

Gordon, K., Chester, M., & Denton, A. (2015). *Getting Active Outdoors*. O.I.A: Sport England.

Harley, G. L., Polk, J. S., North, L. A., & Reeder, P. P. (2011). Application of a cave inventory system to stimulate development of management strategies: The case of west-central Florida. *Journal of Environmental Management, 92*, 2547–2557.

Hewitt, I. (1992). *A Field Investigation and Assessment of the Impact of Man on Virgin Cave Passages and Formations*. Unpublished undergraduate dissertation, Liverpool John Moores University, 102pp.

Hildreth-Werker, V., & Werker, J. eds. (2006). *Cave Conservation and Restoration*. Huntsville, AL: National Speleological Society, 600pp.

Horne, G. (2005). Cave management guidelines for Western Mountain National Parks of Canada. In *Proceedings of the National Cave and Karst Management Symposium* (vol. 53, pp. 53–61). Albuquerque, NM: Speleobooks.

Jasper, J., & Voyles, K. (2005). Using sandblasting to remove Graffiti in Bloomington Cave, Utah. In *2005 National Cave and Karst Management Symposium* (pp. 132–135).

Jones, W. K., Culver, D. C., & Lucas, P. C. (2005). Digging: Guidelines for cavers and resource managers. In *National Cave and Karst Management Symposium* (pp. 88–91), Albuquerque, NM: Speleobooks.

Keeler, R. F., & Bohman, R. (2013). Incorporating cave and karst management into the forest plan revision process of Arizona forests. In *20th National Cave and Karst Management Symposium* (pp. 147–152).

Luo, J., Jiang, T., Lu, G., Wang, L., Wang, J., & Feng, J. (2013). Bat conservation in China: Should protection of subterranean habitats be a priority? *Oryx, 47*, 526–531.

Martin, K. W., Puchette, W. L., Hensley, S. L., & Leslie Jr., D. M. (2000). Internal cave gating as a means of protecting cave-dwelling bat populations in eastern Oklahoma. *Proceedings of the Oklahoma Academy of Science, 80*, 133–137.

Medellin, R. A., Wiederholt, R., & Lopez-Hoffman, L. (2017). Conservation relevance of bat caves for biodiversity and ecosystem services. *Biological Conservation, 211*, 45–50.

Meyer, E., Seale, L. D., Penmar, B., & McClary, A. (2017). The effect of chemical treatments on lampenflora and a collembola indicator species at a popular tour cave in California, USA. *Environmental Management, 459*, 1034–1042.

Mohr, C. E. (1972). The status of threatened species of cave-dwelling bats. *Bulletin of the National Speleological Society, 34*, 33–47.

Montechiano, F., & Giordano, M. (2006). Effect of prolonged dark incubation on pigments and photosynthesis of the cave-dwelling bacterium *Phormidium autumnale* (Oscillatoriales, cyanobacteria). *Phycologia, 45*, 704–710.

Mulec, J., & Kosi, G. (2009). Lampernflora algae and methods of growth control. *Journal of Cave and Karst Studies, 7*, 109–115.

National Speleological Society. (2009). *A Guide to Responsible Caving*. Huntsville, AL: National Speleological Society.

Nelson, E., Hall, J., Randolph-Quinney, P., & Sinclair, A. (2017). Beyond size: The potential of a geometric morphometric analysis of shape and form for the assessment of sex in hand stencils in rock art. *Journal of Archaeological Science, 78*, 202–213.

Nolfi, D. C. (2011). *National Park Service Cave and Karst Resources Management Case Study: Great Smoky Mountains National Park*. Master's theses and specialist projects, Paper 1053, 121pp. Retrieved from https://digitalcommons.wku.edu/theses/1053.

Olson, R. (2002). Control of lamp flora in Mammoth Cave National Park. In T. Hazslinsky (Ed.), *Proceedings of the International Conference on Cave Lighting* (pp. 131–136). Budapest: Hungarian Speleology Society.

Outdoor Foundation. (2016). Outdoor recreation participation in USA. Retrieved from http://www.outdoorfoundation.org/research.participation.2016.topline.html.

Russell, M. J., & McLean, V. L. (2008). Management issues in a Tasmanian tourist cave: Potential microclimatic impacts of cave modifications. *Journal of Environmental Management, 87*, 474–483.

Sargent, N. (1998). *A Study of the Environmental Impact on Sedimentary Mud in Pridhamsleigh Cavern, Devon*. Unpublished undergraduate dissertation, Liverpool John Moores University, 65pp.

Sidaway, R. (1988). *Sport, Recreation and Nature Conservation: Conflict or Cooperation*. Study 32, Sports Council and Countryside Commission, London.

Speakman, J. R., Webb, P. I., & Racey, P. A. (1991). Effects of disturbances on the energy expenditure of hibernating bats. *Journal of Applied Ecology, 28*, 1087–1104.

Stanton, W. I. (1982). Mendip-pressures on its caves and karst. *Transactions of the British Cave Research Association, 9*, 176–183.

Stihler, C. W., & Hall, J. S. (1993). Endangered bat populations in West Virginia caves gated or fenced to reduce human disturbance. *Bat Research News, 34*, 130.

Thomas, G. (1997). *An Investigation into the Conflicting Interests Between Cavers and Bat Workers in North Wales*. Unpublished Outdoor Education undergraduate thesis, Liverpool John Moores University, 89pp.

Tuttle, M. D. (1977). Gating as a measure of protecting cave dwelling bats. In T. Aley & D. Rhodes (Eds.), *Proceedings of the National Cave Management Proceedings* (pp. 77–82). Albuquerque, NM: Speleobooks.

Williams, M. A. (1966). Further investigations into bacterial and algal populations in the caves of South Wales. *International Journal of Speleology, 2*, 389–395.

Wilmut, J. (Ed.). (1972). *Caves and Conservation*. National Caving Association.

Water Sports and Water-Based Recreation

13

Chapter Summary

This chapter first lists some of the different water sport disciplines and then defines those on which the chapter will focus—motorboating/powerboating, canoeing, kayaking, jet skiing, rafting, rowing, sailing, surfing, water skiing, sailing, and windsurfing—distinguishing between motorised and non-motorised activities. It then examines relative and actual participation numbers. The final part of the chapter focuses on specific environmental impacts of water sports: physical impacts to aquatic vegetation, the spread of invasive species, erosion of banks and shores, water pollution and its costs. There is discussion about the impacts of water sports on wildlife as well as the chemical impacts on water sports (heavy metals, hydrocarbons). The final section considers the management of these activities and gives examples of ways in which users can be educated.

13.1 Definitions

The term water sports includes a wide range of activities both in the water and on the water surface. Chapter 14 deals with some of the underwater activities: scuba diving and snorkelling. This chapter focuses on some of the more popular water recreational activities which take place on the water surface. Table 13.1 shows a list of the potential activities, with those in bold to be considered in this chapter.

13.1.1 Non-motorised Water Sports

It is generally agreed that non-motorised watercraft have less impact on the environment than motorised craft. While non-motorised craft have little or no impact on the water over which they pass, there may be impacts on the shore or river bank as well as disturbances to fish, plants, and wildlife.

13.1.1.1 Canoeing

Canoeing is an activity which involves paddling a canoe with a single-bladed paddle (Fig. 13.1A–D). In some parts of Europe, canoeing refers to both canoeing and kayaking (see Sect. 13.1.1.2), with a canoe being called an "Open Canoe" or sometimes, "Canadian Canoe" (after its origins as an ancient mode of transportation, used by *voyageurs* to transport beaver furs across Canada). Canoeing can be combined with other activities such as canoe camping (Fig. 13.1B), or where canoeing is merely a transportation method used to accomplish other activities. Most present-day canoeing is done as or as a part of a sport or recreational activity.

A recreational form of canoeing is canoe camping, the open canoe being very suited to

Table 13.1 List of potential water-based activities which take place on the water surface

Activity name	Description
Barefoot skiing	Water skiing without skis
Boating	The use of small boats manoeuvred without an engine (i.e. with paddles, oars, or poles)
Boat racing/ motorboating/ powerboating	Use of powerboats with engines, often used to participate in races
Bodyboarding	Similar to surfing, but the board is smaller and the person (normally) lies down on the board
Cable skiing	Similar to wakeboarding but with cables for artificial manoeuvring
Canoeing	Canadian or open canoes manoeuvred by a single blade paddle. Normally one person kneeling or two persons seated
Canoe polo	A competitive sport which normally takes place between two teams of five on a pitch which can be set up in swimming pools or any stretch of flat water
Dragon boat racing	Dragon boats are the basis of the team paddling sport of dragon boat racing, a water sport which has its roots in an ancient folk ritual of contending villagers, which has been held for over 2000 years throughout southern China. Teams of 20 paddlers race each other
Fishing	The recreation and sport of catching fish. See Chap. 15
Flyboard	A Flyboard rider stands on a board connected by a long hose to a watercraft. Water is forced under pressure to a pair of boots with jet nozzles underneath which provide thrust for the rider to fly up to 15 m (49 ft) in the air or to dive headlong through the water down to 2.5 m (8 ft). A Flyboard is a brand of hydroflighting device which supplies propulsion to drive the Flyboard into the air to perform a sport known as hydroflying
Flowrider	A water park attraction to simulate the riding of waves in the ocean. In the late 1980s, a patent was taken out for "a wave-forming generator for generating inclined surfaces on a contained body of water." The rider surfs an artificial wave on a small surfboard
Jet skiing	Jet Ski is a proper noun and registered trademark of Kawasaki. The stand-up Kawasaki Jet Ski was the "first commercially successful" personal watercraft in America, having been released in 1972. There is normally one driver and up to two passengers
Kayaking	Kayaking is the use of a kayak for moving across water. It is distinguished from canoeing by the sitting position of the paddler and the number of blades on the paddle. A kayak is a low-to-the-water, canoe-like boat in which the paddler sits facing forward, legs in front, using a double-bladed paddle to pull front-to-back on one side and then the other in rotation[1]. Most kayaks have closed decks, although sit-on-top and inflatable kayaks are growing in popularity as well
Kiteboarding	Kiteboarding is an action sport combining aspects of wakeboarding, snowboarding, windsurfing, surfing, paragliding, skateboarding, and sailing into *one* extreme sport. A kiteboarder harnesses the power of the wind with a large controllable power kite to be propelled across the water, land, or snow. On water, a kiteboard, similar to a wakeboard or a small surfboard, with or without footstraps or bindings, is used
Kitesurfing	Kitesurfing is a style of kiteboarding specific to wave riding, which uses standard surfboards or boards shaped specifically for the purpose. On land, a mountain board or foot-steered buggy is used while skis or snowboards can be used in snow. There are different styles of kiteboarding, including freestyle, freeride, speed, course racing, wakestyle, big air, park, and surfing.[1] In 2012, the number of kitesurfers was estimated by the ISAF and IKA at 1.5 million persons worldwide
Kneeboarding	Kneeboarding is an aquatic sport where the participant is towed on a buoyant, convex, and hydrodynamically shaped board at a planing speed, most often behind a motorboat
Paddleboarding	Paddleboarding participants are propelled by a swimming motion using their arms while lying, kneeling, or standing on a paddleboard or surfboard in the ocean. A derivative of paddleboarding is stand-up paddle surfing and stand-up paddleboarding (SUP). Paddleboarding is usually performed in the open ocean, with the participant paddling and surfing unbroken swells to cross between islands or journey from one coastal area to another
Parasailing	Parasailing, also known as parascending or parakiting, is a recreational kiting activity where a person is towed behind a vehicle (usually a boat) while attached to a specially designed canopy wing that reminds one of a parachute, known as a parasail wing

(continued)

Table 13.1 (continued)

Activity name	Description
Rafting/white-water rafting	Rafting and white-water rafting are recreational outdoor activities which use an inflatable raft to navigate a river or other body of water. This is often done on white water or different degrees of rough water
Rowing	Rowing is the act of propelling a boat using the motion of oars in the water, displacing water, and propelling the boat forward. The difference between paddling and rowing is that rowing requires oars to have a mechanical connection with the boat, while paddles are handheld and have no mechanical connection
Sailing/yachting	Sailing employs the wind—acting on sails, wingsails, or kites—to propel a craft on the surface of the water (sailing ship, sailboat, windsurfer, or kitesurfer), on ice (iceboat) or on land (land yacht) over a chosen course, which is often part of a larger plan of navigation
Sit-down hydrofoiling	The sit-down hydrofoil, first developed in the late 1980s, is a variation on water skiing, a popular water sport. When towed at speed, by a powerful boat or some other device, the board of the hydrofoil "flies" above the water surface and generally avoids contact with it, so the ride is largely unaffected by the wake or chop of the water and is relatively smooth. The air board is a modified hydrofoil where the skier stands up
Skimboarding	Skimboarding (or skimming) is a boardsport in which a skimboard (much like a surfboard but smaller and without fins) is used to glide across the water's surface to meet an incoming breaking wave and ride it back to shore
Skurfing	Water skurfing is a form of water skiing that uses a surfboard or similar board instead of skis. The skurfer is towed behind a motorboat at planing speed with a tow rope similar to that of kneeboarding and wakeboarding. It shares an advantage with kneeboarding in that the motorboat does not require as much speed as it does for water skiing
Surfing	Surfing is a surface water sport in which the wave rider, referred to as a surfer, rides on the forward or deep face of a moving wave, which is usually carrying the surfer towards the shore. Waves suitable for surfing are primarily found in the ocean but can also be found in lakes or in rivers in the form of a standing wave or tidal bore
Wakeboarding	Wakeboarding is a surface water sport which involves riding a wakeboard over the surface of a body of water. The wakeboard is a small, mostly rectangular, thin board with very little displacement and shoe-like bindings mounted to it. The wakeboard is usually towed behind a motorboat, typically at speeds of 30–40 km/h (18–25 mph), depending on the board size, weight, type of tricks, and comfort
Water skiing	Water skiing (also waterskiing or water-skiing) is a surface water sport in which an individual is pulled behind a boat or a cable ski installation over a body of water, skimming the surface on two skis or one ski. The sport requires sufficient area on a smooth stretch of water, one or two skis, a tow boat with tow rope, three people (depending on state boating laws), and a personal flotation device
Windsurfing	Windsurfing is a surface water sport that combines elements of surfing and sailing. It consists of a board usually 2.5 to 3 m long, with displacements typically between 60 and 250 litres, powered by wind on a sail. The rig is connected to the board by a free-rotating universal joint and consists of a mast, boom, and sail. The sail area generally ranges from 2.5 m^2 to 12 m^2 depending on the conditions, the skill of the sailor, the type of windsurfing being undertaken, and the weight of the person windsurfing

carrying large loads (Fig. 13.1B). Other forms include a wide range of canoeing on lakes, slow-moving rivers (Fig. 13.1A), fast-moving rivers (Fig. 13.1C), and even the sea (Fig. 13.1D). Canoe sailing is another strand within the sport, as is canoe poling (Fig. 13.1E) where the canoeist stands in the canoe (slightly back from centre) and propels and steers the craft with a 12 ft. aluminium pole (Fig. 13.1E).

British Canoeing (https://www.britishcanoeing.org.uk/) is the national governing body for paddlesports in the UK. Formerly known as the British Canoe Union (founded in 1936), Canoe England, and GB Canoeing, these bodies have now come together under one unified umbrella organisation for the home nation associations in Scotland (Scottish Canoe Association), Wales (Canoe Wales), and Northern Ireland (Canoe Association Northern Ireland). British Canoeing is responsible for leading and setting the overall framework for all the national associations and includes areas such as coaching,

Fig. 13.1 (A) Tranquil open canoe journey on the River Stour, Southern England. Photo by Tim Stott. (B) Open canoe camping. The canoe is ideal for transporting heavy loads. Photo by Tim Stott. (C) The author solo paddling an open canoe on Grade II water of the River Dee, North Wales. Photo by Clive Palmer. (D) The open canoe can be rigged for sailing. Photo by Tim Stott. (E) In canoe poling the canoeist stands in the canoe (slightly back from centre) and propels and steers the craft with a 12 ft. aluminium pole. The US National Canoe Poling Championships on the Meramec River, Missouri, USA. Photo by Tim Stott

competition, and representing canoeing interests at a UK level.

13.1.1.2 Kayaking

Kayaking is the use of a kayak for moving over water. It is distinguished from canoeing by the sitting position of the paddler and the number of blades on the paddle. A kayak is a low-to-the-water, canoe-like boat in which the paddler sits facing forward, legs in front, using a double-bladed paddle to pull front-to-back on one side and then the other in rotation (Fig. 13.2A–C). Most kayaks have closed decks, although sit-on-top and inflatable kayaks (Fig. 13.2F) are growing in popularity as well.

Kayaks were created thousands of years ago by the Inuit, formerly known as Eskimos, of the northern Arctic regions. They used driftwood and sometimes the skeleton of whale, to construct the frame of the kayak, and animal skin, particularly seal skin, was used to create the body. The main purpose for creating the kayak, which literally translates to "hunter's boat" was for hunting and fishing. Modern kayaks are made from plastic, though some specialist slalom kayaks are still made from glass fibre or Kevlar.

13.1.1.3 Rafting/White-Water Rafting

Rafting and white-water rafting are recreational outdoor activities which use an inflatable raft to navigate a river or other body of water. This is often done on white water or different degrees of rough water (Fig. 13.2D, E). Dealing with risk and the need for teamwork is often a part of the experience. This activity as a leisure sport has become popular since the 1950s, evolving from individuals paddling 3.0 m (10 ft) to 4.3 m (14 ft) rafts with double-bladed paddles or oars to multi-person rafts propelled by single-bladed paddles and steered by a person at the stern or by the use of oars. Rafting on certain sections of rivers is considered an extreme sport, while other sections are not so extreme or difficult. The International Rafting Federation, often referred to as the IRF, is the worldwide body which oversees all aspects of the sport.

13.1.1.4 Rowing

Rowing is the act of propelling a boat using the motion of oars in the water, displacing water, and propelling the boat forward. The difference between paddling and rowing is that rowing requires oars to have a mechanical connection with the boat, while paddles are handheld and have no mechanical connection. In some regions, rear-facing systems are used, while in other places forward-facing systems prevail, especially in crowded areas such as in Venice, Italy, and in Asian and Indonesian rivers and harbours. In another system called sculling, a single oar extending from the stern of the boat is moved back and forth under water somewhat like a fish tail, and quite large boats can be moved.

13.1.1.5 Sailing/Yachting

Sailing employs the wind acting on sails to propel a craft on the surface of the water (sailing ship, sailboat, windsurfer, or kitesurfer), on ice (iceboat), or on land (land yacht) over a chosen course, which is often part of a larger plan of navigation. The forces transmitted via the sails are resisted by forces from the hull, keel, and rudder of a sailing craft. In the twenty-first century, most sailing represents a form of recreation or sport (Fig. 13.3A). Recreational sailing or yachting can be divided into racing (Fig. 13.3B) and cruising (Fig. 13.3C). Cruising can include extended offshore and ocean-crossing trips, coastal sailing within sight of land, and day sailing.

13.1.1.6 Windsurfing

Windsurfing is a surface water sport that combines elements of surfing and sailing. It consists of a board usually 2.5 to 3 m (7–9 ft) long, with displacements typically between 60 and 250 litres, powered by wind on a sail (Fig. 13.3D). The rig is connected to the board by a free-rotating universal joint and consists of a mast, boom, and sail. The sail area generally ranges from 2.5 to 12 m^2 depending on the conditions, the skill of the sailor, the type of windsurfing being undertaken, and the weight of the person. Windsurfing can take place on lakes, reservoirs, estuaries, and the open sea.

Fig. 13.2 (A) Recreational kayaker descends a rapid on the upper Afon Tryweryn, North Wales. Photo by Tim Stott. (B) Recreational kayaker at the same location as part (A) with the kayak almost totally under the water. A spray deck fitted over the cockpit prevents water ingress. Photo by Tim Stott. (C) A recreational white-water kayaker using a high-volume kayak to descent "big water". Photo by Tim Stott. (D) White-water rafting on the upper Afon Tryweryn, North Wales. Photo by Tim Stott. (E) White-water rafting on a wave on the Durance River, France. Note the kayakers in the foreground. Photo by Tim Stott. (F) A typical "beach" on the Durance River, France, in summer gives an impression of the popularity of the sport. Note the red inflatable kayak in the foreground. Photo by Tim Stott

Fig. 13.3 (A) Dinghy sailing in Liverpool Marina. Photo by Tim Stott. (B) Yachts racing on the Mersey Estuary near Liverpool. Photo by Tim Stott. (C) Cruising boats in Liverpool Marina. Photo by Tim Stott. (D) Windsurfing in the inland sea at Valley on Anglesey in North Wales. Photo by Tim Stott. (E) Surfing at Bundoran, Northern Ireland. Photo by Tim Stott

13.1.1.7 Surfing

Surfing is a surface water sport in which the wave rider, referred to as a surfer, rides on the forward or deep face of a moving wave, which is usually carrying the surfer towards the shore (Fig. 13.3E). Waves suitable for surfing are usually found in the ocean but can also be found on rivers as standing waves or tidal bores. However, surfers can also use artificial waves such as those from boat wakes and the waves created in artificial wave pools.

The term surfing refers to the act of riding a wave, regardless of whether the wave is ridden with a board or without a board, and regardless of the stance used. The native peoples of the Pacific, for instance, did so on their belly and knees. The modern-day definition of surfing, however, most often refers to a surfer riding a wave standing up on a surfboard; this is also referred to as stand-up surfing. Another prominent form of surfing is bodyboarding, when a surfer rides a wave on a bodyboard (which is about one third of the length of a surfboard), either lying on their belly, drop knee, or sometimes even standing up on a bodyboard.

13.1.2 Motorised Water Sports

Jet Ski is the brand name of a personal watercraft (PWC) first manufactured by Kawasaki, a Japanese company in the 1970s. The term is often used generically to refer to any type of PWC used mainly for recreation, and it is also used as a verb to describe the use of any type of PWC. A runabout style PWC typically carries up to three people seated in a configuration like a typical bicycle or motorcycle. With the introduction of the Jet Ski, in cooperation with aftermarket companies and enthusiasts, Kawasaki helped in creating the United States Jet Ski Boating Association (USJSBA). In 1982 the name was changed to the International Jet Sports Boating Association (IJSBA).

Water skiing is a surface water sport in which an individual is pulled behind a boat or a cable ski installation over a body of water, skimming the surface on two skis or one ski. The sport requires sufficient area on a smooth stretch of water, one or two skis, a tow boat with tow rope, three people (depending on state boating laws), and a personal flotation device. In addition, the skier must have adequate upper and lower body strength, muscular endurance, and good balance.

A motorboat, speedboat, or powerboat is a boat which is powered by an engine. Some motorboats have an outboard motor installed on the rear (Fig. 13.4), others are fitted with inboard engines. Motorboats vary greatly in size and configuration, from the 4 m, open centre console type to the luxury mega-yachts capable of crossing an ocean.

One thing which all these motorised watercraft have in common is a propeller which, as we shall see later, has the potential to inflict damage to aquatic ecosystems.

13.2 Participation Numbers

In the USA, during the 2016 calendar year, a total of 24,134 online interviews were carried out with a nationwide sample of individuals and households from the US Online Panel of over one million

Fig. 13.4 Small powerboat. Photo by Terry Mitchell

people operated by Synovate/IPSOS (Outdoor Foundation 2017). A total of 11,453 individual and 12,681 household surveys were completed. The total panel is maintained to be representative of the US population for people ages six and older. Over sampling of ethnic groups took place to boost response from typically under responding groups. The 2016 participation survey sample size of 24,134 completed interviews provides a high degree of statistical accuracy.

As can be seen in Table 13.2, the Outdoor Foundation (2017) survey for the USA provided data for eight water sport disciplines. The rank order in terms of the greatest number of participants in 2016 was:

- *Canoeing*—10,046,000 participants in 2016 and a −1.1% decrease over the previous three years.
- *Kayaking* (recreational)—10,017,000 participants in 2016 and a three-year increase of 14.9%.
- *Sailing*—4,095,000 participants in 2016 and a three-year increase of 4.6%.
- *Rafting*—3,428,000 participants in 2016 and a three-year decrease of −10.6%.
- *Stand-up paddling*—3,220,000 participants in 2016 and a massive three-year increase of 61.6%.
- *Kayaking* (sea/touring)—3,124,000 participants in 2016 and a three-year increase of 16.0%.
- *Surfing*—2,793,000 participants in 2016 and a three-year increase of 5.1%.
- *Kayaking* (white water)—2,552,000 participants in 2016 and a three-year increase of 18.9%.
- *Boardsailing/windsurfing*—1,737,000 participants in 2016 and a three-year increase of 31.2%.

If we total all three kayaking disciplines, there was a total of 15,693,000 participants in 2016, making it the most popular of the disciplines in the survey. Stand-up paddling showed by far the greatest three-year increase of 61.6%.

Cordell's (2012) survey showed kayaking (Table 13.3) as having 14,200,000 participants in the 2005–2009 period with a 103.8% increase between the 1999–2001 and 2005–2009 periods. Comparing that with the Outdoor Foundation's (2017) data, it appears that kayaking had continued to increase after Cordell's 2005–2009 survey. Sailing had the second greatest numbers participating within the water sports disciplines with 10,400,000 in the 2005–2009 periods with a decrease of −0.4% between 1999–2001 and the 2005–2009 survey. Rowing had the third greatest numbers participating within the water sports disciplines with 9,400,000 in the 2005–2009 periods with an increase of 8.9% between the 1999–2001 and the 2005–2009 surveys. Surfing had 2,000,000 participants in 2005–2009 (a 46.3% increase between the 1999–2001 and the 2005–2009 surveys), and windsurfing had 600,000 participants in 2005–2009 (a −10.1% decrease between the 1999–2001 and the 2005–2009 surveys).

Bowker et al. (2012) projected changes in total outdoor recreation participants between 2008 and 2060 (Table 13.4). There was an estimated 62 million participants in 2008 taking part in motorised water activities (motorboating, water skiing, PWC), and this was predicted to become 87–112 million by 2060. For non-motorised floating water activities (canoeing, kayaking, and rafting), there was an estimated 40 million participants in 2008 taking part, and this was predicted to increase to 52–65 million by 2060.

Table 13.5 shows the changes over the same period for the total number of outdoor recreation days. There was an estimated 958 million days in 2008 where people took part in motorised water activities (motorboating, water skiing, PWC) and this was predicted to become 1304–1806 million days by 2060. For non-motorised floating water activities (canoeing, kayaking, and rafting), there was an estimated 262 million days in 2008 where people took part, and this was predicted to increase to 338–422 million days by 2060 (Bowker et al. 2012).

In England, data are available from the Sport England's Active People Survey (APS). In terms of water sports, Sport England included rowing, sailing, and canoeing in this survey. Table 13.6 shows that there was an estimated 83,400 rowing participants (0.19% of the population of England)

Table 13.2 Outdoor participation by activity (ages 6+) in the USA, 2006–2016 (The Outdoor Foundation 2017, p. 8)

	2006	2007	2008	2009	2010	2011	2012	2013	2014	2015	2016	3-year change (%)
Adventure racing	725	698	920	1089	1339	1065	2170	2213	2368	2864	2999	35.5
Backpacking overnight >¼ mile from vehicle/home	7076	6637	7867	7647	8349	7095	8771	9069	10,101	10,100	10,151	11.9
Bicycling (BMX)	1655	1887	1904	1811	2369	1547	2175	2168	2350	2690	3104	43.2
Bicycling (mountain/non-paved surface)	6751	6892	7592	7142	7161	6816	7714	8542	8044	8316	8615	0.9
Bicycling (roads/paved surface)	38,457	38,940	38,114	40,140	39,320	40,349	39,232	40,888	39,725	38,280	38,365	−6.2
Birdwatching (more and ¼ mile from home/vehicle)	11,070	13,476	14,399	13,294	13,339	12,794	14,275	14,152	13,179	13,093	11,589	−18.1
Boardsailing/windsurfing	**938**	**1118**	**1307**	**1128**	**1617**	**1151**	**1593**	**1324**	**1562**	**1766**	**1737**	**31.2**
Camping (RV)	16,946	16,168	16,517	17,436	15,865	16,698	15,108	14,556	14,663	14,699	15,855	8.9
Camping (with ¼ mile of home/vehicle)	35,618	31,375	33,686	34,338	30,996	32,925	29,982	29,269	28,660	27,742	26,467	−9.6
Canoeing	**9154**	**9797**	**9935**	**10,058**	**10,553**	**9787**	**9839**	**10,153**	**10,044**	**10,236**	**10,046**	**−1.1**
Climbing (sports/indoor/boulder)	4728	4514	4769	4313	4770	4119	4592	4745	4536	4684	4905	3.4
Climbing (traditional/ice/mountaineering)	1586	2062	2288	1835	2198	1609	2189	2319	2457	2571	2790	20.3
Fishing (fly)	6071	5756	5941	5568	5478	5683	6012	5878	5842	6089	6456	9.8
Fishing (freshwater/other)	43,100	43,859	40,331	40,961	38,860	38,868	39,135	37,796	37,821	37,682	38,121	0.9
Fishing (saltwater)	12,466	14,437	13,804	12,303	11,809	11,983	12,017	11,790	11,817	11,975	12,266	4.0
Hiking (day)	29,863	29,965	32,511	32,572	32,496	34,491	34,545	34,378	36,222	37,232	42,128	22.5
Hunting (bow)	3875	3818	3722	4226	3908	4633	4075	4079	4411	4564	4427	8.5
Hunting (handgun)	2525	2595	2873	2276	2709	2671	3553	3198	3091	3400	3512	9.8
Hunting (rifle)	11,242	10,635	10,344	11,114	10,150	10,807	10,164	9792	10,081	10,778	10,797	10.3
Hunting (shotgun)	8987	8545	8731	8490	8062	8678	8174	7894	8220	8438	8271	4.8
Kayak fishing	n/a	n/a	n/a	n/a	**1044**	**1201**	**1409**	**1798**	**2074**	**2265**	**2371**	**31.8**
Kayaking (recreational)	**4134**	**5070**	**6240**	**6212**	**6465**	**8229**	**8144**	**8716**	**8855**	**9499**	**10,017**	**14.9**
Kayaking (sea/touring)	**1136**	**1485**	**1780**	**1771**	**2144**	**2029**	**2446**	**2694**	**2912**	**3079**	**3124**	**16.0**
Kayaking (white water)	**828**	**1207**	**1242**	**1369**	**1842**	**1546**	**1878**	**2146**	**2351**	**2518**	**2552**	**18.9**
Rafting	**3609**	**3786**	**4226**	**4342**	**3869**	**3725**	**3958**	**3915**	**3924**	**4099**	**4095**	**−10.6**
Running/jogging	38,559	41,064	41,130	43,892	49,408	50,713	52,187	54,188	51,127	48,496	47,384	−12.6
Sailing	**3390**	**3786**	**4226**	**4342**	**3869**	**3725**	**3958**	**3915**	**3924**	**4099**	**4095**	**4.6**
Scuba diving	2965	2965	3216	2723	3153	2579	2982	3174	3145	3274	3111	−2.0
Skateboarding	10,130	8429	7807	7352	6808	5827	6627	6350	6582	6436	6442	1.5
Skiing (alpine/downhill)	n/a	10,362	10,346	10,919	11,504	10,201	8243	8044	8649	9378	9267	12.4
Skiing (cross-country)	n/a	3530	3848	4157	4530	3641	3307	3377	3820	4146	4640	40.3
Skiing (freestyle)	n/a	2817	2711	2950	3647	4318	5357	4007	4564	4465	4640	2.7
Snorkelling	8395	9294	10,296	9358	9305	9318	8011	8700	8752	8874	8717	0.2
Snowboarding	n/a	6841	7159	7421	8196	7579	7351	6418	6785	7676	7602	3.4
Snowshoeing	n/a	2400	2922	3431	3823	4111	4029	3012	3501	3885	3533	−12.3
Stand up paddling	n/a	n/a	n/a	n/a	**1050**	**1242**	**1542**	**1993**	**2751**	**3020**	**3220**	**61.6**
Surfing	**2170**	**2206**	**2607**	**2403**	**2767**	**2195**	**2895**	**2658**	**2721**	**2701**	**2793**	**3.0**
Telemarking (downhill)	n/a	1173	1435	1482	1821	2099	2766	1732	2188	2569	2848	3.0
Trail running	4558	4216	4857	4833	5136	5610	6003	6792	7531	8139	8582	26.4

Note: All participation numbers are in thousands (000)

Table 13.3 Trends in number of people of ages 16 and older participating in recreation activities in the USA, 1999–2001 and 2005–2009 for activities with fewer than 15 million participants from 2005 through 2009 (Source: Cordell 2012, p.40)

	Total participants (*millions*)			Percent participating	Percent change 1999–2001 to 2005–2009
	1994–1995	1999–2001	2005–2009	2005–2009	
Kayaking	3.4	7.0	14.2	6.0	103.8
Mountain climbing	9.0	13.2	12.4	5.3	−5.9
Snowboarding	6.1	9.1	12.2	5.2	33.7
Ice skating outdoors	14.2	13.6	12.0	5.1	−11.5
Snowmobiling	9.6	11.3	10.7	4.5	−5.5
Anadromous fishing	11.0	8.6	10.7	4.5	24.1
Sailing	**12.1**	**10.4**	**10.4**	**4.4**	**−0.4**
Caving	9.5	8.8	10.4	4.4	18.4
Rock climbing	7.5	9.0	9.8	4.2	9.5
Rowing	**10.7**	**8.6**	**9.4**	**4.0**	**8.9**
Orienteering	4.8	3.7	6.2	2.6	−21.7
Cross-country skiing	8.8	7.8	6.1	2.6	−21.7
Migratory bird hunting	5.7	4.9	4.9	2.1	−1.1
Ice fishing	4.8	5.7	4.8	2.1	−15.5
Surfing	2.9	3.2	4.7	2.0	46.3
Snowshoeing	–	4.5	4.1	1.7	−9.4
Scuba diving	–	3.8	3.6	1.5	−5.6
Windsurfing	**2.8**	**1.5**	**1.4**	**0.6**	**−10.1**

Missing data are denoted with "–" and indicate that participation data for that activity were not collected during that time period. Percent change was calculated before rounding
Source: USDA Forest Service (1995) (*n* = 17,217), USDA Forest Service (2001) (*n* = 52,607), and USDA Forest Service (2009) (*n* = 30,398)
Note: The numbers in this table are *annual* participant estimates on data collected during the three time periods
1994–1995 participants based on 201.3 million people of ages 16+ (Woods & Poole Economics, Inc. 2007)
1999–2001 participants based on 214.0 million people of ages 16+ (U.S. Department of Commerce 2000)
2005–2009 participants based on 235.3 million people of ages 16+ (U.S. Department of Commerce 2008)

in the October 2015–October 2016 survey period. For sailing there was an estimated 64,000 participants (0.16% of the population of England) in the October 2005–October 2006 survey period. This had declined to 45,600 (0.10% of the population) by the October 2015–October 2016 survey period. For canoeing there was an estimated 36,500 participants (0.09% of the population of England) in the October 2005–October 2006 survey period. This had increased to 41,900 (0.09% of the population) by the October 2016–October 2066 survey period.

So Table 13.6 shows that the water sports activities of rowing, sailing, and canoeing rank in the lower half of the activities included in the Sport England's APS.

13.3 Environmental Impact

It seems reasonable to examine the impacts of these water sports on the environment in three main areas: physical impacts (wave action, turbidity), biological impacts (on wildlife, fish, invertebrates, plants), chemical impacts (heavy metals, fuel, and oil spillage from engines).

Liddle and Scorgie (1980) reviewed the impacts of recreation on freshwater plants and animals. They made a distinction between water- and shore-based activities and between physical and chemical effects. The impacts of water-based recreation, which result mainly from boating, were considered in terms of wash, turbulence and turbidity, propeller action, direct contact, disturbance to animals,

Table 13.4 Changes in total outdoor recreation participants between 2008 and 2060 across all activities and scenarios (Source: Bowker et al. 2012, p. 28)

Activity[a]	2008 Participants[b] (millions)	2060 Participant range[c] (millions/[percent])	2060 Average participant change[c] (millions)	2060 Participant range[d] (millions/[percent])	2060 Average participant change[d] (millions)
Visiting developed sites					
Developed site use—family gatherings, picnicking, developed camping	194	273–346 [42–77]	116	271–339 [40–75]	112
Visiting interpretive sites—nature centres, zoos, historic sites, prehistoric sites	158	231–294 [48–84]	106	231–289 [46–83]	104
Viewing and photographic nature					
Birding—viewing/photographing birds	82	118–149 [42–76]	53	115–144 [40–76]	47
Nature viewing—viewing, photography, study, or nature gathering related to fauna, flora, or natural settings	190	267–338 [42–76]	114	268–333 [41–75]	112
Backcountry activities					
Challenge activities—caving, mountain biking, mountain climbing, rock climbing	25	38–48 [50–86]	19	37–48 [47–90]	18
Equestrian	17	24–31 [44–87]	11	25–35 [50–110]	13
Hiking—day hiking	79	117–150 [50–88]	55	114–143 [45–82]	50
Visiting primitive areas—backpacking, primitive camping, wilderness	91	120–152 [34–65]	47	119–145 [31–60]	42
Motorised activities					
Motorised off-road use	48	62–75 [29–56]	21	62–76 [28–58]	21
Motorised snow use (snowmobiles)	10	10–13 [10–37]	3	4–10 [(56)–6]	(2.5)[e]
Motorised water use	**62**	**87–112 [41–81]**	**40**	**84–111 [35–78]**	**35**
Consumptive					
Hunting—all types of legal hunting	28	30–34 [8–23]	5	29–34 [5–21]	4
Fishing—anadromous, cold-water, saltwater, warm water	73	92–115 [28–56]	33	89–115 [22–58]	30
Non-motorised winter activities					
Downhill skiing, snowboarding	24	38–54 [58–127]	23	36–54 [50–126]	21
Undeveloped skiing—cross-country, snow-shoeing	8	10–13 [32–67]	4	5–10 [(42)–28]	(1)
Non-motorised water					
Swimming, snorkelling, surfing, diving	144	210–268 [47–85]	99	212–266 [47–85]	99
Floating—canoeing, kayaking, rafting	**40**	**52–65 [30–62]**	**20**	**47–62 [18–56]**	**13**

Source: National Survey of Recreation and the Environment (NSRE) 2005–2009, Versions 1 to 4 (January 2005 to April 2009). $n = 24,073$ (USDA Forest Service 2009)

[a]Activities are individual or activity composites derived from the NSRE. Participants are determined by the product of the average weighted frequency of participation by activity for NSRE data from 2005 to 2009 and the adult (>16) population in the USA during 2008 (235.4 million)

[b]Because of small population and income differences, initial values for 2008 differ across PRA scenarios, thus an average is used for a starting value

[c]Participant range across Resources Planning Act (RPA) scenarios A1B, A2, and B2, without climate considerations

[d]Participant range across RPA scenarios A1B, A2, and B2, each with three selected climate futures

[e]Parentheses denote negative number

13.3 Environmental Impact

Table 13.5 Changes in total outdoor recreation days between 2008 and 2060 across all activities and scenarios (Source: Bowker et al. 2012, p. 29)

Activity[a]	2008 Days[b] (millions)	2060 Days range[c] (millions/[percent])	2060 Average days change[c] (millions)	2060 Days range[d] (millions/[percent])	2060 Average days change[d] (millions)
Visiting developed sites					
Developed site use—family gatherings, picnicking, developed camping	2246	3121–3949 [40–74]	1294	3055–3796 [36–69]	1185
Visiting interpretive sites—nature centres, zoos, historic sites, prehistoric sites	1249	1899–2417 [53–91]	952	1935–2435 [55–95]	988
Viewing and photographic nature					
Birding—viewing/photographing birds	8255	11,680–14,322 [40–74]	4859	10,050–13,313 [36–69]	3764
Nature viewing—viewing, photography, study, or nature gathering related to fauna, flora, or natural settings	32,461	41,805–52,835 [31–61]	14,635	41,550–51,288 [28–58]	13,597
Backcountry activities					
Challenge activities—caving, mountain biking, mountain climbing, rock climbing	121	178–219 [49–83]	4859	179–232 [48–92]	89
Equestrian	263	388–503 [49–92]	196	369–482 [40–83]	166
Hiking—day hiking	1835	2901–3682 [59–98]	1470	2825–3541 [54–93]	1366
Visiting primitive areas—backpacking, primitive camping, wilderness	1239	2046	622	1562–1946 [26–57]	519
Motorised activities					
Motorised off-road use	1053	1264–1532 [21–46]	357	1274–1611 [21–53]	385
Motorised snow use (snowmobiles)	69	74–91 [8–33]	16	23–65 [(6)–(67)]	(27)[e]
Motorised water use	**958**	**1304–1806 [37–90]**	**596**	**1245–1763 [30–84]**	**495**
Consumptive					
Hunting—all types of legal hunting	538	506–576 [(5)–8]	14	494–575 [(8)–7]	(8)
Fishing—anadromous, cold-water, saltwater, warm water	1369	1665–2020 [23–46]	514	1602–1958 [17–41]	397
Non-motorised winter activities					
Downhill skiing, snowboarding	178	274–437 [61–150]	179	258–422 [50–146]	165
Undeveloped skiing—cross-country, snow-shoeing	52	69–87 [35–70]	29	28–64 [(45)–25]	(5)
Non-motorised water					
Swimming, snorkelling, surfing, diving	3476	5037–6429 [46–83]	2446	4396–6257 [42–80]	2298
Floating—canoeing, kayaking, rafting	**262**	**338–422 [30–62]**	**128**	**309–409 [18–56]**	**83**

Source: National Survey of Recreation and the Environment (NSRE) 2005–2009, Versions 1 to 4 (January 2005 to April 2009). n = 24,073 (USDA Forest Service 2009)
[a]Activities are individual or activity composites derived from the NSRE. Participants are determined by the product of the average weighted frequency of participation by activity for NSRE data from 2005 to 2009 and the adult (>16) population in the USA during 2008 (235.4 million)
[b]Because of small population and income differences, initial values for 2008 differ across PRA scenarios, thus an average is used for a starting value
[c]Participant range across Resources Planning Act (RPA) scenarios A1B, A2, and B2, without climate considerations
[d]Participant range across RPA scenarios A1B, A2, and B2, each with three selected climate futures
[e]Parentheses denote negative number

Table 13.6 Once a week participation in funded sports (16 years and over)—Sport England: Active People Survey 10 (October 2015–September 2016) (Source: https://www.sportengland.org/media/11746/1x30_sport_16plus-factsheet_aps10.pdf)

Sport England NGB 13-17Funded sports	Active People Survey 10 (October 2015–September 2016)	
	%	n
Swimming	5.67	2,516,700
Athletics	5.01	2,217,800
Cycling	4.40	1,950,300
Football	4.21	1,844,900
Golf	1.64	729,300
Exercise, movement, & dance	0.98	437,200
Badminton	0.97	425,800
Tennis	0.90	398,100
Equestrian	0.64	282,400
Bowls	1.33	211,900
Squash & racquetball	0.45	199,500
Rugby union	0.46	199,000
Netball	0.42	180,200
Boxing	0.36	159,000
Cricket	0.36	158,500
Basketball	0.35	150,800
Mountaineering	0.25	110,200
Table tennis	0.24	107,100
Angling	0.24	106,200
Snowsport	0.23	99,800
Hockey	0.22	92,700
Weightlifting	0.20	88,100
Rowing	**0.19**	**83,400**
Gymnastics	0.15	65,100
Shooting	0.13	56,600
Sailing	**0.10**	**45,600**
Rugby league	0.10	44,900
Canoeing	**0.09**	**41,900**
Volleyball	0.08	33,800
Archery	0.07	32,400
Taekwondo	0.06	23,900
Judo	0.04	18,900
Rounders	0.03	12,800

Source: Sport England's Active People Survey

pollution from outboard motors and sewage. Those resulting from shore-based activities, such as boat launching and egress (as well as angling discussed in Chap. 15 and swimming), included trampling and associated effects, boat wash erosion, as well as sewage and other chemical impacts.

Aquatic ecosystems, like terrestrial ecosystems, have many parameters that interact to determine water quality. Some of these impacts are direct, occurring on or in the water (Fig. 13.5). Other impacts to water systems are indirect, they come from actions on the shore, land, or watershed adjacent to the water body. Such actions might be due to land use activities like farming, forestry, or industry and lie outside the scope of this chapter.

13.3.1 Physical Impacts of Water Sports

Back in the 1980s, Garrad and Hey (1987) reported that increasing levels of turbidity reported in parts of the Norfolk Broads over the last century had been attributed to algal growth. Their research demonstrated how the resuspension of bed sediments by a single moving boat was possible and how the diurnal variation of boat traffic movement had distinct effects on patterns of suspended sediment concentration and hence turbidity. They concluded that the control of boat speed and frequency therefore had important implications for the management of turbidity levels in the Norfolk Broads, UK. O' Sullivan (1990) examined the effects of recreational usage in the gorges of the Afon Conwy in Snowdonia and noted that it was often difficult to distinguish between the impacts of canoeists and fisherman and other visitors. Spencer (1995) carried out erosion surveys at canoe/kayak access and egress sites on the Afon Llugwy and Afon Tryweryn in Snowdonia and noted a number of examples of erosion which he photographed. Zani (2000) also examined the effect of recreational canoeing on the riparian vegetation of the River Dee in North Wales. He examined sites before and after an organised canoeing event known as the Dee Tour and found on average there to be 70% less vegetation at three popular access and egress sites compared with control sites.

Bradbury et al. (1995) discovered that the wash from high-speed tourist cruise launches caused erosion of the formerly stable banks of

13.3 Environmental Impact

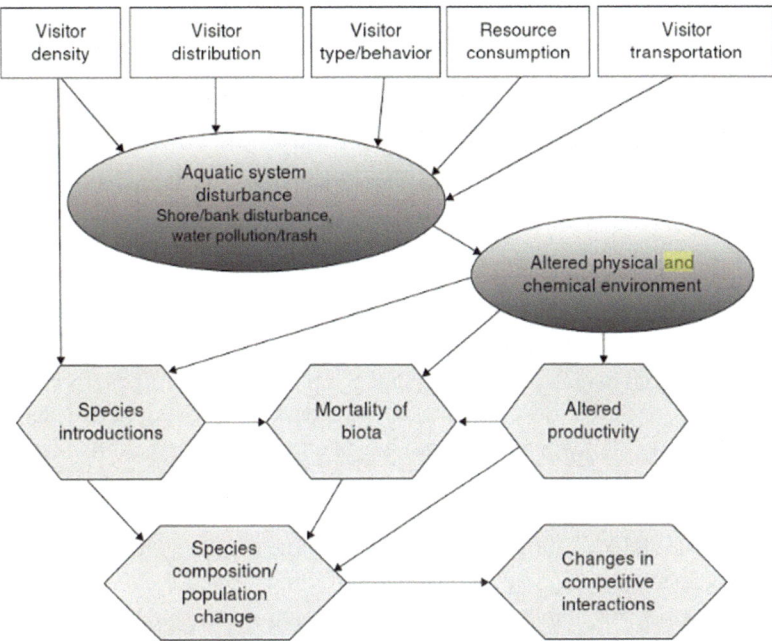

Fig. 13.5 Main impacts of recreation-water resource impacts (after Hammitt et al. 2015; adapted from Monz and Leung 2006)

the lower Gordon River within the Tasmanian Wilderness World Heritage Area. They noted that speed and access restrictions on the operation of commercial cruise vessels had considerably slowed, but not halted erosion, which continued to affect the destabilised banks. Using erosion pins, survey transects, and vegetation quadrats at 48 sites, the mean measured rate of erosion of estuarine banks slowed from 210 to 19 mm/year with the introduction of a 9-knot speed limit. In areas where cruise vessels continued to operate, alluvial banks were eroded at a mean rate of 11 mm/year during the three-year period of the current management regime. Very similar alluvial banks no longer subject to commercial cruise boat traffic eroded at the slower mean rate of 3 mm/year. Sandy levee banks retreated an estimated 10 m maximum during the 10–15 years prior to their study. The mean rate of bank retreat slowed from 112 to 13 mm/year with the exclusion of cruise vessels from the leveed section of the river. Revegetation of the eroded banks was proceeding slowly; however, since the major bank colonisers are very slow growing tree species, they stated that it was likely to be decades until revegetation could contribute substantially to bank stability.

Bishop (2007) showed that waves produced by even low-wash vessels can have a sizeable impact on infaunal assemblages in bottom sediments of sheltered estuaries. Although this impact is widely regarded to be a consequence of wash coarsening sediment grain-size, it may be due to a number of alternative mechanisms which include enhanced turbidity, decreased larval supply, changed resource availability, and/or erosion of animals from the sediment.

With recreational boat usage and ownership in Australia at an all-time high, Ruprecht et al. (2015) investigated the impact that the proliferation of recreational vessels designed and manufactured for the sport of wakeboarding and, more recently, wakesurfing (a popular alternative activity to wakeboarding) were having. Wakeboarding/wakesurfing vessels are designed, through the use and control of ballast and customised trim, to maintain a breaking wave at the optimal operational speed (typically 10 knots for wakesurfing and 19 knots for wakeboarding). Tests were undertaken in a controlled environment (deep water, no currents, controlled boat speeds, repeat runs, etc.) using state-of-the-art measuring equipment, and the results indicated that the wave energy associated with the single maximum wave height for the wakesurf "operating conditions" was approximately four times that of the wakeboard "operating conditions" and twice that of the wakeboard "maximum wave" conditions. Operational wakesurfing was shown to produce significantly different waves to wakeboarding and water skiing. They recommended that these three activities be assessed and managed separately. A common feasible management option would be to restrict those activities to wide parts of the river to allow for natural wave height attenuation. In certain situations, where maximum wave height is a concern, and insufficient distance is available to allow for natural attenuation, management of the sport may result in restricting activities or the implementation of artificial shoreline enhancements (i.e. bank armouring, rip-rap, rock fillets, etc.)

Mujal-Colilles et al. (2017) collected field data in a harbour basin and compared them with analytical formulations for predicting maximum scouring depth due to propeller jets. Spatial data analysis of seven-year biannual bathymetries quantified the evolution of a scouring hole along with the sedimentation process within a harbour basin. The maximum scouring depth was found to be of the order of the propeller diameter with a maximum scouring rate within the first six months of docking manoeuvring.

Gabel et al. (2017) presented a review on the effects of ship-induced waves on the structure, function, and services of aquatic ecosystems based on more than 200 peer-reviewed publications and technical reports. Ship-induced waves act at multiple organisational levels and different spatial and temporal scales. All the abiotic and biotic components of aquatic ecosystems are affected, from the sediment and nutrient budget to the planktonic, benthic, and fish communities. They highlighted how the effects of ship-induced waves cascade through ecosystems and how different effects interact and feed back into the ecosystem finally leading to altered ecosystem services and human health effects.

13.3.2 Biological Impacts of Water Sports

13.3.2.1 Water Quality and Micro-organisms

The Boundary Waters Canoe Area in northern Minnesota is the most heavily used wilderness area in the USA. The majority of its visitors participate in water-based recreational activities that contribute detergents, and sanitary, outboard motor, and solid wastes to the natural chemical budget of the pristine lakes. King and Mace Jr (1974) reported on research which aimed to determine if recreational use caused reduction of water quality in bodies of water located near campsites. Nine parameters of water quality were measured in water near campsites and near unused shorelines for comparison. The parameters measured were dissolved oxygen saturation, temperature, turbidity, hydrogen ion concentration, specific conductivity, nitrate (plus nitrite) concentration, available phosphate concentration, total Kjeldahl nitrogen concentration, and coliform bacteria population. The results show that use of campsites causes highly significant ($\alpha = 0.01$) increases in coliform bacteria populations (Table 13.7) and smaller increases ($\alpha = 0.10$) in available phosphate concentrations in water near the campsites. They suggested that one probable cause of these increases was drainage from the pit toilets located at each campsite.

Fewtrell et al. (1992) noted how there is little quantitative information on the relation between water quality and disease attack rates after recreational activities in freshwater, so they conducted

Table 13.7 Coliform populations for various classes of campsites. University of Minnesota Boundary Waters Canoe Area Campsite Study, 1970 (after Merriam and Smith 1974)

	Number of coliforms/100 ml use categories		
	High	Medium	Low
Campsite	4.61	6.63	5.83
Control	0.28	1.95	4.68
Difference	4.33	4.68	1.15

Note: High-use sites had over 1100 days visitor use; medium-use sites had over 500 days visitor use; low-use sites had less than 300 days total visitor use a prospective cohort study to measure the health effects of white-water and slalom canoeing in two channels with different degrees of microbial contamination. Site A, fed by a lowland river, showed high enterovirus concentrations (arithmetic mean 198 pfu per 10 litre) and moderate faecal coliform concentrations (geometric mean 285/dl); at site B, from an upland impoundment, all samples were free of enteroviruses, and the geometric mean faecal coliform concentration was 22/dl. Between five and seven days after exposure, canoeists using site A had significantly higher incidences of gastrointestinal and upper respiratory symptoms than canoeists using site B or non-exposed controls (spectators). Like seawater bathers, freshwater canoeists can be made ill by sewage contamination. They recommended that the hazard of freshwater may be best measured by counting of viruses rather than bacteria.

Other quantitative risk assessments have estimated health risks of water recreation. One input to risk assessment models is the rate of water ingestion. One published study estimated rates of water ingestion during swimming, but estimates of water ingestion are not available for common limited-contact water recreation activities such as canoeing, fishing, kayaking, motorboating, and rowing. In the summer of 2009, Dorevitch et al. (2011) conducted two related studies to estimate water ingestion during these activities. First, at Chicago area surface waters, survey research methods were utilised to characterise self-reported estimates of water ingestion during canoeing, kayaking, and fishing among 2705 people. Second, at outdoor swimming pools, survey research methods and the analysis of cyanuric acid, a tracer of swimming pool water, were used to characterise water ingestion among 662 people who engaged in a variety of full-contact and limited-contact recreational activities. Data from the swimming study was used to derive translation factors that quantify the volume of self-reported estimates. At surface waters, less than 2% of canoeists and kayakers reported swallowing a teaspoon or more and 0.5% reported swallowing a mouthful or more. Swimmers in a pool were about 25–50 times more likely to report swallowing a teaspoon of water compared to those who participate in limited-contact recreational activities on surface waters. Mean and upper confidence estimates of water ingestion during limited-contact recreation on surface waters are about 3–4 ml and 10–15 ml, respectively. These estimates of water ingestion rates may be useful in modelling the health risks of water recreation.

Phillip et al. (2009) conducted a study to determine the possible influence of recreation on microbiological water quality of a tropical stream. Microbiological water quality was measured at several recreational sites along the stream, and a separate experiment was conducted to look at the effect of sediment resuspension on microbiological water quality. Microbiological quality of the water in the stream was generally poor and varied widely with faecal coliform and *Escherichia coli* levels ranging from 1 to >16,000 and 14 to 9615 organisms 100 ml^{-1}, respectively. Levels of faecal coliforms were higher in the wet (median = 700 organisms 100 ml^{-1}) than the dry (median = 500 organisms 100 ml^{-1}) season while the reverse was true for *E. coli* (median = 300 and 220 organisms 100 ml^{-1} in the wet and dry seasons, respectively). Recreational activity resulted in reduced water quality: sites with recreation had poorer water quality than those without; water quality was generally poorer when there were high numbers of recreational users. Wading resulted in a fourfold increase in mean *E. coli* densities and a threefold increase in total suspended sediments in the overlying water suggesting that the increases were due to suspension of bacteria from the sediments. They concluded that water quality monitoring methodology for assessing recreational water quality should be amended to factor in the effects of wading since environmental strains of bacteria can be pathogenic and thus represent a human health threat.

DeFlorio-Barker et al. (2017) estimated the costs of sporadic gastrointestinal illness associated with surface water recreation. They characterised the disease burden attributable to water recreation using data from two cohort studies using a cost of illness (COI) approach and estimated the largest drivers of the disease burden of water recreation. Comparing data which evaluated swimming and wading in marine and freshwater beaches in six US states, with data which evaluated illness after incidental-contact recreation (boating, canoeing, fishing, kayaking, and rowing) on waterways in the Chicago area, they estimated the cost per case of gastrointestinal illness and costs attributable to water recreation. Data on health care and medication utilisation and missed days of work or leisure were collected and combined with cost data to construct measures of COI. Depending on different assumptions, the cost of gastrointestinal symptoms which were attributable to water recreation were estimated to be $1220 for incidental-contact recreation (range $338–$1681) and $1676 for swimming/wading (range $425–2743) per 1000 recreational users. Lost productivity was found to be a major driver of the estimated COI, accounting for up to 90% of total costs. These estimates suggested that gastrointestinal illness attributed to surface water recreation at urban waterways, lakes, and coastal marine beaches was responsible for costs that should be accounted for when considering the monetary impact of efforts to improve water quality. The COI provides more information than the frequency of illness, as it takes into account disease incidence, health-care utilisation, and lost productivity.

13.3.2.2 Impacts on Plants and the Spread of Invasive Species

Liddle and Scorgie (1980) developed a useful flow chart of the impacts of boats on plants (Fig. 13.6). They saw three main ways in which boats impact plants: by propeller action, from wash, and by direct contact.

The wash created by motorised craft can cause considerable erosion of plant roots. Haslam (1978) investigated the susceptibility of a number of plants to erosion by directing a horizontal jet of water from upstream onto the soil at the base of the plants, and the time taken for the plants to be eroded was noted. On this basis plants were placed into four groups. One of the most resistant, the yellow waterlily (*Nuphar lutea*), has smaller leaves in faster-flowing waters.

In many parts of the world, the spread of invasive aquatic plant species by boats is a major

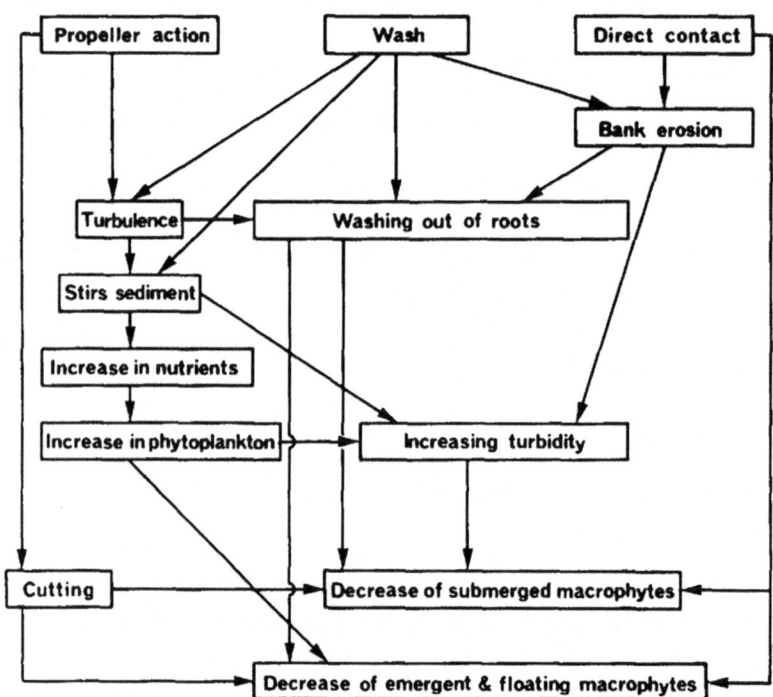

Fig. 13.6 The impacts of boats on plants (after Liddle and Scorgie 1980, p. 185)

problem and is among the most important threats to biodiversity worldwide. For example, aquatic ecosystems in South Africa are prone to invasion by several invasive alien aquatic weeds, most notably, *Eichhornia crassipes* (Mart.) Solms-Laub. (*Pontederiaceae*) (water hyacinth), *Pistia stratiotes* L. (*Araceae*) (water lettuce), *Salvinia molesta* D.S. Mitch. (*Salviniaceae*) (salvinia), *Myriophyllum aquaticum (Vell. Conc.) Verd.* (parrot's feather), and *Azolla filiculoides Lam.* (*Azollaceae*) (red water fern). Hill and Coetzee (2017) reviewed the biological control programme on waterweeds in South Africa and found that there had been significant reductions in the extent of invasions and a return on biodiversity and socio-economic benefits through the control programme. These studies provide justification for the control of widespread and emerging freshwater invasive alien aquatic weeds in South Africa. The long-term management of alien aquatic vegetation relies on the correct implementation of biological control for those species already in the country and the prevention of other species entering South Africa.

Eurasian watermilfoil, an aquatic invasive weed, occurs at a number of sites in western Nevada and northeastern California, including Lake Tahoe. Because Eurasian watermilfoil is easily spread by fragments, transport on boats and boating equipment plays a key role in contaminating new water bodies. This is an important means of the potential spread of this weed throughout key recreational and agricultural areas surrounding Lake Tahoe. Unless the weed is controlled, significant alterations of aquatic ecosystems, with associated degradation of natural resources and economic damages to human uses of those resources, may occur. Eiswerth et al. (2000) used an economic valuation approach known as benefits transfer to estimate the value of a portion of the recreational service flows that society currently enjoys in the Truckee River watershed below Lake Tahoe. The lower-bound estimates of baseline water-based recreation value at a subset of sites in the watershed range from $30 to $45 million/year. Impacts from the continued spread of Eurasian watermilfoil in the watershed could be significant; for example, even a 1% decrease in recreation values would correspond to roughly $500,000/year as a lower bound.

Murphy and Eaton (1983) conducted quantitative surveys of plant growth in British Cruising and Remainder Canals which showed significant associations between community composition, abundance of aquatic macrophytes, and pleasure-boat traffic. They found evidence for a "critical" traffic range, about 2000–4000 movements ha^{-1} m $depth^{-1}$ $year^{-1}$ (4 weeks)$^{-1}$ (my) reducing submerged macrophyte standing crop, perhaps by maintaining high daytime water turbidity during the summer months as there were significant associations between boat traffic density, water turbidity, and summer submerged crop. The seasonal distribution of pleasure-boat traffic appeared to be an important influence on submerged macrophyte community abundance. They noted that the course of macrophyte community development may be largely determined by the stage in the growth season at which traffic in the range 300–600 my is attained. Relationships between the abundance of emergent plants and boat traffic were weaker, and there was no significant association with water turbidity. In 1977 approximately 50% of the canal system carried low pleasure-boat traffic density (<2000 my), 24% had traffic within the critical range (2000–4000 my), and 26% had heavy traffic (>4000 my).

Walsh et al. (2016) evaluated the economic impacts of an invasive species that cascaded through a food web to cause substantial declines in water clarity, a valued ecosystem service. The predatory zooplankton, the spiny water flea (*Bythotrephes longimanus*), invaded the Laurentian Great Lakes in the 1980s and then subsequently spread to inland lakes, including Lake Mendota (Wisconsin), in 2009. In Lake Mendota, *Bythotrephes* reached unparalleled densities compared with in other lakes, decreasing biomass of the grazer *Daphnia pulicaria* and causing a decline in water clarity of nearly 1 m. Time series modeling revealed that the loss in water clarity, valued at US$140 million (US$640 per household), could be reversed by a 71% reduction in phosphorus loading. A phosphorus reduction of this magnitude was estimated to cost between US$86.5 million and US$163 million

(US$430–US$810 per household). Estimates of the economic effects of Great Lakes invasive species may increase considerably if cases of secondary invasions into inland lakes, such as Lake Mendota, are included. Furthermore, they concluded that such extreme cases of economic damages called for increased investment in the prevention and control of invasive species to better maximise the economic benefits of such programmes.

De Ventura et al. (2016) examined how recreational boats that are transported overland could contribute to the dispersal of invasive zebra mussels among lakes in Switzerland. Using a questionnaire sent to registered boat owners, they surveyed properties of transported boats and collected information on self-reported mussel fouling and transport activities of boat owners. They also sampled boat hulls at launching ramps and harbours for biofouling invertebrates. Boats that were kept seasonally or year-round in water were found to have high vector potential with mussel fouling rates of more than 40%. However, only about 6% of boats belonging to these groups were transported overland to other water bodies. Considering that approximately 100,000 recreational boats are registered in Switzerland, they estimated that every year around 1400 boats fouled with mussels were transported overland. Such boats pose a high risk of distributing zebra mussels between water bodies. Their results suggested that there is a considerable risk that recreational boats may spread new fouling species to all navigable water bodies within the study area and speculated that one such species could be the quagga mussel, which has not yet invaded lakes in Switzerland. However, their study has identified the group of high-risk boats so that possible control measures could be targeted at this relatively small group of boat owners.

13.3.2.3 Disturbance to Wildlife

Water-based recreation can result in disturbance to wildlife. Perhaps the best studied group about which the effects of disturbance have been observed is birds. Hockin et al. (1992) noted that human-induced disturbance can have a significant negative effect on breeding success by causing nest abandonment and increased predation. Outside the breeding season, recreation (particularly powerboating, sailing, and coarse fishing on wetlands) reduces the use of sites by birds. Compensatory feeding at night by some species can probably recoup some of the energy losses caused by disturbance. Mosisch and Arthington (1998) presented some data to support this (Table 13.8).

Steven et al. (2011) conducted a review of the impacts of nature-based recreation on birds. Their review of the recreation ecology literature published in English language academic journals identified 69 papers from 1978 to 2010 that examined the effect of these activities on birds. Sixty-one of the papers (88%) found negative impacts, including changes in bird physiology (all 11 papers), immediate behaviour (37 out of 41 papers), as well as changes in abundance (28 out of 33 papers) and reproductive success (28 out of 33 papers). Previous studies are concentrated in a few countries (USA, England, Argentina, and New Zealand), mostly in cool temperate or temperate climatic zones, often in shoreline or wetland habitats, and mostly on insectivore, carnivore, and crustaceovore/molluscivore foraging guilds. They found limited research in some regions with both high bird diversity and nature-based recreation such as mainland Australia, Central America, Asia, and Africa and for popular activities such as mountain bike riding and horse riding. It was clear that non-motorised nature-based recreation has negative impacts on a diversity of birds from a range of habitats in different climatic zones and regions of the world.

The edges of propellers can act like a set of rotating knives. Liddle and Scorgie (1980)

Table 13.8 Breeding densities (pairs/10 km channel) of three common species of English waterbirds in used and disused canals (adapted from Mosisch and Arthington 1998)

Species	Disused canal	Used canal
Little grebe (*Tachybaptus ruficollis*)	5.1	0.2
Coot (*Fulica atra*)	4.7	2.5
Moorhen (*Gallinula chloropus*)	37.8	22.5

reported that a boat with an outboard motor driven through a patch of yellow water lily (*Nuphar lutea*) cut through the petioles, leaving a very jagged end. On a 50 m run, 15 leaves were detached and many more were overturned. Other studies have found that used in shallow water, propellers can cut plants right out of a mud substrate and remove sediment, which can lead to increased turbidity. More recently, Whitfield and Becker (2014) conducted a detailed review of the impacts of recreational motorboats on fishes. They reported that some fish species do not appear to respond behaviourally to the presence of powered outboard engines, for example, lake trout (*Salvelinus namaycush*) in a small Canadian lake did not respond to boat traffic, even during detailed manual tracking of individual fish. However, the passage of boats has been shown to break up fish schools and result in increased energy expenditure as they attempt to move away from the disturbance. Manipulative work examining nest guarding behaviour of longear sunfish found that passing boats caused fish to leave their nests for longer periods than during control times. A recent study on the effects of passing motorboats on the abundance of different sized fish within the main channel of a South African estuary revealed that the 100–300 mm and >500 size classes had no change in their abundance following the passage of boats (Becker et al. 2013). However, the mid-sized fishes (301–500 mm) decreased in abundance, a displacement which was attributed to a number of factors, including noise, bubbles, and the approaching boat itself.

Disturbance effects of motorboats on fishes can be linked to several factors (Fig. 13.7), including noise levels (Slabbekoorn et al. 2010). Since fish have sensitive auditory organs, anthropogenic noise has the potential to cause physiological and behavioural responses (Purser and Radford 2011). In situ recording of powerboat noise spectra indicate that outboard sounds can be detected by species such as cyprinids at a distance of hundreds of metres (Amoser et al. 2004). Noise from boats may also increase fish stress levels or even have a direct impact on the

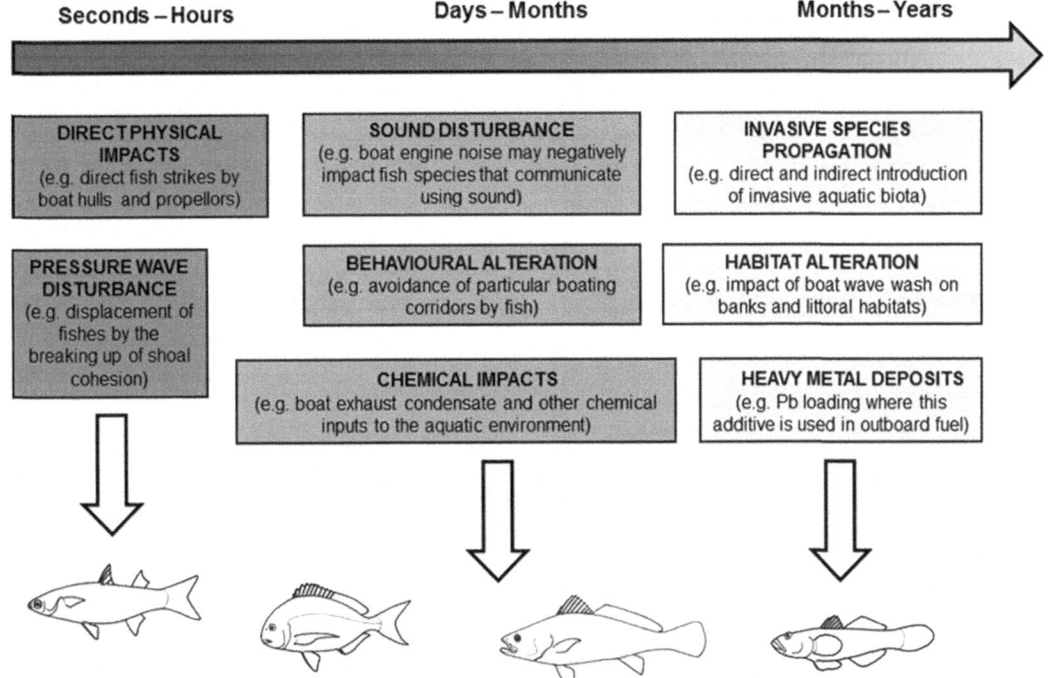

Fig. 13.7 Likely influences and impacts of powerboating activities on fishes and their habitats and the likely time frame over which the impacts may occur (after Whitfield and Becker 2014, p. 25)

breeding behaviour of certain fish. Boat noise has also been shown to adversely affect the territorial behaviour of certain fish species which uses sound production as an effective tool for territorial defence and nest caring.

To date there seems to be very limited work conducted on the environmental effects of PWC or "jet skis," and this seems to be an area for future research. While the air noise pollution of PWC is often complained about by the public (Blomburg et al. 2003), noise emitted by these craft in the water may also be of concern (Erbe 2013). Also, PWC riders tend to change speed and direction far more frequently than those driving typical recreational boats, thus giving rise to unpredictable changes in sound pitch and volume as well as craft direction. Additionally the hull of PWC tends to strike the water surface harder and with greater frequency than typical motorboats, all of which is likely to cause more confusion in nearby fish schools.

13.3.3 Chemical Impacts of Water Sports

13.3.3.1 Heavy Metals

Heavy metal inputs to aquatic environments became a major issue following the industrial revolution, and in the modern era, these pollutants can come from a variety of sources, including boats. In terms of recreational boating, in the recent past, the major sources of heavy metals were antifouling paints and boat exhaust emissions. Secondly, resuspension of sediment bound metals by boat wake, and direct sediment disturbance by boat engine operations in shallow water, have accentuated the problem. Fortunately, considerable progress has been made in reducing toxic metals from paints and lead from petroleum products, thus reducing pollution from these sources. Increasing lead levels in lakes, rivers, and estuaries are perhaps one of the most obvious potential indicators of environmental pollution by outboard motors, particularly where lead is used as an additive to the fuel. Lead in the aquatic environment from exhaust waste is most likely to occur in a relatively insoluble form with lead accumulating in the sediments and being potentially absorbed by certain benthic biota. Boat traffic can also result in the resuspension of heavy metals from polluted sediments. For example, boat traffic in the Deûle River in northern France has been directly linked to the resuspension of sedimentary particles that significantly increased lead and zinc into the overlying water (Superville et al. 2014).

Elevated copper levels in Lake Texoma water were attributed to antifouling-based paint used on boats and high copper levels at specific locations in marinas around the lake appeared to be associated with recreational boat repair activities (An and Kampbell 2003). Leaching of tributyltin-containing antifouling paints used on boats into coastal waters is a major problem for certain invertebrate species (Bhosle et al. 2004) and the enzyme system activities of certain fish. Molluscs appear to be most affected by tributyltin and its degradation products, with fish having low levels of contamination. However, because fish invertebrate prey is negatively affected by tributyltin, it is likely that fish stocks will also be impacted. Fortunately, legislation in many countries has seen these toxic paints being replaced by more environmentally acceptable alternatives.

Eklund and Watermann (2018) used a handheld X-ray fluorescence (XRF) analyser specially calibrated for measurements of metals on plastic boat hulls on leisure boats in Denmark (DK), Finland (FI), and Germany (DE). The results on tin and copper are presented as μg metal/cm^2 and were compared with published data from different parts of Sweden, that is, boats in freshwater, brackish water, and saltwater. The results showed that tin with mean values >50 μg Sn/cm^2 is still found on 42%, 24%, and 23% of the boats in DK, FI, and DE, respectively. The corresponding percentages based on median values are 38%, 22%, and 18% for DK, FI, and DE, respectively. The variation among boats is high with a maximum mean value of 2000 μg Sn/cm^2. As comparison, one layer of an old Tributyltin (TBT) antifouling paint Hempels Hard Racing Superior corresponds to 300 μg Sn/cm^2. The percentage of boats with tin >400 µg Sn/cm² content based on mean values were 10% in DK, 5% in FI, and 1% in DE. The corresponding median values were 9%, 6%, and 1% for DK, FI, and DE, respectively. Copper, >100 µg Cu/cm², was detected on all measured boats in DK and in DE and on all but 3% of the FI boats. One layer of Hempels Mille Xtra corresponds to 4000 µg Cu/cm². The recommendation on the can is to apply two layers. The proportion of boats with higher mean copper values than 8000 µg Cu/cm² was 51, 56, and 61 for boats in DK, FI, and DE, respectively. The proportion based on median values >8000 µg Cu/cm² was 50%, 54%, and 61% for DK, FI, and DE, respectively. The conclusion was that many leisure boats around the Baltic Sea still display or possess antifouling paints containing organotin compounds and that more than half of the boats have more copper than needed for one boat season according to the paint producers. Much of these known toxic compounds will probably be released into the environment and harm the biota.

13.3.3.2 Motorboat Engine Products and Bi-products

Motorboats are usually powered by either diesel or a petroleum and oil-based mixture, both of which are sometimes accidentally spilt into waterways when filling up tanks or servicing engines close to the water. In addition, both types of fuel emit exhaust fumes into or onto the water when the motorboat is underway which can affect fish eggs, larvae, and juveniles, especially in surface waters. Diesel is an important fuel used by both small and large boats in coastal areas and has the potential to influence gene expression in fishes (Mattos et al. 2010). Similarly, petroleum contamination of the surrounding water by small boats was found to negatively influence the health of winter flounder (*Pleuronectes americanus*) in Placentia Bay, Newfoundland (Khan 2003). Accidental spillage of motorboat fuel directly into aquatic ecosystems is a reality and will always remain a water pollution risk.

Laboratory tests conducted in 1960 by English et al. (1963) showed that bluegill sunfish were killed when outboard fuel consumption reached 530 L per million litres of water. However, fish flesh could be tainted by outboard motor exhaust wastes at much lower levels. These tests showed that 90% of persons in a taste panel noted objectionable flavour in fish exposed to cumulative fuel consumption levels of 2.8 L per million litres of water. Lüdermann (1968) found that detrimental changes in the flavour of the flesh of freshwater fishes exposed to the exhaust emissions of outboard motors disappeared after a few days of the fish living in clean freshwater. Carbon monoxide was attributed to fish kills near an outboard testing facility on the Fox River and the suggestion made that such events could to be exacerbated by low river flows and high temperatures (Kempinger et al. 1998). This indicates that there may be potential for carbon monoxide poisoning in areas with very high boat traffic and low flushing rates.

Table 13.9 summarises the major impacts of recreational motorboat activities.

Professor Joy Tivy of Glasgow University (Tivy 1980) wrote a very detailed report on the effect of recreation on freshwater lochs and reservoirs in Scotland which was commissioned and published by the Countryside Commission for Scotland. This comprehensive report examined all aspects of recreational impact on and around the water.

13.4 Management and Education

Future management of water-based recreation must take account of the predicted future increases in participation numbers (as we saw in Sect. 13.2) and climate change.

13.4.1 Managing Physical Impacts

O' Sullivan (1990) examined the effects of recreational usage in the gorges of the Afon Conwy in Snowdonia. He noted that his observations at the time did not present a cause for ecological concern in the gorges and concluded that "positive management plans arrived at through essential dialogue between recreationalists and conservationists are

Table 13.9 Summary of the major findings relating to recreational motorboat activities (adapted from Whitfield and Becker 2014)

Motorboat traffic and direct hits	Evidence of direct hits by boats. Very few studies have quantified fish strikes by boats at various speeds or the fish sizes that are affected. This is an area needing further research
Motorboat traffic and fish behaviour	The effect of motorboat traffic on the behaviour of fish is probably the most studied aspect of boat impacts on fish. Noise emitted from engines may increase stress levels in fishes, and underwater noise has also been linked to disruption in the reproductive behaviour of certain fishes. Noise has been found to influence all fish life history stages, including the larvae. Most studies have been conducted in laboratories, but recent examples from field based studies have provided real data for the testing of hypotheses. Further research is required on fish size-related responses to boat movements, as well as which species are most negatively affected by boat traffic
Heavy metals	Sources of heavy metals in aquatic ecosystems arising from boats include antifouling paints and exhaust emissions, as well as the resuspension of contaminated sediments by boat propeller action and wave wakes. More research is needed to link levels of boating activity to Pb and other metal concentrations in the aquatic environment
Motorboat bi-products	Engine exhaust is the most prominent bi-product of motorboats. Diesel can influence gene expression in fish, while multiple studies have found that other petroleum-based products can adversely affect the health of fish. Carbon monoxide poisoning has been linked to fish kills, and this may be a particular threat in systems with high boat traffic and low flushing rates
Invasive species propagation	Transport of invasive fish species overland from one water body to another is a major issue, with this often being done deliberately. However the inadvertent transport of fish diseases and parasites on/in boats and associated equipment is a topic which has not received research attention and is in need of urgent investigation
Boat infrastructure	Infrastructure which facilitates boating activities such as piers, moorings, ramps, and marinas can impact fish assemblages. Removal of natural habitat to construct infrastructure has the greatest impact, with fish and invertebrate assemblages on man-made structures rarely the same as those found in natural habitats. Research has also been conducted on the negative effects of mooring sites and anchoring chains on seagrass beds. The use of swing mooring has been shown to greatly reduce these impacts
Impacts on aquatic habitats	Moving boats can impact aquatic habitats by increasing turbidity, eroding banks with wave wash, and scouring aquatic macrophyte habitats with boat propellers. Invertebrates in seagrass exposed to boating activity have been found to have lower diversity than control sites, which can have important implications for fish productivity wave wash from boats can be mediated by restricting the speed of boat traffic in sensitive areas

necessary now and are likely to be required in future" (p. 81). Spencer (1995) carried out erosion surveys at canoe/kayak access and egress sites on the Afon Llugwy and Afon Tryweryn in Snowdonia and noted a number of examples of erosion which he photographed and undertook interviews with canoeists who exhibited "negative attitudes towards the environment" (p. iii) and concluded that a code of conduct was needed.

Earlier in this chapter, we reported on the study by Bradbury et al. (1995) which showed that the wash from high-speed tourist cruise launches caused erosion of the formerly stable banks of the lower Gordon River within the Tasmanian Wilderness World Heritage Area. They noted how speed and access restrictions on the operation of commercial cruise vessels had considerably slowed, but not halted erosion, which continued to affect the destabilised banks. Revegetation of the eroded banks was proceeding slowly; however, since the major bank colonisers are very slow growing tree species, they stated that it was likely to be decades until revegetation could contribute substantially to bank stability.

Bonham (1980) reported the results of field trials in the UK which showed that beds of *Phragmites australis*, *Scirpus lacustris*, *Typha angustifolia*, and *Acorus calamus* attenuated ship waves from motorboat wash. They concluded that under suitable conditions of depth, vegetation density and a bed slope of 1 in 4, a bed of any of these species 2 m wide would dissipate almost two thirds of shipwave energy. They suggested that beds be re-established with mixed emergent species in both mixed bed and fen and Broadland Rivers with high motorboat usage would give bank protection and also provide a scarce natural habitat.

13.4.2 Managing Biological/Water Quality Impacts

Sharp et al. (2017) noted how aquatic invasive species (AIS) present a great challenge to ecosystems around the globe, and controlling AIS becomes increasingly difficult when the potential vectors are related to recreational activities. They claimed that an approach combining education and outreach efforts to control AIS may be the best course of action. They therefore designed a survey to measure public perceptions, knowledge of, and attitudes towards AIS, as well as public support for various management actions. Surveys were administered during the summer of 2013 at two boat launches where one launch had active outreach the previous summer and one that did not. A total of 400 surveys were completed with a response rate of 89%. There was support for most proposed management options, and respondents understood the urgency of managing AIS. There was a difference between the launches in how people responded, highlighting that educational programming may need to be tailored for specific recreational uses and recreational settings.

Breen et al. (2017) collected water quality data from Ireland and carried out an on-site survey of waterway users to evaluate how water quality affected trip days demanded for recreational activities. Water quality measures employed in the analysis included Water Framework Directive (WFD) ecological status as well as several physiochemical measures. The analysis found some evidence that higher levels of recreational demand occur at sites with the highest quality metric measures. However, in many of the estimated models, there was no statistical association between the water quality metric (e.g. WFD status, Biological Oxygen Demand (BOD), ammonia, etc.) and the duration of the recreational trip. As most sites considered in the analysis had relatively high levels of water quality, this result possibly suggested that above an unspecified threshold level that water quality is not a significant determinant of recreational trip duration. Model estimates also revealed a relatively high valuation among participants for water-based recreational activity with an estimate of mean willingness to pay equivalent to €204/day.

Figure 8.8 (Chap. 8) shows visitor signs alerting water users to the spread of invasive species.

Hussner et al. (2017) noted how introduced invasive alien aquatic plants (IAAPs) threaten ecosystems due to their excessive growth and have both ecological and economic impacts. To minimise these impacts, effective management of IAAPs is required according to national or international laws and regulations (e.g. the new EU regulation 1143/2014). Prevention of the introduction of IAAPs is considered the most cost-effective management option. If/when prevention fails, early detection and rapid response increase the likelihood of eradication of the IAAPs and can minimise ongoing management costs. For effective weed control (eradication and/or reduction), a variety of management techniques may be used. The goal or outcome of management interventions may vary depending on the site (i.e. a single waterbody or a region with multiple waterbodies) and the feasibility of achieving the goal with the tools or methods available. Broadly defined management goals fall into three different categories of containment, reduction or nuisance control, and eradication. Management of IAAP utilises a range of control methods, either alone or in combination, to achieve a successful outcome. Hussner et al. reviewed the biological, chemical, and mechanical control methods for IAAPs, with a focus on the temperate and subtropical regions of the world and provided a management diagram illustrating the relationships between the state of the ecosystem, the management goals, outcomes, and tools.

13.4.3 Managing for Climate Change

There have been much interest in the effect that climate change in future might have on water resources in many parts of the world (e.g. Arnell

1998; La Jeunesse et al. 2016; Wang et al. 2016) but less specifically on how and whether water-based recreation will be affected and how it will need to adapt.

Faccioli et al. (2015) noted that climate change will further exacerbate wetland deterioration, especially in the Mediterranean region. On the one side, they claimed that it will accelerate the decline in the populations and species of plants and animals, resulting in an impoverishment of biological abundance. On the other hand, it will also promote biotic homogenisation, resulting in a loss of species' diversity. In this context, they stated that different climate change adaptation policies could be designed: those oriented to recovering species' abundance and those aimed at restoring species' diversity.

Based on the awareness that knowledge about visitors' preferences is crucial to better inform policy makers and secure wetlands' public use and conservation, Faccioli et al. assessed the recreational benefits of different adaptation options through a choice experiment study carried out in S'Albufera wetland (Mallorca). Their results showed that visitors display positive preferences for an increase in both species' abundance and diversity, although they assigned a higher value to the latter, thus suggesting a higher social acceptability of policies pursuing wetlands' differentiation. This finding acquires special relevance not only for adaptation management in wetlands but also for tourism planning, as most visitors to S'Albufera are tourists. Thus, given the growing competition to attract visitors and the increasing demand for high environmental quality and unique experiences, promoting wetlands' differentiation could be a good strategy to gain competitive advantage over other wetland areas and tourism destinations.

13.4.4 Concluding Comments

Clearly there are a number of examples of research which indicates that water-based recreation can have damaging effects on the environment. These can be physical such as in-stream macrophytes being cut by outboard engine propellers, wave wash accelerating bank erosion, and scouring the bed thereby increasing turbidity. Biological impacts include the spread of invasive species, the contamination of waterbodies by sewage and other products brought in by recreationists and disturbance to wildlife. The third general set of impacts is concerned with the chemical changes to water brought about by antifouling paints on the hulls of recreational craft and spills and leakage of oils and fuels.

It is probably reasonable to conclude therefore that the environmental impacts of motorised recreation craft are far greater than for non-motorised ones.

Conclusions

1. The term water sports includes a wide range of activities both in the water and on the water surface. Chapter 14 dealt with some of the underwater activities: scuba diving and snorkelling. This chapter focussed on some of the more popular water recreational activities which take place on the water surface, and these can be divided into the non-motorised (canoeing, kayaking, rafting, rowing, sailing, surfing, and windsurfing) and motorised (jet skiing, motorboating/powerboating, water skiing).
2. The Outdoor Foundation (2017) survey for the USA provided data for eight water sport disciplines. The rank order in terms of the greatest number of participants in 2016 was:
    - *Canoeing*—10,046,000 participants in 2016 and a −1.1% decrease over the previous three years.
    - *Kayaking* (recreational)—10,017,000 participants in 2016 and a three-year increase of 14.9%.
    - *Sailing*—4,095,000 participants in 2016 and a three-year increase of 4.6%.

- *Rafting*—3,428,000 participants in 2016 and a three-year decrease of −10.6%.
- *Stand-up paddling*—3,220,000 participants in 2016 and a massive three-year increase of 61.6%.
- *Kayaking (sea/touring)*—3,124,000 participants in 2016 and a three-year increase of 16.0%.
- *Surfing*—2,793,000 participants in 2016 and a three-year increase of 5.1%.
- *Kayaking (white water)*—2,552,000 participants in 2016 and a three-year increase of 18.9%.
- *Boardsailing/windsurfing*—1,737,000 participants in 2016 and a three-year increase of 31.2%.

If we total all three kayaking disciplines, there were a total of 15,693,000 participants in 2016, making it the most popular of the disciplines in the survey. Stand-up paddling showed by far the greatest three-year increase of 61.6%.

3. Bowker et al. (2012) projected changes in total outdoor recreation participants between 2008 and 2060. There was an estimated 62 million participants in 2008 taking part in motorised water activities (motorboating, water skiing, PWC), and this was predicted to become 87–112 million by 2060. For non-motorised floating water activities (canoeing, kayaking, and rafting), there was an estimated 40 million participants in 2008 taking part, and this was predicted to increase to 52–65 million by 2060. So water-based recreation is projected to increase over the next four decades at least.
4. Water-based recreation can have damaging effects on the environment. Physical impacts include damage to in-stream macrophytes by cutting by outboard engine propellers, wave wash accelerating bank erosion, and scouring the bed thereby increasing turbidity.
5. Biological impacts of water-based recreation include the spread of invasive species, the contamination of waterbodies by sewage and other products brought in by recreationists, and disturbance to wildlife.
6. Water-based recreation can cause chemical changes to water brought about by antifouling paints on the hulls of recreational craft and spills and leakage of oils and fuels.
7. It is probably reasonable to conclude therefore, that the environmental impacts of motorised recreation craft are far greater than for non-motorised ones.
8. There are numerous examples of how these impacts can be managed. For example, controlling the speeds at which watercraft move reduces the impact of wave action and its effect on bank erosion. Planting certain reed species on gently sloping banks can ameliorate the effects of wave action generated by boats. Educating boat users (e.g. by placing information signs at water access points) has been shown to control the spread of invasive species. Changes to the law regarding the use of certain antifouling paints used to protect boat hulls have resulted in reductions in heavy metal concentrations in water and improved water quality.

References

Amoser, S., Wysocki, L. E., & Ladich, F. (2004). Noise emission during the first powerboat race in an alpine lake and potential impact on fish communities. *The Journal of the Acoustical Society of America, 116*, 3797–3798.

An, Y. J., & Kampbell, D. H. (2003). Total, dissolved, and bioavailable metals at Lake Texoma marinas. *Environmental Pollution, 122*(2), 253–259.

Arnell, N. W. (1998). Climate change and water resources in Britain. *Climatic Change, 39*(1), 83–110.

Becker, A., Whitfield, A. K., Cowley, P. D., Järnegren, J., & Næsje, T. F. (2013). Does boat traffic cause displacement in estuarine fish? *Marine Pollution Bulletin, 75*, 168–173.

Bhosle, N. B., Garg, A., Jadhav, S., Harjee, R., Sawant, S. S., Venkat, K., et al. (2004). Butyltins in water, biofilm, animals and sediments of the west coast of India. *Chemosphere, 57*(8), 897–907.

Bishop, M. J. (2007). Impacts of boat-generated waves on macroinfauna: Towards a mechanistic understanding. *Journal of Experimental Marine Biology and Ecology, 343*(2), 187–196.

Blomburg, L. D., Schomer, P. D., & Wood, E. W. (2003). The interest of the general public in a national noise policy. *Noise Control Engineering Journal, 51*(3), 172–175.

Bonham, A. J. (1980). *Bank Protection Using Emergent Plants against Boat Wash in Rivers and Canals*. Report, No. IT 206. Wallingford: Hydraulics Research Station.

Bowker, J. M., Askew, A. E., Cordell, H. K., Betz, C. J., Zarnoch, S. J., & Seymour, L. (2012). *Outdoor Recreation Participation in the United States—Projections to 2060: A Technical Document Supporting the Forest Service 2010 RPA Assessment*. Ashville: Southern Research Station. Retrieved from www.srs.fs.usda.gov.

Bradbury, J., Cullen, P., Dixon, G., & Pemberton, M. (1995). Monitoring and management of streambank erosion and natural revegetation on the lower Gordon River, Tasmanian wilderness world heritage area, Australia. *Environmental Management, 19*(2), 259.

Breen, B., Curtis, J. A., & Hynes, S. (2017). *Recreational Use of Public Waterways and the Impact of Water Quality (No. 552)*. ESRI working paper.

Cordell, K. (2012). *Outdoor Recreation Trends and Futures: A Technical Document Supporting the Forest Service 2010 RPA Assessment*. Ashville: Southern Research Station. Retrieved from www.srs.fs.usda.gov.

De Ventura, L., Weissert, N., Tobias, R., Kopp, K., & Jokela, J. (2016). Overland transport of recreational boats as a spreading vector of zebra mussel *Dreissena polymorpha*. *Biological Invasions, 18*(5), 1451–1466.

DeFlorio-Barker, S., Wade, T. J., Jones, R. M., Friedman, L. S., Wing, C., & Dorevitch, S. (2017). Estimated costs of sporadic gastrointestinal illness associated with surface water recreation: A combined analysis of data from NEEAR and CHEERS studies. *Environmental Health Perspectives, 125*(2), 215.

Dorevitch, S., Panthi, S., Huang, Y., Li, H., Michalek, A. M., Pratap, P., et al. (2011). Water ingestion during water recreation. *Water Research, 45*(5), 2020–2028.

Eiswerth, M. E., Donaldson, S. G., & Johnson, W. S. (2000). Potential environmental impacts and economic damages of Eurasian watermilfoil (*Myriophyllum spicatum*) in western Nevada and northeastern California. *Weed Technology, 14*(3), 511–518.

Eklund, B., & Watermann, B. (2018). Persistence of TBT and copper in excess on leisure boat hulls around the Baltic Sea. *Environmental Science and Pollution Research, 25*(15), 14595–14605.

English, J. N., McDermott, G. N., & Henderson, D. (1963). Pollution effects of outboard motor exhaust—Laboratory studies. *Journal of the Water Pollution Control Federation, 35*(7), 923.

Erbe, C. (2013). Underwater noise of small personal watercraft (jet skis). *The Journal of the Acoustical Society of America, 133*(4), EL326–EL330.

Faccioli, M., Font, A. R., & Figuerola, C. M. T. (2015). Valuing the recreational benefits of wetland adaptation to climate change: A trade-off between species' abundance and diversity. *Environmental Management, 55*(3), 550–563.

Fewtrell, L., Jones, F., Kay, D., Wyer, M. D., Godfree, A. F., & Salmon, B. L. (1992). Health effects of whitewater canoeing. *The Lancet, 339*(8809), 1587–1589.

Gabel, F., Lorenz, S., & Stoll, S. (2017). Effects of ship-induced waves on aquatic ecosystems. *Science of the Total Environment, 601*, 926–939.

Garrad, P. N., & Hey, R. D. (1987). Boat traffic, sediment resuspension and turbidity in a Broadland river. *Journal of Hydrology, 95*(3–4), 289–297.

Hammitt, W. E., Cole, D. N., & Monz, C. A. (2015). *Wildland Recreation: Ecology and Management*. John Wiley & Sons.

Haslam, S. M. (1978). *River Plants*. London.

Hill, M. P., & Coetzee, J. (2017). The biological control of aquatic weeds in South Africa: Current status and future challenges. *Bothalia-African Biodiversity & Conservation, 47*(2), 1–12.

Hockin, D., Ounsted, M., Gorman, M., Hill, D., Keller, V., & Barker, M. A. (1992). Examination of the effects of disturbance on birds with reference to its importance in ecological assessments. *Journal of Environmental Management, 36*(4), 253–286.

Hussner, A., Stiers, I., Verhofstad, M. J. J. M., Bakker, E. S., Grutters, B. M. C., Haury, J., et al. (2017). Management and control methods of invasive alien freshwater aquatic plants: A review. *Aquatic Botany, 136*, 112–137.

Kempinger, J. J., Otis, K. J., & Ball, J. R. (1998). Fish kills in the Fox River, Wisconsin, attributed to carbon monoxide from marine engines. *Transactions of the American Fisheries Society, 127*, 669–672.

Khan, R. A. (2003). Health of flatfish from localities in Placentia Bay, Newfoundland, contaminated with petroleum and PCBs. *Archives of Environmental Contamination and Toxicology, 44*(4), 485–492.

King, J. G., & Mace Jr., A. C. (1974). Effects of recreation on water quality. *Journal (Water Pollution Control Federation)*, 2453–2459.

La Jeunesse, I., Cirelli, C., Aubin, D., Larrue, C., Sellami, H., Afifi, S., et al. (2016). Is climate change a threat for water uses in the Mediterranean region? Results from a survey at local scale. *Science of the Total Environment, 543*, 981–996.

Liddle, M. J., & Scorgie, H. R. A. (1980). The effects of recreation on freshwater plants and animals: A review. *Biological Conservation, 17*(3), 183–206.

Lüdermann, D. (1968). Water pollution by outboard motors and its effects on fauna and flora. *Helgoländer Wiss. Meeresun, 17*(1–4), 356–369.

Mattos, J. J., Siebert, M. N., Luchmann, K. H., Granucci, N., Dorrington, T., Stoco, P. H., et al. (2010). Differential gene expression in *Poecilia vivipara* exposed to diesel oil water accommodated fraction. *Marine Environmental Research, 69*(Suppl. 1), S31–S33.

Merriam, L. C., & Smith, C. K. (1974). Visitor impact on newly developed campsites in the boundary waters canoe area. *Journal of Forestry, 72*(10), 627–630.

Monz, C., & Leung, Y. F. (2006). Meaningful measures: Developing indicators of visitor impact in the national park service inventory and monitoring program. In *The George Wright Forum* (Vol. 23, No. 2, pp. 17–27). George Wright Society.

Mosisch, T. D., & Arthington, A. H. (1998). The impacts of power boating and water skiing on lakes and reservoirs. *Lakes & Reservoirs: Research & Management, 3*(1), 1–17.

Mujal-Colilles, A., Gironella, X., Sanchez-Arcilla, A., Puig Polo, C., & Garcia-Leon, M. (2017). Erosion caused by propeller jets in a low energy harbour basin. *Journal of Hydraulic Research, 55*(1), 121–128.

Murphy, K. J., & Eaton, J. W. (1983). Effects of pleasure-boat traffic on macrophyte growth in canals. *Journal of Applied Ecology, 20*, 713–729.

O' Sullivan, P. (1990). *The Effect of Recreational Usage, in Particular Canoeing, and Agricultural Input upon the Gorges of the Afon Conwy, North Wales*. Unpublished undergraduate dissertation, Liverpool John Moores University.

Phillip, D. A. T., Antoine, P., Cooper, V., Francis, L., Mangal, E., Seepersad, N., et al. (2009). Impact of recreation on recreational water quality of a small tropical stream. *Journal of Environmental Monitoring, 11*(6), 1192–1198.

Purser, J., & Radford, A. N. (2011). Acoustic noise induces attention shifts and reduces foraging performance in three-spined sticklebacks (*Gasterosteus aculeatus*). *PLoS One, 6*, e17478.

Ruprecht, J. E., Glamore, W. C., Coghlan, I. R., & Flocard, F. (2015). Wakesurfing: Some wakes are more equal than others. In *Australasian Coasts & Ports Conference 2015: 22nd Australasian Coastal and Ocean Engineering Conference and the 15th Australasian Port and Harbour Conference* (p. 779). Engineers Australia and IPENZ.

Sharp, R. L., Cleckner, L. B., & DePillo, S. (2017). The impact of on-site educational outreach on recreational users' perceptions of aquatic invasive species and their management. *Environmental Education Research, 23*(8), 1200–1210.

Slabbekoorn, H., Bouton, N., van Opzeeland, I., Coers, A., ten Cate, C., & Popper, A. N. (2010). A noisy spring: The impact of globally rising underwater sound levels on fish. *Trends in Ecological Evolution, 25*, 419–427.

Spencer, M. (1995). *Environmental Impact by Canoeing on the Afon Llugwy and Afon Tryweryn*. Unpublished undergraduate dissertation, Liverpool John Moores University.

Steven, R., Pickering, C., & Castley, J. G. (2011). A review of the impacts of nature based recreation on birds. *Journal of Environmental Management, 92*(10), 2287–2294.

Superville, P. J., Prygiel, E., Magnier, A., Lesven, L., Gao, Y., Baeyens, W., et al. (2014). Daily variations of Zn and Pb concentrations in the Deûle River in relation to the resuspension of heavily polluted sediments. *Science of the Total Environment, 470–471*, 600–607.

The Outdoor Foundation. (2017). *Outdoor Participation Topline Report 2017*. Washington, DC: The Outdoor Foundation. Retrieved from www.outdoorfoundation.org.

Tivy, J. (1980). *The Effect of Recreation on Freshwater Lochs and Reservoirs in Scotland*. Perth: Countryside Commission for Scotland.

Walsh, J. R., Carpenter, S. R., & Vander Zanden, M. J. (2016). Invasive species triggers a massive loss of ecosystem services through a trophic cascade. *Proceedings of the National Academy of Sciences, 113*(15), 4081–4085.

Wang, X. J., Zhang, J. Y., Shahid, S., Guan, E. H., Wu, Y. X., Gao, J., et al. (2016). Adaptation to climate change impacts on water demand. *Mitigation and Adaptation Strategies for Global Change, 21*(1), 81–99.

Whitfield, A. K., & Becker, A. (2014). Impacts of recreational motorboats on fishes: A review. *Marine Pollution Bulletin, 83*(1), 24–31.

Zani, J. (2000). *The Effect of Recreational Canoeing on the Riparian Vegetation of the River Dee*. Unpublished undergraduate dissertation, Liverpool John Moores University.

Recreational Scuba Diving and Snorkelling

14

Chapter Summary

In this chapter, recreation diving is defined, and the numbers involved and impacts discussed. However, it is difficult to quantify the impact of any one of many stressors on coral reefs independently of the others. There is often a lack of scientific understanding of the interrelationships which limits the success of efforts to effectively create policy and regulations preventing reef decline. The impacts include direct trampling by reef walking, from the effects of pontoons (installation, anchor damage, fish feeding), direct impacts from diver behaviour including sediments raised, the effects on the corals (including effects of sunscreens), impacts on fish communities and kelp forests. The ways of managing activities to reduce the impacts are discussed, including reducing use levels, modifying diver behaviour, establishing underwater diving and sculpture trails, artificial reefs, fee charging changes, and diver involvement in conservation projects.

14.1 Introduction

Recreational scuba diving and snorkelling take place mainly in the warm and sunny climates of the tropics, in shallow seas around various types of coral reefs. These coral reefs are built mainly from stony corals, consisting of polyps which are clustered in colonies and secrete calcium carbonate exoskeletons which support and protect their bodies. These polyps belong to a group of animals, Cnidaria, which also include jellyfish and sea anemones. The reefs are biologically very rich and contain 25% of all known marine species, and they are important in tourism, fisheries, and shoreline protection. However, scuba diving and snorkelling can also provide recreation in lakes, flooded quarries, at wreck sites throughout the world, in more temperate seas of South Africa and Australia, in cenotes and sinkholes of the Yucatan in Mexico, and even in the polar regions (Garrod 2008).

Early studies suggested that diver activity generated minor damage compared to hurricanes and natural sources (e.g. Tilmant and Schmahl 1981; Tilmant 1987; Talge 1991). Tilmant and Schmahl (1981) conducted a three-year study in Biscayne National Park (Florida) and concluded that natural damage was more prevalent than anthropogenic damage but found a significant correlation between reef use and physical damage, suggesting increased use would result in greater damage.

14.2 Types of Recreational Diving

There are several categories of recreational diving:

14.2.1 Recreational Scuba Diving

This is diving for the purpose of leisure and enjoyment using a self-contained underwater breathing apparatus which is carried by the diver who is therefore not connected to the surface. The diver uses compressed air as the breathing mixture and confines themselves to depths of under 40 m which does not require a decompression stop to surface and so, if necessary, divers can ascend to the surface without stopping. The dive is usually not solo but either with a buddy, who can share air, or give other assistance if necessary, as long as they remain close, or more likely with a dive leader and several other divers in a team. This activity is underpinned by an internationally recognised and standardised approach to skills development and training.

14.2.2 Technical Diving

This is diving that exceeds recreational scuba diving limits but is not engaged in for profit. It is still considered recreational diving but the nature of the diving and type of equipment used exceed the boundaries of recreational scuba diving and so, for example, can use gas mixtures other than air, such as nitrox or trimix (Fig. 14.1). It includes cave diving which can include access to cenotes from the sea in Mexico, ice diving in very cold water and under an ice surface layer, deep penetration wreck diving and very deep diving over 40 m.

14.2.3 Wreck Diving

Divers are constantly in pursuit of new challenges, and wreck diving can give an exciting alternative which is scuba diving on artificial reefs based on shipwrecks which can take diving traffic away from coral reefs. In many cases ships have been sunk for the sole purpose of creating an artificial reef for wreck diving, and the trend continues to grow with the development of many types of artificial reef. However, many wrecks have historical or war grave significance, and four types of diver impacts on shipwrecks have been identified by Edney (2006): the removal of artefacts as souvenirs or personal mementos by divers; disturbance to wreck sites associated with this activity; direct contact with wrecks and protective marine growth and concretions by divers and their equipment; exhaled air bubbles trapped inside wrecks from divers penetrating wrecks. These exhaled air bubbles can also accelerate corrosion and affect the stability and longevity of a wreck, first by damaging the layer of marine growth, then by setting vertical currents in motion that remove the protective layer of rust. A third way by which exhaled air bubbles may accelerate the corrosion rate is through increasing the supply of oxygen. The impact of exhaled bubbles resulting in increased corrosion has not been quantified but is widely acknowledged within the recreational diving community as an impact of diving. Edney and Howard (2013) review wreck diving as a recreational diving pursuit.

14.2.4 Snorkelling

Snorkelling has been less studied than scuba diving because most tourists stay on the sea surface when snorkelling, wearing a diving mask and using a snorkel (a breathing tube) and usually wearing swimfins, to view the shallow underwater environment. Hence they are less likely to impact on the reef. However, Allison (1996) reported that most

Fig. 14.1 Technical diver using mixed gases to dive to 60 m. Credit: OAR/National Undersea Research Program (NURP); Univ. of North Carolina—Wilmington. Photo by D. Kesling

damage in the Maldives occurred when snorkellers kicked, or stood on, coral colonies. Management of snorkelling so that sites were deep enough that people could not stand (i.e. >2 m deep) would reduce damage from this effect. In an examination of seven studies of snorkeller damage at reef pontoons on the Great Barrier Reef, Nelson and Mapstone (1998) reported that there were no statistically significant changes in coral cover at the seven snorkelling sites. In fact at five of the seven snorkel sites, coral cover increased during the study (3% to 13% cover per year). Only one of four sites studied demonstrated a significant difference between control and impact sites in the amount of damaged corals. On the intensively used pontoon sites on the Great Barrier Reef, snorkellers are generally restricted to sites within close distance to the pontoon, partly for the safety of the visitors. These areas are generally around 50 m by 50 m, marked with ropes and floats. Even on the most heavily visited reefs, any damage which occurred is likely to be restricted to a very small proportion of the reef.

Snorkel trails have been proposed as a way to promote interpretation of the marine environment to tourists and to restrict impacts of snorkellers to small areas. Plathong et al. (2000) examined the effects of snorkel trails and found that even at low levels of use (15 snorkellers per week), a difference could be detected in the number of broken corals between the trails and control sites. They report that the change occurred within a month of the opening of the trail to snorkellers but that after the initial period, the amount of damage stabilised. They raise the interesting question of whether it is best to concentrate snorkellers and their damage to a small area, as occurs with trails, or whether it is best to spread the impacts over a wider area. It is suggested that short briefings, careful site selection, establishment of floating rest stations, and periodic rotation of trails might be useful management strategies.

14.2.5 Swimming with Cetaceans

Since 1991, there has been a small but increasing industry based on encounters with dwarf minke whales for a two-month season on the Ribbon Reef area between Port Douglas and Lizard Island, Great Barrier Reef (Arnold and Birtles 1999). Visitors on live-aboard charter vessels interact with the whales in the water, with the whales approaching snorkellers who hold a rope tethered to the boat. The whales and the small industry based on them have been studied since 1996 by Arnold and Birtles (1999), who have developed a research programme and industry code of practice in collaboration with the Marine Park Authority and the industry. Recommended practices to minimise any negative impacts on the whales include the requirement for tourists to be well briefed and the need for participants never to swim towards the whales. In 2000, the Great Barrier Reef Marine Park Authority (GBRMPA) adopted a policy on whale and dolphin conservation that underlines whale watching and other management practices for cetaceans in the Marine Park (GBRMPA 2000).

14.3 Benefits of Dive Tourism

Dive tourism is important in many parts of the world such as South East Asia, the Caribbean, the Red Sea, many islands in the Pacific, and parts of Australia because it can bring economic benefits, improves the quality of life in the destination, fosters community pride, allows cultural exchange, reduces over-exploitation, and promotes conservation. In some areas it has been promoted as an essential part of the tourism industry as in Malaysia and Thailand, and it is an important component of the world tourism industry (Abidin and Mohamed 2014; Brander et al. 2007; Dimmock and Musa 2015). Hence any degradation of coral reefs in particular (Shiviani 2007) could lead to dissatisfaction by the tourists, negative impacts on local tourism businesses, and a decline in biodiversity. There are also many challenges confronting the dive tourism industry unrelated to diving per se, such as the global economic downturn since 2008 and the vagaries of currency exchange rates, socio-political instability in Egypt, natural disasters such as hurricanes and typhoons in the Caribbean and South East Asia, tsunami in South East Asia, and the impacts of climate change.

14.4 Estimates of the Numbers Involved in These Recreation Pursuits

It is useful to know if there is a growth in recreational diving as the more divers there are, the more the impacts are likely to be. However, because it is not a well-regulated industry, there is no accurate estimated worldwide number, yet despite the economic downturn, one estimate by Garrod and Gössling (2008) suggested that there were about 28 million active participants in scuba diving. Another estimate suggested a rapid growth in the number of worldwide certifications at around 1 million/year but only a global figure of around 23 million. However, other figures suggest the number of new divers per year is much lower at around 225,000, with an annual attrition rate estimated to be 50–80%.

In Australia the number of club divers in 2010 was 28,400, whilst domestic tourists in 2014 numbered 156,000 divers and the international tourists 279,655, to give a grand total of around 464,000. The figures for the USA are much higher, and it is estimated that there are between 2.7 and 3.5 million active scuba divers and about 11 million snorkellers in the USA, with around 20 million snorkellers worldwide (Diving Equipment and Marketing Association 2014). Cordell's (2012) and the Outdoor Recreation Topline for Participation (2016) estimates for the USA are given in Table 14.1.

Table 14.1 Estimates for scuba diving and snorkelling (participants 16+) in the USA (adapted from Cordell 2012 and the Topline Survey, 2016)

	1994–1995	1999–2001	2005–2009	% participating
Numbers in millions scuba diving		3.8	3.6	1.5, a decrease of −5.6%
Annual days in millions		71.3	70.9	A decrease of −0.6%
Snorkelling	16.2	13.6	15.2	6.5, an increase of +11.8%
Topline survey (2016)	**2006**	**2010**	**2016**	**Three-year change**
Participants 6+ in millions				
Scuba diving	2.965	3.153	3.111	−2.6%
Snorkelling	8.395	9.305	8.717	+0.2%

There are major regional differences in the USA as would be expected, and between 2005 and 2013, the newly certified scuba divers numbered 182,395 in California, 128,633 in Florida, and 100,550 in Texas. In 2016 there were 433,000 scuba divers in the USA between 18 and 24 years old, but the figures seem to have oscillated between a high of 623,000 in 2006 and a low of 275,000 in 2011 so it is difficult to know whether there will be a future increase in recreational diving based on these figures.

14.5 Direct and Indirect Impacts of Recreational Diving

There can be both direct and indirect effects of recreation divers. These include the following effects:

14.5.1 Direct Trampling by Reef Walking

Reef walking by significant number of tourists is practised at a few locations, for example, on Heron Island, Low Islands, and Hardy Reef on the Great Barrier Reef (Dineson and Oliver 1997). Studies of the impact of reef walking have concluded that repeated passes break a significant number of corals but that there is no detectable difference in coral cover in areas subject to reef walking than in control sites (Liddle and Kay 1987; Kay and Liddle 1989; Hawkins and Roberts 1994). Corals recover from the breakage of fragments by regrowing branches, and some fragments are capable of establishing new colonies. However, Woodland and Hooper (1977) had shown that high levels of human activity can have a negative effect on the health of coral reefs. Their experiment showed that trampling reefs reduced live coral cover from 41% to 8% after only 18 people walked over the coral. This was confirmed by Rogers and Cox's research (2003) in Hawaii where coral transplantation was used to evaluate the response of corals to trampling by determining the growth and mortality at sites that ranged along a gradient of human use. There was a clear progression of coral survivorship along the gradient as survivorship dropped from 70% at the low impact site to 55% at the medium impact site. At the high impact site, total loss was reported after only 8 months, which was estimated as the equivalent to under 20,000 total visitors, or 63 people in the water/hour. Where transplanted corals survived, there was no difference in growth which was assumed to be due to control of activities of people in the water at those sites.

Nevertheless, on a reef-wide scale, impacts of reef walking seem negligible. Management of reef walking includes providing information on coral fragility to tourists, avoiding areas with fragile branching corals, and keeping to obvious tracks where possible. However, the effect of trampling on the medium-sized, harpacticoid copepod assemblage (benthic types of crustacean) inhabiting turfs on a coral reef was investigated by Sarmento and Santos (2012) in Porto de Galinhas (NE Brazil). Reef formation is near the beach in one of Brazil's main tourist destinations. Two areas were compared: one protected and one subject to intensive tourism. The densities of total Harpacticoida and of the most abundant species showed strong reductions in the trampled area. An analysis of covariance revealed that the loss of phytal habitat (dense seaweed coverage) was not the main source of density reductions indicating that trampling affected the animals directly. In addition multivariate analysis showed that there were differences in the structure of the harpacticoid assemblages between the areas. Of 43 identified species, 12 were detected by the indicator species analysis as being indicators of the protected or trampled areas. Moreover, species richness was reduced in the area open to tourism. At least 25 species are new for science and of these 20 were more abundant or occurred only in the protected area while five were more abundant or occurred only in the trampled area. These results highlight the possibility of local extinction of still-unknown species as one of the potential consequences of trampling coral reefs.

14.5.2 Effects of Pontoons

Pontoons are moored offshore up to 60 km from the coast in areas of the Great Barrier Reef where there are few coral cays to provide a base for up to 400 day visitors each day (Fig. 14.2). They were intended in part to relieve tourist pressure on the few accessible coral cays (Inglis 1997). Tourist pontoons were first installed in the 1980s and were simple low-cost structures with improvised mooring systems, comprising chains attached to miscellaneous concrete and steel anchors (Inglis 1997; Kapitze 1999). Vandrezee (1996) estimated that 50% of tourists to the GBR travelled on day trips to a moored pontoon. Following escalating demand for pontoons and other structures in the Marine Park, a "no-structures" subzone covering 22% of the Cairns section of the Marine Park was declared in 1992 to limit the number of sites at which pontoons were permitted (Dineson and Oliver 1997).

Several accidents, including the sinking of the "Fantasy Island" structure in 1988 and the breaking of a pontoon from its moorings during Cyclone Justin in 1997, intensified pressure to optimise pontoon mooring design (Kapitze 1999). Guidelines now include recommendations on design of moorings, anchors, and the pontoon body, as well as procedures for siting pontoons and their installation and maintenance.

Concerns about the impacts on the reef communities of pontoon structures resulted in a series of environmental monitoring programmes which have been a required condition of pontoon permit approval since 1989 (reviewed by Inglis 1997; Nelson and Mapstone 1998). Nelson and Mapstone's detailed analysis of 11 pontoon monitoring studies reported that early pontoons had an impact on benthic communities under the pontoons as a result of shading and movement of the mooring chains. The response to these early studies was to site pontoons over sandy areas and away from living corals and to improve the technology of mooring systems. More recent monitoring has demonstrated few significant impacts on coral or fish communities by operating pontoons (Nelson and Mapstone 1998). There is strong motivation for tourism operators to implement practices to protect the environment near pontoons, because of the limited number of suitable sites and the high cost of moving the pontoons should the reef be damaged.

Fig. 14.2 Diving platform or pontoon, Agincourt Reef, Great Barrier Reef, Queensland. Note the buoys marking the limits for diving. Photo by Bob Linsdell. www.panoramio.com/photo96044723

14.5.3 Fish Feeding and Pontoons

Sweatman (1996) reviewed information about fish feeding associated with tourist pontoons, following concerns that aggregation of predatory fish would deplete local fish populations. It was found that the fish species studied aggregate under natural conditions and no evidence was found of any impacts of the fish on prey or competitor species. The study also found that fish respond to human signals, with fish dispersing away from the pontoons outside the times when the tourist boats were present. The fish are apparently attracted to the pontoon by feeding, and the primary management concern is that the amount of food is limited and the quality of the food is appropriate for the species. There is also potential for aggression by the fish, and as a result all marine wildlife feeding has been banned in Florida and a permit is required for fish feeding activities within the Great Barrier Reef Marine Park. The practice of feeding food scraps has been discontinued, and a limit on the amount of raw marine product or fish pellets to 1 kg per site per day has been applied.

14.5.4 Direct Impact by Mooring Installation

Installation of moorings has the potential to cause minor damage to the surrounding areas. The presence of moorings also concentrates divers and other activities to a restricted area. Studies have shown higher diver damage around moorings than in control sites. However, there is agreement that the presence of moorings greatly reduced anchor damage (Dineson and Oliver 1997), and the GBRMPA management policy is to provide public moorings and require private moorings as part of permit conditions in heavily used sites.

14.5.5 Direct Impact by Anchor Damage

Coral reefs are highly susceptible to vessel-based damage because they are often located in very shallow water and the coral has slow growth rates which reduce recovery (Shiviani 2007). Reefs large enough to accommodate a high number of vessels often suffer degradation from anchors, and anchoring is the most commonly and thoroughly studied type of vessel damage to coral reefs. It has been demonstrated to cause considerable and long-lasting damage to coral communities (Dixon et al. 1993; Tratalos and Austin 2001). Anchors can cause damage during setting, while at anchor, and during retrieval. Corals can be broken, fragmented, or detached as the anchor is dropped to the substratum. Once set, further disturbance to the benthos is frequently associated with the anchor's chain dragging across the substratum or entangling the reef structure used. Large anchors and chain have been observed to predominantly affect the reef's lower slope and smaller reef anchors and associated chain, or rope primarily affecting the reef's crest (Dinsdale and Harriot 2004). This anchor damage on coral colonies has been recognised as problematic for several decades. For instance, in the Dry Tortugas, Davis (1977) found that up to 20% of an *Acropora cervicornis* (staghorn coral) zone had been destroyed by anchors. Rogers (1990) reported that 14% of vessels in the Virgin Islands National Park anchored on coral reef habitat and over a quarter of these vessels had some impact on corals. Increased frequencies of injured coral colonies were seen on intensely anchored sites (Rogers 1990; Davis 1977; Hawkins and Roberts 1994; Schleyer and Tomalin 2000). Jameson et al. (1999) compared four high-use coral reefs in the Egyptian Red Sea and found higher levels of broken coral and rubble compared with rates of natural damage reported in the literature, and Dustan and Halas (1987) recorded higher numbers of fragmented coral at Carysfort Reef (Florida Keys), which had high intensities of boating compared to nearby lower-use reefs. Lutz (2006) evaluated the condition of 315 shallow-water coral colonies from 49 sites in the upper Florida Keys and determined that nearly 60% of the sites and 80% of the coral heads showed vessel-based damage. Lutz also reported that the presence of mooring buoys did not affect the frequency of damage incidences but instead, sites near metropolitan areas and high vessel use were the most heavily impacted. Anchoring of large numbers of recreation and charter vessels was recognised as causing damage, particularly to the fringing reefs in the Whitsunday Islands and offshore from Cairns (GBRMPA) (Dineson and Oliver 1997).

Anchors can damage corals in a variety of ways including abrasion of tissue and skeletons, death to portions of the colony, fragmentation, and detachment from the substratum (Dinsdale and Harriot 2004; Giglio et al. 2017). The colony morphology often determines its susceptibility to various types of injury (Hall 2001). Branching species are more prone to physical damage (Liddle and Kay 1987), while massive and encrusting species are more vulnerable to overgrowth by algae (Hall 2001). A coral's resistance to damage depends on the intensity and duration of the damage, the geomorphology and depth of the reef zone (Connell 1997). Physical destruction may not necessarily kill coral colonies, but even partial mortality may favour infestation by pathogens and reduce the growth and reproductive potential of individuals (Hall 2001; Chabanet et al. 2005). The broken surfaces of corals will often serve as a substratum for algae and other organisms, which may infect coral tissue and further damage the colony (Bak et al. 1977; Riegl and Velimirov 1991). While anchoring impacts may have been lessened by mooring buoy installations in countries like Australia and the USA, anchoring on coral reefs remains a problem in developing countries where a majority of the world's coral reefs are located (Shiviani 2007).

Even though there are laws in place that protect coral reef resources from anchor damage (e.g. in Florida the Coral Reef Protection Act (CRPA)), law enforcement does not yet have the ability to hold a recreation boater responsible for anchoring on coral. While the officer may have location data which strongly suggests that the vessel is anchored on coral, they still need to have visible proof of such, which would require them to dive on each anchor suspected of impacting corals. Even if they are able to visually establish that the vessel is anchored on coral, the officer then has to prove intent. Intent is a mental attitude with which an individual acts, and therefore it cannot ordinarily be directly proved but must be inferred from surrounding facts and circumstances, which is difficult to do.

The only option to provide reef protection through effective (and efficient) regulatory enforcement is to establish "no-anchor" zones, so that the need to establish intent and the need to link specific people and their actions to reef damage is eliminated. Once an area has been designated as a no-anchor zone, a buoy system will need to be adopted and installed to delineate the restricted area. The system should be recognisable and similar to what has already been established within a location, for example, like in the Florida Keys National Marine Sanctuary and associated national parks. Consistency between buoy systems and reef markings is essential. A large-scale public education campaign has to be initiated, and outreach materials should include signage at boat ramps, marinas, and fuel docks and local maps with the no-anchor zones clearly labelled. The nautical paper and electronic chart providers should be engaged to include the no-anchor zones into their database. Additionally, coral reef habitat and no-anchor zone layers could be added to the Google Earth/Ocean software as another way of informing the public.

14.5.6 Other Boating Impacts

Other boating impacts relate to release of sewage and other waste water and littering. Discharge of waste from ships has been regulated by requirements for storage tanks in ships over 10 m and limits to areas where boats can discharge wastes (Dineson and Oliver 1997; Edwards 1997). Littering in the GBRMPA is illegal and is considered to be a relatively minor problem, best addressed by public education (Edwards 1997).

14.6 Direct Scuba Diving Impacts

14.6.1 Direct Impact by Divers (See Fig. 14.3 and Table 14.2)

Scuba diving may result in the deterioration of benthic communities, because divers can easily damage marine organisms through physical contact through touching, holding with their hands, their body, through dangling equipment, and fin contacts (Rouphael and Inglis 1995, 1997; Tratalos and Austin 2001; Zakai and Chadwick-Furman 2002; Uyarra and Côté 2007). Although the damage produced by indi-

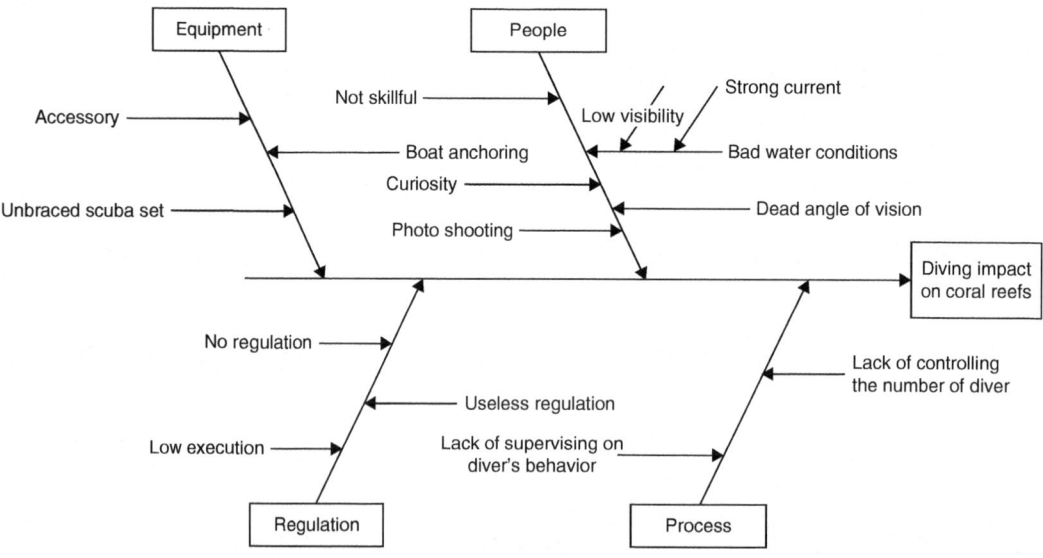

Fig. 14.3 Scuba diving impacts on coral reefs (after Abidin and Mohamed 2014)

viduals is usually minor, there is some evidence that the cumulative effects of the disturbances can cause significant localised destruction of sensitive organisms (Garrabou et al. 1998; Hawkins et al. 1999; Plathong et al. 2000). Two types of diver damage to stony corals (Fig. 14.4) have generally been recorded: (1) skeletal breakage, defined as fractured hard skeleton, and (2) tissue abrasion, defined as damaged tissue that exposes the underlying intact coral skeleton. These types of anthropogenic disturbances may appear relatively minor compared to natural disturbances, such as hurricanes; however, human impacts may significantly affect the recovery process of a reef, particularly if they are long term (Connell 1997). Furthermore, chronic and low-level perturbations may cause more damage to the reefs in the long term than discrete and highly destructive events, because the former do not allow sufficient time for recovery (Davis 1977; Dustan and Halas 1987; Tilmant 1987; Chabanet et al. 2005). Even with seemingly minor perturbations such as mooring emplacements, boating, snorkelling, and diving, impacted reef communities may take several years to return. Measures at the coral-colony scale potentially provide the earliest warning of possible deterioration, while measures on the community scale may better indicate the magnitude and ecological importance of the disturbances (Underwood and Peterson 1988).

Table 14.2 Diver-induced damage compared (adapted from many authors)

Impact per dive	Place
1. 2.5–5.5 contacts with corals per 10 min of diving, 1.7 breakage of corals and raised 9.4 sediment clouds per 60-min dive at 4–8 m	Eilat, Israel
2. 19 breakage of corals	Thailand
3. 14.65 contacts with the substratum or 5.7 contacts per 10 min, 5.86 contacts with coral per dive	Hong Kong (Chung et al. 2013)
4. 4.9–33.1 contacts with corals with 0.6–1.9 coral breakages per 30 min of diving	Eastern Australia
5. 0.95–3.17 contacts with corals per 10 min	Rock Islands, Palau
6. 35 contacts with the substratum; 2.45 causing damage	St. Lucia
7. 3.3 contacts and 0.96 coral touch per 10 min	Florida Keys

14.6.2 Effects of Diver Behaviour

The strategy of directly observing scuba divers to quantify their interactions with corals and the reef was first reported by Talge (1991). At Looe Key

Fig. 14.4 Stony coral: table coral of genus Acropora at French Frigate Shoals, North-western Hawaiian Islands. These types of corals are fragile and can easily be broken by divers. Photo by Yumi Yasutake (NOAA)

National Marine Sanctuary (Florida), inexperienced divers were observed most often interacting with corals by inadvertently kicking them with their fins or by using corals to push themselves away from the reef. More experienced divers, who can better control their buoyancy, have a lesser impact than inexperienced ones. A major finding was that only a small percentage of divers were responsible for the majority of the human-coral interactions and were mostly due to inexperienced divers with poor buoyancy control or experienced divers engaged in specific activities, such as photography (Fig. 14.1). In fact most studies have found that a limited subset of divers were responsible for most of the damage observed, prompting research to discover other relationships between diver demographics and interactions with the reef. For example, Rouphael and Inglis (1995) reported that 17% of divers were observed to break corals during a dive, but only 4% of the divers were responsible for more than 75% of recorded damage. Divers without gloves also had fewer interactions with the reef than those wearing gloves. These conclusions provided specific suggestions to reduce diver impact, including more emphasis on proper buoyancy control. Subsequent studies also sought to determine patterns to identify diver impact reduction strategies, and diver training was an obvious starting point. Zakai and Chadwick-Furman (2002) reported exceptionally high rates of damage to corals at dive training sites, with up to 100% of all corals broken in quadrats at the most heavily used site. They found that during a typical hour-long dive at 4–8 meters of depth, each diver broke 1.7 ± 4.9 corals and raised 9.40 ± 11.90 sediment clouds onto the reef ($n = 251$ divers).

However, not all experience-related observations have been consistent. Roberts and Harriot (1994) found that inexperienced divers (i.e. those that had completed <100 dives) may be more likely to damage the reef, although a later study found no such trend (Harriot et al. 1997). Similarly, Rouphael and Inglis (2001) found no relationship between the level of diving experience and the number of times divers contacted the reef. Some studies have noted that male divers caused less damage per contact than females.

In the Worachananant et al. (2008) study, observation participants were primarily female (77%), while females made up only 45% of subjects observed by Talge (1991) and 44% by Rouphael and Inglis (2001) leading to a potential sampling bias. Talge (1991) noted that a high proportion of inexperienced female divers stayed well above the reef, while inexperienced males were more likely to be negatively buoyant. It has also been found that most damage occurred during the early part of a dive whilst the divers were adjusting their equipment and that there was more damage during night dives than on those during the day.

Several studies have also supported the observations of Talge (1991) who revealed the relationship between diver damage and underwater photography. Medio et al. (1997) noted that divers using cameras (estimated at 27%) were responsible for 72% of contacts. Rouphael and Inglis (2001) found that amateur photographers did not damage corals more frequently than divers without cameras, although underwater photographers with specialised equipment were the most damaging of all divers observed. In a study in St. Lucia, Barker and Roberts (2004) also found significantly higher reef contact by divers with cameras. Serour (2004) found that during the Red Sea's tourist season, scuba divers averaged 1.3 contacts for every 10 minutes of diving, including 0.9 contacts with live coral. Photographers made up 7.2% of observed divers and were responsible for 67% of coral breakage, percentages very similar to those reported by Medio et al. (1997). Serour (2004) found that during months of low visitation, photographers accounted for 17% of divers and were responsible for 80% of the contacts that caused of coral breakage. Throughout the study, divers with underwater cameras were frequently observed negatively buoyant, using the reef to stabilise themselves while photographing. Luna et al. (2009) suggested that control of instruments used at the most fragile sites by divers, such as cameras or diving lights, may be a good measure for controlling damage by scuba diving, because carrying any element causes divers to have a greater interaction with the environment than when divers carry nothing. This finding, in terms of photographers, has been observed in other studies, but the effect of carrying diving lights had not been evaluated before. The photographers being observed tended to adopt the most comfortable and best position to avoid movement and to obtain better images. They then cause damage by anchoring themselves at the bottom, using their knees, fins, and elbows. When carrying a lantern, divers exhibit a particular behaviour, looking for small holes, cracks, caves, or animals to illuminate, and they often disregard their buoyancy or fail to keep their equipment or body off the seabed.

In a study in Bonaire, Uyarra and Côté (2007) found that divers came into contact with corals more often when viewing seahorses or frogfish compared to diving outside the vicinity of these sought-after species. The authors hypothesised that the increased rate of contact was related to the concealed or hidden nature and benthic habit of these fishes and the need for divers to come close to the substratum for observation. When divers were observing frogfish and seahorses, their contact rate increased 45-fold, and they also contacted corals for longer periods of time, both accidentally and intentionally.

The use of a muck stick (a handheld stainless steel or aluminium rod approximately 30 cm in length) as a means for a diver to stabilise themselves whilst underwater is a controversial practice within the scuba diving industry and one which has been banned in the Red Sea. A concern amongst representatives of the diving industry is the use of muck sticks to manipulate animals unnecessarily, for example, pushing animals out of holes for better viewing, stressing animals to show customers their stress behaviour (e.g. an octopus changing colour), and physically breaking hard coral to be used in photographs. Proponents of their usage, however, suggest that they may help prevent reef contact or reduce the level of damaging contact. The data of Roche et al. (2016) found that divers carrying a muck stick contacted the reef more than those who did not, but muck sticks caused the lowest proportion of obviously damaging contacts of body and equipment parts which were observed to contact the reef.

Branching corals are more likely to be damaged by divers than other morphologies. Rouphael and Inglis (1997) reported high numbers of breaks caused by divers on reefs with high cover of branching corals. They found that up to 45% of the qualified scuba divers break coral colonies but noted that the amount of damage per diver was generally small. Overall reef topography appeared to be unimportant in determining the impact rates (Rouphael and Inglis 1997).

Meyer and Holland (2008) found that recreation impacts on coral reef habitats were found to be relatively low in number: 71% of boat-based snorkellers had no contact with the substrate, compared to only 3.5% of shore-based snorkellers and scuba divers. Boat-based diving also had less impact per dive than shore-based diving. Overall, of the 1340 substrate contacts recorded, only 0.7% showed obvious substrate damage, and only 5% of observed contacts with live coral resulted in noticeable damage. However, Worachananant et al. (2008) in Thailand found that 93% of divers observed came into contact with the reef during a 10-minute observation period, averaging 97 contacts per hour of diving. In 66% of cases, divers damaged coral at least once during the 10-minute period, averaging 19 coral breakages per hour of diving. Photographers came into contact with corals more often than other divers, causing more damage per dive, but the damage was less per contact.

Luna et al.'s (2009) study in SE Spain also identified underwater photographers as making contact with the seabed more frequently than divers without cameras. During the observation period, 97% of divers observed had some interaction with the benthos, with an average 41.2 contacts per 10 minutes per diver. Diver's hands were found to be the part of the body that made most impacts, as has been observed in previous studies (Zakai and Chadwick-Furman 2002; Barker and Roberts 2004; Uyarra and Côté 2007). Pre-dive briefings and underwater intervention by a dive leader were also found to be effective at reducing the average impact of divers, as has been previously reported (Medio et al. 1997; Barker and Roberts 2004; Uyarra and Côté 2007). Luna et al. (2009) suggested that identifying the different factors that describe diver behaviour and their environmental effects may help managers to develop more effective training procedures, pre-dive briefings, and site regulations, to prevent or reduce the incidence of damaging behaviours. For this reason, an evaluation of the relationships between factors that could influence underwater behaviour showed the existence of several intrinsic diver and dive factors that may influence the effects of scuba diving. Contact frequency with the seabed is strongly influenced by diver profile and immersion characteristics. By following divers, it became clear that nearly all (96.7%) made some contact during the 10 min immersion, causing potentially serious damage to the environment.

It is estimated that in a 45-min immersion, each recreation diver had more than 60 contacts removing algae, 8 contacts with fragile organisms, and 14 contacts that result in the trampling of organisms. Most contacts were caused by flapping and contact with fins, confirming similar results from the Red Sea (Zakai and Chadwick-Furman 2002), Australia (Roberts and Harriot 1994; Harriot et al. 1997; Rouphael and Inglis 2001), and the Caribbean (Barker and Roberts 2004; Uyarra and Côté 2007), which attribute most diver damage to the effect of fins. The hands were the part of the body that made most impacts, as also observed for divers in the Red Sea (Zakai and Chadwick-Furman 2002) and the Caribbean (Barker and Roberts 2004; Uyarra and Côté 2007). Most contacts appeared to be unintentional and caused by poor swimming technique and incorrect weighting, factors that, in general, indicate a poor diving proficiency. There were gender differences in the contact frequency for flapping and contact with any part of the body, women causing fewer impacts than men. In general, a male diver is less cautious in his underwater behaviour, tending to be more adventurous and more likely to take risks than women, a relationship also shown by other studies of environmental attitudes and the behaviour of male and female divers (Rouphael and Inglis 2001). Another explanation for this result is that men ignore pre-dive instructions on safety and environmental behaviour advice compared to women and have a more independent attitude.

The effect of diver experience on the number of impacts has been documented in some areas (Roberts and Harriot 1994), but not in others (Harriot et al. 1997; Rouphael and Inglis 2001), but these differences could be because of differences in the definitions. Diving experience should be measured through three variables: level of the diving certificate, the total number of dives, and the number of years diving. Both total number of dives and the years of diving were associated with environmental impact: less damage was caused by the more experienced divers, those with a greater number of years diving and the total number of dives. The diving certificate level did not show this association, so it does not appear to reflect diving experience. A diver can obtain a higher certification by taking an advanced course, but this does not mean that the person becomes more proficient. Divers can even become diving instructors with fewer than 100 dives. Dive training certificates are lifetime qualifications and do not require periodic renewal or dive proficiency testing. Therefore, diver training level may not be a good indicator of current diving skills, and this factor is not sufficient to determine whether a diver is qualified for diving at a site. Briefings before the dive and underwater intervention by a dive leader were highly effective at reducing the average impact of flapping and contact by any part of the body. These differences in dive behaviour were more obvious for intentional contacts by any part of the body, confirming that deliberate contacts may be reduced by the implementation of simple measures by the diving centres. Attending a briefing emphasising the importance of buoyancy control and careful action (educational tools) increases the environmental awareness of recreation divers and might reduce diver damage at dive sites. Moreover, the use of dive leaders during dives is clearly an effective tool in minimising scuba divers' physical impact on their environment, because dive leaders can take measures when they see divers behaving inappropriately. For this reason, smaller dive groups tend to be better, dive leaders being able to supervise all members of the group adequately (Barker and Roberts 2004). In any case, smaller groups are preferred by the divers themselves.

Medio et al. (1997) demonstrated that providing divers with a 45-minute pre-dive educational briefing on the fragility of corals reduced contact with reef substrate by more than 80%. However, according to Barker and Roberts (2004), a one sentence addition to pre-dive briefings about not touching the reef did not reduce diver contact rates, although intervention by a dive leader was found to be very effective.

In the area investigated by Luna et al. (2009) in NE Spain, there is an advanced coralligenous community composed of many sessile, filter-feeding, long-lived organisms with fragile skeletons and slow rates of growth. The risk of long-term degradation should be determined by the impact rate and the speed with which it is repaired (Rouphael and Inglis 1997). The problem is that organisms living in Mediterranean coralligenous communities are not adapted to severe disturbance, and their recovery after moderate pressure might be difficult (Garrabou et al. 1998). The sustainability of diving activity at particular sites depends on both the number of divers accessing the sites and the capacity of the ecosystem to regenerate and recover from any damage incurred (Harriot et al. 1997). Monitoring programmes need to be established to detect environmental changes at dive sites before diving impact levels become critical and, perhaps, irreparable.

The scale of the impacts can be seen from Poonian et al.'s (2010) estimate that recreation divers may be responsible for 589,000 ± 117,000 coral contacts and approximately 400 coral breakages annually from 49,378 dives of 40 minutes duration at German Channel, Palau (one of the Caroline islands in Micronesia). This could have a long-term effect on coral health at this site, and the visitation rates far exceed the 4000–6000 (Hawkins and Roberts 1994), or the 7000 (Schleyer and Tomalin 2000) dives/year/site considered to be a reliable rule of thumb to estimate the carrying capacity for scuba divers, although this would depend on the biophysical character of the site.

14.6.3 Impacts from Sediments Raised by Divers

Different ways can be used to measure diver-induced sediment disturbance. Barker (2003) measured how many times sediment was raised by different parts of a diver, and her findings showed that flippers and hands were the parts of a diver's body that most frequently raised sediment. However, since Barker (2003) presented her findings in a diagram and did not state the average figures for raising sediment by each diver, it is not possible to make a detailed comparison with other results. On the other hand, Zakai and Chadwick-Furman (2002) measured sediment disturbance by the number of sediment clouds raised, a term that was not defined in their study. Furthermore, there was no discussion on the adverse impact of diver-induced sediment disturbance in their paper. Hassler and Ott (2008), however, were able to measure sedimentation rates with the use of sediment traps. They concluded that sedimentation rates were higher at diving site entrances but decrease as the distance from the entrance increases. Even when measured, since different units of measurements were used by different studies, it was difficult to compare the findings. Neil (1990) carried out an experiment on the Heron Reef flat (Great Barrier Reef) to assess the potential impact of sediment resuspension and subsequent deposition by reef walkers. Suspended sediment concentration up to 0.93 mg/litre was measured and deposition up to 38 mg/cm, and, if frequently repeated, it was thought that these levels were capable of causing moderate stress to reef-flat corals. The result is though that diver-induced sediment disturbance remains something of an information lacuna in diving tourism impact studies, but it seems likely that sediment mantling of the coral organisms will have a detrimental effect, by asphyxia, reduction in growth rates of the coral skeleton, and abrasion. The coral growth can be altered too when precipitation of calcium carbonate is changed by a lack of available light due to the suspended sediment, and corals have to use excess energy to remove sediment from their polyps. Erftemeyer et al. (2012) in a detailed review of the effects of sediment disturbances on corals suggested that the duration that corals can survive high turbidity ranges from several days (sensitive species) to at least 5–6 weeks (tolerant species). Increased sedimentation can cause smothering and burial of coral polyps, shading, tissue necrosis, and population explosions of bacteria in coral mucus. Fine sediments tend to have greater effects on corals than coarse sediments. Turbidity and sedimentation also reduce the recruitment, survival, and settlement of coral larvae. Maximum sedimentation rates that can be tolerated by different corals range from <10 mg/cm^2/d^1 to >400 mg/cm^2/d^1. The durations that corals can survive high sedimentation rates range from <24 h for sensitive species to a few weeks (>4 weeks of high sedimentation or >14 days complete burial) for very tolerant species.

14.6.4 Impacts on the Corals

In the Red Sea, Riegl and Velimirov (1991) found coral breakage to be the most common form of damage at high-use coral reef sites. They also noted damage most frequently within 10 m of the surface, where most human interaction with the reef (snorkelling, scuba diving, and anchoring) takes place. In Egypt's Red Sea, Hawkins and Roberts (1994) reported finding significantly more damaged corals at heavily dived sites, recording broken coral colonies at up to ten times higher frequencies than lightly dived reefs. More recently, Hassler and Ott (2008) also reported that reefs in the Red Sea that were subjected to intense levels of scuba diving showed a significantly higher number of damaged and broken corals and lower coral cover. On the reef crest, at dived sites, over half of coral colonies were damaged and 27% were broken, with branching coral species making up over 95% of broken corals (Hassler and Ott 2008). That study also noted that diver-related sedimentation rates decreased further from the dive site entrance, indicating poor buoyancy control was common at the beginning of observed dives, consistent with previous studies of diver-related sedimentation by Barker and Roberts (2004).

In a study in Bonaire, Dixon et al. (1993) determined that high dive use areas showed lower percent coral cover while species diversity was higher at lower use sites. A direct relationship between coral damage and distance from a mooring buoy was also documented, giving rise to the concept of a site's diver carrying capacity based on these findings (Dixon et al. 1993). In Australia, Rouphael and Inglis (1995) concluded that a diver's lack of "environmental awareness" contributed to a greater number of impacts on the reef. Hawkins et al. (1999) explored the possible effects of diver-related damage to coral reef-fish communities but ultimately did not detect any significant changes.

However, a higher number of abraded corals were observed in higher-use areas (Hawkins et al. 1999). In Grand Cayman, Tratalos and Austin (2001) concluded that diving has had significant impacts on heavily used dive sites, as a reduction in coral cover and an increased amount of dead coral were documented as diver visitation increased. Comparative benthic surveys have typically compared long-established dive sites with relatively undived reefs, allowing researchers to elucidate the impacts of high levels of human use on coral reefs. Most studies documented that well-established dive sites had greater numbers of broken and damaged coral colonies than undived reefs nearby.

However, some authors have concluded that the impact of divers may be more related to their experience and behaviour as we have already outlined than just their number. Dive sites also become popular for a range of biological and physical attractions which may not often be seen on other local reefs (Shivlani and Suman 2000; Rouphael and Inglis 2001). For these reasons, researchers may fail to distinguish the impacts of scuba diving from existing differences inherent between reefs. Nevertheless, if resource managers, dive operators, and dive educators have increased knowledge of diver behaviour, more effective strategies to protect reef resources can be developed.

14.6.4.1 Coral Species and Physical Damage

The connection between coral-colony structure and damage resistance was first detailed by Charles Darwin in 1874, who contrasted the exceedingly strong honeycombed mass, which generally assumes a circular form of *Porites* and *Millepora* colonies, which dominated the exposed edges of the Cocos-Keeling Atoll, with the brittle and thinly branched species inside the protected lagoon. This was because the structure of *Porites* and *Millepora* seemed able to resist the fury of the breakers (Darwin 1874). Corals found in areas of greater wave energy typically have a higher skeletal strength (Chamberlain and Graus 1975; Chamberlain 1978). Chamberlain (1978), using strength testing, later determined that the strength of dry, dead coral material was statistically similar to the strength of living corals. Liddle and Kay (1987) studied the effects of trampling on reef-flat coral colonies and found massive species to be significantly more resistant to physical damage than branching species. They also found that size, morphology, porosity, and density all contribute to a coral's resistance to breakage (Liddle and Kay 1987).

On South African reefs, Reigl and Reigl (1996) found that the likelihood of coral damage is related to its growth form, with open arborescent coral species found to be easily damaged by physical disturbances. Censuses of those fragments showed low chances of survival, especially in areas of high wave action (Riegl and Riegl 1996). However, in Hawaii, it was found survivorship for large coral fragments to be higher than 70%, compared to 5% to 70% for small fragments. Species differences were also found to affect survival rates of fragments, with *Montipora capitata* and *Porites lobata* fragments having lower survival rates than *Porites compressa* and *Pocillopora meandrina*, two other dominant Hawaiian corals.

Direct damage caused by divers (e.g. skeletal breakage) may not be the main source of damage to hard corals in intensively dived areas, and physical damage caused by divers may increase their susceptibility to disease and therefore reduce their resilience to the physical damage. It has been shown by Bak and Criens (1981) that experimentally fragmented *Acropora cervicornis* and *A. palmata* were more likely to be infected and killed by disease than unfragmented corals in control plots. So indirect impacts of recreational diving on stony corals and other organisms are complex (Guzner et al. 2010) and should not be overlooked. Indirect impacts (e.g. tissue lesions) are usually minor impacts on, or damage to, corals that develop subsequently more significant damage. For instance, tissue lesions and compromised physiological conditions together may increase vulnerability of hard corals to predatory attacks (e.g. Guzner et al. (2010) showed the effects of predation by the corallivorous gastropod *Drupella cornus*). Algal colonisation of corals may soon follow tissue damage with algae competing for space with corals and acting as sediment traps, which hinder coral recovery (Hall 2001). One factor emphasised by Scott et al. (2017) affecting coral colonies after physical damage is that it does attract and create a point of entry for corallivores, which may form aggregations inside, or around the wounds (Potkamp et al. 2017). In addition, an increase of corallivore abundance has been linked to a range of other stress-inducing events (Plass-Johnson et al. 2015), and some of these could also catalyse a secondary or indirect effect of overuse through scuba diving. However, at the Koh Tao site (Gulf of Thailand), the relative abundance of *Drupella* showed no significant correlation to the observations of coral health and disease as shown by Lamb et al. (2014). Little is understood about the causes and implications of *Drupella* snail population increases, except that multiple factors interact to create the conditions for outbreaks (McClanahan 1994), and more robust data and studies are needed to fully understand these interactions. However, *Drupella* snails have been shown to be a highly effective transmission vector for brown band disease, as well as white syndromes and SDR (shut-down-reaction) and black band disease (Antonius and Riegl 1997). It is currently unclear whether *Drupella* snails promote coral disease outbreaks or are merely attracted to them. What does seem clear from the work of Lamb et al. (2014) is that there is a correlation between high and low dive use sites and coral health at Koh Tao (Thailand). They compared the predominance of four coral diseases and eight other indicators of coral health. The mean prevalence of healthy corals at low-use sites (79%) was nearly twice than at high-use sites (45%). They also found that there was a three times increase in coral disease at high intensities as well as significant increases in sponge overgrowth, physical injury, tissue necrosis from sediment and non-normally pigmented coral tissues. Injured corals were more susceptible to skeletal eroding band disease only at high-intensity sites which suggested that additional stressors associated with use intensity facilitated disease development. Sediment necrosis of coral tissues was strongly associated with the prevalence of white syndrome, a devastating group of diseases across all sites, but there were no significant differences in mean levels of coral growth anomalies or black band disease between high- and low-use sites.

When the abundance of *Drupella* snails was plotted against the descriptions of high- and low-use sites as designated by Hein et al. (2015), the abundance of *Drupella* snails did not show significant differences, which does not agree with results from Eilat (Gulf of Aqaba), where higher abundances were found in high-use than in low-use areas (Guzner et al. 2010).

Scott et al. (2017) concluded that the present findings suggest that although there is no significant correlation between corallivore population densities and diving pressure, *Acanthaster* densities have remained unchanged and those of *Drupella* have significantly increased throughout the study period, but few studies and no management reports for Koh Tao have addressed the issue of coral predation. However, the ongoing outbreak of *Drupella* snails and the occurrence of some significantly high local densities of *Acanthaster* found warrant inclusion in all future assessments of reef threats, decline, or management on the island, and elsewhere. More accurate global data on the population trends and dynamics of corallivores which potentially contribute significantly to reef decline can further improve understanding and management of coral reefs in the face of all the other stressors like human population growth, development, natural disasters, and climate change.

Lyons et al. (2015) suggested that Bonaire's benthic communities differed between heavy and light diving traffic areas both in structural complexity and in benthic assemblage. There was lower rugosity at heavy diver traffic than light diver traffic areas. This is probably the first time that diving impact has been shown to directly reduce structural complexity of coral reefs, although previous studies have demonstrated that an increase in diving traffic reduces survivorship of upright organisms (e.g. Coma et al. 2004), which certainly reduces structural complexity. The average rugosity of all sites in the Lyons et al. study was 2.06, which is greater than the Caribbean-wide average of 1.2 (Alvarez-Filip et al. 2009) but lower than values from Bonaire's shallow reefs in the late 1970s, which were as high as 3.62. Given that structural complexity is important to ecologically important mobile organisms that inhabit reefs, it is concerning that there has already been a large loss of structural complexity on Bonaire's shallow reefs and that diving traffic had an effect on the abundance of some benthic substrata but not on others. Similar to the results of two previous studies on the effect of diving in Bonaire, it was found that divers are having an effect on massive corals (Dixon et al. 1993; Hawkins et al. 1999). Lyons et al. (2015) also found lower abundances of branching and leafy corals at heavy diver traffic locations, whereas Dixon et al. (1993) and Hawkins et al. (1999) found that branching corals were more abundant in diving areas than no-diving reserves on Bonaire. They concluded that diving impact favoured branching and leafy species by giving these "weedy" species a competitive advantage over massive corals. In addition, they found that gorgonians were more abundant at heavy diver traffic areas than light diver traffic areas. Lyons et al. (2015) found that while gorgonians and sponges were correlated with heavy diver traffic locations in their multivariate analysis, there was no significant difference in their abundances between heavy and light diving traffic sites. The explanations for why their results differ from those of Dixon et al. (1993) and Hawkins et al. (1999) were that the surveys of Dixon et al. (1993) were conducted in 1991 and Hawkins et al. (1999) in 1994. Since that time the number of scuba divers in Bonaire has more than doubled from 17,000 in 1991 (Dixon et al. 1993) to 36,444 divers in 2014. Thus, benthic organisms such as branching stony corals and gorgonians that once benefited from the disturbance caused by divers (possibly due to reduced competition with stony corals) may now be negatively affected by the greater amounts of diver traffic. Secondly, other global and local anthropogenic stressors have had increasing impacts on Bonaire's reefs. The population of Bonaire has risen from 10,000 in 1991 to 15,666 in 2011, and consequently nutrient pollution has increased in Bonaire. Moreover, Bonaire's reefs were damaged by Hurricane Lenny in 1999, Omar in 2008, and a moderate bleaching event in 2010. The importance of synergistic effects of multiple stressors on marine systems is becoming increasingly apparent (Nyström et al. 2000) and the physical disturbance caused by divers may act synergistically with these other stressors.

14.6.4.2 Effects of Sunscreens and Insect Repellents on Coral Growth

Over the last 20 years, massive coral bleaching (i.e. loss of symbiotic zooxanthellae hosted within scleractinian corals) has increased dramatically, both in frequency and spatial extent. This phenomenon has been associated with positive temperature anomalies, excess ultraviolet (UV) radiation or altered available photosynthetic radiation, and the presence of bacterial pathogens and pollutants. Production and consumption of personal care products like insect repellents and cosmetic sun products are increasing worldwide, reaching unexpected levels, with potentially important consequences on environmental contamination. The release of these products is also linked with the rapid expansion of tourism in marine coastal areas, particularly in the tropics. Chemical compounds contained in sunscreens and other personal care products have been demonstrated to reach detectable levels in both fresh and seawater systems, and these compounds are expected to be potentially harmful for the environment. Hence, the use of sunscreen products is now banned in a few popular tourist destinations, for example, in marine ecoparks in Mexico (XCaret and Xel-Ha). It has been shown that sunscreens are lipophilic (the ability of a chemical compound to dissolve in fats, oils, and lipids), their UV filters can bioaccumulate in aquatic animals and cause effects similar to those reported for other xenobiotic compounds (a chemical compound foreign to a living organism). Paraben preservatives and some UV absorbers contained in sunscreens have estrogenic activity, and it has been demonstrated that several sunscreen agents may undergo photodegradation, resulting in the transformation of these agents into toxic by-products. It has also been demonstrated that sunscreens have an impact on marine bacterioplankton, but there is no scientific evidence for their impact on coral reefs. To evaluate the potential impact of sunscreen ingredients on hard corals and their symbiotic algae, Danovaro et al. (2008) conducted several independent in situ studies with the addition of different concentrations of sunscreens to different species of *Acropora* (one of the most common hard-coral genera), *Stylophora pistillata*, and *Millepora complanata*. In all replicates and at all sampling sites, sunscreen addition even in very low quantities (i.e. 10 μL/L) resulted in the release of large amounts of coral mucus (composed of zooxanthellae and coral tissue) within 18–48 hr. and complete bleaching of hard corals within 96 hr. Different sunscreen brands, protective factors, and concentrations were compared, and all treatments caused bleaching of hard corals, although the rates of bleaching were faster when larger quantities were used. Untreated coral branches of 3–6 cm used as controls did not show any change during the entire duration of the experiments. Bleaching was faster in systems subjected to higher temperature, suggesting synergistic effects with this variable.

According to estimates discussed in their paper, Danovaro et al. (2008) believe that up to

10% of the world's coral reefs would be threatened by sunscreen-induced, coral bleaching. To add to this worrying story, Downs et al. (2016) investigated the effects of oxybenzone (benzophenone-3), a sunscreen ultraviolet filter, on *Stylophora pistillata* coral larvae (planulae), as well as its toxicity in vitro to coral cells from this and six other coral species. The planulae were transformed from a motile state to a deformed, sessile condition, and they showed an increased rate of coral bleaching in response to increased concentrations of oxybenzone. This chemical is also a geno-toxicant to corals showing a positive relationship between DNA-AP lesions and increased oxybenzone concentrations, and it is a skeletal endocrine disruptor. It induced ossification of the planulae, encasing the entire planulae in its own skeleton making them unable to float with the currents and disperse. This work shows that oxybenzone poses a hazard to coral reef conservation and threatens the resiliency of coral reefs to climate change. The oxybenzone leaches the coral of its nutrients and bleaches it white, and it is thought maybe to also disrupt the development of fish and other wildlife. The US National Park Service in South Florida, Hawaii, US Virgin Islands, and American Samoa recommend using "reef-friendly" sunscreens made with titanium oxide, or zinc oxide, or do not use these products at all when recreational diving but use wetsuits and rash guards.

14.7 Effects of Recreational Diving on Fish Communities

Hawkins et al. (1999) in Bonaire (Caribbean) found that comparisons between fish abundance in high- versus low-use dive areas showed a very small significant difference only for groupers, but when the fish sizes were converted to biomass, there were no significant differences between areas of diving intensity for any of the fish families examined. Nor was there any significant difference in total number of species observed/counted at either of the depths measured between high- and low-intensity areas. Reserves though had significantly higher-quality habitats than the dived sites, with higher coral cover, greater structural complexity, and less sand and bare rock, which resulted in differences in fish communities. There were significantly more snappers, surgeon fish, and overall fish numbers in the reserves than the dive sites. Surprisingly the biomass of predators was greatest in the reserves and not the dive sites which might have been expected because of divers feeding fish.

In NE Brazil two very similar studied reefs sustained similar fish assemblages with the most speciose and abundant families being Pomacentridae, Scaridae, Haemulidae, Acanthuridae, and Labridae (Medeiros et al. 2007). These families illustrate the typical reef-dwelling fishes, which are very common on the north-eastern Brazilian coast. However, despite these similarities, major differences were detected between the reefs (Fig. 14.5). The most striking disparity concerns the extremely high abundance of the sergeant major (*Abudefduf saxatilis*) at Picãozinho, which accounted for almost two thirds of all individuals recorded on this reef and was over six times more abundant than at Quebra Quilhas. During tourist presence at Picãozinho, this species was conspicuously the most affected fish, becoming attracted by the external sources of food provided by tourists. In fact, a remarkable change in the behaviour of this species has been observed since tourist arrivals at the reef, and changes in their overall abundance may also be observed when comparing between the human-frequented and unfrequented periods. Feeding by tourists has the potential to directly alter the behaviour of many fish species, which are attracted by this external source of food (Sweatman 1996) and eventually the fish may display aggressive behaviours. It is very likely that over the past decades, uncontrolled tourism may have strongly benefited the individuals of *A. saxatilis* and, as a consequence, the overall abundance of this species has dramatically increased.

Human trampling is another activity that may supply the fish with other sources of food in the studied reef. When humans trample the substrate, especially in algal-dominated areas, burrowing invertebrates become exposed and attract potential invertivores and carnivores which are nearby. Consequently, this activity also has the potential

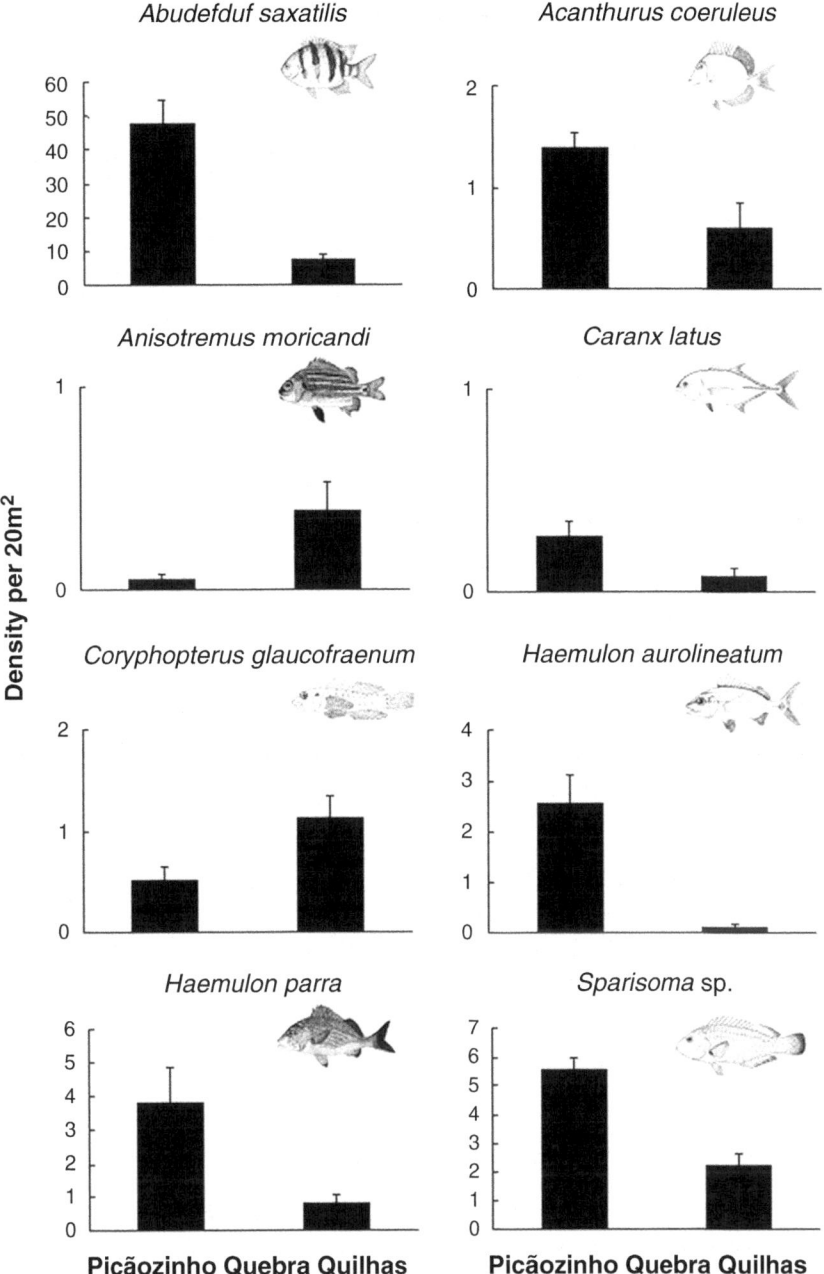

Fig. 14.5 Density of fishes (per 20 m^2) (mean ± SE) at Picãozinho and Quebra Quilhas. Student's t test showed significant differences in fish abundance between reefs ($P < 0.05$). Note that different scales were used (adapted from Medeiros et al. 2007)

to alter fish behaviour considerably. Furthermore, the higher abundances of juveniles of *Acanthurus coeruleus* and *Sparisoma* sp. at Picãozinho suggest that this reef is under a higher level of human interference, since high abundance of herbivores in marine ecosystems may indicate a sign of degradation due to the higher biomass of algae. Although the above-mentioned species have benefited by these activities, a discrepancy in the population levels within a fish assemblage is considered a sign of historical environmental change, which is caused or, at least, enhanced by human activities. These discrepancies were also noticed elsewhere (Milazzo et al. 2002) and have been related to recreation activities benefiting one or few species, with harsh consequences on others. Although in reefs worldwide, protected or non-protected, it is common to observe a pattern shift from common abundant species to a progressive decline in rarer species (Magurran 1996), there has not been any undisturbed reef investigated in Brazil, where one single species dominates over all others by 75%. Nevertheless, although only two species were less abundant at Picãozinho, most other species did not seem to be negatively affected by the recreation activities and their abundances did not differ from those of the control reef. However, the increased population size of only a few species indicates a significant shift in fish assemblage structure at Picãozinho. Thus, the observed structure of both reefs suggests that, besides tourism at Picãozinho, other external factors may have contributed to the current patterns observed. Today, large predators in NE Brazil are restricted to deeper reefs as a consequence of a long history of overfishing that took place in the shallow reefs of this area. Thus, it is reasonable to consider that fishing has also played a major role in shaping the assemblage structure on these shallow reefs by altering the reef trophic dynamics over the decades. The removal of large predators is considered to have a major influence on the trophic structure of a reef, often leading to an increase in the population of their prey, which, in turn, influences the whole base of the food web.

The spiny lobster (*Panulirus argus*) fishery in Florida is closed during the spawning season (March–July) except for a two-day recreation "mini-season" for sport divers in July, several days prior to the opening of the commercial fishing season. In Monroe County, recreation fishers, who possess a valid saltwater fishing licence with crawfish stamp, are allowed to harvest six lobsters per day, each with a minimum carapace length of 76.2 mm (3.0 inches). During these two days, approximately 50,000 people attempt to catch lobster, and the number of boats visiting the reef has been estimated to be up to 900 times higher than during the regular lobster season. Hartman (2012) found an increase in benthic damage incident at impact sites in three surveys conducted after the mini-season, whilst no significant changes occurred at the control sites. This suggested that detectable benthic damage associated with lobster fishing occurred during the mini-season and that this was at least partly caused by diver impacts whilst searching for and capturing lobsters. The cause of this damage is likely to be the result of the additional gear that the divers bring such as gloves, a 92 cm tickle stick, a hand net, a lobster gauge, and a lobster bag, all of which make buoyancy control much more difficult. By actively searching for and attempting to capture spiny lobster which are cryptic and hide, maintaining close proximity to the reef, the lobster-fishing diver damages the benthos at higher rates than other recreation divers.

In a study in South African coral reefs, the densities of the top-level predator *Epinephelus tukula* (the potato cod or grouper) were in decline in an intensive dive site, and the diving activities may be causing competitive disturbance to this species. At the same time, there was a higher abundance of herbivorous fish juveniles, such as *Acanthurus coeruleus* and *Sparisoma* species. The higher abundance of these species may indicate a sign of coral degradation due to the higher biomass of algae. So we see here that the changes caused by one species from diving have an effect on others and the fish assemblage structure can be drastically modified.

14.8 Impact of Recreation Divers on Kelp Forests

As we have seen, most of the impacts noted by researchers caused by recreation divers have been observed in tropical waters, and the effects have been on corals. However, Schaeffer and Foster (1998) noted diver disturbance in Monterey Bay (California) which is well known for its diverse subtidal giant kelp (*Macrocystis pyrifera*) forests, along a relatively small section (2.8 km) of coast which was estimated to see 65,000 diver days annually. Forty-two divers were observed in 1997, and the disturbance they caused was noted (number of contacts with the bottom, number of algal blades detached, and the number of animals touched) in 30 minutes. The average diver would unintentionally contact the bottom 10 times, stir up sediment 83 times, and detach 2.6 algal blades. The divers would entangle themselves in giant kelp fronds and then break the fronds to get free; they would get the fronds wrapped around their fins or tank valves without them realising, and continued movement would usually break the fronds. If the estimate of the diver number is correct in one year, the diver contact with the bottom would be 2.8 million times; they would touch animals 260,000 times and detach 130,000 algal blades. Other disturbances not quantified were noted during diver entry and exits, from diver boat anchors or the detachment of giant kelp surface fronds by diver boat propellers and divers attempting to swim on the surface through these fronds. Although the effects on the giant kelp population and structure were not taken further, it was assumed that there would be major effects over a period of time.

14.9 Management of Scuba Diving and Snorkelling

14.9.1 Introduction

Ensuring the sustainable future of scuba diving and snorkelling involves a difficult set of problems to solve which requires an understanding of the issues which result from a diver's desire to maximise their experiences, the tourism industry efforts to enable these experiences whilst achieving profitable commercial goals, the host community's needs and priorities, and the absolutely imperative set of goals to preserve pristine marine environments and conservation values in the long term (Dimmock and Musa 2015). The key is the marine environment in which all the stakeholders are dependent. What is required is effective management to protect ecological and cultural values to ensure the sustainable use of the resources in the broader, difficult world climate system, with global warming, greater numbers of natural hazard events, and ocean acidification. However, there seem to be two broad sets of management approaches: to manage effectively the physical environment, including the provision of artificial reefs and to modify diver behaviour, for example, by education, enforcing codes of conduct and regulations, and charging fees.

14.9.2 Management of the Divers' Physical Environment

The obvious management approach would be to reduce the level of use at certain sites, especially where the recreation carrying capacity has been exceeded, particularly in areas where there is a high abundance of stony corals.

It might be possible to rest some sites for all diving activity for a period of time, or there could be a blanket ban, or another method would be to ban access to some sites for diver groups who are known to cause most damage, such as novices or photographers. Restricting diver numbers is based on the concept of a reef dive site's "carrying capacity"; a level beyond which diving impacts become readily apparent. This has been reported to vary between 5000 and 6000 dives per year (Hawkins et al. 1999) to up to 7000 dives per year (Schleyer and Tomalin 2000). Restriction of scuba diving equipment has focused on banning the use of accessories believed to increase reef contacts within marine protected areas, such as gloves, muck sticks, or underwater cameras; however, such regulations are often unpopular within the scuba diving community (Poonian et al. 2010). An alternative would be that these diver categories and equipment users could be restricted to certain sites thought to be more resilient to damage. However, the main problem with these kinds of approaches is the policing of the sites and the divers which would be very difficult, or impossible, without a permit system.

A similar management approach is the establishment of no-anchor zones or anchoring at the mooring buoys only and restricting scuba exiting points to specific areas of the reef, and this is more easily managed.

It might be possible to establish underwater diving trails which are the only sites to be used and is one way to restrict recreational diving and concentrate diving impacts within specific locations (e.g. Ríos-Jara et al. 2013; Rouphael and Inglis 2001). One such example comes from Isabel Island off the Mexican Pacific coast where Ríos-Jara et al. (2013) proposed two management strategies: (a) the creation and use of underwater trails and (b) the estimation of the specific tourism carrying capacity (TCC) for each trail. Six underwater trails were selected at sites that presented elements of biological, geological, and scenic interest. The methodology to estimate the TCC was based on the physical and biological conditions for each site, the infrastructure and equipment available, and the characteristics of the service providers and administrators of the park. The TCC values ranged between 1252 and 1642 dives/year/trail, with a total of 8597 dives/year for all six trails. These numbers are much higher than the actual numbers of recreation divers to the island (about 1000 dives/year), but there is a need for adequate preventative management if the dive sites are to maintain their aesthetic appeal and biological characteristics in the future. Many other dive locations could establish similar trails.

14.9.3 The Development of Artificial Reefs and Marine Sculpture or Art Trails

These have been developed in many diving tourism locations worldwide and are useful because they take away divers from real reefs and so limit the impacts of scuba diving (Shani et al. 2011). An artificial reef is a structure that is colonised by plant and animal communities which resemble those occurring naturally on a coral reef. The structures can include scuttled ships and aircraft and can be composed of many types of material (Fig. 14.6), some of which may be convenient to the site location and others long distances, like the sludge tanker brought from New York to form part of the basis of Shipwreck Park, Pompano (Florida). There have been complaints on aesthetic grounds that artificial reefs do not resemble real reefs, especially during their early stages of development, but van Treek and Schumacher (1998) suggested creating artificial reefs that imitate the physical appearance of natural reefs. They suggested that it was possible to draw calcium minerals from seawater by in situ electrolysis and create a new calcium carbonate substrate akin to limestone, and this can be generated to coat a flexible metal matrix, such as chicken wire, which could be shaped like natural boulders and transplanted with corals. However, it appears there has been some problems with the technology where it has been tried out, but there are many other possibilities like the EcoReefs, which are artificial reefs using ceramic modules that mimic branching corals. These can be installed in close-packed arrays to create a dense, spatially complex habitat, and they have been used, for example, at Bunaken National Park (Indonesia) and El Nido Marine Park in the Philippines. The Reef Ball Foundation too has been very successful in establishing artificial reefs and has conducted over 3500 projects worldwide and placed over half a million reef balls. These reef balls are produced in marine concrete, engineered to last over 500 years, and they have been tested successfully and cannot overturn (Fig. 14.7). They are created with preformed attachment holes which accept standardised coral fragment plugs, and they have rough surfaces to enhance natural coral settlement so that corals can be transported already in place on the reef balls or propagated in the reef location. Studies have shown that recreation divers would prefer artificial reef sites which are located relatively close to shore and accessible; sites which are under 30 m deep; sites that contain a sunken ship or shipwrecks, especially a large naval ship or aircraft rather than more generic or amorphous forms, such as concrete blocks or pipes; the use of mooring buoys on the artificial reefs which reduces the anchoring damage; the artificial reefs to be designed so that there can be zones within existing reefs where specific recreation activities can exclusively take place so there are no conflicts between types of user (Milon 1989; Ditton et al. 2002; Perhol-Finkel et al. 2006; Stolk et al. 2007). It has been found that the artificial reefs when

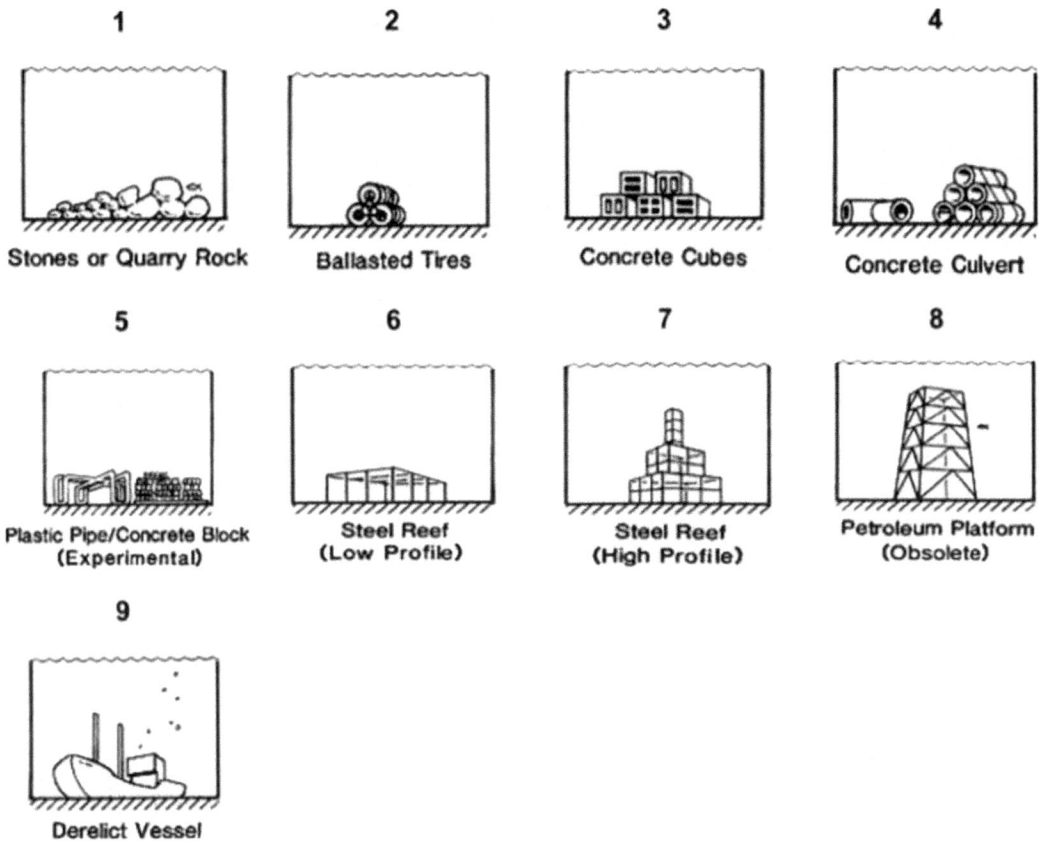

Fig. 14.6 Materials commonly used for artificial reefs (adapted from Stolk et al. 2007)

established are as good as, or better than, natural reefs in terms of fish recruitment and overall biodiversity.

In a similar fashion, underwater art and museums have been created at many sites (see examples from Fig. 14.8A, B and the growth of coral), such as Musha Cay (Bahamas), Molinere (Grenada), Manchones Reef (Mexico) and Punta Nizuc (Cancun, Mexico), and Museo Atlántico, Playa Blanca, Lanzarote (Canary Islands). This has occurred mainly due to the artist Jason deCaires Taylor who has produced statues constructed from marine-grade concrete at depths of 4–9 m, with a rough texture to allow coral larvae to gain a strong foothold and gaps and crevices in the sculptures to ensure that fish can swim in and out and located, where possible, downstream of a thriving reef in order to catch coral larvae and other marine organisms as they float by. Both the artificial reefs and the underwater sculpture parks are a way of trying to alleviate the problem of coral decline by providing a structure which allows more areas to be colonised by corals, by providing protection for small fish from predators and by providing alternative, interesting dive sites with good biodiversity viewing, so reducing the pressure on natural coral reefs and helping to aid coral reef recovery and resilience. In this they have been successful because in Koh Tao island (Thailand) from diver surveys 21% of the total dives annually were on an artificial reef (Nicholls 2013), and Belhassen et al. (2017), who targeted an online survey at divers at Eilat (Red Sea), found that 35% of dives took place on artificial reefs. It has been suggested though that the artificial sites are more appropriate for dive training, and it was proposed by Nicholls (2013) that all dives involving novice divers should be on an artificial reef and that the divers should be only allowed on natural

Fig. 14.7 Reef balls, Lake Pontchartrain Basin. Source: Louisiana Sea Grant College Program, Louisiana State University, USA

reefs when their buoyancy is to an acceptable standard that they will not cause much damage to corals on natural reefs. However, for this to happen, the artificial reefs in Koh Tao, for example, must become larger, be more interesting, and have more marine life, and there must be more of them close to the dive resorts. As Harris (2009) suggested snorkel and scuba diving trails could be incorporated into artificial reefs as has happened in Antigua on reef ball foundations. It has also been found that visitors are willing to pay a significant amount of money to view marine life, especially turtles on artificial reefs in Barbados (Kirkbride-Smith et al. 2016). These user fees could provide a considerable income source to help reef conservation.

14.9.4 Modification of Diver Behaviour

It has been shown earlier in this chapter that much of the impact on reefs has been caused by poor diver practices and there needs to be more responsible underwater behaviour (Fatt and Musa 2013) so it seems necessary to incorporate more diver and diver guide education into the recreational diving industry to try and improve this aspect. Whilst the diver certification programme is a start here, it would be possible to improve the training of dive guides and incorporate more environmental education so that they pass on this knowledge to their customers.

Offering divers this extra environmental education may reduce the frequency of contacts by divers. Pre-dive briefings are one of the most effective educational tools for introducing divers to conservation and ecological awareness. Pre-dive briefings focused on environmental awareness, diver impacts, and conservation concepts significantly decrease the amount of contact divers have with the benthos (Medio et al. 1997; Townsend 2000; Camp and Fraser 2012; Krieger and Chadwick 2013; Toyoshima and Nadaoka 2015), in some cases by as much as 70% (Medio et al. 1997; Townsend 2000). Briefings that give divers knowledge about coral reef ecology (Meyer 2002) and methods for minimising their impacts while diving are most effective (Belknap 2008), and it could be an excellent management practice to make this pre-diving briefing with underwater demonstrations, particularly of buoyancy control, compulsory. Lyons et al. (2015) suggested that because Bonaire already has in place a mandatory introduction and orientation dive, they suggested that the education should stress the ecological and economic importance of the reefs, the fragility of the system, the evidence of diver impacts in Bonaire, and the ways personal impacts can be reduced. Beyond information gained from dive briefings, Meyer (2002) found that divers with a stronger understanding of coral reef ecology had lower contact rates with the benthos. Thus, affiliation with a coral reef conservation programme (e.g. the Green Fins programme; NOAA Blue Star programme, Reef Environmental Education Foundation, Padi (Professional Association of Diving Instructors) AWARE, Coral Restoration Foundation) could provide effective information dissemination via educational pamphlets,

Fig. 14.8 (A) Artificial reef sculptures by Jason deCaires Taylor. (B) Reef sculptures, Playa del Carmen, Lanzarote. Photo by D. Huddart

updated online information on coral conservation, and coral identification cards and also benefits the dive shop as an attractive marketing tool (Camp and Fraser 2012). Dive operators could be more proactive in promoting this environmental education which should include the selection of the best diving sites to minimise damage and to suit the standards of their clientele and to focus on minimum impact diving techniques.

Dive leader supervision and intervention underwater would intuitively appear to be linked to rates of reef contact, and when examined, the willingness of dive guides to intervene in correcting diver behaviour underwater has been found to significantly reduce diver contact rates (Barker and Roberts 2004). This is part of the education of these leaders to provide this intervention, and the leaders of dives should ensure that during the first 10 minutes of each dive there will be more monitoring/intervention and be spent in areas with minimal stony coral cover and special attention should be paid to camera users and divers at night.

Adoption of environmentally responsible practices like the Green Fins or Blue Star programmes seems a sensible way to incorporate this environmental education into the industry.

Green Fins is a conservation initiative that works with dive centres to promote a set of standards for environmentally sustainable, dive tourism through the implementation of the Green Fins Code of Conduct and a robust assessment system to monitor compliance. It was initiated in 2004 by the United Nations Environmental Programme (UNEP) under the Regional Seas programme as part of the effort to increase public awareness and management practices that will benefit the conservation of coral reefs and reduce unsustainable tourism practices. It is overseen by the Coral Reef Unit of the UNEP based in Bangkok in collaboration with the UK charity The Reef-World Foundation. The Code of Conduct consists of 15 points (see Fig. 14.9 for the details); members receive training and the tools to promote environmental education and awareness. It has created a network of Green Fins-certified, dive centres to tackle environmental threats (Harvey 2015, 2017). Hunt et al. (2013) suggest that findings from the Philippines demonstrate through an evaluation that the programme is an effective way of providing a comprehensive and sustainable management tool for the protection of the marine ecosystem and may significantly reduce the recreational diving impacts on the reefs. They suggest that this type of programme should be implemented worldwide.

Roche et al. (2016) provide evidence that the effective implementation of environmentally responsible practices, via programmes designed to reduce diving impacts, may translate to reduced diver reef contacts. The finding of low overall diver reef contact rates in their study, comparable to other locations worldwide where environmentally responsible dives take place, gives additional support for the effectiveness of the Green Fins approach. Differences observed between high and low Green Fins compliance dive operators indicate that levels of engagement within a dive impact minimisation programme can influence the number of reef contacts made by divers. Many diver characteristics which might intuitively be expected to impact reef contact rates, such as level of qualification and overall experience, were not influencing factors in this study, and high versus low levels of Green Fins compliance did not influence the number of interventions made by dive guides underwater. Roche et al. (2016) suggest that dive operator's behaviours and attitudes towards conservation measures are more important factors in influencing diver reef contact rates.

There are other established programmes which range from those with a primarily educational focus, such as Padi AWARE, Blue the Dive in the USA, and REEF survey courses, to regional programmes with a policy background, such as the NOAA Blue Star charter within the Florida Keys (Camp and Fraser 2012). In 1992 the Project Aware Foundation registered as an independent non-profit organisation in the USA to engage divers and activities around the world in conservation and became an established UK charity in 1999 and received formal non-profit status in Australia in 2002 (http://www.projectaware.org). The Blue Star charter recognises tour operators who are committed to promoting responsible and sustainable diving and snorkelling practices to reduce the impact of these practices on coral reefs in the Florida Keys. This education of both divers and diving operators is necessary to help in the conservation of coral reefs. A review of environmental codes of conduct for entry-level divers and how they should be educated and trained, as this is where many problems lie, is given in Johansen and Koster (2012) and Johansen (2013).

GREEN FINS

- No stepping on Coral
- No stirring the sediment
- No touching or chasing marine life
- No feeding fish
- No littering
- Don't buy souvenir of coral and marine life
- Do not support shark finning
- No spear fishing
- Do not anchor on coral reefs
- Do not collect dead or alive marine life
- No gloves
- Wear life jacket when snorkelling
- Use mooring buoys
- Report environmental violations
- Join in conservation projects

www.greenfins.net

UNEP

Fig. 14.9 Green Fins information leaflet. Permission from Green Fins organisation

14.9.5 Charging Fees or Differential Fees

An effort to concentrate divers in areas away from high stony coral abundance might help reduce their impact, considering that it has been found that stony corals are prone to the effects of scuba divers while gorgonians and sponges are relatively resistant or resilient. Economic incentives to dive in those areas (e.g.

cheaper marine park admission tags to those that only dive those sites) could reduce the amount of diving in sites with a large abundance of stony corals that are prone to diver impacts. Green and Donnelly (2003) in a survey of marine protected areas found that only 25% containing coral reefs charged divers on entry, or they charged very low fees. They thought the potential for generating income had not been fully realised and more conservation could be achieved if higher fees were applied. In fact this has happened and in South East Asia 70% of marine parks have a diver use fee. However, these user fees are generally substantially lower than willingness to pay estimates (Pascoe et al. 2014), and therefore the potential fee income that could be used for management could be substantially increased. The number of dives that would occur if increases took place is estimated not to be greatly affected which would be part of the reason for the increases but the ability of managers to conserve the reefs would benefit. As the diver tourism industry is so important in countries like Thailand and Malaysia, it is always a delicate balance between loss to the industry by fee increases and conservation of the reefs. Emang et al. (2016) considered that it is natural to consider the potential of expanding the "user pays" principle at Sipadam (Malaysia), and they suggested an upward revised pricing structure to increase funding for reef conservation. Lower fees for local divers are already in place in some countries like Thailand, for example, and this seems the correct policy. In Guam Grafeld et al. (2016) found that divers strongly prefer ecosystems with greater ecological health and that they would be willing to pay for projects that would improve reef conditions. This collecting of fees from divers for coral reef management seems sensible and also creates a feeling amongst divers that they are partners in the conservation efforts. In Barbados Kirkbride-Smith et al. (2016) found that snorkellers would be willing to pay a significant fee for both natural and artificial reef use and that these fees could be used to help manage and maintain both types of reefs.

14.9.6 Recreation Divers As Part of the Conservation Effort

The value of recreation divers and snorkellers being involved in scientific/conservation projects is that it has educational value as it leads to an individual's increased awareness of scientific and conservation issues. They have too the opportunity to contribute to the environmental cause in a practical way, and the massive amount of data that can be generated can be important, generated at a low cost. Projects grounded on data collected by scuba divers in coral reef monitoring have been ongoing with good results for many years, such as the Coral Watch (www.coralwatch.org) and Reef Check (www.reefcheck.org) projects where recreation scuba divers, diving centres, and scientists have been involved in big monitoring projects. Two examples from the Italian sector of the Mediterranean Sea illustrate the potential value. Goffredo et al. (2004) offered Italian recreation scuba divers precisely these types of opportunity where they could participate in the first study of the geographical and ecological distribution of sea horses in the Mediterranean (The Mediterranean Hippocampus Mission). Here 2536 divers spent 6077 diving hours, coupled with information from 8827 questionnaires. It was evident that volunteers (citizen diver scientists) could be used to collect data that were intrinsically difficult to obtain. However, difficulties can arise related to the quality and validity of the data generated, but it has been shown that with appropriate recruitment and training, volunteer-collected data are qualitatively equivalent to those collected by professionals (e.g. see Pattengill-Semmens and Semmens 2003). In the Mediterranean sea horse project, the data were reliable because (a) volunteers were assisted during data collection by dive guides and instructors who had attended workshops and received training on the project's objectives and methodology; (b) sea horse identification was not difficult as there were clear morphological differences between the two species of interest; (c) information requested on the questionnaire, such as dive location, depth, dive time, and habitat are routinely recorded in the diver's personal dive-logs;

and (d) the data were remarkably consistent across the years of the project, which indicated a strong degree of reliability. It would have taken a professional researcher 20 years to collect the data that were collected over the three-year project by the volunteers at a cost of over US$ 1.365 million. The second study's aim by Lorenz et al. (2011) was to collect data related to the geographical, bathymetric distribution and health status of *Corallium rubrum* (Mediterranean Red Gold) which is a slow-growing gorgonian, endemic to the Mediterranean Sea. In this project 616 questionnaires were obtained from 12 sites by 16 diving centres and 3 scuba associations. Both projects could be used as models for biodiversity monitoring.

14.9.7 Activity Standards, Park Rules and Regulations

All recreation scuba divers and snorkellers must follow any adventure activity standards that are in place. For example, in Australia each state developed their own standards for these two activities. In Western Australia there are the Adventure Activity Standards of Recreational Scuba Diving version 1.0 published in July 2008 and the Snorkelling and Wildlife Swims Standards, version 1.2, published in October 2009. However, the Australian Adventure Activity Standard is currently being developed which is a national approach to safety outdoors by establishing best practice guidelines used to manage risk and safety across a wide range of outdoor adventure activities, including scuba diving and snorkelling, although these standards are not legally binding.

Any rules and regulations which are in place in any marine park or National Park must also be adhered to, such as the Red Sea Diving Regulations (www.projectoceanvision.com/redsearegulations.htm); the Bonaire Marine Park Rules (www.bonairepro.com/blog/scuba-diving/rules-you-should-know-for-bonaire-marine-park), the Koror Marine park, Malaysia (Davis and Kearns 2005); and the Marine Park Rules for Cozumel (Mexico) (https://thisiscozumel.com/things-to-do/dive/368-marine-park-rules).

14.9.8 Other Damaging Impacts on Coral Reefs

There are many other damaging effects on coral reefs which have nothing to do with recreational diving, but they are much more important in terms of environmental degradation. These include global warming and climate change (Hoey et al. 2016; Hughes et al. 2017 and Hoegh-Guldberg et al. 2017) which have been linked to coral bleaching, ocean acidification (Anthony et al. 2016), hurricanes/typhoons, tsunami, and coral diseases. There are also invasive species of fish and population blooms in certain organisms which can have a marked effect on the biodiversity of the reefs. For example, there has been the invasion of the Indo-Pacific Lionfish to reefs across the western Atlantic, the Caribbean, and the Gulf of Mexico (Co et al. 2012) and outbreaks of crown-of-thorns starfish (Kayal et al. 2012) which have been linked to increasing nutrients in the ocean from anthropogenic activity.

> **Concluding Remarks**
>
> It is usually very difficult to quantify the impact of any one of the many stressors on a coral reef ecosystem independently of the others. There is also often a lack of scientific understanding of the interrelationships which can limit the success of efforts to effectively create policy and regulate uses to prevent reef decline and manage the reefs. Moreover Wongthong and Harvey (2014) suggest that what is needed to manage reef-based, dive tourism should be an attempt to integrate the management and governance of integrated coastal management and sustainable tourism development, with a voluntary management strategy and a community-orientated approach. Each locality or area must select or adapt an appropriate set of strategies for its own needs based on the biophysical, socio-cultural, and managerial settings. This is before we even consider the much more global effects, such as global warming, coral bleaching, ocean acidification, and the damage cause by tsunamis and hurricanes/typhoon.

References

Abidin, S. Z. Z., & Mohamed, B. (2014). A review of SCUBA Diving impacts and implication for Coral Reefs Conservation and Tourism Management. In *SHS Web of Conferences 12*, 01093, https://doi.org/10.1051/shsconf/20141201093.

Allison, W. R. (1996). Snorkeler damage to reef corals in the Maldive Islands. *Coral Reefs, 15*, 215–218.

Alvarez-Filip, L., Dulvy, N. K., Gill, J. A., Côté, I. M., & Watkinson, A. R. (2009). Flattening of Caribbean coral reefs: Region-wide decline in architectural complexity. *Proceedings of the Royal Society, Biological Science, 276*, 3019–3025.

Anthony, K. R. N., Kline, D. N., Diaz-Pulido, G., Dove, S., & Hoegh-Guldberg, O. (2016). Ocean acidification causes bleaching and productivity loss in coral reef builders. *PNAS, 105*, 17442–17446.

Antonius, A., & Riegl, B. (1997). A possible link between coral diseases and a corallivorous snail (*Drupella cornus*) outbreak in the Red Sea. *Atoll Research Bulletin, 447*, 1–9.

Arnold, P. W., & Birtles, R. A. (1999). *Towards Sustainable Management of the Developing Dwarf Minke Whale Tourism Industry in Northern Queensland*. CRC Reef Research Centre Technical Report No. 27. CRC Reef Research Centre, Townsville, 30pp.

Bak, R. P., Brouns, J., & Heys, F. (1977). Regeneration and aspects of spatial competition in the scleractinian corals Agaricia agaricites and Montastrea annularis. In *Proceedings of the 3rd International Coral Reef Symposium 2* (pp. 143–148).

Bak, R. P. M., & Criens, S. R. (1981). Survival after fragmentation of colonies of *Madracis mirabilis*, *Acropora palmata*, and *A. cerviconis*. *Atoll Research Bulletin, 441*, 1–9.

Barker, N. H. L. (2003). *Ecological and Socio-economic Importance of Dive and Snorkel Tourism in St. Lucia, West Indies*. PhD thesis, University of York, 220pp.

Barker, N. H. L., & Roberts, C. M. (2004). Scuba diver behaviour and the management of diving impacts on coral reefs. *Biological Conservation, 120*, 481–489.

Belhassen, Y., Rousseau, M., Tynyakov, J., & Shashar, N. (2017). Evaluating the attractiveness and effectiveness of artificial coral reefs as a recreational ecosystem service. *Journal of Environmental Management, 203*, 448–456.

Belknap, J. (2008). *A Study of the Relationship between Conservation Education and Scuba Diver Behaviour in the Flower Garden Banks National Marine Sanctuary*. PhD thesis, Texas A&M University, 120pp.

Brander, L. M., Beukering, P. V., & Cesar, H. S. J. (2007). The recreational value of coral reefs; a meta-analysis. *Ecological Economics, 63*, 208–218.

Camp, E., & Fraser, D. (2012). Influence of conservation education dive briefings as a management tool on the timing and nature of recreational SCUBA diving impacts on coral reefs. *Ocean Coast Management, 61*, 30–37.

Chabanet, P., Adjeroud, M., Andrefouet, S., Bozec, Y. M., Ferraris, J., Garcia-Charton, J. A., et al. (2005). Human-induced physical disturbances and their indicators on coral reef habitats: A multi-scale approach. *Aquatic Living and Resources, 18*, 215–230.

Chamberlain, J. A. (1978). Mechanical properties of coral skeleton: Compressive strength and its adaptive significance. *Paleobiology, 4*, 419–435.

Chamberlain, J. A., & Graus, R. R. (1975). Water flow and hydromechanical adaptations of branched reef corals. *Bulletin of Marine Science, 25*, 112–125.

Chung, S. S., Au, A., & Qiu, J. (2013). Understanding the underwater behaviour of scuba divers in Hong Kong. *Environmental Management, 51*, 824–837.

Co, I. M., Green, S. J., Akins, J. L., & Malijokovic, A. (2012). Invasive lionfish drive Atlantic coral reef fish declines. *PLoS One, 7*, 1–4.

Coma, R., Pola, E., Ribes, M., & Zababala, M. (2004). Long-term assessment of temperate octocoral mortality patterns, protected vs. unprotected areas. *Ecological Applications, 14*, 1466–1478.

Connell, J. H. (1997). Disturbance and recovery of coral assemblages. *Coral Reefs, 16*, 101–113.

Cordell, H.K. (2012). *Outdoor Recreation: Trends and Futures a Technical Document Supporting the Forest Service 2010 RPA Assessment*. General Technical Report SRS-150. Asheville, NC: USDA Forest Service, Southern Research Station.

Danovaro, R., Bongiornia, L., Corinaldesi, C., Gionvandli, D., Dantiani, E., Astolfi, P., et al. (2008). Sunscreens cause coral bleaching by promoting viral infections. *Environmental Health Perspectives, 116*, 441–447.

Darwin, C. (1874). *The Structure and Distribution of Coral Reefs*. London: Smith, Elder and Company.

Davis, G. E. (1977). Anchor damage to a coral reef on the coast of Florida. *Biological Conservation, 11*, 29–34.

Davis, P. Z. R., & Kearns, C. M. (2005). *Koror State Government Marine Tour Guide Certification Program Manual* (2nd ed.), Koror State Department of Conservation and Law Enforcement and the Coral Reef Research Foundation, Koror, 172pp.

Dimmock, K., & Musa, G. (2015). Scuba diving tourism system: A framework for collaborative management and sustainability. *Marine Policy, 54*, 52–58.

Dineson, Z., & Oliver, J. (1997). Tourism impacts. In D. R. Wachenfeld, J. Oliver, & K. Davis (Eds.), *State of the Great Barrier Reef World Heritage Area, Workshop 1995* (pp. 414–427). GBRMPA Workshop Series No. 23. Townsville, Australia.

Dinsdale, E. A., & Harriot, V. J. (2004). Assessing anchor damage on coral reefs: A case study in selection of environmental indicators. *Environmental Management, 33*, 126–139.

Ditton, R. B., Osburn, H. R., Baker, T. L., & Thailing, C. E. (2002). Demographics, attitudes, and reef management practices of sport divers in offshore Texas waters. *ICES Journal of Marine Science, 59*, 186–191.

Diving Equipment and Marketing Association. (2014). *Diver Study*. San Diego, CA.

References

Dixon, J. A., Fallon-Scura, L., & van't Hof, T. (1993). Meeting ecological and economic goals: Marine parks in the Caribbean. *Ambio, 22*, 117–125.

Downs, C. A., Kramarsky-Winter, E., Segal, R., Fauth, J., Knutson, S., Bronstein, O., et al. (2016). Toxicopathological effects of the sunscreen UV filter, Oxybenzone (Benzophenone-3) on coral Planulae and cultured primary cells and its environmental contamination in Hawaii and the US Virgin Islands. *Archives of Environmental Contamination and Toxicology, 70*, 265–288.

Dustan, P., & Halas, J. C. (1987). Changes in reef-coral community of Carysfort Reef. Key Largo, Florida, 1974–1982. *Coral Reefs, 6*, 91–106.

Edney, J. (2006). Impacts of recreational scuba diving on ship-wrecks in Australia and the Pacific. A review. *Micronesian Journal of the Humanities and Social Sciences, 5*, 201–233.

Edney, J., & Howard, J. (2013). Review: Wreck diving. In K. Dimmock & G. Musa (Eds.), *SCUBA Diving Tourism* (pp. 52–56). London and New York: Routledge.

Edwards, M. (1997). Concerns and potential threats to the ecologically sustainable use of the Cairns section of the Great Barrier Reef Marine Park by the marine tourism industry. In *Reef Tourism 2005*, Cairns, 41pp.

Emang, D., Lundhede, T. H., & Thorsen, B. J. (2016). Funding conservation through use and potentials for price discrimination among scuba divers at Sipadan, Malaysia. *Journal of Environmental Management, 182*, 436–445.

Erftemeyer, P. L. A., Riegl, B., Hoeksema, B. W., & Todd, P. A. (2012). Environmental impacts of dredging and other sediment disturbances on corals: A review. *Marine Pollution Bulletin, 64*, 1737–1765.

Fatt, O. T., & Musa, G. (2013). Responsible underwater behaviour. In K. Dimmock & G. Musa (Eds.), *SCUBA Diving Tourism* (pp. 117–133). London and New York: Routledge.

Garrabou, J., Sala, E., Arcas, A., & Zabala, M. (1998). The impact of diving on rocky sublittoral communities: A case study of a bryozoan population. *Conservation Biology, 12*, 302–312.

Garrod, B. (2008). Market segments and tourist typologies for diving tourism. In B. Garrod & S. Gössling (Eds.), *New Frontiers in Marine Tourism: Diving Experiences, Sustainability, Management* (pp. 31–39). Amsterdam: Elsevier.

Garrod, B., & Gössling, S. (2008). Introduction. In B. Garrod & S. Gössling (Eds.), *New Frontiers in Marine Tourism: Diving Experiences, Sustainability, Management* (pp. 3–28). Amsterdam: Elsevier.

Giglio, V. J., Ternes, M. L. F., Mendes, T. C., Cordeiro, C. A. M. M., & Ferreira, C. E. L. (2017). Anchoring damages to benthic organisms in a subtropical dive hotspot. *Journal of Coastal Conservation, 21*, 311–316.

Goffredo, S., Piccinetti, C., & Zaccanti, F. (2004). Volunteers in marine conservation monitoring: A study of the distribution of sea horses carried out in collaboration with recreational scuba divers. *Conservation Biology, 18*, 1492–1503.

Grafeld, S., Oleson, K., Barnes, M., Peng, M., Chan, C., & Weijerman, M. (2016). Divers willingness to pay for improved coral reef conditions in Guam: An untapped source of funding for management and conservation? *Ecological Economics, 128*, 202–213.

Great Barrier Reef Marine Park Authority. (2000). Whale and Dolphin conservation in the GBRMPA. Retrieved from http://www.gnrmpa.gov/au/corp.site/info.services/publications/whale-dolphin/index.

Green, E., & Donnelly, R. (2003). Recreational scuba diving in Caribbean marine protected areas: Do the users pay? *Ambio, 32*, 140–144.

Guzner, B., Novplansky, A., Shalit, O., & Chadwick, N. E. (2010). Indirect impacts of recreational scuba diving: Patterns of growth and predation in branching stony corals. *Bulletin of Marine Science, 86*, 727–742.

Hall, V. R. (2001). The response of *Acropora hyacinthus* and *Monitpora tuberculosa* to three different types of colony damage: Scraping injury, tissue mortality and breakage. *Journal of Experimental Marine Biology and Ecology, 264*, 209–223.

Harriot, V. J., David, D., & Banks, S. A. (1997). Recreational diving and its impact in marine protected areas in eastern Australia. *Ambio, 26*, 173–179.

Harris, L. E. (2009). Artificial reefs for ecosystem restoration and coastal Erosion protection with aquaculture and recreational amenities. *Reef Journal, 1*, 235–246.

Hartman, M. L. (2012). *Assessment of Diver Impact during the Spiny Lobster Sport Season, Florida Keys, USA*. MSc thesis, University of South Florida, 75pp (Graduate theses and dissertations). Retrieved from http://scholarcommons.usf.edu/etd/4330.

Harvey, C. (2015). Green Fins: An approach to managing a sustainable diving industry. Retrieved from http://panorama.solutions/en/solutions/green-fins-an-approach-tomanaging-sustainable-divingindustry.

Harvey, C. (2017). Green Fins: A tool for reducing direct impacts of diving and tourism industries. Retrieved from http://Vimeo.com/225451965.

Hassler, H., & Ott, J. A. (2008). Diving down the reefs? Intensive diving tourism threatens the reefs of the northern Red Sea. *Marine Pollution Bulletin, 56*, 1788–1794.

Hawkins, J. P., & Roberts, C. M. (1994). The growth of coastal tourism in the Red Sea: Present and future effects on coral reefs. *Ambio, 23*, 503–508.

Hawkins, J. P., Roberts, C. M., Van't, T., de Meyer, K., Tratalos, J., & Adam, C. (1999). Effects of recreational scuba diving on Caribbean coral and fish communities. *Conservation Biology, 13*, 888–897.

Hein, M. Y., Lamb, J. B., Scott, C., & Willis, B. L. (2015). Assessing baseline levels of coral health in a newly established marine protected area in a global scuba diving hotspot. *Marine Environmental Research, 103*, 56–65.

Hoegh-Guldberg, O., Poloczanska, E. S., Skirving, W., & Dove, S. (2017). Reef ecosystems under climate change and ocean acidification. *Frontiers in Marine Science, 29*. https://doi.org/10.3389/fmars.2017.0158.

Hoey, A. S., Howells, E., Johansen, J. L., Hobbs, J.-P. A., Messner, V., McGowan, D. M., et al. (2016). Recent advances in understanding the effects of climate change on coral reefs. *Diversity, 8*(2), 12. https://doi.org/10.3390/d8020012.

Hughes, T. P., Kerry, J. T., Alvarez-Moriega, M., Alvarez-Romero, J. G., Anderson, K. D., Baird, A. H., et al. (2017). Global warming and recurrent mass bleaching of corals. *Nature, 543*, 373–377.

Hunt, C. V., Harvey, J. J., Johnson, V., & Pongsuwan, N. (2013). The Green Fins approach for monitoring and promoting environmentally sustainable scuba diving operations in South East Asia. *Ocean and Coastal Management, 78*, 35–44.

Inglis, G. J. (1997). Science and tourism infrastructure on the Great Barrier Reef: learning from experience or just "muddling through"? In *Proceedings of Great Barrier Reef Science Use and Management* (pp. 319–333).

Jameson, S. C., Ammar, M. S. A., Sandalla, E., Mostafa, H. M., & Riegl, B. (1999). A coral damage index and its application to diving sites in the Egyptian Red Sea. *Coral Reefs, 18*, 333–339.

Johansen, K. M. (2013). SCUBA diving education and training. In K. Dimmock & G. Musa (Eds.), *SCUBA Diving Tourism* (Chap. 5). London and New York: Routledge.

Johansen, K. M., & Koster, R. L. (2012). Forming scuba diving environmental codes of conduct: What entry level divers are taught in their first certification course. *Tourism in Marine Environments, 8*, 61–77.

Kapitze, I. R. (1999). Sustainable infrastructure development on the Great Barrier Reef. In *Coasts and Ports '99, Challenges and Directions in the New Century. Proceedings of 14th Australasian Coastal and Ocean Engineering Conference and the 7th Australasian Port and Harbour Conference*.

Kay, A., & Liddle, M. (1989). Impact of human trampling in different zones of a coral reef flat. *Environmental Management, 13*, 509–520.

Kayal, M., Vercelloni, J., Lison de Loma, T., Bosserelle, P., Chancerelle, Y., Geoffrey, S., et al. (2012). Predator crown-of-thorns starfish (*Ancathaster planci*) outbreak. Mass mortality of corals and cascading effects on reef fish and benthic communities. *PLoS One, 7*(10), e 47363.

Kirkbride-Smith, A. E., Wheeler, P. M., & Johnson, M. L. (2016). Artificial reefs and marine protected areas: A study in willingness to pay to access Folkestone marine reserve, Barbados, West Indies. *Peer Journal, 4*, e2175. https://doi.org/10.7717/peerj.2175.

Krieger, J. R., & Chadwick, N. E. (2013). Recreational diving impacts and the use of pre-dive briefings as a management strategy on Florida coral reefs. *Journal of Coastal Conservation, 17*, 179–189.

Lamb, J. B., True, J. D., Piromvaragorn, S., & Willis, B. L. (2014). Scuba diving damage and intensity of tourist activities increases coral disease prevalence. *Biological Conservation, 178*, 89–96.

Liddle, M., & Kay, A. (1987). Resistance, survival and recovery of trampled corals on the Great Barrier Reef. *Biological Conservation, 42*, 1–18.

Lorenz, B., Ilavia, V., Sergio, R., Stefano, S., & Giovanni, S. (2011). Involvement of recreational scuba divers in emblematic species monitoring: The case of Mediterranean red coral (*Corallium rubrum*). *Journal for Nature Conservation, 19*, 312–318.

Luna, B., Pérez, C. V., & Sanchez-Lizaso, J. (2009). Benthic impacts of recreational divers in a Mediterranean marine protected area. *ICES Journal of Marine Science, 66*, 517–523.

Lutz, S. J. (2006). A thousand cuts? An assessment of small-boat grounding damage to shallow corals of Florida keys. In W. F. Precht (Ed.), *Coral Reef Restoration Handbook* (pp. 25–38). Boca Raton, FL: CRC Press.

Lyons, P. J., Arboleda, E., Benkwitt, C. E., Davis, B., Gleason, M., Howe, C., et al. (2015). The effect of recreational SCUBA divers on the structural complexity and benthic assemblage of a Caribbean coral reef. *Biodiversity and Conservation, 24*, 3491–3504.

Magurran, A. E. (1996). *Ecological Diversity and Its Measurement*. Princeton, NJ: Princeton University Press, 192pp.

McClanahan, T. R. (1994). Coral-eating snail *Drupella cornus* population increases in Kenyan coral reef lagoons. *Marine Ecology Progress Series, 115*, 131–137.

Medeiros, R. P., Grempel, R. G., Souza, A. T., Tlarri, M. I., & Sampaio, C. L. (2007). Effects of recreational activities on the fish assemblage structure in a northeastern Brazilian reef. Panamjas. *Pan-American Journal of Aquatic Sciences, 2*, 288–300.

Medio, D., Ormond, R., & Pearson, M. (1997). Effect of briefings on rates of damage to corals by scuba divers. *Biological Conservation, 79*, 91–95.

Meyer, C. E., & Holland, K. (2008). Spatial dynamics and substrate impacts of recreational snorkelers and SCUBA divers in Hawaiian marine protected areas. *Journal of Coastal Conservation, 12*, 209–216.

Meyer, L. A. (2002). *Recreation Specialization and Environmental Behaviors: An Exploratory Analysis among Scuba Divers*. MSc thesis, University of Florida, 122pp.

Milazzo, M., Cheniello, R., Badalamenti, F., Camarda, R., & Reggio, S. (2002). The impact of human recreation activities in marine protected areas: What lessons should be learnt in the Mediterranean Sea? *Marine Ecology, 23*, 280–290.

Milon, J. W. (1989). Artificial marine habitat characteristics and participation behaviour by sport anglers and divers. *Bulletin of Marine Science, 44*, 853–862.

Neil, D. (1990). Potential for coral stress due to sediment resuspension and deposition by reef walkers. *Biological Conservation, 52*, 221–227.

Nelson, V. M., & Mapstone, B. D. (1998). *A Review of Environmental Impact Monitoring of Pontoon Installations in the Great Barrier Reef Marine Park*. CRC Reef Research Centre Technical Report No. 13, CRC Reef Research Centre, Townsville, 85pp.

References

Nicholls, R. S. (2013). *Effectiveness of Artificial Reef as Alternative Dive Sites to Reduce Diving Pressure on Natural Coral Reefs, a Case Study of Koh Tao, Thailand.* BSc thesis, Department of Forestry and Conservation, University of Cumbria, 42pp.

Nyström, M., Folke, C., & Moberg, F. (2000). Coral reef disturbance and resilience in a human-dominated environment. *Trends in Ecology and Evolution, 15*, 413–417.

Pascoe, S., Doshi, A., Thébaud, O., Thomas, C. R., Schultenberg, H. Z., Heron, S. F., et al. (2014). Estimating the potential impact of entry fees for marine parks on dive tourism in South East Asia. *Marine Policy, 47*, 147–152.

Pattengill-Semmens, C. V., & Semmens, B. X. (2003). Conservation and management application of the reef volunteer fish monitoring program. *Environmental Monitoring and Assessment, 81*, 43–50.

Perhol-Finkel, S., Shashar, N., & Benayanu, Y. (2006). Can artificial reefs mimic natural reef communities? The roles of structural features and age. *Marine Environmental Research, 61*, 121–135.

Plass-Johnson, J. G., Schwieder, H., Heiden, J., Weiand, L., Wild, C., Jompa, J., et al. (2015). A recent outbreak of crown-of thorns starfish (*Acanthaster planci*) in the Spermonde archipelago, Indonesia. *Regional Environmental Change, 15*, 1157–1162.

Plathong, S., Inglis, G. J., & Huber, M. E. (2000). Effects of self-guided snorkeling trails in a tropical marine park. *Conservation Biology, 14*, 1821–1830.

Poonian, C., Davis, P. Z. R., & McNaughton, C. K. (2010). Recreational divers on Palauan coral reefs and options for management. *Pacific Science, 64*, 557–565.

Potkamp, G., Vermeij, M. J. A., & Hoeksema, B. W. (2017). Host-dependent variation in density of coralliverous snails (*Coralliophila* spp.) at Curaçao, southern Caribbean. *Marine Biodiversity, 47*, 91–99.

Riegl, B., & Riegl, A. (1996). Studies on coral community structure and damage as a basis for zoning marine reserves. *Biological Conservation, 77*, 269–277.

Riegl, B., & Velimirov, B. (1991). How many damaged corals in Red Sea reef systems? A quantitative survey. *Hydrobiologia, 216/217*, 249–256.

Roberts, L., & Harriot, V. J. (1994). Recreational scuba diving and its potential for environmental impact in a marine reserve. In O. Bellwood, H. Choat, & N. Saxena (Eds.), *Recent Advances in Marine Science and Technology, '94* (pp. 695–704). Townsville, Australia.

Roche, R. C., Harvey, C. V., Harvey, J. J., Kavanagh, A. P., McDonald, M., Stein-Rostaing, V. R., et al. (2016). Recreational diving impacts on coral reefs and the adoption of environmentally responsible practices within the SCUBA diving industry. *Environmental Management, 58*, 107–116.

Rogers, C. S. (1990). Responses of coral reef organisms to sedimentation. *Marine Ecology Progress Series, 62*, 185–202.

Rogers, K. S., & Cox, E. F. (2003). The effects of human trampling on Hawaiian corals along a gradient of human use. *Biological Conservation, 112*, 383–389.

Rouphael, B. A., & Inglis, G. J. (1995). *The Effects of Qualified Recreational Scuba Divers on Coral Reefs.* Technical Report No. 4, CRC Reef Research Centre, Townsville, Australia.

Rouphael, B. A., & Inglis, G. J. (1997). Impacts of recreational scuba diving at sites with different reef topographies. *Biological Conservation, 82*, 329–336.

Rouphael, B. A., & Inglis, G. J. (2001). "Take only photographs and leave only footprints"? : An experimental study of the impacts of underwater photographers on coral reef dive sites. *Biological Conservation, 199*, 281–287.

Ríos-Jara, E., Galván-Villa, C. M., Rodriguez-Zaragosa, F. A., Lopez-Uriate, E., & Muñon-Fernández, V. T. (2013). The tourism carrying capacity of underwater trails in Isabel Island National Park, Mexico. *Environmental Management, 52*, 335–347.

Sarmento, V. C., & Santos, P. J. P. (2012). Trampling on coral reefs: Tourism effects on harpacticoid copepods. *Coral Reefs, 31*, 135–146.

Schaeffer, T. M., & Foster, M. S. (1998). *Diver Disturbance in Kelp Forests.* Research Technical Report, Monterey Bay National Marine Sanctuary. Retrieved from http://montereybay.noaa/research/techreports/diver_report/html.

Schleyer, M. H., & Tomalin, B. J. (2000). Damage on south African coral reefs and an assessment of their sustainable diving capacity using a fisheries approach. *Bulletin of Marine Science, 67*, 1025–1042.

Scott, C. M., Mehrotra, R., Hein, M. Y., Moerland, M. S., & Hoeksema, B. W. (2017). Population dynamics of corallivores (Drupella and Acanthaster) on coral reefs of Koh Tao, a diving destination in the Gulf of Thailand. *Raffles Bulletin of Zoology, 65*, 68–79.

Serour, R. K. (2004). *An Environmental Economic Assessment of the Impacts of Recreational Scuba Diving on Coral Reef Systems in Hurghada, the Red Sea, Egypt.* MSc thesis, University of Maryland, 63pp.

Shani, A., Polak, O., & Shashar, N. (2011). Artificial reefs and mass marine ecotourism. *Tourism Geographies: An International Journal of Tourism, Space and Environment, 14*, 361–382.

Shivlani, M. (2007). *A Literature Review of Sources and Effects on Non-extractive Stressors to Coral Reef Ecosystems.* South East Florida Coral Reef Initiative, Final Report, Project 19, Phase 1, 18pp.

Shivlani, M., & Suman, D. (2000). Dive operator use patterns in the designated no-take zones of the Florida Keys National Maritime Sanctuary (FKNMS). *Environmental Management, 25*, 647–659.

Stolk, P., Markwell, K., & Jenkins, J. M. (2007). Artificial reefs and recreational scuba diving resources: A critical review of research. *Journal of Sustainable Tourism, 15*, 331–350.

Sweatman, H. P. A. (1996). *Impact of Tourist Pontoons on Fish Assemblages on the Great Barrier Reef.* Technical Report 5, CRC Reef Research, Centre, Townsville, Australia.

Talge, H. (1991). *Impact of Recreational Divers on Scleractinian Corals of the Florida Keys*. MSc thesis, University of South Florida, St. Petersburg, USA, 92pp.

Tilmant, J. T. (1987). Impacts of recreational activities on coral reefs. In B. Salvat (Ed.), *Human Impacts on Coral Reefs: Facts and Recommendations* (pp. 551–556). French Polynesia: Antenne Museum EPHE.

Tilmant, J. T., & Schmahl, G. P. (1981). A comparative analysis of coral damage on recreationally used reefs within Biscayne National Park, Florida. In *Proceedings of the Fourth International Coral Reef Symposium* (pp. 187–192), University of the Philippines, Quezon City, Philippines.

Townsend, C. (2000). *The Effects of Environmental Education on the Behaviour of SCUBA Divers: A Case Study from the British Virgin Islands*. MSc thesis, University of Greenwich, 116pp.

Toyoshima, J., & Nadaoka, K. (2015). Importance of environmental briefing and buoyancy control on reducing negative impacts of SCUBA diving on coral reefs. *Ocean and Coastal Management, 116*, 20–26.

Tratalos, J. A., & Austin, T. J. (2001). Impacts of recreational SCUBA diving on coral communities of the Caribbean island of grand Cayman. *Biological Conservation, 102*, 67–75.

Underwood, A. J., & Peterson, E. H. (1988). Towards an ecological framework for investigating pollution. *Marine Ecology Progress Series, 46*, 227–234.

Uyarra, M. C., & Côté, I. M. (2007). The quest for cryptic creatures: Impacts of species-focused recreational diving on corals. *Biological Conservation, 136*, 77–84.

van Treek, P., & Schumacher, H. (1998). Mass diving tourism—A new dimension calls for new management approaches. *Marine Pollution Bulletin, 37*, 499–504.

Vandrezee, M. (1996). Managing tourism use in Australia's GBRMPA. In *Proceedings of World Congress on Coastal and Marine Tourism*, Hawaii.

Wongthong, P., & Harvey, N. (2014). Integrated coastal management and sustainable tourism: A case study of the reef-based SCUBA dive industry from Thailand. *Ocean and Coastal Management, 95*, 138–146.

Woodland, D. J., & Hooper, N. A. (1977). The effect of human trampling on coral reefs. *Biological Conservation, 11*, 1–4.

Worachananant, S., Carter, R. W., Hockings, M., & Reopanichkul, P. (2008). Managing the impacts of SCUBA divers on Thailand's coral reefs. *Journal of Sustainable Tourism, 16*, 645–663.

Zakai, D., & Chadwick-Furman, N. E. (2002). Impacts of intensive recreational diving on reef corals at Eilat, northern Red Sea. *Biological Conservation, 105*, 179–187.

Recreational Fishing 15

Chapter Summary

The definitions of recreation fishing and the numbers involved are discussed in this chapter. The direct impacts on fish stocks, on endangered fish species through trophy fishing, on size selection and fish community structure, including truncation of age and size structure, are reviewed. There is a loss of genetic diversity and evolutionary changes. Discards and catch-and-release impacts and the effects of invasive, non-native species are discussed. There are a series of indirect impacts such as habitat disturbance, walking tracks, off-road vehicles, effects on wildlife, loss of fishing gear, boat strikes, nutrient impact, pollution, plastics and pathogen transmission. The management of recreation fishing impacts such as the use of marine protection areas, best practice guidelines and codes of conduct, and the education of practitioners, including mandatory programmes, are reviewed.

15.1 Introduction

Overfishing throughout the world is a major ecological problem with much reduced fish stocks and individual fish species threatened with extinction. There is no doubt that commercial fishing has played a major role in this problem and it is clear there have been dramatic effects from commercial fisheries on marine fish stocks and marine ecosystems, but what seems to have been ignored to some extent is the potential role of the other major fishery sector to contribute to this crisis: recreational fishing. Furthermore, previous analyses have focused exclusively on marine environments, with little consideration of the role of freshwater fisheries. However, Cooke and Cowx (2004) outline the reasons why this has happened and suggest that failure to recognise the contribution of recreational fishing to fishery declines, environmental degradation, and ecosystem alterations places ecologically and economically important resources at risk. So we will see that the same issues that have led to commercial fisheries concerns have the equivalent and sometimes magnified impacts in recreation fisheries. The over-exploitation of fish stocks has also affected traditional activities such as small-scale artisanal fishing in some parts of the world by reducing the availability of catches. Artisanal fishing is usually operated by relatively small vessels typically fishing three nautical miles from the coast in, for example, the Mediterranean, with considerable cultural and historical significance, but is now declining in many areas with a downward trend in the number of vessels, licences, catches, and net revenues. The interacting impacts of artisanal and recreation fisheries have been described by Prato et al. (2016).

In this chapter it is a hope to outline aspects of these ecological problems by considering the environmental impacts of recreational fishing, some of the management techniques and solutions and approaches to educating the recreational fishing community about related topics.

15.2 Definition of Recreational Fishing

Recreational fishing is where aquatic animals that do not constitute the individual's primary resource to meet nutritional needs and are not generally sold or otherwise traded on export, domestic, or black markets are caught. This definition is sufficiently broad to include other animals beyond fish (e.g. invertebrates such as lobster and crabs); it avoids pointing to individual motivations (fun, sport, enjoyment, thrill of the catch, social bonding), does not discriminate against particular methods of fish capture (e.g. recreation rod and line angling vs. recreation gill netting, which is an important recreational fishing activity in some countries), does not preclude the catch being taken for personal consumption (as long as the catch does not become the primary resource to meet essential physiological needs), does not discriminate against non-Western cultures, but does discriminate commercial and purely subsistence fishing (artisanal) from recreational fishing. It is acknowledged that the unambiguous demarcation between pure recreation fisheries and pure subsistence fisheries is impossible because many recreation fishers have strong subsistence-like incentives to harvest fish. However, using fishing activity to generate resources for livelihoods marks a clear differentiation between recreation fisheries and pure subsistence fisheries, and, as a rule, recreation fishers have the capacity to substitute the products of their fishing experience by other products to meet nutritional needs. Globally, angling is by far the most common recreational fishing technique, which is why recreational fishing is often used synonymously with (recreation) angling.

Hence recreation or sportfishing is defined as fishing for pleasure, as opposed to commercial fishing for income or subsistence fishing for survival. Angling is typically conducted with a rod, reel, and line with a baited hook, lure, or fly attached. Some recreational fishing is conducted with a spear, net, or bow and arrows. In addition to finfish, recreation fishers collect crustaceans by net or trap; molluscs by hand, rake, or shovel; and frogs and turtles by net. Fishing may occur from the shore of the water body, by wading in shallow waters (Fig. 15.1), or from watercraft ranging in size from large multipassenger, live-aboard ocean-going ships, to single-passenger kayaks and other small boats.

15.2.1 Types of Recreational Fishing

The main forms of recreational fishing according to the National Survey of Fishing (2011) are shoreline (49.9%), boat (48.3%), riverbank (43.0%), and kayak fishing (3.9%). However, the terminology can be confusing. For example, coarse fishing is a term used in the UK and Ireland for fishing for game fish (like barbel, carp, pike, perch, roach, bream) which are not salmonids whilst game fishing is angling for freshwater salmonids (particularly salmon, trout, and char) using a fly-fishing technique (Fig. 15.1). However, there is no taxonomic basis for the distinction between coarse and game fish. Sportfishing includes fly fishing (Fig. 15.1), coarse and game fishing, and if it takes place offshore, for fish like marlin, tuna, sailfish, and shark, it might be called big-game fishing (Fig. 15.2). Tailrace fishing is angling immediately below natural, or man-made dams, or where there are restrictions in water flow on rivers and canals. There are other types of fishing like ice fishing which are carried out by very low numbers of recreation fishers.

15.3 Numbers of Recreation Fishers

Estimates for recreation fishers worldwide vary between 220 and 700 million (FAO212, World Bank 2012), but accurate figures are not easy to collect and different ages are used in different surveys which make comparisons difficult. In Australia there were 3.36 million in 2001 or

Fig. 15.1 Fly fishing on River Sava, Bohinjka, Slovenia. Photo by Ziga

Fig. 15.2 Trophy fishing for striped blue marlin, caught off Cabo San Lucas, Baja California. Photo by Kate Crandell

19.5% of the population, although in Western Australia this figure was estimated to be one third of the population, whilst in Canada in 2005 over 3.2 million adults bought licences which was about one in every ten adults. This number though was as high as one third of the population in Newfoundland and Labrador. The trends show that the numbers had decreased by 2% during the period 1995–2005 whilst the total days fishing declined, but the number of days fished per angler stayed the same (Hoffman 2009). In the USA Cordell (2012) suggested that the numbers had fallen 15% from 1996 to 2006, although 30 million of the 229 million citizens over 16 years of age (i.e. one in every eight) went fishing in the latter year. This was made up of 25.4 million freshwater and 7.7 million saltwater anglers.

In the Topline Survey (2017) for the USA, fishing was the second most popular activity with 14.6% of the over 25 year olds participating (31.5 million), whilst in terms of frequency of participation, fishing was third overall with 40.1 average outings and a total of 628 million. Figures were also gathered for 6–24 year olds where fishing was the third most popular activity with 19.5% participating (15.6 million), a total of 16.1 average outings per person and a total of 252.4 million. The two types of fishing which showed growth were kayak fishing up to 38% in the period 2013–2016 to 2.371 million participants and fly fishing which had a 6% growth 2015–2016. The figures for both saltwater and freshwater had declined slightly from 2006 to 2016. In 2013 almost 46 million Americans fished, or 15.8% of the population aged 6+ (Outdoor Foundation 2014) and for the first time since 2010 this was a loss of 1.2 million. However in 2017 there was a net gain of 1.5 million, with a total for freshwater 38.1 million, saltwater 12.3 million, and fly fishing 6.5 million. There are regional variations too in the USA with figures declining for the Great Lakes and increasing for Colorado where there was a 36.4% participation rate in 2011, with over 26 million activity days. In the estimates to 2060 produced by Bowker and Askew (2012), the fishing participation rate is expected to fall from 30.9% to around 28% of the population, although in total numbers the estimates are a growth between 27% and 56% because of population growth, depending on a series of factors, including climate change. The projected increase in fishing days/year is likely to exceed 200 million.

Overall on a global basis, rates for recreational fishing are variable and can exceed 45% of the population for some Scandinavian countries, like Finland, and every tenth European Union citizen goes fishing, with the global average about 11% (Arlinghaus and Cooke 2009). In Australia 20% of residents aged 5+ in 2015 took part in recreational fishing at least once within the last year, 20% in freshwater, and 35% in estuarine ecosystems.

In the UK it is estimated that the figure for recreational fishing is 4.2 million, about 9% of the population in England and Wales (Simpson and Mawle 2010), although in Wales the figure was between 10% and 12% between 2008 and 2014 (Wales Outdoor Recreation Survey 2015). Arkenford (2014) though put the total at 1.135 million (2.1%) for 16+ for the UK whilst according to the Environment Agency there were 1.4 million rod licences sold to freshwater anglers in England and Wales (2010–2011). This figure is rather less than the 2.3 million estimate for freshwater anglers during 2009 (Simpson and Mawle 2010). For 2012 Armstrong et al. (2013) estimated over 1.08 million sea anglers for the UK, a figure much less than the 1.9 million of sea anglers estimated by Simpson and Mawle (2010).

15.4 Direct Impacts of Recreational Fishing

Due to this large population of recreation fishers, it is no real surprise that there are a whole raft of direct and indirect impacts of this fishing on the fish populations, on the ecosystems in which they live, and on the overall environment. These range from effects on fish stocks, through evolutionary changes, disease and pathogen transmission to pollution. However, each mode of fishing, such as shore or boat fishing, is implicated in a variety of ecological impacts that are specific to each one.

15.4.1 Effects on Fish Stocks

In terms of biomass, recreational fishing has been estimated to take up to 12% of global fish catches (Cooke and Cowx 2004) which does not seem a large figure, and there seem to be few documented declines to fish stocks in recreation fisheries. However, Post et al. (2002) documented four in Canada that showed evidence of dramatic declines which were attributed to recreational fishing: lake trout, walleye, northern pike, and rainbow trout. These were largely unnoticed by the fisheries managers and it seems

15.4 Direct Impacts of Recreational Fishing

Table 15.1 Comparative catches of species shared by recreation and commercial fishers in various Australian studies

Location	Species	Recreational catch (tons/year)	Commercial catch
SE Queensland	Snapper	148	50
Metropolitan Adelaide waters	King George whiting	48.5	15.4
Fraser Island (Queensland)	Tailor	180	25–55
Richmond and Clarence Rivers (New South Wales)	Yellowfin bream, dusky flathead	70	54
Pumicestone Passage (Queensland)	Yellowfin bream, dusky flathead, and sand whiting	43.1	0
Leschenault Estuary (Western Australia)	Blue swimmer crab	45.7	2.8
Eastern Gulf of Shark Bay (Western Australia)	Snapper	100	3
Greater Metropolitan Perth	Tailor	651	7
Port Phillip Bay (Victoria)	Mixed inshore species including snapper and King George whiting	469	482

that this may well be widespread in recreation fisheries, and it was concluded that recreation and commercial fisheries were not inherently different, with both having the potential to affect fisheries negatively. In Australia in the eastern Gulf of Shark Bay (Western Australia), the biomass of snapper was estimated to be only 2–10% of the original virgin stock and that recreational fishing was thought to be the main cause. This can be seen from Table 15.1 where the comparative catches of species shared by recreation and commercial fishers are shown. In North America the NOAA (2009) landings data show 13.3 million pounds of red drum (*Sciaenops ocellatus*) caught compared to only 200,000 pounds by their commercial counterparts. The discrepancies may not be that large, and, for example, recreation fishers took 319 t of the West Australia Dhufish in 2005/2006 compared to commercial fishers who took 163.9 t, 66% of the catch. However, this fish is typically long-living and slow breeding and highly susceptible to overfishing. Where the harvest rates exceed sustainable levels of a target species and affect the abundance and size structure, this is called growth overfishing, and with continued or extreme overfishing, this can affect recruitment (recruitment overfishing). This can affect biodiversity and whole ecosystems. The exploitation rates (i.e. the fraction of the fish in a population at a given time) that is caught and removed during a particular time interval, for example, a year by angling, are highly variable and can range from <10% to > 80% and thus can be substantial but depends on all kinds of factors like regulations, angling gear, and angling effort. In the Mediterranean, Font and Lloret (2014) illustrated that often the size of individual fish caught is below the minimum landing size, which is illegal. Furthermore they found that when comparing the minimum landing size of 17 species targeted to their corresponding size at maturity, they found that only four species had a mean that was greater than their size at maturity which raises questions about the sustainability of the fisheries. This problem of retention of juvenile fish by anglers seems common, and McPhee et al. (2002) documented several examples from Australia, the USA, and South Africa.

15.4.2 Large Species Range

The harvest includes a wide species range, for example, McPhee et al. (2002) gave figures of 201 different taxa by boat anglers in New South Wales, 194 species in S.E. Queensland, and 170 species in Biscayne National Park, Florida. Even small recreation catches though may impact on fish stocks because of their life-history characteristics, like slow growth rate, small population sizes, and restricted ranges.

15.4.3 Endangered Fish Species and Trophy Fishing

These declines in fish stocks are rarely considered a real threat or even halted when endangered species are targeted, and in some cases anglers are drawn to fish because they are rare and endangered species. This seems especially to be the case in trophy fishing where anglers target the largest individuals of a species with the goal of catching a "world record" sized fish which are certified as such by the International Game Fish Association (IGFA). However, they have to be weighed at an official IGFA station which requires transport and killing the fish. This can be an ecological problem because for many species the number of offspring is based on how big they are, and by removing the biggest individual fish this has a disproportionate impact on the population dynamics of a species. For example, removing a single 61 cm long red snapper is equivalent to removing over 200 of 41 cm fish from the population. Larger mothers have higher energy reserves and are able to invest more resources into each individual larva. For example, the black rockfish (*Sebastes melanops*) larvae from larger mothers have larger globules of oil than those from smaller mothers which is a feature associated with growth rates three times as high and survival twice as high as larvae from smaller mothers. As fish exhibit infinite growth, the larger fish are also older and more experienced. Many fish species also show sexual dimorphism where females are larger than the males so that many gravid females are often the largest individual fish within a population. Due to the prestige mentality, it is obvious that despite a low probability of catching a record-sized fish, anglers will land near record fish for weighing. With these factors in mind where species have reduced populations removing the largest, or near largest, individuals can impede a population's recovery or contribute to its decline. Unfortunately in the IGFA World Record List for 1222 species, there are 858 species considered threatened with extinction by the IUCN Red List. Shiffman et al. (2014) discuss these issues and suggest a simple solution: the IGFA should stop issuing records that implicitly require killing the fish for all International Union for Conservation (IUCN) Red List of Threatened Species which would immediately reduce the fishing pressure on the largest individuals of the most vulnerable species. This would still allow anglers to target over 93% of species that records have been issued for.

15.4.4 Size Selection and Fish Community Structure

As many fishing techniques are size-selective, changes in the size structure of populations should be expected. Decreases in mean size of the target fish and reductions in abundance of the larger fish are widely reported as fishing increases. Size-selective fishing will affect different species in different ways because species have different life-history traits so that species with late maturity and slow growth towards a larger maximum size are affected more by size-selective fishing than small, fast-growing species with early maturity. Species composition of fish communities therefore should change and smaller, fast-growing species should dominate the biomass. As this affects a number of life-history traits which are at least partially heritable, it should be expected that the exploited populations should evolve in response to harvesting.

Selective removal of larger fish may affect their predators or prey and is one process that may cause the size distribution of biota within an ecosystem to differ from that predicted by models. In multispecies communities, species with short lifespans and rapid population growth which have early maturity and channel a large proportion of their resources into reproductive activities are likely to respond to fishing rapidly. As long as fishing intensity and recruitment are balanced, they can be fished sustainably at younger ages and higher mortality levels. Slower-growing species, with a later maturity and larger size, are likely to be vulnerable to intensive fishing, despite having more naturally stable population sizes which in the unexploited state are buffered by numerous age classes against recruitment failure of individual cohorts. The larger and late maturing species are more susceptible to exploitation whilst the life histories of smaller species may enable them to sustain higher instantaneous mortality rates than larger species. They may also suffer lower fishery mortality simply because they are less desirable and less accessible targets in a size-selective fishery. So we can see that changes in fish community structure result from combined effects of differential fish mortality and the variable susceptibility of species with different life histories.

Size-selective fishing can have a marked impact on fish population, sex ratios, and artificially curtail reproductive lifespans. The relative fish fecundity increases as they grow, and so a population of a given biomass will have a greater potential fecundity when composed of larger rather than smaller individuals.

15.4.5 Consequences of High Exploitation Rates and Selectivity

Angling therefore adds a further trophic level to aquatic ecosystems, and anglers can be regarded as keystone predators in aquatic ecosystems. The fishing mortality can be rather high for particular, highly valued and sought-after fish species, for example, salmonids, and within these species, larger-size classes are positively selected. The combination of high exploitation rates and pronounced selectivity may have some direct and indirect effects on exploited fish populations.

15.5 Direct Consequences

High fishing mortality has been shown to repeatedly influence fish population dynamics and to contribute to the collapse of recreationally exploited fish populations. The reasons that may play a role are depensatory mechanisms, truncation of age and size structure, loss of genetic variability, and evolutionary changes. These reasons are discussed in detail in Lewin et al. (2006).

15.5.1 Depensation Instead of Compensation

Mechanisms of compensation are central to most classical fisheries biological concepts such as surplus production and maximum sustainable yield. It assumes that compensatory effects (e.g. enhanced growth rate, enhanced fecundity, enhanced juvenile survival) arise through attenuating intraspecific interactions and food competition when fishing reduces the abundance of the target. The compensatory potential of fish populations, however, is a matter of debate, and the relative strength and frequency of density-dependent population regulation depend on environmental conditions and life-history strategies, and there is growing evidence that there are some limits for compensatory responses. Some mechanisms, collectively referred to as depensatory responses, may counteract compensation if the population size is reduced below a specific threshold. After reaching such a threshold, group dynamics and cooperative interactions might be impaired, which compromise mating success, foraging, or anti-predator strategies. Furthermore, large-bodied fish can exert a top-down control on smaller species that are competitors or predators of their own progeny. Reducing the abundance of large piscivores may relax smaller prey species from top-down control and impair the fish potential for compensatory responses once the population is fished down under a threshold level. This might occur when the prey of the fish achieve a competitive advantage over the young of the piscivores. Also environmental stochasticity and genetic mechanisms such as drift and inbreeding may have a stronger influence on small populations and may impair their compensatory potential. Depensatory effects capture the positive relationship between the per capita population growth rate and population density at low population sizes, which increases the per capita mortality probability of intensively exploited fish populations at low population abundances.

15.5.2 Truncation of Age and Size Structure

Size-selective angling may not only reduce the biomass but also truncates the age and size distributions in the targeted fish population. The removal of large individuals may increase the growth rates of juvenile fishes if competition for food is relaxed at lowered population abundances. However, because the fish size correlates with many reproductive traits, the selective removal of most of the large individuals will affect the reproductive capacity of the exploited fish population despite compensatory growth of surviving individuals. Older fishes often have a higher hatching success than first-time spawners which may be attributed to a variety of factors, such as egg size and quality, or ideal spawning time. In many marine and freshwater fishes, larger age, size, or weight results in the production of larger eggs. The egg size, influenced by maternal effects (condition factor, weight, size, or age at maturity), positively correlates with offspring survival. For example, large salmonid eggs have higher survival rates than smaller eggs, when the concentration of dissolved oxygen is low.

Old and large fish also increase their reproduction success in breeding competition because they have higher competitive abilities which enable them to obtain better spawning sites or, in the case of salmonids, to dig deeper redds. Moreover, the fecundity in fish exponentially increases with age and size. Larger fish produce more eggs simply because of geometric constraints but also because they provide a greater proportion of energy stores to egg production. The fish age also influences the spawning time. Younger and smaller fish may start later with spawning, because they emerge from the winter with lower lipid reserves than larger individuals and the need to acquire sufficient energy reserves may delay spawning. The spawning time influences the recruitment, as the larval survival depends highly on the coincidence of larval production and peak zooplankton production. An earlier birthdate may enhance the survival of the progeny, presumably as a result of a longer growing season. For fishes showing age-related temporal spawning, the removal of old age classes will shorten the spawning season and can result in recruitment failure in years when successful recruitment depends on early spawning time.

Furthermore, it has been shown for some fish species, they are capable of social learning from more experienced and sometimes older individuals concerning anti-predator behaviour, migration and orientation, mate choice, foraging, and communication behaviour (for a review, see Brown and Laland 2003). In addition, the recruitment in populations is not only influenced by cannibalism or intraspecific competition for food or space among fish of the same size. Older fish may also contribute to the regulation as is assumed for pike and various salmonid species.

15.5.3 Loss of Genetic Diversity

The genetic variability plays a crucial role in the survival of species and is essential for their potential for successfully evolving in response to short- and long-term environmental changes. This aspect is of crucial importance, especially in freshwater populations. Local populations of freshwater fishes are genetically more divergent than those of marine species and are more susceptible to the loss of genetic variability, in particular, small and isolated freshwater populations that are confronted with a high selective mortality.

Many fish species targeted by recreation fishers have a spatially phylogeographic structure defined by evolutionary history, demographic processes, the level of gene flow, and genetically based adaptations to the local environment, which is detectable on different spatial scale. The biodiversity at the level of discrete populations ensures their adaptive potential and the resilience against environmental changes of a species and plays a critical role in keeping fisheries sustainable. In particular populations living in an uncommon or variable habitat constitute an important part of the evolutionary legacy. The reduction of population densities can lead to the loss of populations which obviously results in a loss of genes, or gene combinations. In addition, demographic bottlenecks are expected to reduce the number of rare alleles by genetic drift and inbreeding.

A loss of genetic variation may further be caused by the skewing of the sex ratio as a result of the selective removal of male or female individuals from a population. In addition, the removal of the largest individuals may lower the genetic variability. In general, the loss of genetic diversity and allelic richness decreases the adaptive potential and lowers the long-term fitness of populations. In addition, there is some evidence that the genetic diversity on the level of individual organisms can provide fitness benefits and may increase disease resistance.

15.5.4 Evolutionary Changes Due to Selective Angling

As many commercially exploited fish stocks declined and failed to recover even after exploitation ceased, there is growing concern that heavy and selective exploitation over decades results not just in demographic consequences for targeted and non-targeted fish species but may have led to detrimental evolutionary changes in some life-history characters. The changes of life-history parameters in response to fishing are well known, and the possibility that the fishery may inevitably change exploited fish stocks has been discussed for decades, but there is great difficulty in determining whether the change of life-history traits reflects phenotypic variability or is caused by genetic changes. However, the prerequisites for evolutionary changes in fish population in response to recreational fishing such as local adaptation, heritable population variation, and a high and selective fishing mortality exist. The perception that evolution is a very slow process has been challenged by studies on fish species demonstrating that, under an appropriate life history and a sufficient strong selection pressure, a so-called contemporary evolution can occur in comparatively short time periods and change production-relevant life-history traits, such as age and size at maturation, growth rate, and annual reproductive investment.

Also behavioural traits might undergo a selection in response to fishing as behavioural individuality has a genetic basis. It has been demonstrated that angling creates a selection for avoidance behaviour. Behavioural traits are usually determined by genes of more than one locus. Consequently, behavioural traits that are related to the vulnerability to angling can be correlated with other characteristics, such as metabolic rates and parental care. A selection against aggressive behaviour may reduce the fitness of the surviving population. The aggressiveness of some nesting male fish correlated positively with the quantity of eggs in a male's nest. Consequently, the males with the greatest potential to contribute to annual recruitment were those fished and removed by anglers.

To sum up, angling may have the potential to cause an evolution in some life-history traits. Angling may select for, or against, certain life-history traits provided that the fishing mortality is high and the survivors represent genotypes that are less vulnerable to the force of mortality and then proliferate in subsequent generations. A prediction of the effects on life- history evolution though is difficult, because the effects depend on the multiple interactions within the aquatic ecosystem, and it has been argued that evolutionary change induced by fishing is slow and therefore unimportant to fisheries management. Yet Conover and Munch (2002) concluded that fisheries-induced selection from a simulated harvest from a hypothetical fishery resulted over four generations in the removal of large individuals. So changes in size-related, life-history traits can influence population persistence and yield.

The higher the rate of fishing mortality and the higher the number of generations over those that a population has been fished, the greater the probability that genetic responses occur. The outcomes of the selection are not necessarily positive, neither from the populations nor from the angler's point of view. Genotypes that survive fishing pressure may be less than optimal with respect to natural selection, and this may prevent a recovery of a population, even after the fisheries have ceased. Sutter et al. (2012) nevertheless show that size-selective fishing or even a just elevated level of fishing mortality has the potential to induce rapid evolutionary change in a range of production-related traits in fish populations. Using males from two lines of large-mouth bass (*Micropterus salmoides*) selectively bred over three generations for either high or low vulnerability to angling as a model system, they show that the trait vulnerability to angling positively correlates with aggression, intensity of parental care, and reproductive fitness. This experimental research has demonstrated that angling vulnerability is heritable in large-mouth bass and is correlated with elevated resting metabolic rates (RMR) and higher fitness.

However, whether such differences are present in wild populations is unclear. Hessenauer et al. (2015) sought to quantify differences in RMR among replicated exploited and unexploited populations of large-mouth bass. They collected large-mouth bass from two Connecticut drinking water reservoirs unexploited by anglers for almost a century and two exploited lakes, then transported and reared them in the same pond. Field RMR of individuals from each population was quantified using intermittent-flow respirometry. Individuals from unexploited reservoirs had a significantly higher mean RMR (6%) than individuals from exploited populations. These findings are consistent with expectations derived from artificial selection by angling on large-mouth bass, suggesting that recreation angling may act as an evolutionary force influencing the metabolic rates of fishes in the wild. Reduced RMR as a result of fisheries-induced evolution may have ecosystem-level effects on energy demand and be common in exploited recreation populations globally.

Recreation angling therefore selectively captures individuals with the highest potential for reproductive fitness. This suggests that selective removal of the fittest individuals likely occurs in many fisheries that target species engaged in parental care. As a result depending on the ecological context, angling-induced selection may have negative consequences for recruitment within wild populations of large-mouth bass and possibly other populations of exploited species in which behavioural patterns that determine fitness, such as aggression or parental care, also affect their vulnerability to fishing gear.

15.5.5 Discards or By-Catch from Recreational Fishing

There is substantial discarding of fish which include unwanted species, juveniles, and catch limits, and it has been estimated that the rates vary between 30% and 76% from various studies in the USA and Australia, and Cooke and Cowx (2004) estimated that 60% were returned overall. There are also management rules or legislation that demands the release of all captured fish or of fishes of protected size or species.

Voluntary catch-and-release (CR) behaviours where anglers release fish because it is the modus operandi or for ethical, conservation, or sporting reasons (Policansky 2002) are also a characteristic of recreation fisheries, which may contribute to the view that recreational fishing is benign relative to commercial fishing. In some fisheries, voluntary release rates can reach nearly 100%, such as in the coarse fisheries of Western Europe or elitist resources such as bonefish (Policansky 2002). However, an unknown proportion of fish captured by anglers and released under the assumption that they will survive die post-release which brings us on to a consideration of CR effects.

15.5.6 Catch-and-Release Impacts

There are two perspectives on CR. Some people see fishing solely as a means of catching fish and consider that there is no purpose to catch a fish other than pleasure, and so CR is an unethical fishing practice which can cause distress and physical damage to the fish. Another view is that CR is ethical and a conservative approach to sustaining recreation fisheries and preferable to catch-and-kill. However, CR can be an effective practice in offsetting angling-induced impacts to individual fish and their populations and can encourage the biological, economic, and social sustainability of a fishery (Bartholomew and Bohnsack 2005). Numbers involved can be high. In 2011, the National Marine Fisheries Service estimated that while 393,193 striped bass were harvested in New Jersey, over 900,000 were released (NMFS 2012).

In CR angling strategy is the assumption that fish experience low mortality and minimal sublethal effects and that they are released in good condition and will return to the common population, with negligible effects on lifetime fitness.

Sometimes this is the case but mortality is highly dependent on species and a range of other factors like water depth, temperature, and type of tackle used. It can vary from 0% to 95%. Poor CR practices can cause physical injury and physiological stress to fish. In fact, according to a review by Cooke et al. (2002) in virtually all CR fisheries, some proportion of released fish die as a result of being captured, while others experience sublethal effects, such as injury, physiological disturbance, behavioural alterations, and fitness impairments.

In a review of the impacts of CR practice on striped bass (*Morone saxatilis*), Tiedemann and Danylchuk (2012) found that despite the best intentions of anglers practising CR angling, the mortality rate associated with this practice was not trivial. The Atlantic States Marine Fisheries Commission currently applies an 8% hooking mortality rate for striped bass caught and released by recreation anglers in saltwater ecosystems (ASMFC 2011). This mortality rate is based on the results of a study on mortality of hooked and released striped bass in a saltwater impoundment in Massachusetts by Diodati and Richards (1996). Applying this mortality rate to estimates of striped bass caught and released annually in New Jersey yields the annual discard mortality estimate of 72,366 out of 904,576 released. This mortality rate may, in part, be due to a general lack of understanding among anglers as to how CR techniques can physically injure and physiologically stress fish. Many anglers make the assumption that all fish released survive the experience since they observe that the fish appear relatively unharmed and swim away, with dead fish rarely resurfacing but in fact the behaviour of released fish is a poor indicator of whether fish live or die. Muoneke and Childress (1994) reported that fish that appear to be healthy when they are released may exhibit post-release injuries or stress caused by angling and handling and actually experience mortality sometime after release.

Angled striped bass may experience stress for a variety of reasons. The exercise induced by angling is the first cause of physiological stress response. Environmental factors can then exacerbate the rate of stress during angling, including water temperature, air temperature, and salinity.

A number of studies have documented the fact that stress and stress-related mortality in striped bass caught-and-released is temperature dependent. In general, as water temperatures rise above the bass' optimum temperature range, angler-induced stress increases, along with increased potential for post-release mortality. Therefore, it can be assumed that the warmer the water temperature, the longer it will take for a bass to recover from a fight or factor related to stress and stress-induced mortality. It has been documented that high mortality of striped bass released in freshwater occurs during the warm summer months, with air temperature when fish were landed and handled as the most important factor related to mortality. Any abrupt or substantial temperature increase experienced by angled striped bass, even for a brief period of time, can also add physiological disruption to that caused by fighting during angling. This is especially important during hot weather, when there are large differences between water and air temperatures. In general, research has shown that environmental stress and stress-induced mortality of caught-and-released striped bass is potentially higher in freshwater ecosystems (Diodati and Richards 1996). In marine waters, salinity appears to help moderate physiological imbalances associated with stress. This is an important consideration for anglers participating in the coastal striped bass fishery, as fish caught in low-salinity areas, such as the upper portions of estuaries, are more likely to experience stress during CR than bass taken in high-salinity waters.

The tackle type including the number and style of hooks and the type of bait used are all factors that can affect anatomical hooking location and the likelihood of physical injury to organs and tissue from hook wounds. Anatomical location of hook wounds has been found to be one of the most important factors influencing survival rates for released striped bass and, in general, mortality is highest if the wound site includes a vital organ. For example, a fish hooked in the jaw stands a much better chance of survival than a fish that is hooked in the gills, oesophagus or stomach. Diodati and Richards (1996) reported that the odds of death for gut-hooked fish were almost six times the odds of death for fish hooked in the lip. Deep hooking was the single most important factor that caused death of striped bass caught and released in studies by the Maryland Department of Natural Resources. Between 1996 and 2000, nearly 1300 striped bass were used in their CR studies and they estimated a 17-times higher chance of dying if a striped bass is deep hooked rather than shallow hooked (MD DNR 2010). Deep hooking in striped bass is often higher with live baits or natural baits than with artificial baits, because fish often swallow hooks and baits more deeply which increases the possibility of injury and mortality.

Many studies have reported this lower mortality and injury to striped bass when fish are angled with artificial baits. However, while lures generally hook fish in the jaw or mouth, they can also present problems. For example, large plugs rigged with multiple treble hooks can cause injury to a bass since the free hooks often swing around and catch in the fish's gills or eyes. Treble hooks may also require an inordinate amount of time for removal.

To counter these concerns, it is often recommended that anglers replace treble hooks on plugs and metal lures with single hooks. The IGFA endorsed the idea of replacing treble hooks with single hooks to facilitate easy dehooking and faster release of fish (IGFA 2011). However, even single hooks can cause problems with hook removal if they are barbed hooks, and it is recommended crushing hook barbs, or using barbless hooks on plugs and lures to facilitate easy hook removal and reduce handling time and hooking injuries.

Aside from physical injury from deep hooking, physiological stress from fighting is another important factor that can result in angler-induced mortality of striped bass caught and released by recreation anglers. Fish that struggle intensely for prolonged periods of time during angling become exhausted. When fish are angled to exhaustion, lactic acid builds up in the tissues of the fish from muscle function. Increased levels of lactic acid can lead to a situation known as acidosis, and exhausted fish may reach a point where physiological imbalance, muscle failure, or death is possible. Therefore, the longer a fish is fought, potentially the less likely it is to survive after release. A fish that is landed quickly has a better chance of survival after release than one that has been exhausted by a lengthy fight.

The inability of striped bass to recover from physiological stress incurred during capture can disrupt normal feeding patterns, increase vulnerability to attack from predators, and reduce the striped bass' ability to fight off diseases and parasites or heal wounds caused by hooks.

Stress and the potential for post-release mortality also increase dramatically if fish are mishandled. Landing, handling, and release methods all may further exacerbate stress and result in post-release mortality. For example, the longer a fish is kept out of the water, the lower its chances for post-release survival, especially if it has endured a prolonged fight (see Cooke and Suski 2005).

Therefore mortality can be attributed to physical damage from hooking injuries or to physiological stress associated with stress-inducing hypoxia due to lifting the fish out of the water for hook removal or photography. To evaluate the effects of air exposure and angling-induced exhaustive exercise on released grayling condition, Lennox et al. (2016) observed the blood physiology and reflexes after angling and air exposure in fish from the subarctic River Lakselva (Norway). Blood samples were drawn 30 min after angling and analysed for lactate abions, glucose, sodium ions, and pH. Reflex impairment was determined with orientation and tail grab reflex action assessment immediately after landing, after air exposure, and after 30 min holding. Blood physiology did not indicate an exacerbating effect of air exposure relative to just angling-induced exercise but significant and prolonged reflex impairment associated with the 120 s air exposure interval.

The conclusion was that anglers must take care to minimise air exposure to adhere to best handling practice and this was under 10s for European grayling at summer water temperatures. Similarly Bower et al. (2016) angled blue-finned mahseer using a range of bait/lure types and angling and exposure times. There were no cases of mortality observed, and the rates of moderate to major injury were low with 91% hooked in the mouth. More extreme physiological disturbances (blood lactate, glucose, pH) were associated with longer angling times. 33% exhibited at least one form of reflex impairment so that the conclusion here was that these fish were fairly robust to CR but that anglers should avoid unnecessarily long fight times and minimise air exposure to decrease the likelihood of sublethal effects that could contribute to post-release mortality.

However, Pope and Wilde (2004) found no effect on largemouth bass caught on plastic grubs. There was no difference in weight gain between caught and uncaught fish over a 40-day angling and recovery period which was much less than the 22% and 38% reported from previous studies. It seems that angling mortality varies with anatomical location in which the fish is hooked. All of the fish in this study were hooked in locations for which mortality is generally low (under 2%). Angling mortality was twice as great among fish that bleed from hooking wounds (only 3% in this sample). It is clear that CR mortality varies substantially among species and a number of observations suggest that sublethal effects are at least equally variable. This can be shown too from Thomé-Souza et al.'s work (2014) on the sustainability of sportfishing for peacock bass (Cichla spp.) in the Brazilian Amazon, where they found that fish caught through CR fishing had mean mortalities of only between 2.3 and 5.2% for three species. The fish lengths were between 4–26 and 79 cm but only fish under 42 cm died. They found that hooking was more dangerous to fish if it occurs in the throat or gills and no fish died with injuries to the lip, jaw, or mouth region.

Tiedemann and Danylchuk (2012) summarised the existing literature to develop five general trends that could be adopted for species for which no data are currently available: (1) minimise angling duration, (2) minimise air exposure,(3) avoid angling during extremes in water temperature, (4) use barbless hooks and artificial lures/flies, and (5) refrain from angling fish during the reproductive period. These generalities provide some level of protection to all species, but do have limitations. Therefore, a goal of conservation science and fisheries management should be the creation of species-specific guidelines for CR.

15.6 Impact of Invasive, Non-native Species

Inland waters in particular are often enhanced through stocking with introduced species, and this practice has been ongoing for a long time. Salmonids have had an impact on a variety of native fauna through predation and competition and have been implicated in reducing the diversity of macro-invertebrate assemblages and population declines and/or the reduction and fragmentation of the ranges of several species endemic to Australia and New Zealand. Predation by salmonids is also considered to have played a major role in the decline of the critically endangered spotted tree frog and possibly other frog species in South Eastern Australia, whilst the populations of at least two bird species in New Zealand (the crested grebe and the blue duck) have had their prey availability reduced by salmonids.

In the extensive fishless lake and stream habitats in the montane ecosystems of North America, salmonid fish have been introduced since the mid-1800s, and the fish stocking has been undertaken by the various agencies responsible for the management of fish and wildlife since the mid-1900s. Trout stocking has stopped in most national parks in western North America during the 1970s and 1980s but continues in other protected areas such as those managed by the United States Forest Service. These stocking programmes have transformed the formerly fishless aquatic ecosystems so that of the estimated 1600 naturally fishless lakes in the Western USA, 60% of all lakes and 95% of larger, deeper lakes now contain non-native trout and over 7000 mountain lakes are regularly stocked with trout, usually with the use of aircraft. This has been controversial as the fish introductions have dramatically altered the native vertebrate and invertebrate communities, often leading to the dying out of native fish, amphibians, zooplankton, and benthic macroinvertebrates (McGarvie Hirner and Cox, 2007). In a review of this process by Knapp et al. (2001), it has been shown that the spread of introduced trout from headwater lakes has had a disproportionately larger effect on native fishes than introductions lower in the drainage systems and that in many river basins the remaining populations of native fish are concentrated in headwater refugia, where they are already protected by natural barriers from introduced species that are already established at lower elevations. It has also been shown that in the formerly fishless, oligotrophic lakes, there is increased phosphorus and this new nutrient source results in increased algal biomass and production. Further it has been shown that the abundance of all life stages of long-toed salamanders and spotted frogs was lower in lakes containing non-native trout than in those that remained fishless.

However, although it is generally assumed that the introduction of a non-native species to an area with numerous endemic species would be deleterious to the recipient ecosystem, in several shallow New Zealand lakes, it was found that the addition of non-native, brown and rainbow trout did not alter the lakes' original invertebrate composition on any appreciable level (Wissinger et al. 2006). This is in contradiction to findings in North America and Europe which showed dramatic invertebrate compositional shifts after trout introduction. This illustrates that environmental characteristics of the lakes can determine whether the invasive species is detrimental or not and shows that the result of the invasion depends on more than invader characteristics.

Nevertheless the round gobies in the USA, which are bottom-dwelling fish introduced into the Great Lakes from Central Eurasia via the ballast water of ocean-going, cargo ships, show food chain predation (Fig. 15.3). They have competed successfully with native, bottom-dwelling fish, like the sculpins and darters. There have been major reductions in the local populations of these species where round gobies are established, and this impacts the food chain of recreationally important fish, like the smallmouth bass and the walleye. Whilst there is direct predation of darters and other small fish, they also feed on the

Fig. 15.3 Round goby (*Neogobius melanostomus*). Accidentally introduced into Great Lakes probably by ballast water transfer from cargo shipping. Originally from Sea of Marmara, Black Sea, Sea of Azov, and rivers of Crimea and Caucasus. Source: National Digital Library of the US Fish and Wildlife Service. Photographer: Eric Engebretson

eggs and fry of lake trout. Moreover they eat large quantities of zebra mussels, another invader which has been successful. Whilst this might be thought a good thing, there is a problem, because as filter feeders the mussels consume toxins/contaminants and as the gobies are preyed on by sportfish this can cause a direct transfer of the contaminants to these fish, such as the small mouth and rock bass, walleyes, and brown trout. The round gobies are nuisance competition too, and they fish aggressively and take bait from hooks. Anglers in the Detroit area have reported that at times they can catch only gobies when they are fishing for walleye.

Another problem fish is the northern snakehead (*Channa argus*) which is a top predator invasive species which is native to parts of China and possibly Korea and Russia and disrupts the natural aquatic feeding structure (Fig. 15.4). It was found in California in 1997 and is now established in many states on the east coast of the USA. Its presence is thought to have resulted from aquarium owners discarding unwanted exotic captive species into local waterways, although it could have been to create a local food source for recreational fishing. Many native species have been outcompeted for food sources:

Fig. 15.4 Northern snakehead (*Channa argus*). This is a top-level predator which poses a major threat to freshwater fish. Originally native to China, Russia, and North and South Korea but introduced to other regions where it is a major invasive species. Photo by Brian Gratwicke

fish, crustaceans, small amphibians, reptiles, and even some birds and mammals. During the spawning season and after the young are born, they can become very aggressive towards trespassing species.

Burgin (2017) reviews the introduction of the carp into Australia where since the mid-1800s it has become the most abundant large freshwater fish in South East Australia and the catch is around twice as great as any other species, with over 2 million annually. The impact of this species has been exacerbated due to the hybridisation of two taxa with the subsequent emergence of the vigorous Boolarra strain and the associated large increase in carp numbers. The impacts have been destruction of aquatic plants and an associated increase in water turbidity which result in reduced prey availability and further impacts on native fish that require sight, for example, for foraging. The carp tend to outcompete native species because of their greater abundance, high fecundity, robustness, and tolerance of a wide range of aquatic environments.

15.7 Indirect Impacts of Recreational Fishing

There are many, diverse, indirect impacts of recreational fishing that are not associated with the exploitation and harvesting of fish which have the potential to be detrimental to the long-term sustainability of both freshwater and nearshore ecosystems which now need to be considered, such as habitat disturbance, wildlife disturbance, or loss of fishing gear.

15.7.1 Disturbance of Habitats

Recreation fishermen have access to nearshore and littoral habitats which are of crucial importance for many fish species, and as interfaces between terrestrial and aquatic ecosystems, littoral zones fulfil a variety of physical and ecological functions. They delay or prevent the transport of nutrients to lakes from eroding upland soils, enhance local energy and nutrient availability fostering a higher biological productivity, support material and energy cycles and a variety of life-history strategies (Lewin et al. 2006). The diverse riparian zone processes affect biodiversity, reproduction, feeding, and predator-prey interactions. Woody debris and submerged or emerged macrophytes are refuge or feeding habitats for juvenile fish and invertebrates, and macrophytes serve as spawning substrate for phytophilic species (species associated with plants). At the same time, the alteration of littoral habitats by human activities has resulted in a loss of refuge habitats and resource heterogeneity which has affected fish communities by changing the fish species richness, biomass, growth rates, and the spatial distribution of fish.

15.7.1.1 Walking Tracks (See Burgin 2017)

Anglers can affect littoral habitats if they make paths to gain access to the water and walk parallel to the shoreline. A medium or heavy use of pathways and shores can change, or destroy, the natural plant communities of freshwater or marine littoral habitats. Anglers may also cut bank vegetation and remove submerged vegetation at the beginning of the fishing season. The removal of the aquatic plants and shoreline vegetation can affect phytoplankton development, invertebrates, fishes, and birds, enhance erosion processes, and change nutrient fluxes.

15.7.1.2 Impacts of Off-Road Vehicles

Tracks from the use of such vehicles could cause indirect impacts especially when used near the water's edge, for example, with the movement of watercraft and/or trailers and the carrying of fishing gear to the lake shore. Compared with other recreation activities, the effects of angling on the aquatic vegetation may be of minor importance but can still have ecological consequences.

15.7.1.3 Wading Associated with Instream Angling

Significant differences in egg and larvae development have been found where anglers waded. A single wading on shallow, salmonid-spawning habitats during the period before the hatching killed 43% of eggs and fry, while a twice-daily wading killed up to 96%.

15.7.2 Disturbance of Wildlife

Nearly all activities carried out on the shores are potentially disturbing wildlife who live in littoral areas or are sitting on the surface. The disturbances associated with recreational fishing originate mainly from direct contact, sound, and sight. Above all, water birds are closely associated with littoral and shoreline habitats. Therefore, the research on human disturbance of wildlife in aquatic ecosystems concentrates mainly on water birds. Quan et al. (2002) demonstrated that species richness and abundance on a highly exploited lake were correlated with human disturbance and not to habitat quality. Human disturbances, especially those caused by recreation activities, can affect distribution, species richness, and abundance of waterbirds by disturbing overwintering, resting and feeding, and reproduction (e.g. the prelaying phase and the egg and chick phase). The disturbance of feeding may be more pronounced if the feeding is restricted to certain places or time periods and can result in adults having insufficient time to fulfil their own energy demands and those of their chicks. The disturbances to the nesting birds can result in higher rates of non-hatching and abandonment, in the exposure of eggs to predators, or unfavourable environmental conditions, such as solar radiation, or thermal stress and may therefore decrease breeding success. Compared to other land-based activities such as bird watching, walking, or picnicking, shore angling is considered to have serious impacts on water birds, since anglers often use vehicles to gain access to the angling sites and remain there for long periods. Furthermore, they frequently show long periods of inactivity, interspersed with short periods of rapid movements. Liddle and Scorgie (1980) cite some studies which showed that activities by anglers substantially decreased the breeding success and breeding stocks of different water bird species. Anglers had a similar effect as boats on water birds, creating an area around them within which birds did not venture. According to this, Sudmann et al. (1996) observed a reproduction failure of breeding waterbirds in a reservoir during years when angling took place. The reproduction improved after termination of angling.

The avoidance and redistribution in response to human disturbances are species-specific.

Species that do not avoid disturbance may be affected by disturbance even more seriously, if they are forced to tolerate the disturbance in case suitable other alternative habitats are lacking. Other bird species may show an adaptation to recreation disturbances. The great crested grebes left their nest at shorter distances to approaching rowing boats, presumably as an adaptation to recreation activities. However, the short flight distances were disadvantageous, as the birds did not cover their eggs before leaving, so that the clutch was not protected from predation.

Few studies deal with the angling-associated, indirect disturbance of taxa other than birds. Angling may disturb otter (*Lutra lutra*) populations if there is a lack of sufficient refuges, such as dense woodland structures along river banks, especially if anglers prefer remaining cut-back trees and stumps as fishing sites.

15.8 Plastic in Various Forms

Plastic degrades extremely slowly or not at all and gets into the food chain. For example, Possato et al. (2011) noted that in an investigation into three catfish species in a tropical estuary in North East Brazil that individual catfish had ingested plastics (*Cathorops spixii* 18%, *C. agassizii* 33%, and *Sciades herzbergii* 18% of individuals). Nylon fragments from fishing activity played a major role in this contamination. Sigler (2014) also noted that small pieces of plastic had been found in fish stomachs and that it is of concern because these fragments may facilitate the transport of absorbed pollutants to predators within the food chain. Raison et al. (2014) further report that as the popularity and use of soft plastic lures (SPLs) by recreation anglers have increased recently, there are anecdotal reports of them being found in the digestive tract of a variety of fish species and in aquatic environments. Fieldwork was carried out in Charleston Lake (Eastern Ontario), a system known to have a SPL problem based on angler's reports, in lake trout and smallmouth bass. Snorkel surveys revealed about 80/km of shoreline/yr had SPLs and when immersed in water at two temperatures of 4 °C and 21 °C they showed little evidence of decomposition. Anglers interviewed reported 18% had found at least one ingested SPL when cleaning lake trout but when the lake trout were samples by gill net and smallmouth bass by rod and reel, there was only 2.2% and 3.4%, respectively, affected. What is needed is angler education to rig SPLs so that they are less likely to be lost during fishing, and the tackle industry needs to develop SPLs less likely to be pulled off by fish and/or degrade quickly. Meanwhile Skaggs and Allen (2016) reported that studies showing the occurrence or abundance of SPL ingestion by wild fish populations are rare but that there may be inconsistency in diet reporting. Hence the degree to which SPLs are ingested across fish species remains unknown. Studies showing the effects of SPL ingestion are also rare, but it is possible from one study that consumption of SPLs caused reduction in body weight and condition in brook trout (Danner et al. 2009). Regulations to restrict use of SPLs in order to protect fish populations and fisheries currently would not be based on good scientific proof of impacts, despite the consideration of a ban from Maine Department of Inland Fisheries and Wildlife in 2014.

15.9 Fishing Gear Loss

Attempts at broad-scale quantification of marine litter enable only a crude approximation of Abandoned Lost and Discarded Fishing Gear (ALDFG) which comprise less than 10% of global marine litter by volume, with land-based sources being the predominant cause of marine debris in coastal areas and merchant shipping, the key sea-based source of litter. The impacts of ALDFG can be:

- *Continued catch of target and non-target species.* This is called "ghost fishing." The state of the gear at the point of loss is important. For example, lost nets may operate at maximum fishing efficiency and will thus have high ghost-fishing catches and, if well anchored, be slow to collapse. Some abandoned or lost gear may collapse immediately and have lower initial fishing efficiencies, unless they become snagged on rock, coral, or wrecks where they are held in a fixed fishing position. Discarded gear or parts thereof would also have a low fishing efficiency. Fish that are killed in nets may also attract scavengers that are then caught in the nets, resulting in cyclical catching by the fishing gear.
- *Interactions with threatened/endangered species.* ALDFG, especially when made of persistent synthetic material, can impact marine fauna, such as sea birds, turtles, seals, or cetaceans through entanglement or ingestion. For example, a census by the Australian Seabird Rescue Group in the Richmond River (New South Wales) showed that of the 108 resident pelicans, 37 were suffering injuries from being entangled or hooked by fishing tackle and a later survey of Australian pelicans showed that 92% of human-induced injuries were from entanglement. Wells et al. (1998) concluded that the number of deaths or serious injuries to bottlenose dolphins in Florida from recreational fishing line entanglement exceeded that from the region's commercial fishing. Entanglement is generally considered far more likely a cause of mortality than ingestion, and, although rec-

reational fishing gear losses may be important, it is the commercial fishing gear that really causes the biggest impacts. While it is an important commercial gear, hook and line is also used by a large number of recreation and subsistence fishers, and therefore losses, especially within shallow inshore waters, may be very high. In the Florida Keys, it was reported that the debris type causing the greatest degree of damage was hook-and-line gear (68%), especially monofilament line (58%), and that it accounted for the majority of damage to branching gorgonians (69% of damage), fire coral (83%), sponges (64%), and colonial zoanthids (77%). This indicated that a gorgonian sponge-dominated reef would be more susceptible to damage from lost hook-and-line gear than coral-dominated reefs.

Asoh et al. (2004) assessed the extent of damage from monofilament fishing lines as a cause of cauliflower coral (*Pocillopora meandrina*) damage or death in fished and unfished areas in Hawaii. They found a positive linear relationship between the proportion of colonies entangled with fishing lines and the proportion of dead or damaged corals which indicated a negative impact on the health and survival of *P. meandrina* colonies.

15.10 Boat Strike and Boat Traffic Impacts

Wolter and Arlinghaus (2003) categorised the effects of boat traffic on fish, most notably on fish larvae, into direct and indirect stressors. Direct effects were caused by the physical forces generated from moving boats, directly related to fish mortality (propeller action, waves, wash waves, and dewatering). Indirect effects result from stress, disturbances which prevent fish from feeding or nest guarding, dislodgement of eggs or larvae, an increase of turbidity, or a loss of macrophytes following wave action. Their review deals mainly with the impacts of commercial navigation in waterways, and some effects may be restricted to large vessels. However, wave effects may also result from recreation boat traffic. Experimental studies have shown that shear stress can increase the mortality of eggs and larvae from different fish species, but given the smaller size of the boats used by recreation fishermen, impacts of shear stress, stranding, and dewatering following wave action may be less important. But obstructing nest-guarding behaviour and dislodgement and redistribution of eggs and larvae may affect the fitness of fish populations. Small recreation boats travelling at slow speed near the nests drove males of longear sunfish (*Lepomis megalotis*) from the nests, thus increasing the likelihood of egg predation. Boats moving at higher speeds increased the turbidity and therefore the possibility of the predation success. The passage of even a single paddle or motorboat over low-density organic mud can lead to a resuspension of sediments and if the frequency of boat traffic is sufficient during the season, it may result in an increase of turbidity. An increase in turbidity beyond the natural level may have physiological effects (gill trauma, sublethal stress response) and behavioural effects (avoidance, predator-prey interactions). Additionally, the increased turbidity may contribute to a loss of macrophytes in littoral habitats, and macrophytes serve as colonisation substrate for various species and as feeding and refuge habitat for juvenile fish.

Motorboat traffic in rivers, lakes, and along the coastline results in the emission of inorganic and organic compounds into the water and into the air near the surface, which is toxic to zooplankton and fishes. Also in marine ecosystems, the engine emissions from outboard motors can contribute to the surface microlayer, and the toxic substances on the air-water interface can significantly affect the survival and development of early life-history stages of marine fishes and other surface-dwelling organisms. However, even if it is not possible to quantify the effects of boat traffic linked exclusively to recreational fishing, given a substantial level of boating activity, there could be some negative effects on the aquatic environment or fish stocks, whereas the effects depend on motor type, travelling speed, bottom structure of the ecosystem, or slope of the shoreline. However boat strike is the single biggest cause of marine turtle mortality in Queensland and between 12.8% and 48.5% sampled had injuries consistent with propeller strike, and in Florida it has been reported as a significant source of manatee mortality.

15.11 Nutrient Input

Groundbaiting or chumming is widely practised in freshwater by some anglers to attract fish, such as the cyprinids bream, carp, or tench to the angling site (Arlinghaus and Mehner 2003). Groundbaiting up to a certain limit can be effective in increasing the carrying capacity of the fishery and the catch of cyprinid fishes. Higher groundbaiting rates can negatively affect the catch though, and the existence of an upper limit may result from negative impact of not consumed baits on the water quality and invertebrate community. Groundbaiting over the entire fishing season may lead to significant changes in the benthic invertebrate community. The rapid breakdown of cereal baits by microbial activity resulted in a high oxygen consumption of the sediment, and presumably, as a result of the alteration of the microbial and chemical conditions through the decay of uneaten baits, the densities of naididae, cyclopoidae, and cladocera decreased. Only tubificidae showed no reductions in density. In general, the lack of oxygen on the sediment surface can result in diminished decomposition rates, causing an accumulation of organic surplus. The decay of this organic matter can enhance the ammonium flux from the sediment and initiate the redox-dependent release of iron-bound phosphorus, therefore contributing to the internal nutrient loading. At the same time, nutrients especially phosphorus from the egestion or excretion of fish after having fed these baits or from uneaten baits may substantially contribute to anthropogenic eutrophication and therefore either directly or indirectly enhance primary production and algal growth (Arlinghaus and Mehner 2003). On the other hand, the angler harvest can counterbalance the nutrient input from groundbaiting. However, such a harvest rate may be unrealistic as many specialised anglers mainly practise CR fishing. However, the contribution of groundbaiting to an anthropogenic eutrophication is strongly dependent on local conditions. Water depth, trophic state, effective nutrient load and loading history, water retention time, as well as fisheries-connected factors (harvest rates, digestibility, and nutrient composition) affect the impact of the groundbaiting on the water body. Small, shallow, oligotrophic lakes with long water retention times, high angler densities, and low harvest rates may be sensitive to groundbaiting (Arlinghaus and Mehner 2003). Preservatives that may leach from commercial baits have received little scientific attention. The impacts of effluent discharge from shore-based recreation facilities like detergents and chemical toilet discharge can cause extensive pollution, but those associated with recreational fishing are an unknown fraction of the total (Burgin, 2017).

15.12 Exotic Species of Bait and Bait Gathering Effects

With the popularity of recreational fishing, the demands for live bait rise. Some studies on marine coastal habitats have shown that bait digging can locally influence the littoral fauna and size structure of harvested, benthic organisms. Some of the species intensively used have a role in structuring the bottom communities. Therefore, an intensive harvest affects not only the harvested species but other components of the fauna, as well as bacteria and algae. For example, such cascading effects can result from an intensive collection of sandprawns (*Callianassa kraussi*). Ghost shrimps (*Trypaea australiensis*), a popular bait species in Australia, changed the distribution and abundance of other benthic taxa (polychaetes, amphipods, soldier crabs). In addition, the reduction of benthic organisms (cockles or worms) may potentially affect the behaviour and foraging success of higher trophic level species, such as shorebirds. The bait digging or pumping and the associated trampling can involve a considerable disturbance to the sediment and affect taxa, sensitive to disturbance of the sediment structure.

There is some evidence that the intensive bait digging for lugworm (*Arenicola marina*) and ragworm (*Nereis diversicolor*) reduced the abundance of cockles (*Cerastoderma edule*). The digging can lead to a burial of many cockles and to a surface exposure of some other species. In addition, bait collecting can affect not only the biological but also the physical and chemical sediment parameters. Bait pumping and trampling changed the porosity, organic carbon content and redox potential of the sediment and increased the chlorophyll concentration. There are also indications that the perturbation of the sediment through intensive digging influences bioavailability and uptake of heavy metals (lead and cadmium) by polychaetes.

The intertidal and subtidal boulders on rocky shores exhibit a diverse assemblage of sessile and mobile fauna. Consequently, a frequent sampling can cause short-term effects on the abundances of sessile organisms. For example, the collection of mussels used as baits by anglers significantly reduced cover, density, biomass, and size of the mussels (*Mytilus californianus*) on rocky shores even during a period of high natural disturbance.

The use of exotic species as bait can be a threat to the coastal ecosystem, and the introduction of exotic species resulting from the release of certain baits has been well documented. Also to keep them alive and moist, the live bait can be packed with living substrates, like algae which can then be discarded. These can contain living organisms such as small crustaceans, snails, and worms which may establish themselves in a new ecosystem. It has also been shown that live or dead bait can transfer viruses that can significantly affect wild fish stocks. An example here is viral haemorrhagic septicaemia virus (VHSv which was found in Canada's Lake St. Clair in 2003 and has spread throughout the Great Lakes). The virus affects a wide variety of fish species and has killed 28 freshwater species since 2006, yet more than 50 species may be susceptible. It rapidly leads to internal bleeding and haemorrhaging from open sores (Fig. 15.5). It seems to be spread through fishing bait, and bait dealers around the Great Lakes have to certify that the fish they use as bait are disease-free.

So we can see that the invertebrate harvest for bait constitutes an important component of angling's ecological footprint and the concern can be illustrated by a new law approved in 2006 in Croatia which prohibited fishing with live bait.

Fig. 15.5 Gizzard shad with VHSv, a deadly infectious disease which causes bleeding. It affects over 50 freshwater and marine fish species in the northern hemisphere. Source: https://open.nim.nih.gov/detailedresult.php?img=PMC3386630

15.13 Pathogen Transmission

Virus strains have been spread from Asia to North America, for example, the whirling disease, a parasitic infection affecting young trout and salmon which enters the fish's head and, due to pressure, makes the fish swim erratically. In Australia native species such as the Macquarie perch are vulnerable to a fatal infection by the epizootic haematopoietic virus which may be carried by introduced redfin and trout. Through fish introductions such as carp, there has been established the Asian fish tapeworm which infects native fish. In Europe nearly 100 known pathogens originating from a wide range of taxa have been introduced to European freshwaters, and, although aquaculture was likely to be the main source, recreational fishing was also a pathway.

15.14 Inadvertent Overland Dispersal of Non-native Plants

The movement of vehicles, boats, and trailers between freshwater bodies has the potential to support dispersal of organisms, including the non-native invasive aquatic weeds like alligator weed and salvinia, which seem to have been transported long distances in Australia. These types of plants could result in hydrological changes as they can form dense stands that blanket the water, affect other species of plant, and ultimately can impact on water chemistry and quality, faunal and floral diversity, and ultimately freshwater fisheries. The movement of such fishing-related vehicles in North America seems to have played a role in the continued dispersal of invasive aquatic species and appears to be a vector in weed transmission.

15.15 Management of Recreational Fishing Impacts

There are a whole range of organisations which have at least as part of their brief to attempt to manage some of the impacts of recreational fishing that have just been described. These range from voluntary codes of practice, fishing clubs, to state and government laws to regulate fishing. Various types of zones and protection areas have also been established to regulate recreational fishing. The traditional regulatory options imposed by government agencies, such as harvest and gear restrictions, represent the standard in recreation fisheries management, at least in developed countries, but there are other methods discussed by Cooke et al. (2013) which hold out great promise because they involve the recreation anglers themselves in the practices, and Cooke et al. (2014) suggest that these types of informal institutions/practices may be as effective as formal regulations when addressing fishing management issues.

For example, there is a Standard for National Environmental Assessment of Tournament Fishing set up by national organisations, such as NEATFish organised by RecFish Australia (www.neatfish.com), where the participation is voluntary, but the Standard must be adhered to by organisations of all tournaments which claim certification under NEATFish (Neatfish 2009). It is based on a five-star model which classifies fishing tournaments on their environmental, social, and economic impacts. In its development it drew on two national initiatives, the National Code of Practice for Recreational and Sport Fishing and the National Strategy for the Survival of Released Line Caught Fish. RecFish Australia consulted widely with stakeholders in the recreational fishing industry, including national fishing agencies, recreation fishers themselves, state and government fisheries authorities, research organisations, and consultants in natural resource management. The purpose is to recognise that tournaments have a potential impact on fish stocks at specific locations and on particular species. Other issues such as fish welfare in CR tournaments needed to be considered. The Environmental Assessment includes outcomes that there are no adverse impacts on the sustainability of fish stocks, a minimisation of detrimental impacts on the environment, and the provision of useful data to fisheries research and management.

Some of the alternatives include the use of angler education programmes that attempt to evoke voluntary changes in angler behaviour, resulting in the emergence of voluntarily motivated, resource-conserving, informal institutions. These "softer" approaches to aquatic stewardship and fisheries management can be developed in cooperation with stakeholders and in many cases are led by avid anglers and angling groups. Examples of such measures include voluntary sanctuaries, informally enforced seasonal closures, personal daily bag limits, self-imposed constraints on gear, development of entirely live-release fisheries, and adoption of fish and aquatic ecosystem conservation-oriented gear and release practices. Education efforts that provide anglers with knowledge on best practices and empower them to modify their behaviour hold great promise to meet formal management goals and objectives but seem to be underutilised relative to formal regulations. Cooke et al. (2013) highlight the benefits and challenges of relying on informal institutions as alternatives to traditional regulatory options but informal institutions that protect resources and help overfished stocks recover hold great promise in both developed and developing countries, particularly when there is a single stakeholder group or when the capacity to enforce traditional regulations or to invest in stock assessments is limited. Informal institutions may help make formal institutions more effective or can even be alternatives to costly institutions that depend on enforcement to be effective.

However, sometimes regulations are needed, for example, in Quebec a regulation prohibited the targeted angling of all redhorse and suckers in regions where the endangered copper redhorse (*Moxostoma hubbsi*) was present. This was because it was extremely difficult to distinguish this species from other redhorse and sucker species. An educational campaign alone was deemed not likely to succeed so a more sweeping ban was thought to be needed.

Veiga et al. (2013) in Portugal suggested that the existence of fishing regulations is a good starting point for effective management, despite the lack of acceptance and detailed knowledge of the regulations in place by fishers based on almost 1300 interviews before and after the 2006 restrictions to control recreational fishing harvests. This may result in lack of compliance and hinder the success of recreational fishing regulations in Portugal. This was also the case in British Columbia where Lancaster et al. (2015) found a lack of knowledge related to rockfish conservation measures, and they suggested that public outreach and an educational campaign was necessary. Alós et al. (2009) suggested the use of shrimp as bait as opposed to worms to reduce the catch of undersized fish and the incidences of deep hooking. Here managing bait type might complement standard harvest regulations and facilitate more sustainable exploitation rates.

Bag or creel limits aiming to regulate the harvest of individual anglers per fishing event, or angling day, are widely used and can successfully limit the angler effort (Beard et al. 2003). However, bag limits may not be sufficient to limit total harvest (Cox et al. 2002) which may be related to the fact that they may restrict the harvest by the individual anglers but often do not restrict neither the amount of anglers nor the total harvest. In addition, bag limits may affect only the catch of the experienced anglers, because many anglers do not catch their bag limits. High bag limits may increase the attractiveness of a lake to the anglers and may set a target. As the angling satisfaction is linked to the catch and influences the management preferences of anglers, the dissatisfaction following unrealistic expectations may reduce the acceptance of sustainable management practices, such as habitat management (Arlinghaus and Mehner 2005). If the angler effort varies with the bag limits, independent of quality or density of the exploited fish population, high bag limits may fail to protect the fish population. In case the bag limits are higher than the biological capability of the fisheries, lowering the bag limits is often suggested to prevent over-exploitation. However, although a lowering of bag limits may be meaningful from a biological point of view and may work educationally altering the perceptions on fishing success and reminding anglers that their resource is not unlimited, some aspects may counteract the effects of lower bag limits. Low bag limits may lead to the replacement of small fish with larger fish after the bag limit is reached.

Quotas on the total recreation catch are thought to be impractical but currently New Jersey anglers are permitted to harvest two fish per day with a minimum size of 28 inches and an additional fish at a minimum size of 28 inches, if the angler obtains a bonus permit from the New Jersey Division of Fish and Wildlife (NJDEP 2012).

Widespread restoration of habitats is probably not a realistic aim when we consider the range and extent of environmental impacts that have occurred in lake and river wetland habitats and in the nearshore marine environments, but Lewin et al. (2006) discuss the possibilities. The best way forward here would be voluntary involvement of the recreational fishing community on small-scale restoration schemes. This has been stressed by McPhee (2017) when outlining the importance of recreation fisheries in Australian urban coastal cities. Recreation fishers can be important drivers for improvements to urban coastal environments that are subjected to cumulative stressors. Typical fisheries management frameworks and management objectives are not optimal for recreation fisheries and, in particular, urban recreation fisheries. Although a number of specific traditional fisheries management tools, such as minimum legal sizes and gear restrictions, remain relevant, they are insufficient however for the full benefits of recreational fishing to be realised. There are important issues that affect the fishery which are outside the traditional fisheries management frameworks. Major Australian coastal cities should have specific Urban Fisheries Management Plans that recognise the specific issues associated with urban recreation fisheries. These plans should coordinate within and between levels of government and have clear management objectives relevant to urban fisheries. These plans need to incorporate opportunities where relevant for habitat restoration, or habitat creation, as well as necessary infrastructure support which can enhance the recreational fishing experience. Urban recreation fisheries represent a substantial catalyst for habitat restoration activities in particular, which can have wider benefits for aquatic conservation. Stock enhancement is also a relevant potential tool for urban fisheries management. Citizen science opportunities are significant within the scope of urban recreational fisheries and are a chance for stakeholders to take greater stewardship of the local resource and collect valuable monitoring information in a cost-effective manner. Overall, Urban Fisheries Management Plans are a substantial opportunity to make fisheries management more holistic and more focused on end-user requirements, without compromising the resource base.

Sutinen and Johnston (2003) have outlined how angling management organisations could be important in the integration of the recreation sector into fishery management, but because of the multiplicity of regulations related to recreational fishing in particular, this does not seem to have occurred. As Sanchirico et al. (2010) describe, there needs to be comprehensive planning, dominant-use zones, and user rights, rather than the 321 regulations related to recreational fishing and 226 spatially explicit regulations in state and federal waters on the Californian coast (1 January 2005). Within these figures catch-limit regulations that apply to a single species, such as bag limits or boat trip limits, are not included in these figures because they are so numerous. They do not combine to create a coordinated set of interlocking regulations but rather they are a cluster of single species and single gear-type regulations that have little relationship to other regulations. There is similar fragmentation in the regulations in Massachusetts so a much more integrated set of coastal and inland fisheries plans seems to be the next stage for future fishery management. In fact as long ago as 2002 Dayton et al. suggested that the way to resolve the recreational fishing management problems was to adopt a proactive, precautionary management regime based on planning and marine zoning. This has only slowly been developed.

So far marine protected areas (MPAs) have been established as the most immediate and most effective means to conserve threatened marine ecosystems, and they have been established in highly productive coastal environments, like estuaries and reefs, many in the same habitats frequented by recreation anglers. These areas are associated with and relative to unprotected areas, increased species diversity, biomass, organism size, and density. They serve as a protection for vulnerable species and habitats and can export biomass to surrounding waters. The most common approach is to prohibit all extractive or consumptive activities that result in the harvest of organisms. They are known as "no take" MPAs and the reserves that permit recreational fishing show differences in population structure and abundance from those where no fishing is permitted (Cooke and Cowx 2004). These types of MPAs circumvent the problem of non-compliance by anglers with management measures because infringements are clearly visible (McPhee et al. 2002). They have not been popular in Australia compared with South Africa, and McPhee et al. (2002) suggested several reasons for this.

As an example of what can happen by removing recreational fishing from MPAs is the Poor Knights Islands marine reserve in New Zealand. This was closed to commercial fishing in 1981, but by 1998 it was clear that there were not significant benefits for protecting snapper populations and the reserve was made fully protected. The result was that the snapper population increased to 14 times their previous abundance in just 5 years. There was noticed that there were seasonal variation in snapper numbers too which means that some large fish were migrating to the surrounding waters where they were accessible to fishers. The environmental impact of the depletion of large fish had been that sea urchins had thrived and reduced the kelp cover over the rocks. This was thought normal until the marine reserve was established, and within a few years renewed populations of large snapper and crayfish had eaten many of the urchins, allowing the kelp to regrow and increase the productivity of these coastal waters. Similarly at Rottnest Island (Western Australia) when a marine reserve was established, the density of lobsters was about 34 times higher, and the density of lobsters above the minimum legal size was around 50 times higher than in other areas around the island, where recreational fishing was allowed. The mean carapace length, the total biomass, and egg production of lobsters in the reserve were significantly higher than in the adjacent fished areas. An alternative approach would be to restrict recreational fishing to particular zones and/or to mandate CR or to permit fishing for certain species in MPAs that can be clearly targeted and do not involve substantial by-catch. More detailed discussion can be found in Cooke et al. (2006). The usefulness of partial MPAs that implement some form of fisheries management regulations, but do not ban fishing and the take of fish entirely, has been questioned due to its perceived limited conservation benefits. However, Alós and Arlinghaus (2013) provide empirical data demonstrating fish conservation benefits of partial MPAs when the stocks in question are mainly exploited by recreation angling. They studied a multispecies recreation fishery from the Balearic Islands (Mediterranean Sea) comparing three kinds of spatially close, managed areas. The implementation of a partial MPA decreased the fishing pressure attracted, and the protected areas hosted greater abundances and larger-sized fish compared to areas of open access. Possibly the greatest conservation benefit of partial MPA resulted from the reduced fishing effort attracted, likely as a result of aversion of anglers to use areas where some form of management is affecting the recreation experience. In addition, the constraints on artisanal fishing may also have contributed to the conservation benefits found. Depending on the right social and ecological context, partial MPA may therefore work as expected.

Another strategy might be to provide local communities and stakeholders like anglers and guides, with a bigger role in determining the goals for MPAs. In some cases only recreational fishing is involved, and an approach is the establishment of voluntary sanctuaries where community stakeholders promote sustainable fisheries through education (Suski et al. 2002). Another approach would be to have zones that prohibit or mandate certain fishing gear, such as barbless hooks or live bait. If clearly defined and combined with an education programme, such regulations can be clearly enforced and need testing in MPAs (Smallwood and Beckley, 2012).

The Magnuson-Stevens Fishery Conservation and Management Act sets strict, scientifically adjusted, annual catch limits on US commercial, charter, and recreation fisheries in order to sustain saltwater fish stocks. It is seen as a model of fishery management globally. However, a new piece of legislation going through the American senate at the moment may effectively deregulate saltwater fishing to a large degree. This is officially known as the Modernizing Recreational Fisheries Management Act of 2017 (Ortolani, 2017). It has been praised by sportsmen, boating and outdoor organisations, but it has also drawn strong opposition from conservationists and some commercial and charter fishermen, and critics say that the new act would muddy the waters between federal and state management and allows political and economic considerations to override science in management decisions. There appears to be a major loophole in that annual recreation catch limits would be no longer required for stocks whose fishing rates were being maintained below their federal target and annual catch limits would be removed for fisheries in which overfishing is not occurring. Currently the status of US fisheries is annually assessed by the NOAA which tracks 474 stocks or stock complexes where fish are grouped for management. In 2016 the overfishing list included 30 stocks and stated that 444 stocks were not at present overfished. Under this Act those stocks would lose their current requirement for science-based, sustainable annual limits on catch for recreational fishing. When the NOAA set the shortest recreation snapper season ever in the Gulf of Mexico (three days in June 2017), recreation fishers lobbied the Trump administration. The Commerce Department issued a rule that permits overfishing of red snapper in the Gulf of Mexico by private anglers while acknowledging that it will delay the stock rebuilding schedule for the species by six years. This overrules good science and could eventually reduce fishing opportunities.

15.16 Education Related to Recreational Fishing Impacts

15.16.1 Trade Sector and Recreation Fisheries Conservation

The recreation angling community is composed of diverse stakeholders, including the trade sector responsible for the manufacturing, distribution, and sales of tackle, boats, and clothing, angler-based travel, revenue-generating popular media, and angling services. Through marketing and promotion, fishing companies compete for customers by convincing anglers as to what success means when they go fishing. If the angling trade can influence the social norms in the recreation angling community, then this could hold true for norms related to the conservation of recreationally targeted fishes and their habitats. Danylchuk et al. (2016) questioned whether individuals working within the fishing trade are adequately informed about best practices for recreation fisheries conservation, since these perceptions could, in turn, influence the values portrayed in the marketing and promotion of fishing. They surveyed fishing trade employees during five industry and consumer shows to evaluate their perceptions about recreation fisheries conservation and where they believe their consumers learn about these issues. Across events, respondents believed that commercial fishing and habitat loss were the greatest threats to recreation fisheries. Specific to the angling event, physical injury when handling (e.g. during hook removal) and duration of the fight were selected as having the greatest impacts on fish, with between 74% and 91% of respondents indicating that they felt impacts were species-specific. Respondents believed that their customers received information on best practices and conservation predominantly from peer-to-peer interactions, social media, and fishing magazines. They also indicated that one of the primary roles of the angling trade when it comes to recreation fisheries conservation is to convey best practices in marketing and promotion. Overall, the trade sector appears to be an important mechanism for reaching anglers, yet more work is needed to ensure that the conservation information they share is consistent with science-based, best practice and this needs to be pursued in the future.

15.16.2 Mandatory Education Programmes

Associated with licencing in MPAs, this has been suggested as it may be an effective strategy for ensuring that anglers understand the purpose of the MPA, how to minimise CR mortality, and how to minimise their footprint on the environment (Cooke et al. 2006), and it has been successfully used in Germany and Switzerland related to the licencing process. So formal course training should be required to obtain fishing licences prior to granting permission to fish for endangered species (Cooke et al. 2014). It could also be a mandatory requirement to hire fishing guides/charter captains when targeting endangered species. This might ensure proper handling, compliance with regulations, and the date and control of the fishing effort as long of course as the guides are adequately trained in, and committed to, conservation of best practices. This would require fishing guide certification programmes and could be an effective way of fostering respect for endangered fish (Cooke et al. 2014).

15.16.3 Best Practices Guidelines for Catch-and-Release in General and, As an Example, Guidelines for Striped Bass Catch-and-Release

There has been much research related to best practices that anglers should carry out in this vital aspect of fish conservation and fish welfare, and it has been suggested that what needs to be developed are techniques which are species-specific (Cooke and Suski 2005; Cooke and Shramm 2007; Nguyen et al. 2013; Gagne et al. 2017).

15.16.3.1 Techniques to Increase Survival of Released Fish

Anglers control many factors that can exacerbate stress imparted on striped bass that are caught and intended to be released (Tiedemann and Danylchuk 2012). For example, when fishing for striped bass, anglers should use appropriate weight-class tackle that allows fish to be brought in quickly to reduce exhaustion and minimise stress. Other angler-controlled factors include terminal tackle type, playing time, landing, handling and unhooking techniques, and release methods.

15.16.3.2 Terminal Tackle Type

Terminal tackle type, including the number and style of hooks and the type of bait used, can affect anatomical hooking location and the likelihood of physical injury to organs and tissue from hook wounds. Two types of hooks that are known to reduce injury and mortality of released striped bass are barbless hooks and circle hooks. In addition, corrodible, non-stainless steel hooks are encouraged. When fishing with plugs and lures with multiple treble hooks, removing one or two sets of hooks or replacing them with single hook should be considered. Single, barbless hooks are even better, as they reduce tissue damage and handling stress because they can be quickly and easily removed. In general, then fishermen should use single barbless hooks whenever possible, or crimp, bend, file, or flatten the barbs on the hooks to make them easier to remove. When fishing with natural or live bait, non-offset circle hooks should be used to minimise gut hooking and the chance of lethal wounding of striped bass to be released. The unique shape and hook point location of a circle hook ensures that when a fish takes a bait and continues to swim, or make a turn, the hook pulls until the point catches the fish in the corner of the mouth. This causes minimal damage, reduces the chance of lethal wounding, and makes it easier to unhook and quickly release the fish. Even if a circle hook is swallowed by the fish, it will slide out of the stomach when the fish moves off with the bait. As the line is pulled through the fish's mouth, it guides the hook around the jaw where it locks in place.

15.16.3.3 Playing Time

The longer a fish fights, the higher the stress level and greater the chance for exhaustion and physiological disturbance which reduces the chance of survival after release. When a strike is felt, the hook should be set quickly. Setting the hook as soon as you feel a strike will help prevent the fish from taking the hook deep in its throat where it may cause internal organ damage and be hard, or impossible, to remove. Once a fish is hooked, it should be landed quickly, rather than playing it to exhaustion. A fish brought to the boat or shore quickly has a much better chance of survival after release than one that has been exhausted by a lengthy fight.

15.16.3.4 Landing and Handling Techniques

If at all possible, striped bass should be kept in the water while hooks are removed. If a fish must be removed from the water to unhook it, always try to minimise the amount of time it is kept out of the water, handle the fish as little as possible, and release it quickly. Avoid using gaffs to land striped bass that are going to be released. In a jetty situation, if a gaff must be used, gaff fish in the jaw or corner of the mouth only.

When using a landing net, use a net with small mesh made out of rubber, knotless nylon, or other soft nonabrasive material rather than a large mesh polypropylene landing net. These materials remove less slime and reduce potential wounding. Make sure the net basket is shallow and of sufficient circumference so that it does not bend the fish severely. If a fish must be removed from the water, refrain from holding the fish in a vertical position to avoid displacing or stressing internal organs. If you are bringing a striper onboard a boat using a lip gripper or other landing device to hold the fish while you remove the hook, grab the fish by the lower jaw. However, do not lift the fish clear of the water with the gripper to avoid placing the fish's entire body weight on the jaw. Hold fish horizontally by firmly gripping the lower jaw with one hand and gently supporting its weight under the belly with the palm of the other hand. Once a striper is landed, keep it from thrashing around and injuring itself. Stripers can be calmed down by covering their eyes and head with a wet rag, or towel, or by turning them on their back. When unhooking a striped bass, handle fish carefully using wet hands, wet cotton gloves, or a wet towel to minimise removal of the fish's protective mucous layer. Striped bass have a protective mucous layer that prevents disease and infection from entering through the skin. The more a fish is handled, the more of this protective slime that is removed. Avoid touching or injuring the eyes. Never touch the gills or insert your hand into a gill cover to hold a striper as this will damage the gills and impair the fish's ability to breath.

15.16.3.5 Unhooking Techniques

Striped bass should be unhooked quickly and carefully in the water whenever possible, to reduce stress and the potential for injury or post-release mortality, especially when air temperature is much higher than water temperature. Do not tear tissue when removing the hook. If a hook is embedded in a fish's throat or difficult to remove by hand, use a proper dehooking tool for hook removal, such as long-nosed pliers, hemostats (forceps), or a commercially available hook removal tool. Do not forcefully remove the hook if you cannot see it or if it appears that you may cause greater harm to the fish by attempting to remove the hook when a fish is hooked deep in the throat, or stomach, or hooked in the gills. Cut the leader as close to the eye of the hook as possible and leave the hook in the fish. There is evidence that fish are capable of rejecting, expelling, or encapsulating hooks by secreting an inert matrix of calcified cellular material.

15.16.3.6 Release Methods

Fish in good condition should be quickly and gently returned to the water head first in an upright position. Fish that are stressed by the fight, or handling and unhooking, should be revived prior to release. Exhausted fish can be revived by holding them head first into the current, or direction of the seas, in the swimming position with one hand under the tail and the other under the fish's belly, or by grasping its jaw between the thumb and forefinger. Gently move the fish to get water flowing through the mouth and over the gills. Use a figure-8 pattern to always keep the fish moving forward and never move the fish backwards. When the fish is revived, let it swim away on its own. Do not let the fish go until it clamps down on the thumb or is able to swim strongly and freely out of the grasp.

There is growing interest in educating anglers on CR best practices, yet there is little information on whether angler education programmes yield measurable improvements in fish condition and survival. As such, Delle Palme et al. (2016) conducted a study focused on mixed-gender youth groups (aged 8–10) and contrasted three levels of training intervention. Treatment 1 training had no mention of CR best practices. Treatments 2 and 3 training involved visual aids to illustrate best practices, while Treatment 3 added a hands-on demonstration. When caught by the most highly trained participants, fish experienced the least amount of air exposure but were handled for longer periods, as trained anglers were more careful. Higher levels of training led to a higher likelihood that anglers wet their hands and used a bucket filled with water while handling fish but all treatment groups yielded similar incidences of deep hooking and bleeding. Overall, mortality (initial and after ~12 h) was low across all treatments. These findings suggest that a short (~20 min) fishing workshop can transfer information on CR practices, at least in the short term, that can lead to some improved conditions for angler-caught fish. It is unclear the extent to which this information is retained in the long term, or how different target populations or training strategies might influence knowledge transfer and adoption and thus biological outcomes. With growing interest in sharing CR best practices with anglers, it was suggested that there is a need for additional research on outreach strategies to ensure that such efforts are effective and yield meaningful benefits to fish welfare and conservation, and it has been suggested by Cooke et al. (2017) that anglers need to be involved much more in the science and practice of CR science.

15.17 Behavioural Response of Anglers to Management Actions

A primary concern in managing recreation fisheries is the behavioural response of anglers to management actions. Use restrictions on public sport fisheries are often necessary because the demand for a superior fishing experience in terms of catch puts pressure on fish populations and the sustainability of aquatic systems. Efficient fisheries management requires that agencies be able to anticipate angler reactions to new fishing regulations considering they ultimately alter the attractiveness of the affected fishing opportunity (Beardmore et al. 2011). Furthermore, fisheries agencies may also be concerned about the impact of regulations on participation because most of their revenues are derived from licence and equipment sales. Klatt et al. (2014) present a model of anglers' reactions to regulations designed to slow the spread of an aquatic infectious disease. This is a critical issue for fisheries managers because aquatic diseases (and, similarly, invasive species) tend to suppress catch rates by increasing the mortality and altering the behaviour of fish. Furthermore, anglers that travel between different lakes and rivers become an important vector through which aquatic diseases can spread and thereby trigger reductions in the fishing quality of an entire region. Yet there is little research that addresses disease regulations in fisheries.

The results demonstrated that anglers significantly alter their behaviour at the site choice and participation levels in response to a new disease and its regulations. Specifically, it was found that disease regulations implemented by the Michigan Department of Natural Resources to slow the spread of VHSv have had an impact on angler behaviour for areas where the virus is present and most heavily regulated. Anglers were less likely to visit a site considered to be VHSv positive and subject to bait use restrictions and more likely to choose a site free of disease regulations. This suggests that the VHSv regulations have been successful in reducing the opportunities for the disease to be spread by anglers. Furthermore, there is evidence that the presence of these regulations affected how anglers fished the Great Lakes through their choice of bait, but not through boat use.

To be clear, there is no absolute clarity explicitly distinguishing among two possible effects driving these results: the influence of VHSv on resource quality and the influence of the disease management zone restrictions on angler actions per se. Moreover, the extent to which the regulations have prevented damage to the fishery by limiting the spread of VHSv cannot be measured.

In Europe research and policy debates point to the need to increase efforts to rehabilitate or restore habitat structure and function, at the expense of a traditional recreation fisheries management approach which is to intensively stock fish.

Arlinghaus and Mehner (2005) consider that rehabilitation of habitat on larger scales can be considered as the most sustainable recreation fisheries management strategy. This particularly applies in densely populated countries such as Germany, where most aquatic ecosystems have experienced anthropogenic disturbances dating back several centuries. Although habitat-orientated recreation fisheries management offers solutions to many management problems, major advances in research and training, restructuring of institutions, and support from all stakeholders, including the public, are needed. Anglers are amongst the key players in this shifting fisheries management policy as they are often users and at the same time managers in Central Europe, but other stakeholders are clearly equally important, like water management authorities or land owners. Arlinghaus and Mehner (2005) suggest that multiple factors are responsible for anglers being orientated towards a more sustainable habitat management as opposed to a less sustainable stocking management approach. These factors offer insights into paramount variables that might be targeted by managers. Obviously, the most promising way is to increase the pro-ecological values and attitudes of anglers, for example, by appropriate education outreaches. The most efficient ones will include anglers in habitat management project design, implementation, and evaluation, which ensure that anglers directly experience potentially positive effects of habitat improvements or alternatively possible negative effects of habitat modifications. Personal experiences might be judged as one of the most effective ways of environmental education.

There are also situations where continuous stocking seems appropriate, for example, in artificial fisheries, or to preserve fish at risk of extinction. One paramount factor that emerged in their analysis was that to satisfy anglers, what was essential was catching fish to meet catch expectations. The most straightforward implication might be that angler satisfaction should be enhanced by increasing the catch quality, which in turn would increase the probability of habitat management as opposed to stocking to be supported by anglers. This might most easily be achieved by simply improving the effectiveness of traditional inland fisheries practices, such as appropriately planned harvest regulations, closed seasons, or promotion of CR practices. However, increasing the fish stock quality as perceived by the anglers may not always increase the catch quality for individual anglers due to increased angling effort/mortality directed at the recovering water with unlimited access. Some access restrictions may therefore be needed in specific vulnerable fisheries because high-quality (here equalled with high catch rates) angling is often only found in waters where: (1) high cost/time required to access the fishery (e.g. remote waters without driving routes) exclude anglers or (2) access and effort is strictly controlled by private or local interests. Angling effort to indirectly increase stock abundance, catch quality, angler satisfaction, and support for habitat management may be controlled by lottery or licence rotating systems, individual transferable effort or access quotas, protected areas, or high access costs (e.g. time or money). This procedure is already being pursued in some of the highest-quality recreation fisheries, which is particularly feasible in private property fisheries that characterise large parts of Central Europe.

15.18 Voluntary Codes of Practice, Codes of Conduct, and Angler's Codes of Ethics (Like the Fly Fishers Code of Ethics, 2002)

There are many such codes which are available, both nationally (e.g. RecFish Australia 2014), and regionally (like the Code of Conduct for Recreational Fishing in the Kimberley, Western Australia, 2012). Other voluntary codes of practice exist like in the UK where the National Angling Alliance produced a Code of Conduct for Coarse Anglers covering aspects like the care of the environment, general behaviour, tackle and fish handling. The Nordic Angler Association established a code for recreation angling covering the whole of Scandinavia, including Iceland. European Codes cover recreational fishing (FAO 2008) and recreational fishing and invasive alien species (Owen 2013). The main problem with all these codes is that whilst they are available on the internet, the average recreation fisher is unlikely to spend time delving into these codes and there needs to be better ways of establishing educational outreach to anglers with the information that they contain.

Anglers can help by:

- Promoting ethical behaviour in the use of aquatic resources.
- Never disposing trash, waste, or plastics into the ocean. Avoiding spilling and never dumping any pollutants on land or in the water.
- Recycling rubbish, including worn-out lines, leaders, and hooks.

- Limiting the use of boats and vehicles to approved areas thus avoiding sensitive marine habitats.
- Volunteering for beach clean-ups and wetland restorations.
- Starting their own fishing line recycling programme if there is no convenient location in their community.
- Participating in community natural resources planning efforts.
- Getting involved in protecting essential fish habitat.
- Reporting pollution problems to local, state, or federal authorities. They can also follow the Angler's Code of Ethics which is modified from the US National Marine Fisheries Service's code.
- Demonstrating and promoting, through education and practice, ethical behaviour in the use of aquatic resources.
- Valuing and respecting the aquatic environment and all living things. Avoiding spilling and never dumping pollutants, such as gas or oil.
- Keeping fishing sites litter-free.
- Purchasing and keeping current their fishing licence, if necessary. If they are exempt, they may still purchase a licence as a way to contribute to conservation.
- Treating other anglers, boaters, and property owners with courtesy and respect.
- Respecting property rights, and never trespassing on private lands or waters.
- Keeping no more fish than needed for consumption and never wastefully discarding fish.
- Carefully handling and releasing alive all fish that are unwanted or prohibited by regulation.
- Using tackle and techniques that minimise harm to fish when CR angling.

Concluding Remarks

It is clear that the way forward with recreation fisheries management is to involve the participants much more in this process with a hands-on approach and to find the best ways of educational outreach, with important scientific knowledge related to fisheries so that the recreational fishing community becomes better informed. This includes the wider recreational fishing industry too and the tourism-related aspects of fishing holidays, guiding, and charter trips. Participation in recreational fishing creates one of the strongest social and political constituencies for environmental education and the conservation of aquatic resources. Planning of the coastal ecosystems worldwide must include recreational fishing as one important component, and the protection of both coastal and inland wetlands must be extended. One of the biggest future issues this century is the likely influence of climate change, and it will have effects on recreational fishing and there will need to be further research into this topic. Jones et al. (2013) suggested that the US cold-water fisheries were highly vulnerable to climate change through loss of suitable habitat. With a low-emission trajectory, it was estimated that there would be 18% less habitat by 2100, but under high emissions the cold-water fish habitat will disappear from most of New England, the Appalachians, and Upper Midwest by that year. Other stressors before then may be affected by increased temperature and precipitation variability which in some areas will make streams and lakes unsuitable for fish populations before climatic effects occur so that there will be changes in nutrient budgets, changes in run-off and riparian vegetation. Gilbert and Smith (2016) say that so far research on the effects of climate change on recreational fishing has been conducted at a coarse scale making specific impacts of climate change uncertain. Meanwhile there has been a special issue of Fisheries Magazine (issue 41) in July 2016 on the Effects of Climate Change on North American Inland Fisheries, and rather more specifically Clark et al. (2017) propose that wild populations in a warming climate may become skewed towards low-performance phenotypes which would have ramifications for predator-prey interactions and community dynamics and influence fish evolution in the future.

References

Alós, J., & Arlinghaus, R. (2013). Impacts of partial marine protected areas on coastal fish communities exploited by recreational angling. *Fisheries Research, 137*, 88–96.

Alós, J., Arlinghaus, R., Palmer, M., March, D., & Alvarez, I. (2009). The influence of type of natural bait on fish catches and hooking location in a mixed-species marine recreational fishery, with implications for management. *Fisheries Research, 97*, 270–277. https://doi.org/10.1016/j.fishres.2009.03.003.

Arkenford Ltd. (2014). *Watersports Participation Survey 2013*. Retrieved from http://www.britishmarine.co.uk/pdf/WatersportsReview2013ExecutiveSummary.pdf.

Arlinghaus, R., & Cooke, S. J. (2009). Recreational fisheries: Socioeconomic importance, conservation issues and management challenges. In B. Dickson, J. Hutton, & W. M. Adams (Eds.), *Recreational Hunting, Conservation and Rural Livelihoods: Science and Practice* (pp. 39–58). Chichester: Wiley-Blackwell.

Arlinghaus, R., & Mehner, T. (2003). Socio-economic characterization of specialized common carp (*Cyprinus carpio* L.) anglers in Germany, and implications for inland fisheries management and eutrophication control. *Fisheries Research, 61*, 19–33.

Arlinghaus, R., & Mehner, T. (2005). Determinants of management preferences of recreational anglers in Germany: Habitat management versus fish stocking. *Limnologica—Ecology and Management of Inland Waters, 35*, 2–17.

Armstrong, M., Brown, A. Hargreaves, J., Hyder, K., Pilgrim-Morrison, S., Munday, M., et al. 2013. Sea Angling 2012—A survey of recreational sea angling activity and economic value in England. Retrieved from www.marinemanagement.org.uk/seaangling/documents/finalreport.pdf.

Asoh, K., Yoshikawa, T., Kosaki, R., & Marschall, E. A. (2004). Damage to cauliflower coral by monofilament fishing lines in Hawaii. *Conservation Biology, 18*, 1645–1650.

Atlantic States Marine Fisheries Commission. (2011). Atlantic States Marine Fisheries Commission striped bass stock assessment update, November 2011.

Bartholomew, A., & Bohnsack, J. A. (2005). A review of catch-and-release angling mortality with implications for no-take reserves. *Reviews in Fish Biology and Fisheries, 15*, 129–154.

Beard, T. D., Cox, S. P., & Carpenter, S. R. (2003). Impacts of daily bag limit reductions on angler effort in Wisconsin walleye lakes. *North American Journal of Fisheries Management, 23*, 1283–1293.

Beardmore, S., Dorow, M., Haiden, W., & Arlinghaus, R. (2011). The elasticity of fishing effort response and harvest outcomes to altered regulatory policies in eel (*Anguilla anguilla*) recreational angling. *Fisheries Research, 110*, 136–148.

Bower, S. D., Danylchuk, A. J., Raghavan, R., Danylchuk, S. C., & Pinder, A. C. (2016). Rapid Assessment of the physiological impacts caused by catch-and-release angling on blue-finned mahseer (Tor sp.) of the Cauvery River, India. *Fisheries Management and Ecology, 23*, 208–217.

Bowker, J. M., & Askew, A. E. (2012). US Outdoor Recreation Participation Projections to 2060. In H. K. Cordell (Ed.), *Outdoor Recreation Trends and Futures: A Technical Document Supporting the Forest Service 2010, RPA Assessment* (pp. 105–124). General Technical Report SRS-150. Asheville, NC: USDA Forest Service, Southern Research Station.

Brown, C., & Laland, K. N. (2003). Social learning in fisheries: A review. *Fish and Fisheries, 4*, 280–288.

Burgin, S. (2017). Indirect consequences of recreational fishing in freshwater ecosystems: An exploration from an Australian perspective. *Sustainability, 9*, 280. https://doi.org/10.3390/su9020280.

Clark, T. D., Messmer, V., Tonin, A. J., Hoey, A. V., & Pratchett, M. S. (2017). *Rising Temperatures may Drive Fishing-Induced Selection of Low Performance Phenotypes*. Scientific Reports 7, 40571. https://doi.org/10.1038/srep40571.

Conover, D. O., & Munch, S. B. (2002). Sustaining fisheries yields over evolutionary time scales. *Science, 297*, 94–96.

Cooke, S. J., & Cowx, I. G. (2004). The role of recreational fishing in global fish crisis. *Bioscience, 543*, 857–859.

Cooke, S. J., Danylchuk, A. J., Danylchuk, S. E., Suski, C. D., & Goldberg, T. L. (2006). Is catch-and-release-recreational angling compatible with no-take marine protected areas? *Ocean and Coastal Management, 49*, 342–354.

Cooke, S. J., Hogan, Z. S., Butcher, P. A., Stokesbury, M. J. W., Raghavan, R., Gallagher, A. J., et al. (2014). Angling for endangered fish: Conservation problem or conservation action? *Fish and Fisheries, 17*, 249–265.

Cooke, S. J., Palensky, L., & Danylchuk, A. J. (2017). Inserting the angler into catch-and-release angling science and practice. *Fisheries Research, 186*, 599–600.

Cooke, S. J., Schreer, J. E., Dunmall, K. M., & Philipp, P. (2002). Strategies for quantifying sub-lethal effects of marine catch-and-release angling—Insights from novel freshwater applications. *American Fisheries Society Symposium, 30*, 121–134.

Cooke, S. J., & Shramm, H. L. (2007). Catch-and-release science and its applications to conservation and management of recreational fisheries. *Fisheries Management and Ecology, 14*, 73–79.

Cooke, S. J., & Suski, C. D. (2005). Do we need species-specific guidelines for catch-and-release recreational angling to effectively conserve diverse fishery resources? *Biodiversity and Conservation, 14*, 1195–1209.

Cooke, S. J., Suski, C. D., Arlinghaus, R., & Danylchuk, A. (2013). Voluntary institutions and behaviours as alternatives to formal regulations in recreational fisheries management. *Fish and Fisheries, 14*, 439–457.

Cordell, H. K. (Ed.). (2012). *Outdoor Trends and Futures: A Technical Document Supporting the Forest Service 2010 RPS Assessment*. General Technical Report SRS-150. Asheville, NC: USDA Forest Service, Southern Research Station, 167pp.

Cox, S. P., Beard, T. D., & Walters, C. (2002). Harvest control in open-access sport fisheries: Hot rod or asleep at the reel? *Bulletin of Marine Science, 70*, 749–761.

Danner, G. R., Chacko, J., & Brautigern, F. (2009). Voluntary ingestion of soft plastic lures affects brook trout growth in the laboratory. *North American Journal of Fisheries Management, 29*, 352–360.

Danylchuk, A. J., Tiedemann, J., & Cooke, S. J. (2016). Perceptions of recreational fisheries conservation within the fishing industry: Knowledge gaps and learning opportunities identified at east coast trade shows in the United States. *Fisheries Research*. https://doi.org/10.1016/j.fishres.2016.05.015.

Dayton, P. K., Thrush, S., & Coleman, F. C. (2002). *Ecological Effects of Fishing in Marine Ecosystems of the United States*. Arlington, VA: Pew Oceans Commission, 44pp.

Delle Palme, C. A., Nguyen, V. M., Gutowsky, L. F. G., & Cooke, S. J. (2016). Do fishing education programs effectively transfer "catch-and-release" best practices to youth anglers yielding measurable improvements in fish condition and survival? *Knowledge and Management of Aquatic Ecosystems, 417*, 42. https://doi.org/10.1051/kmae/2016029.

Diodati, P., & Richards, R. A. (1996). Mortality of striped bass hooked and released in salt water. *Transactions of the American Fisheries Society, 125*, 300–307.

FAO. (2008). *EIFAC Code of Practice for Recreational Fisheries*. FAO of the United Nations European Inland Fisheries Advisory Committee. EIFAC Occasional Paper 42, Rome, 54pp.

Fly Fishers International. (2002). *Federation of Fly Fishers Code of Angling Ethics*. Fly Fishers International.

Font, T., & Lloret, J. (2014). Biological and ecological impacts derived from recreational fishing in Mediterranean coastal areas. *Reviews in Fisheries Science and Aquaculture, 22*, 73–85.

Gagne, T. O., Ovitz, L., Griffin, L. P., Brownscombe, J. W., Cooke, S. J., & Danylchuk, A. J. (2017). Evaluating the consequences of catch-and-release angling on golden dorado (*Salminus brasiliensis*) in Salta, Argentina. *Fisheries Research, 186*, 625–633.

Gilbert, L., & Smith, J. W. (2016). *The impact of Climate Change on Inland Recreational Fishing*. Institute of Outdoor Recreation and Tourism, Utah State University. IORT-PR-2016-01, 3pp.

Hessenauer, J.-M., Vokoun, J. C., Suski, C. D., David, J., Jacobs, R., & O'Donnell, E. (2015). Differences in the metabolic rates of exploited and unexploited fish populations: A signature of recreational fisheries induced evolution? *PLoS One, 10*(6). https://doi.org/10.1371/journal.pone0128336.

Hoffman, N. (2009). *Gone Fishing: A Profile of Recreational Fishing in Canada*. Statistics Canada.

IGFA. (2011). IGFA school of sportfishing tips and techniques in modifying lures. *International Angler, 74*(5).

Jones, R., Travis, C., Rodgers, C., Lazar, B., English, E., Lipton, J., et al. (2013). Climate change impacts on freshwater recreational fishing in the United States. *Mitigation and Adaptation Strategies for Global Change, 18*, 731–758.

Klatt. J., Lupi, F., & Melstrom, R. (2014). A natural experiment identifying disease regulation effects on recreational fishing. In *Agriculture and Applied Economics Association's Annual Meeting*, Minneapolis. Retrieved from http://purl.umn.edu/170580.

Knapp, R. A., Corn, P. S., & Schindler, D. E. (2001). The introduction of normative fish into wilderness lakes: Good intentions, conflicting mandates, and unintended consequences. *Ecosystems, 4*, 275–278.

Lancaster, D., Dearden, P., & Ban, N. C. (2015). Drivers of recreational fisher compliance in temperate conservation areas: A study of rockfish conservation areas in British Columbia, Canada. *Global Ecology and Conservation, 4*, 645–657.

Lennox, R. J., Mayer, I., Havn, T. B., Johansen, M. R., Whoriskey, K., Cooke, S. J., et al. (2016). Effects of recreational angling and air exposure on the physiological status and reflex impairment of European grayling (*Thymallus thymallus*). *Boreal Environment Research, 21*, 461–470.

Lewin, W.-C., Arlinghaus, R., & Mehner, T. (2006). Documented and potential biological impacts of recreational fishing: Insight for management and conservation. *Reviews in Fisheries Science, 14*, 305–367.

Liddle, M. J., & Scorgie, H. R. A. (1980). The effects of recreation on freshwater plants and animals: A review. *Biological Conservation, 17*, 183–206.

Maryland Department of Natural Resources. (2010). Deep hooking is the single most important factor causing sport angled fish to die. MDDNR Fisheries Service. Retrieved from http://www.dnr.state.md.us/fisheries/recreational/circlehookFisheriesHomePage.html.

McGarvie Hirner, J. L., & Cox, S. P. (2007). Effects of rainbow trout (*Oncorhynchus mykiss*) on amphibians in productive recreational fishing lakes of British Columbia. *Canadian Journal of Fisheries Aquatic Science, 64*, 1770–1780.

McPhee, D. (2017). Urban recreational fisheries in the Australian coastal zone: The sustainability challenge. *Sustainability, 9*, 422. https://doi.org/10.3390/su9030422.

McPhee, D. P., Leadbitter, D., & Skilleter, G. A. (2002). Swallowing the bait: Is recreational fishing in Australia ecologically sustainable? *Pacific Conservation Biology, 8*, 40–51.

Muoneke, M. I., & Childress, W. M. (1994). Hooking mortality: A review for recreational fisheries. *Reviews in Fisheries Science, 2*, 123–156.

National Marine Fisheries Service. (2012). *Atlantic States Marine Fisheries Commission 2011*.

National Survey of Fishing, Hunting and Wildlife Association. (2011). US Department of the Interior, US Fish and Wildlife Service, US Department of Commerce, US Census Bureau 2011. US National Survey of Fishing 2011, Washington, DC, 161pp.

NEATFish. (2009). *A Standard for National Environmental Assessment of Tournament Fishing*. Neatfish, 55pp.

Nguyen, V. M., Rudd, M. A., Hinch, S. G., & Cooke, S. J. (2013). Recreational anglers' attitudes, beliefs and behaviours related to catch-and-release practices of Pacific salmon in British Columbia. *Journal of Environmental Management, 128*, 852–865.

NJDEP. (2012). *The Striped Bass Bonus Program*. New Jersey Department of Environmental Protection, Division of Fish and Wildlife. Retrieved from https://www.njfishandwildlife.com/bonusbas.htm

NOAA. (2009). *Fisheries of the United States 2009*. Current Fishery Statistics No. 2009, Final report. Silver Spring, MD.

Ortolani, G. (2017). Modern fish act: Boon to recreational fishing or risk to US fishery? Retrieved from http://news.mongabayu.com/by/giovanni-ortolani/.

Outdoor Foundation. (2014). *Special Report on Fishing*. Recreational Boating and Fishing Foundation, The Outdoor Foundation. 63pp.

Owen, M. G. (2013). *European Code of Conduct on Recreational Fishing and Invasive Alien Species*. Strasbourg: Council of Europe, 35pp.

Policansky, D. (2002). Catch-and-release recreational fishing—A historical perspective. In T. J. Picher & C. E. Hollingworth (Eds.), *Recreational Fisheries: Ecological, Economic and Social Evaluation*. Oxford: Blackwell Science.

Pope, K. L., & Wilde, G. R. (2004). Effect of catch-and release angling on growth of largemouth bass (*Micropterus salmoides*). *Fisheries Management and Ecology, 11*, 39–44.

Possato, F. E., Barletta, M., Costa, M. F., Ivar do Sul, J. A., & Dantas, D. V. (2011). Plastic debris ingestion by marine catfish: An unexpected fisheries impact. *Marine Pollution, 62*, 1098–1102.

Post, J. R., Sullivan, M., Cox, S., Lester, N. P., Walters, C. J., Parkinson, E. A., et al. (2002). Canada's recreational fishery: The invisible collapse? *Fisheries, 27*, 6–17.

Prato, G., Barrier, C., Francour, P., Cappanera, V., Markantonaton, V., Guidetti, P., & et al. (2016). Assessing interacting impacts of artisanal and recreational fisheries in a small Marine Protected Area (Portifino), NW Mediterranean Sea. *Ecosphere 7*(12), 18pp. e01601. https://doi.org/10.1002/ecs2.1601.

Quan, R.-C., Wen, X., & Yang, X. (2002). Effects of human activities on migratory waterbirds at Lashihai Lake, China. *Biological Conservation, 108*, 273–279.

Raison, T., Nagrodski, A., Susk, C. D., & Cooke, S. J. (2014). Exploring the potential effects of lost or discarded soft plastic fishing lures on fish and the environment. *Water Air Soil Pollution, 225*. https://doi.org/10.1007/s11270-014-1809-1.

Recfish Australia. (2014). *A National Code of Practice for Recreational and Sport Fishing*. Recfish Australia.

Sanchirico, J. N., Eagle, J., Palumbi, S., & Thompson Jr., B. H. (2010). Comprehensive planning, dominant-use zones and user-rights: A new era in ocean governance. *Bulletin of Marine Science, 86*, 1–13.

Shiffman, D., Gallagher, A. J., Wester, J., MacDonald, C. C., Thaler, A. D., Cooke, S. J., et al. (2014). Trophy fishing for species threatened with extinction: A way forward building on a history of conservation. *Marine Policy, 50*, 318–322.

Sigler, M. (2014). The effects of plastic pollution on aquatic wildlife: Current situation and future solutions. *Water Air Soil Pollution, 225*. https://doi.org/10.1007/s11270-104-2184-6.

Simpson, D., & Mawle, G. W. (2010). *Public Attitudes to Angling 2010*. Bristol: Environment Agency.

Skaggs, J., & Allen, M. S. (2016). Data needs to assess effects of soft plastic lure ingestion on fish populations. American Fisheries Society, January 11, Essays, *Fisheries Magazine News*.

Smallwood, C. B., & Beckley, L. E. (2012). Spatial distribution and zoning compliance of recreational fishing in Ningaloo Marine Park, North-Western Australia. *Fisheries Research, 125–126*, 40–50.

Sudmann, S. R., Distelrath, F., Mayer, B. C., & Bernert, P. (1996). Auswirkungen der Einstellung des Angelsports auf den Brutvogelbestand am südlichedn Teil des Althein Bienen-Praest. *Natur und Landschaft, 12*, 536–540.

Suski, C. D., Phelan, F. J. S., Kubacki, M. R., & Philipp, D. P. (2002). The use of community-based sanctuaries for protecting smallmouth bass and largemouth bass from angling. *American Fisheries Society Symposium, 2002*(31), 371–378.

Sutinen, J. G., & Johnston, R. J. (2003). Angling management organizations: Integrating the recreational sector into fishery management. *Marine Policy, 27*, 471–487.

Sutter, D. A. H., Suski, C. D., Philipps, D. P., Kleforth, T., Wahl, D. H., Kersten, P., et al. (2012). Recreational fishing selectively captures individuals with the highest fitness potential. *PNAS, 109*(51), 20960–20965.

Thomé-Souza, M. J. F., Maceina, M. J., Forsberg, B. R., Marshall, B. G., & Carvalho, A. L. (2014). Peacock bass mortality associated with catch-and-release sport fishing in the Negro River, Amazonas state, Brazil. *Acta Amazonica, 44*, 527–532.

Tiedemann, J., & Danylchuk, A. (2012). Assessing impacts of catch and release practices on striped bass (*Morone saxatalis*). Implications for Conservation and Management, 15pp. Retrieved from http://www.monmouth.edu/uploadedFiles/Resources/Urban_Coast_Institute/Best20%PracticesStriped%20BassCatchandRelease.

Topline Survey. (2017). *Outdoor Recreation Participation*. Outdoor Foundation, Washington, 12pp.

Veiga, P., Pita, C., Leite, L., Ribeiro, J., Ditton, R. B., Gonçalves, J. M. S., et al. (2013). From a traditionally open access fishery to modern restrictions: Portuguese anglers' perceptions about newly implemented recreational regulations. *Marine Policy, 40*, 53–63.

Wales Outdoor Recreation Survey. (2015). *Wales Outdoor Recreation Survey 2014 Final Report*. Cardiff: Natural Resources Wales, 108pp.

Wells, R. S., Hoffman, S., & Moors, T. L. (1998). Entanglement and mortality of bottlenose dolphins, *Tursiops truncates* in recreational fishing gear in Florida. *Fisheries Bulletin, 96*, 647–650.

Wissinger, S., McIntosh, A., & Greig, H. S. (2006). Impacts of introducing brown and rainbow trout on benthic invertebrate communities in shallow New Zealand lakes. *Freshwater Biology, 51*, 2009–2028.

Wolter, C., & Arlinghaus, R. (2003). Navigation impacts on freshwater fish assemblages. The ecological relevance of swimming performance. *Reviews of Fish Biology and Fisheries, 13*, 63–89.

World Bank. (2012). *Hidden Harvest—The Global Contribution of Capture Fisheries*. Report No. 66469, Washington, DC, 92pp.

Expeditions

16

Chapter Summary

This chapter first defines what constitutes an expedition and then gives examples of expeditions with different purposes and modes of travel and concludes that many use camping (discussed in Chap. 8) as their primary accommodation. It then examines the history of overseas expeditions (e.g. the South Pole and Everest expeditions) before examining participation numbers. The final part of the chapter focuses on specific environmental impacts of expeditions in four areas: movement and access, campsites on local communities, and the impacts of expedition fieldwork. The final section considers the management of these impacts and gives examples of ways in which expedition organisers can minimise the impact of their expeditions.

16.1 Definitions

What constitutes an expedition? A journey undertaken by a group of people with a particular purpose, especially that of exploration, research, or war. There are no readily available criteria by which an expedition can be defined or classified. The number of people involved might be one criterion. This can range from a single expeditioner to groups in excess of 100. Stott et al. (2015) in a review of youth expeditions defined an expedition, for the purposes of their literature search, as having a "duration exceeding 14 days, self-propelled, and was based overseas or out-of-state" (p. 197).

The environment in which the expedition takes place could be another way to classify or define an expedition. For example, the British Exploring Society, based in the UK, which has organised overseas youth expeditions since 1932, uses four environments to categorise their expeditions: polar, mountain, desert, or jungle.

Expeditions may have different objectives or purposes:

To cross a continent:

1. Various Antarctic expeditions were led by Shackleton, Scott, Amundsen, and Mawson to reach the South Pole in the early 1900s.
2. The Lewis and Clark Expedition from May 1804 to September 1806 was the first American expedition to cross what is now the western portion of the USA.
3. Robert Peary led various expeditions to the Arctic in the late nineteenth and early twentieth centuries, claiming to have reached the geographic North Pole with his expedition on 6 April 1909; Wally Herbert became the first man fully recognised for walking to the North Pole in 1969, on the 60th anniversary of

© The Author(s) 2019
D. Huddart, T. Stott, *Outdoor Recreation*, https://doi.org/10.1007/978-3-319-97758-4_16

Robert Peary's famous, but disputed, expedition.

4. The Burke and Wills expedition of 1860–1861 had the objective of crossing Australia from Melbourne in the south to the Gulf of Carpentaria in the north.

To climb a mountain:

5. Various expeditions were undertaken to climb Everest—in 1920 by Mallory and, the first which was proved to have reached the summit, in 1953 by Hilary and Tenzing.

To cross a desert:

6. Bertram Thomas was the first European to cross the Rub' al Khali (the Empty Quarter) in Oman in 1930–1931 which was crossed again by Wilfred Thesiger in 1946 and 1948 and by Mark Evans in 2017.

To navigate a river/jungle:

7. John Blashford-Snell led the expeditions which made the first descent of the Blue Nile (in 1968) and a complete navigation of the Congo River (1974–1975).

To circumnavigate the world:

8. Mark Beaumont has cycled around the world (Beaumont 2011) and holds the record for cycling his 18,000-mile (29,000 km) route, completed on 18 September 2017, having taken less than 79 days; Robin Knox-Johnston was the first person to sail single-handed, unassisted, and non-stop around the world in 1969.

Another way in which expeditions can be categorised is by means of their mode of travel. Not all, as defined by Stott et al. (2015), are necessarily self-propelled. Just taking the short list of expeditions above, Table 16.1 categorises them based on the modes of travel.

While perhaps a more recent style of expedition, a number of people have used bicycles to undertake expeditions. For example, Mark Beaumont has cycled around the world (Beaumont 2011) and holds the record for cycling his 18,000-mile (29,000 km) route, completed on 18 September 2017, having taken less than 79 days.

Most expeditions use camping as their primary accommodation. The impacts of camping are discussed in detail in Chap. 8.

Table 16.1 Types of expeditions

Mode of expedition travel	Chapter in this book in which this is reviewed	Examples (number of expedition in the list in Sect. 16.1)
On foot	Chap. 2	1, 2, 4, 5, 6
On ski	Chap. 11	1, 3
By self-propelled boat, yacht, raft, canoe, or kayak	Chap. 13	7, 8
With animals (horseback, camels, ponies, mules, dogs)	Chap. 9	1, 2, 3, 4, 5, 6
Off-road vehicles (snowmobile, quadbike, motorbike, 4WD Land Rover)	Chap. 5	1
Motorised boat	Chap. 13	7
Bicycle	Chap. 7	8

16.2 History of Overseas Expeditions

Expeditions are almost certainly as old as humankind. Archaeological evidence clearly shows that our ancestors travelled widely on foot and overseas on rafts and various types of boats as far back as 60,000 or 70,000 years ago. More recently, expeditions in the UK have a history that can be traced back to exploration for geographical purposes. These expeditions can be linked to characters such as Scott, Shackleton, Watkins, and Herbert in the polar regions and Younghusband and Hilary in the Himalaya. A brief consideration on Scott's famous Antarctic expeditions might be useful to communicate the huge scale and commitment of an expedition of that type.

Robert Falcon Scott (1868–1912) was a British Royal Naval officer and explorer who led two expeditions to the Antarctic regions: the Discovery Expedition, 1901–1904, and the ill-fated Terra Nova Expedition, 1910–1913. Scott took along a large team of scientists, and his ship became the most completely equipped vessel for scientific purposes in polar regions. The scientific crew included meteorologists, hydrologists, zoologists, glaciologists, biologists, and geologists, all under control of Dr. E. A. Wilson, the Chief Scientist. During this second venture, Scott led a party of five who reached the South Pole on 17 January 1912, to find that they had been preceded by Roald Amundsen's Norwegian party in an unsought "race for the Pole." On their return journey, Scott and his four comrades all perished because of a combination of exhaustion, hunger, and extreme cold. The bodies of Scott, Wilson, and Bowers were discovered the following spring in their tent some 12 miles from One Ton Depot. Surgeon E. L. Atkinson RN of the recovery party concluded: "We recovered all their gear and dug out the sledge with their belongings on it. Amongst these were 35 lb. of very important geological specimens which had been collected on the moraines of the Beardmore Glacier; at Doctor Wilson's request they had stuck to these up to the very end, even when disaster stared them in the face and they knew that the specimens were so much weight added to what they had to pull." A total of 1919 rock specimens from the expedition are housed at the Natural History Museum today.

Among Scott's party was a Surgeon Commander George Murray Levick who, as well as being one of the expedition's medical doctors, also became a member of the Eastern party which after a brief meeting with Amundsen was to become the Northern party and occupied Evans Cove for summer fieldwork. As a result of impenetrable ice, they were not picked up by boat and forced to overwinter. After tents were ravaged by blizzards, their only hope was to dig a cave in the largest snow patch they could find. Through the winter they cooked primarily with a seal blubber stove and ate seal meat. After surviving the winter, they were able to take a photo of themselves (Fig. 16.1) impregnated with seal blubber oil.

After returning from the expedition, Levick served in the Navy in the First World War and then became a doctor in London where he must have reflected on his expedition experiences. There were expeditions leaving the UK on a regular basis, most notably from some older universities such as those led by Gino Watkins from Cambridge to Greenland (later to lead the British Arctic Air

Fig. 16.1 These six men, the northern party of Captain Scott's last expedition, stand outside the entrance to the snow hole in which they have just spent the 1911–1912 Antarctic Winter in darkness. The low spring sun allows the zoologist and photographer of the party, Surgeon George Murray Levick RN (second from right), to take this picture. Their clothing and hair were impregnated with seal blubber because all their cooking, mostly of seal meat, was carried out over a seal blubber stove. Source: British Schools Exploring Society archive, Royal Geographical Society

Route Expedition in 1930–1931 and the following year died in East Greenland on a hunting expedition), but there were no opportunities for young people at school to obtain adventure experiences abroad. Levick saw a need for tough, demanding challenges and in 1932 took eight boys to Finland with basic equipment for a cost of £30 per boy. In 1933 he founded the Public Schools Exploring Society (PSES) and continued to lead expeditions (growing in size each year) which has today evolved to become the British Exploring Society (http://www.britishexploring.org/).

Interestingly it was not until 1978 when the next youth expedition organisation was formed with similar aims: Operation Drake and then in 1984 Operation Raleigh, since 1991 known as Raleigh International (https://raleighinternational.org/). The 1980s and 1990s saw the beginning of many similar organisations with a variety of aims, operating both in the commercial and charitable sectors with an obvious wide range of aims and objectives. Most expedition providers offer some combination of adventurous activities, science work, and community projects for time periods varying from 3 weeks to 12 months. Some providers work directly with individuals while others operate through schools, education authorities, and youth organisations. Expeditions are staffed by a wide range of qualified personnel including professional outdoor leaders, scientists and researchers, educators, and outdoor enthusiasts, and personnel may be paid staff or volunteers or a combination of both. In addition, expeditions are increasingly connected to offer components of other certifying organisations such as the Duke of Edinburgh Award scheme and the John Muir Award.

16.3 Participation Numbers

Travel and overseas experiences, particularly those involving some form of outdoor education, are regarded by many young people, parents, university admissions, and employers as somehow beneficial to a young person's development. Expeditions have been used in the UK as an educational tool since 1932 when the Public Schools Exploring Society ran their first expedition to Finland.

While gap years and expeditions are slightly different (as the former often incorporates the latter, but not vice versa), no specific statistics are available on the numbers of people engaged in expeditions from the UK each year. Jones (2004), however, estimated that 250,000–350,000 Britons between 16 and 25 years old were taking a gap year annually. In 2008 Rowe reported that "the gap year market is valued at £2.2 billion in the UK and globally at £5 billion. It's one of the fastest growing travel sectors of the 21st century, and the prediction is for the global gap year market to grow to £11billion by 2010" (p. 47).

It is also worth noting the development of British Standard 8848 (specification for the provision of visits, fieldwork, expeditions, and adventurous activities outside the UK) in concert with the Learning Outside the Classroom (LOtC) quality badge scheme (underpinned by the Expedition Providers Association)—further indications of the growth in numbers of people travelling overseas on expeditions and gap years.

Figure 16.2 shows the number of planned and executed expeditions on the Royal Geographical Society (RGS) Expeditions Database, 1964–2018 (https://www.rgs.org/in-the-field/rgs-fieldwork-database/, accessed 05/04/18).

According to the RGS Expeditions Database, between 1964 and 2018 an average of 229 expeditions per year was planned and an average of 177 returned. These expeditions span all continents and all types of expeditions. By accessing the database, it is possible to select for categories of expedition and 114 categories are represented. We are not certain whether these numbers are an accurate reflection of the total number of expeditions which were happening. It's not clear what proportion of all expeditions actually bother to put their plans/reports on the RGS database. It is possible that most of them are there because they had been awarded grants from the RGS (and so were required to upload their plans and reports), so the trend may just reflect RGS grant funding rather than a real decline in number of expeditions since the 1990s.

16.4 Environmental Impact

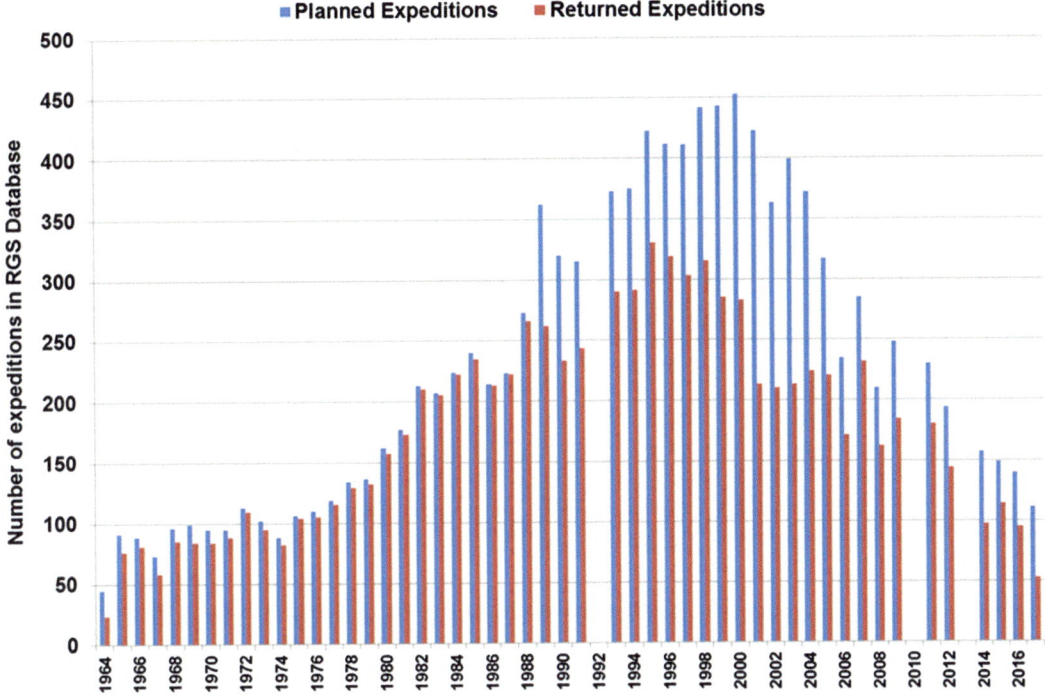

Fig. 16.2 The number of planned and executed expeditions on the Royal Geographical Society Expeditions Database, 1964–2018 (https://www.rgs.org/in-the-field/rgs-fieldwork-database/, accessed 05/04/18)

16.4 Environmental Impact

"A wheel mark in the desert lasts for decades. A footprint in the Arctic takes years to fade. Yet the expeditions which make these marks may further our knowledge of the world in which we live, helping us to conserve it."

(Footprints Forever, *Geographical Magazine*, 1991)

In 2008 the British Ecological Society (BES) and the Young Explorers' Trust (YET) (British Ecological Society/Young Explorers' Trust 2008) published *Environmental Responsibility for Expeditions: A Guide to Good Practice*. In their guide they identified four main areas in which expeditions can impact the environment. These were (1) reducing the impact of movement and access, (2) reducing the impact of campsites, (3) promoting good community relations, and (4) responsible fieldwork. These are considered next.

16.4.1 The Impact of Movement and Access

Travelling to and from an expedition area and moving around during the expedition have potential to have significant environmental impacts. Most overseas expeditions use air travel, and this arguably results in the biggest environmental impact of all. However, if the carbon costs of expeditions are compared with those of business travel or academics flying to conferences, Allison et al. (2011) would argue that they are far more justifiable. If long haul flights are used, then clearly larger and longer expeditions are perhaps better when the carbon emissions per person per day are calculated. However, BES/YET (2008) argue that the carbon costs of an expedition must be offset against the value of the expedition in terms of what it achieves. So it is important that the expedition is well planned and executed so that the benefits gained justify the environmental cost of long-distance travel.

The frequency and type of movement to/from project sites or centres of activity during an expedition can impact the local environment. While Chaps. 2 and 3 discussed the impacts of walking/running on soil and vegetation, Gellatly et al. (1986) conducted an interesting set of trampling experiments while on a British Exploring Society expedition on the Lyngen Peninsula in Arctic Norway. The effects of regular trampling by members of a large expedition in an area of arctic heath were assessed over a six-week period and again the following summer. Characteristic visual changes included the reduction in vegetation cover and an increase in the width, depth, and extent of lateral erosion. Trampling increased soil compaction and bulk density which in turn influenced levels of soil moisture and porosity. Levels of compaction increased with recreational intensity and partial stripping of the surface organic horizon led to a reduction in organic soil material. Their study highlighted the importance of interactive forces such as surface roughness, drainage, and natural obstacles. An assessment of the recovery of damaged sites nearly a year later led to recommendations for more awareness of the potential degradation and fragility of this environment under continued heavy recreational pressure by visitors.

Many expeditions use pack animals to move equipment and supplies to, from, and around the expedition area. Barros and Pickering (2015) conducted a manipulative experiment to assess damage to alpine meadows by pack animals and hikers in the Aconcagua Provincial Park, Andes, Argentina. They recorded vegetation height, overall cover, cover of dominant species and species richness immediately after and two weeks after different numbers of passes (0, 25, 100, and 300) by hikers or pack animals in an experiment, using a randomised block design. They found that pack animals had two to three times the impact of hiking on the meadows, with greater reductions in plant height, the cover of one of the dominant sedges and declines in overall vegetation cover after 300 passes. Impacts of pack animals were also apparent at lower levels of use than for hikers. These differences occurred despite the meadow community having relatively high resistance to trampling due to the traits of one of the dominant sedges (*Carex gayana*). They concluded that pack animals caused more damage than hikers to the alpine meadow, but the scale of the difference in short-term impacts depended upon the characteristics of the plant community, the amount of use, and the vegetation parameters measured. Use of the meadows by hikers and pack animals should be minimised as these meadows are scarce and have high conservation values.

In another study, Cousquer and Allison (2012) examined mountain guide's and expedition leader's ethical responsibilities towards pack animals on expedition. They noted that in the absence of motorised transport, the mule's ability to carry heavy loads over difficult mountainous terrain was exploited. However, they found that the nature of the contract between the leader and the mule was far from clear and the leader's responsibilities towards pack animals could be easily overlooked. They discussed the industry's failure to recognise its responsibilities to pack animals. Chapter 9 discusses horseback riding and its ecological impacts.

During a British Exploring Society expedition in Gipsdalen, Svalbard, in 1995, Kate Eldon and Alan Swan (Eldon and Swan 1993) carried out an experiment which aimed to establish the amount of damage inflicted by walkers on the tundra vegetation and to determine whether the type of footwear had any effect on the damage caused. Two adjacent 10 m "paths" were delimited on the tundra (Fig. 16.3A, left), and the % cover, number, and height of flower stalks (*Dryas octopetala*) were recorded in randomly located fixed quadrats. Each path was walked first 50, then 100 times, then a further 100, Path A with wellington boots and path B with stiff-soled mountain boots. Figure 16.3A shows that % vegetation and mean stalk height of *Dryas octopetala* both declined with the number of passes, though the decline on the track trampled by mountain boots was slightly faster than on the one trampled by wellington boots.

These results may have implications for the best type of footwear to use to reduce damage. Many trekkers now wear light approach shoes or trainers to trek into mountains, keeping their

16.4 Environmental Impact

	Wellington boots	Mountain boots
% vegetation cover at start	36	54
% vegetation cover after 100 walks	23	35
DIFFERENCE	**-13**	**-19**
% vegetation cover after 200 walks	13	22
DIFFERENCE compared with start	**-23**	**-32**
Mean height of stalks at start (mm)	43	39
Mean height of stalks after 100 walks	27	29
DIFFERENCE	**-16**	**-10**
Mean height of stalks after 200 walks	16	14
DIFFERENCE compared with start	**-27**	**-25**

Fig. 16.3 (A) Expedition footwear trampling experiment on tundra in Svalbard (Eldon and Swan 1993). Photo by Tim Stott. (B) Crampons used by mountaineering expeditions can leave scratches on rock, which on popular routes can leave a permanent scar. Photo by Tim Stott

heavy mountain boots for snow, ice, glaciers, and moraines where there is little or no vegetation to damage. However, the wearing of crampons (Fig. 16.3B) on mixed ground (i.e. rock, moraine, and ice/snow) can lead to permanent scratches on rock which, on some popular routes, can easily be seen as a scar and can detract from the wilderness feeling that some walkers and climbers seek.

Some expeditions use vehicles and the impacts of off-road vehicles in wilderness areas are discussed in detail in Chap. 6. These impacts include damage to vegetation and soils, exhaust emissions, potential oil and fuel leakage, and noise and disturbance to wildlife. Vehicles driven on snow (snowmobiles) can compact the snow, and we know that compacted snow takes longer to melt (see Chap. 11: Snow sports) and can damage underlying vegetation.

Many expeditions will use boats as a means to access remote coastlines or islands. Boats can be a very efficient and effective way to ship large amounts of heavy equipment and supplies (Fig. 16.4A–D).

However, as discussed in Chap. 13, there are also negative impacts associated with propeller action (cutting plants or disturbing bottom sediments in shallow water), wave action, oil and fuel spillage, and noise/disturbance to wildlife. However, when used carefully and correctly, in

Fig. 16.4 (A) Bringing in supplies by boat for the 2009 British Exploring Society expedition to Tasermiut Fjord, SW Greenland. Photo by Tim Stott. (B) Supplies for the 2009 British Exploring Society expedition to Tasermiut Fjord, SW Greenland, were brought on this raft. Photo by Tim Stott. (C) Loading supplies onto the Langoysund in Longyearbyen for the 2001 British Exploring Society expedition in Svalbard. Photo by Tim Stott. (D) Small inflatable zodiacs with an outboard engine are popular for use on expeditions to moving equipment and people. Photo by Tim Stott. (E) Cruise ships started to visit Longyearbyen, Svalbard, in the 1990s. Photo by Tim Stott

certain situations small boats can be one of the most environmentally sound forms of transport. However, whether one would consider travel on large vessels such as the cruise ship shown in Fig. 16.4E, as an "expedition" is debatable. While some of the passengers may see it as an adventure, they are hardly self-propelled. However, in the past there has been a serious proposal to run a large (50+ people) British Exploring Society expedition to Bylot Island in the Canadian Arctic and to have the whole expedition based on the ship (moored offshore) and to ferry the expedition

16.4 Environmental Impact

Fig. 16.5 (A) Small expedition camp site in Zara Valley, Ladakh. Photo by Tim Stott. (B) Large British Exploring Society Expedition base camp site in Gipsdalen, Svalbard. The whole expedition only camped here for two nights at the beginning of the six-week expedition. They then moved away in six smaller groups. Photo by Tim Stott. (C) British Exploring Society Expedition base camp site on a storm beach at Brucebyen on Isfjord, Svalbard, in 2001. This site would score 1 using Sørbel et al. (1990) terrain vulnerability classification system. There would be minimal impact from camping as there is no soil or vegetation cover. Photo by Tim Stott. (D) British Exploring Society Expedition campsite on dry river bed in Ladakh in 2013. This site would score 1 using Sørbel et al. (1990) terrain vulnerability classification system. Camping here will have minimal impact as there is no soil or vegetation cover to be damaged. Photo by Tim Stott

members each day onto the island for their science and adventurous activities. To date this idea has not been tested, but it may be a good way of avoiding the environmental impact of camping in large numbers as described in the next section.

Brida and Zapata (2009) discussed the economic, socio-cultural, and environmental impacts of cruise tourism, which in the Caribbean region increased from three million in 1980 to more than 25 million in 2007, and Butt (2007) reported on the impact of cruise ship-generated waste in Southampton. Winser (2004) section 6 offers some useful advice on transport for expeditions and fieldwork.

16.4.2 Expedition Campsites

Chapter 8: "Camping, Wild Camping, Snow Holing, and Bothies" discussed the general impacts of camping on soils and vegetation, wildlife, and water resources. Clearly these same impacts will apply to all expeditions which use camping as the main form of accommodation. Campsites on small expeditions (Fig. 16.5A), however, will differ in their impact from those where there are perhaps 50 or more people in the same place at the same time (e.g. in an expedition base camp, Fig. 16.5B), whilst permanent base camps or bothies (as also discussed in Chap. 8) will present a different range of problems to transient overnight camps.

Drystan Jones conducted an environmental impact study of expedition base camp sites on the East Coast of Oscar II Land, Svalbard (Jones 1997). In summer of 1996, as a member of a large British Exploring Society (BES) expedition, he visited base camps used by BES expeditions in 1987, 1990, and 1993. He was therefore able to compare sites which had been used as expedition base camps nine, six, and three years before his visits, as well as the 1996 base campsite. Each base camp site was paired with a nearby undisturbed "control site." He recorded the number of plant species, % cover, and vegetation height as well as compaction of soil. All four expeditions had changed the base camp areas they occupied. Valuable studies in this field have been carried out by Leif Sørbel and co-workers at the Department of Physical Geography, University of Oslo (Sørbel et al. 1990). They developed a terrain vulnerability classification system which Jones used in his study. Sørbel et al. defined the following classes:

1. *Invulnerable areas*. Examples are active alluvial plains, fans, and tidal shores.
2. *Moderately vulnerable areas*. Dry, well-drained areas with a discontinuous vegetation cover.
3. *Vulnerable areas*. Characterised by continuous vegetation cover, often fine-grained material and relatively high ground moisture.
4. *Very vulnerable areas*. Wear easily causes further erosion. Areas are characterised by fine material, moisture saturation, and continuous, thick vegetation cover, often combined with inclination and proximity to drainage ways.
5. *Areas of conservational value*. Localities which contain biotopes, landforms, or other features which are found to be particularly valuable and therefore should be protected from disturbances.

Jones then used Sørbel's terrain vulnerability scale to describe each of the base campsites. The 1987 site scored 2–3 on the terrain vulnerability classification; it had lost its original cover of *Cassiope tetragona* which he described as a loss to the aesthetic quality of the area. The 1990 site scored 3 on the terrain vulnerability classification and in 1996 supported a different community type to that of the control. The terrain was generally flattened throughout camp area, and the ground was damper (which he ascribed to the thaw of permafrost). The 1993 site scored 2 on the terrain vulnerability classification and showed less obvious change due to the more discontinuous nature of original cover; he observed some scarring of coarse material (which he did not measure). The 1996 site scored 1 on the terrain vulnerability classification. Examples of expedition camps on this type of terrain are shown in Fig. 16.5C, D. Using this site the expedition caused little or no impact on the vegetation or underlying terrain of the base camp—though there were "paths" leading from the area where vegetation had been damaged. He concluded that after nine years the 1987 site was still visibly different to the surrounding area and was occupied by pioneering vegetation species.

In addition to damage to vegetation and soils, which as we have seen, can be minimised by careful selection of campsites, another problem faced by large expeditions is waste disposal, particularly that of human waste. Ideally all human waste would be carried out (as described in the Cairngorm Poo project in Chap. 8), but this is not always practical on a large expedition with 50+ people who are away in wilderness areas for up to six weeks. An appropriate method of dealing with human waste must be adopted, and there are a number of options ranging from group pits to individual burying depending on the number of people, the duration of the camp, and the nature of the location. The generally accepted best practice is, where possible, to dig pits for the latrines and to bury the waste. Clearly this introduces large quantities of nutrients into the normally nutrient poor soils of these wilderness areas. For expeditions based on coastlines, disposal of human waste into saltwater may be an option, but this needs careful consideration and assessment for each situation. Tidal currents, wind, turbulence, nearby habitation, and wildlife all need to be considered in this event. It may come down to a question of concentration vs. dispersal. All toilet paper and sanitary products should be brought out or burnt, depending on any local codes of practice. Latrines should be located at least 75 m away from water courses to avoid contamination of the water supply with coliform bacteria. When washing clothes or bodies, there is the potential to contaminate water, and the BET/YET (2008) guidelines give advice on how expeditions can minimise these impacts.

Kuniyal (2002) noted that biotic pressure due to expeditions, trekking, tourism, and transhumance practices by the shepherds is continuously increasing in mountain areas. Practices like indiscriminate throwing of wastes, leaving behind self-generated wastes and emission of poisonous gases from unattended wastes, cutting of trees like *Rhododendron spp.* (for fuelwood), introduction of hybrid sheep to replace indigenous ones, extraction of invaluable and endangered medicinal plants, and reduction in wildlife because of illegal hunting and poaching (for meat, skin, and medicine) have adversely affected the expedition areas. Kuniyal's study which was conducted on one of the expeditions to the Pindari Valley of Indian Himalayas showed that 61% non-biodegradable waste problem could be resolved by reuse (39%) and recycling (21%), but all the waste needed to be brought back by the visitors from expedition/trekking areas to the road heads for easy transportation to places where it can be reused, recycled, or new products discovered with innovative recovery initiatives.

Mount Everest, the highest mountain on our planet, was first climbed in 1953 and has since become a magnet for mountaineers. Since Sir Edmund Hillary and Tenzing Norgay reached the summit on 29 May 1953, as part of the British expedition led by Lord John Hunt, the British Mountaineering Council (2018) stated that there had been 6871 ascents of Everest by 4042 different climbers (up until February 2014), meaning that some climbers, most of them Sherpas, have reached the top multiple times. Two Sherpas, Apa and Phurba Tashi, held the record for the most ascents—21. Kenton Cool holds the British record for multiple ascents, having reached the summit 11 times, 2 of them within a week in 2007.

Sadly Everest shows the signs of over seven decades of these climbers' quests to stand on the roof of the world. Bishop and Naumann (1996) reported on how climbers would find trash on the mountain in the form of old tents, fixed ropes, used oxygen bottles, human waste, tins, glass, paper, and other garbage left behind by expeditions. Overall, 265 people have died on Everest, between 1922 and 2014. Because it is virtually impossible to rescue ill or injured climbers in what is called the "death zone" (above 8000 m), a substantial number of dead bodies are also left high on the mountain. Since the 1990s, there has been a raising of awareness of this problem, and a number of attempts are being made to clean up on Everest (Ken 2000; Nuwer 2015). *The Guardian* newspaper (2017) reported that the government of Nepal and Everest expedition organisers had launched a clean-up operation at 21,000 ft. to remove rubbish. Sherpas and other climbers were given canvas bags each capable of holding 80 kg (176 lbs) of waste to place at different elevations on Mount Everest. Once full, the bags were winched by helicopters and flown down the mountain. Removing the sacks by air means Sherpa guides do not have to risk carrying heavy loads of waste through the treacherous Khumbu Icefall to the base camp. The operation used helicopters that would ordinarily return empty after dumping climbing ropes on the site. Recreational climbers were being urged to pick up any rubbish along their route, while Sherpas who carried equipment up the mountain for their clients are paid extra— US$2 per kg—to return with bags of trash.

16.4.3 Impact on Local Communities

Most expeditions interact in some way with local inhabitants, and all expeditions, regardless of their objectives, will need to establish some kind of relationship with their host country. To a certain extent, expeditions will change the communities they aim to experience; cultural exchange is not possible without some erosion of cultural differences.

Allison and Beames (2010) discussed the issue of cultural sensitivity and environmental responsibility on expeditions. They noted that critics have identified some problematic aspects of certain practices on youth expeditions, including cultural sensitivity, the use of drugs, and the environmental costs associated with young people travelling outside of their home country (Allison and Higgins 2002). Expedition groups that did not show appropriate cultural sensitivity when travelling in developing nations such as those who do not cover themselves suitably and wear short and sleeveless tops in Muslim countries are criticised. Is flying a group of 50+ young people across the world justifiable (Allison and Higgins 2002) in a time when air travel is now becoming widely accepted as a contributor to global climate change? It seems that many operators and participants are convinced that they must visit lands far away, despite sometimes knowing little of their homeland. This point is contentious and has been responded to by the YET which has convincingly argued that the benefits outweigh the costs. This debate will no doubt gain more momentum if climate change (warming, increased storminess) continues in the future.

Expeditions should be sensitive to their impact on local communities, acknowledge that they are privileged visitors to a host country, and recognise that cultural sensitivities may impose constraints on their activities.

16.4.4 Impact of Expedition Fieldwork

Many expeditions carry out some form of fieldwork either for research or as a means of educating the members of the expedition (Stott 2010). The intention is often to find out more about a particular area so that the findings can, in some way, benefit the scientific community and humankind in general or may even be of direct benefit to the host country or local community. In some cases, the findings might result in the local area being managed more effectively.

Earthwatch (2018) provides citizens with the opportunity to work alongside leading scientists to combat some of the planet's most pressing environmental issues. Expeditions run projects concerned with research on wildlife and ecosystems, climate change, archaeology and culture, and ocean health. Likewise the British Exploring Society offer expeditions for young people which encompass science and adventure. Figure 16.6 shows a field site in Greenland where BES expedition carried out a river study for over a month (Stott et al. 2014) in August 2009. The photograph shows the site next to the river where a group of 12 young expeditioners assisted a professor in sampling the river for four weeks. The trampling impact around the tent, used for shelter when sampling through the night, can clearly be seen. However, it would have been very difficult to have completed the study without some damage to the riparian vegetation, but the outcome was a publication in an international academic journal which may contribute in some way to worldwide understanding of how rivers work.

If expeditions to such places are planned and executed carefully, the likely benefit of the expedition and its research should justify any adverse impact on the environment. As with community projects and other cultural interaction on expeditions, if it is possible to involve the host country or local scientists in the expedition fieldwork, this could lead to the research having greater impact and/or longer-term benefits. However, if the expedition is primarily educational (e.g. a youth expedition) rather than a research expedition, it may be more difficult to justify the use of a particularly sensitive habitat for the fieldwork. Winser (2004) section 5 offers some useful advice for organisers of field research projects on expeditions.

16.5 Management and Education

Publications such as the RGS' Expedition Handbook (Winser 2004) offer a great deal of sound advice for planning, organising, and managing expeditions. The RGS offers a great deal of advice and training for expeditions and fieldwork (Royal Geographical Society 2018) which cover all aspects of expeditions from initial planning stage to the reporting after it's over. Stott et al. (2013) argued that of the many benefits

16.5 Management and Education

Fig. 16.6 River in SW Greenland where a group of 12 young expeditioners assisted a professor in sampling the river for four weeks. The trampling impact around the tent, used for shelter when sampling through the night, can clearly be seen. Photo by Tim Stott

young people gain from taking part in an expedition, one is the real life experiences and powerful lasting memories (see Allison et al in review) which will foster positive attitudes towards wilderness environments for the rest of their lives.

In 2008 the BES and the YET (British Ecological Society/Young Explorers' Trust 2008) published *Environmental Responsibility for Expeditions: A Guide to Good Practice*. The publication is not a comprehensive handbook on reducing environmental impacts. Much has been written on the techniques of minimal impact camping, using vehicles on expeditions, working with local communities, and on specific fieldwork techniques. The BES/YET guide is designed to prompt thinking during the initial planning of an expedition. Expedition leaders are expected to consider each of the points made in this guide and bear them in mind when deciding where to locate campsites, identifying access routes, choosing fieldwork projects, and so on. The checklists at the end (Table 16.2) are for photocopying for the benefit of leaders during planning and for use as an audit during screening by the two organisations.

Before an expedition will be considered for grants or approval by either organisation, it must be able to show that the group have an adequate plan to limit their environmental impact and that they will be able to implement that plan in the field. They must be able to demonstrate the steps they intend to take in order to keep the impact of their activities on the local environment to a minimum. It is important that a group ethos is developed so that all members of a group take responsibility for the consequences of their actions, not just the leaders.

In a study focused on university students, Harper et al. (2017) reported that expedition participants believed that the expeditions provided real benefits to the communities visited. Organisations like World Challenge Expeditions

Table 16.2 Checklists provided by British Ecological Society/Young Explorers' Trust (2008, pp. 11–13)

1 Access and movement

1.1 Are you keeping to established routes/tracks? If not, how are you establishing your own routes?
1.2 Have you chosen the least damaging routes to and from campsites or sites of activity? Will these routes be varied to reduce impact?
1.3 Are all the expedition members aware of techniques for minimising movement impact? What training will be provided?
1.4 Are you making the minimum number of journeys during the expedition?
1.5 Do you have the resources and expertise to minimise damage to sensitive areas, for example, walkway construction?
1.6 Will you be using vehicles or powerboats? How will you limit noise and damage?
1.7 What arrangements will you have for refuelling, and how will you avoid leakage and spillage?
1.8 Will you be using pack animals? What will their impact be on the surroundings?
1.9 Will you have a designated movements/transport officer?

2 Campsites

2.1 Do you have adequate information about the environmental sensitivity of campsite locations?
2.2 What will the conditions be like during the season you are visiting, and will they require any special care?
2.3 Who owns the land and how will you obtain permission to camp, if required?
2.4 How much interaction can you expect with any local inhabitants around the campsite?
2.5 Can you justify the location of your campsites in terms of minimum impact?
2.6 On reaching potential campsites, what is your procedure for checking for sensitive areas, nesting sites, animal drinking access, and so on?
2.7 Who will manage the campsite and monitor environmental impact?
2.8 How will you minimise trampling and/or tent damage around the campsite?
2.9 How will you cook? If you are intending using open fires, do you know how much fuelwood is available, and can you justify using it?
2.10 How will you minimise the damage caused by using open fires?
2.11 What is your waste management plan?
2.12 What steps will you take to reduce the expedition's waste?
2.13 How do you intend to deal with human waste, and is this the most appropriate method for the situation?

(continued)

16.5 Management and Education

Table 16.2 (continued)

3 Local communities

3.1 Is the local community aware of the expedition and its aims, and what evidence do you have that it is supportive?
3.2 Do you have all appropriate permissions for working in the area?
3.3 Who will liaise with the local community, and who is your contact in the community?
3.4 If you intend carrying out a community project, how have you ensured that it is supported by the community and that it is appropriate and sustainable?
3.5 How will expedition members be briefed about maintaining good community relations?
3.6 Have you considered all local options for supply of goods and services? If you are not using local sources, can you justify this?
3.7 Are you certain that you are not using crucial local resources?
3.8 What is your policy regarding requests for medical treatment?
3.9 How will you avoid compromising the safety of any local participants?
3.10 Will the expedition be self-sufficient in terms of evacuation and/or rescue?
3.11 Do you plan to leave any equipment or materials with the local community? How will you ensure that this will be distributed fairly and utilised appropriately?
3.12 Will there be any follow-up with the local community after the expedition?

4 Fieldwork

4.1 What permissions or permits do you require to carry out your fieldwork?
4.2 What liaison have you had with host country scientists or institutions, and who is your point of contact?
4.3 Do you have adequate information about the environmental sensitivity of fieldwork sites?
4.4 What will the conditions be like during the season you are visiting, and will they require any special care?
4.5 How sensitive is the site to trampling, use of vehicles, collecting samples, and so on?
4.6 Are there any cultural sensitivities you need to be aware of in relation to your fieldwork?
4.7 Has any work of a similar nature already been carried out in the area, and if so, can you justify any further similar work?
4.8 What are your sampling techniques, and will they result in any adverse environmental impact?
4.9 If necessary, how will you minimise the impact of repeated visits to the same site during the expedition?
4.10 Will you be collecting any material, and if so, why is this necessary?
4.11 What fieldwork expertise exists in your team, and/or what training will expedition members have in the necessary fieldwork techniques?
4.12 Who will monitor your impact on the environment during the expedition?
4.13 How will your results be disseminated, and what arrangements do you have for ensuring that the results are communicated to the host country?

(World Challenge Expeditions 2018), which sends around 350 expeditions overseas each year, build in a community project to all its expeditions. This is intended to "give something back" to the host country. It may be helping to dig a well, build part of a school, or paint a community hall. Some expedition providers try to revisit the same regions each year or develop a three-year rolling programme, so that established links can be renewed and developed to maximise the benefits to the host community. Many expedition providers (like World Challenge) are increasingly employing guides and leaders from the host countries. In this way, expeditions are far less likely to inadvertently negatively impact the country they are visiting. Local guides will know the best places to visit and will understand cultural sensitivities. Expeditions should be sensitive to their impact on local communities, acknowledge that they are privileged visitors to a host country, and recognise that cultural sensitivities may impose constraints on their activities.

Many expeditions carry out some form of fieldwork. The BES/YET Guide (2008) offers plenty of sound advice on how to plan, execute, and minimise any environmental impact. Expeditions which encounter rare and threatened habitats which are especially vulnerable to damage must be planned and executed particularly carefully. The likely benefit of the expedition and its research must fully justify any adverse impact on the environment. The possible impact of any subsequent expeditions must also be taken into account.

Geneletti and Dawa (2009) explained how mountain tourism in developing countries was becoming a growing environmental concern due to extreme seasonality, lack of suitable infrastructures and planning, and interference with fragile ecosystems and protected areas. Their study aimed to assess the adverse environmental impacts of tourism, and in particular of trekking-related activities, in Ladakh, Indian Himalaya. Their approach was based on the use of Geographical Information System (GIS) modelling and remote sensing imageries to cope with the lack of data that affect the region. First, stressors associated with trekking and environmental receptors potentially affected were identified. Subsequently, a baseline study on stressors (trail use, waste dumping, camping, pack animal grazing, and off-road driving) and receptors (soil, water, wildlife, vegetation) was conducted through fieldwork, data collection, and data processing supported by GIS. Finally, impacts were modelled by considering the intensity of the stressors and the vulnerability and the value of the receptors. The results were spatially aggregated into watershed units and combined to generate composite impact maps. The study concluded that the most affected watersheds were located in the central and south eastern part of Ladakh, along some of the most visited trails and within the Hemis and the Tsokar and Tsomoriri national parks. This example of a modern approach to understand patterns of tourism-induced environmental degradation is exciting and could be used to support mitigation interventions, as well as in the development of sustainable tourism policies.

Conclusions

1. An expedition is defined as a journey undertaken by a group of people with a particular purpose—especially that of exploration, research, or war. There are no readily available criteria by which an expedition can be defined or classified. The number of people involved might be one criterion. This can range from a single expeditioner to groups in excess of 100. The environment in which the expedition takes place could be another way to classify or define an expedition, for example, polar, mountain, desert, or jungle.
2. Expeditions use various modes of travel which include: on foot, on ski, with animals (dogs, horses, ponies, mules, camels), self-propelled boat (canoe, kayak, raft, yatch), off-road vehicles (snowmobile, quadbike, motorbike, 4WD Land Rover), motorised boat, and bicycle.

3. Expeditions are as old as humankind. Archaeological evidence clearly shows that our ancestors travelled widely on foot and overseas on rafts and various types of boats as far back as 60,000 or 70,000 years ago.
4. There has been a growth in organised expeditions since the first youth expedition provider, the Public Schools Exploring Society (now British Exploring Society), was established in 1932. The number of expeditions on the RGS Expeditions Database, 1964–2018, shows an increase from 1964 to the mid-1990s, followed by a decline.
5. The environmental impact of expeditions can be considered in four categories:
   - Impact of movement and access
   - Expedition campsites
   - Impact on local communities
   - Impact of expedition fieldwork
6. The BES and the YET (2008) published *Environmental Responsibility for Expeditions: A Guide to Good Practice* which is an excellent guide with checklists which expedition leaders can use to minimise their impact.
7. By organising youth expeditions and taking young people into wilderness areas for extended periods, some would argue that the benefits outweigh the costs to the environment and local host community. Taking part in an expedition might be one of the most powerful learning experiences available, leaving memories which can last a lifetime.

References

Allison, P., & Beames, S. (2010). The changing geographies of overseas expeditions. *International Journal of Wilderness, 16*(3), 35–42.

Allison, P., & Higgins, P. (2002). Ethical adventures: Can we justify overseas youth expeditions in the name of education? *Australian Journal of Outdoor Education, 6*(2), 22–26.

Allison, P., Stott, T. A., Felter, J., & Beames, S. (2011). Overseas youth expeditions. In M. Berry & C. Hodgson (Eds.), *Adventure Education: An Introduction* (pp. 187–205). London: Routledge.

Allison, P., Stott, T. A., Palmer, C., Jose, M. J., & Fraser, K. (in review). Forty years on: Just how 'life changing' are school expeditions? *Journal of Experiential Education*.

Barros, A., & Pickering, C. M. (2015). Impacts of experimental trampling by hikers and pack animals on a high-altitude alpine sedge meadow in the Andes. *Plant Ecology and Diversity, 8*(2), 265–276.

Beaumont, M. (2011). *The Man Who Cycled The World*. London: Transworld Publishers.

Bishop, B., & Naumann, C. (1996). Mount Everest: Reclamation of the world's highest junk yard. *Mountain Research and Development, 16*(3), 323–327.

Brida, J. G., & Zapata, S. (2009). Cruise tourism: Economic, socio-cultural and environmental impacts. *International Journal of Leisure and Tourism Marketing, 1*(3), 205–226.

British Ecological Society/Young Explorers' Trust. (2008). *Environmental Responsibility for Expeditions: A Guide to Good Practice* (Ed. Mark Smith, 2nd ed.).

British Mountaineering Council. (2018). Retrieved April 8, 2018, from https://www.thebmc.co.uk/everest-facts-and-figures.

Butt, N. (2007). The impact of cruise ship generated waste on home ports and ports of call: A study of Southampton. *Marine Policy, 31*(5), 591–598.

Cousquer, G., & Allison, P. (2012). Ethical responsibilities towards expedition pack animals: The mountain guide's and expedition leader's ethical responsibilities towards pack animals on expedition. *Annals of Tourism Research, 39*(4), 1839–1858.

Earthwatch. (2018). Retrieved April 8, 2018, from http://earthwatch.org/Expeditions.

Eldon, K., & Swan, A. (1993). Trampling effects on Tundra. In T. A. Stott (Ed.), *The Science Report of the British Schools Exploring Society Expedition to Svalbard (Summer 1993)*, BES Expeditions, London. British Exploring Society. Retrieved from http://www.britishexploring.org/.

Gellatly, A. F., Whalley, W. B., & Gordon, J. E. (1986). Footpath deterioration in the Lyngen Peninsula, North Norway. *Mountain Research and Development, 6*, 167–176.

Geneletti, D., & Dawa, D. (2009). Environmental impact assessment of mountain tourism in developing regions: A study in Ladakh, Indian Himalaya. *Environmental Impact Assessment Review, 29*(4), 229–242.

Guardian Newspaper. (2017). Retrieved April 8, 2018, from https://www.theguardian.com/world/2017/mar/29/climbers-prepare-clean-up-mission-mount-everest-nepal-waste.

Harper, L. R., Downie, J. R., Muir, M., & White, S. A. (2017). What can expeditions do for students… and for science? An investigation into the impact of University of Glasgow Exploration Society expeditions. *Journal of Biological Education, 51*(1), 3–16.

Jones, A. (2004). *Review of Gap Year Provision*. Research Report RR555. London: Department for Education & Skills.

Jones, D. (1997). *An Environmental Impact Study of Expeditions to the East Coast of Oscar II Land, Svalbard*. Unpublished undergraduate dissertation, Liverpool John Moores University.

Ken, N. (2000). Cleaning up Mount Everest. *Japan Quarterly, 47*(4), 45.

Kuniyal, J. C. (2002). Mountain expeditions: Minimising the impact. *Environmental Impact Assessment Review, 22*(6), 561–581.

Nuwer, R. (2015). Cleaning up on Everest. *New Scientist, 228*(3043), 29.

Royal Geographical Society. (2018). Retrieved April 8, 2018, from https://www.rgs.org/in-the-field/advice-training/.

Sørbel, L., Høgvard, K., & Toigensbakk, J. (1990). Geomorphology and Quarternary Geology of Gipsdalen, Svalbard. In *Environmental Atlas, Gipsdalen, Svalbard*. Report No. 61, Norsk Polarinstitutt, Oslo.

Stott, T. A. (2010). Science on expeditions. In S. Beames (Ed.), *Understanding Educational Expeditions* (pp. 45–53). Rotterdam: Sense Publishing. ISBN 978-94-6091-123-1.

Stott, T., Allison, P., Felter, J., & Beames, S. (2015). Personal development on youth expeditions: A literature review and thematic analysis. *Leisure Studies, 34*(2), 197–229.

Stott, T. A., Allison, P., & Von Wald, K. (2013). Learning outcomes of young people on a Greenland expedition: Assessing the educational value of adventure tourism. In S. Taylor, P. Varley, & T. Johnston (Eds.), *Adventure Tourism: Meanings, Experience and Learning* (pp. 148–160). London: Routledge.

Stott, T., Nuttall, A. M., & Biggs, E. (2014). Observed run-off and suspended sediment dynamics from a minor glacierized basin in south-west Greenland. *Geografisk Tidsskrift-Danish Journal of Geography, 114*(2), 93–108.

Winser, S. (ed.) (2004). *RGS-IBG Expedition Handbook*. Profile Books, 502pp. ISBN I 86197 0447. £16.99 including UK postage. PDFs of chapters freely available at: http://www.rgs.org/OurWork/Publications/EAC+publications/Expedition+Handbook/Expedition+Handbook.htm.

World Challenge Expeditions. (2018). Retrieved April 8, 2018, from https://weareworldchallenge.com/uk.

Overall Summary 17

Chapter Summary
The direct impacts of outdoor recreation are summarised in this chapter. Because of future increase in this recreation, which is estimated, both recreation management and environmental impact management are necessary. Generalisations related to recreation ecology research, in particular disturbance-use relationships, are discussed, and the need for further research is emphasised.

17.1 Introduction

It is well known that all outdoor activities create changes in the environment where they take place, and we have demonstrated these in the previous chapters. In this final chapter, there is an attempt to summarise the direct impacts of outdoor recreation and to look ahead to the likely future impacts in this century by establishing the possible trends in outdoor recreation. Outdoor recreation inevitably causes degradation of the natural environment, even under low use, and ecologically this impact is usually negative. However, due to the increased popularity of outdoor recreation and because of these environmental resource impacts, both recreation management and environmental impact management are necessary. This is because the elimination of recreation in the majority of areas is not possible or desirable, except perhaps for conservation reasons in specialist nature reserves. To counteract, or at least minimise, the changes caused by the activities, there is a need to educate the outdoor practitioners, and an attempt is made to summarise the best methods. This means that management processes and systems are crucial in the attempt to limit the environmental changes that are caused by outdoor recreation and some of these have been considered in this book. It seems likely that there are greater current and likely future impacts on the environment than those related to outdoor recreation activities, such as the impact of long-distance air travel to carry out the various outdoor activities, climate change, marine pollution, and geomorphic change. Nevertheless, despite the relatively minimal impacts of outdoor recreation activities, they still cause damage to the environment which needs to be minimised where possible.

17.2 Numbers Involved in Outdoor Recreation, the Relative Importance of the Various Activities, and Likely Future Trends

First it is necessary to summarise the overall numbers involved in the various outdoor activities, and we have already attempted this in each chapter but it is extremely difficult to get accurate figures. However in this section we are attempting to see the relative importance of the various activities because the more people taking part in an activity, the more likely it is that there will be environmental impacts. We will use the American participation in outdoor recreation activities as a basis for this as it is the best documented, over a long time period. The participation has increased each decade since 1960 (Cordell et al. 2008), the total numbers and the number of days of participation have increased too, and Hammitt et al. (2015) offer a detailed discussion on the growth of outdoor recreation in their Chap. 7 on trends in wildland recreation. Up to 1986 the annual growth rates for total use appeared to have decreased in all wilderness areas. Prior to the mid-1960s, there were two figure increases annually, but this was lowered to a 1% increase in the early 1980s, and wilderness use declined on a per acre basis between 1975 and 1986. This decline occurred in both the Forest Service wildernesses and the National Park Service areas.

However since that period of decline or at least stabilisation, the recreational use has increased (Cole 1996), and there has been a major increase in acreage use intensity from 0.32 recreational visitor days/acre in 1965 to 0.4 visitor days/acre in 1979 but a decline to 0.18 in 1994. This latter decline was mainly because in 1980 56 million acres of wilderness were designated in Alaska and therefore if this state is excluded (where in 1994 visitor days were only 0.02) recreational use intensity is as high as it has ever been at around 0.4 recreational visitor days/acre and there was a use increase in nearly every wilderness area during the 1990s. Recent growth since then is particularly high in various National Park Service areas where there are double-digit annual growth rates as had occurred several decades ago. The authors have personally witnessed the impact and management issues this has caused in areas like Yosemite, Zion, and the Arches National Parks.

The growth trends for some of the activities considered in this book in terms of total numbers and the total annual days from 1982/1983 to 2005/2009 are illustrated in Table 17.1.

The figures in terms of the percentage of the population participating for several of the activities are illustrated in Table 17.2 from 1982/1983 through to 2008/2009.

Here it can be seen that walking for pleasure, day hiking, and backpacking have all increased and are the most popular activities. The biggest increase was viewing or photographing wild birds, probably driven by advances in digital camera and phone technology, whilst there has been a steady decline in hunting and fishing and

Table 17.1 Outdoor recreation activity numbers and total number of participation days in the USA (after Cordell et al. 2008); in millions of participants and days

Activity	Total annual days 2005/2009	1982/1983	1994/1995	1999/2001	2005/2009
Walking for pleasure	20,928	91.9	138.5	175.6	200
Day hiking		24.3	53.6	69.1	79.7
Viewing or photographing birds	8275	20.8	54.3	68.5	35.7
Primitive camping	249.6	17.3	31.4	33.1	34.2
Backpacking	235.8	8.7	17.0	21.5	23.2
Fishing	816.3		48.0	49.8	56.0
Canoeing	106.7	13.9	19.2	23.0	22.8
Horseback riding	262.1	15.6	20.7	19.8	21.5
Downhill skiing		10.4	22.8	17.4	15.9
Cross-country skiing	35.9	5.2	8.8	7.8	6.1
Snorkelling	70.9		16.2	13.6	15.2
Scuba diving	22.7			3.8	3.6
Caving	19.5		9.5	8.8.	10.4
Rock climbing	44.7		7.5	9.0	9.8
Orienteering			4.8	3.7	3.6
Snowmobiling	77.4	5.2	9.6	11.3	10.7

Table 17.2 American participation in certain outdoor recreation activities (after Cordell et al. 2008), figures in % of population participating

1982/1983	1999–2001	2005–2009	% change	Activity
53.0	82.4	84.1	111.3	Walking for pleasure
14.0	32.4	32.6	209.9	Day hiking
34.0	34.2	33.8	32.2	Fishing
12.0	31.8	34.9	287.0	Viewing/photographing birds
50.0	56.7	55.1	46.4	Visiting nature centres and so on
8.0	11.5	12.4	105.8	Canoeing
9.0	9.7	9.7	43.6	Horseback riding
5.0	10.4	9.8	160.9	Backpacking

Table 17.3 Future estimated demand for some outdoor activities between 2020 and 2040 (from Cordell et al. 2008) as a percentage of 1987 demand

1987 trips in millions	2020	2030	2040	Activity
69.5	146	162	174	Wildlife observation and photography
26.0	196	230	255	Backpacking
91.2	198	244	293	Day hiking
266.5	146	164	177	Walking for pleasure
63.2	160	177	190	Horseback riding

a decline in cross-country skiing and snowmobiling, with the latter two likely to decrease further with the effects of climate change. Orienteering has declined in the USA but continues to be extremely popular in Scandinavia, whilst there has been a phenomenal growth in geocaching not captured in the American surveys, whilst specialist activities like caving and rock climbing have marginally increased in popularity, although the latter's popularity may be because of the growth in artificial climbing walls. Canoeing seems to have increased in popularity but is thought to have a limited impact on the environment, whilst snorkelling and scuba diving have declined in popularity.

According to Cordell (2012) the five activities projected to grow the most in the USA in terms of number of participants are developed skiing (68–147%), undeveloped skiing (55–106%), challenge activities which are mountain and rock climbing and caving (50–86%), horseback riding (44–87%), and motorised water activities (41–81%). The activities with the lowest growth in participant numbers are visiting primitive areas (33–65%), motorised off-road activities (29–56%), motorised snow activities (25–61%), hunting (8–23%), fishing (27–56%), and so-called floating activities, such as canoeing, kayaking, and rafting (30–62%). While activities currently having high participation levels may not show large percentage increases in participant numbers, even small percentage increases in already highly popular activities can mean quite large increases in participants and therefore potentially much greater environmental impacts. Figures taken from Cordell et al. (2008) for this future demand up to 2040 are shown in Table 17.3.

The future participation and use for many of the activities considered in this book are considered for a longer time period up to 2060 in Table 17.4, from a baseline in 2008, taken from Bowker et al. (2012).

It is clear from all this data that outdoor recreation impacts on the environment will continue to grow this century because of population growth and that the major impacts will occur because of the popularity of walking for pleasure, day hiking, backpacking, and activities like nature viewing, all of which cause trampling pressure on the vegetation and soils. Some activities like horseback riding and off-road vehicle driving can cause major damage to the terrain and water quality because of the specialist nature of the

Table 17.4 Future participation in selected outdoor recreation activities up to 2060 (after Bowker et al. 2012): numbers in millions and % increase based on three possible scenarios outlined in that report

Activity	Numbers 2008	Possible numbers 2060	Total days, % increase	Estimated potential increase
Challenge activity (mountain and rock climbing, caving)	25.1	46.4	81–88	120 million to 219 million 81–90%
Nature viewing	189.4	307	60–63	32,303 million to 46,648 million 39–44%
Horseback riding	16.39	30.57	87–110	262 million to 503 million
Day hiking	78.3	134.4	63–72	1826 million to 3330 million 74–78%
Primitive area visitation and use (primitive camping, backpacking, and visiting wilderness)	90.2	141.0	49–56	1233 million to 1909 million 41–55%
Off-road driving	47.9	75.0	55–58	1048 million to 1532 million 46–53%
Snowmobiling	9.44	12.99	6–56	68.4 million to 91.0 million 6–67%
Hunting	27.9	34.2	13–23	535 million to 576 million 2–8%
Fishing	72.7	110.6	44–54	1363 million to 1965 million, 25–44%
Skiing (downhill and snowboarding)	23.7	54.2	123–127	171 million to 437 million 142–150%
Canoeing, kayaking, and rafting	39.8	64.6	26–62	261 million to 422 million 24–62%

activity with heavy animals or vehicles. Fishing remains a very popular but specialist activity, with particular environmental impacts of its own.

Whilst most of this discussion related to trends has been from the USA because of the long and detailed datasets available from the Outdoor Recreation Resources Review Commission surveys published since 1960, similar trends with some regional variations are obvious in Europe (Bell et al. 2007) and within individual countries like England and Wales. Walking and hiking remain in pole position in most countries whilst studying and enjoying nature is important. Some of the regional differences include the importance of cross-country skiing in Norway and Finland where between 40 and 50% of the population take part and the picking of berries and wild mushrooms in both those countries. There is however always a problem in comparing data because of the ways in which the data has been collected and presented.

17.3 Research into the Recreation Impacts (Recreation Ecology) and the Land Management of These Impacts

The number of scientists who would be considered as recreation ecologists on a world scale is relatively few, and the research that has been completed has been concentrated mainly in a few geographical areas, especially in the USA, Australia, and Western Europe. Much of the research has been carried out by an even smaller group of these scientists, and they have been supported by specialist research institutions and universities, such as, for example, the International Centre for Ecotourism Research at Griffith University (Queensland); the Aldo Leopold Wilderness Research Institute (Missoula, Montana); the Department of Parks, Recreation and Tourism Management at North Carolina State University; the Department of Forestry,

Virginia Tech (Blacksburg); and the Department of Environment and Society at Utah State University. The research has also often been on particular ecosystems and on particular species, so there has been much research, for example, on trampling pressure by recreational walkers/hikers in temperate grassland ecosystems and on wilderness campsites in the USA. Buckley (2013) suggests that much of the research has been localised, where direct effects are easily observed and where measurements of various parameters are not difficult to make as in the classic trampling pressure experiments (Cole and Bayfield 1993). Where the effects are less obvious, diffuse, delayed, and the effects difficult to detect, like where there might be the introduction of pathogens on clothing or by horses or the disruption to animal energetics, reproduction, and behaviour by recreationists, these have been less frequently researched. However, recently sophisticated research has been carried out using remote telemetry to monitor hormones and heart rate in birds disturbed by recreationists, but these are expensive to set up and have been relatively uncommon.

The trampling pressure research on the other hand has been inexpensive to carry out and it has been relatively straightforward to measure the vegetation and soil parameters, and as a result there have been many such studies in different terrains, vegetation types, and with different recreation impactors, such as walkers, horses, mountain bikes, off-road vehicles, and even llamas. The stress-response curves of vegetation cover and soil loss with this trampling are curvilinear and asymptotic. When the results with known levels of use were examined, the results and the model of the relationship between use and impact indicate that on previously undisturbed sites, even small increases in the amount of initial use resulted in dramatic increases in impacts (Cole 1981; Hill and Pickering 2006). Hence, where use levels are low, small differences in the amount of use can lead to substantial differences in impact. It was believed that this type of curve applied to all forms of recreation impact. However, this is now known to be incorrect.

Alternatively, where use levels are high, additional impact has proportionally less effect. This relationship underpins strategies that seek to confine use to designated trails and sites in areas popular for nature-based tourism and recreation. This are commonly called "confinement" strategies, with the assumption that once a site is extensively disturbed, impacts will not change considerably, despite substantial increases in use. Although the literature suggests some management challenges with confinement strategies, for example, in Cole et al. (2008), they can assist in limiting the total area impacted by recreationists (Hammitt and Cole 1998; Hammitt et al. 2015; Newsome et al. 2012). However, in contrast to the curvilinear model, studies by Cole and Monz (2004) and Growcock (2005) observed a sigmoidal response to use, as opposed to the more commonly reported single asymptote at the top of the curve. Sigmoidal models are useful in various ecological applications, for example, in Kuznar (2002), and the use of this model in recreation disturbance was suggested as a possibility in some earlier work on the subject by Liddle (1975) and Cole (1992). This has practical implications for managing areas with dispersed low levels of use, particularly on non-vegetated substrates or trampling-resistant vegetation. In these cases, it may be more likely to effectively manage low trampling levels and limit ecological change than the curvilinear response suggests. Growcock's (2005) work further suggests that along the stress-response curve, different impacts may be more pronounced at different levels of use which means that plant physiological stress can precede mechanical damage and loss of plant cover. Also, research indicates that more sensitive techniques for assessing change may be required than those previously used in trampling studies. Recent work addressing this issue has modelled the shape of the relationship with more replicates and levels of use and where the disturbance-use relationship was closer to linear than curvilinear (Hill and Pickering 2006).

Highly resistant substrates can obviously display a different relationship between use and impact than that observed for vegetated areas. These types of substrates can occur naturally such as on exposed rock outcrops, or on rocky shores, or can result from overuse which has eliminated completely the vegetation and compacted soils, or might be created by management via the maintenance of trails and sites with resistant substrates, such gravel and rock. Concentrating use on these hardened surfaces reduces ecological change, and sites with these hard surfaces will exhibit a flat relationship with increasing use (no slope). A different relationship between use and impact on a hardened surface though was seen in a study by Milazzo et al. (2004) who examined the effects of human trampling on the tolerance of algal communities on rocky shorelines. Here impacts were more subtle than those in some terrestrial ecosystems, and some relationships may more closely resemble the curvilinear relationship than what has been typically assumed. So despite what was mentioned earlier, the impact of increasing recreation trampling on soil has been found to be highly variable and not necessarily curvilinear. From the research literature, several use-soil-loss relationships have emerged, including linear and exponential relationships. Where surfaces are hard, the relationship is flat, but where soils are soft and deep, as in valley-floor, alpine humus soils, the relationship can be linear and steep. Furthermore, as Liddle (1997) described, recreation disturbance is often the "trigger" for soil erosion by damaging the protective vegetation and litter cover, resulting in direct mechanical disturbance to exposed soil horizons and subsequent erosion from wind and water, leading to rapid soil loss.

When it comes to wildlife, one widely reported generalisation regarding increased recreation use and impact is birds taking flight when approached by recreationists (Buckley 2004; Steven et al. 2011). Sometimes described as a "flight or fight" response (Knight and Gutzwiller 1995), this kind of behaviour results in a step relationship between use and impact, with a sudden reaction from the pre-disturbance steady state as numbers increase, or visitors approach the animals. Buckley (2013) gives the example of the often repeated experiments when the researcher walks towards feeding shorebirds and measurements are made of how close these birds can be approached before they take off. This gives a stress-response curve with the vertical step from no flight to flight. It can also induce complete avoidance of areas that are more intensively used by people, resulting in decreased animal diversity close to high-use sites (Buckley 2004; Steven et al. 2011; Newsome et al. 2012).

The work reported by Monz et al. (2013) advances recreation ecology theory by summarising these new generalisations about the relationship between use and impacts, enhancing both future research opportunities and improving management. The original use-impact relationship stands as one of the few well-developed research generalisations in this field. This is mainly due to research that focused on easily observable ecological responses, such as changes in vegetation cover and the use of rapid assessment techniques, rather than more sensitive and difficult measurements. However, as we have seen, more recent research has suggested several alternative models for some ecological responses. Future research could more directly model the use-response relationship through more sensitive methods of measurement and improved experimental designs. The alternative models proposed also highlight the need for caution while employing standard recreation dispersal and confinement management strategies. Although successful in minimising the proliferation of certain impacts, such as vegetation loss in many environments which often are the most obvious and common impacts, these approaches may have unintended consequences, especially in situations where impacts do not "level off" with increased use.

Monz et al. (2013) emphasise that increasing intensity of use is just one important factor influencing outdoor recreation impacts. Management decisions should not be solely based on any one factor like limiting use numbers. For example, visitor behaviour, degree of site hardening and maintenance, trail design, and environmental durability all affect the severity and extent of impacts and may be more important than use level in some circumstances. Moreover, recreation area management involves integrating ecological, social/cultural, and managerial components to achieve the best possible overall use outcome. This suggests that recreation managers cannot simply accept that all ecological impacts follow a curvilinear response and then assume that a confinement strategy will minimise unacceptable changes. We have seen that the job of the recreation resource manager is not easy as impacts can be complex, interrelated, and not equally distributed in time and space. The impacts can vary by the type of environment and by the type of use, and they are seldom directly related to the amount of recreation use in a linear way. What is required from the land manager is that after they understand the impacts and their spatial and temporal patterns in relation to recreation use, they develop a Limits of Acceptable Change managerial system where conditions can be manipulated so that impacts are kept within limits. At the same time, recreation opportunities are preserved. However, there has always been tension between managing land in national parks and other protected areas for conservation and managing them for recreation. The land manager has to walk a tightrope to balance the use. In the past often the land manager made decisions in the absence of much scientific evidence, simply because of the lack of research and the fact that the managers did not fully understand the ecology of their areas. There is less excuse today and land managers need to know how ecological systems operate, how ecological impacts occur, and how the changes can be monitored. They need to know how much change is allowable or necessary, desirable, or acceptable in any area. This means that management objectives need to be set which determine the level at which an impact becomes problematic and needs action from the land management. It is necessary to monitor changes on a regular basis so that trends can be seen and then recreation use and the sites themselves can be changed. Initially it is probably easier to try and change the recreationists' behaviour by giving information on the sensitivity of the resources and how to achieve minimum impact behaviour from individuals and groups. It may be possible to change use to more durable locations or redistribute users to either concentrate or disperse them depending on the management objectives. It may be best though to concentrate the use on already impacted sites rather than to spread the impacts, that is, to develop "honeypot sites." Management of the sites to include increasing the durability of the trail or campsite and other means of rehabilitating the sites may be needed as well as visitor control. Change though is inevitable but the aim should be to keep the recreation areas in as natural a condition as possible, or as managerially possible. Yet what is meant by natural and how do we define this? For example, the uplands of the Lake District, Pennines, or Wales in the UK have changed dramatically with the effects of sheep pastoralism since medieval times, and this has gone hand in hand with climate change, including the Medieval Warm Period and the Little Ice Age when the climate has been warmer, wetter, and colder at various times. What is ecologically natural is difficult to define because ecosystem change is going on most of the time and ecosystems are rarely in equilibrium and respond at different time scales to a wide range of disturbances, including climate change.

As Buckley (2013) suggests there are some other perspectives worth noting when it comes to the effects of recreation impacts on ecosystems. For example, some impact parameters are more ecologically important than others so that whole populations are more important than individuals and mortality is more important than a change in behaviour. There have been situations where disturbance of the breeding colony of a rare bird species can lead to the loss of an entire year's young. It is usual that species are more susceptible to impact during reproduction or periods of energetic or nutrient stress so susceptibility can depend on the timing of the disturbance. Rather than just the intensity of the disturbance, the interval between disturbance and the timing of repeated disturbance can be important too. For example, repeated disturbance of feeding by migratory birds on their way to overwintering or breeding grounds can cause significant mortality. The recovery too is important and some species take longer to recover than others. For example, the algae and lichens making up the cryptogrammic crusts of desert soils take much longer to recover from trampling than grasslands or estuarine mudflats. Some impacts can be self-limiting whilst others are self-propagating, so that, for example, plant cover loss can lead to soil erosion whilst the introduction of weed seeds, pathogen propagules, or insect pests can change whole ecosystems.

It is clear that there is still the need for much more scientific research by ecologists in a wide range of ecosystems on the effects of recreation impacts across a wider range of outdoor recreation. They can then pass on the results of their research to the land managers for appropriate action (e.g. Hadwen et al. 2010). This research should be over a sufficient time period for the results to be worthwhile, and the scientists should have sufficient research funds to carry out this research using the most up-to-date techniques. This research is needed because relative to the demand for recreation and the demands of the adventure tourism market the research has been small scale up to now. With future increase in numbers across most recreation pursuits as we have noted earlier, there must be appropriate management based on scientific evidence. This is difficult because managers face more recreationists, usually with no more money to manage them and often less. Sometimes it may be possible to make new land available for recreation but this may not always be possible. A characteristic of recreation impacts is that they are usually concentrated in certain locations like mountains, lakes, rivers, and other scenic attractions like waterfalls. Substantial impacts might only though be under 1% of the total land area so it would seem sensible to continue to concentrate campsites, nature trails, and visitor centres in these locations. Site hardening of the trails and campsites reduces the impacts. Quotas and other regulations and restrictions in some areas may be needed like we have seen in some caves; in the Zion National Park canyons and the climbs in Yosemite National Park. Marketing, or not advertising, certain areas may allow the possibility of moving recreation to different areas.

In this book we have concentrated on the environmental impacts of the main outdoor recreation pursuits, and there is no duplication of the work of Buckley (2006) who considered detailed case studies across a range of outdoor recreation activities supplied by commercial adventure tourism companies. However, these tour companies create environmental impacts just like individuals, or independent groups, and we have not distinguished between the two main types of outdoor recreation in terms of their environmental impact.

Overall Conclusions

The concluding remarks to each chapter are as follows and many apply to more than one of the chapters. In Chap. 2 we see that recreation impacts can occur on all types of recreation land but when vegetation loss occurs there is often parallel soil erosion, especially in mountains where there are severe environmental conditions and moorlands with peat soils so that deep gully erosion can occur. Where "honeypots" occur the carrying capacity of the land is often far exceeded, and it may be more effective to educate and regulate the recreation walkers so that the land is sustainably managed. Nevertheless, there is plenty of recreation resource available for recreation walking.

In Chap. 3 we saw that due to the relatively recent increase in popularity of mountain marathons and adventure racing, there is a need to undertake more research and further develop policies to sustainably manage mountain marathons, adventure racing, and mountain tours.

In Chap. 4 we see that there needs to be a further fostering of indoor climbing walls, artificial climbing competitions, and artificial ice climbing. The latter is especially important with future climate change and an inevitable decline in the outdoor resource availability for this recreation activity. Education and appropriate environmental stewardship seem to be the key to conserving the natural climbing environment. Outreach, education, and appropriate site management are the way forward, for example, through Action Fund projects in the USA and the British Mountaineering Council initiatives in the UK. There have been doubts too as to how much damage the activity causes the natural rock and compared with natural processes, such as rockfall or glacial erosion, the impacts appear trivial.

There are also doubts as to the impacts of climbers on natural vegetation, and further research is needed here. Seasonal climbing restrictions have been successful for maintaining breeding bird populations, especially raptors and other factors like marine pollution or overfishing are likely to be having bigger impacts on sea-bird populations than climbing as again there appear to be well-observed seasonal restrictions on sea cliffs. Education and outreach have produced a much more environmentally aware climbing population who has a greater concern for the stewardship of the mountain and crag environment.

In Chap. 5 it has been observed that there are relatively few participants taking part in gorge or canyon recreation. However, in the UK there has been floral loss of a rare association of plants due to the impacts of mainly outdoor centre groups, although through education of the centre staff this probably has now declined. In the Greater Blue Mountains of Australia, the numbers of participants seems lower than originally thought, they are concentrated in only a few locations, and their impacts on stream invertebrates and water quality appear negligible. In the classic canyon country of the southwest USA, there has been rock damage by ropes and tree damage from anchors, but overall it looks like management regulations, Codes of Conduct, and Ethics and Education of the practitioners and commercial company guides currently maintain an overall sustainable environment in most gorges and canyons.

In Chap. 6 there is an example of the use of modern research methods using airborne remote sensing and Geographical Information Systems to collect data on off-road vehicles which has informed planning and management decision-making. Guides for snowmobile tours have suggested that the tourists may even become spokespersons for sustainable tourism because of the tour that they had experienced, although whether this is really true is debatable.

In Chap. 7 it is shown that purpose-built mountain biking centres and bases have helped enormously to confine the impacts as mountain biking has caused major damage to vegetation, soils, water quality, and wildlife disturbance outside these centres.

In Chap. 8 we see that camping impacts again cause major damage to vegetation, soils, water quality, and wildlife, especially black bears in the USA, in a range of environments and over different time scales. Additionally there can be introduction of weed propagules, the spread of exotic species, and soil-borne pathogens.

In Chap. 9 the high impact of recreation horseback riding on the environment, its increasing importance as a recreation activity, and an established conflict with other recreation users is outlined. This means that there needs to be an emphasis on best-practice management which is imperative, including education, changing user behaviour, and the use of codes of conduct. It may also be necessary to prohibit this activity in certain areas where significant conservation and biodiversity issues may take precedence.

In Chap. 10 it is shown that the idea that orienteering causes significant disturbance to wildlife and damage to flora has not been proven by scientific evidence and this activity seems to have a low ecological impact. The impact of geocaching which has increased markedly in the last few years and its impacts need to be carefully researched, although both activities are though to promote environmental stewardship.

In Chap. 11 it is shown that the impacts of ski resort development include changes in vegetation, changes in bird and arthropod populations and significant soil erosion. There are impacts too from snow grooming and artificial snowmaking which uses a lot of water and can cause change in plant species composition. Climate change will result in major changes in the ski industry, and there will be closures, although the high-altitude resorts will be least affected.

In Chap. 12 it is suggested that the long time scale for caves to develop means that once damaged they will never recover. The aim has to be to minimise the damage. The number of cavers is relatively small but the physical environment and the specialised cave fauna and flora can suffer change by the activity. Controlling access is a major management technique, more so than in many other outdoor pursuits, and there are alternatives to the activity, like mines and artificial caves, that can be considered.

In Chap. 13 the huge variety of watercrafts which potentially have impacts on the environment are considered, before making the important distinction between motorised (with propellers) and non-motorised types. The latter are shown to generally have far greater environmental impacts. These impacts include physical impacts to aquatic vegetation; the spread of invasive species; erosion of banks and shores; water pollution and its costs. There is discussion about the impacts of water sports on wildlife as well as the chemical impacts on water sports (heavy metals, hydrocarbons). The final section considers the management of these activities and gives examples of ways in which users can be educated.

In Chap. 14 it is shown that the impacts of scuba diving and snorkelling are often difficult to quantify for any one stressor on a coral reef ecosystem independently of others, and there are interrelationships which are not fully understood. There are many global effects too like global warming, coral bleaching, ocean acidification, and damage by tsunamis and hurricanes which can cause major change to the reefs. It has been suggested that the best approach should be to integrate management and governance in integrated coastal management and sustainable tourism development in each area of reef, with a voluntary management strategy and a community-orientated approach.

In Chap. 15 it is suggested that recreational fishing management needs much more involvement with participants and that best ways for educational outreach should be found, with important scientific knowledge related to fisheries, so that the participants are better informed and can act on the information. Climate change impacts need further research as the effects on recreational fishing are uncertain but may have major implications and even influence future fish evolution.

Finally, Chap. 16 considers expeditions, starting by giving examples of expeditions with different purposes and modes of travel and concludes that many use camping as their primary accommodation. Specific environmental impacts of expeditions are considered in four areas: movement and access, campsites, on local communities, and the impacts of expedition fieldwork. The final section considers the management of these impacts and gives examples of ways in which expedition organisers can minimise the impact of their expeditions.

References

Bell, S. L., Tyrväcken, L., Sievänen, T., Pröbstl, U., & Simpson, M. (2007). Outdoor recreation and nature tourism: A European perspective. *Landscape Research Living Reviews*. Retrieved from http://www.Livingreviews.org/lrlr-2007-2.

Bowker, J. M., Askew, A. E., Cordell, H. K., Betz, C. J., Zarnock, S. J., & Seymour, L. (2012). *Outdoor Recreation Participation in the United States—Projections to 2060: A Technical Document Supporting the Forest Service 2010 RPA Assessment*. General Technical Report SRS-160, US Department of Agriculture Forest Service, Southern Research Station, Asheville, NC, 34pp.

Buckley, R. (2004). Environmental impacts of off-highway vehicles. In R. Buckley (Ed.), *Environmental Impacts of Ecotourism* (Chap. 5, pp. 83–98). Wallingford, Oxfordshire: CABI Publishing.

Buckley, R. (2006). *Adventure Tourism*. Wallingford: CABI, 515pp.

Buckley, R. (2013). Next steps in recreational ecology. *Frontiers in Ecology and the Environment, 11*, 399.

Cole, D. N. (1981). Managing ecological impacts at wilderness campsites and evaluation of techniques. *Journal of Forestry, 79*, 86–89.

Cole, D. N. (1992). Modeling wilderness campsites: Factors that influence amount of impact. *Environmental Management, 16*, 255–264.

Cole, D. N. (1996). *Wilderness Recreation Use Trends, 1965 through 1994*. USDA Forest Service Research Paper INT-RP-448.

Cole, D. N., & Bayfield, N. G. (1993). Recreational trampling of vegetation: Standard experimental procedures. *Biological Conservation, 63*, 209–215.

Cole, D. N., Foti, P., & Brown, M. (2008). Twenty years of change on campsites in the backcountry of Grand Canyon National Park. *Environmental Management, 41*, 959–970.

Cole, D. N., & Monz, C. A. (2004). Spatial patterns of recreation impact on experimental campsites. *Journal of Environmental Management, 70*, 73–84.

Cordell, H. K. (2012). *Outdoor Recreation Trends and Futures: A Technical Document Supporting the Forest Service 2010 RPA Assessment*. General Technical Report SRS-150, US Department of Agriculture Forest Service, Southern Research Station, Asheville, NC, 167pp.

Cordell, H. K., Betz, C. J., & Green, G. T. (2008). Nature-based outdoor recreation trends and wilderness. *International Journal of Wilderness, 14*, 7–13.

Growcock, A. J. W. (2005). *Impacts of Camping and Trampling on Australian Alpine and Subalpine Vegetation*. PhD dissertation, Griffith University, Gold Coast, Queensland, 288pp.

Hadwen, W. L., Arthington, A. H., & Boon, P. I. (2010). *Guidelines for Design and Implementation of Monitoring Programs to Assess Visitor Impacts in and Around Aquatic Ecosystems within Protected Areas*. Gold Coast, QLD: Sustainable Tourism Cooperative Research Centre Press, 46pp.

Hammitt, W. E., & Cole, D. N. (1998). *Wildland Recreation: Ecology and Management* (2nd ed.). Chichester and New York: Wiley, 376pp.

Hammitt, W. E., Cole, D. N., & Monz, C.A. (2015). *Wildland Recreation, Ecology and Management* (3rd ed.). Chichester: Wiley Blackwell, 313pp.

Hill, W., & Pickering, C. M. (2006). Vegetation associated with different walking track types in the Kosciuszko alpine area, Australia. *Journal of Environmental Management, 90*, 1305–1312.

Knight, R. L., & Gutzwiller, K. J. (1995). *Wildlife and Recreationists: Coexistence through Management and Research*. Washington, DC: Island Press, 369pp.

Kuznar, L. A. (2002). Evolutionary applications of risk sensitivity models to socially stratified species: Comparison of sigmoid, concave, and linear functions. *Evolution and Human Behaviour, 23*, 265–280.

Liddle, M. J. (1975). A selective review of the ecological effects of humans trampling on natural ecosystems. *Biological Conservation, 8*, 251–255.

Liddle, M. J. (1997). *Recreation Ecology: The Ecological Impact of Outdoor Recreation and Ecotourism*. London: Chapman and Hall, 639pp.

Milazzo, M. F., Badalamenti, F., Riggio, S., & Chemelo, R. (2004). Patterns of algal recovery and small-scale effects of canopy removal as a result of human trampling on a Mediterranean rocky shallow community. *Biological Conservation, 117*, 191–202.

Monz, C. A., Pickering, C. M., & Hadwen, W. L. (2013). Recent advances in recreational ecology and the implications of different relationships between recreation use and ecological impacts. *Frontiers in Ecology and the Environment, 11*, 441–446.

Newsome, D., Moore, S. A., & Dowling, R. K. (2012). *Natural Area Tourism: Ecology, Impacts, and Management* (2nd ed.). Clevedon, UK: Channel View Books, 457pp.

Steven, R., Pickering, C. M., & Castley, G. (2011). A review of the impacts of nature based recreation on birds. *Journal of Environmental Management, 92*, 2287–2294.

Index

A

Abseil points, 78
Acanthaster, 376
Accelerated erosion, 78, 221, 242, 281
Access agreements, 34, 98, 313–314, 319, 327
Access Fund, 93, 96, 103
Acclimatisation, 57, 61
Acidosis, 406
Adder, 252
Adirondacks, 31, 87
Adopt-a-Gorge scheme, 121, 129
Adventure racing, 55–70, 173, 454
Adventure Racing World Series (ARWS), 60, 61, 70
Afon Tryweryn, 336, 344
Aggregate paths, 42, 45–48
Aid canyoneering, 113
Aid climbing, 76, 89, 90
Airborne remote sensing, 152, 455
Algal blooms, 29
All-terrain vehicle (ATV), 28, 131–160
The Alpine Climb Scale, 64
Alpine saxifrage, 83
Alpine skiing, 267, 271, 273, 277, 278
Amblers, 17
American National Standards Institute (ANSI), 132, 133
American oystercatcher, 146
Amundsen, Roald, 429, 431
Anchoring, 46, 76, 80, 90, 92, 97, 98, 100, 101, 103–105, 113, 125, 126, 128, 129, 354, 365–368, 371, 374, 381, 382, 411, 455
Angling, 344, 396, 399, 401, 403–407, 410, 413–416, 418, 419, 423, 424
Antelope ground squirrel, 203
Antifouling, 352–353, 356, 357
Apical meristems, 25
Aquatic invasive species (AIS), 205, 355
Aquatic vegetation damage, 456
Arctic fox, 148, 149
Arctic-alpine species, 82, 95, 115
Armour, 178
Arthropods, 144, 235, 236, 286, 288, 289, 295, 456
Artificial caves, 321, 324–327, 456
Artificial reefs, 362, 381–385, 388
Artificial snowmaking, 284, 285, 288–290, 294, 295, 456
Artisanal fishing, 395, 418
Asian fish tapeworm, 414

B

Backpacking, 17, 21, 44, 188, 448, 449
Bag limits, 415–417
Bait digging, 413, 414
Bald eagles, 175
Bank armouring, 346
Bank erosion, 29, 356, 357
Barefoot skiing, 332
Basal rosette, 25
Basalt lava tubes, 301
Bat roosts, 303, 310
Beaumont, Mark, 430
Bedouin, 190
Belay ledges, 78
Benzene-toluene-xylene (BTX), 291
Big-game fishing, 396
Biodiversity reservoirs, 301
Biomass, 26, 29, 220, 232–234, 287, 288, 295, 306, 349, 378–380, 398–401, 408, 409, 414, 417, 418
Biophysical change, 1
The Birketts, 59, 70
Bison, 175, 319
Bivouac, 104, 188, 189
Black band disease, 375, 376
Black bear, 203, 209, 212, 455
Blanket peat, 44
Blashford-Snell, John, 430
Bluegill sunfish, 353
BMX, 61, 70, 164, 167, 170, 171, 183
Boardwalk, 40, 43, 46, 48
Boat strike, 412–413
Boat wash, 344, 354
Bodyboarding, 332
Bolts, 75, 88, 90, 94, 97, 98, 100, 101, 103, 104, 113, 123, 125, 126, 128, 129, 305, 318
Booking, 100, 129, 205, 313, 315, 327
Bothy, 194, 196, 201–203, 209, 211, 212
Bothy Code, 209, 210
Bouldering, 74, 76, 77, 93, 104–105
Braking, 172, 173

Branching corals, 365, 371, 374, 376, 382
British Ecological Society (BES), 433, 438, 440–445
Bryophytes, 42, 81–84, 88, 115, 118, 120, 283, 305
Buffer distance, 34
Buoyancy control, 370, 372, 374, 380, 384
Burke and Wills, 430
Bushwacking, 17

C

Camping, 7, 18, 20, 59, 60, 66, 103, 104, 164, 183, 187–212, 216, 217, 228, 229, 241, 277, 305, 313, 315, 331, 334, 430, 437–439, 441, 444, 455, 456
Campsites, 7, 27, 29, 79, 104, 113, 188, 189, 191, 198–201, 205–207, 211, 212, 229, 234, 241, 242, 346, 433, 437–439, 445, 451, 453, 454, 456
Canoeing, 188, 277, 332, 335, 339, 341, 344, 347, 348, 356, 357, 449
Canyoneering management plan, 126, 127
Capercaillie, 34, 260
Carabids, 288, 289
Carabiners, 75, 76
Caravan, 188, 190, 191, 196, 212
Carbide, 31, 304, 326
Carbon dioxide, 306
Carbon monoxide, 353, 354
Carp, 396, 409, 413, 414
Carrion crows, 288
Carrying capacity, 5–7, 34, 37, 38, 49, 79, 257, 310, 373, 374, 381, 413, 454
Catch-and-release (CR), 404–407, 420–422
Cave adoption, 319, 327
Cave conservation, 301, 315, 322–325
Cave conservation codes, 314, 323
Cave conservation plans, 323, 324, 327
Cave diving, 304, 308, 362
Cave fauna, 304, 307–311, 319–321, 327, 456
Caving suits, 305
Cereal baits, 413
Chalk, 88–89, 94, 97, 98, 101, 103, 105
Chamaephytes, 27
Chamois, 175
Chatter marks, 94
Chemical binders, 41, 241
Chipping, 31, 45–47, 88, 92–94, 101, 105, 127
CITO event cache, 253
Clean-up and restoration, 127–129, 424
Cliff goldenrod, 82
Climbers Pact, 103
Climbsite, 79
Clogwyn Du'r Arddu, 95, 96, 102
Closures, 6, 9, 10, 87, 88, 97, 98, 103, 106, 122, 175, 199, 205, 292, 415, 456
Cnidaria, 361
Code of conduct, 121, 128, 168, 239, 240, 243, 386
Coefficient of floristic dissimilarity, 30
Coliform bacteria, 29, 201, 202, 346, 439
Coliforms, 201, 202, 290, 346, 347
Colorado chipmunks, 203
Colorado Plateau, 126
Compensation, 401
Confinement strategies, 451, 453
Conservation, 5, 27, 33, 34, 37, 38, 66, 73, 84, 92, 98–100, 102, 115, 118, 121, 122, 148, 150, 151, 153, 160, 178, 179, 183, 215, 216, 230, 238, 241, 242, 244, 245, 259, 262, 285, 289, 290, 293, 303, 308, 314–316, 318, 320–324, 327, 356, 363, 378, 381, 384–386, 388–389, 404, 407, 416–420, 422, 424, 434, 447, 453, 455
Coot, 350
Copper, 352, 353, 416
Coral bleaching, 377, 378, 389, 456
Coral breakage, 370, 371, 373, 374
Coral reefs, 13, 361–368, 371, 374, 376–378, 380, 382–384, 386, 388, 389, 456
Corallivorous gastropod, 375
The Corbetts, 58, 70
Corms, 26
Cost of illness (COI), 347
Crash pads, 74, 77, 93, 103
Crayfish, 123, 418
Cross-country, 61, 70, 134, 163, 180, 183, 215, 267, 273, 274, 287, 293–295
Cross-country (XC) mountain bike, 169, 449
Cruise ship, 436, 437
Cryptogrammic crusts, 454
Cryptophytic buds, 26
Cushion-plants, 146, 147
Cyanobacteria, 305, 322

D

DDT, 86, 87
de Saussure, Horace-Bénédict, 57
Deadman anchors, 126
Deep hooking, 406, 416, 421
Deer, 23, 34, 36, 141, 175, 230, 233, 257–259, 262, 288
Deer mice, 203, 230
Depensation, 401
Desert bighorn sheep, 175
Desert cottontails, 203
Desert racing, 131–133, 159
Devil's Tower, 87
Devotional trails, 18
Diesel, 145, 147, 353, 354
Dirt jumping, 170
Disc brakes, 135, 163
Dissolved oxygen, 29, 346, 402
Diver behaviour, 369–375, 381, 384–387
Diversionary feeding, 209, 211
Diversity index, 30
Diving trails, 382, 384
Dolomite, 83, 270, 301
The Donalds, 58, 70
Downhill, 31, 40, 142, 163, 169, 171–173, 178, 180, 183, 222, 267, 273, 274, 286, 291, 293, 294
Downhill (DH) bike, 169–171, 183
Dragon boats, 332
Drilling, 44, 97, 126, 304

Driving, 67, 68, 131, 132, 135, 139, 141–143, 145, 146, 153–156, 158, 160, 171, 205, 236, 258, 259, 352, 422, 423, 444, 449
Drupella snails, 375, 376
Dry tooling, 74, 75, 97
Dune bashing, 132, 159
Durance River, 336
Dye testing, 305
Dyke intrusions, 114

E
EarthCache, 254
Earthwatch, 440
Ecobolts, 305
Eco-centric, 293
Economic activities, 291
EcoReefs, 382
Eco-regional networks, 6
The eight-thousanders, 58, 70
Elk, 233, 258–259
Endemic, 82, 237, 289, 301, 307, 308, 320, 327, 389, 407, 408
Endozoochory, 227
Endurance conservation, 318
Endurance riding, 217
Enduro, 132–134, 153, 169
Enduro/All-Mountain (AM) bike, 169
Enterovirus, 347
Environmental changes, 13, 29, 288, 310, 373, 380, 402, 447
Environmental concerns, 69, 70, 103, 257, 262, 444
Environmental education, 255, 384–386, 423, 424
Environmental stewardship, 255–256, 263, 454, 456
Epilithic lichens, 81
Epizoochory, 227
Erosion, 8, 13, 20–23, 26–29, 31, 35, 39–41, 43–45, 47, 48, 60, 67, 69, 78, 79, 90–95, 100, 104–106, 111, 114, 117, 118, 120–124, 129, 142–144, 151, 160, 171, 172, 176–178, 205, 219–221, 223, 225, 234, 237, 239, 242, 244, 255, 281–283, 286–288, 294, 295, 303, 305, 344, 345, 348, 354, 356, 357, 409, 434, 438, 439, 452, 454–456
Erosion Index, 118, 120
Erosion pins, 345
Escherichia coli (*E. coli*), 29, 202, 225, 226, 290, 347
Eurasian watermilfoil, 206, 349
Event cache, 253
Everest, 57, 430, 439
Exhaust fumes, 291, 353
Expedition, 56, 57, 60, 61, 70, 179, 188, 199, 206, 429–445, 456
Expedition fieldwork, 440, 445, 456

F
Faecal coliforms, 202, 206, 209, 290, 347
Fat bike, 164
Fault line, 114
Faunal Sanctuaries, 320
Feral horses, 23, 227
Field experiments on impacts, 223–225
Fingerling brook trout, 148, 290
Fires, 27, 104, 200, 442
Fish communities, 346, 365, 374, 378–380, 400–401, 409
Fish feeding, 366
Fish stocks, 352, 395, 398–400, 403, 407, 412, 414, 415, 418, 423
Fixed anchors, 90, 92, 97, 100, 103, 125–127
Flash flooding, 111
Flexboard, 46, 48
Flight response, 33, 34, 36, 175
Flowstone, 303, 304, 317, 322
Flushing distance, 175, 258
Flyboard, 332
Footpath erosion, 22–23, 48, 78, 90, 307
Forest-based mountain biking, 179–182
Formal and informal trails, 9–12, 18–20, 177, 221, 222
Four–wheel-drive (FWD), 132, 135, 137, 159, 221, 430, 444
Freeride (FR) mountain bike, 170
Free soloing, 75
Freestyle, 133, 134, 164, 182, 267, 270, 273, 274, 277, 294, 332
Front pointing, 94
Front suspension, 168–171
Fruticose lichen, 82
Full suspension, 135, 164, 168, 169
The Furths, 58, 70

G
Game fishing, 396
Gardening, 78, 101, 102, 126
Gastropods, 84–85, 88, 93, 307, 375
Gating, 312, 315, 316, 321, 327
Genetic drift, 402
Genetic variability, 401, 402
Geocaching, 249–263, 449, 456
Geocoin, 254
Geographical Information Systems (GIS), 152, 444, 455
Geojute, 40
Geological variability, 111, 114
Geophytes, 27
Geotextiles, 40–42, 45, 46, 178, 244
Geotrail, 254
Ghost crab, 145, 146
"Ghost" fishing, 411
Ghosting, 113
Giardia lamblia, 29
Glamping, 188, 212
Gleyed soil, 28
Gluing, 93, 126
Golden plover, 35
Gorge Code of Conduct, 121, 128
Gorge rotation, 129
Graffiti removal, 127, 319
The Grahams, 58, 59, 70

Grasshoppers, 289
Grass skiing, 273
Green Fins, 384, 386, 387
Green guides, 85, 102
Grizzly bears, 175
Groundbaiting, 413
Group size, 97, 103, 126, 127, 129, 146, 175, 244
Growth forms, 25, 26, 375
Growth overfishing, 399
Guano, 309
Guild density, 33
Gully erosion, 44, 287, 454
Gullying, 23, 28
Gypsum, 301

H
Habituation, 34, 36, 211
Hammock, 188, 189
Hard tail, 168, 171
Harpacticoida, 365
Heather, 24, 26, 30, 79, 116, 119, 283
Heavy metals, 226, 341, 352–354, 357, 414, 456
Helicopter, 43, 46, 267, 439
Helophytes, 27
Hemicryptophytes, 27
Herbert, Wally, 429, 430
The Hewitts, 58, 70
Hibernacula, 309
Highlining, 126
Hiking pole impacts, 30–31
Hillary, Edmund, 439
Hoar frost, 95
Holocene, 114
Honeycomb weathering, 92
Honeypots, 38, 49, 316, 453, 454, 120
Hooking, 94, 405–407, 420
Horse riding code of conduct, 239
Horse trekking, 217, 226
Human faeces, 202, 206
Human waste, 103, 125, 127, 199, 202, 207, 209, 290, 305, 326, 438, 439, 442
Hydrocarbons, 147, 148, 290, 291, 456
Hydroelectric schemes, 122
Hydrofoiling, 333
Hydrogen peroxide, 322
Hydroseeding, 42, 44, 47, 48, 286
Hypoxia, 406

I
Idwal Slabs, 90, 91, 100
Igloo, 190–194, 201, 212
Impact penetrometer, 29
Inbreeding, 401, 402
Infiltration rate, 28, 198
Insect repellents, 377–378
Interannual variability, 291
Intercalary, 25

International Mountain Biking Association (IMBA), 168, 171, 172, 178, 183
International Ski Federation (FIS), 267, 273, 294
Invasive alien aquatic plants (IAAPs), 355
Invasive aquatic weeds, 415
Invasive species, 26, 142, 205, 206, 228, 234, 348–350, 354–357, 389, 408, 409, 422, 456
Inventory, 4–6, 32, 97, 237, 311, 312

J
Jet skiing, 332, 356

K
Kangaroo rat, 203
Kayaking, 277, 331, 332, 335, 339, 347, 348, 356, 357, 449
Kelp forests, 380–381
Keystone predator, 401
Kiteboarder, 332
Kitesurfer, 332, 333, 335
Kneeboarding, 332
Knox-Johnston, Robin, 430

L
Lake trout, 351, 398, 408, 411
Lampenflora, 305, 306, 322
Landslide, 287
Leaf litter, 10, 24, 221, 326
Leave No Trace, 7, 9, 12–13, 103, 176
 practices and ethics, 12
LeTrec, 217
Letterboxing, 249–263
Lewis and Clark, 429
Liaison groups, 98, 100, 319
Libraries, 256
Lichens, 25, 31, 78, 80–85, 88, 89, 119, 126, 146, 257, 283, 454
Lichen species density, 81
Life-forms, 27, 37, 82
Light pollution, 305, 306
Limestone, 45, 47, 82–86, 89, 113, 301, 302, 306, 326, 382
Limestone gorges, 114
Limits of Acceptable Change (LAC), 6, 8, 178, 198, 211, 212, 453
Litter, 8, 12, 28, 31, 47, 65, 67, 69, 101, 105, 118, 121, 125, 128, 152, 178, 198, 203, 219, 220, 238, 253, 284, 295, 304, 307, 411, 452
Littering, 67, 122, 305, 368
Little grebe, 350
Little Tryfan, 92
Local communities, 68–70, 182, 418, 439–441, 443–445, 456
Logbook, 251, 253, 254, 263
Longear sunfish, 351, 412
The Longmynd Hike, 56, 70
The Lowe Alpine Mountain Marathon (LAMM), 55, 70

M

Macro-invertebrates, 123, 129, 226, 235, 407
Malham, 86
Mallory, George, 430
Management plans, 6, 96–98, 100, 106, 122, 123, 126, 129, 151, 312, 323, 353, 442
Manure, 8, 220–222, 225–234, 237, 238, 243
The Marilyns, 58, 59
Marine cave species, 308
Marine protected areas (MPAs), 381, 388, 417–419
Maritime species, 115
Marmot Dark Mountains, 55, 65, 70
Memorandums of understanding, 98
Meramec River, 334
Mesh elements, 41
Micro-topography, 82, 84, 224
Minimum impact caving, 320, 326–327
Mites, 37
Mixed climbing, 61, 74, 93–95
Monetary loss, 291
Monitoring, 3–9, 21, 27, 32, 44, 97, 98, 117, 121, 124, 125, 147, 205, 222, 225, 226, 234, 237, 240, 242–245, 259, 291, 309, 311, 320, 324, 347, 365, 373, 386, 388, 389
Moorhen, 350
Moorland species, 78, 116
Mosses and liverworts, 78, 115, 116, 118
Motocross, 133–134, 153, 159, 164
Motor boating, 339, 347, 357
Motor home, 212
Motorised water sport, 338
Mountain bike (MTB), 163–164, 168, 169, 171–173, 178–180, 182, 183, 249, 350
Mountain cross/"Four-cross," 171
Mountain goats, 146
Mountain marathon, 55–70, 454
Mountain rescue teams, 65, 67, 68
Mountain tours, 55–70, 187, 454
Mountain Training UK, 209
The Mourne Mountain Marathon, 55, 70
Movement and access, 433–437, 445, 456
Muck stick, 371, 381
Mulch mats, 41
Mule deer, 34, 36, 175, 230
Multi-cache, 253
Murray Levick, George, 431
Mystery cache, 253

N

National Park Service (NPS), 2–8, 27, 97, 100, 103, 125, 126, 310–313, 323, 378, 448
National Three Peaks Challenge, 67–69
National Trails, 17, 18, 21
NAVSTAR GPS, 251
Naylor, Joss, 59
Nesting birds, 85, 260–261, 410
Noise, 32, 33, 67, 69, 103, 141, 145, 147–149, 153, 160, 174, 238, 326, 351, 352, 435
Noise restrictions, 148

Non-motorised water sports, 331–337
Non-native species spread, 11, 82, 232, 407–409
Nordic skiing, 267–269, 277, 287, 294
Norgay, Tenzing, 439
Nutrient cycling, 28, 234, 286
Nuts, 76, 89, 90, 104
The Nuttalls, 58

O

Off-road vehicle (ORV), 131–132, 134–138, 141–146, 150–153, 159, 160, 174, 203, 205, 410, 435, 444, 449, 451, 455
Offset caches, 253
Online reservation system, 126
Opilionids, 289
Organotin compounds, 353
Original Mountain Marathon (OMM), 55, 59, 64, 70
Overfishing, 380, 395, 399, 418, 419, 455
Overland Track, 205–207

P

Pack animals, 235, 434, 444
Packrafts, 124
Packstock, 8, 28, 29, 217, 218, 226, 228, 236, 238, 244
Packstock grazing, 234–236
Paddleboarding, 332
Parasailing, 332
Pathogen-free catchments, 205
Pathogenic bacteria, 285
Pathogens, 199, 205, 206, 212, 219, 234, 236, 237, 242, 367, 377, 398, 414, 451, 454, 455
Peary, Robert, 429, 430
Peat pipes, 45
Pennine Way, 18, 23, 35, 43–45
Peregrine falcon, 80, 86, 87, 97
Permit system, 100, 126, 242, 321, 381
Petroglyphs, 124
Petrol, 291
Phanerophytes, 27
Phenology, 287, 288, 295
Phototrophs, 305, 322
Physiological stress, 405, 406, 451
Phytophthora cinnamomi, 236
Pine marten, 87
Pinnacles National Park, 87
Piste, 267, 272, 283–287, 289, 294, 295
Piste caterpillar, 283
Pitons, 74, 76, 88, 90, 94, 101, 104, 125, 126
Plant biomass, 29, 232
Plant resilience, 26
Plastics, 13, 411, 423
Pleistocene, 82, 114
Polar bears, 149, 150, 190
Pollution, 13, 34, 67, 73, 85, 122, 145, 147–149, 151, 160, 211, 305, 306, 344, 352, 353, 377, 398, 413, 424, 447, 455, 456
Pontoons, 363, 365, 366
Potholes, 113

Potholing, 299
Power boating, 350, 351
Powered water craft (PWC), 338, 352
Prairie falcons, 87
Pre-dive briefings, 371–373, 384
Preservational zones, 127
Primary production, 23, 413
Pronghorn antelope, 174
Propeller action, 341, 348, 412, 435
Protection, 12, 27, 39, 41, 42, 74, 78, 89–92, 94, 98, 99, 101, 102, 104, 113, 134, 139, 141, 143, 188, 190, 211, 257, 290, 311, 312, 319, 320, 323, 355, 361, 368, 383, 386, 407, 415, 417, 424
Ptarmigan, 288, 294
Public right of way, 22
Puddling, 28
Pygmy shrew, 289

Q

Quinzhees, 190–193, 201, 212

R

Radio-location devices, 259
Rafting, 56, 66, 70, 203, 204, 333, 335–336, 339, 356, 357, 449
Raleigh International, 432
Ramblers, 17
Raptors, 85–87, 98, 455
Raunkiaer life-forms, 27
Ravines, 114
Recreational Opportunity Spectrum, 178
Recreation ecology, 1–13, 21, 350, 450–454
Recreation vehicle (RV), 164, 183, 188, 196, 212
Recruitment overfishing, 399
Red grouse, 34, 288
Redstart, 261
Reef balls, 382, 384
Reef walking, 364–365
Reflex impairment, 406, 407
Reindeer, 134, 148, 149, 288
Relict plants, 114
Remote telemetry, 451
Resin acids, 79
Resource protection, 3, 5, 6, 8, 12, 20, 96, 98, 127
Response distances, 35, 150
Resting metabolic rates (RMR), 404
Resurgence, 303
Rigid, 23, 25, 26, 168–170
Ring barking, 79
Riparian vegetation, 29, 344, 424, 440
River bank erosion, 29
River Dee, 334, 344
River Stour, 334
River tracing, 113
Rock crawling, 131, 132, 139, 159
Rockfall, 13, 73, 88, 106, 127, 305, 455
Rock polish, 92–93
Rogaining, 249

Root exposure, 8, 26
Round gobies, 408
Roundworms and threadworms, 37
Rowing, 333, 335, 339, 341, 347, 348, 356, 410
Royal Geographical Society (RGS), 431, 432, 440, 445

S

Sacrificial caves, 300, 316–317, 327
Sacrificial gorges, 129
Sailing, 61, 332–335, 337, 339, 341, 350, 356
Salts, 22, 285, 352, 438
Sandblasting, 310, 319
Saunders Lakeland Mountain Marathon (SLMM), 55, 59, 70
Scott, Robert Falcon, 430, 431
The Scottish Munros, 58, 70
The SCOTT Snowdonia Trail Marathon, 55, 70
Scramblers, 17, 122
Scree slopes, 78
Seasonal restrictions, 98–99, 455
Sediment disturbances, 352, 373, 374
Sediment production, 22
Seeding, 26, 41, 44, 47, 48, 283
Sergeant major, 378
The Seven Summits, 57–58, 70
Sewage, 344, 347, 356, 357, 368
Shackleton, Earnest, 429, 430
Shawangunks, 93
Shore erosion, 456
Show caves, 299, 300, 305, 306, 313, 314, 322, 323, 325
Shrew, 288, 289
Signs, 6, 10, 12, 19, 26, 36, 38, 125, 150, 205, 206, 209, 254, 258, 304, 321, 322, 355, 357, 379, 380, 439
Site management, 6, 9–11, 44, 106, 242, 455
Size-selective fishing, 400, 403
Skidding, 172
Ski-doo, 134, 152
Skiing, 34, 56, 61, 70, 233, 267–295, 332, 333, 338, 339, 356, 357, 449, 450
Ski jumping, 267, 273
Ski lift, 21, 169, 267, 281, 283, 288
Skimboarding, 333
Ski mountaineering, 269, 287, 294
Skurfing, 333
Slalom, 132, 163, 171, 183, 273, 335, 347
Slopestyle, 170, 171, 270, 273
Snorkelling, 331, 356, 361, 364, 389, 456
Snow boarding, 272
Snow bunting, 288
Snowcat, 267, 283
Snow cave, 190–193, 201, 212
Snow density, 284, 287, 295
Snow grooming, 283, 284, 287, 290, 295, 456
Snow hole, 196, 198, 207, 431
Snowmobile, 134, 137–138, 141, 146–151, 153, 159, 174, 267, 283, 284, 290, 291, 444, 455
Snow scooter, 134
Snow-shoeing, 267–295

Index

Sodium hypochlorite, 322
Soil compaction, 26, 28, 29, 39, 172, 173, 201, 203, 219–221, 234, 237, 255, 287, 434
Soil crusts, 26, 92
Solidry, 45
Soil loss, 28, 78, 173, 176, 451, 452
Southern Sandstone, 92, 105
South Pole, 429, 431
Species diversity, 26, 30, 47, 81, 82, 85, 146, 285, 295, 356, 374, 417
Speleology, 299
Speleothems, 303, 305, 306, 319, 322
Spelunking, 299
Spiders, 288, 289, 307, 308, 320, 326
Spiny lobster, 380
Sport climbing, 75, 77, 79, 90, 93, 97, 98
Springtails, 37, 304, 322
Stalactite, 303, 304, 325
Standard Positioning Service (SPS), 251
Star system, 100
Starts, 18, 25, 56, 57, 65, 69, 70, 87, 95, 101, 131, 204, 216, 226, 243, 253, 261, 270, 303, 314, 370, 384, 402, 416, 424, 456
Stewardship, 3, 5, 12, 13, 98, 105, 106, 255–256, 263, 312, 323, 415, 417, 455, 456
Stolons, 26
Stonechat, 261
Stone pitching, 46–47
Stony corals, 361, 369, 375, 377, 381, 386–388
Stress-response curves, 451, 452
Striations, 94
Sump drainage, 304
Sunscreens, 29, 377–378
Surface moulding, 42
Surfing, 270, 332, 333, 335, 337, 339, 356, 357
Suspended sediment concentration, 344, 373
Sustainable tourism, 153, 389, 444, 455, 456
Swift-water canyoneering, 113
Swiss International Mountain Marathon, 55, 70

T
Tailrace fishing, 396
Tarpaulin, 188, 189, 228
Technical diving, 362
Telemark skiing, 268–270
Tent, 55, 59, 187–189, 199–201, 203, 205, 212, 228, 229, 431, 439–442
Tethering areas, 219, 226, 232, 242
Therophytes, 27
Thesiger, Wilfred, 430
Thomas, Bertram, 430
Three Peaks Project, 39, 41, 45–48
Thru' trails and thru'-hiking, 18
Tin, 352, 353
Top rope climbing, 76
Torquing, 94, 132
Tourism carrying capacity (TCC), 382
Traditional summer climbing, 78–84
Trail bike, 169

Trailer tent, 188, 212
Trail proliferation, 223
Trail riding, 134, 135, 216, 217
Tramping, 18
Trample dosage, 30
Trample-resistant species, 24
Trample-tolerant species, 24
Trampling, 8, 9, 11, 21–30, 32, 37–39, 44, 47, 48, 78, 83, 93, 95, 101, 106, 118, 119, 121, 122, 173, 176, 200, 201, 203, 205, 220–225, 234, 237, 255–257, 259, 262, 303, 320, 326, 344, 364–365, 372, 375, 378, 413, 414, 434, 440–443, 449, 451, 452, 454
Trampling experiment, 201, 434, 435
Trampling injury, 26
Tramplometers, 30
Transplanting, 44
Travelways, 255
Tread lightly, 69, 142, 256
Tread Lightly booklet, 101
Tree density, 80
Tree pipit, 261
Trials bike, 170
Triglav National Park Management Plan, 127
Troglobites, 308
Trophy fishing, 397, 400
Trundling, 78
Tsunami, 13, 363, 389, 456
Turbidity, 28, 122, 142, 220, 225, 341, 344–346, 349, 351, 354, 356, 357, 374, 409, 412
Turf placement, 94

U
Umbilicate lichen, 81
Underwater art, 383

V
Vascular plants, 81–84, 88, 237, 283
Vegetation damage, 30, 78, 94, 178, 201, 222, 257, 281, 295
Viral haemorrhagic septicaemia (VHS), 414
Virtual cache, 254
Visitor carrying capacity, 6–7
Visitor education, 11
Visitor use management (VUM), 7, 8
Vole, 288, 289

W
The Wainwrights, 58–59, 70
Wakeboarding, 332, 333, 345, 346
Wakestyle, 332
Warthogs, 94
Waste water, 368
Water crossings, 177, 225–226
Waterfalls, 74, 93, 111–114, 118, 121, 125, 254, 317, 325, 454
Water Framework Directive (WFD), 355

Water ingestion, 347
Water pollution, 353, 456
Water quality, 29, 32, 123, 129, 142, 151, 220, 222, 225, 226, 290, 308, 344, 346–348, 355, 357, 413, 449, 455
Water resources, 177–178, 183, 198, 201–202, 206–209, 212, 278, 289–291, 294, 345, 355, 437
Water skiing, 332, 333, 338, 339, 356, 357
Watkins, Gino, 430, 431
Weed spreading, 227–232
Wheatear, 252, 261, 288
Wheel spinning, 172
Wherigo cache, 253
Whirling disease, 414
White nose syndrome, 309, 310, 321, 327
White syndrome, 375, 376
Wilderness areas, 2, 3, 29, 37, 114, 126, 142, 145, 178, 187, 200, 217, 228, 229, 234, 238, 241, 242, 244, 256, 346, 435, 438, 445, 448
Wildlife disturbance, 66, 175, 178, 409, 455
Windsurfing, 168, 332, 333, 335, 337, 339, 356, 357

Winter climbing, 74, 77, 93–96, 102
Winter flounder, 353
Woodlark, 260
Woodrats, 203
Wood warbler, 261
World Challenge Expeditions, 441
Wreck diving, 362

Y

Yellow whitlow grass, 83
The Yorkshire Three Peaks, 69, 70
Yosemite, 29, 49, 74, 87, 88, 90, 226, 235, 448
Young Explorer's Trust (YET), 433, 439–445

Z

Zebra mussels, 350, 408
Zodiac, 436
Zoning off, 318–319

Printed by Printforce, the Netherlands